McKean
MW01026369

GALACTIC ASTRONOMY

Second Edition

GALACTIC ASTRONOMY

Structure and Kinematics

DIMITRI MIHALAS
High Altitude Observatory
National Center for Atmospheric Research

JAMES BINNEY
Magdalen College
Oxford University

W. H. FREEMAN AND COMPANY
San Francisco

Project Editor: Larry Olsen
Copy Editor: Ruth Cottrell
Designer: Sharon H. Smith
Production Coordinator: Bill Murdock
Illustration Coordinator: Cheryl Nufer
Artist: Victor Royer
Compositor: Syntax International
Printer and Binder: The Maple-Vail Book Manufacturing Group

Cover: NGC 5272, a globular star cluster in Canes Venatici. Courtesy of the Hale Observatories. (Palomar Observatory, California Institute of Technology.) Back cover: A relatively loose globular cluster. (Lick Observatory photograph.)

Library of Congress Cataloging in Publication Data

Mihalas, Dimitri, 1939–
Galactic astronomy.
Includes bibliographies and index.
1. Milky Way. 2. Galaxies. I. Binney, James, 1950– . II. Title.
QB857.7.M53 1981 523.1′13 81-1612
ISBN 0-7167-1280-6 AACR2

Printed in the United States of America

1 2 3 4 5 6 7 8 9

To W. W. Morgan and J. H. Oort,
and to the memory of
W. Baade, E. P. Hubble, J. C. Kapteyn,
H. Shapley, and R. J. Trumpler,
pioneers in the study of galactic structure

Contents

Preface

The first edition of this book, written in 1967, has been out of print for over five years. Despite numerous requests for a new impression, it was clear to the authors that the original book had become hopelessly out of date and that the amount of information concerning the nature of galaxies had grown so rapidly during the intervening decade that a simple revision would not be adequate, but instead a completely new work would have to be written. The matter would have rested there had it not been for Professor Jeremiah Ostriker of Princeton University, who insisted that a new version of the book had to be produced, and who helped plan the present work.

Work on the present book began early in 1978 when the senior author was on sabbatical leave from High Altitude Observatory to the Department of Astrophysics at Oxford University and the Department of Physics and Astronomy at University College, London. It soon became evident that our plans could not be accomodated within a single volume of reasonable size. Accordingly, the second edition will be produced in two volumes, the first volume dealing with the structure and kinematics of galaxies, and the second volume dealing with the dynamics of galaxies and clusters of galaxies.

As a major extension of the scope of the first edition, we now treat not only our own Galaxy but external galaxies as well. Throughout the second edition, we shall attempt to exploit the complementarity of the information provided by studies of our Galaxy and of external systems. The former studies provide detailed information about a localized region in the disk of a typical spiral galaxy, but they allow us to draw only fragmentary conclusions about the large-scale structure of the Galaxy as a whole. The latter studies provide much lower-resolution information, but they allow us to see galaxies in their entirety and to study the enormous variety in their morphology.

Galactic astronomy today must certainly be one of the most exciting and rapidly developing fields of physical science. A sequence of major technical developments that extends back two decades or more (and shows no sign of ending) has vastly expanded the quality and scope of the information available to us concerning the structure and evolution of galaxies and clusters of galaxies. Along with this information explosion has grown a considerable body of theory. Some of these theories have stood the tests of time and further observation; others are speculative and often fragmentary, and the field today is in an almost explosive state of development.

Our awareness that we are not describing a mature and static field of knowledge has affected our labors in two ways. First, we have tried to distinguish as clearly as possible between bare observational *facts* and possible *interpretations*, however reasonable the latter may appear to be. We have learned much about galaxies, but there are quite a number of stubbornly puzzling observations that indicate the possibility of serious gaps in our present information or flaws in our present reasoning. Problems of this kind will be cleared up only if we scrupulously separate certain fact from plausible speculation and if we constantly bear in mind that the "results" of observations are often model-dependent *interpretations* of the numbers actually furnished by the observing apparatus.

Second, in many places we have tried to give an account of the historical development of our understanding of the nature of galaxies, which surely is one of the major intellectual adventures of mankind. We hope that our sketch of the formation of the seminal ideas, of some of the controversies and false leads surrounding them, and of moments when great discoveries have illuminated the whole landscape in a clearer and brighter light will be not only of historical interest but also will assist the reader to imagine what relationship our present ideas might have to the underlying truth and to understand what types of mistakes and misconceptions may cut us off from a consistent picture of things. We have cited, wherever possible, landmark papers by the masters in the field. Although we recognize that the literature will be too large for students to read all of the important studies in a semester or two, we nonetheless urge them to seek out some of these great papers to re-experience for themselves the excitement of these major discoveries; in so doing, they will be preparing themselves well for many profound developments undoubtedly yet to come.

As one reads these volumes, the feeling will probably grow that, just as was the case in the study of the structure of our own Galaxy fifty years ago, before the presence of interstellar absorption was recognized, we are at present missing some very fundamental points in the whole picture. We expect that the enigmas that now confront us will vanish suddenly when our eyes are opened by major new discoveries, and we expect that a text written on this subject, say, twenty years from now, will look much different from those we are writing today. We shall feel sufficiently rewarded if it is our readers who are instrumental in advancing the subject and writing the next generation of texts.

Dimitri Mihalas wishes to express his deepest gratitude to Professors D. E. Blackwell and M. J. Seaton for making possible his sabbatical visit to

England and for the innumerable kindnesses they showed him while he was there. Further, he thanks Dr. G. Newkirk of HAO and Drs. J. Firor and F. P. Bretherton of NCAR for granting him leave to work in an area so far afield from his usual interests and duties. James Binney owes a debt of gratitude to the President and Fellows of Magdalen College, Oxford University, for their material and moral support and to Professor R. J. Elliot for the hospitality of the Department of Theoretical Physics, Oxford University, during the writing of this first volume.

Further, we wish to thank those who helped in the writing of this book. First, we thank Dr. J. P. Ostriker for help in planning the scope and structure of the book and for unfailingly good advice about, and criticism of, what we have written. Second, we thank Drs. I. R. King and P. O. Vandervoort for reading the book in its entirety and for offering numerous helpful comments and criticisms. Next we thank Drs. G. T. Bath, D. Carter, W. W. Morgan, R. Sancisi, W. F. van Altena, and S. van den Bergh for reading various chapters and offering corrections and comments. Any errors that remain after the conscientious efforts of these colleagues are solely our own responsibility. In addition, we thank Drs. F. Bertola, A. Bosma, D. Carter, G. de Vaucouleurs, J. Kormendy, W. W. Morgan, A. H. Rots, S. E. Strom, and P. C. van der Kruit for generously providing illustrative material, and the late Mr. J. W. Tapscott of Yerkes Observatory for his skill in preparing a satisfactory print for Figure 1-6. We particularly thank Dr. F. Hohl for supplying us with the computer-generated plots used on the opening page of each chapter. Finally, we thank Paulina Franz for her patience and skill in producing a superb typescript, and Kathlyn Auer for preparing the index.

Dimitri Mihalas
Sunspot, New Mexico

James Binney
Princeton, New Jersey

GALACTIC ASTRONOMY

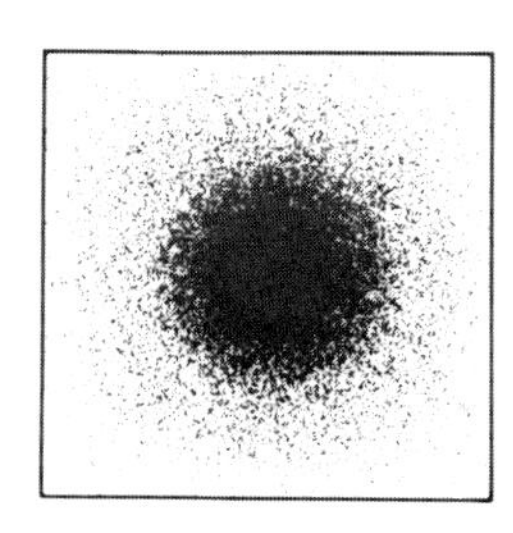

1 The Galaxy: An Overview

Our Sun is located in a stellar system called the *Milky Way Galaxy*. We know today that our Galaxy* is similar to countless other galaxies, and we now recognize that galaxies are the major structural units of the Universe. Our goals in this book will be to learn what galaxies are and how they reached their present state. We shall attempt to infer the distribution and kinematics of the material in galaxies directly from observations, and then we shall attempt to obtain insight into the dynamics and evolution of galaxies through the application of known physical laws.

These goals are not straightforward and easily accomplished. The growth of our conception—even to its present incomplete level—of the nature of our Galaxy and its place in the Universe must be viewed as one of mankind's major intellectual accomplishments. Difficulties arise for several reasons. First, astronomy is an observational science—we cannot perform experiments, adjust the circumstances in which we find our objects of study, or change our vantage point. Often we find that, because of our location in the plane of our Galaxy, the range of our observations is limited, and our ability to deduce necessary information is seriously impaired. Frequently the observational data are fragmentary or are of insufficient accuracy; thus the answers to essential questions are often more vague than we should like them to be.

Astronomers have gotten considerable insight into the nature of our Galaxy from observations of distant galaxies, which reveal large-scale features that are completely hidden from us in our own Galaxy. In a real sense, studies of our Galaxy and of others go hand-in-hand, because the detailed

---

* A note about nomenclature: In this book we shall write *Galaxy* whenever we mean our own (Milky Way) Galaxy and *galaxy* for any galaxy in general. Similarly, we shall use the adjective *galactic* to denote objects in our Galaxy and *galaxian* for objects in galaxies in general.

information we can deduce about the structure of our local neighborhood in our Galaxy is complemented by the large-scale information provided by observations of distant galaxies. Yet, it was only fifty years ago that it became unequivocally clear that our Galaxy *is* a galaxy, similar to the *spiral nebulae*, as they were then called, and that the spiral nebulae are major systems comparable to our own. Only then did it become possible to exploit the complementarity provided by these two types of observations.

Galactic astronomy today is one of the most integrative branches of astronomy, drawing from practically every other area of modern astronomy and astrophysics. By its nature, it is sometimes difficult to explain, because often several disparate pieces of evidence bear upon a single point, and because frequently the reasoning follows long and complicated chains of arguments. Often we shall find that different sets of observations are enmeshed in an almost bewildering thicket of interrelations. Sometimes we reach a dilemma because two lines of reasoning should ideally be developed in parallel, but to do so is didactically impossible. In our discussion, we shall attempt to follow a definite logical route and to provide cross-references among related points; nevertheless, in the end students will profit most if they will reread this book at least once.

In this chapter, we shall give first a brief historical sketch of some of the major developments in our conceptual picture of our Galaxy. We shall make no attempt to be comprehensive, but instead we refer the reader to the excellent histories contained in the books *Man Discovers the Galaxy* (**B3**), *Astronomy of the 20th Century* (**S9**), and *The Discovery of Our Galaxy* (**W1**) for the details of this fascinating story. The point of our discussion will be to highlight the emergence of key ideas and to show both how recently our understanding of our Galaxy has developed and how radical changes in our picture have occurred virtually overnight as the result of new discoveries. We hope that, if nothing else, this introduction will whet the student's appetite for the major discoveries and conceptual changes yet to come. We hope also that it will show clearly a characteristic of this field—that time and time again the application of a new observational technique or a more powerful instrument has led to major advances in our knowledge.

The second section of this chapter will give a sketch of our Galaxy as a whole, describe its present form, kinematics, and dynamics, and outline a possible process by which it formed. We shall not attempt to justify the statements made (that being the task of the remainder of the book), nor shall we hesitate to introduce terms without detailed definition beyond that implied by the context. Our goal here is to develop a broad view of how various parts of our Galaxy are interrelated, in the hope that this will provide students with a kind of map to orient further study in the following chapters; one might find it helpful to refer back to this section from time to time in what follows.

1-1. THE GROWTH OF OUR CONCEPTION OF OUR GALAXY

As we look out into the nighttime sky, we see an enormous number of stars with a large range of apparent brightnesses. The brightest stars, those easily visible to the naked eye, are fairly uniformly distributed over the entire sky. On a clear, dark night we can see, in addition, the *Milky Way*, a faint band of light, cut by a dark rift, stretching around the sky; the faint glow is the whole impression produced visually by our Galaxy, and the dark band is caused by obscuring dust, which severely limits our ability to see distant parts of the system.

Scientific study of the physical nature of our Galaxy dates from 1610, when Galileo turned his telescope to the Milky Way and discovered that it could be resolved into "innumerable" faint stars. Henceforth it was realized that the diffuse light from the Milky Way could no longer be attributed to some kind of luminous "celestial fluid" but instead originated from vast numbers of unresolved stars; in short, the Milky Way was discovered to be a *stellar system.*

By the middle of the eighteenth century, Thomas Wright and Immanuel Kant had offered a description of our Galaxy as consisting of a *disk* of stars in which the Sun is immersed (see Figure 1-1). On such a picture we would observe nearby stars scattered fairly uniformly over most of the sky, and, in those directions that happen to lie in the plane of the disk, we would also observe the light from a great number of distant stars. Kant further remarked that our Galaxy might not be unique and that many similar disklike systems, which he called *island universes*, might be distributed throughout space at enormous distances from our own system. He even went so far as to suggest that the small, faint, nebulous, elliptical patches on the sky, then called *elliptical nebulae*, were these island universes seen at various angles to the line of sight. But these ideas, however appealing, had no hard core of scientific evidence to support them and were little more than philosophical speculation (albeit correct in this instance).

Further empirical evidence was brought to bear on the problem by William Herschel's work at the end of the eighteenth century. Herschel built telescopes that were giants for their time, and he used them to study both our own Galaxy and other stellar systems. As one of his major scientific

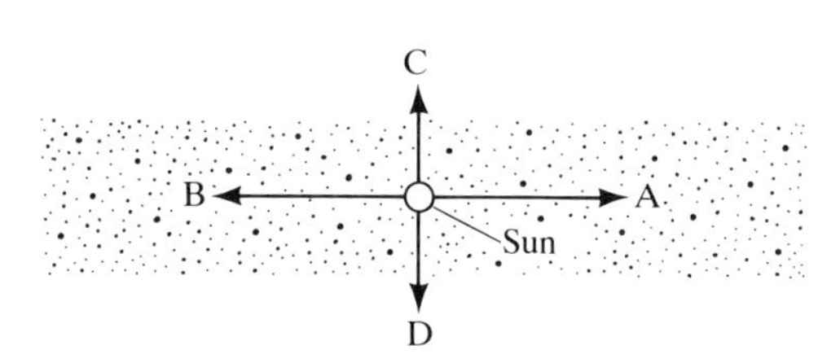

Figure 1-1. Schematic representation of our Galaxy as a disk-shaped stellar system containing the Sun. Note that, from the Sun's position, many more stars would be observed in directions *A* and *B* than in directions *C* and *D*.

research undertakings, he attempted to determine the shape of our Galaxy by a technique he called *star-gauging*. To this end, he counted the number of stars that he could observe to successive limits of apparent brightness in about 700 different regions of the sky. Then, by assuming that all stars have approximately the same absolute brightness and are distributed uniformly in space, that their apparent brightness falls as the inverse square of their distances, and that he could observe to the borders of the system, he used his counts to deduce the relative dimensions of our Galaxy. He concluded that the Sun lies near the center of a flattened, roughly elliptical system that extends about five times farther in the plane of the Milky Way than in the direction perpendicular to that plane.

In addition, Herschel compiled an extensive catalog of nebulae, which was extended to include southern-hemisphere objects by his son, John Herschel. Herschel believed that most of his nebulae were in fact stellar systems similar to our own Galaxy and that, with sufficiently powerful instruments and refined techniques, they would be resolved into individual stars. He also realized that some nebulae, such as the Orion nebula and the *planetary nebulae*, are *not* stellar systems but are composed of "a shining fluid of a nature totally unknown to us." Thus the distinction between true nebulae (glowing gaseous clouds) and unresolved stellar systems had been recognized, but it was not until the spectroscopic work of Huggins and others in the nineteenth century that the nature of any particular unresolved object could be decided unambiguously by observation. Even after Herschel's work, there was still no known way to determine the distances to nebulae and thus to decide whether they were extragalactic "island universes," comparable in size to our own Galaxy but located at great distances from it, or only relatively minor systems contained wholly within our Galaxy.

By the middle of the nineteenth century, yet another interesting observation emerged. Using a gigantic telescope with a mirror 72 inches in diameter (not to be surpassed until the completion of the 100-inch Mount Wilson telescope in 1917), William Parsons, Third Earl of Rosse, discovered that several of Herschel's nebulae showed a *spiral* structure. Furthermore, with the aid of his more powerful instrument, he was able to discern individual stars (more probably clusters of giant emission nebulae) in objects that Herschel was unable to resolve. Parsons' observations added support to the view that the nebulae were indeed "island universes" (galaxies) external to our own, and they led to a new hypothesis: The very shape of the spiral nebulae suggests that they *rotate* about an axis perpendicular to the plane containing the spiral whorl. But verification that spiral galaxies do indeed rotate had to wait until 1914, when V. Slipher presented direct proof by means of spectroscopic observation of the Doppler shifts produced by the rotation of a number of these systems.

At the end of the nineteenth century, the development of astronomical photography opened fabulous new possibilities for astronomical research. With its cumulative light-gathering ability, the photographic plate can reach light levels far too low to be accessible to visual observation; moreover, photographs can record thousands to millions of individual stellar images on a single plate. The time was right for a new, quantitative discussion of the structure of our Galaxy, and this discussion was initiated in the opening years of the twentieth century by H. von Seeliger, J. C. Kapteyn, and P. J. van Rhijn.

Kapteyn conceived a plan to study some 200 *Selected Areas* distributed carefully over the sky, and he enlisted international cooperation among astronomers to obtain the necessary plates, to make star counts, brightness estimates, and spectroscopic classifications, and to measure proper motions and radial velocities. From an analysis of the proper-motion data, Kapteyn and van Rhijn were able to estimate average distances for stars at various apparent brightness levels, and, from an analysis of the star-count data, they inferred the distribution of stars in space. In their analysis, it was again assumed that the apparent brightness of a star falls off as the inverse square of its distance, that is, that interstellar space is completely transparent. As we mentioned before and shall return to later, this was a serious mistake, for in fact there is strong absorption of starlight by *interstellar material* in the galactic plane.

The final picture that emerged from Kapteyn's work (**K1**), (**K2**) is commonly referred to as the *Kapteyn Universe*; it depicts our Galaxy as a flattened spheroidal system of modest size, roughly five times longer in the galactic plane (that is, the plane of the Milky Way) than in the direction perpendicular to the plane. The same qualitative picture had been derived by Herschel, but Kapteyn added a *scale* to the system and quantitative estimates of the variation of star density within it. In the final model, the Sun was located slightly out of the galactic plane at a distance of about 650 parsecs (pc) away from the center of the system (1 pc = 3.26 light-years = 3.1×10^{13} km). The star density decreased uniformly away from the center of the system (see Figure 1-2), dropping to half its central value at a distance of about 800 pc in the galactic plane and 150 pc in the direction of the galactic pole. The corresponding distances for a decrease to 10% of the central density were about 2800 pc and 550 pc respectively, and, for a decrease to 1%, about 8500 pc and 1700 pc.

Certainly one of the most uncomfortable features of the Kapteyn Universe was its strong heliocentric flavor. Why, after all, should the Sun be so near the center of our Galaxy (or of the Universe, if the nebulae are not regarded as separate systems)? Kapteyn fully realized that an alternative explanation of the data was possible: *If* there was an *absorbing medium* in interstellar space, then the light from distant stars would suffer extra dimming; if this dimming were incorrectly interpreted as a distance effect, then the stars would be erroneously placed at too-large distances, leading to an artificial systematic falloff of the star density in all directions away from the observer. Kapteyn had, in fact, spent considerable effort over many years searching carefully for such absorption effects, but he was unable to find any convincing

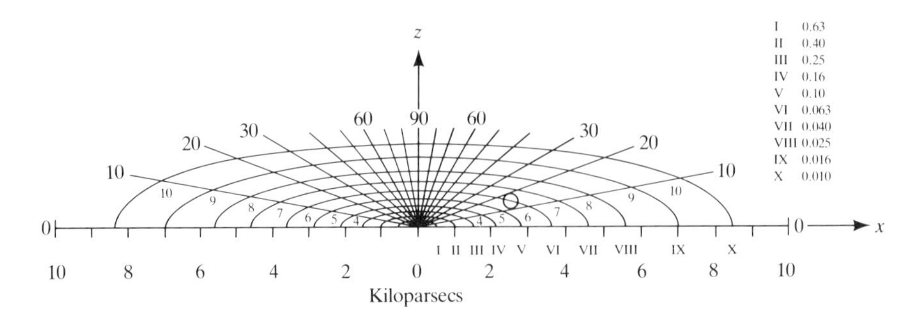

Figure 1-2. The Kapteyn Universe. The x axis lies in the galactic plane and is labeled with distances in kiloparsecs from the center of the system. The z axis points toward the galactic pole; radial lines denote cones of constant galactic latitude. The position of the Sun is indicated by the circle. This representation of the system is divided into ellipsoidal shells with a plane: pole axial ratio of 5:1; curves of constant star density are labeled with Roman numerals, and the densities on these curves, in units of the central density, are given in the table at the right edge of the figure. [From (**K1**), copyright © 1922, University of Chicago Press.]

evidence for them. He therefore concluded that interstellar space is essentially transparent, an opinion shared by virtually all eminent astronomers of the time. Kapteyn was thus forced to the conclusion that our Galaxy is of the form shown in Figure 1-2.

Even before Kapteyn's final model was formally published (his results were widely known in the astronomical community before their publication), it was seriously challenged by a radically different picture developed by H. Shapley in a classic series of papers published between 1915 and 1919 (**S2**), (**S3**), (**S4**), (**S5**), (**S6**). At the Mount Wilson Observatory, Shapley had undertaken a detailed observational study of *globular clusters*, which are compact spherical systems containing from 10^5 to 10^6 stars. Because of their great brightness and distinctive appearance, these objects can be identified and observed at very great distances from the Sun. Furthermore, many of these systems are found at large distances from the galactic plane, and hence their light is not dimmed much by the absorbing material in the plane. Shapley pointed out that, while globular clusters are distributed uniformly above and below the galactic plane, they are not uniformly distributed in longitude around the plane; instead they show a marked concentration toward the direction of the great star clouds in Sagittarius (see Figure 1-3). Shapley argued that such massive systems must be a major structural element of our Galaxy and that it should be reasonable to suppose them to be distributed symmetrically around the galactic center. If that is the case, then their asymmetric apparent distribution must be interpreted as implying that the Sun is *not* located near the center of our Galaxy but is actually quite far from the center. Using distances estimated from the apparent brightness

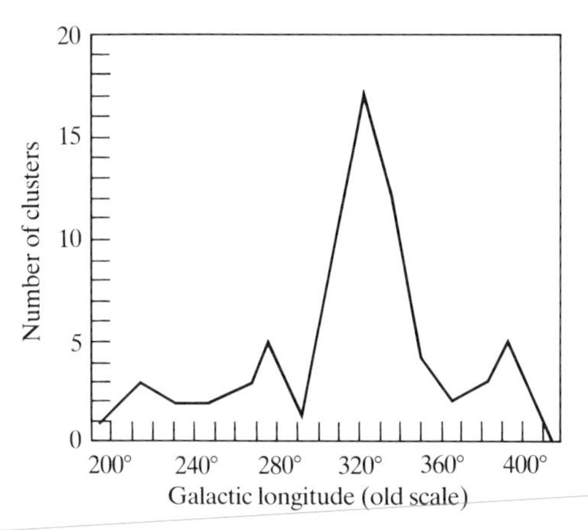

Figure 1-3. Distribution of globular clusters in galactic longitude as observed by Shapley. Note that many more clusters are observed near longitude 325°, in the direction of the Sagittarius star clouds. The position of the galactic center is at about 327° on the old longitude scale used for the abscissa. [From (**S2**), copyright © 1918, University of Chicago Press.]

of variable stars (of known intrinsic brightness) observed in the clusters and from the size and brightness of each cluster as a whole, Shapley estimated that the Sun must be about 15,000 pc or 15 kiloparsecs (kpc) from the galactic center (see Figure 1-4). Today, our best estimate of the Sun's distance from the galactic center is about 9 kpc.

Shapley's bold conclusion that the Sun lies far out toward the edge of our Galaxy has been proved correct by all subsequent investigations. The revolutionary change it wrought in our picture of our Galaxy has often been likened to the Copernican shift from a geocentric to a heliocentric picture of the solar system. Shapley also estimated that the most distant clusters in his sample were of the order of 70 kpc away, and he concluded that the full diameter of the outermost reaches of our Galaxy is of the order of 100 kpc, about a factor of 10 larger than the Kapteyn Universe! Actually, Shapley's estimate of the size of our Galaxy was somewhat too large because he neglected interstellar absorption, as had Kapteyn, and hence he overestimated the distances of some of the clusters. Today we would estimate a diameter of the order of 50 kpc for the disk of our Galaxy.

In retrospect, Shapley was extraordinarily lucky in his choice of objects, the majority of which happen to lie far from the galactic plane and hence suffer little from the effects of interstellar absorption. Shapley himself pointed out that he could find no clusters within ± 1300 pc of the galactic plane. To explain the absence of clusters in the plane, he argued that they would be disrupted by strong gravitational forces acting there. We know today that his failure to observe such clusters is to be explained by the presence of strong interstellar absorption in the galactic disk, which prevents distant clusters in the plane from being observed. Similarly, the spiral nebulae (external galaxies) were found in great numbers near the galactic poles but not near the galactic plane. This irregular zone within $\pm 10°$ latitude above and below the galactic plane is called the *zone of avoidance*. Had these two

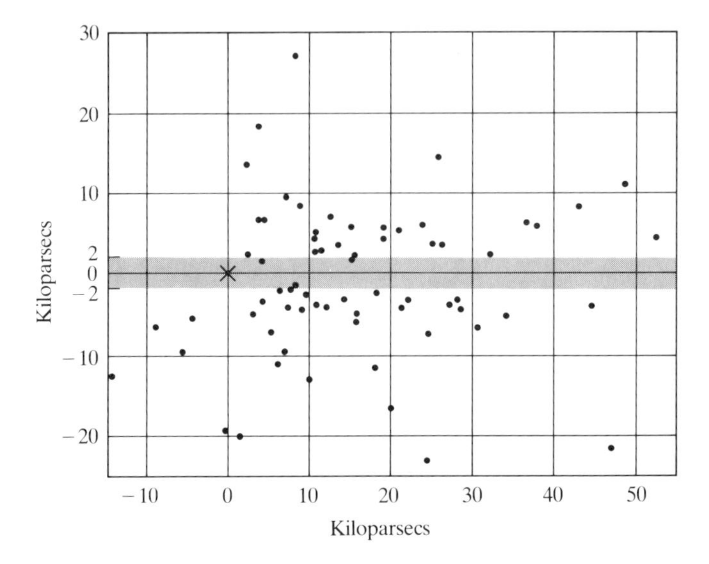

Figure 1-4. Cross section of our Galaxy through the Sun and the galactic center, showing projections of the positions of the globular clusters observed by Shapley. The position of the Sun is marked by a cross. The center of the system of globular clusters lies about 13,000 pc from the position of the Sun. Note that no globular clusters are found within ± 1300 pc of the galactic plane. [From (**S2**), copyright © 1918, University of Chicago Press.]

types of observations been interpreted properly, they would have demonstrated convincingly why the Kapteyn model was wrong. As it was, Shapley's conclusion that there was no interstellar absorption simply served to affirm Kapteyn's model and left no alternative explanation for Kapteyn's observations. These two conflicting models stood in stark contrast to one another, and astronomers were forced to choose one or the other in the face of discordant evidence.

Not all astronomers accepted Shapley's ideas; indeed, it is probably accurate to say that the majority supported Kapteyn's star-count model. One of Shapley's strongest critics was H. D. Curtis of the Lick Observatory, a leader in the study of spiral nebulae. Curtis was convinced that the spirals were systems external to our Galaxy, but he believed our Galaxy to be a system the size of the Kapteyn Universe. In April 1920, Curtis and Shapley met at the National Academy of Sciences in what has since been called astronomy's *great debate.* On this occasion, two primary problems were discussed: (1) the size of our Galaxy and the distance scale within it, and (2) the distances of the spiral nebulae and, by implication, the question of whether or not they are extragalactic systems.

Shapley's views and reasoning concerning the first issue have already been described. Curtis attacked Shapley's picture of our Galaxy on the basis of rather technical arguments concerning distance determinations. With the advantage of hindsight, we know today that several of Curtis' arguments were faulty, being based in part on the incorrect assumption that the red stars observed by Shapley were dwarfs of the kind found in the solar neighborhood and hence intrinsically faint, whereas in actuality they were intrinsically bright giants. On balance, Shapley's position on the first question posed for the debate was basically correct.

Concerning the second question, Curtis advanced the view that the spiral nebulae are galaxies like our own, lying at distances ranging from 150 kpc for the Great Nebula in Andromeda (the spiral galaxy M31) to 3000 kpc for the most distant systems; Shapley held the opinion that the spirals are relatively near and are not comparable in size to our Galaxy. It is interesting that Shapley was as conservative (and wrong) on the second question as he was radical (and correct) on the first. To support his view, Curtis offered several cogent arguments:

1. The average apparent brightness of novae in the spiral nebulae, when compared to novae in our own Galaxy, indicates distances of 150 kpc (or more) for these systems.
2. At a distance of 150 kpc, the Andromeda nebula would be comparable in size to Kapteyn's estimate of the size of our Galaxy.
3. The radial velocities of the spiral nebulae are much larger than those of any objects known to be associated with our Galaxy, hence the spirals must be separate systems because they would not be bound dynamically by a system of the size of the Kapteyn Universe. Moreover, the spirals do not show measurable proper motions across the line of sight, despite their great radial velocities. They must therefore be very distant, unless we are to assume that their velocity vectors all happen to point radially away from the Sun, which is very unlikely.
4. A spiral, when viewed edge-on, generally shows a band of absorbing material in its disk projected against its bright nucleus (see Figure 1-7). By analogy, our own Galaxy should possess such a band, which would explain the zone of avoidance.

On his side, Shapley had two strong arguments concerning the second question of the debate:

1. If the spirals are of the same size as our Galaxy (100 kpc, according to Shapley), then a system such as the Andromeda nebula would be at so large a distance from our Galaxy that its novae would be intrinsically much brighter than those in our Galaxy.

2. Measurements of proper motions in spiral nebulae by A. van Maanen indicated that the spirals rotate with an angular motion of 0".02 per year. For any plausible rotation velocity, this implies that the spiral nebulae must be close. For example, if the rotation velocities are 200 km s^{-1}, the nebulae must be within 50 kpc (*inside* Shapley's model of our Galaxy). At the distances required by the view that the spirals are separate galaxies, the velocities implied by van Maanen's proper motions would be ridiculously large.

In retrospect, we know that Curtis' evidence was more correct than Shapley's, although his arguments were not flawless. Thus, in the conflicting assertions about the novae, Curtis had underestimated their intrinsic brightness, and Shapley had overestimated the size of our Galaxy. In his second point, Curtis had (ironically) underestimated the distance to the Andromeda galaxy by about a factor of five. He was right that it is indeed external to our Galaxy, but at its correct distance its size is comparable to Shapley's estimate of our Galaxy's size and is much larger than the Kapteyn Universe. Curtis was basically correct in his third argument, and in the fourth he was remarkably prescient. Yet, there is a logical dilemma in his last argument, for if there *is* heavy absorption in the galactic plane, then the reasoning leading to the Kapteyn Universe can no longer be supported.

The argument that probably had the greatest influence in forming the opinion of the astronomical community at large was Shapley's point about van Maanen's proper motion measurements. Van Maanen had an excellent reputation as a meticulous observer and enjoyed the general respect of the community; hence his results were almost universally accepted as correct. Unfortunately, they were purely and simply spurious—the rotational proper motions of spiral galaxies are in fact too small to measure, even by the most accurate present-day techniques. But definitive proof that van Maanen's work was in error did not emerge for another fifteen years after the Curtis–Shapley debate. Even today we cannot reconstruct unequivocally the causes of the errors in his results [see (**B3**) for an interesting analysis of this matter]. The great debate itself settled neither of the questions it addressed, and opinions remained sharply divided. It did, however, bring the issues into sharp focus, and it stimulated a great deal of thought.

The question of the nature of the spiral nebulae was settled within five years by E. Hubble. By 1923, Hubble, working with the recently completed 100-inch telescope at Mount Wilson, had found that the outer portions of the disks of two nearby spiral galaxies (M31 and M33) could be resolved into swarms of images indistinguishable from those of stars. If these were assumed to be stars of an intrinsic brightness comparable to that of the most luminous stars in our Galaxy, then these systems had to be at enormous distances and were certainly extragalactic. The evidence was not conclusive, however, because the images could have been those of star clusters (which would nevertheless have implied still larger distances) or of dense knots of nebulosity. (Some "stars" identified by Hubble were, in fact, later found to be gigantic H II regions—ionized hydrogen nebulae.) The main point is that there was at first no definitive way to establish the intrinsic brightnesses of the objects that Hubble had detected. The situation changed radically in late 1923 when Hubble discovered *Cepheid variables* in M31. The intrinsic brightness of these stars can be estimated accurately from their distinctive light variations if it is assumed that they are identical to galactic Cepheids. In this way, Hubble obtained [see **(H2)**, **(H3)**] a value of about 300 kpc for the distance to M31 and M33. In a masterly sequence of papers [see, for example, **(H4)**, **(H5)**, **(H6)**] that culminated in his book *The Realm of the Nebulae* **(H8)**, Hubble established once and for all the true nature of spiral galaxies.

The question of the nature of our Galaxy was settled almost simultaneously by considerations of its kinematics and dynamics. In 1926, B. Lindblad affirmed Shapley's location for the galactic center and developed a mathematical model for the rotation of our Galaxy about an axis through its center **(L1)**, **(L2)**, **(L3)**. He proposed that our Galaxy might consist of a number of subsystems, each of which was symmetric about the rotation axis of the whole system and rotated about this axis with some characteristic speed. Each subsystem had a characteristic degree of flattening and internal velocity dispersion, the most slowly rotating systems being the least flattened and having the largest internal (nonrotational) velocity dispersions.

In support of a large size for our Galaxy, Lindblad advanced an independent and telling argument against the Kapteyn Universe. He pointed out that the total mass calculated from Kapteyn's model produced a gravitational field too weak to retain the globular clusters and RR Lyrae variables as gravitationally bound members of the system. These objects had been observed to have velocities of about 250 km s^{-1} with respect to the Sun, which is much larger than the escape velocity of the Kapteyn Universe. But inasmuch as both RR Lyrae stars and globular clusters are, in fact, found in large numbers, one concludes that either they are formed at a very rapid rate to replace those that escape, or they are really permanent members of our Galaxy and are bound by stronger gravitational forces than those predicted from the Kapteyn model. Because the globular clusters are themselves so massive, it seemed extremely unlikely that they could be formed quickly enough to replace their loss. Lindblad was therefore led to the latter alternative, which implies a much larger and more massive Galaxy than that envisaged by Kapteyn. (It is interesting to note that recent developments in our understanding of galactic dynamics imply a still larger and much more massive Galaxy—by perhaps as much as a factor of ten above current estimates. These theoretical arguments are supported by the existence of at least one globular cluster that would be escaping from our Galaxy if conventional estimates of its mass are correct.)

Lindblad further proposed that the globular clusters might constitute a subsystem nearly at rest (that is, not rotating) relative to our Galaxy as a

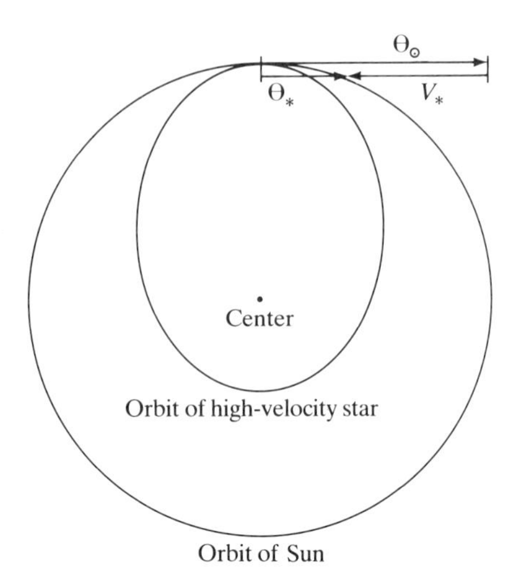

Figure 1-5. The Sun moves on a nearly circular orbit with a velocity of revolution $\Theta_\odot$ around the galactic center. The velocity of revolution Θ_* around the galactic center of a "high-velocity" star, such as shown in the figure, is much smaller. Therefore, the high-velocity star moves on an eccentric orbit around the center, and when it is at the Sun's distance from the center, it has a large negative tangential velocity V_* with respect to the Sun.

whole and that the Sun travels on a circular orbit around the galactic center with a velocity of the order of 200 to 300 km s^{-1}. Because a typical value for the observed dispersion in stellar velocities in the solar neighborhood is only 30 km s^{-1} (excluding the high-velocity stars), he argued that all the low-velocity stars had essentially the same energy of motion as the Sun and that they also traveled on nearly circular orbits around the galactic center.

In 1927 and 1928, J. H. Oort extended Lindblad's ideas and showed [see (**O1**), (**O2**)] that several predictions directly verifiable by observation were possible. First, Oort pointed out that a number of features in the velocity distribution of the so-called *high-velocity stars*, which were so puzzling when viewed in terms of Kapteyn's model, could be understood quite naturally on the basis of Lindblad's model. Oort noted that stars moving at speeds less than a certain characteristic value relative to the Sun have a symmetric velocity distribution, whereas at higher relative speeds the distribution suddenly becomes quite asymmetric. The nature of the asymmetry is such that the high-velocity stars tend to lag behind the Sun in a direction that is perpendicular to the direction of the galactic center as identified by Shapley and Lindblad. This is precisely what would be expected if the Sun belongs to a rapidly rotating system and moves on a circular orbit around the galactic center while the high-velocity stars belong to a system that rotates much more slowly around the center. Such stars would lag behind the sun in the solar neighborhood, and, because they do not have sufficiently large tangential velocities to maintain them on circular orbits against the gravitational forces exerted on them by our Galaxy, their orbits carry them in toward the galactic center (see Figure 1-5). Relative to the galactic center, these stars are not fundamentally high-velocity stars at all; rather, as a group, they are *low rotation-velocity* (or low angular-momentum) objects. Nor are they stars that are escaping rapidly from the system; rather, many of them are actually more tightly bound to our Galaxy than the Sun is. As we shall see in later chapters, some of these stars have almost no rotation-velocity component, and therefore they move on almost radial orbits with respect to the galactic center. Some even move on *retrograde* orbits; that is, they revolve backward around the center as compared to the direction of revolution of the Sun.

Next, Oort showed how a rough estimate of the total mass of our Galaxy and its spatial distribution relative to the galactic center could be made. The very abrupt change in the nature of the velocity distribution at a speed of about 65 km s^{-1} relative to the Sun and, in particular, the fact that no stars were observed to have relative speeds larger than this value in the direction of galactic rotation suggested to Oort that such stars might escape from our Galaxy. If we take a rotation velocity $\Theta_\odot$ of about 250 km s^{-1} (see Chapter 6), then Oort's conjecture implies that the escape velocity Θ_{esc} is about 315 km s^{-1}. From the size of Θ_{esc}, one can obtain an estimate of the total mass of our Galaxy interior to the Sun. The result turns out to be much larger than can be explained with the Kapteyn model, but it is compatible with the larger system conceived by Shapley and Lindblad. As we shall see in later chapters, Oort's estimate of the escape velocity was too low, and our Galaxy is more massive than he estimated. This fact, of course, merely strengthens his conclusion about the inadequacy of the mass contained in the Kapteyn Universe.

Finally, Oort developed a complete kinematical theory of the *differential rotation* (that is, faster angular rotation near the center and slower near the edge) of our Galaxy, and its predictions were consistent with observations. This theory proved unequivocally the basic correctness of the picture that our Galaxy is a large, rotating system with the Sun located far from its center.

Thus, by 1927, the Kapteyn Universe had become history, and our Galaxy had become a peer of the spiral nebulae. The great debate was over. Yet, two nagging questions remained: (1) What was to be made of van Maanen's evidence for rotational proper motions in the spiral galaxies? (2) Why did the star-count analyses give such an erroneous picture of our Galaxy?

Most astronomers had long been convinced of the distances to the spiral galaxies by Hubble's evidence. Consequently, they proceeded to forget or ignore van Maanen's earlier work. Van Maanen's measurements still stood virtually unchallenged in the literature until 1935, when Hubble published a paper (**H7**) containing an independent set of proper-motion measurements of the spiral galaxies M33, M51, M81, and M101 made by himself, Baade, and Nicholson. The new measurements were based on many of the plates used earlier by van Maanen as well as on new plates spanning a much larger time interval than that embraced by van Maanen's data. If the motions claimed by van Maanen had been real, they would have been even more readily detected in the new material. The results were completely negative; van Maanen's measurements had been spurious. Thus was demolished one of the primary objections against the position that the spiral nebulae are extragalactic.

As we mentioned before, most astronomers working on the problem of the structure of our Galaxy (including Kapteyn himself) were fully aware that, if there was significant absorption of starlight in interstellar space, then the stellar space densities derived from star counts would show a spurious decline at large distances. There had long been evidence for clouds of absorbing material in space; for example, the dark rift in the Milky Way is plainly visible to the naked eye. Numerous other dark patches and lanes showed clearly on photographs of the Milky Way obtained early in the twentieth century by E. E. Barnard. These areas on the sky seem to be totally devoid of stars, even when examined with the largest available telescopes. It was early recognized that it is exceedingly improbable that these dark areas could be real gaps in the stellar system, for gaps would imply the existence of numerous tunnel-shaped openings in our Galaxy, all of which are aligned along lines of sight passing through the solar system. By 1920, there was no doubt that large dark clouds of absorbing material (*interstellar dust*) exist in our Galaxy. In addition, there was evidence for the presence of *interstellar gas*. Spectroscopic observations of bright nebulae by W. Huggins in the late nineteenth century showed that some of them are luminous bodies of gas. In 1904, J. Hartmann discovered that the lines of Ca II in the spectrum of the double star δ Ori are *stationary*; that is, they have a constant wavelength and do not periodically oscillate as do all other lines in the spectrum. It was soon realized that the Ca II lines must arise in diffuse absorbing clouds of gas located along the line of sight to the star.

Nonetheless, at the time of the Curtis–Shapley debate, there was still no conclusive evidence for the existence of a pervasive general *absorbing medium*, extending throughout the entire system, which would be capable of dimming the light from all stars observed. The strong hint provided by the zone of avoidance continued to be overlooked. However, its significance was recognized by J. Oort in 1927 (**O1**). Astronomers had realized for some time that any likely scattering process from small interstellar dust grains would be *selective* (would tend to dim light more at shorter wavelengths than at longer wavelengths). A *reddening* of starlight would thus be produced along with absorption, just as Rayleigh scattering in the Earth's atmosphere affects sunlight. Such effects had, in fact, been sought by Kapteyn, but without success. His lack of success was probably caused by inadequate precision in photometry and insufficient information about the intrinsic brightnesses and colors of stars in his sample.

Even though strong evidence for the existence of a general absorption in the plane of our Galaxy was obtained in 1929 from studies of the distribution of early-type stars by C. Schalen, the first absolutely irrefutable proof was published in 1930 by R. J. Trumpler (**T1**). Trumpler made an exhaustive study of *galactic clusters* (star clusters found in the galactic plane), obtaining spectral types and making brightness and color measurements for cluster members. By plotting diagrams of apparent brightness versus spectral type for cluster stars, and by comparing these with the then-known correlation of intrinsic brightness versus spectral type for nearby stars (the *Hertzsprung-Russell diagram*), he could estimate cluster distances on the assumption that there was no interstellar absorption. At the same time, he measured the angular diameters of the clusters. If we assume that all clusters have nearly the same *linear* diameters, then their *angular* diameters should be inversely proportional to their distances from us.

Trumpler found that there was a systematic discrepancy between the distances inferred from the clusters' Hertzsprung-Russell diagrams and their angular diameters. The angular sizes of the fainter (and therefore presumably more distant) clusters were too large as compared with their predicted sizes from the hypothesis of equal linear sizes. Discarding the unlikely hypothesis that clusters grow progressively larger with increasing distance in all directions from the Sun, Trumpler argued instead that the angular diameters gave a true indication of cluster distances. He proposed that the effect he had discovered was to be explained by a progressive dimming of the light of the distant clusters (which would make them appear to be yet more distant) by the absorption of starlight in the interstellar medium in the galactic plane. He estimated that this absorption produces a dimming of 0.7 stellar magnitudes per kiloparsec (that is, a reduction in intensity by about a factor of two per kiloparsec, over and above the inverse-square falloff) in all directions in the galactic plane. To clinch the argument, he showed conclusively that the colors of stars of a given spectral type tended to become progressively redder with increasing distance, an effect that had been sought earlier by Kapteyn. We shall see in later chapters that this interstellar reddening provides the basis for methods of determining the amount of interstellar absorption that the light from any particular star has suffered. The amount of general absorption found by Trumpler was sufficient to explain the (spurious) stellar density falloff in the Kapteyn Universe, and the star-count data were thus finally reconciled with Shapley's picture of our Galaxy.

W. Baade opened a whole new aspect of the study of galaxies in 1944 when he resolved the nucleus of the spiral galaxy M31, its two companions M32 and NGC 205, and the elliptical galaxies NGC 147 and NGC 185 into stars on red-sensitive plates taken with the 100-inch Mount Wilson telescope during the wartime blackout of Los Angeles (**B1**), (**B2**). Baade realized that the brightest stars in these spheroidal systems must be *red giants* and hence of a totally different character from the luminous blue supergiants found within spiral arms. He therefore suggested that stars in a galaxy could be categorized into two distinctive *stellar populations*. Baade described *Population I* as those objects closely associated with *spiral arms*. Some examples are luminous, young hot stars (spectral types O and B), Cepheid variables, interstellar dust lanes, ionized hydrogen regions (emission nebulae), and clusters similar to the galactic clusters in our own Galaxy. The arrays produced by these stars in a Hertzsprung-Russell diagram (or in a plot of intrinsic brightness versus intrinsic color, called a *color-magnitude diagram*) resemble those found by Trumpler for galactic clusters. Baade described *Population II* as being composed of those objects found in the *spheroidal components* of the galaxies—that is, in the nuclear bulge, in the halo, and in globular clusters. These stars have a Hertzsprung-Russell diagram quite different from diagrams of the galactic clusters, but one that is similar to Shapley's diagrams for the globular clusters in our Galaxy.

The notion of stellar populations has proved to be very fruitful in a variety of contexts, and it has led to important advances in our understanding of the structure and evolution of galaxies. For example, after Baade had delineated the Population I objects most closely associated with spiral structure, those objects could be selected as tracers of spiral structure in our Galaxy. In the early 1950s, Morgan and his associates carried out several studies **(M1)**, **(M2)** of young clusters and associations and of the most luminous stars known to be spiral-arm tracers. They thus were able to map the spiral arms of our Galaxy in the neighborhood of the Sun.

The concept of stellar populations also provided key insights into the scheme of stellar evolution. In the two decades following Baade's discovery, there was a rapid development of both observational and theoretical methods for studying stellar structure, evolution, and star formation. On the observational side, the postwar availability of sensitive photomultipliers revolutionized astronomical photometry, and it became possible to construct color-magnitude diagrams of hitherto unachievable accuracy for star clusters. On the theoretical side, both the rapid growth of theoretical insight into the basic physics of energy generation and energy transport in stars and the availability of high-speed electronic computers made feasible, for the first time, the detailed computation of evolution tracks for individual stars in a color-magnitude diagram. Thus a physical interpretation of observed cluster color-magnitude diagrams became possible.

The analysis of cluster color-magnitude diagrams showed that they could be understood in terms of the *age* and *composition* of stars in a cluster. It was found that the Population II stars (for example, globular-cluster stars) are all old, having ages nearly equal to the estimated age of the Universe itself. In contrast, Population I objects show a wide range of ages, some galactic clusters being almost as old as the globular clusters, whereas others are forming new stars at the present time. The key composition parameter was found to be the abundance of elements heavier than hydrogen and helium. These elements are customarily called *metals* by astronomers, and, despite the fact that it is a misnomer, we shall employ that term in this book.

Detailed spectroscopic analyses showed that Population I stars have metal abundances similar to those in the Sun, and Population II stars in the galactic halo (globular clusters and halo subdwarfs) are metal deficient by about a factor of one hundred relative to the Sun. Because the halo Population II stars are found to be so old, it follows that their chemical composition must nearly reflect that of the primordial material formed in the initial explosive creation event of the Universe (the *big bang*) implied by Hubble's discovery in 1929 of the large-scale expansion of the Universe. [We shall not attempt to describe the historical development of the concept of the *expanding Universe* here; see (**B3**) or (**S9**).] Because the metal abundance of the material now forming into stars is so much greater than that of very old stars, it was realized that elements heavier than helium must have been created by *nucleosynthesis* in the interiors of stars and then recycled back into the interstellar medium. Plausible mechanisms for recycling the material are provided by the observed existence of both quiescent stellar mass-loss (stellar winds) and supernova explosions. Work to enlarge our knowledge of stellar evolution and stellar populations and their relation to the structure and evolution of our Galaxy as a whole proceeds apace today.

The year 1944 also marked the beginning of a different set of developments of profound importance to our understanding of our Galaxy and of galaxies in general. In that year, H. C. van de Hulst predicted the existence of 21-cm radio emission in a spectral line of neutral hydrogen in the interstellar medium. It was known from the work of K. G. Jansky in 1932 that our Galaxy is a source of *continuum radio-wave emission*. By 1940, G. Reber showed that the maximum emission came from the galactic center, and in 1944 he reported two other strong sources in Cassiopeia and Cygnus. (The former was later identified as a *supernova remnant*, and the latter as an *extragalactic radio source*.) The *line radiation* predicted by van de Hulst offered two important advantages—it would be observable from our entire Galaxy, and it would make possible radial-velocity measurements through Doppler shifts of the wavelength of the line.

In 1951, the 21-cm line was actually observed by H. I. Ewen and E. M. Purcell at Harvard, by W. N. Christiansen in Sydney, and by C. A. Muller and J. H. Oort in the Netherlands. Over the next fifteen years, groups in the Netherlands and Australia made detailed measurements of the rotation speed of the hydrogen around the galactic center as a function of distance from the center (key data for dynamical studies) and of the distribution of neutral atomic hydrogen (H I) in the galactic plane. The continuing development of radio astronomy has revealed many other important features of our Galaxy, for example, nonthermal emission from relativistic particles and surprisingly complex molecules in the interstellar medium.

A major advance occurred with the discovery of the *cosmic microwave background* in 1965 by A. A. Penzias and R. W. Wilson (**P1**), which provided conclusive evidence favoring a big-bang model of the Universe over rival theories. This radiation was interpreted by R. Dicke, P. J. Peebles, P. Roll, and D. Wilkinson (**D1**) as being the relic of the fireball accompanying the initial cosmic explosion. Its existence had actually been predicted in the late 1940s by G. Gamow and his collaborators, but it was not observable with techniques then available, and their prediction had been virtually forgotten.

The study of our Galaxy and external galaxies by radio observations is made difficult by the relatively low resolution afforded by even the largest radio dishes, because of the long wavelengths of radio waves. This difficulty led to the development of ingenious *interferometer* techniques to provide higher resolution; these techniques in turn have been elaborated into *aperture synthesis* techniques, which use arrays of moderate-sized radio telescopes arranged in such a way as to provide high sensitivity and resolution. At present, extremely interesting information about the distribution and kinematics of hydrogen in external galaxies is being gathered by several such systems; noteworthy examples are the Westerbork array in the Netherlands and the Cambridge array in England.

The current major development is the recently built Very Large Array, a part of the U.S. National Radio Astronomy Observatory. This complex, which occupies hundreds of square kilometers in New Mexico, provides hitherto unachievable clarity and sensitivity in our view of galaxies at radio wavelengths, and it will undoubtedly produce results of the most far-reaching significance to galactic astronomy.

In optical studies of galaxies, great strides have been made through the development of *image tubes* and two-dimensional *solid-state detectors* of various kinds. These instruments offer much-improved sensitivity and information-gathering capacity for observations of faint objects, and with them we can obtain galaxian spectra for radial-velocity measurements and astrophysical analyses as well as do precision surface photometry for much fainter objects than we could in the past. A major improvement in our observational ability will result when the *Space Telescope* (2.4 m in diameter) is put into orbit. In addition to permitting observations in the ultraviolet, this instrument will yield a factor-of-ten increase in resolving power over ground-based observations, and it will thus offer unprecedentedly sharp views of galaxies. This improvement in resolving power will allow the discrimination of point sources (for example, stars) at light levels a factor of one hundred fainter (and hence to a factor-of-ten greater distance) than is presently possible. This development should revolutionize our knowledge of the stellar content of star clusters and galaxies and be of great value to studies of the central regions of galaxies.

The successful operation of the Space Telescope will, however, by no means supersede the use of large ground-based telescopes for extragalactic astronomy because, for studies of the faint outer regions of galaxies, it is more important to have a large light-collecting area than to have high spatial resolving power. Unfortunately, about half of the light that makes the dark night sky so depressingly bright (from the point of view of the extragalactic astronomer) is extraterrestrial in origin and therefore will plague observations made with the Space Telescope. Nevertheless, all branches of extragalactic astronomy will certainly receive important impetus from Space Telescope observations, complemented by ground-based observations using a new generation of very large aperture *multiple mirror telescopes*, such as the one recently commissioned in Arizona.

Finally, it is worth mentioning that, at the present time, great strides are also being made in developing an understanding of the way galaxies are formed and how they evolve, both as individuals and in clusters. A particularly interesting issue that has emerged from dynamical studies of galaxies is whether they do or do not contain large amounts of essentially nonluminous matter, which, according to some lines of evidence, may even constitute the bulk of the mass in galaxies. This question is still unsettled [see (**O3**) and (**B5**)], and, in terms of its importance and intrinsic interest, it could well form the basis for a contemporary version of the great debate. When a definitive answer finally emerges, it may herald a new era in our conception of the nature of our Galaxy and of the distribution of matter in the Universe.

1-2. A PORTRAIT OF OUR GALAXY

The basic conclusion reached in §1-1 is that our Galaxy is a spiral galaxy. Thus, from a suitable vantage point in space outside the system, it would resemble many of the other spiral galaxies we can observe at large distances. Much of the evidence available today indicates that our Galaxy can be considered to be approximately of type Sb in the galaxian classification system developed by Hubble (see §5-1). We would thus suppose that it should look something like galaxies of that type (for example, M31, NGC 2841, NGC 3031, NGC 4565, and NGC 7331) as shown in the *Hubble Atlas* (**S1**) and in Figure 5-5; this conjecture is supported by comparing Figures 1-6 and 1-7. In Figure 1-6, we have a full-sky photograph of our Galaxy taken with the Henyey–Greenstein camera by Code and Houck (**C1**) in infrared light, which penetrates the interstellar absorption much better than light of shorter wavelengths. In Figure 1-7, we have a photograph of the edge-on Sb galaxy NGC 891. In both cases, we can clearly see the prominent nuclear bulge surrounded by an extensive, flat, dust-laden disk. The resemblance is striking when allowance is made for the fact that our Galaxy is being observed from a position within the disk of the system, whereas NGC 891 is being viewed from a position completely outside the system and in a different wave band. Of course, the fact that we cannot obtain the same type of overall picture of our Galaxy as is used to classify other systems makes any attempt at classification of our Galaxy very uncertain, but on the basis of present evidence, it seems likely that our Galaxy is a spiral system of type Sb or Sc.

In the broadest terms, our Galaxy comprises two main structural elements—a *spheroidal component* and a *disk*. Each of these contains quite different and characteristic stellar and nonstellar populations, and they have different compositions, kinematic and dynamic properties, and evolutionary histories. The spheroidal component can be thought of as being an approximately axially symmetric system that, for expository convenience, can be

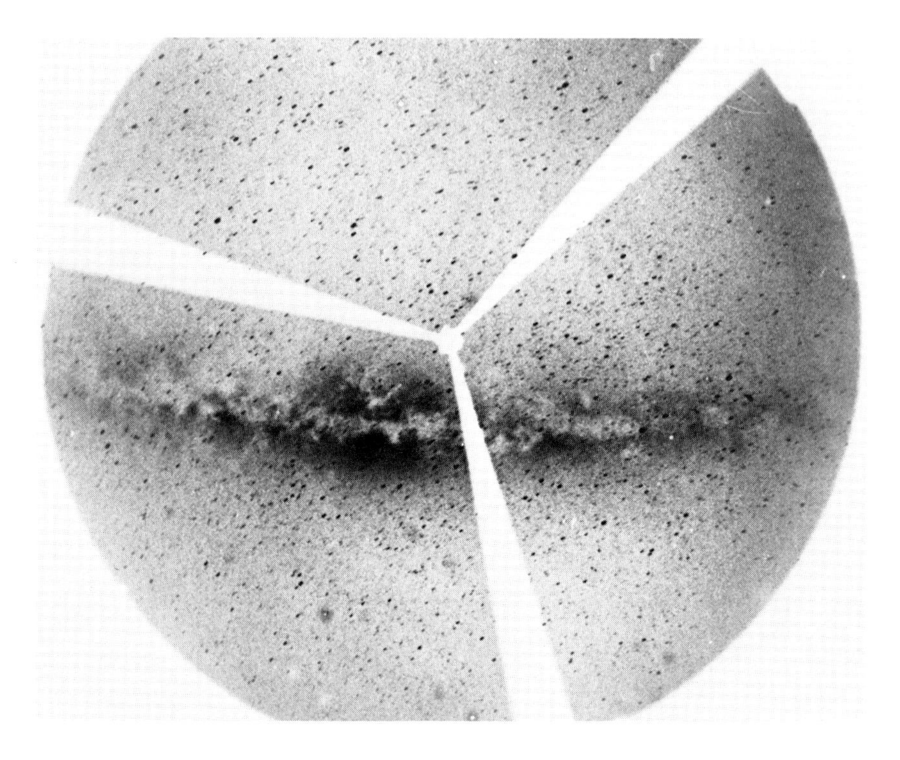

Figure 1-6. Full-sky photograph of our Galaxy taken by Code and Houck with the Henyey-Greenstein camera on infrared film (**C1**). Note the nuclear bulge, disk, and dust lanes; compare with the photograph of NGC 891 shown in Figure 1-7. [Yerkes Observatory photograph, courtesy Yerkes Observatory.]

divided into several subcomponents, ranging from the *nucleus* (the innermost part of our Galaxy, having a size of ~ 3 pc) to the *bulge* (of the order of 3 kpc in radius), through an *intermediate* component, out into an extensive *halo* that may extend to radii of 30 kpc or more. Each of these subsystems has a different degree of flattening as well as physical and chemical properties that vary in a fairly systematic way with increasing radius. These subsystems are often assumed to be axially symmetric, although there is now some evidence that this picture may be oversimplified. The disk is an extremely thin (about 200 pc thick), flat system extending in the galactic plane from the center to radii of perhaps 25 or 30 kpc. The Sun is located in the disk at a distance of about 9 kpc from the galactic center. An impression of the spatial characteristics of the disk–bulge–halo systems can be obtained from Figure 1-8, which compares a deep photograph of NGC 3115, showing the halo and globular cluster system, with a less deeply exposed photograph, showing mainly the bulge and vestiges of a disk. (We should point out that this galaxy

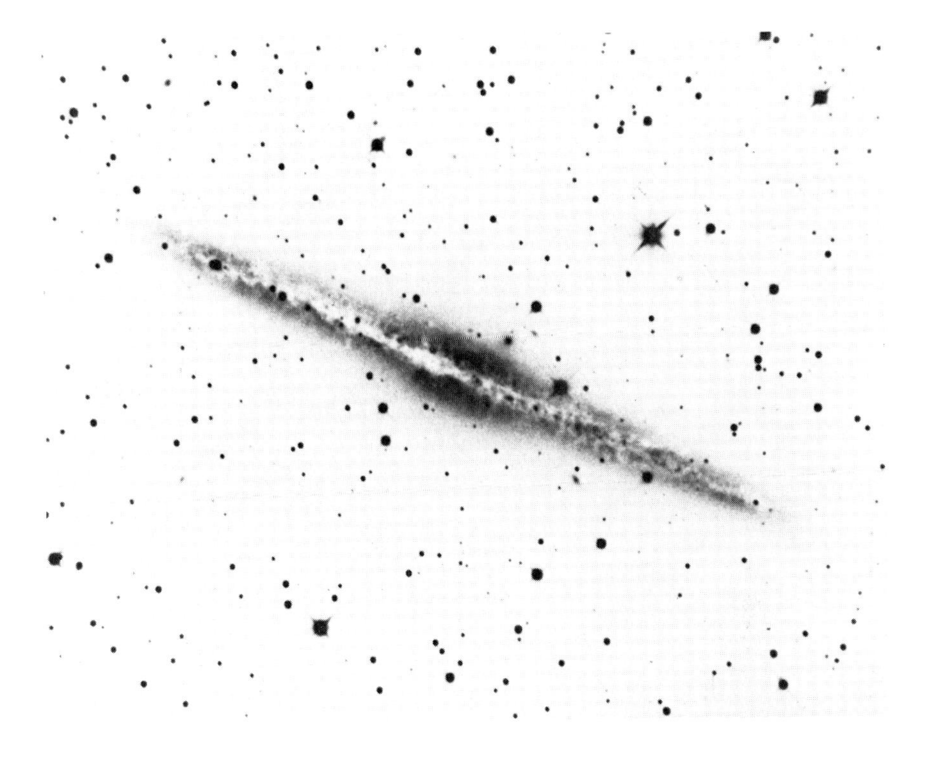

Figure 1-7. Photograph of the edge-on Sb spiral galaxy NGC 891 taken with the 60-inch telescope of the Mount Wilson Observatory on November 23 and 24, 1916. Note the striking resemblance of this object to our Galaxy as shown in Figure 1-6. [Yerkes Observatory photograph, courtesy Yerkes Observatory.]

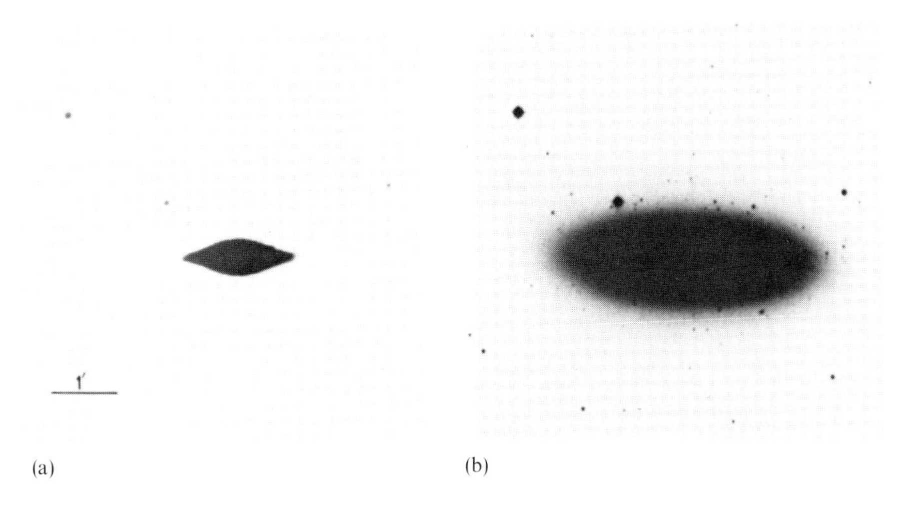

Figure 1-8. Comparison of (a) shallow and (b) deep photographs of the S0 galaxy NGC 3115 taken with the Kitt Peak National Observatory 4-meter telescope in *U* and *B* light, respectively. The nuclear bulge and vestigal disk are shown in part (a); part (b) shows the halo component and the system of globular clusters. [From (**S8**), by permission. Copyright © 1977 by the American Astronomical Society.]

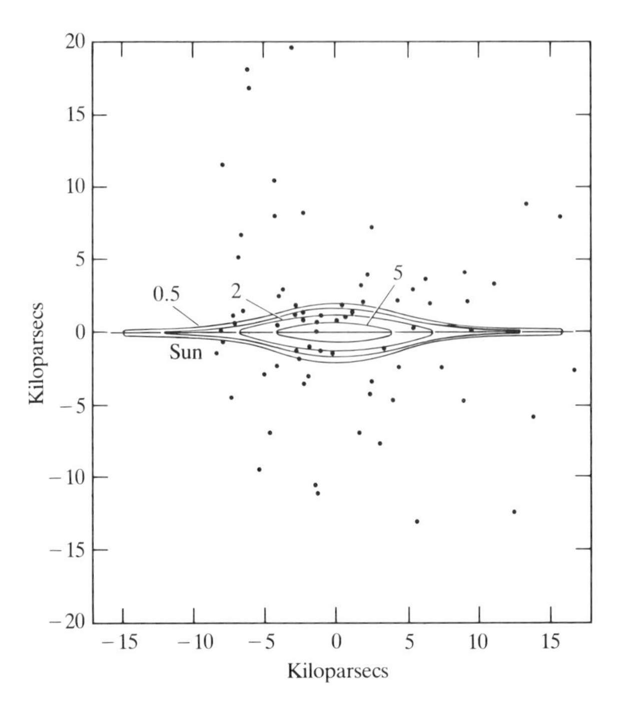

Figure 1-9. A model of the mass distribution in our Galaxy. The contours show curves of equal density measured in units of the density near the Sun. The diagram does not show the steep increase in density near the galactic center. The dots show the projected positions of globular clusters. [From (**B4**, 454), by permission. Copyright © 1965 by the University of Chicago.]

is a somewhat different type than our own Galaxy, which must therefore look somewhat different.) A highly schematic view of the galactic disk and bulge is given in Figure 1-9, which shows a model of our Galaxy derived from a dynamical study. The contours show lines of equal mass density (omitting, however, the rapid rise near the center) and the positions of globular clusters projected onto the plane containing the axis of the system and the Sun.

The characteristic objects in the spheroidal component are stars and globular clusters. The spheroidal component contains little if any dust and gas. Typical stellar representatives are the metal-poor *subdwarfs* and *RR Lyrae variables*; the latter are easy to identify even at large distances because of their distinctive light variation, and hence they provide an ideal group of objects to use in determining the space distribution of stars within the spheroidal systems. From such determinations, we find a very rapid rise in density toward the galactic center. This rise implies a strong central concentration of mass in our Galaxy, a conclusion which applies, however, only to stars that we can observe. The situation could be quite different for the hypothetical dark population discussed later.

Spectroscopic analyses of both globular clusters and individual field stars indicate a general radial composition gradient within the spheroidal component, ranging from quite metal poor in the outermost halo (deficient by a factor of 10^2 or more relative to solar) to metal rich in the nuclear region (that is, solar or slightly above-solar metal content). This gradient suggests that it is appropriate to subdivide the spheroidal-component stars, which are the Population II stars as defined by Baade, into a radial sequence of populations, which we shall designate as *bulge-component stars*, *intermediate spheroidal-component stars*, and *halo-component stars*. Obviously, other subdivisions may be possible as more information becomes available, but the ones we have chosen are satisfactory to describe the presently available data.

Each of the aforementioned populations fills the entire volume from the galactic center out to some maximum distance; that is, bulge-component stars are found away from the center only out to some characteristic outer radius of the bulge, whereas halo stars are found in that volume and also out to still greater distances from the center. Thus, within a given volume, there will be a mixture of populations, which changes progressively from position to position within the spheroidal component.

In addition to the relatively easily observed spheroidal-component objects mentioned thus far, there are several different lines of dynamical evidence that point to the presence of a distribution of objects of unknown nature throughout the halo (and perhaps beyond?), whose total mass is large enough to play a significant role in galactic dynamics. Because these objects have not yet been conclusively identified observationally [see **(H1)**, **(S7)** for preliminary evidence], we shall refer to them as the *dark population*, for they obviously emit very little light (per unit mass) compared with normal stars. We are not absolutely certain at present that these objects exist, despite the appealing arguments that can be adduced in support of the hypothesis. Even if we do admit their existence, we do not know the nature of the objects, though several possibilities can be suggested. Among them are very faint red dwarf stars; cool degenerate dwarfs, neutron stars, and black holes (all terminal remains of an ancient stellar population); or substellar masses below the limit needed to produce thermonuclear energy release (for example, Jupiter-sized masses or below). The existence and nature of the dark population are problems of tremendous importance and current interest, and they will be solved only by astute and persistent observational and theoretical work.

The characteristic constituents of the disk include: interstellar dust and gas, which give rise to interstellar absorption, reflection nebulae, and emission nebulae (or H II regions); young metal-rich stars, which have recently formed in stellar associations and galactic clusters in the spiral arms and which have a clumpy distribution; and older metal-rich stars, which have a smooth distribution throughout the disk. The dust, gas, and young stars, which we shall call the *spiral-arm population*, correspond to Baade's Population I. We shall call the smoothly distributed stars the *disk population*. The radial-scale length over which properties such as star density, metal richness, and so on change appreciably is much greater for the disk component than for the spheroidal component. Therefore, the variation of star density and metal richness from the galactic center to beyond the solar neighborhood is much greater in the spheroidal component than in the disk component. In fact, most of the observable disk has a relatively high metallicity, which lies within a factor of ten of that of the innermost parts of the spheroidal component. This suggests that the disk is made entirely of material already processed in an earlier generation of stars.

In the range from 3 kpc to 15 kpc away from the galactic center, the density of interstellar gas is roughly proportional to the star density. Within about 3 kpc of the center, there is a sharp drop in gas density, and, in this region, observations show strongly noncircular motions of the gas relative to the center. One interpretation of the central drop in the gas density is that the gas has been swept out of the central region of our Galaxy; similar low-density regions are observed near the centers of some other spiral galaxies. In external spiral galaxies, the H I component in the spiral arms is usually observed to extend to much larger radii than the stellar component; the same may well be true for our Galaxy.

The cause of the spiral structure of our Galaxy is not, at present, well understood. One current theory seeks to explain it as a spiral density wave that propagates self-consistently around the disk in such a way that the density enhancements and rarefactions associated with the wave pattern produce just the gravitational forces that are required to sustain the shape and motion of the wave.

As will be discussed later, the disk is known to be in a state of rapid rotation. The density-wave theory of spiral structure predicts that the whole spiral wave pattern rotates coherently more slowly than the material in the bulk of the disk. Hence, we expect the gas in the disk to overtake the wave pattern from behind and run through it. As the gas encounters the density concentration in the wave, it is compressed. This compression could possibly lead to relatively rapid star formation downstream in the flow, as is apparently indicated by observation in some external galaxies (see §8-5). Although this picture is appealing, attempts to formulate it mathematically have encountered formidable technical obstacles. The problem of the formulation of a completely satisfactory theory of spiral structure remains one of the most important challenges in theoretical astrophysics.

Our whole Galaxy is, of course, in a state of motion. The mass distribution in our Galaxy produces gravitational forces that give rise to accelerations, and hence motions, of both stars and gas. The forces that act on stars are essentially purely gravitational, whereas the gaseous component is subject to both gravitational and gas-dynamic forces (and possibly also magnetic forces in certain situations). The stars respond mainly to the smooth gravitational field of the system as a whole, and they are only weakly perturbed by chance encounters (except in regions of extremely high star density, for example, the centers of globular clusters). Stellar gravitational encounters are elastic, and the kinetic energy of the interacting stars is conserved. In contrast, gas clouds have large mutual collision cross sections, and their motions are strongly controlled by intercloud collisions; these collisions are strongly inelastic and dissipative, with the kinetic energy of the clouds being lost into internal gas excitation and subsequent radiation of photons.

The spheroidal component of our Galaxy has a very slow rate of rotation around the center. Many spheroidal-component stars have angular momenta that are much too small for them to move on circular orbits around the galactic center. Hence they move on highly eccentric, nearly radial orbits (that is, "in-and-out" or "plunging" orbits). As a consequence, they traverse large ranges of distance from the galactic center; for example, the halo-component stars observed in the solar neighborhood typically move from minimum radii of less than 4 kpc out to maximum radii of the order of 20 kpc or more. Within any given volume of space, the stars all have individual *residual velocities* relative to the *mean velocity* of that volume. For spheroidal-component stars, these residual velocities are very large. Because the dispersion of the residual-velocity distribution for spheroidal-component stars is actually larger than the systemic rotation speed, some of the stars have tangential velocity components (that is, perpendicular to the radius vector from the center) that are negative (in the direction opposite to the rotation) and hence they move on retrograde orbits around the center. Furthermore, because they can have large residual velocities perpendicular to the galactic plane, the spheroidal-component stars can move on orbits that are highly inclined to the plane and take them large distances from it.

In the galactic disk, the situation is quite different. The interstellar gas cannot move freely in the radial direction because of the large collision cross sections of the interstellar clouds. The gas is therefore constrained to move on almost perfectly circular orbits around the galactic center. The material in the disk has a high angular momentum per unit mass (as is appropriate to circular motion), and the disk as a whole is in a state of rapid *differential rotation*; that is, the inner parts of the disk have a higher angular rotation rate than the outer parts, as is the case for the planets in the solar system. The gas has a very small internal velocity dispersion (~ 10 km/s) compared to the rotation speed (~ 250 km s^{-1} in the vicinity of the Sun). Thus, when spiral-arm stars form from the gas in the disk, they also move on virtually circular orbits around the galactic center with little if any motion out of the galactic plane. If we consider only the forces arising from the smooth galactic potential, then the orbits of individual disk stars can be well described in terms of a *harmonic oscillation* perpendicular to the plane and an *epicyclic motion* in the plane relative to the *local standard of rest*, a hypothetical point that moves on a perfectly circular orbit around the galactic center.

As time passes, a cluster or association of stars tends to be torn apart by the shear of the differential rotation of the disk and by the evaporation of individual stars that have gained enough energy to escape the binding forces exerted by the other cluster members. The energy necessary for a star to escape by evaporation can be gained in individual stellar encounters, by encounters of the cluster with interstellar clouds, or by successive passages of the cluster through the spiral-wave pattern. Individual cluster stars thus tend to be dispersed into the disk, and the random velocities within groups of disk stars, relative to the local standard of rest, tend to increase with age. As a result of the mechanisms just mentioned, older and older groups of disk stars diffuse both *in* the galactic plane, to orbits of increasingly large eccentricities, and *perpendicular to* the plane, to larger and larger distances above and below it. We can, in fact, identify a sequence of disk population groups, each with a highly characteristic set of physical and kinematic properties.

How did our Galaxy arrive at its present state? Although we are not yet in a position to answer this question definitively in every detail, a few basic features of the processes of formation and evolution of our Galaxy have begun to emerge fairly clearly from recent research. We shall sketch here just briefly a *possible* history of our Galaxy; alternative scenarios for both our own and other galaxies will be dealt with in greater detail in Chapter 19. We have already noted that our Galaxy has two major structural parts, the disk and the spheroidal component. According to present theoretical ideas, these are formed separately in a two-stage process, in which the physical nature and dynamical behavior of the material is somewhat different in each step.

About 10^{10} years ago, the Universe began in the big bang, which initiated the cosmic expansion we observe today. In an early phase of the expansion, primeval hydrogen and helium were created with a relative H:He number abundance of about 10:1. At some point in the expansion of this material (perhaps very early), inhomogeneities developed, and they produced density fluctuations in the pregalaxian material. In some regions of above-average density, gravity was able first to slow to a halt and then to reverse the original cosmic expansion. We call these collapsing regions *protogalaxies*.

It is not clear at present whether the gas in the protogalaxy for our Galaxy then collapsed as one great cloud or was first converted more or less completely into stars. To explain the presence of heavy elements (with about 1/1000 to 1/100 of the solar abundance) in even the most metal-deficient stars, it may be necessary to argue that at least some of the primordial gas condensed into stars at a very early stage in the formation of our Galaxy and was recycled back into the interstellar medium. In any case, we have reason to believe that gas-dynamic processes played a central role in the early history of the system.

The spheroidal component of our Galaxy may have been formed when the gas in the protogalaxy underwent a rapid *radial collapse*, accompanied by rapid star formation and a simultaneous metal enrichment of the collapsing material. A radial metal-abundance gradient was set up in this process, as collapsing protogalaxian material became progressively more metal-enriched while it streamed toward the center through a background of already formed stars. Later generations of stars that formed from this material merely reflect the resulting metal-abundance gradient. In such a picture, both the final central metal richness of the spheroidal component and the abundance gradient within the spheroidal component depend on the rate of star formation in the early stages of the collapse. If there is very efficient star formation in these stages, then the system will have a fairly high metal abundance throughout, whereas less efficient star formation implies a steeper abundance gradient and rather low metal abundances at large radii.

The present rate of systemic rotation of the spheroidal component was fixed by the angular momentum of the infalling protogalaxian material. Because the distribution of the residual velocities of the halo stars appears to be roughly of Maxwellian form (which for a gas characterizes the final equilibrium distribution of particles that interact dynamically), we infer that this system of stars must have undergone a strong *dynamical relaxation*. There are several possible mechanisms that could have produced this relaxed state. For example, the form of the velocity-distribution function and its large dispersion could simply be those appropriate to a near dynamical equilibrium of the gas from which these stars formed, with the potential from a very massive halo of dark-population objects. Or, the relaxed velocity distribution could have been produced by rapid large fluctuations of the gravitational forces within the system during the collapse itself. In this *violent relaxation* process, stars move on orbits along which energy is not conserved, and their orbital properties become randomly mixed in such a way that their velocity distribution approximates the Maxwellian distribution. Alternatively, there could have been a *hierarchical relaxation*, in which stars in the halo were relaxed during an early phase of formation of a huge system of globular clusters, which in turn were largely obliterated during the formation of our Galaxy. In this scheme, the present globular clusters represent only a small remnant of the original group. It is an observed fact that there are now many more high-velocity, low-metal stars outside globular clusters than inside them.

The formation of the disk apparently proceeded rather differently. The condensation of material into the disk occurred only after the collapse of the spheroidal component, with a relatively low rate of star formation. During the formation of the disk, gas of higher and higher angular momentum settled into the galactic plane at ever-increasing radii from the galactic center. This material may have collapsed radially in the plane, conserving angular momentum in the process. But, as a result of collisions among gas clouds, energy was dissipated, and the material was forced into circular orbits around the center. Because the material ended up much closer to the center than where it originated, it will have, at any position in the disk, a higher specific angular momentum than spheroidal-component stars at the same distance from the galactic center. The infall of outlying gas into the outermost parts of the disk may take place over an appreciable part of the lifetime of a spiral galaxy, and it may well be occurring even today at the outer fringes of our Galaxy.

Even the oldest known disk stars have near-solar metal content (that is, within a factor of, say, three to five of the solar value). The entire disk is metal rich, with only a modest gradient increasing toward the galactic center. These facts suggest that much of the material that condensed into the disk was recycled material that had been shed by previous generations of stars (perhaps evolved spheroidal-component stars). Indeed, detailed computations of the chemical evolution of the material in the galactic plane show (**O4**) that the relative paucity of metal-poor stars in the disk in the solar neighborhood cannot be understood unless either the initial metal abundance of disk material was nearly solar or there was a continuous inflow of material into the disk from outside. Models that start with a disk of primordial metal-poor material and do not envisage continuous infall of gas shed from evolving halo stars either produce many too many metal-poor stars or violate other physically plausible constraints.

Finally, after the disk settled, a spiral pattern was set up, and perhaps a bar developed at the center of our Galaxy. The spiral structure (the individual arms of which may be only transient features, even if the overall pattern persists through many rotations) may be thought of as a device by which angular momentum is passed from the mass near the center of our Galaxy, via the gravitational field of the arms, to outlying gas and stars. This angular-momentum transfer enables the innermost matter to sink deeper into our Galaxy's potential well, thus releasing gravitational energy. Some of the energy released in this way produces an increase in the random velocities of disk stars, and some produces shock waves in the material of gas clouds. These shock waves have two important effects—they induce dense regions to collapse and fragment into new stars, and they cause less dense regions to radiate strongly at infrared and radio wavelengths.

The existence (and effect) of a bar at the center of our Galaxy is more controversial. Certainly, many other spiral galaxies do have more or less prominent central bars, and the observations of disturbed gas motions near the galactic center are suggestive of such a phenomenon in our own Galaxy as well. There are also dynamical reasons for believing a bar should form in that part of our Galaxy that rotates approximately like a solid body. If there is a bar at the galactic center, it will push much of the central gas out into a ring near its edge while feeding other gas in toward the nucleus itself. It is an observed fact that much of the gas in our Galaxy is concentrated in a ring of giant molecular clouds about 4 kpc from the galactic center, and the galactic center itself is the seat of a great deal of very ill-understood activity, which almost certainly involves substantial quantities of dense gas. However, at this point, one steps over the limit of actual understanding of galaxies into the perplexing and fascinating problems that arise when one attempts to understand the extraordinary phenomena that power quasars and radio galaxies, and in this book we shall have to stop just short of that mark. Others must tell that tale at another time; but to do so they will need to draw heavily on the knowledge presented in the following pages.

REFERENCES

(B1) Baade, W. 1944. *Astrophys. J.* **100**:137.

(B2) Baade, W. 1944. *Astrophys. J.* **100**:147.

(B3) Berendzen, R., Hart, R., and Seely, D. 1976. *Man Discovers the Galaxies.* (New York: Science History Pub.).

(B4) Blaauw, A. and Schmidt, M. (eds.). 1965. *Galactic Structure.* (Chicago: University of Chicago Press).

(B5) Burbidge, G. 1975. *Astrophys. J. Letters.* **196**:L7.

(C1) Code, A. D. and Houck, T. E. 1955. *Astrophys. J.* **121**:533.

(D1) Dicke, R. H., Peebles, P. J. E., Roll, P. G., and Wilkinson, D. T. 1965. *Astrophys. J.* **142**:414.

(H1) Hegyi, D. J. and Gerber, G. L. 1977. *Astrophys. J. Letters.* **218**:L7.

(H2) Hubble, E. P. 1925. *Observatory.* **48**:139.

(H3) Hubble, E. P. 1925. *Popular Astron.* **33**:252.

(H4) Hubble, E. P. 1926. *Astrophys. J.* **63**:236.

(H5) Hubble, E. P. 1926. *Astrophys. J.* **63**:281.

(H6) Hubble, E. P. 1929. *Astrophys. J.* **69**:103.

(H7) Hubble, E. P. 1935. *Astrophys. J.* **81**:334.

(H8) Hubble, E. P. 1936. *The Realm of the Nebulae.* (New Haven: Yale University Press).

(K1) Kapteyn, J. C. 1922. *Astrophys. J.* **55**:302.

(K2) Kapteyn, J. C. and van Rhijn, P. J. 1920. *Astrophys. J.* **52**:23.

(L1) Lindblad, B. 1926. *Arkiv für Mat. Astron. och Fysik.* **19B**:No. 7.

(L2) Lindblad, B. 1927. *Arkiv für Mat. Astron. och Fysik.* **20A**:No. 17.

(L3) Lindblad, B. 1927. *Mon. Not. Roy. Astron. Soc.* **87**:553.

(M1) Morgan, W. W., Sharpless, S., and Osterbrock, D. E. 1952. *Astron. J.* **57**:3.

(M2) Morgan, W. W., Whitford, A. E., and Code, A. D. 1953. *Astrophys. J.* **118**:318.

(O1) Oort, J. H. 1927. *Bull. Astron. Inst. Netherlands.* **3**:275.

(O2) Oort, J. H. 1928. *Bull. Astron. Inst. Netherlands.* **4**:269.

(O3) Ostriker, J. P., Peebles, P. J. E., and Yahil, A. 1974. *Astrophys. J. Letters.* **193**:L1.

(O4) Ostriker, J. P. and Thuan, T. X. 1975. *Astrophys. J.* **202**:353.

(P1) Penzias, A. A. and Wilson, R. W. 1965. *Astrophys. J.* **142**:419.

(S1) Sandage, A. 1961. *The Hubble Atlas of Galaxies.* (Washington: Carnegie Institution).

(S2) Shapley, H. 1918. *Astrophys. J.* **48**:154.

(S3) Shapley, H. 1918. *Pub. Astron. Soc. Pacific.* **30**:42.

(S4) Shapley, H. 1919. *Astrophys. J.* **49**:249.

(S5) Shapley, H. 1919. *Astrophys. J.* **49**:311.

(S6) Shapley, H. 1919. *Astrophys. J.* **50**:107.

(S7) Spinrad, H., Ostriker, J. P., Stone, R. P. S., Chiu, L. G., and Bruzual, G. A. 1978. *Astrophys. J.* **225**:56.

(S8) Strom, K. M., Strom, S. E., Jensen, E. G., Moller, J., Thompson, L. A., and Thuan, T. X. 1977. *Astrophys. J.* **212**:335.

(S9) Struve, O. and Zebergs, V. 1962. *Astronomy of the 20th Century.* (New York: Macmillan).

(T1) Trumpler, R. J. 1930. *Lick Obs. Bull.* **14**:154.

(W1) Whitney, C. A. 1971. *The Discovery of Our Galaxy.* (New York: A. Knopf).

2 Astronomical Background

Our basic observations of astronomical objects yield information about their positions, their motions along and across the line of sight, and the distribution of the energy emitted in their spectra as a function of frequency and time. Essentially everything else we wish to know about astronomical bodies must follow from deductions and inferences made from analyses of these data, using the laws of physics established in terrestrial laboratories. It is actually rather remarkable that modern astrophysics has derived as much information as it has concerning the properties of stars and galaxies, despite the severe limitation (compared to other sciences) of having at its disposal only the photons emitted casually by the objects under study.

The determination of positions and transverse motions lies in the domain of *positional astronomy*, or *astrometry*, whereas radial velocities and energy distributions are obtained by *spectroscopic* and *photometric* measurements. The relevant measurements can be made with conventional optical telescopes, with radio telescopes, or with space-borne instruments that can reach certain spectral regions inaccessible from the Earth's surface (for example, ultraviolet or X ray).

In principle, the acquisition of the data is fairly straightforward, but in practice the measurements require careful techniques, using delicate and complex equipment, often followed by an elaborate set of reductions to eliminate extraneous effects. It is beyond the scope of this book to discuss either the equipment or the reduction procedures involved, especially as there are a number of excellent books dealing with positional astronomy [for example, (**M1**), (**S6**, Chapters 1–6), (**S3**), and (**V1**)] and astronomical spectroscopy and photometry [for example, (**G2**), (**H2**) and (**S6**, Chapters 9, 11, 13)]. In this chapter, we shall attempt only to describe the kinds of astrometric, photometric, and spectroscopic data that we shall call upon in our work. The interpretation of these data, and their application to the development of an understanding of galaxies, will occupy much of the remainder of this book.

2-1. POSITIONS AND COORDINATE SYSTEMS

Positional astronomy is the oldest branch of astronomy. Catalogs of the locations of astronomical objects on the sky have come down to us from antiquity. It is probably fair to say that astrometry was the central problem of observational astronomy through the late nineteenth and early twentieth century. Since that time, astrophysical observations have attracted more interest and generated more excitement than astrometry, to the extent that students often have the (mistaken) impression that positional astronomy is somewhat old fashioned and perhaps even unimportant. Such an opinion is grievously wrong, not only because the data obtained by astrometric measurements lie at the foundation of much of astrophysics, but also because in recent years the subject has undergone a renaissance, for two reasons. First, new observational techniques offer hitherto unattainable accuracy, and second, technological advances (such as computers) can be applied to the acquisition of observations, to the control of automatic measuring engines, and to the reduction of data. Even more important, the issues that positional astronomy deals with are anything but trivial. Among them is the task of establishing, in practical terms, an *inertial reference frame*. This matter is one of profound importance to a wide range of problems, including the navigation of space vehicles, the dynamics of the solar system, the large-scale dynamics of matter in the Universe, and theoretical questions about the physical nature of gravity and inertia (see Chapter 11). In short, astrometry is now a lively subject in which exciting results of great importance are likely to emerge in the near future.

The Equatorial System

Any mention of the position of an object implies that a coordinate system has been chosen and established observationally. The fundamental coordinate system for observations made from the surface of the Earth is the *equatorial system*. The Earth rotates eastward on its axis once a day, and, as a consequence, the sky appears to rotate westward about the Earth. The extension of the Earth's axis to the *celestial sphere* (an imaginary sphere of infinite radius centered on the earth) defines the *north* and *south celestial poles* (NCP and SCP), and the extension of the Earth's equatorial plane determines the *celestial equator* (see Figure 2-1). The equator is a *great circle*, that is, a circle on the celestial sphere defined by the intersection of a plane passing through the sphere's center with its surface.

For each observer on the Earth's surface, the direction of gravity fixes the direction of the *local vertical*; the point at which the extended vertical line intersects the celestial sphere is the *zenith*. The *zenith distance* z of a star is its angular distance from the zenith. The *horizon* is the great circle whose pole is the zenith (and thus lies 90° from the zenith). The angular distance

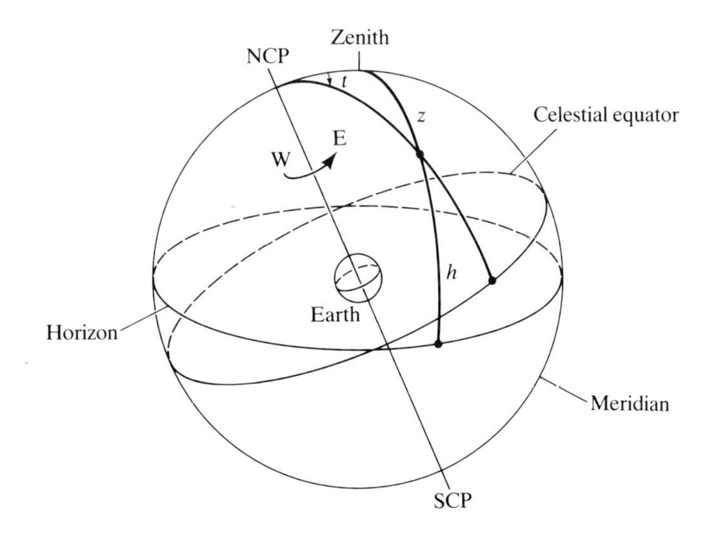

Figure 2-1. The celestial sphere, showing the celestial poles, celestial equator, meridian, zenith point, and horizon. The hour angle of a star is the angle t, its zenith distance is the arc z, and its altitude is the arc h.

of an object above the horizon is its *altitude*. The great circle passing through the celestial poles and the zenith is the *meridian*. The *hour angle* t of an object is the angle between the meridian and its *hour circle*, which is the great circle through the poles and the object. Hour angle is defined to increase westward.

Angles on the celestial sphere can always be expressed in units of degrees, minutes, and seconds of arc. Because its operational definition involves a time measurement, the hour angle is often expressed in terms of hours, minutes, and seconds of time. The conversions from time to angular measure are $24^h = 360°$, $1^h = 15°$, $1^m = 15'$, and $1^s = 15''$.

As the Earth revolves annually in its orbit around the Sun, the Sun appears to revolve in the opposite sensc around the Earth on the celestial sphere on a path called the ecliptic. The *ecliptic* is the great circle defined by the intersection of the Earth's orbital plane with the celestial sphere. The Earth's rotation axis is inclined away from the normal to its orbit by an angle of about 23° 27′; hence the ecliptic is also inclined to the celestial equator by this angle, which is called the *obliquity of the ecliptic*. The ecliptic and the celestial equator intersect at two points (separated by 180°) called the *vernal* and *autumnal equinoxes*. The Sun passes through the vernal equinox on approximately March 21, moving from south to north of the celestial equator. About six months later, it passes through the autumnal equinox traveling from north to south. The time interval between successive returns of the Sun to the vernal equinox is called the *tropical year* (see Figure 2-2).

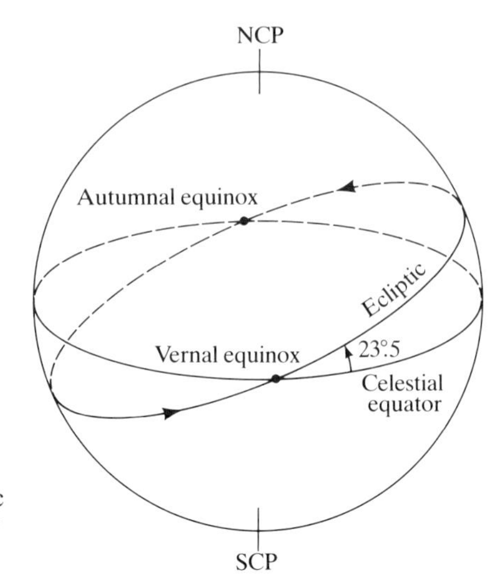

Figure 2-2. The relationship between the celestial equator, the ecliptic, and the equinoxes. The angle between the ecliptic and the equator is called the obliquity of the ecliptic.

The longitudelike coordinate of the equatorial system is called the *right ascension* α; the zero point of α is taken to be the vernal equinox. Thus the angular distance measured eastward along the celestial equator from the vernal equinox to the point of intersection of the equator and the hour circle of a star is the star's right ascension (see Figure 2-3). As was true for the hour angle, right ascension can be expressed either in angular units or in time units. The latitudelike coordinate, called the *declination* δ, is the angular distance measured from the equator along a star's hour circle to

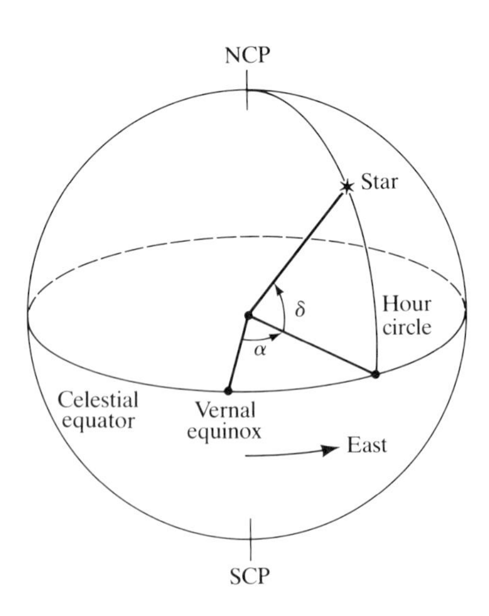

Figure 2-3. The position of a star in the equatorial system is specified by its right ascension α and its declination δ.

the star, and it is positive northward and negative southward. For example, the NCP is at $\delta = +90^\circ$; the SCP is at $\delta = -90^\circ$; the declination of the celestial equator is 0°.

The altitude of the NCP above the horizon equals the observer's *latitude* on the surface of the Earth. The latitude can be determined by measuring the zenith distance of upper and lower meridian passage of a circumpolar star. The declinations of other stars then follow from knowledge of the observer's latitude and the zenith distance of each star at meridian passage.

Time measurement is based ultimately on the observed position of the Sun. The obliquity of the ecliptic is derived by observing the maximum and minimum declination of the Sun at meridian passage throughout a year. Knowledge of the obliquity and of the Sun's observed declination at a given meridian passage yields the right ascension of the Sun (or, equivalently, the hour angle of the vernal equinox) from standard relations of spherical trigonometry. *Apparent solar time* is then given by the hour angle of the actual Sun plus 12 hours, so that noon occurs when the Sun crosses the meridian above the horizon.

Apparent solar time is nonuniform because (1) the Earth moves around the Sun on a noncircular (elliptical) orbit with nonuniform velocity (which implies that the apparent motion of the Sun along the ecliptic is nonuniform), and (2) the ecliptic is inclined to the equator (which implies that the Sun's projected rate along the equator would be nonuniform even if it moved uniformly along the ecliptic). Therefore, a fictitious *mean Sun* is defined such that it moves at a uniform rate along the celestial equator; the motion of the mean Sun defines *mean solar time*, which is our practical *clock time.*

The time interval between successive meridian passages of the mean Sun is a *mean solar day*. Because the Sun appears to move eastward on the sky as a result of the Earth's orbital motion (and completes a revolution around the equator in a year), the mean solar day is approximately 3^m56^s (that is, 24h/365.25 days) longer than the *sidereal day*, which is defined as the time interval between two successive meridian passages of the vernal equinox. The difference between apparent solar time and mean solar time, which can be computed from knowledge of the elements of the Earth's orbit and the obliquity of the ecliptic, is called the *equation of time.*

Given the Sun's right ascension at apparent noon and the equation of time, we know both the mean solar time and the hour angle of the vernal equinox at that instant of mean solar time. Then, given an accurately uniform clock (running either at the mean solar or the sidereal rate), we can easily infer the hour angle of the vernal equinox, hence the right ascension of the meridian, at subsequent times. Therefore, by noting the time of meridian passage of a star, we determine its right ascension. In short, the right ascensions of stars follow, in effect, by comparison of the times of meridian passage of the Sun and stars via a clock, which serves as an interpolation device to connect daytime observations of the Sun to nighttime observations of the stars.

Thus far, we have described the equatorial system as though it were fixed in space. In fact, it is not; rather, it moves relative to a true inertial frame owing to the effects of *precession* and *nutation*, which cause the direction of the Earth's rotation axis—the celestial pole—to move on a complicated path on the celestial sphere, and hence cause the whole coordinate system based on that pole to change in an involved way as a function of time.

The dominant precession effect arises because the Earth is not perfectly spherical but is approximately an oblate spheroid (that is, it has an equatorial bulge). This allows the Sun and Moon to exert gravitational torques on the Earth, which cause the Earth's rotation axis to drift on a cone around the normal to its orbital plane (the pole of the ecliptic). The pole of the celestial equator therefore describes a small circle of radius $23°27'$ around the pole of the ecliptic in a period of about 25,725 years. Accordingly, the vernal equinox moves slowly westward along the ecliptic, which implies a slow change in the right ascension and declination of every object in the sky. This secular (that is, cumulative) motion is called the *luni-solar precession.*

In addition, the Earth's orbit is perturbed by gravitational interactions with other planets, and this fact leads to changes in the shape of the orbit and in the orientation of its plane (the plane of the ecliptic) in space. These *planetary precession* effects are much smaller than luni-solar precession. The sum of the luni-solar and planetary precessions is called the *general precession*, which amounts to an angular motion of the equinox of $50''.25$ per years.

In addition to the slow secular effect of precession, the celestial pole undergoes a relatively short time-scale periodic oscillation around its mean position, called nutation. The Moon's orbit is inclined by about $5°$ to the plane of the ecliptic; the forces exerted by the Sun on the Moon cause the pole of the Moon's orbit to drift along a small circle of radius $5°$ around the pole of the ecliptic. This leads to a westward *regression* of the line of nodes of the Moon's orbit in a period of about 18.6 years, and the resulting periodic change in direction of the forces exerted by the Moon on the Earth produces nutation. The combination of precession and nutation causes the pole to follow a wavy path around a circle on the celestial sphere.

Because the position of the celestial pole changes with time, the numerical values of the position coordinates α and δ of a given object constantly change (even though the object may be fixed "absolutely" in space). Therefore, when we state an object's position in the equatorial system, we must also specify the *epoch* of that position. The epoch tells, in effect, the position of the pole relative to which the position is measured.

Positional astronomers thus face the following challenge: not only must they determine (with high precision) the positions of objects within a coordinate system capable of observational realization, but they must determine the motion of the system itself! In principle, this can be done directly from *fundamental* right ascensions and declinations determined along the lines sketched out in this section. In practice, there is yet another complication.

The stars themselves actually move as a result of random motions and their revolution around the center of our Galaxy, and these movements lead to measurable changes in stellar positions called *proper motions*. As we shall see in §2-2 and in Chapters 6, 7, and 8, proper motions provide vital information about the kinematics of our Galaxy.

In the present context, to untangle the combined systematic effects of precession, solar motion, and differential galactic rotation (see Chapters 6 and 8) from the individual motions of the stars themselves, we must resort to a hypothesis: we assume that stellar proper motions are truly random and that, if we use a large enough sample of stars, the proper motions will average to zero, so that any residual changes in observed position can then be attributed to the systematic effects just listed. In this way it is possible to derive the precession constants empirically, as was done more than eighty years ago by S. Newcomb, whose values (with subsequent corrections) for these constants have been used in practically all later work. The great danger that lurks here must be stated explicitly. The proper motion of a star is derived by determining the change in its position between two widely separated epochs in time, but these positions must first be reduced to a single coordinate system (that is, to a common position of the celestial pole) by correcting for the effects of precession. Therefore, any error that exists (for whatever reason) in the precession constants used to make this reduction will necessarily produce a false change in position. Hence this error will make a spurious contribution to the derived proper motion and thus ultimately will lead to distortions of our picture of the kinematic state of our Galaxy.

Thanks to recent improvements in estimates of planetary masses, the planetary precession, which can be calculated by celestial mechanics, is now known to a high precision. But luni-solar precession cannot be calculated accurately because of inadequacies of our knowledge of the internal mass distribution of the Earth. Therefore, it must be determined empirically. It is now known that the adopted value for luni-solar precession *is* slightly in error and that a corrected value should be used in future work. Unfortunately, a definitive value of the correction to be made has not yet been determined. It is quite possible, therefore, that stellar proper motions on the fundamental meridian-circle system do contain systematic errors, perhaps of significant size. What is urgently needed now is a precise method for determining precession, independent of any hypothesis about stellar motions. We shall return to this point again, for recent developments in *radio astrometry* offer promise of providing such a method.

As already outlined, the classical techniques for measuring *fundamental positions* employ a variety of instruments, including meridian circles, transits, and astrolabes, along with accurate clocks. These techniques are typically used to measure the time and zenith distance of the meridian passage of a star. The instruments used must be carefully calibrated to eliminate errors of alignment, collimation, flexure, graduation of circles, and so on. Advances in technology have greatly improved the accuracy and ease with which unwanted instrumental effects can be eliminated. Furthermore, corrections must be made for effects such as *atmospheric refraction*, which systematically increases an object's observed altitude, and *aberration* (both annual and diurnal), which results from the Earth's revolution and rotation and the finite velocity of light and which displaces the apparent position of a star from its true position. In addition, the images of stars are never perfect points; they are either blurred or fluctuate randomly in position because of atmospheric *seeing* effects that are produced by variations in atmospheric properties along the path of the ray from star to observer. All together, the difficulties just described lead to standard errors in fundamental positions of $\pm 0\overset{''}{.}15$ in the best catalogs. This is an internal precision; absolute external (systematic) errors can be, and in some cases are known to be, significantly larger.

Relative positions can be measured on photographic plates with respect to stars whose positions have already been determined fundamentally. Historically, most astrometric plates were taken with long-focus refractors. But recently, high-quality results have been obtained using Schmidt telescopes and specially designed astrometric reflectors. These instruments have the advantages of wide field and greater light-gathering power. In reducing photographic measurements, one must make allowance for the projection of the celestial sphere onto the plane of the plate and for a number of instrumental effects (for example, optical aberrations, alignments, and so on), which may depend on the color and brightness of a star, in order to solve for the scale and orientation of the plate with respect to reference stars whose positions are considered known. One problem encountered in measuring plates is the growth of images caused by scattering in the photographic emulsion; the images of bright stars are sometimes so large as to preclude accurate measurement. Moreover, guiding errors and seeing effects can produce systematic differences between the photocenters of faint and bright stars on the plate. A typical standard error for a relative position measured on a single plate is $\pm 0\overset{''}{.}02$, and, by combining several plates from a number of nights, positions in the best cases have been determined to $\pm 0\overset{''}{.}005$.

In the past, most fundamental position measurements with meridian circles have been limited to relatively bright stars, typically with apparent magnitudes (see §2-5) $m \lesssim 7.5$. Recently, these observations have been pushed to $m \approx 9.0$, and, by use of photoelectric detection techniques, astronomers hope to reach much fainter stars in the near future. This work is important both because of the intrinsic interest in positions and motions of faint stars and because it will help to tie into the fundamental system both the new photographic measures of faint stars made with large reflectors and surveys using faint galaxies as a reference frame (see §2-2). At present, it is difficult to bridge the gap between the bright stars used in meridian-circle work and the faint objects used in photographic work without making systematic errors.

Two stimulating new developments in positional astronomy are the possibilities offered by measurements from space vehicles and the emergence of radio astrometry. In space, the problems of atmospheric refraction and seeing vanish, the accessible spectrum is extended into the ultraviolet and infrared, and there is no net gravity force to produce instrumental flexure. It has been estimated (**G1**, 277) that observations from a small, general-purpose satellite (for example, TD-1) could yield positions accurate to $\pm 0''.01$, and observations from a large instrument, such as the Space Telescope, or from special-purpose astrometric satellites could yield positions accurate to $\pm 0''.002$ (**G1**, 283), (**M6**, 361).

The prospects for radio astrometry are equally exciting (**C1**). In the radio spectral region, atmospheric refraction again practically vanishes. The interferometric technique employed yields fundamental declinations and right ascensions (with an arbitrary zero-point) directly. A crucial advantage of this approach is that large angular separations can be measured as accurately as small ones, and hence high internal consistency is easily attained over the whole sky. In contrast, it is extremely difficult to achieve such consistency by optical methods (except from space data). Radio methods are now as precise as optical methods, and they are likely to improve. Most of the radio sources observed are external galaxies so distant that their proper motions can safely be assumed to be zero. Thus radio astrometry offers the possibility of a definitive determination of precession, untainted by any kinematic hypothesis. Because most of the objects observed in radio work are not stars, special steps must be taken to connect the radio data back into the fundamental stellar system. Some of the observed sources are compact. If these can be identified with compact optical counterparts, a reliable connection can be made with the optical system. Also, a few of the sources actually are stars. When very precise optical positions are determined for these stars, the two systems can again be connected reliably. Even though some practical problems remain, and even though much observational work must yet be done, it is likely that definitive precession constants and extremely accurate positions will soon emerge from combined radio and optical astrometric work.

The Galactic System

Although the equatorial system constitutes the fundamental observable system, it is clearly geocentric, and, as such, provides an inappropriate viewpoint for problems of galactic structure and dynamics. It is useful, therefore, to set up a *galactic system* of coordinates that has a direct physical connection with the structure of our Galaxy. The *galactic equator* is chosen to be the great circle that most closely approximates the plane of the Milky Way. This plane is inclined at an angle of $62^\circ 36'$ to the celestial equator. The north pole of the galactic system is located at $\alpha_{1950} = 12^h 49^m$ and

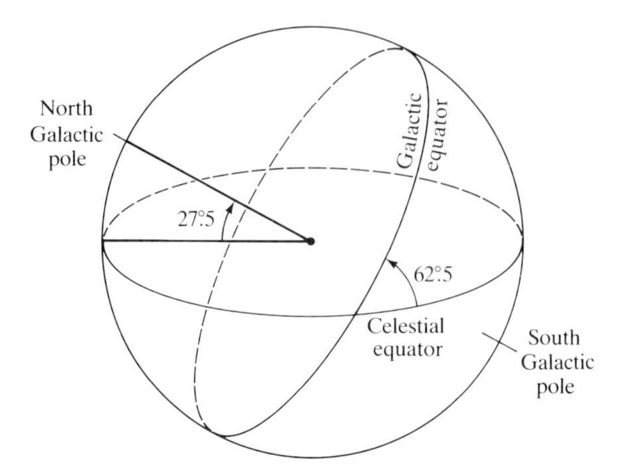

Figure 2-4. The galactic equator is the great circle whose plane most closely approximates the plane of the Milky Way. It is inclined at an angle of about 62°.5 to the celestial equator.

$\delta_{1950} = +27°24'$. The *galactic latitude b* is the angular distance along a great circle perpendicular to the galactic equator; for example, the north galactic pole is at $b = +90°$, and the south galactic pole is at $b = -90°$ (see Figure 2-4).

Actually, there are two systems of galactic coordinates. The old system $(\ell^{\mathrm{I}}, b^{\mathrm{I}})$ measures *galactic longitude* ℓ^{I} from one of the points where the galactic equator intersects the celestial equator. This choice is arbitrary and without physical significance. The new system $(\ell^{\mathrm{II}}, b^{\mathrm{II}})$ uses a slightly different pole (as given above) and measures galactic longitude ℓ^{II} with respect to a point almost in the direction to the galactic center. (The compact radio and infrared source believed to mark the position of the galactic nucleus actually lies about 5′ away from this point.) Thus, in the new system, the direction to the galactic center is $\ell^{\mathrm{II}} = 0°$, $b^{\mathrm{II}} = 0°$, with $\alpha_{1950} = 17^{\mathrm{h}}42^{\mathrm{m}}.4$ and $\delta_{1950} = -28°55'$, whereas, in the old system, the galactic center was located at $\ell^{\mathrm{I}} = 327°41'$, $b^{\mathrm{I}} = -1°24'$. We shall use the new system throughout this book and dispense henceforth with the superscript II. Conversions of coordinates between the equatorial and galactic systems (and between the two galactic systems) are easily effected with standard formulae of spherical trigonometry [see, for example, (**S3**, Chapter 1)].

2-2. PROPER MOTIONS

Because of their intrinsic motions in space, the stars change position with respect to one another on the sky. The component of a star's motion across the line of sight, relative to the Sun, produces an angular rate of change in position called the *proper motion*. This angular rate of change is directly proportional to the star's linear velocity perpendicular to the line of sight and inversely proportional to its distance (convenient units will be specified in §3-1). Therefore, if we can measure a star's proper motion and estimate its distance, we thereby determine its *transverse velocity*—one component of its *space velocity* relative to the Sun (the other component is the radial velocity; see §2-4). The space motions of stars can be analyzed (see Chapters 6, 7, and 8) to give information about the distribution of stellar velocities about the center of our Galaxy. Hence proper motions are an essential component in the development of our knowledge of the kinematics, and thus, ultimately, the dynamics, of our Galaxy. It is for this reason that so much concern is focused on their accuracy, and so much attention lavished on their determination.

Proper motion is a vector quantity and has both a magnitude μ, typically measured in seconds of arc per year, and a direction, which may be specified by its *position angle* θ (the angle between the direction to the NCP and the direction of motion, measured positive from north to east). In the equatorial system, the motion is usually resolved into two components: $\mu_\delta = \mu \cos \theta$, which is perpendicular to the celestial equator, and $\mu_\alpha \cos \delta = \mu \sin \theta$, which is parallel to the celestial equator. Here, μ_α is the annual rate of change in right ascension (often expressed in seconds of *time* instead of seconds of *arc*), and the factor $\cos \delta$ comes from the convergence of hour circles to the celestial poles.

Proper motions are usually tiny; the largest known is that of Barnard's star, which has a motion of $10''.3$ per year. Typical measured motions are only a few hundredths of a second of arc per year. In view of the errors in position measurements quoted in §2-1, it is clear that many years must elapse before these small motions accumulate to a measurable effect. Thus, to determine fundamental proper motions, one compares meridian-circle positions obtained at epochs separated by twenty to fifty years. The observed differences in position consist of (1) changes produced by precession, nutation, solar motion, and differential galactic rotation; (2) the accumulated proper motion; and (3) errors of measurement. If the precession constants were known precisely, then proper motions (plus errors) would follow immediately. Ignoring the possible systematic errors discussed in §2-1, the standard errors (accidental errors only) in the highest-quality catalogs such as FK4 (see Table 2-6) are stated to be $\pm 0''.002$ per year.

By measuring positions of stars on photographic plates taken at widely separated times (twenty years or more), we can obtain *relative* proper motions with respect to a predetermined set of reference objects on the plate. If the reference objects are stars, then the relative motions must be reduced to *absolute* motions by applying a correction for the average proper motions of the reference stars themselves. In the event that these are stars with known fundamental proper motions, a direct reduction to the meridian-circle system is possible. The accuracies achieved in this way are typically only $\pm 0''.005$ to $\pm 0''.010$ per year, principally because there will generally be only a few stars on the plate with reliable fundamental positions and motions. Furthermore, both the accidental and systematic errors in the reference-star motions are necessarily propagated into the final absolute motions derived from the relative motions. In some cases, one measures motions of bright, nearby stars relative to very faint (and presumably distant) comparison stars in the field. In such cases, often only a statistical correction (which depends on the galactic coordinates of the field and the magnitude of the reference stars) is applied to reduce the relative motions to absolute motions. Obviously this procedure is less precise, but it is often very useful for survey work, and it must be used whenever the fundamental motions of the reference stars are unknown. Photographic proper-motion programs have received enormous impetus in recent years from the development of automatic measuring engines and digital computers, which make elaborate reduction procedures feasible. It is now possible to conduct programs that would have been unmanageable a few years ago.

An entirely different approach to measuring absolute proper motions has been taken in programs under way at Lick Observatory and Pulkovo Observatory (**V2**). These surveys are based on plates separated by long time intervals, and they use faint external galaxies as the reference system. Because galaxies lie at vast distances, we know that their possible proper motions are orders of magnitude below our measurement threshold and thus can be taken to be zero. Hence, for practical purposes, these galaxies define an inertial reference frame. The proper motions derived in these programs are therefore truly absolute, and they are free of those systematic errors present in the fundamental (meridian-circle) system. At present, preliminary results from these programs are beginning to become available. The values of μ_δ from both observatories agree with one another and with the fundamental (FK4) system. In contrast, the values of $\mu_\alpha \cos \delta$ found in the two programs disagree with one another by a systematic difference of $0''.01$ per year, and neither agrees with the fundamental system. The source of the discrepancy is not yet known, but some astronomers favor the Lick results because the Pulkovo measurements employ far fewer reference galaxies and may therefore be less secure. It is reasonable to expect that this temporary problem will soon be resolved and that we will then have a set of reliable absolute proper motions measured in a truly inertial frame.

Space observations also hold promise for proper-motion measurements. Because position determinations from space observatories can be very precise, accurate proper motion measurements can be made in shorter time intervals than those needed for ground-based observations. For example, it is estimated (**G1**, 277) that data from a small general-purpose satellite such as TD-1 can yield proper motions accurate to $\pm 0''.005$ per year in two years of observing. From special-purpose astrometric satellites, one should be able to obtain proper motions accurate to $\pm 0''.002$ in three years (**M6**, 361). With the Space Telescope, it should be possible (**G1**, 283) to obtain proper motions good to $\pm 0''.001$ in three years (not fifty), and if the lifetime of the instrument is twenty years, then accuracies of $\pm 0\overset{''}{.}002$ are possible based on only one early and one late observation. The latter figures would permit measurements of internal motions in a number of star clusters, and these measurements would significantly improve our understanding of cluster dynamics (see Chapters 14 and 18).

Although accurate measurement of proper motions is exacting and sometimes tedious work, it cannot be emphasized too strongly that it is of central importance to the development of a picture of the structure, kinematics, and dynamics of our Galaxy

2-3. PARALLAX

As the Earth revolves around the Sun, the vantage point from which we view the stars continually changes. Thus their apparent directions also change slightly. Consider Figure 2-5, and suppose the Earth is at E_1; the direction to a star S is then along the line E_1S. Six months later, when the Earth is at E_2, diametrically opposite the Sun, the star will appear in the direction E_2S. During the course of the year, the apparent position of a star traces out an elliptical path called the *parallactic ellipse*. The lines E_1S and E_2S contain an angle at S that is defined to be twice the *parallax* π of the star. If r is the radius of the Earth's orbit, and d is the distance from the Sun to the star, then, because $d \gg r$, π is a small angle, and

$$\frac{r}{d} = \tan \pi \approx \pi \text{ rad} \tag{2-1}$$

If we convert to seconds of arc,

$$\pi'' = 206{,}265\pi \text{ rad} \tag{2-2}$$

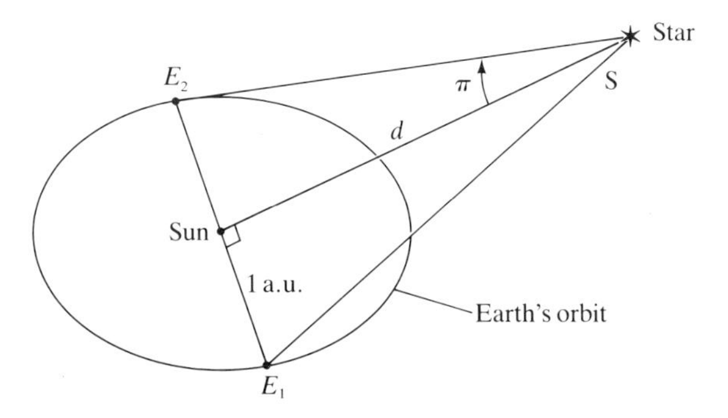

Figure 2-5. The parallax π of a star is the angle subtended by the star at 1 a.u.

and set r equal to one *astronomical unit* (a.u.), then we have

$$d = \frac{206{,}265}{\pi''} \text{ a.u.} \tag{2-3}$$

We now define another unit of distance, the *parsec* (pc), as the distance at which a star would have a parallax of $1''$. Then 1 pc = 206,265 a.u. = 3.086×10^{13} km = 3.26 light-years, and the distance in parsecs to a star with observed parallax π'' is

$$d = \frac{1}{\pi''} \text{ pc} \tag{2-4}$$

For galactic-structure work, it is convenient to use the *kiloparsec* (kpc) = 10^3 pc as a distance unit, and, for the discussion of distances between galaxies and in the Universe at large, the *megaparsec* (Mpc) = 10^6 pc. As is obvious from equation (2-4), the greater the distance to the star, the smaller is its parallax. This fact makes it possible to measure the *trigonometric parallaxes* of the nearer stars against the background of very distant stars.

Because the stars are far away, their parallaxes are small. For several hundred years, numerous attempts to measure them to "prove" the heliocentric theory of the solar system were frustrated because the available observational techniques were not accurate enough to detect parallax shifts, even though they *are* there. It might be good to keep this historical example in mind when considering contemporary attempts, often still unsuccessful, to decide important questions about the structure of our Galaxy and the Universe by means of difficult observations. It was not until 1838 that F.W. Bessel finally determined the parallax of the star 61 Cygni as $0''.29$. With present-day equipment, parallaxes are routinely measured out to distances of the order of 50 pc (and sometimes beyond, but with great uncertainty).

In principle the method for determining parallax is simple. A photograph of star S is taken when the Earth is at E_1, and another is taken when it is at E_2. The position of star S is measured on each photograph with respect to faint background stars (which are presumably very distant), and the difference in these positions is then twice the *relative parallax*. Naturally, the procedure is more complicated in practice. Many photographs are required, great care must be taken in making the measurements, and an elaborate set of reductions must be made [see (**M1**), (**S3**), (**V1**) for details]. The reference stars do not actually lie at infinite distances, and they must therefore have finite (if small) parallaxes themselves. They will thus move on parallactic ellipses similar to that of star S (but, ideally, of much smaller amplitude), and a correction must be applied to reduce the measured relative parallax to an *absolute parallax*, which refers to a frame truly fixed in space. The reduction to absolute parallax is usually made by applying a statistical correction, which depends on the brightness of the reference stars and the galactic coordinates of the field. Alternatively, absolute parallaxes could, in principle, be measured with respect to objects known in advance to have negligibly small parallaxes, for example, compact external galaxies or stars that are known on astrophysical grounds (from their spectra) to be intrinsically bright but apparently faint because they lie at enormous distances.

Because of the large effort required to determine parallaxes, observers will often attempt to select stars that have a high probability of yielding a measurable result, unless there is some overriding astrophysical reason for including a particular star in their program. A common way of doing this is to choose stars that have large proper motions because, for a given limit on their space velocities, the nearest stars will generally show the largest proper motions. Although this is a reasonable (if not infallible) strategy, it introduces a *selection effect* that leads to a statistical bias in the group of stars known to be nearby, and the sample of such stars is thus not complete. We shall return to this point in Chapter 4 in our discussion of the stellar composition of our Galaxy in the solar neighborhood.

All trigonometric parallaxes determined before about 1963 are collected in the *Yale Parallax Catalog* (see Table 2-6); more recent results are scattered through the literature. The largest known parallax is 0".75 for α Centauri, which places it at a distance of 1.3 pc. The second largest is 0".55 for Barnard's star—the largest proper-motion star—which is at a distance of 1.8 pc. The typical error in a parallax determination has long been about $\pm 0''.01$, but modern work with improved techniques now achieves errors of about $\pm 0''.005$. In a few cases, the use of a very large number of plates and the most modern methods allows the errors to be reduced to $\pm 0''.002$. Given the typical errors for modern parallax work, we see that the distance of a star at 20 pc ($\pi'' = 0''.05$) is uncertain by about 10%, which implies a 20% uncertainty in the determination of its intrinsic brightness (see §2-6). The fractional errors at larger distances quickly become unacceptably large, hence detailed astrophysical studies based on trigonometric parallaxes are generally confined to stars within 20 to 25 pc. Stars within these distance limits have been collected into special catalogs (see Table 2-6).

There are about 1900 stars known to be within 25 pc, and about 1300 within 20 pc, of which only perhaps 900 really have distances known to an accuracy of 10%. There are 52 individual stars with parallaxes $\geq 0''.20$ (that is, $d \leq 5$ pc). Many of these are members of *visual binaries* (see §3-1), and the number of distinct systems of one or more stars within 5 pc is 39, so that the visible *multiplicity ratio* is 1.3. If one accounts for the fact that some of these stars are known to be *spectroscopic binaries* (see §3-1) and that some are suspected to have invisible companions, the total number of stars within the 39 systems rises to 61, and the multiplicity ratio is 1.5. Estimates of the completeness of the parallax data suggest that we have identified only about 35% of the stars actually within a sphere of a 20-pc radius and that the degree of incompleteness is worst for intrinsically faint stars.

The nearby stars provide a sample of a typical population of stars in the disk of our Galaxy. A striking feature of this sample is that *most of the stars in it are intrinsically fainter than the Sun*, which shows that stars of low luminosity are quite common. Of the stars within 5 pc of the Sun, only one (Sirius) is intrinsically brighter, and only one other (α Cen) is as bright. A sample selected according to apparent brightness (such as found in the *Yale Bright Star Catalog*) is radically different in character; it contains a preponderance of intrinsically bright stars which, though actually rare, can easily be seen to large distances. We shall return to this point in Chapter 4.

To determine the intrinsic brightness of a star, we must know its distance (see §2-6). Thus the limitation that trigonometric parallaxes are reliable only for distances $\lesssim 25$ pc is a severe one, for there are many types of stars of great astrophysical importance that are simply too rare to be found within this volume. To calibrate the intrinsic brightness of such stars, we must develop alternative methods for estimating their distances. Some of these methods are geometric in nature and hence are in some sense as fundamental as trigonometric measures (though in most cases they invoke additional hypotheses). But, for many stellar types, the calibrations can be made only by comparing stars of unknown intrinsic brightnesses with stars whose brightnesses have already been calibrated geometrically and with which they are known to be physically associated (for example, those in binaries and clusters). We shall return to these points in Chapter 3; for the present, we wish only to stress that trigonometric parallaxes for nearby stars play a key role in fixing the intrinsic brightness of practically every other kind of astrophysical object and thus ultimately in setting the distance scale for the entire Universe.

As was true for positions and proper motions, parallax measurements from space observatories may yield a considerable improvement in accuracy. With observations from astrometric satellites, it may be possible to achieve errors of $\pm 0\overset{''}{.}003$ (**M6**, 361), and from the Space Telescope, errors of only $\pm 0\overset{''}{.}001$ (**G1**, 283). If this could be done, then we could survey reliably to at least five times the present distance limit, that is, through a volume 125 times as large. Among other things, it would then be possible to obtain reliable distances for stars in the Hyades cluster (the importance of which is discussed in §3-1) and for subdwarfs (see §3-5).

2-4. RADIAL VELOCITIES

When a source of radiation moves toward or away from an observer, the observed wavelengths of photons will be different from their emitted wavelengths. To the lowest order in v/c, the wavelength difference $\Delta\lambda$ is given by the Doppler formula

$$\Delta\lambda = \frac{v_R}{c}\lambda_0 \tag{2-5}$$

where c is the velocity of light, λ_0 is the wavelength of the radiation in the rest frame of the source, and v_R is its *radial velocity* (that is, its velocity along the line of sight to the observer). In astronomical work, the sign of v_R is taken to be positive if the source moves away from the observer and negative if it approaches. Thus $\Delta\lambda$ for a receding source is positive, and the observed spectrum is *redshifted* relative to its rest wavelength; for an approaching source, $\Delta\lambda$ is negative and the observed spectrum is *blueshifted.*

When we examine the spectrum of a star, we can identify the observed spectrum lines with those produced by various chemical elements in definite states of excitation and ionization; from terrestrial experiments, we know the rest wavelengths of these lines. Now suppose that we photograph the spectrum of a star and that, on the same plate, we also photograph the spectrum of a laboratory source at rest (typically an arc or a discharge tube). The laboratory spectrum provides a reference scale of wavelengths directly on the plate, so by measuring the positions of the stellar lines with respect to those in the comparison spectrum, we can determine their wavelength shift and hence the radial velocity of the star with respect to the spectrograph. After correcting for any components of the Earth's orbital velocity (30 km s^{-1}) and rotation velocity (0.5 km s^{-1}) along the line of sight, we obtain the *heliocentric radial velocity* of the star (see **S6**, Chapter 7 for additional discussion).

Stellar radial velocities are needed if we wish to know the space velocity of stars with respect to the Sun. They are thus essential to our understanding of the kinematics of our Galaxy. Some stars are observed to have variable radial velocities. These may be caused by the orbital motion of a star around a companion (see §3-2) or by pulsation of the star (see §3-7). In either case, the radial velocity data can be analyzed to yield astrophysically important information.

Radial velocity measurements are usually made on high-dispersion spectrograms in order to obtain high accuracy. For example, suppose we wish to determine v_R to within ± 1 km s^{-1} (the error quoted for a high-quality velocity in the *Mount Wilson Catalog of Radial Velocities*; see Table 2-6); then, at λ4500 Å, we must measure $\Delta\lambda$ to ± 0.015 Å, or to within ± 3 microns on a plate with a dispersion of 5 Å/mm! A complication that enters here is that some stellar lines are actually *blends* of several features, and slight adjustments, depending on spectral type, must be made in the rest wavelengths adopted for the stellar lines. In general terms, the best accuracy is obtained for stars of near-solar temperature and cooler, which usually have spectral lines that are sharp and easy to measure. For stars that are hotter than the Sun, the lines are often broad and diffuse, owing to the effects of pressure broadening and stellar rotation; the uncertainties of measurement are then correspondingly larger. Recently, a photoelectric method for measuring radial velocities has been developed **(G3)**, **(G4)**, **(G5)**. This approach gives very high accuracy (± 0.2 km s^{-1}) for bright stars and good accuracy (± 1 km s^{-1}) for stars that were hitherto much too faint for standard photographic techniques, such as individual members of globular clusters.

2-5. STELLAR SPECTRA

The richest source of information about stars are their spectra. The recovery and exploitation of this information has long been one of the central problems of astrophysics. The two main approaches to this problem are based on different philosophies, use complementary methods, and aim at meeting different needs. We shall categorize them as *spectrum analysis* and *spectral classification.* In this section, we shall deal mainly with the latter, but it is worthwhile to consider both in broad terms for the sake of orientation.

When we examine the light from a star with a spectrograph, we observe a bright *continuum* of radiation, upon which are superimposed *absorption lines* (and, occasionally, *emission lines*). The distribution of energy in the continuum, and both the profiles (fractional depth into the continuum as a function of wavelength) and strengths (integrated absorption) of the lines are determined by the physical conditions in the stellar atmosphere—temperatures, densities, velocity fields, magnetic fields, element abundances, and so on. Hence, by a suitable theoretical analysis of these features, we can derive a detailed physical picture of the structure and composition of the outer layers of stars.

The basic data required for a spectrum analysis are quantitative descriptions of line strengths and profiles and descriptions of the frequency variation of the continuum. These data are obtained from *spectrophotometric* measurements. The continuum information can be obtained from *color indices* (see §2-6) or from absolute energy distributions (see §2-7); the line data are obtained from high-dispersion spectra. The basic philosophy of the procedure is to describe the spectrum with a small number of numerical indices and then, by a theoretical calculation using established physical laws, to determine the physical conditions in the stellar atmosphere that are required to match these numerical values. The procedure is complicated, and we shall not discuss it here [see (**M2**)], although we shall use the results of such work in following chapters, particularly Chapter 3. In practice, these analyses are very time consuming and are restricted to relatively bright stars for which suitable spectra can be obtained.

In spectral classification, one attempts, from a study of the morphology of the spectrum, to group together stars with very similar physical characteristics. In principle, one may use the whole spectrum; in practice, present-day classification is based on only the part of the spectrum transmitted through the Earth's atmosphere. One first chooses one or more sets of labels (*dimensions*) with which to categorize the spectrum. The standard system now in use (the MK system) employs two sets of labels; that is, it is *two dimensional.* For each choice of labels, one declares the name of a standard star (or stars) whose spectrum defines what is meant by that particular classification. Successive choices of labels to which standards are assigned establish classification "boxes," and the variation of the nature of the spectrum from box to box is fixed by the properties of the standards in those boxes. Each classification box corresponds to a unique *spectral type* of the system. When the whole system is defined, one classifies a star by finding the standard whose spectrum it most closely matches (or interpolates between standards, if necessary). In this way, each star is put into one of the boxes of the system. We shall see later that similar systems are used for classifying star clusters, galaxies, and clusters of galaxies.

If an astute choice of dimensions has been made, and if the classification process itself is done with care using well-defined criteria, then (1) a group of stars having the same type will, in fact, be nearly identical to one another in those physical properties that produce variations of the classification criteria (but not necessarily in all properties), and (2) the groups with different types will be distinguishable from one another in a significant way. After the system has been defined and the classification carried out, one then calibrates the system in terms of physical parameters, such as stellar temperatures, luminosities, compositions, and so on. Quantitative estimates of these parameters are derived for each spectral type by performing a detailed spectrum analysis on a typical member of that type. At that point, one can say that, if some star has a certain spectral type, then the temperature, luminosity, or other property appropriate to that type can be assigned to it without further analysis. The spectral type thus gives a concise description of both the spectrum and the physical properties of a star.

In practice, classification is normally done by visual inspection of moderate-dispersion spectrograms, and therefore it can be carried out for large numbers of stars. It is also possible to classify huge numbers of stars with less accuracy on low-dispersion (objective-prism) plates, by the empirical development of criteria that can be seen at low dispersion but yet uniquely characterize a particular classification box (or boxes).

Pioneering work in spectral classification was done in the 1860s by A. Secchi, who divided stars into four broad *spectral classes*. Parallel efforts were made about the same time by W. Huggins and H. C. Vogel. The first great steps toward our present system were made at Harvard College Observatory in 1890. Under the direction of E. C. Pickering, Williamina P. Fleming published a catalog of 10,000 stars grouped into a system of spectral classes denoted by the letters A, B, C, and so on. In 1888, Antonia C. Maury, without benefit of astrophysical data (which was almost nonexistent at that time), rearranged these spectral classes into the order that has been used ever since, solely by studying the progression of line patterns observed in the spectra. Subsequently, Annie J. Cannon introduced decimal subdivisions of the spectral classes, and, during the interval from 1911 to 1924 (with extensions in 1949), she compiled results for hundreds of thousands of stars in the classical *Henry Draper Catalog* (HD)—see Table 2-6. From the observed variation of ratios of line strengths of successive ionization stages of the chemical elements, and from photometric data, this *spectral sequence* O, B, A, F, G, K, M was later recognized to be primarily a temperature sequence, listed here in order of decreasing temperature. At the cool end of the sequence, additional classes R, N, and S were added to describe stars in the K-M temperature range that have markedly different compositions. Finally, with the advent of M. Saha's ionization theory in 1920 (**S1**), (**S2**), quantitative analysis became possible, and in 1925 a comprehensive theoretical interpretation of the Harvard spectral sequence appeared in Cecilia Payne's book *Stellar Atmospheres* (**P1**).

With the work of E. Hertzsprung and H. N. Russell in the years from 1905 to 1913, it became evident that stars of a given spectral class could have vastly different luminosities (see §3-5), and this fact implied that they could have markedly different radii and atmospheric densities. The most common, fainter stars are called *dwarfs* (or *main-sequence* stars); brighter, larger stars are called *giants*; and the brightest, largest stars are called *supergiants*. The effects of differing envelope sizes are reflected directly in changes in the spectrum, and therefore a second parameter is required to describe the spectrum completely. A start in this direction was made in the 1890s by Antonia Maury at Harvard, who added the symbols a, b, or c to some spectral types; we now know that her c class corresponds to supergiants and the others correspond to dwarfs. Much later, a fully two-dimensional system—the *MK system*—was developed by W. W. Morgan and P. C. Keenan (**M5**), who added a *luminosity class* as the second classification parameter. In physical terms, this parameter reflects stellar envelope size and atmospheric density and hence correlates with stellar surface-gravity. From the point of view of galactic structure, the luminosity class is closely correlated with stellar luminosity. The MK system, with subsequent revisions and extensions [(**J1**), (**S6**, Chapter 8), (**A1**), (**M3**), (**M4**)], has become the standard classification system in use today.

The basic precepts of the MK system are:

1. It is empirical; only directly observable features of the spectrum are used to determine a star's classification.
2. It is based on homogeneous material. It uses well-widened spectra with dispersion of around 60–130 Å/mm, which gives high enough resolution to provide sensitive criteria but low enough dispersion to allow one to reach faint stars, and hence ones at large distances in our Galaxy.
3. It is defined by standards. Thus the classification system is autonomous, in the sense that it remains unchanged even when the interpretation of the classes in terms of physical conditions in the stars changes, as models of stellar structure are refined. Furthermore, observers using different spectrographs and dispersions can classify on the same system simply by reobserving the standard stars with their own particular equipment.

The spectral classes of the MK system are essentially those of the Harvard sequence, and some of the principal spectral features characterizing each of

Table 2-1. Principal Characteristics of Spectral Classes

Spectral class	Spectral features
O	He II lines visible; lines from highly ionized species, for example, C III, N III, O III, Si IV; H lines relatively weak; strong uv continuum
B	He I lines strong, attain maximum at B2; He II lines absent; H lines stronger; lower ions for example, C II, O II, Si III
A	H lines attain maximum strength at A0 and decrease toward later types; Mg II, Si II strong; Ca II weak and increasing in strength
F	H weaker, Ca II stronger; lines of neutral atoms and first ions of metals appear prominently
G	Solar-type spectra; Ca II lines extremely strong; neutral metals prominent, ions weaker; G band (CH) strong; H lines weakening
K	Neutral metallic lines dominate; H quite weak; molecular bands (CH, CN) developing; continuum weak in blue
M	Strong molecular bands, particularly TiO; some neutral lines, for example, Ca I, quite strong; red continua
C	Carbon stars; strong bands of carbon compounds C_2, CN, CO; TiO absent; temperatures in range of classes K and M
S	Heavy-element stars; bands of ZrO, YO, LaO; neutral atoms strong as in classes K and M; overlaps these classes in temperature range

these classes are listed in Table 2-1. The luminosity classes, and the stars to which they pertain, are listed in Table 2-2. The characteristics mentioned in these lists are only illustrative; the system is *defined* by standard stars. The complete spectral type is specified by both the spectral class and the luminosity class of a star as determined by comparison with the standards. In Chapter 3, we shall give detailed tables of the stellar physical properties (for example, temperatures and luminosities) that have been associated with MK spectral types through astrophysical calibrations.

Spectral classes are subdivided into decimal subclasses, running from 0 at the hot end through 9 at the cool end: for example, B0, B1, B2, . . . , B9;

Table 2-2. MK Luminosity-Class Designations

Ia-0	Most extreme supergiants
Ia	Luminous supergiants
Iab	Moderate supergiants
Ib	Less-luminous supergiants
II	Bright giants
III	Normal giants
IV	Subgiants
V	Dwarfs

Table 2-3. Distribution of Stellar Types in HD Catalog ($V \leq 8.5$)

Spectral class	Percent
O	1
B	10
A	22
F	19
G	14
K	31
M	3

A0, A1, A2, . . . , A9; F0, F1, F2, . . . , F9, and so on. The luminosity classes are usually not subdivided except for supergiants. Examples of spectral types are: Sun (G2V), ε Ori (B0Ia), α Lyr (A0V), α Tau (K5III). Stars hotter than the Sun (classes O, B, A, F) are commonly called *early types*, and solar-type and cooler stars (classes G, K, M) are called *late types*. (These designations are archaic remnants of an obsolete scheme of stellar evolution and are devoid of physical significance, but they are universally used by astronomers.) Although it is explicitly two-dimensional, the MK system implies the existence of and the need for formal consideration of a third (or more) dimension in localized regions of the system in order to describe, say, the *weak-lined stars* (for example, the subdwarf HD 140283 or the variable star RR Lyrae) or *peculiar stars* (see the following discussion).

Perhaps 90% or more of all stars can be classified with standard MK spectral types. The relative numbers of stars in different spectral classes in the HD Catalog are listed in Table 2-3, and a similar distribution for MK types is given in Table 2-4. Both of these tables refer to apparently bright stars, and they do not reflect true space densities, which are discussed in Chapter 4.

Additional spectral classes are used for certain unusual stars: Class C (which replaces Harvard classes R and N) for carbon stars; S for heavy-

Table 2-4. Distribution of MK Types of Apparently Bright Stars (%)

Spectral class \ Luminosity class	V	IV	III	II	I
O, B	10	3	6	2	3
A, F	14	3	5	1	4
G, K, M	1	4	25	6	4

metal stars; and WC and WN (or sometimes WR) for Wolf–Rayet stars, which are early-type stars that show strong, broad emission lines. Additional letters may be appended to spectral types to denote special characteristics; for example, p(peculiar), e(emission lines), f(He II and N III emission in O stars), and n(broad lines). Some stars have special notations. Some examples are the *peculiar A stars* (Ap), which show strong lines of certain elements (Si, Mn, Cr, Sr, Eu) and have strong magnetic fields, and the *metallic-line A stars* (Am), which show abnormally strong metal lines and weak Ca II. The prefix w or D is often used to denote *white dwarfs* (wA, wG, DA, and so on), and the prefix sd denotes *subdwarfs*, extreme metal-poor stars (for example, sdG). In some work, particularly at low dispersion, it is not possible to derive accurate MK luminosity classes, although it is possible nevertheless to distinguish among dwarfs, giants, and supergiants; this is of interest because we can then assign intrinsic brightnesses to the stars. In such cases, the prefixes d, g, and c are often used to denote these three groups (for example, dK, gK, cA).

The standard stars that define the MK system are given in (**M5**), (**J1**), (**A1**), (**M3**), and (**M4**). A fairly large number of stars have been classified on this system, and extensive lists can be found in the literature (see Table 2-6). Recently, a repeat of the HD Catalog has been started at the University of Michigan Observatory, using extremely high-quality objective prism plates. This survey will yield almost a full two-dimensional classification, very nearly on the MK system, for most stars in the HD catalog. When completed, it will have enormous value for studies of galactic structure.

2-6. MAGNITUDES AND COLORS

In addition to measuring the positions and motions of stars, we can also measure their light output. Every astronomical body emits radiation over much of the electromagnetic spectrum. Ideally, we should like to measure the complete spectral distribution of this radiation and to determine the energy received by an observer in terms of the flux per unit frequency interval f_ν in ergs cm^{-2} s^{-1} hz^{-1} or the flux per unit wavelength interval f_λ in ergs cm^{-2} s^{-1} Å^{-1} over the entire spectrum. The measurement of such *absolute energy distributions* (see §2-7) is difficult, for two reasons. First, a determination of the absolute response of the observing equipment must be made. Second, electromagnetic radiation can penetrate to the earth's surface only in certain limited ranges of the spectrum. Three major windows of atmospheric transmission exist: (1) the *optical window*, roughly in the range 3200 Å $\lesssim \lambda \lesssim$ 7500 Å; (2) the *infrared window*, from about 0.75μ to 20μ, in which atmospheric absorption bands alternate with transmission bands; (3) the *radio window*, which transmits roughly from a few millimeters to 50 m. (We shall defer further discussion of radio measurements until Chapter 8.) Access to the X-ray and ultraviolet (uv) regions is gained from satellites, and access to the submillimeter region is gained from high-altitude aircraft and balloons.

For many purposes, we do not need to find the complete energy distribution of a star but merely the total energy received by a detector in some definite range of wavelengths, that is, the integrated radiation flux f, measured in units of ergs $\text{cm}^{-2}\ \text{s}^{-1}$, contained in a wavelength band $\Delta\lambda$ set by the instrumental response. In particular, we can use detectors such as photographic plates and photoelectric cells to measure the apparent brightnesses of stars in various bands in the optical region of the spectrum; this procedure is called *astronomical photometry*.

Apparent Magnitudes

The apparent brightnesses of stars are expressed in terms of their apparent magnitudes. The ancient Greek astronomers divided stars into six magnitude groups judged to be separated by equal steps in brightness. The brightest stars were of the first magnitude, and the faintest that could be seen by the naked eye were of the sixth magnitude. From subsequent physiological studies, it was learned that equal steps of brightness sensed by the eye correspond fairly well to equal ratios of radiant energy; that is, the response of the eye to stimulus by light is essentially logarithmic. Thus, if m_1 and m_2 denote the magnitudes assigned to stars with energy fluxes f_1 and f_2, then

$$m_1 - m_2 = -k \log_{10}\left(\frac{f_1}{f_2}\right) \tag{2-6}$$

where the minus sign is chosen so as to assign smaller numerical values to brighter stars.

Photometric studies in the nineteenth century showed that sixth-magnitude stars are about a hundred times fainter than first-magnitude stars. Hence, following the suggestion of N. Pogson, the magnitude system was defined such that a difference of 5 mag corresponds exactly to a factor of one hundred in the ratio of radiation fluxes. Thus, for $(f_1/f_2) = 100$, $m_2 - m_1 = 5$. Hence, in equation (2-6), $k = 2.5$, and, in general,

$$m_1 - m_2 = -2.5 \log_{10}\left(\frac{f_1}{f_2}\right) \tag{2-7}$$

or

$$\frac{f_1}{f_2} = 10^{-0.4(m_1 - m_2)} \tag{2-8}$$

Notice that if $(f_1/f_2) = 1 + \Delta f$, where $\Delta f \ll 1$, then $\Delta m \equiv m_2 - m_1$ is given by $\Delta m \approx 1.086\ \Delta f$; that is, the magnitude difference (when small) between two objects is about equal to the fractional difference in their relative brightness. It follows from equation (2-7) that a magnitude difference of 1 mag corresponds to a flux ratio of $(100)^{1/5} \approx 2.512$, and 2.5 mag corresponds to a flux ratio of 10.

The precise definition of the observed energy flux f has thus far been left rather vague; we take it to be

$$f \equiv \int_0^\infty f_\nu^o T_\nu R_\nu F_\nu \, d\nu = \int_0^\infty f_\lambda^o T_\lambda R_\lambda F_\lambda \, d\lambda \tag{2-9}$$

where f_λ^o denotes the stellar flux incident outside the Earth's atmosphere; T_λ is the transmission of the atmosphere; R_λ is the efficiency of the telescope-spectrograph-receiver system, that is, $R_\lambda =$ (instrumental response/incident energy)$_\lambda$; and F_λ is the transmission of a filter, which can be put into the system to isolate a particular range of wavelengths. The effects of atmospheric transmission can be eliminated, as described in the following paragraph. The instrumental efficiency is a composite of the reflectivity of telescope mirrors (or transmission of lenses), the efficiency of the spectrograph, and the sensitivity of the photon receiver itself. While mirror reflectivities can be measured fairly easily, the other two factors just mentioned are extremely difficult to determine accurately a priori. In practice, therefore, the system must be calibrated by measuring its response to a source whose absolute energy output is known with precision (see §2-7). Finally, the filter transmissions are readily determined and can, in fact, be chosen at will to measure the energy contained in definite wavelength intervals. Most photometric systems employ several different filters, and the filter band used must always be stated when giving an apparent magnitude.

The transmission of the atmosphere is proportional to $\exp(-a)$, where a is the column density of air, or *air mass*, along the line of sight. For a plane-parallel atmosphere, $a \propto \sec z$, where z is the zenith distance of the object being observed. Thus *atmospheric extinction* produces a dimming of starlight according to the relation

$$m(z) = m(z = 0) + k \sec z \tag{2-10}$$

where $m(z)$ is the observed stellar magnitude at zenith distance z. We can correct for atmospheric extinction (at wavelengths where the atmosphere is not totally opaque) by observing the magnitude of a star at several values of z. A fit to these data then yields the constant k in equation (2-10) and allows us to correct not only to unit air mass at $z = 0$ but also to extrapolate to zero air mass (that is, "$\sec z = 0$") and thus obtain the magnitude outside the Earth's atmosphere [see (**H2**, Chapter 8) for details].

Until the 1950s, most astronomical photometry was done photographically with two systems: the blue-violet-sensitive *international photographic system* giving magnitudes m_{pg}, and the *photovisual system*, m_{pv}, whose wavelength sensitivity simulates that of the eye. Unfortunately, the photographic plate has a nonlinear response to different levels of light intensity, and the relation between incident intensity and photographic darkening must be determined empirically by calibrating each plate. Furthermore, the dynamic response range of a plate is only about a factor of twenty in intensity; to cover larger ranges, a set of several plates must be used. Given such properties of the detector, it becomes very difficult to do absolute photometry and to extend a magnitude scale over a wide range of intensity without making large systematic errors. Many of the results in the older photographic photometric catalogs (see Table 2-6) are, unfortunately, seriously affected by such errors. On the other hand, if some stars on the plate have accurately known brightnesses, then it is usually possible to make differential measurements to ± 0.1 mag, which is adequate for many purposes. In the best cases, such differential measurements may be accurate to ± 0.03 mag.

Astronomical photometry was revolutionized with the advent of photoelectric photometers. These instruments are strictly linear; they have (with suitable auxiliary electronics) an enormous dynamic range; and they are capable of factor-of-ten better precision than photographic plates, yielding magnitudes accurate to ± 0.01 mag and magnitude differences often accurate to ± 0.002 mag. At this level of accuracy, color-magnitude diagrams of star clusters become sensitive diagnostic tools for the study of stellar evolution (see §3-6).

The standard photometric system today is the *ultraviolet-blue-visual* (*UBV*) *system* of H. L. Johnson and W. W. Morgan (**J1**), (**S6**, Chapter 11), which employs a photometer with a 1P21 photomultiplier tube on a telescope with two aluminum-coated mirrors and three filters, whose characteristics are listed in Table 2-5. Magnitudes in this system are denoted by the capital letters designating the filter; thus V magnitudes are now the standard visual magnitudes. The system is defined by standard stars whose magnitudes and colors are listed in the references cited. Magnitudes and colors observed with other telescopes and photometers (similar to, but never exactly the same as the original equipment) using *UBV* filters can always be transformed precisely back to the original *UBV* system via the standards. Examples of V magnitudes are: Sun, $V = -26.74$; Sirius (apparently brightest star), $V = -1.45$; Vega, $V = +0.04$; faintest stars measured, $V \approx 23$. The total range from the Sun to the faintest measurable stars is about 50 mag or a ratio of 10^{20} in apparent brightness! A recent photoelectric photometric catalog (see Table 2-6) lists magnitudes for about 20,000 stars, and many other lists can be found in the literature.

The methods of photoelectric and photographic photometry are essentially complementary, and, taken together, they make an effective team. The photoelectric method is well suited for precision measurements of individual objects; it can therefore provide, for example, accurate light curves for variable stars (see §3-7). Because the measurements can be made on only one star at a time, this approach is unsuitable if we need magnitudes for thousands of stars in a field. In contrast, we can obtain images of enormous

Table 2-5. Filter Characteristics of Astronomical Photometry Systems

System	Filter	λ_0	$\Delta\lambda_{1/2}$
UBV (Johnson-Morgan)	*U*	3650 Å	700 Å
	B	4400 Å	1000 Å
	V	5500 Å	900 Å
Six-color (Stebbins-Whitford-Kron)	*U*	3550 Å	500 Å
	V	4200 Å	800 Å
	B	4900 Å	800 Å
	G	5700 Å	800 Å
	R	7200 Å	1800 Å
	I	10,300 Å	1800 Å
Infrared (Johnson)	*R*	7000 Å	2200 Å
	I	8800 Å	2400 Å
	J	1.25μ	0.38μ
	K	2.2μ	0.48μ
	L	3.4μ	0.70μ
	M	5.0μ	1.2μ
	N	10.4μ	5.7μ
$uvby\beta$ (Strömgren-Crawford)	*u*	3500 Å	340 Å
	v	4100 Å	200 Å
	b	4700 Å	160 Å
	y	5500 Å	240 Å
	β	4860 Å	30 Å, 150 Å

numbers of stars simultaneously on a single photographic plate, but accurate photometry can be done only differentially, relative to known standards. Therefore, we combine the methods by measuring the magnitudes of a moderate number of stars in a field photoelectrically to high precision and then using these stars as absolute reference standards in subsequent photographic measurements of magnitudes for large numbers of stars on plates of the same field. In this way, we exploit both the high intrinsic precision of photoelectric photometry and the enormous information capacity of photographs to determine relatively accurate magnitudes (with typical errors of a few hundredths of a magnitude) for large numbers of stars, say, within a cluster.

Absolute Magnitudes

The energy flux we receive at the Earth from a star depends on both its intrinsic brightness and its distance. If F is the flux received when the star is at distance D, the flux f that would be received if it were at some other distance d is given by the inverse square law

$$f = \left(\frac{D}{d}\right)^2 F \tag{2-11}$$

Obviously, the farther away a star is, the fainter it will appear, and to obtain information about the relative intrinsic brightnesses of stars, we must account for differences in their distances from us. We therefore define the *absolute magnitude* M to be the apparent magnitude a star would have if it were located at some standard distance D. From equations (2-7) and (2-11), we see that

$$m - M = -2.5 \log\left(\frac{f}{F}\right) = 5 \log\left(\frac{d}{D}\right) \tag{2-12}$$

The standard distance D is always taken to be 10 pc, so if d is measured in parsecs, then

$$m - M = 5 \log d - 5 \tag{2-13}$$

The quantity $(m - M)$ is called the *distance modulus* of a star. If we know m and d, we can immediately correct for the nonstandard distance of the star and reduce the apparent magnitude m to the absolute magnitude M via equation (2-13). Conversely, if we know m and M, we can infer d.

Absolute magnitudes are normally derived from visual apparent magnitudes and are denoted M_V. It should be noted that the absolute magnitude is *not* a direct measure of the total energy output (*luminosity*) of a star, but only of the energy in the V band. To measure total energy output, we use so-called *bolometric magnitudes*, M_{bol}, which will be discussed in §3-4.

One of the important practical problems of galactic astronomy is the determination of M_V for each MK spectral type. As we shall see in §3-1, a variety of techniques must be employed to effect this calibration, and the results of these procedures will be summarized in §3-5. Once the relation $M_V = f(\text{spectral type})$ is known, an observation of a star's apparent magnitude and its spectral type yields an estimate of its distance. This method of estimating distance is referred to as the *spectroscopic parallax* method.

Note in passing that the distance from the Earth to the Sun is 1 a.u. = (1/206,265) pc. Thus we know immediately that the distance modulus of the Sun is -31.57 mag and therefore that the absolute magnitude of the Sun is $M_V(\odot) = +4.83$. We shall see in Chapter 3 that the Sun is a star of rather average intrinsic brightness—the most luminous stars are about 10^6 times brighter, and the least luminous are about 10^4 times fainter.

Finally, we must caution that, up to this point, we have tacitly assumed that the light from stars suffers no absorption on its journey to the earth; *this is not the case in reality*. In fact, there is material between the stars, the *interstellar medium*, which absorbs stellar radiation, thus causing the stars to appear dimmer than they would from distance effects alone and hence increasing their apparent magnitudes. This absorption is usually expressed in magnitude units. If there are A magnitudes of interstellar absorption, then equation (2-13) must be rewritten as

$$m = M + 5 \log d - 5 + A \tag{2-14}$$

In practice, we must always correct apparent magnitudes for the (often large) effects of interstellar absorption, or else they will yield spuriously large estimates of stellar distances.

Color Indices

Suppose we have a photometric system with several filter bands at different wavelengths. Then, by taking the difference in magnitudes measured in two different bands, we can form a *color index*, or *color*. That is, if A and B denote two different filters, we write

$$(C.I.)_{AB} \equiv m_A - m_B = \text{const.} - 2.5 \log \frac{\int_0^\infty S_\lambda(A) f_\lambda^0 \, d\lambda}{\int_0^\infty S_\lambda(B) f_\lambda^0 \, d\lambda} \tag{2-15}$$

where S_λ denotes the combined telescope-receiver-filter sensitivity. A color index is usually written using the letters that denote the different filters involved, that is, $(A - B)$ for the hypothetical example just given or $(B - V)$ and $(U - B)$ for the standard UBV system. As is clear from equation (2-15), a color index essentially measures the ratio of stellar flux between two characteristic wavelengths (the *effective wavelengths* of the filters). In general, an arbitrary normalization factor and hence an arbitrary zero-point constant are present. By convention, color-index zero-points are chosen so that an average A0 star has the same magnitude at all wavelengths; thus, for example, $(B - V) = (U - B) \equiv 0$ for A0 stars.

The detailed shape of a stellar energy distribution is determined by a few basic physical parameters such as the temperature, surface gravity, and chemical composition of the star. By an astute choice of filters, we can isolate and measure features in the energy distribution that are sensitive to the values of these variables. Then, by a calibration procedure, we can derive correlations between the chosen color indices and the physical parameters so that, in the end, we can use observed colors as diagnostic tools to infer physical properties of stellar atmospheres. For example, it is known both from observation and from theory that the wavelength of maximum emission in stellar energy distributions decreases from the infrared for M stars, to the visible for solar-type stars, into the blue for A stars, and out into the ultraviolet for OB stars. That is, the coolest stars *look* red, solar stars look yellow, and early-type stars look blue. A color index that measures the flux ratio between a short and a long wavelength [for example, $(B - V)$ in the *UBV* system] thus provides a measure of the stellar color (hence the name) and therefore yields information about the atmospheric temperature. As another example, it is known that, in stellar spectra, we can find some spectral bands that contain large numbers of absorption lines whose strengths reflect the abundances of heavier elements (and hence the atmospheric composition) and other bands that are relatively line free. Suitably chosen color indices using such bands can thus give information about the absorption-line strengths (*line-blanketing*) in the continuum of a star and hence give information about its composition. As a final example, it is observed that there is a sharp drop in continuum intensity in early-type stars at wavelengths shorter than about $\lambda 3700$ Å. This drop arises from the sudden onset of continuum absorption from the $n = 2$ level of hydrogen, and it is called the *Balmer jump*. Suitably chosen filters can give an index [$(U - B)$ in the *UBV* system] that in effect measures the size of the Balmer jump, which, it turns out, is a sensitive function of stellar temperature for O–A stars and of surface gravity for A–F stars.

A large number of photometric systems have been devised. In choosing filters for a system, one must balance the desire to reach faint stars against the desire to achieve crisp spectral resolution (by excluding, with narrow passbands, most of the photons received from a given star). Practical compromises between these two opposing considerations have led to the development of systems that are *wide-band* with filter bandwidths $\Delta\lambda \gtrsim 300$ Å, *intermediate-band* with $100 \text{ Å} \lesssim \Delta\lambda \lesssim 300$ Å, or *narrow-band* with $\Delta\lambda \lesssim 100$ Å.

The properties of four widely used photometric systems are summarized in Table 2-5. The first three are wide-band systems, and the last is an intermediate-band system. For each filter, the table lists the wavelength λ_0 of peak transmission and $\Delta\lambda_{1/2}$, the (full) half-intensity width.

The *UBV* system is generally regarded as standard. It is closely coupled to the MK system of spectral types and is well suited for observing even the faintest stars. Two color indices, $(B - V)$ and $(U - B)$, are defined in this system: $(U - B)$ gives essentially a measure of the amount of ultraviolet radiation emitted by a star, and $(B - V)$ measures essentially its temperature, being negative for very blue (hot) stars and positive for red (cool) stars. For example, $(B - V)$ is 0.00 for Vega, +0.65 for the Sun, and +1.81 for Antares. In Chapter 3, we shall give extensive tabulations of stellar properties correlated with particular values of $(U - B)$ and $(B - V)$. The Stebbins–Whitford–Kron six-color system **(S4)**, **(S5)** gives more spectral information than the *UBV* system and extends into the infrared. It is valuable for studies of cool stars and of line-blanketing. The Johnson infrared system is an expansion of the *UBV* system into the far infrared, and it is probably the standard infrared system at present. It is well suited for studies of cool stars. A large number of color indices can be formed using the ten bands, *U*, *B*,

$V, \ldots, L, M, N$. The Strömgren system (**S6**, Chapter 9), (**S7**) has four intermediate-band filters, *uvby*, and a narrow-band pair, β, measuring the Hβ-line and adjacent continuum. The usual color indices employed in this system are $(b - y)$, $(u - b)$, $c_1 \equiv (u - v) - (v - b)$, and $m_1 \equiv (v - b) - (b - y)$. Both $(b - y)$ and $(u - b)$ serve as temperature indicators; c_1 is a temperature indicator for O–A stars and a luminosity indicator for A–F stars; β is a luminosity indicator for O–A stars and a temperature indicator for A–G stars; m_1 is a line-blanketing (composition) indicator for F–G stars and a line-blanketing or peculiarity indicator for A stars.

Finally, it must be remembered that the effects of interstellar absorption on colors have been ignored in this discussion. Because the interstellar medium scatters light more efficiently at short wavelengths than at long wavelengths, transmitted starlight is not only dimmed by interstellar absorption but is also *reddened.* It is therefore necessary to correct observed colors for interstellar reddening in order to derive *intrinsic* colors. As we shall see in §3-8, it is fortunate that interstellar reddening exists, for it is possible to devise combinations of color indices that allow us to determine the amount of reddening present and hence to correct for both reddening and absorption.

2-7. ABSOLUTE ENERGY DISTRIBUTIONS

Color indices provide a good deal of useful information about stellar energy distributions; but, as was mentioned earlier, we should ideally like to know the detailed variation of the energy flux per unit wavelength (or frequency) interval throughout the entire spectrum. Having these data, we can apply the theory of stellar atmospheres and analyze them to infer physical characteristics of stars [see (**M2**, §7-4)]. Furthermore, we can use absolute energy distributions to obtain bolometric magnitudes and stellar temperatures (see §3-4).

The basic problem in this work is the measurement of the absolute efficiency of the telescope-spectrograph-receiver system, that is, the determination of the amount of energy that must be put into the system in order to produce a unit instrumental response. The only practical way to derive this efficiency is to use the instrument to observe a source whose absolute energy distribution is known in advance, for then we know the amount of incident energy that produces the measured responses. There are only two types of sources for which we can specify the rate of energy emission from unchallengeable theory: (1) a *blackbody* source at a known temperature, the energy distribution of which is given by the *Planck function* $B_\lambda(T)$, and (2) a *synchrotron-radiation* source, from which the emission by relativistic electrons can be calculated as a function of their energy. The procedure is to observe such an absolute reference source with a given telescope-spectrometer system and then, with the same system, to observe a standard star (usually Vega).

Observations of the absolute reference source calibrate the instrumental efficiency at each wavelength and thus provide the factor required to convert the observed responses to the stellar radiation into an absolute energy emission from the star (after correction, of course, for atmospheric extinction).

Thanks to the advent of space observations and to persistent effort by several astronomers in developing the necessary apparatus and techniques, great progress has been made in the last decade in measuring absolute energy distributions. Accurate measurements of the absolute energy distribution of Vega on the range 3300 Å $\leq \lambda \leq$ 10,800 Å (**H1**) and of α Vir, η UMa, and α Leo on the range 1370 Å $\leq \lambda \leq$ 2920 Å (**B1**) are now available; the latter data were obtained from rocket-borne spectrometers. The ultraviolet data are given directly in ergs cm^{-2} s^{-1} per unit wavelength interval. The Vega distribution is usually given in terms of a relative absolute energy distribution, written in *monochromatic magnitudes*,

$$m_\nu\left(\frac{1}{\lambda}\right) \equiv -2.5 \log\left[\frac{f_\nu(1/\lambda)}{f_\nu(1/\lambda = 1.8\mu^{-1})}\right] \tag{2-16}$$

plus the absolute flux at $(1/\lambda) = 1.8\mu^{-1}$ (that is, λ5556 Å), for which the present best estimate is $f_\lambda = 3.39 \times 10^{-9}$ ergs cm^{-2} s^{-1} Å^{-1}, or $f_\nu = 3.50 \times 10^{-20}$ ergs cm^{-2} s^{-1} hz^{-1}, or $N_\lambda = 948$ photons cm^{-2} s^{-1} Å^{-1}. Here $f_\nu = f_\lambda|d\lambda/d\nu| = (\lambda^2/c)f_\lambda$, and $N_\lambda = f_\lambda/(h\nu)$. This format is chosen because relative monochromatic magnitudes are easier to determine, and probably more accurately known, than the absolute normalization flux. They are therefore likely to remain unchanged even if the absolute flux requires future adjustment.

Once the absolute energy distribution of even a single star is known precisely, it can henceforth be used as a reference standard directly on the sky. Observations of the absolute energy distributions of other stars are then relatively simple—one merely measures program stars and the reference star with the same equipment, corrects for differences in atmospheric extinction, and then uses the known absolute distribution of the standard to obtain absolute distributions for the program stars. Catalogs of spectrophotometric data for hundreds of stars of various spectral types are now available (see Table 2-6). These data are indispensable for performing astrophysical analyses both of individual stars and of the integrated energy distributions from stellar systems (clusters and galaxies), which are composites of contributions from many different spectral types.

2-8. ASTRONOMICAL CATALOGS AND ATLASES

A list of astronomical catalogs is given in Table 2-6. These catalogs contain information about positions, motions, spectra, magnitudes, and colors of stars and similar data for galaxies and clusters of galaxies. In the table, a

Table 2-6. Some Important Astronomical Catalogs

Catalog	Reference	Principal content	Number of objects
	Stellar survey catalogs		
Bonner Durchmusterung (BD)	*Beob. Bonn Obs.*, **3**, **4**, **5**, 1860	Position and magnitude (plus atlas)	320,000
Southern Durchmusterung (BD)	*Beob. Bonn Obs.*, **8**, 1886	Position and magnitude (plus atlas)	130,000
Cordoba Durchmusterung (CD)	*Result. Nat. Obs. Argentina*, **16**–**18**, **21**a, **21**b, 1892	Position and magnitude (plus atlas)	580,000
Cape Photographic Durchmusterung (CPD)	*Ann. Cape Obs.*, **3**, **4**, **5**, 1896	Position and magnitude	450,000
Carte du Ciel	Various observatory publications	Position and magnitude	$\sim 10^7$
EBL_2	Hamburg Sternwarte, 1936	Proper motion for BD, CD, and CPD stars	95,000
Faint Blue Star Survey (LB)	Luyten et al., Minneapolis, 1954–1969	Position, proper motion, magnitude, color	11,500
Large Proper Motion Survey (LTT, LFT)	Luyten, Minneapolis, 1955–1961	Proper motions $\gtrsim 0''.2$/year	$\sim$20,000
Bruce Proper Motion Survey	Luyten, Minneapolis, 1963	Proper motion	98,000
48″ Schmidt Proper Motion Survey	Luyten, Minneapolis, 1963–1971	Proper motion	$\sim$50,000
Lowell Proper Motion Survey	Giclas et al., Lowell Obs., 1971	Proper motion	9,000
South Galactic Pole Survey	Luyten and La Bonte, Minneapolis, 1973	Proper motion	7,000
North Galactic Pole Survey	Luyten, Minneapolis, 1976	Proper motion	11,000
Large Proper Motion Survey (LHS)	Luyten, Minneapolis, 1976	Proper motion $\gtrsim 0''.5$/year	6,000

Table 2-6. *(continued)*

Catalog	Reference	Principal content	Number of objects
	Stellar precision catalogs		
AGK_1	Astron. Gesellschaft, Leipzig and Paris, 1890–1924	Position	~500,000
Yale Catalog (reobservation of AGK_1)	*Yale Obs. Trans.*, **4**–**31**, 1925–1971	Position and proper motion	~210,000
Cape Zone Catalog	Jones, Cape Obs., 1927, 1936	Position, proper motion, magnitude	~21,000
General Catalog	Boss, *Carnegie Inst. Wash. Pub. No. 486*, 1936	Position and proper motion	33,000
Cape Catalog (Epochs 1925, 1950)	Jackson et al., 1949–1953	Position and proper motion	23,000
N30	*Astr. Papers, Amer. Ephemeris*, **13**, part III, 1952	Position and proper motion	5,300
AGK_2	Hamburg Sternwarte, 1950–1958	Position	82,000
FK4	Astron. Rechen-Inst., Heidelberg, 1963	Position and proper motion	1,500
Smithsonian Star Catalog	Smithsonian Ap. Obs., 1966	Position, proper motion, magnitude, spectral type	260,000
AGK_3	Hamburg Sternwarte, 1975	Position and proper motion	183,000
	Selected areas		
Harvard-Groningen Selected Areas	Pickering, Kapteyn, and van Rhijn, *Harvard Ann.*, **101**–**103**, 1920	Magnitude, 206 areas ($15' \times 15'$)	250,000
Mount Wilson Selected Areas	Seares, Kapteyn, and van Rhijn, *Carnegie Inst. Wash. Pub. No. 402*, 1930	Magnitude, 139 areas	68,000

Radcliffe Catalog of Proper Motions	Knox-Shaw and Barrett, Oxford Univ. Press, 1934	Proper motion, Selected areas 1–115	~30,000
Bergerdorf Spectral Durchmusterung	Schwassman and van Rhijn, Hamburg Sternwarte, 1935–1953	Spectral type and magnitude, Selected areas 1–115	174,000
Potsdam Spectral Durchmusterung	Becker and Brück, *Pub. Ap. Obs. Potsdam*, **27**, **28**, 1929–1938	Spectral type and magnitude, Selected areas 116–206	50,000
	Nearby, bright, and high-velocity stars		
Yale Parallax Catalog	Jenkins, Yale Univ. Obs., 1963	Parallax	6,000
Catalog of Nearby Stars	Gleise, *Veröff. Rechen Inst. Heidelberg*, **3**, No. 22, 1969	Position, proper motion, radial velocity, distance, magnitude, color for stars within 20 pc	900
Catalog of Stars within 25 pc of Sun	Woolley et al., *Roy. Obs. Ann. No. 5*, 1970	Position, proper motion, radial velocity, space velocity, distance, magnitude, color	1,700
Catalog of Stars within 5 pc of the Sun	van de Kamp, *Ann. Rev. Astr. Ap.*, **9**, 103, 1971	Position, proper motion, radial velocity, distance, magnitude	45
Yale Bright Star Catalog (BS)	Hoffleit, Yale Univ. Obs., 1964	Position, proper motion, radial velocity, spectral type, color, magnitude, parallax	9,000
Catalog of High-Velocity Stars	Roman, *Ap. J. Supp.*, **2**, 198, 1955	Position, magnitude, color, spectra, proper motion, space motion	600
Space Velocity Catalog	Eggen, *Roy. Obs. Bull. No. 51*, 1962	Magnitude, color, spectral type, radial velocity, proper motion, space velocity	3,500
Catalog of High-Velocity Stars	Eggen, *Roy. Obs. Bull. No. 84*, 1964	Magnitude, color, distance, space motion	660

Table 2-6. *(continued)*

Catalog	Reference	Principal content	Number of objects
	Spectral types		
Henry Draper Catalog (HD)	Cannon and Pickering, *Harvard Obs. Ann.*, **91–99**, 1918–1924	Spectral class and magnitude	225,000
Henry Draper Extension (HDE)	Cannon, *Harvard Obs. Ann.*, **100**, 1925–1936; **112**, 1949	Spectral class and magnitude	134,000
Luminous Stars in the Northern Milky Way	Hardorp et al., Hamburg Sternwarte, 1959–1965	Spectral and luminosity class, magnitude	~8,000
Luminous Stars in the Southern Milky Way	Stephenson and Sanduleak, *Pub. Warner and Swasey Obs.*, **1**, No. 1, 1971	Spectral and luminosity class, magnitude	5,000
Catalog of Early-Type Emission-Line Stars	Wackerling, *Mem. Roy. Astr. Soc.*, **73**, 153, 1970	Spectral type, magnitude	11,000
Catalog of Stellar Spectra Classified in the MK System	Jaschek et al., LaPlata Obs., 1964	MK spectral type and magnitude	~20,000
MK Spectral Classifications (additions to LaPlata Catalog)	Buscombe, Northwestern Univ., 1974–1977	MK spectral type, *UBV* photometry	~45,000
Univ. of Michigan Catalog of Two-Dimensional Spectral Types (Reclassification of HD)	Houk and Cowley, U. of Mich., 1975–present	MK spectral type	35,000 to date

		Photometry	
UBVRIJKL Photometry of Bright Stars	Johnson et al., *Commun. Lunar and Plan. Lab.*, **4**, 99, 1966	*UBV* and infrared magnitudes	~1,500
Photoelectric Catalog	Blanco et al., *Pub. U.S. Nav. Obs.*, **21**, 1968	*UBV* magnitudes and colors	21,000
Two Micron Survey	Neugebaur and Leighton, *NASA SP-3047*, 1969	Infrared magnitude	5,000
Six-Color Catalog	Kron et al., *Pub. U.S. Nav. Obs.*, **20**, Part V, 1972	Six-color magnitudes (*UVBGRI*)	1,300
Celescope Catalog of *UV* Stellar Observations	Davis et al., Smithsonian Ap. Obs., 1973	Ultraviolet stellar magnitudes	5,100
uvbyβ Photoelectric Photometric Catalog	Lindemann and Hauck, *Astr. and Ap. Supp.*, **11**, 119, 1973	$(b - y)$, m_1, c_1, β index	7,600
O Star Catalog	Goy, *Astr. and Ap. Supp.*, **26**, 273, 1976	*UBV* magnitudes, spectral types, polarization	800
Spectrophotometric Catalog	Breger, *Ap. J. Supp.*, **32**, 7, 1976	Absolute energy distributions	940
UV Bright-Star Spectrophotometric Catalog	European Space Agency, ESA SR-27, 1976	UV spectral energy distributions	1,350
General Catalog of *UBV* Photometry	Mermilliod and Nicolet, *Astr. and Ap. Supp.*, **29**, 259, 1977	*UBV* magnitudes	53,000
		Radial velocity	
General Catalog of Radial Velocities (Lick)	Moore, *Pub. Lick Obs.*, **18**, 1932; 20, 1949	Radial velocity	6,700
General Catalog of Radial Velocities (Mount Wilson)	Wilson, *Carnegie Inst. Wash. Pub. No. 601*, 1953	Radial velocity	15,000
Bibliography of Stellar Radial Velocities	Abt. and Biggs, Kitt Peak Nat. Obs., 1972	Radial velocity	25,000
Catalog of Individual Mount Wilson Radial Velocities	Abt, *Ap. J. Supp.*, **19**, 387, 1970; **26**, 365, 1973	Radial velocity	7,200

Table 2-6. (*continued*)

Catalog	Reference	Principal content	Number of objects
	Double stars		
General Catalog of Double Stars	Burnham, *Carnegie Inst. Wash. Pub. No. 5*, 1906	Double star observations	13,500
New General Catalog of Double Stars	Aitken, *Carnegie Inst. Wash. Pub. No. 417*, 1932	Double star observations and orbits	17,200
Catalog General de Etoiles Doubles	Jonckheere, *Pub. Obs. de Marseille*, 1962	Double star observations	3,400
Catalog of Visual Binary Orbits	Worley, *Pub. U.S. Nav. Obs.*, **18**, Part III, 1963	Double star orbits	540
Index Catalog of Visual Double Stars	Jeffers, et al., *Lick Obs. Pub.*, **21**, 1963	Double star observations	64,200
	Star clusters and associations		
Photometry of Stars in Galactic Clusters	Hoag et al., *Pub. U.S. Nav. Obs.*, **17**, 347, 1961	*UBV* magnitudes, color-magnitude diagrams	70 clusters
Atlas of Color-Magnitude Diagrams for Clusters and Associations	Barkhatova and Syrovoy, Acad. Sci. USSR, Moscow 1963	*UBV* color-magnitude and color-color diagrams	320 clusters
Catalog of Star Clusters and Associations	Alter et al., Akad. Kiado, Budapest, 1970	Star-cluster data, literature review	1055 open clusters, 125 globular clusters, 70 associations

Atlas of Open-Cluster Color-Magnitude Diagrams	Hagen, *Pub. David Dunlap Obs.*, **4**, 1970	*UBV* color-magnitude diagrams	190 clusters
Atlas of Globular-Cluster Color-Magnitude Diagrams	Alcaino, Univ. Catolica-Chile, 1973	*UBV* color-magnitude diagrams, photographs	44 clusters
Catalog of *UBV* Photometry of Globular-Cluster Stars	Philip et al., *Dudley Obs. Report No. 11*, 1976	*UBV* magnitudes, spectral types	115 clusters 37,000 stars
Catalog of *UBV* Photometry and MK Spectral Types in Open Clusters	Mermilliod, *Astr. and Ap. Supp.*, **24**, 159, 1976	*UBV* magnitudes, MK spectral types	200 clusters 10,800 stars
	Variable stars		
General Catalog of Variable Stars	Kukarkin et al., Sternberg Obs., Moscow 1969–1976 (3rd ed. plus supplements)	Position, magnitude, variability type	26,000
Catalog of Supernovae	Karpowicz and Rudnicki, Warsaw Univ. Press, 1968	Data and literature review	~200
	Nebulae		
Catalog of Dark Nebulae	Lynds, *Ap. J. Supp.*, **7**, 1, 1962	Position, area, opacity	1,800
Catalog of Bright Nebulae	Lynds, *Ap. J. Supp.*, **12**, 163, 1965	Position, size, color, brightness	1,100
Catalog of Galactic Planetary Nebulae	Perek and Kohoutek, Czech. Acad. Sci., Prague, 1967	Position, dimensions, magnitude, velocity, spectrum	1,000

Table 2-6. *(continued)*

Catalog	Reference	Principal content	Number of objects
		Radio sources	
Master List of Radio Sources	Dixon, *Ap. J. Supp.*, **20**, 1970	Position and flux; list of catalogs	12,000
		Galaxies and clusters of galaxies	
New General Catalog (NGC)	Dreyer, *Mem. R.A.S.*, **44**, 1, 1888	Positions and descriptions of nebulae, galaxies, and star clusters	7,800
Index Catalog (IC; two additions to NGC)	Dreyer, *Mem. R.A.S.*, **51**, 185, 1895; **59**, 105, 1908	Positions and descriptions of nebulae, galaxies, and star clusters	6,900
Survey of Galaxies Brighter than 13th Magnitude	Shapley and Ames, *Harvard Obs. Ann.*, **88**, 41, 1932	Positions, magnitudes, and forms of bright galaxies	1,250
Catalog of Rich Clusters of Galaxies	Abell, *Ap. J. Supp.*, **3**, 211, 1958	Positions, distances, richnesses of clusters of galaxies	2,700
Catalog of Galaxies and Clusters of Galaxies	Zwicky et al., Calif. Inst. of Technology, 1961–1968	Positions, magnitudes, radial velocities of galaxies, and clusters of galaxies	31,300
Morphological Catalog of Galaxies	Vorontsov-Velyaminov et al., Sternberg Obs., Moscow, 1962–1968	Positions, sizes, forms of galaxies, esp. multiple and interacting systems	30,000
Second Reference Catalog of Bright Galaxies	de Vaucouleurs, de Vaucouleurs, and Corwin. Univ. of Texas Press, 1976	Position, magnitude, colors, classification, radial velocity	2,600
Catalog of Compact and Post-Eruptive Galaxies	Zwicky, Guemlingen, Switzerland, 1971	Positions and descriptions	3,700
Revised New General Catalog of Nonstellar Objects	Sulentic and Tifft, U. of Ariz. Press, 1973	Position, description, magnitude	7,800
Uppsala General Catalog of Galaxies	Nilson, Uppsala Obs., 1973	Position, classification, magnitude, radial velocity	13,000

Table 2-7. Galaxy Atlases

Atlas	Reference	Instrument	Principal content	Number of objects
Hubble Atlas of Galaxies	Sandage, *Carnegie Inst. Wash. Pub. No. 618*, 1961	Mount Wilson 60″, 100″, Palomar 200″ reflectors	Descriptions, classification, photographs of galaxies	180
Atlas of Peculiar Galaxies	Arp, Calif. Inst. Technology, 1966	Palomar 200″ reflector	Photographs of peculiar and interacting galaxies	340
Isophotometric and Photographic Atlas of Peculiar Galaxies	Schanberg, *Ap. J. Supp.*, **26**, 115, 1973	24″ Curtis Schmidt	Photographs and isophotes of interacting and peculiar galaxies	90
Atlas of Dust and H II Regions in Galaxies	Lynds, *Ap. J. Supp.*, **28**, 391, 1974	Arizona 2.3-meter reflector	Photographs and maps of dust and H II regions in galaxies	40
Berkeley Low-Latitude H I Survey	Weaver and Williams, *Astr. and Ap. Supp.*, **8**, 1, 1973; **17**, 1, 1974	Hat Creek 85′ radio telescope	21-cm line profiles and contour maps, $\|b\| \leq 30^\circ$	–
Survey of 21-cm Line Radiation	Heiles and Habing, *Astr. and Ap. Supp.*, **14**, 1, 1974	Hat Creek 85′ radio telescope	21-cm contour maps, $\|b\| \geq 10^\circ$	–

distinction is made between *precision catalogs*, which give high-accuracy results for a limited number of objects, and *survey catalogs*, which contain results of moderate precision for a large number of objects. Table 2-7 contains a short list of atlases showing pictorially or graphically the structure of our Galaxy and of other galaxies. Table 2-8 lists some of the more important photographic sky surveys showing stars and nebulae in our Galaxy and galaxies and clusters of galaxies outside our own.

Taken together, these catalogs and atlases form a rich mine of information, which our theories on the nature of galaxies seek to order and explain. They also provide the standards against which we must measure the usefulness of our speculations concerning the structure of the Universe. In a sense, they form the backbone of galactic astronomy, and it is almost impossible to overestimate the great service that has been rendered to astronomy by those who have so painstakingly assembled these data. Any effort expended to examine and become familiar with these materials will be rewarded with deepened insight into the nature of galaxies. Further, here are found the raw

Table 2-8. Sky Surveys

Survey	Telescope	Region	Approximate limiting magnitude
Franklin-Adams Survey, 1911	10″ Astrograph	$-90^\circ \leq \delta \leq 90^\circ$	15
Barnard Atlas of Milky Way, 1927	Bruce 10″ Astrograph	Milky Way	–
Ross-Calvert Atlas of Milky Way, 1936	Ross 5″ Astrograph	Milky Way	–
Palomar-National Geographic Society Sky Survey, 1960	Palomar 48″ Schmidt	$-33^\circ \leq \delta \leq 90^\circ$	21
Atlas of Hα Emission in Southern Milky Way, 1960	Mt. Stromlo 8″ Schmidt	Milky Way	–
Lick Observatory Sky Atlas, 1967	Ross 5″ Astrograph	$-37^\circ \leq \delta \leq 90^\circ$	16
The Large Magellenic Cloud, 1967	ADH 32″ Schmidt	Large Magellenic Cloud	17.5
ESO Atlas of Southern Milky Way, 1969	Boyden 10″ Astrograph	Milky Way	16
Vehrenberg Atlas Stellarium, 1970	12-cm Astrograph	$-90^\circ \leq \delta \leq 90^\circ$	14
Canterbury Sky Atlas, 1972	Ross 5″ Astrograph	$-90^\circ \leq \delta \leq -22^\circ$	16
ESO-SRC Southern Sky Survey, 1978 (in progress)	U.K. 48″ Schmidt	$-90^\circ \leq \delta \leq -22^\circ$	23

materials needed to mount an attack on a great variety of original research problems. For example, the information contained in the *Palomar Sky Survey* and the *ESO-SRC Southern Sky Survey* has barely been tapped, and much important work that can be based on this information remains to be done. A large number of significant and penetrating questions that can be posed concerning the physical structure and behavior of galaxies can be answered using already existing data.

Only a few of the most representative catalogs and atlases are included in our lists, and there are many others worthy of attention. Many useful references can be found in the *Astronomische Jahresbericht* or in *Astronomy and Astrophysics Abstracts.* In fact, as a result of improved data-acquisition and data-reduction techniques, there has been in recent years a virtual explosion of papers containing large sets of important observational results. It has become increasingly difficult for astronomers to keep abreast of this information outside of a fairly narrow range of specialization. One approach to dealing with this problem has been the establishment of data centers—such as the Strasbourg Stellar Data Center—to collect, digest, and evaluate critically various kinds of astronomical data and to provide them to astronomers upon request. Perhaps catalogs and atlases in their present form are becoming obsolete and will be replaced in the future by information stored in computer memories.

REFERENCES

(**A1**) Abt., H.A., Meinel, A. B., Morgan, W. W., and Tapscott, J. 1969. *An Atlas of Low-Disperson Grating Stellar Spectra.* (Kitt Peak National Observatory and Yerkes Observatory).

(**B1**) Bless, R. C., Code, A. D., and Fairchild, E. T. 1976. *Astrophys. J.* **203**:410.

(**C1**) Counselman, C. C. 1976. *Ann. Rev. Astron. and Astrophys.* **14**:197.

(**G1**) Gliese, W., Murray, C. A., and Tucker, R. H. (eds.). 1974. *New Problems in Astrometry.* (Dordrecht: D. Reidel).

(**G2**) Golay, M. 1974. *Introduction to Astronomical Photometry.* (Dordrecht: D. Reidel).

(**G3**) Griffin, R. F. 1967. *Astrophys. J.* **148**:465.

(**G4**) Griffin, R. F. 1971. *Mon. Not. Roy. Astron. Soc.* **155**:1.

(**G5**) Griffin, R. F. 1975. *Mon. Not. Roy. Astron. Soc.* **171**:407.

(**H1**) Hayes, D. and Latham, D. 1975. *Astrophys. J.* **197**:593.

(**H2**) Hiltner, W. A. (ed.). 1962. *Astronomical Techniques.* (Chicago: University of Chicago Press).

(**J1**) Johnson, H. L. and Morgan, W. W. 1953. *Astrophys. J.* **117**:313.

(**M1**) McNally, D. 1974. *Positional Astronomy.* (London: Muller Educational).

(**M2**) Mihalas, D. 1978. *Stellar Atmospheres*. 2nd ed. (San Francisco: W. H. Freeman).

(**M3**) Morgan, W. W., Abt, H. A., and Tapscott, J. W. 1978. *Revised MK Spectral Atlas for Stars Earlier Than the Sun*. (Kitt Peak National Observatory and Yerkes Observatory).

(**M4**) Morgan, W. W. and Keenan, P. C. 1973. *Ann. Rev. Astron. and Astrophys.* **11**:29.

(**M5**) Morgan, W. W., Keenan, P. C., and Kellman, E. 1943. *An Atlas of Stellar Spectra*. (Chicago: University of Chicago Press).

(**M6**) Müller, E. A. (ed.). 1977. *Highlights of Astronomy*. XVIth General Assembly of the IAU. Vol. **4**, part 1. (Dordrecht: D. Reidel).

(**P1**) Payne, C. H. 1925. *Stellar Atmospheres*. (Cambridge: Harvard College Observatory).

(**S1**) Saha, M. 1920. *Phil. Mag.* **40**:472.

(**S2**) Saha, M. 1921. *Proc. Roy. Soc. (London)*. **A99**:135.

(**S3**) Smart, W. M. 1977. *Textbook on Spherical Astronomy*. (Cambridge: Cambridge University Press).

(**S4**) Stebbins, J. and Kron, G. E. 1956. *Astrophys. J.* **123**:440.

(**S5**) Stebbins, J. and Whitford, A. E. 1943. *Astrophys. J.* **98**:20.

(**S6**) Strand, K. A. (ed.). 1963. *Basic Astronomical Data*. (Chicago: University of Chicago Press).

(**S7**) Strömgren, B. 1966. *Ann. Rev. Astron. and Astrophys.* **4**:433.

(**V1**) van de Kamp, P. 1967. *Principles of Astrometry*. (San Francisco: W. H. Freeman).

(**V2**) Vasilevskis, S. 1973. *Vistas in Astron.* **15**:145.

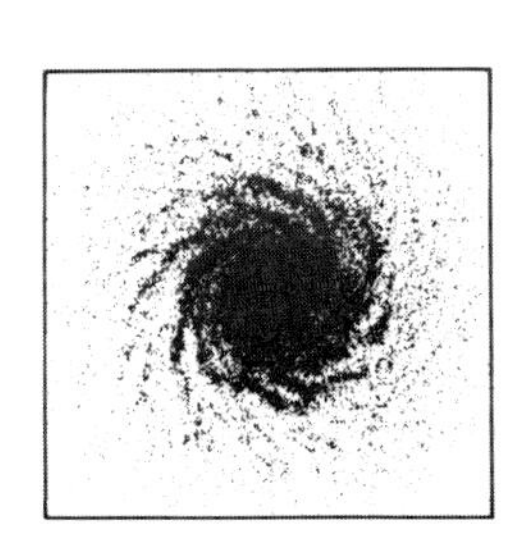

3 Physical Properties of Stars and the Interstellar Medium

Stars and interstellar matter are the two primary constituents of our Galaxy, and a thorough knowledge of their physical properties is essential for the development of a picture of galactic structure, dynamics, and evolution. In the first four sections of this chapter, we shall describe the methods used to deduce the basic physical properties of stars. We shall emphasize fundamental methods that give the results as directly as possible; these provide calibrations that are essentially free from the assumptions that enter into most secondary methods, which can sometimes be used to infer a particular property of a star if its other properties are presumed to be normal. We next summarize (in §3-5 and §3-6) some of the systematic interrelations among stellar properties (masses, radii, and luminosities) and then interpret these (in §3-7 through §3-10) in terms of the theory of stellar structure and evolution. The resulting information can be used, in turn, to discuss the chemical and dynamical evolution of our Galaxy as a whole. Finally, in the last section, we discuss the effects of absorption and reddening on starlight by interstellar material.

3-1. STELLAR DISTANCES

The primary reason we wish to know the distance to a star is that we can then determine its absolute magnitude from equation (2-13), ignoring, for the present, interstellar absorption. If we can do this for a sufficient number of stars, then we can calibrate stellar luminosities as a function of MK spectral type. Once this has been done, we can turn the procedure around and use the apparent magnitudes of stars (corrected for interstellar absorption) along with their now-known absolute magnitudes to find their distances, and hence analyze the distribution of stars in space to build a picture of some of the major structural features of our Galaxy. An interesting broad discussion of these topics can be found in (**H12**).

There are only two direct (that is, wholly geometric) methods by which stellar distances can be determined. We shall now consider these in turn.

Direct Parallaxes

The *trigonometric parallax* method, which is, in the final analysis, the most fundamental method, was described in Chapter 2. We noted that, at present, we can obtain reliable distances (errors less than 10%) only for stars closer than about 20 pc from the Sun. Unfortunately, many spectral types do not have a single representative within the 20-pc sphere centered on the Sun, and this sample is restricted almost exclusively to faint, late-type stars.

Moving Clusters

The second geometric method by which we can determine stellar distances is the moving-cluster method. We sometimes observe a well-defined group of stars whose individual proper motions appear to converge on a point (or, more realistically, within a small region) on the sky, as illustrated schematically in Figure 3-1. Such a group is called a *moving cluster*. The interpretation of this phenomenon is that the members of the cluster share a common motion in space and that the apparent *convergence point* lies in the direction of this motion; that is, we are observing the direction of the point at infinity at which the parallel paths of motion appear to intersect. Thus the angular distance λ between a given cluster star and the convergence point is also the angle between the space-velocity vector of that star and the line of sight from the Sun to it. If the hypothesis of parallel motions is strictly true—that is, if the random motions of stars relative to the cluster motion are negligible and the cluster does not expand, contract, or rotate as a whole—and if the

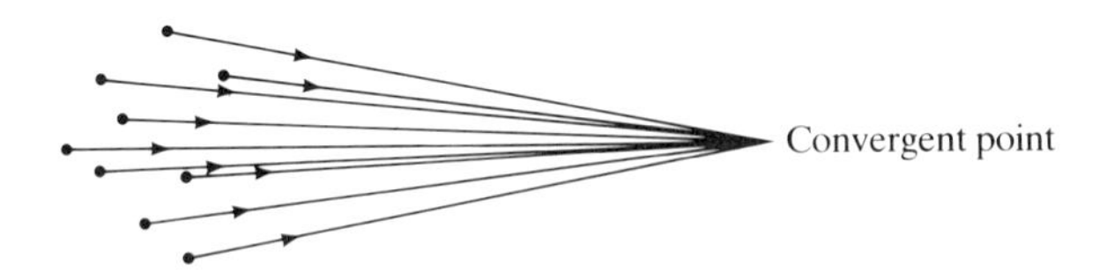

Figure 3-1. Schematic diagram of a moving cluster. When the proper motions of the cluster stars are extended, they appear to intersect at a point of convergence on the sky.

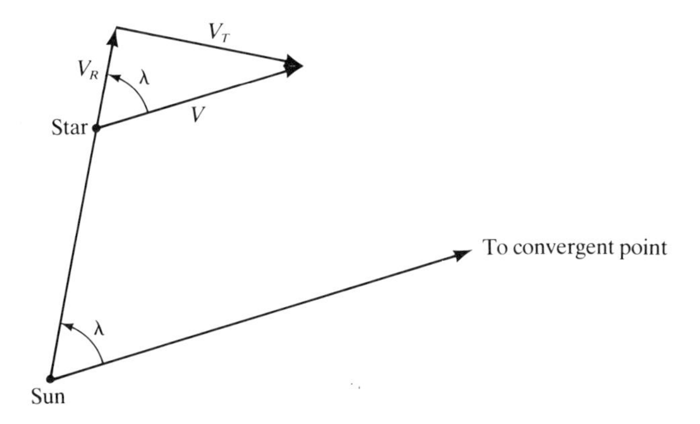

Figure 3-2. The space motion of a star in a moving cluster is assumed to be parallel to the line of sight from the Sun to the apparent convergent point of the cluster. Then the angle λ is both the angle between the positions of the star and of the convergent point on the sky, and the angle between the star's space-velocity vector and the Sun–star line of sight.

proper motions can be measured accurately and are free from systematic errors, then the distances to cluster members can be determined from their observed proper motions and radial velocities.

Suppose v is the *space velocity* of a star relative to the Sun. Its radial velocity is $v_R = v \cos \lambda$, and its tangential velocity across the line of sight is $v_T = v \sin \lambda$ (see Figure 3-2). Thus

$$v_T = v_R \tan \lambda \tag{3-1}$$

which gives v_T in terms of directly observable quantities. A star's tangential velocity gives rise to its proper motion μ. If μ is expressed in seconds of arc per year, v_T in kilometers per second, and the star's distance d in parsecs, then

$$\mu\,(''\text{per year}) = \frac{v_T \text{ km s}^{-1}}{d \text{ pc}} \times \frac{(206{,}265'' \text{ per rad}) \times (3.16 \times 10^7 \text{ s year}^{-1})}{(206{,}265 \text{ a.u. pc}^{-1}) \times (1.497 \times 10^8 \text{ km a.u.}^{-1})}$$

$$= \frac{v_T}{4.74d} \tag{3-2}$$

Thus, eliminating v_T via equation (3-1), we find the star's distance

$$d(\text{pc}) = \frac{v_R \tan \lambda}{4.74\mu''} \tag{3-3}$$

or, equivalently, its parallax

$$\pi'' = \frac{4.74\mu''}{v_R \tan \lambda} \tag{3-4}$$

Using equation (3-3), we can obtain the distance (or parallax) of each cluster member from its observed proper motion, radial velocity, and angular distance from the convergent point.

In practice, the procedure is somewhat more complicated than that just outlined. Details are described in (**H1**) and the references cited therein. The method works best when the angular size of the cluster is sufficiently large that the proper motions of the individual cluster stars are oriented in substantially different directions and thus define a convergent (or divergent) point accurately. Furthermore, the cluster must be close enough that the proper motions are large enough to be measured accurately, the angle λ must be such that the space motion makes a significant contribution to both v_R and v_T, and finally, cluster members must be reliably identifiable so that foreground and background field stars can be excluded.

The moving-cluster method has been applied to a number of clusters, among them (1) the Hyades, which contains about 200 stars at an average distance $d \approx 46$ pc (**H1**), (**H2**), (**V1**); (2) the Ursa Major group, which contains about 60 stars at an average distance of about 24 pc (**E1**), (**E2**); and (3) the Scorpio-Centaurus group, which contains about 100 stars at an average distance of about 170 pc (**B6**). Distance data for the stars in these clusters are of enormous importance because they include types of stars whose distances, and hence whose absolute magnitudes, cannot be measured by the trigonometric parallax method. These particular clusters define the position of much of the main sequence in the Hertzsprung–Russell diagram (see §3-5) and are thus fundamental to the calibration of the main-sequence fitting procedure.

In fact, it is not an exaggeration to say that the distance to the Hyades sets the scale for essentially all galactic and extragalactic distance measurements. It was, therefore, disconcerting when P. W. Hodge and G. Wallerstein (**H19**) pointed out that the Hyades distance modulus of 3.1 mag found from the moving-cluster method was in disagreement with the average result from a variety of other determinations [see the summary in (**V1**)]. This raised the question of whether the method is actually valid for the Hyades. The problem was resolved when new proper motions, measured with respect to external galaxies, became available. All methods now agree quite closely, giving a distance modulus of 3.30 ± 0.05 mag (**H2**), (**V1**).

Other Methods

In addition to the two basically geometric methods just described, there are several less direct approaches to stellar distance determination. Among the most important of these are:

1. The secular and statistical parallax methods (see §6-6). The secular parallax method is, in effect, an application of the moving-cluster method to stars of a single type (a restriction made so that we can assume all the stars have the same absolute magnitude) in a volume big enough to contain a large sample. The method invokes kinematic hypotheses about the random motions of the stars, and it presumes that the solar motion relative to the group can be found.
2. Dynamical parallaxes and parallaxes from spectroscopic-visual binaries (see §3-2). The former method invokes a dynamical law and requires that the masses of the stars in a binary be normal for their spectral type. The latter method is more fundamental, but it applies only to a very special group of stars.
3. The spectroscopic parallax method (see §3-5). Here one estimates distances using the absolute magnitudes associated with each spectral type by means of calibrations made in binaries and clusters. The basic assumption made in this method is that all stars of a given MK type do in fact have the same absolute magnitude, as is expected on both theoretical and empirical grounds. In practice, this method is the most important tool for distance estimation in galactic-structure research.

3-2. STELLAR MASSES

The mass of a star is its most basic physical attribute. For a given chemical composition, the star's mass essentially determines its structure and evolution. All fundamental determinations of stellar masses are based on an application of *Kepler's third law* to the orbits of binary stars. This law states that

$$G(\mathcal{M}_1 + \mathcal{M}_2)P^2 = 4\pi^2 a^3 \tag{3-5}$$

where $\mathcal{M}_1$ and $\mathcal{M}_2$ are the masses of two bodies in mutual revolution on a relative orbit with semimajor axis a and period P, and G is the Newtonian gravitation constant (see Chapter 13).

Kepler's third law applies both to planets revolving around the Sun and to stars revolving around one another. Indeed, we can use the Earth's motion around the Sun to make a convenient choice of units for binary stars. Because $\mathcal{M}_\odot$, the mass of the Sun, is much larger than $\mathcal{M}_\oplus$, the mass of the Earth, we can write $\mathcal{M}_\odot + \mathcal{M}_\oplus \approx \mathcal{M}_\odot$ to a high degree of approximation. Therefore,

$$G\mathcal{M}_\odot P_\oplus^2 = 4\pi^2 a_\oplus^3 \tag{3-6}$$

where $P_\oplus$ is the period of revolution of the Earth around the Sun (a year), and $a_\oplus$ is the mean distance from the Earth to the Sun (an astronomical unit). Taking the ratio of equation (3-5) to (3-6), the numerical constants cancel, and we can write

$$\frac{(\mathcal{M}_1 + \mathcal{M}_2)P^2}{\mathcal{M}_\odot P_\oplus^2} = \frac{a^3}{a_\oplus^3} \tag{3-7}$$

so that, if we express masses in solar masses, periods in years, and distances in astronomical units, we have simply

$$(\mathcal{M}_1 + \mathcal{M}_2)P^2 = a^3 \tag{3-8}$$

Further, by recalling equation (2-3), we see that we can also write

$$a\,(\text{a.u.}) = \frac{a''}{206{,}265} \times d\,(\text{pc}) = \frac{a''}{206{,}265} \times \frac{206{,}265}{\pi''}$$

or

$$a\,(\text{a.u.}) = \frac{a''}{\pi''} \tag{3-9}$$

where a'' is the observed semimajor axis in seconds of arc, and π'' is the parallax of the star, also in seconds of arc.

As we mentioned in Chapter 2, binary and multiple stars are quite common. At least 50% of the "stars" within 5 pc are actually double or multiple, so that at least 60% of the individual stars are members of such systems. Binary stars are classified into several broad groups. The categories we shall deal with are (1) *visual binaries*, in which the individual components can be seen directly, (2) *spectroscopic binaries*, which reveal their orbital motion by shifts in the wavelengths of their spectral lines, and (3) *eclipsing binaries*, which show evidence of orbital motion by periodic variations in their apparent brightnesses. We can obtain different kinds of astrophysical information from each of these groups (which are not mutually exclusive) as is discussed here and in §3-3. An excellent general discussion of binary and multiple stars can be found in (**B3**).

The Mass of the Sun

To convert stellar masses expressed in solar units to physical units, we obviously must know the mass of the Sun in grams. We first determine the ratio of the mass of the Sun to the mass of the Earth by applying equations

(3-5) and (3-6) to the orbit of the Moon around the Earth. Thus

$$G(\mathscr{M}_\oplus + \mathscr{M}_{☾})P_{☾}^2 = 4\pi^2 a_{☾}^3 \tag{3-10}$$

so that

$$\frac{(\mathscr{M}_\oplus + \mathscr{M}_{☾})P_{☾}^2}{\mathscr{M}_\odot P_\oplus^2} = \frac{a_{☾}^3}{a_\oplus^3} \tag{3-11}$$

or

$$\frac{\mathscr{M}_\odot}{\mathscr{M}_\oplus} = \left(\frac{P_{☾}}{P_\oplus}\right)^2 \left(\frac{a_\oplus}{a_{☾}}\right)^3 \left(1 + \frac{\mathscr{M}_{☾}}{\mathscr{M}_\oplus}\right) \tag{3-12}$$

All the terms on the right-hand side are known from observation, including the ratio $(\mathscr{M}_{☾}/\mathscr{M}_\oplus)$, which is obtained by measuring the parallax in the observed position of a nearby planet produced by the Earth's motion around the center of mass of the Earth–Moon system during the course of a month [see (**A1**, Chapter 8)]. Using the result $(\mathscr{M}_{☾}/\mathscr{M}_\oplus) = 1/81.3$ and inserting the appropriate values for $(P_{☾}/P_\oplus)$ and $(a_\oplus/a_{☾})$ into equation (3-12), we find $(\mathscr{M}_\odot/\mathscr{M}_\oplus) = 332{,}945$. The mass of the Earth can be measured directly [see (**A1**, Chapter 5)] and is found to be 5.98×10^{27} g; hence we derive $\mathscr{M}_\odot = 1.99 \times 10^{33}$ g.

Visual Binaries

When stars are examined telescopically, relatively close pairs are frequently found. Some of these stars are mere *optical doubles*, that is, two stars that happen to lie, by coincidence, along nearly the same line of sight. Herschel searched for optical doubles in the hope of measuring the parallax of the nearer star relative to the farther one, and, in 1781, he recognized that most of the systems he had found were actually *physical binaries* that interact dynamically and therefore exhibit a relative orbital motion. Careful measurements of the *separation* and *position angle* of the two stars make it possible to determine the *apparent relative orbit* (the projection of the *true relative orbit* onto the plane of the sky) of one star around the other. The true orbit is an ellipse, and one can show that the projected orbit is also an ellipse.

Because the two components of a visual binary must be resolved, we tend to select systems that are widely separated in linear distance. This separation implies that they will have long periods, and in fact many known visual binaries have completed only a fraction of an orbital revolution in the entire time they have been observed (sometimes a century or more). The theoretical resolving power (in seconds of arc) of a diffraction-limited telescope of diameter D (in meters) is given by

$$\theta'' = \frac{0''.13(\lambda/5000)}{D} \tag{3-13}$$

where λ is the wavelength of the observed light in Ångstroms. In principle, the 5-meter Hale reflector can resolve a pair separated by $0''.03$; in practice, blurring by *atmospheric seeing* caused by turbulence in the Earth's atmosphere prevents measurement of binary-star separations smaller than about $0''.15$. By use of interferometric techniques (see §3-3), smaller separations can be measured in favorable cases.

Considerable gains will be realized when observations can be made above the Earth's atmosphere with the Space Telescope, because seeing effects will then be eliminated and observations can be made in the ultraviolet ($\lambda \approx$ 1000 Å). An order-of-magnitude improvement in the limit of resolution should then be obtained. With this improvement, we shall be able to resolve binary systems of much shorter periods. Indeed, it will be possible to resolve some stars that are presently considered spectroscopic binaries. The combination of the visual orbit with spectroscopic data leads to an enormous increase in the amount of information that we can deduce about those systems.

From a geometric analysis of the apparent orbit [for details see (**H17**, Chapter 22), (**M4**, Chapter 12), or (**S19**, Chapter 14)], one can obtain the *elements* that describe the shape and orientation of the true relative orbit of a visual binary, in particular, the semimajor axis a'', in seconds of arc. Now, if we know the distance to the system and hence π'', then we can convert the observed value of a'' to linear measure via equation (3-9), and, knowing P, we can find the sum of the masses $(\mathscr{M}_1 + \mathscr{M}_2)$ in solar units from Kepler's third law, equation (3-8). In practice, the lack of reliable parallaxes, which enter as the third power in the expression for $(\mathscr{M}_1 + \mathscr{M}_2)$, poses a very severe difficulty for the determination of accurate stellar masses.

To find the individual masses $\mathscr{M}_1$ and $\mathscr{M}_2$, we need to know not only the relative orbit of the two stars but also the *absolute orbit* of each of the components around their common center of mass. The center of mass of a binary system moves along a nearly rectilinear trajectory. The ratio of the two masses is easily determined as the inverse of the ratio of the amplitudes of the individual motions relative to the straight-line path of the center of mass.

Reliable masses are available for only about fifty visual binaries because of the difficulties of obtaining accurate distances and accurate apparent orbits. (Easily resolved systems usually have long periods, and hence only a fraction of their orbit has been observed, whereas systems with short periods often are at the limit of resolution.) Lists of well-determined visual-binary masses and a discussion of the practical difficulties are given in (**S24**, Chapter 15) and (**E4**).

We have just noted that masses found from visual binaries depend sensitively on the parallax of the system. This fact suggests that it can be advantageous to turn the procedure around and use it to estimate the parallax instead of the stellar masses. Thus, if a reasonably good guess can be made for $\mathcal{M}_1$ and $\mathcal{M}_2$, say, by assuming that the star's masses are normal for their spectral types, then we can determine a *dynamical parallax*

$$\pi'' = \frac{a''}{(\mathcal{M}_1 + \mathcal{M}_2)^{1/3} P^{2/3}} \tag{3-14}$$

Because only the cube root of the total mass enters in this formula, even a crude estimate of $\mathcal{M}_1$ and $\mathcal{M}_2$ will yield π'' with good precision. Once we have an estimate of π'', we can estimate the absolute magnitude of each star and hence its luminosity class. We can then obtain a more refined estimate of $\mathcal{M}_1$ and $\mathcal{M}_2$ from the *mass-luminosity relation* (see §3-5), if the stars are main-sequence stars, or from the correlation of stellar mass with MK spectral type. With improved values for the masses, a better value for π'' can be derived from equation (3-14). If necessary, the procedure may be iterated and yields rapid convergence.

Distances obtained from the dynamical parallax method are not fundamental as they are based on the assumption that the stars in the binary are normal. Nevertheless, they are often very useful estimates and occasionally provide an important check on other methods. For example, the dynamical parallaxes of binaries in the Hyades were discordant with the older moving-cluster distance modulus but in agreement with other methods. Hence they lent support to the suggestion that the moving-cluster modulus was in error.

Spectroscopic Binaries

When we examine the spectra of a large number of stars, we occasionally find one in which (1) the lines split into two components, each of which shows a periodic back-and-forth wavelength shift, or (2) the spectrum is a blend from two distinct spectral types and both sets of lines show periodic wavelength shifts, or (3) the spectrum is single but shows a periodic wavelength shift relative to an absolute standard, such as a comparison spectrum or an interstellar line. These stars are actually binary stars whose components are too close to be resolved visually. As the stars revolve around their common center of mass, their motions along the line of sight produce periodic Doppler shifts in their observed spectra. Relative to the center of mass, one star approaches the Earth while the other recedes from it, leading to case (1) if both stars have almost the same spectral type and absolute magnitude, or case (2) if they have different types but are of comparable brightness. Such stars are called *double-line spectroscopic binaries*. A *single-line spectroscopic binary* [case (3)] occurs when one component is so much brighter than the other (typically, $\Delta m \gtrsim 2.5$ mag) that it swamps the spectrum of the secondary and renders it invisible. For a given Δm, it is generally easier to detect both spectra if the two components are of very different spectral types (for example, B and M).

Generally, spectroscopic binaries have short periods and small separations because these imply large mutual velocities, and hence they tend to favor the discovery of the system. The periods may be as short as $0.^{d}1$, in which case the stars may be in *contact* with one another, or they can be as long as 7500^{d}, in which case the stars may be so widely separated as to be resolved as visual binaries. Separations typically are from about 3×10^{-3} a.u. to about 10 a.u.

From an analysis of the radial velocity of one or both of the components as a function of time, one can determine the elements of the binary orbit [see (**H17**, Chapter 23), (**M4**, Chapter 12), or (**S19**, Chapter 14) for detailed methods]. The orbital plane is inclined to the plane of the sky by some angle i, which in general is unknown and cannot be determined from the spectroscopic data alone, because the observed radial velocity v_R yields only the projection of the orbital velocity v along the line of sight; that is, $v_R = v \sin i$. As we shall see, this fact limits the information we can obtain about the system unless i can be determined some other way.

Individual masses can be determined only for the favorable case of a double-line binary, so we shall henceforth restrict our attention to this case exclusively. Suppose for simplicity that the orbits of both stars are circular. Then, from the observed radial velocities, we can immediately determine the projected radii (in absolute units, that is, kilometers or astronomical units) of the two orbits:

$$a_1 \sin i = \frac{(v_1 \sin i)\, P}{2\pi} \tag{3-15a}$$

and

$$a_2 \sin i = \frac{(v_2 \sin i)\, P}{2\pi} \tag{3-15b}$$

and hence the mass ratio

$$\frac{\mathscr{M}_1}{\mathscr{M}_2} = \frac{a_2}{a_1} = \frac{a_2 \sin i}{a_1 \sin i} \tag{3-16}$$

Because we do not know the semimajor axis, $a \equiv a_1 + a_2$, but only its projection $a \sin i = (a_1 \sin i) + (a_2 \sin i)$, we cannot determine $(\mathscr{M}_1 + \mathscr{M}_2)$

from Kepler's third law but only

$$(\mathcal{M}_1 + \mathcal{M}_2)\sin^3 i = \frac{(a \sin i)^3}{P^2} \tag{3-17}$$

Thus, from the analysis of the spectroscopic orbit, we can find $\mathcal{M}_1 \sin^3 i$ and $\mathcal{M}_2 \sin^3 i$ separately, but we obtain $\mathcal{M}_1$ and $\mathcal{M}_2$ only if we can determine the inclination by some other means.

Because we know that $\sin^3 i$ must always be less than or equal to unity, from spectroscopic data alone we can determine only a lower bound on the masses of the binary components. A statistical correction for inclination effects can be made by using the result that $\langle \sin^3 i \rangle = 0.59$ for orbital planes oriented randomly with respect to the plane of the sky. This procedure is open to question, however, as selection effects obviously favor the discovery of systems with large values of $\sin i$ because these have larger velocities for a given orbital velocity.

The masses of both stars in a binary can be obtained unambiguously in the event that the system is either an *eclipsing spectroscopic binary* or a *visual spectroscopic binary*. In the former case, if the stars are reasonably well separated, the occurrence of eclipses implies that the orbital plane lies nearly in the line of sight, and hence we know that $i \approx 90°$ and $\sin i \approx 1$. In *contact binaries*, eclipses can still occur for inclinations significantly smaller than 90°, but even for these systems an analysis of high-quality photometric data enables one to determine i and hence the masses $\mathcal{M}_1$ and $\mathcal{M}_2$.

The second case in which both the visual orbit of the binary and the radial velocity of each star can be measured is extremely favorable. Examples of such systems are α Aur and α Cen. Unfortunately, these systems are rare because, if the stars are widely enough separated to be a visual binary, then their physical separation is generally so large that the period is very long, and their radial velocities unobservably small. From the analysis of the visual orbit, we can determine i and hence the masses $\mathcal{M}_1$ and $\mathcal{M}_2$ of the individual stars from the spectroscopic orbit. Furthermore, from the visual data we know a'', the angular size of the orbit, and from the spectroscopic data we find the linear size a (a.u.) [see equations (3-15a and b)]. Hence we can determine the parallax π'' directly from equation (3-9). Distances obtained by this method are as fundamental as those found by trigonometric measurements or from moving clusters. As we mentioned earlier, an exciting advance will occur when observations can be made with the Space Telescope, because then we will be able to resolve as visual binaries many short-period systems that are now observable only as spectroscopic binaries. In this way, we will obtain accurate masses and distances for a large number of very interesting systems.

In practice, the analysis of spectroscopic binaries is often fraught with difficulty. This is particularly true in contact systems and interacting systems in which mass is being exchanged between the two stars. Up to the present time, there are only about two dozen reliably determined sets of masses, mostly for stars on or near the main sequence. Summaries of the data and critical evaluations of their accuracy are given in (**L3**), (**P8**), and (**G9**, 13).

3-3. STELLAR RADII

The *radius* of a star is a parameter of great importance because (1) for a given mass, it sets the stellar surface gravity and mean density (which have important implications in terms of stellar structure and evolution—see §3-7), and (2) for a given luminosity, it determines the effective temperature of the stellar atmosphere (see §3-4) and hence the appearance of the star's spectrum. To measure radii, we can either (1) measure a star's angular diameter (directly, interferometrically, or by means of lunar occultations), which, when combined with the star's distance, gives its physical size in kilometers or (2) use observations of the light curves and orbital velocities in an eclipsing spectroscopic binary to obtain the component stars' radii in physical units, independent of knowledge of the system's distance.

Direct Angular Measurement

If we can resolve the disk of a star, we can measure its angular diameter directly. At present, this can be done for precisely one star: the Sun. The Sun has an angular diameter of $1919''.3 = 0.009305$ rad, which implies that $R_\odot = 0.004652$ a.u. $= 6.96 \times 10^5$ km.

The angular diameters of all other stars are much smaller than the limit set by seeing effects in the Earth's atmosphere, and hence they cannot be observed directly. For example, a main-sequence star like the Sun seen at a distance of 2 pc would have an angular diameter of only $0''.005$. Some stars, the red giants (see §3-5), have radii of the order of a few hundred solar radii, but typically they lie at distances of 10 to 20 pc. Hence they still have angular diameters less than $0''.05$. Therefore, for the measurement of stellar angular diameters, we must at present use other techniques, described next. However, when the Space Telescope becomes operational, it should be possible to resolve stellar disks of the order of $0''.02$ in diameter, and this will yield direct angular-diameter measurements for a large sample of stars.

Interferometric Measurement

Angular diameters smaller than the limit for direct measurement can be determined by *interferometric techniques*. These techniques can also be used to measure binary-star separations that are too small to be resolved visually. Three different approaches have been developed thus far.

Phase Interferometry Around 1920, A. A. Michelson and F. G. Pease measured the angular diameters of several nearby stars by observing the interference patterns produced in an interferometer formed by mounting a 25-foot track bearing two small mirrors (equivalent to the slits of a Young's double-slit experiment) on the 100-inch Mount Wilson telescope. The visibility of the fringes set up when this system is directed at a star of angular diameter θ'' declines as the distance D between the entrance mirrors is increased. The visibility effectively vanishes when D is given by equation (3-13). In this way, Michelson and Pease obtained the angular diameters of about ten red giants. These angular diameters, when combined with distances, gave stellar radii in physical units. Subsequent attempts by Pease to use a 50-foot interferometer failed because of mechanical flexure of the instrument and the limitations, set by atmospheric seeing, on the distance by which two detectors can be separated and still retain adequate phase coherence. For ground-based observations, $0''.01$ is probably the practical limit for this method, but if observations could be made from space, much smaller limits could be reached.

Intensity Interferometry In the mid 1950s, a new interferometric technique was developed by R. H. Brown and R. Q. Twiss. This technique exploits the fact that the fluctuations in the intensities of the signals received from a star by two instruments separated by distance D are correlated in a way that depends on D and the angular diameter of the star. A clear discussion of this method (and also of phase interferometry and the lunar-occultation method) is given in (**B10**). In intensity interferometry, phase coherence is no longer required, and it becomes possible to use separate telescopes and large baselines. An interferometer consisting of two 6.5-meter mosaic "light-buckets" separated by baselines from 10 to about 190 meters operated for several years at Narrabri, Australia, and produced angular-diameter measurements for about thirty stars (**B11**), (**B12**).

Angular diameters as small as 5×10^{-4} seconds of arc have been measured with this technique. However, this method is restricted to early-type stars, which (with present equipment) must be brighter than $V \approx 2.5$. The list of such stars observable from the southern hemisphere has already been exhausted, and new results will emerge only from northern-hemisphere observations or from the use of a larger instrument that will reach fainter stars. The present data include a wide variety of O5–F5 stars of luminosity classes I–V.

Speckle Interferometry In 1970, A. Labeyrie (**L1**) pointed out that the image of a star at the focus of a large telescope consists of a broad, time-fluctuating *speckle pattern.* This pattern comprises innumerable stellar images, each the size of the Airy disk (that is, the diffraction-limited image), which spread out and move about randomly as a result of seeing-induced phase fluctuations in the wave front received by the telescope. By a suitable Fourier analysis of the speckle pattern, it is possible to recover an estimate of the angular size of each elementary image down to the theoretical limit of resolution of the telescope. This method has been applied using the Hale 200-inch telescope [see **(G2)**] and yields reliable angular diameters down to about 0.″02. In most respects, it is an effective substitute for the Michelson technique. A review of this method and its accomplishments is given in **(L2)**.

Lunar Occultations

As the Moon (or a planet) moves on the sky, it occasionally passes in front of a star and *occults* it. Typically, the light from a star is extinguished within a few milliseconds. To the extent that we can idealize a star as a point source and the Moon's disk as a semi-infinite plane, we expect to observe fringes in a *Fresnel diffraction pattern*, which is produced as the moon passes in front of the star. These fringes cause a fluctuation in intensity that can be recorded with a fast-response photoelectric photometer. Because a star is not actually a point source but has a finite (if small) angular diameter, the observed diffraction pattern will differ slightly from a perfect point-source pattern, and, by a careful analysis of the data, one can use the differences to make an estimate of the star's angular diameter [see **(B10)**, **(N1)**, and **(N2)** for a detailed description of the method].

This method can be used to measure angular diameters down to about 0.″002 as well as separations of binary stars that are too close for visual resolution (two diffraction patterns are observed). Numerous results are reported in **(D6)**, **(M2)**, **(R2)**, **(W4)**, and related papers. The method is obviously restricted to stars near the plane of the ecliptic, and it is limited in part by imperfect knowledge of the irregularities of the Moon's limb.

Eclipsing Binaries

It sometimes happens that the orbital plane of a binary system is observed nearly edge-on, so that, as the stars revolve around their common center of mass, they pass in front of and behind one another (see Figure 3-3) and hence produce *eclipses*. There will then be periodic variations in the light received from the system; a plot of its apparent magnitude as a function of time is called its *light curve*. The first eclipsing binary to be discovered was Algol (β Per), which was noticed to be variable in 1670 and explained physically by J. Goodricke in 1782. Most known eclipsing systems have short periods (90% less than 10^d), but a few have very long periods (for examples ε Aur, $P \approx 27$ years). The probability of discovering a long-period eclipsing system

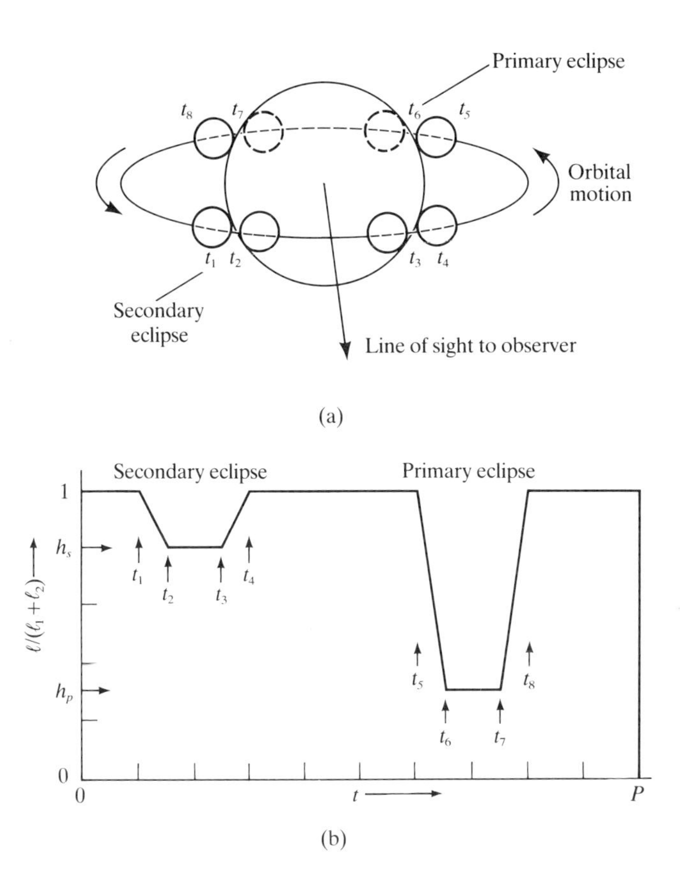

Figure 3-3. If the orbital plane of a binary star happens to be nearly edge-on as seen from the Earth, then eclipses will occur as one of the components passes in front of, or behind, the other. The eclipse geometry is shown in part (a), and a schematic light curve for the case that the smaller star has the greater surface brightness is shown in part (b).

is small, both because the orbital plane of a widely separated system must be inclined almost exactly 90° to the plane of the sky and because the eclipses occupy only a tiny fraction of the orbital period.

The *primary eclipse* occurs when the star having the higher surface brightness (hence higher effective temperature—see §3-4) is eclipsed, and the *secondary eclipse* occurs when the component having the lower surface brightness is eclipsed. These phenomena are illustrated in Figure 3-3 for the case where the smaller star is an early-type (hot) dwarf and the larger star a late-type (cool) giant.

The simplest case to analyze occurs if the orbits are circular, the orbital plane is inclined at 90° to the plane of the sky, and the eclipses are *total* (primary eclipse in Figure 3-3) and *annular* (secondary eclipse in Figure 3-3). Let d_1 be the diameter of the larger star, and d_2 the diameter of the smaller. These diameters in units of the orbital radius a can be determined from the duration of the eclipses. For example, if $t_1, \ldots, t_4$ denote times of first contact, . . . , fourth contact during the secondary eclipse, then

$$\frac{d_1}{2\pi a} = \frac{(t_3 - t_1) + (t_4 - t_2)}{2P} \tag{3-18a}$$

and

$$\frac{d_2}{2\pi a} = \frac{(t_2 - t_1) + (t_4 - t_3)}{2P} \tag{3-18b}$$

where for convenience we have assumed that $d_1 \ll a$ and $d_2 \ll a$. Similar formulae can be written for the primary eclipse, and they provide a check on the results.

Furthermore, it is clear that, if ℓ_1 and ℓ_2 are the apparent brightness of the larger and smaller stars, respectively, and h_p is the residual brightness of the system at primary eclipse in units of the total brightness outside of eclipse, then for the case illustrated in Figure 3-3,

$$\frac{\ell_1}{\ell_1 + \ell_2} = h_p \tag{3-19a}$$

and

$$\frac{\ell_2}{\ell_1 + \ell_2} = 1 - h_p \tag{3-19b}$$

Again, a check on the results can be obtained from the secondary eclipse because then we know that

$$\frac{\ell_1(1 - k^2) + \ell_2}{\ell_1 + \ell_2} = h_s \tag{3-20a}$$

and

$$\frac{k^2\ell_1}{\ell_1 + \ell_2} = 1 - h_s \tag{3-20b}$$

where h_s is the residual brightness at secondary eclipse and $k^2 \equiv (d_2/d_1)^2$; k^2 should be found to be consistent with the results of equations (3-18).

If, in addition to measuring the light curve, we can also measure the radial velocities of the two stars (that is, the star is an eclipsing spectroscopic binary), then we can determine the orbital radius in physical units (kilometers) as $a = (v_{\max} P/2\pi)$, where $v_{\max}$ is the observed maximum relative radial velocity of the two components. We can then reexpress the stellar radii in kilometers. In this favorable case, we determine both masses (from Kepler's third law) and radii (from the eclipses) for the stars.

More generally, a detailed analysis of the light curve yields complete orbital elements for the system, including parameters that specify the shape and orientation of the orbit (in particular the inclination i) and that describe the radii and surface-brightness distributions of the two component stars [see (**H17**, Chapter 24), (**M4**, Chapter 12), or (**S19**, Chapter 14) for methods]. In practice, the analysis is difficult because the relationships between the stellar and orbital parameters and the properties of the observed light curve are complex. The quality of the results depends sensitively on the accuracy of the data. Moreover, numerous physical complications can occur. For example, the eclipses may be only partial, the orbits may be elliptical, the stars may be deformed (oblate or even ellipsoidal), the light curves may contain *reflection effects* (light from one star illuminating the other), or *mass exchange* may occur in the system. Most of these complications occur in close binaries, which are therefore often extremely troublesome to analyze [see (**K4**) for details]. In view of the complexity of the problem, it is perhaps not surprising that accurate results are available for only a small number of systems [see (**G9**, 13), (**L3**), (**P8**), (**S24**, Chapter 15)].

Astrophysical Estimates

For many astrophysically interesting stellar types, none of the direct methods just described can be applied, and yet even an approximate estimate of the radii of these stars (for example, white dwarfs) can be of enormous importance for an understanding of stellar structure and evolution. In such cases we can estimate radii by means of Stefan's law [see equation (3-25)] if we know a star's luminosity (from its apparent magnitude and distance) and its effective temperature (from spectroscopic evidence). Such estimates, although not fundamental measurements, can nevertheless be quite accurate and useful.

3-4. ANALYSIS OF STELLAR SPECTRA

From a theoretical analysis of the line and continuum spectrum of a star, we can deduce the physical structure and chemical composition of its atmosphere and predict the frequency variation of its emitted radiation. We shall discuss here only a few important concepts and results, and we refer the reader to (**A4**), (**G6**), or (**M13**) for details of methods.

Blackbody Radiation

The *specific intensity* emitted at frequency ν by a *perfect radiator* (or *blackbody*) at absolute temperature T is given by *Planck's law*

$$B_\nu(T) = \frac{2h\nu^3/c^2}{e^{h\nu/kT} - 1} \text{ ergs cm}^{-2}\text{ s}^{-1}\text{ hz}^{-1}\text{ ster}^{-1} \tag{3-21}$$

where h is Planck's constant, k is Boltzmann's constant, and c is the velocity of light. The energy emitted outward in all directions by a unit area of a blackbody in unit time and into unit frequency interval is the *monochromatic flux*

$$\begin{aligned}\mathscr{F}_{\nu(\mathrm{BB})} &= \oint B_\nu(T)\mu\, d\omega = B_\nu(T) \int_0^{2\pi} d\phi \int_0^1 \mu\, d\mu \\ &= \pi B_\nu(T) \text{ ergs cm}^{-2}\text{ s}^{-1}\text{ hz}^{-1}\end{aligned} \tag{3-22}$$

where $d\omega = d\phi\, d\mu$ is an element of solid angle, and $\mu \equiv \cos\theta$, where θ is the angle between the normal to the surface and an emerging ray [see (**M13**, §1-3)]. In equation (3-22), we used the fact that radiation emerging from a perfect radiator (for example, a small hole in the wall of an oven in thermal equilibrium) is isotropic. The total energy, summed over all frequencies, emitted per unit time per unit area by a blackbody is given by the *integrated flux*

$$\mathscr{F}_{(\mathrm{BB})} = \pi \int_0^\infty B_\nu(T)\, d\nu = \sigma T^4 \text{ ergs cm}^{-2}\text{ s}^{-1} \tag{3-23}$$

where σ is the *Stefan–Boltzmann constant*. Equation (3-23) is *Stefan's law*.

Monochromatic fluxes emitted by blackbodies at temperatures characteristic of stellar atmospheres are shown in Figure 3-4. As is evident in the plot, as the temperature rises, the wavelength of maximum intensity decreases, and the amount of radiation at short wavelengths rises very rapidly. Clearly, a hot radiator will have a bluer color than a cool radiator. If the intensity is measured per unit wavelength interval, the wavelength at which the emission is maximum is $\lambda_{\max} = (2.898 \times 10^7)/T$, where λ is in Ångstroms and T is in degrees Kelvin.

Now, if a star of radius R, and hence surface area $4\pi R^2$, emitted energy exactly like a blackbody of temperature T, then its *luminosity*, that is, its total radiative energy output, would be

$$L_{\mathrm{BB}} = 4\pi R^2 \sigma T^4 \tag{3-24}$$

But, of course, real stars are not blackbodies, and the actual emergent-energy distribution of most spectral types is not Planckian but generally deviates rather strongly from a blackbody curve. We can nevertheless define

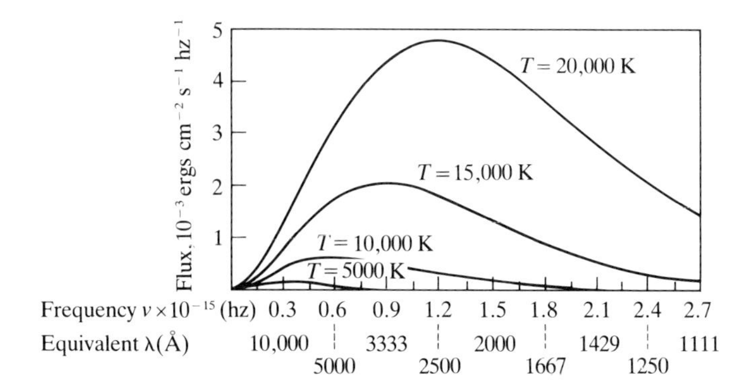

Figure 3-4. Flux emitted by a perfect radiator (blackbody) at temperature T in ergs cm^{-2} s^{-1} hz^{-1}. Note that, as the temperature rises, the point of maximum emission shifts to higher frequencies (shorter wavelengths), and the total energy emitted increases rapidly.

a temperature T_{eff}, called the *effective temperature*, such that, for a star of luminosity L and radius R, a generalization of equation (3-24) does apply; that is, we choose T_{eff} so that

$$L \equiv 4\pi R^2 \sigma T_{\text{eff}}^4 \tag{3-25}$$

The effective temperature not only specifies the integrated flux emitted per unit area by the star, namely

$$\mathscr{F} = \sigma T_{\text{eff}}^4 \tag{3-26}$$

but it also is a genuinely representative temperature of the material in the stellar atmosphere. Insofar as this is true, it follows from Planck's law that we can expect stars with low effective temperatures to emit most of their energy in the infrared and to appear red to the eye (or a photometer) and stars with high effective temperatures to emit most of their energy in the ultraviolet and to appear blue. How we actually determine T_{eff} for real stars is discussed next.

Stellar Atmospheres Theory

From a theoretical point of view, the structure of a stellar atmosphere, and therefore the spectral energy distribution emitted by a star, is determined essentially by three parameters: (1) the effective temperature, which specifies the amount of energy passing outward through each square centimeter of the material; (2) the *surface gravity* $g = G\mathcal{M}/R^2$, which fixes the pressure stratification in the atmosphere; and (3) the *chemical composition* of the material, which determines the frequency variation of the opacity of the stellar material. We shall see in §3-7 that the structure of a star as a whole is basically determined by its mass, chemical composition, and age. In particular, these parameters essentially fix a star's radius and luminosity and hence its effective temperature and surface gravity. Thus the atmospheric structure and the emergent spectrum of a star are uniquely determined by its mass, chemical composition, and age at each instant in its evolutionary history. Reciprocally, an analysis of a star's spectrum by detailed theoretical modeling yields its effective temperature, gravity, and chemical composition and hence some information about the structure of the star. However, it is important to note that, while a star's interior structure fixes its atmospheric structure uniquely, the converse is not true. In fact, if we have determined T_{eff} and g for a star spectroscopically, we really know only its luminosity per unit mass

$$\frac{L}{\mathcal{M}} = (4\pi G\sigma)\frac{T_{\text{eff}}^4}{g} \qquad (3\text{-}27)$$

Only if we can deduce L, $\mathcal{M}$, or R by some independent means can we determine all three of the basic parameters individually.

To make a theoretical model of a stellar atmosphere, we choose a chemical composition, an effective temperature, and a surface gravity. We then demand that pressure forces balance and that energy be conserved. These constraints specify the temperature and density structure of the atmosphere. Once the structure of the atmosphere is known, the *equation of transfer* can be solved to calculate the emergent radiation field. The distribution of radiation as a function of wavelength is fixed by continuum and line opacities, and an accurate simulation of a stellar energy distribution must account for the *line-blanketing effects* of thousands to millions of absorption lines in the spectrum. Recently, it has become possible to treat these effects realistically and to compute reliable theoretical emergent-energy distributions and photometric indices.

Effective Temperatures and Surface Gravities

From the point of view of galactic astronomy, the primary goal of stellar spectrum analysis is to provide estimates of the effective temperatures and surface gravities of stars of each MK type. We consider effective temperatures first.

Effective Temperatures There are two different empirical methods by which T_{eff} can be found. First, it is clear from equation (3-25) that we can find T_{eff}

if we know L and R for a star. If we know the distance to a star, we can find its absolute visual magnitude M_V from its apparent magnitude V. If, in addition, we can estimate the star's *bolometric correction* (see below) from its spectral type, then we can find $(L/L_\odot)$ from equation (3-31). Finally, if we know R, either because we can measure the star's angular diameter (and by hypothesis we already know its distance) or because it is a spectroscopic eclipsing binary, then we find T_{eff}. Notice that $T_{\text{eff}} \propto L^{1/4}R^{-1/2}$, so that even modest accuracies for L and R suffice to yield a reasonably accurate effective temperature.

If we can measure the angular diameter θ and the complete absolute energy distribution of a star, we can apply a more powerful empirical method that does not require the distance to be known. If f_λ is the absolute monochromatic flux in ergs cm^{-2} s^{-1} Å^{-1} received outside the Earth's atmosphere, then $\mathscr{F}_\lambda$, the absolute monochromatic flux at the surface of the star, is given by [see equation (2-11)]

$$f_\lambda = \left(\frac{R}{D}\right)^2 \mathscr{F}_\lambda = \tfrac{1}{4}\theta^2 \mathscr{F}_\lambda \qquad (3\text{-}28)$$

where R and D are the star's radius and distance, respectively (neither of which need be known individually). Assuming that f_λ is known for all wavelengths, we then can find the integrated flux $\mathscr{F} = \int_0^\infty \mathscr{F}_\lambda \, d\lambda$, and hence T_{eff} via equation (3-26). Again, T_{eff} is well determined even if f_λ and θ are only modestly accurate.

Our assumption that f_λ is known completely implies that we have an accurate absolute calibration of the receiver and that we can eliminate the effects of absorption by both the Earth's atmosphere and the interstellar medium. For the visible-wavelength range, we can use ground-based data corrected empirically for atmospheric extinction; in the infrared, we can measure the radiation received through transmitting windows and interpolate between them. But, at other wavelengths, the Earth's atmosphere is opaque. Hence, to measure in the far infrared, we must observe from balloons or high-altitude aircraft. In the ultraviolet, we must observe from space vehicles. Most stars for which θ can be measured reliably are fairly close and therefore suffer only a small absorption by the interstellar medium for $\lambda > 912$ Å. These effects can be eliminated empirically (see §3-11). The interstellar medium is completely opaque for $\lambda \lesssim 912$ Å (the threshold for absorption by atomic hydrogen in its ground state). Thus, for very hot stars, the contribution of the very short-wavelength part of the spectrum to the total flux must be estimated from theoretical models.

The classic application of this method is to the Sun, which has an easily measured angular diameter and a fairly well-known absolute energy distribution. Using the available data, one finds $T_{\text{eff}}(\odot) = 5770$ K. Furthermore, because $R_\odot$ is known directly, we immediately find $L_\odot = 3.826 \times 10^{33}$ ergs s^{-1} from equation (3-25). Until recently, the only other stars to which this approach could be applied were late-type giants having known angular diameters, using absolutely calibrated broadband photometry (**J2**). These results are of great importance because theoretical models for the atmospheres of such stars are too primitive to give reliable effective temperatures. Thanks to the development of the Narrabri intensity interferometer and the Orbiting Astronomical Observatory (OAO-2), it became possible to apply the method to early-type stars as well. A reliable empirical effective-temperature scale for dwarfs on the whole range O–G, and for a few giants and supergiants of various types, is now available [see (**C13**) for a comprehensive discussion of this work]. Numerical results are given in Table 3-5.

While the empirical approaches just described yield a reliable effective-temperature scale for much of the main sequence, there are many types of stars of great astrophysical importance to which they cannot yet be applied. For these stars, we must turn to theory. The basic method is to compute a grid of theoretical models for several effective temperatures and surface gravities (with an appropriate chemical composition) and then to compare predicted spectral features with those observed. Temperature-sensitive parameters can be obtained from, say, ratios of line strengths for two successive ionization stages of an element or from such features as continuum slopes and discontinuities. In general, one uses several criteria simultaneously, and, in favorable cases, the best-fit solution yields reasonably precise values for both T_{eff} and $\log g$. It should be stressed, however, that, unlike the empirical temperatures based on observed total fluxes, these temperatures are only model parameters and are not necessarily properly related to L and R via equation (3-25).

For most stars we do not have enough spectrophotometric information to permit a detailed analysis of the kind just described, and we are forced to work with the limited information contained in magnitudes and colors. In principle, one could hope to use the empirical results to derive a relation $T_{\text{eff}} = f(B - V)$, but in practice the reliable determinations are too sparse. We must therefore eke out the calibration by computing theoretical color indices from model atmospheres. For any color index $(a - b)$ we can write from equation (2-15)

$$(a - b) = k_{ab} - 2.5 \log_{10} \frac{\int_0^\infty f_\lambda S_\lambda(a)\, d\lambda}{\int_0^\infty f_\lambda S_\lambda(b)\, d\lambda} \tag{3-29}$$

where f_λ is the absolute flux at wavelength λ, S_λ denotes the receiver response in the appropriate band, and k_{ab} is an arbitrary constant that must be evaluated. The constant k_{ab} is found by applying equation (3-29) to stars having measured colors $(a - b)$ and known absolute flux distributions f_λ. Once k_{ab} has been determined, equation (3-29) can be applied to the fluxes predicted by models to find a color index for each model for which T_{eff} is known.

Surface Gravities Fundamental surface gravities can be determined for all stars of known $\mathscr{M}$ and R, for example, the Sun and eclipsing spectroscopic binaries. For other stars, we must estimate log g theoretically, using model atmospheres. Typically, we analyze pressure-sensitive lines (of, say, hydrogen) or ionization equilibria, which are sensitive to both pressure and temperature. A difficulty with this approach is that, if there are important dynamical effects operating in the atmosphere (such as pulsation or a stellar wind), then the value of g derived spectroscopically may not actually equal the surface gravity implied by $\mathscr{M}$ and R.

Bolometric Corrections

The absolute magnitude of a star in any definite photometric band, say M_V, obviously provides a measure of only a fraction of the total radiant energy output of the star. To make allowance for the energy emitted at wavelengths outside of the photometric band observed and thereby obtain the luminosity of a star, we define the *bolometric magnitude* as the magnitude that would be measured with an ideal bolometer that absorbs radiation at all wavelengths with perfect efficiency. We then define the *bolometric correction* (*B.C.*) to be such that

$$M_{\text{bol}} \equiv M_V + B.C. \tag{3-30}$$

The luminosity of a star in solar units can now be expressed as

$$\begin{aligned}\log\left(\frac{L}{L_\odot}\right) &= 0.4[M_{\text{bol}}(\odot) - M_{\text{bol}}(*)] \\ &= 0.4[M_V(\odot) - M_V(*) + B.C.(\odot) - B.C.(*)]\end{aligned} \tag{3-31}$$

To find bolometric corrections, one must know the distribution of a star's emitted flux over the entire spectrum. Until recently, this could be done reliably using empirical data only for a few spectral classes (in particular, for the Sun, which emits almost all its energy at wavelengths observable from the ground). However, by using the ultraviolet and infrared data now available, one can determine empirical bolometric corrections for a wide range of spectral types.

Considering the definitions of M_V and M_{bol}, equation (3-30) can be rewritten as

$$B.C. = 2.5 \log \frac{\int_0^\infty f_\lambda S_\lambda(V)\, d\lambda}{\int_0^\infty f_\lambda\, d\lambda} + C_1 \tag{3-32}$$

where $S_\lambda(V)$ is photometer response for the V band (**M7**), or alternatively as

$$B.C. = -2.5 \log\left(\int_0^\infty f_\lambda \, d\lambda\right) - V + C_2 \tag{3-33}$$

Each of these expressions has certain advantages [see the discussion in (**C13**)]. The constant C_1 in equation (3-32) contains an arbitrary zero-point. This is usually fixed by setting the value of the bolometric correction of the Sun. The conventional value derived in early work (**K7**), (**P7**) on the bolometric-correction scale is $B.C.(\odot) = -0.07$ mag [which implies that $M_{\text{bol}}(\odot) = +4.76$]. When this value is used, one finds $C_1 = +0.95 \pm 0.01$ mag (**C13**). To find the constant C_2 in equation (3-33), calculated values of $2.5 \log[\int_0^\infty f_\lambda S_\lambda(V) \, d\lambda]$ are correlated with observed values of V for stars with known f. This calculation yields (**C13**)

$$V = -2.5 \log\left[\int_0^\infty f_\lambda S_\lambda(V) \, d\lambda\right] - 12.47 \tag{3-34}$$

Combining equations (3-32) and (3-34), we have

$$B.C. = -\left[V + 2.5 \log\left(\int_0^\infty f_\lambda \, d\lambda\right) + 11.52\right] \tag{3-35}$$

When empirical fluxes are unavailable, theoretical model-atmosphere fluxes may be used in equation (3-32). Again, unless line-blanketing is treated adequately, the results will contain serious errors. One will tend to overestimate the bolometric correction (that is, find too negative a value) from model fluxes. Numerical values of the bolometric corrections for various spectral types are listed in Table 3-5.

Chemical Compositions

The chemical composition of stellar material is determined from quantitative analyses of stellar spectra. In favorable cases, the analysis can be based on a physical model of the atmosphere and a detailed *spectrum synthesis* of both line profiles and line strengths (equivalent widths). By matching the calculated spectrum to the observational data, one can estimate the star's effective temperature and gravity and the abundance of various elements relative to hydrogen (normally the dominant constituent). In most cases, high-quality line profiles are unavailable, and the analysis must be based on equivalent widths alone using a *curve of growth*, which gives the strength of a line as a function of the number of absorbing atoms in the material.

Many stars of interest to galactic-structure research are so faint that the data needed for even a curve-of-growth analysis are unobtainable, and one must estimate the abundances of all the elements heavier than helium (the "metals") from photometric indices that measure the fraction of the energy absorbed by spectrum lines in specially chosen bands. A number of systems can be used (see §3-6). These are calibrated by observing stars for which abundances have been determined spectroscopically.

Hundreds of spectroscopic analyses of stellar spectra have been performed, and they have yielded a number of extremely important conclusions:

1. Most disk stars and nebulae in the solar neighborhood are composed of a single, standard mixture of elements. In these objects, the abundances (*by number*) of various elements relative to hydrogen are as follows: helium, 0.10; carbon, 3×10^{-4}; nitrogen, 10^{-4}; oxygen, 6×10^{-4}; neon, 10^{-4}; magnesium, 3×10^{-5}; silicon, 3×10^{-5}; and iron, 4×10^{-5}. The variations in these numbers from one disk star to another are relatively minor, with the metals fluctuating (usually together) relative to the solar abundances by perhaps a factor of two up, and by a factor of from three to five down. In the theory of stellar structure, it is customary to use the parameters (X, Y, Z), which give the fractional abundances *by weight* of hydrogen, helium, and everything else, respectively. For the solar mixture, $X = 0.70$, $Y = 0.28$, and $Z = 0.02$.
2. In contrast, the locally observed stars in the spheroidal component are found to have total metal abundances that range from a factor of 10 to a factor of 10^3 below solar; however, the relative distribution of individual elements within this group [measured as $n(\text{element})/n(\text{Fe})$] is roughly the same as in solar material. With few exceptions, the spheroidal-component stars are too cool to excite lines of He I in their observable spectrum, hence the helium abundance of these stars cannot be determined spectroscopically. Helium lines can, in fact, be seen in the spectra of some *horizontal-branch stars* (see §3-6) that are hot enough to show them, but these stars are in an advanced evolutionary state and seem to have peculiar atmospheric compositions. We shall see in §3-7 and §3-10 that it now appears from several lines of evidence that, for spheroidal-component stars, $Y \approx 0.25 \pm 0.03$; that is, their helium content is nearly the same as that of disk stars.
3. The chemical abundances in both disk and spheroidal-component stars show moderate variations with spatial position in our Galaxy, strong correlations with stellar kinematics, and a probable correlation with age. Valuable inferences about galactic evolution can be drawn from these results, as will be discussed in later chapters.
4. From evolutionary analyses, we find that the spheroidal-component stars are all very old and that disk stars range from very young to

moderately old. We thus conclude that the element abundances in the halo stars are close to those of primeval cosmic material, whereas the disk stars have been enriched in the heavier elements. We therefore infer that hydrogen and helium were produced in the cosmic big bang in essentially their present-day ratio. The heavier elements are the products of *nucleosynthesis* in early generations of stars, the material from which is recycled into the interstellar medium by supernova and noncatastrophic mass-loss processes and thence into subsequent generations of stars. We shall return to these points in §3-9 and §3-10.

5. A small number of stars show atmospheric abundance anomalies, which are usually related to structural properties of the atmosphere or associated with a particular evolutionary state. Examples are (a) helium-rich stars showing very strong He lines; (b) helium-deficient stars (among them, the horizontal-branch stars); (c) peculiar A stars (Ap), which have enhanced abundances of certain metals and rare earths in atmospheres containing intense magnetic fields; (d) metallic-line A stars (Am) whose metal lines are too strong compared to their hydrogen lines, while their Ca II lines are too weak; (e) carbon stars, carbon-rich red giants; (f) S stars, showing anomalously high abundances of Zr, Y, La, and such unstable elements as technetium; and (g) Wolf–Rayet stars, divided into types WC and WN, which show strong lines of carbon and nitrogen, respectively.

   Some of these stars, such as those in categories (e) and (f), are in very advanced evolutionary phases and have atmospheric abundances that evidently have been altered by mixing with internal material that has undergone thermonuclear processing. In other cases, the anomalous abundances are the result of the operation of specific physical mechanisms in the atmosphere alone (for example, gravitational settling by diffusion) and are not characteristic of the composition of the star as a whole. In still other cases, anomalous atmospheric abundances may be artifacts of the models used to interpret stellar spectra. More sophisticated models (which include, for example, the effects of magnetic fields and stellar rotation) might well be able to account for the observed line strengths in terms of normal abundances.

3-5. SYSTEMATICS OF STELLAR PROPERTIES: SPIRAL-ARM AND DISK STARS

Thus far in this chapter, we have discussed stars as essentially isolated individuals and have described how their physical properties can be determined. We now inquire into the nature and extent of correlations among these properties. We shall find that very important and distinctive systematic relationships do exist, both for spiral-arm and disk-component stars in the solar neighborhood (discussed in this section) and for the spheroidal-component stars observed both in the solar neighborhood as halo stars and at large distances as members of globular clusters or the galactic bulge (discussed in §3-6).

The Hertzsprung-Russell Diagram

Unquestionably the most important correlation among stellar properties discovered to date is the *Hertzsprung-Russell diagram* (*H-R diagram*), which was developed independently by the Danish astronomer E. Hertzsprung in 1911 and the American astronomer H. N. Russell in 1913. As we shall see in this and the following five sections, the H-R diagram has been of profound importance to the development of our understanding of stellar evolution. It provides both one of the most stringent tests of evolution theories and one of the most incisive tools for exploring the evolutionary history of our Galaxy as a whole. We remind the reader that all discussion of interstellar reddening and absorption effects is deferred to §3-11.

Types of H-R Diagrams The H-R diagram displays the relationship between stellar spectral types and luminosities in a two-dimensional plot, such as that shown in Figure 3-5. Each point in that plot represents a nearby field star of known distance. In its original form, the H-R diagram was a plot of absolute visual magnitude versus spectral class, but other variants are now more commonly used. Of these, the most useful observational form is the *color-magnitude* (*CM*) *diagram*, which is a plot of a magnitude (either absolute, for stars of known distances, or apparent, for stars known to be all at the same distance—for example, within a cluster) versus a color index. The choice most commonly used is V magnitude versus $(B - V)$. An advantage of CM diagrams over the original H-R diagram is that the spectral classification procedure groups stars into discrete categories and thereby introduces a clumpiness into the diagram (easily seen in Figure 3-5) which obliterates fine details. In contrast, CM diagrams yield smooth distributions that allow us to exploit fully the inherent high precision of photoelectric photometry and thus discriminate fine features easily. Furthermore, accurate colors can often be measured for stars that are much too faint for spectral classification. Insofar as stars of a given MK spectral type have a unique color, these two observational diagrams are equivalent.

The most convenient form of the H-R diagram for evolutionary analyses is a plot of $\log (L/L_\odot)$ versus T_{eff}; this is usually called the *theoretical H-R diagram*. Insofar as all stars of a given MK spectral type have the same luminosity and effective temperature, the theoretical H-R diagram is equivalent to either form of observational H-R diagram just described. But, of

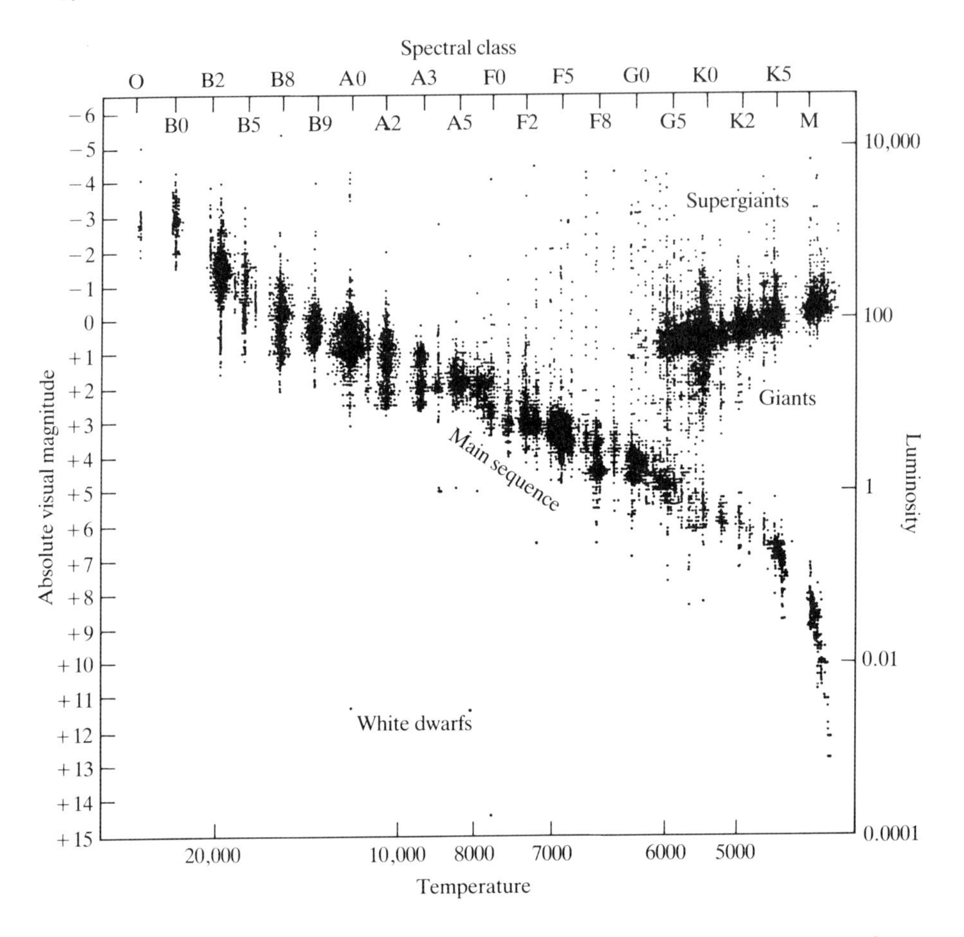

Figure 3-5. The Hertzsprung-Russell diagram, in its classical form, for nearby stars of known distance. The great majority of stars in a unit volume of space fall along the *main sequence*. Also shown are the positions of the *giant*, *supergiant*, and *white-dwarf* sequences. [From Otto Struve, *Stellar Evolution*. Copyright 1950 © 1978 by Princeton University Press, Fig. 2, p. 32. Reprinted by permission of Princeton University Press.]

course, to make a correspondence between the theoretical H-R diagram and a CM diagram, we must know the relations $T_{\text{eff}} = f(\text{MK spectral type})$ and $B.C. = g(\text{MK spectral type})$ with good accuracy. This *mapping problem* between the two sets of diagrams has long been, and remains today, one of the most troublesome practical obstacles encountered in theoretical evolutionary analyses of observed H-R diagrams.

Basic Morphology As is instantly obvious from even a casual inspection of Figure 3-5, stars are not scattered at random in the H-R diagram. Instead, they fall into several distinctive groups along well-defined sequences. The vast majority of all stars are found to lie along the *main sequence*, which stretches from luminous, hot, blue, O stars to faint, cool, red, M stars; these stars are called *dwarfs* (MK luminosity class V). The Sun is a main-sequence star. The next most prominent sequence is the *giant branch*, which stretches from spectral type G0, at a luminosity about thirty times brighter than main-sequence stars of the same type, toward cooler and brighter stars. These stars are the *red giants*; they correspond to MK luminosity class III. Sprinkled across the top of the diagram are a few extremely bright stars at most (but not all) spectral types; these are the *supergiants* (MK luminosity class I). Similarly, about 10 mag below the main sequence, we find a few faint, hot stars known as *white dwarfs*. The region between the main sequence and the giant branch that is almost devoid of stars is known as the *Hertzsprung gap*.

H-R Diagrams of Clusters and Associations In the galactic plane, we observe numerous physical aggregations of 10^2 to 10^3 stars called *galactic* (or *open*) *clusters* and *associations*. A typical galactic cluster is a rather irregular, loosely concentrated group of stars in a volume having a radius of the order of 10 pc. Galactic clusters have a wide range of star densities, from about 0.25 star pc^{-3} for the Hyades to about 10^3 stars pc^{-3} at the centers of the richest clusters. These figures are to be compared to the density of stars in the field in the solar neighborhood, which is of the order of 0.1 star pc^{-3}. The richest clusters can be detected at large distances (the limit being set by interstellar absorption), whereas the poorer clusters can be discriminated from random fluctuations in the field only if they are nearby. About 10^3 galactic clusters are known. The total masses of galactic clusters lie in the range of 100 to $3000\mathscr{M}_\odot$, and their integrated brightness can be as high as $M_V \approx -10$, though a more typical value is $M_V \approx -5$. Galactic clusters contain a wide variety of stellar types and span a wide range of ages, a point that we shall discuss later.

An association is an extremely loose and irregular group of stars. It has a low space density (perhaps 100 stars in a volume of 100-pc radius) and is identifiable mainly because it is a noticeable (if weak) concentration of relatively rare stars of a distinctive type. For instance, *O associations* are aggregations rich in O stars (for example, I Ori) and *T associations* are rich in T Tauri variables (for example, Per T2). About 100 associations are known. The space densities within associations are so low that they cannot be dynamically bound, and they are actually dissolving into the field on a relatively short time scale (a few million years); they often show definite expansional motions. The stars in associations are very young: O stars have lifetimes of only about a million years, and T Tauri stars are still in a stage of contraction onto the main sequence. Evidently, associations are mere transient members of the spiral-arm population, and they are continually forming, disintegrating and dispersing into the field.

In broad terms, the morphology of a typical galactic-cluster CM diagram is similar to that of common field stars, showing the same general sequences.

But, because cluster stars are all essentially at the same distance from the Sun, and hence have the same distance modulus and suffer similar amounts of interstellar absorption and reddening, it is possible to establish observationally the features in their CM diagrams clearly and crisply. Virtually all galactic-cluster CM diagrams have a well-defined main sequence, which extends to some specific upper limit of brightness. As we shall see later, this truncation of the upper end of the main sequence is a result of stellar evolution—the period of time that a star can spend on the main sequence, its *main-sequence lifetime*, decreases markedly from the faint late-types to the luminous early-types. Hence, in a cluster of a given age, all stars brighter than some critical limit will have evolved off the main sequence. Thus, while the very youngest clusters (for example, NGC 2362 in Figure 3-6) can contain O stars, progressively older clusters will contain no stars earlier than type B, A, F, and so on. The CM diagrams of young clusters typically have a main sequence extending to type O or B, a sprinkling of supergiants across a wide Hertzsprung gap, and a concentration of M supergiants. The CM diagrams of intermediate-age clusters, such as M11 or M41, show a main sequence terminating at late B stars or A stars, a well-developed red-giant branch separated from the main sequence by a narrower Hertzsprung gap, and no supergiants. The oldest clusters, such as M67 and NGC 188, show a main sequence only up to F stars and then a continuous *subgiant branch* (no Hertzsprung gap) extending into a moderately luminous giant branch. White dwarfs are commonly found in clusters that are close enough for them to be detected (for example, the Hyades). In a very few clusters, a Cepheid variable is found in the Hertzsprung gap (see §3-8 for the significance of this fact).

The Spectral Type Versus Absolute-Magnitude Relation

Starting from the fact that stars fall on definite sequences in the H-R diagram, one can derive a great deal of information about stellar properties by comparing cluster CM diagrams. In particular, if we hypothesize that the main sequence has a unique locus in the CM diagram, then the observed main sequences (corrected for interstellar reddening) of various clusters should differ only because the clusters have different distance moduli. We can therefore fit the main sequences of cluster CM diagrams together by shifting them up or down in apparent magnitude at fixed color. Then, by fitting all clusters to a cluster of known distance (in practice, the Hyades), we can associate an absolute magnitude with every star in each diagram and hence calibrate the absolute magnitudes of a wide variety of MK spectral types.

In fitting together the main sequences of different clusters, one must make use of the results of stellar-evolution theory (§3-7), which show that, as stars age, they move away from the main sequence up and to the right in the H-R diagram (that is, they become more luminous and redder). Thus, to obtain the *unevolved* or *zero-age main sequence* (*ZAMS*), one fits the lower envelope

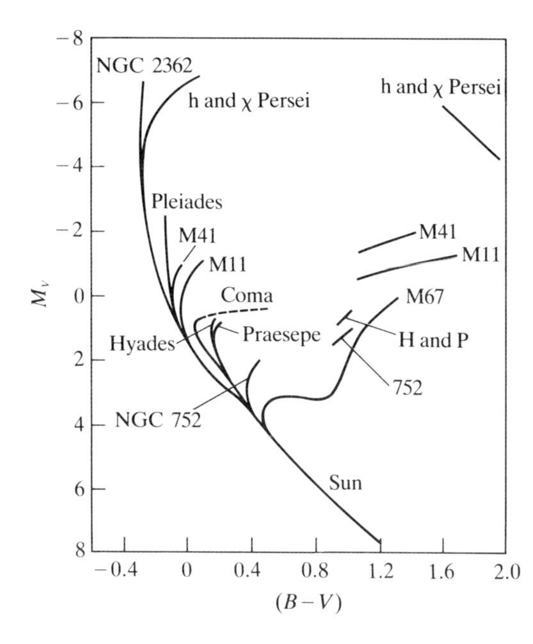

Figure 3-6. Color-magnitude diagram for several representative galactic clusters. NGC 2362 is the youngest, and M67 is the oldest cluster shown in the diagram. [Adapted from (**S1**), by permission. Copyright © 1957 by the University of Chicago.]

of the observed main sequences. To avoid all evolutionary effects, the fitting should be performed for stars at least 3 mag below the point where the main sequence terminates in a given cluster. In practice, to construct the complete ZAMS, one has to patch together pieces of the main sequences of several individual clusters. Details of how this was actually done are given by Sandage (**S1**), Johnson and Iriarte (**J4**), and Blaauw (**B7**, Chapter 20). The result of this work is shown in Figure 3-6.

In the procedure just described, it has been tacitly assumed that all stars have the same chemical composition and that complicating effects, such as rotation and magnetic fields, are unimportant. The composition is especially important both because it fixes the intrinsic brightness of the main sequence at a given color and because stars with different abundances will suffer differing amounts of line-blanketing in their measured colors. Both of these effects must be taken into account. This can be done by using the results of theoretical computations that give the position of the main sequence for different compositions, and theoretical and semiempirical blanketing corrections for the colors. For galactic clusters, the spread in metal abundances relative to the Sun is small, and the resulting differential line-blanketing effects from cluster to cluster are minor (although not negligible).

A more serious problem is that of finding a genuinely unevolved portion of the main sequence in any given cluster. In some cases, the observations may not extend to stars that are faint enough to assure that we have reached the unevolved main sequence. In the youngest clusters, the luminous stars

Table 3-1. The Zero-Age Main Sequence (ZAMS)

$(B-V)_0$	M_V	$(B-V)_0$	M_V
−0.30	−3.50	0.30	2.80
−0.25	−2.30	0.40	3.35
−0.20	−1.30	0.50	4.05
−0.15	−0.50	0.60	4.60
−0.10	0.30	0.70	5.20
−0.05	0.90	0.80	5.70
0.00	1.30	0.90	6.10
0.05	1.55	1.00	6.60
0.10	1.80	1.10	7.00
0.20	2.25	1.20	7.45
		1.30	7.90

SOURCE: Adapted from (**S24**, 216)

may already have begun to evolve away from the main sequence while less-massive stars are still contracting onto it. In this event, the cluster may not have a truly unevolved part to its main sequence. Despite these difficulties, various independent determinations of the ZAMS are in good agreement with one another, and the empirical ZAMS given in Table 3-1 seems well established. Because the ultimate reference standard in the cluster-fitting procedure is the Hyades, the numbers in the table have been made 0.2 mag brighter than those given in the sources cited to allow for the increase in the accepted value of the Hyades distance modulus from 3.1 mag to 3.3 mag.

Once cluster CM diagrams have been matched together, one can read absolute magnitudes for stars of various spectral types. Combining these data with those obtained from other methods, one can finally assemble a calibration of M_V for almost all MK spectral types. A typical compilation of results is shown in Table 3-2. It should be mentioned that the absolute magnitudes tabulated for each type are *averages* for stars of that type. Thus the main sequence given in the table includes an admixture of somewhat evolved stars, and it will therefore necessarily be brighter than the ZAMS for stars earlier than type G. For example, the ZAMS for O stars lies almost a magnitude below the main sequence given in Table 3-2. The final calibration given in Table 3-2 is rather eclectic, and the results for different types are not all of equal reliability [see (**B7**, Chapter 20) for an assessment of accuracies]. The main techniques used for the calibration of various stellar types can be summarized as follows:

1. supergiants: CM diagram fitting;
2. O–A stars: CM diagram fitting, secular and statistical parallaxes (see §6-6);
3. F–M dwarfs: trigonometric parallaxes, moving-cluster method;

Table 3-2. Mean Absolute Visual Magnitude, M_V, Versus MK Spectral Type.

Spectral class	Luminosity class						
	V	IV	III	II	Ib	Ia	Ia0
O5	−5.6						
O9	−4.8	−5.3	−5.7	−6.0	−6.1	−6.2	
B0	−4.3	−4.8	−5.0	−5.4	−5.8	−6.2	−8.1
B5	−1.0	−1.8	−2.3	−4.4	−5.7	−7.0	−8.3
A0	0.7	0.2	−0.4	−3.0	−5.2	−7.1	−8.4
A5	1.9	1.4	0.3	−2.7	−4.8	−7.7	−8.5
F0	2.5	1.9	0.8	−2.4	−4.7	−8.5	−8.7
F5	3.3	2.1	1.2	−2.3	−4.6	−8.2	−8.8
G0	4.4	2.8	0.9	−2.1	−4.6	−8.0	−9.0
G5	5.2	3.0	0.5	−2.1	−4.6	−8.0	
K0	5.9	3.1	0.6	−2.2	−4.5	−8.0	
K5	7.3		−0.2	−2.3	−4.5	−8.0	
M0	8.8		−0.4	−2.3	−4.6	−7.5	
M2	10.0		−0.6	−2.4	−4.8	−7.0	
M5	12.8		−0.8				

SOURCE: Adapted from (**A3**, 200), (**S24**, 401)

4. F–M giants: moving-cluster method, CM diagram fitting, secular and statistical parallaxes.

Special calibration techniques may also be helpful in specific cases. For example, absolute magnitudes can be estimated for several supergiants that are components of noninteracting binaries containing another star of known absolute magnitude (for example, a dwarf), using the fact that the difference in apparent magnitudes of the two components equals the difference in their absolute magnitudes. Implicit in this method is the assumption that both stars in the system are normal and unaffected by the presence of the companion.

Finally, absolute magnitudes for white dwarfs have been determined from trigonometric parallaxes and their membership in clusters and binaries. The characteristic ranges of absolute magnitude for the various white-dwarf spectral classes are: DB, $M_V \approx +10$ to $+11$; DA, $M_V \approx +11$ to $+12$; DF, $M_V \approx +13$ to $+14$; DG, $M_V \approx +14$ to $+15$; and DM, $M_V \approx +15$.

Because of the wide variety of methods used to put together the calibration given in Table 3-2, the effect of the recent change in the Hyades distance modulus cannot be judged without reworking the original data; hence the numbers in the table are the same as in the original sources cited. This absolute-magnitude calibration is of tremendous importance to galactic-structure research, for it allows us to assign a distance to any star of known type and apparent magnitude and thus provides the foundation for the analysis of the space distribution of stars in the solar neighborhood (see §4-2 and §4-3).

The Spectral Type Versus Color, the Color–Color, and Spectral Type Versus Effective-Temperature Relations

The MK spectral type of a star is a concise description of the morphology of its line-spectrum. Inasmuch as both the line and continuous spectrum emitted are determined by the physical structure of a star's atmosphere, we

Table 3-3. The Spectral Type Versus Color Relation

Main sequence (luminosity class V)					
Spectral class	$(B-V)$	$(U-B)$	Spectral class	$(B-V)$	$(U-B)$
O5	−0.32	−1.15	B9	−0.06	−0.19
O6	−0.32	−1.14	A0	0.00	0.00
O7	−0.32	−1.14	A5	0.15	0.09
O8	−0.31	−1.13	F0	0.29	0.04
O9	−0.31	−1.12	F5	0.42	−0.01
B0	−0.30	−1.08	G0	0.58	0.05
B1	−0.26	−0.93	G5	0.69	0.20
B2	−0.24	−0.86	K0	0.85	0.47
B3	−0.20	−0.71	K5	1.16	1.09
B5	−0.16	−0.56	M0	1.42	1.25
B7	−0.12	−0.42	M5	1.61	1.22
B8	−0.09	−0.30			

Giants (luminosity class III)					
Spectral class	$(B-V)$	$(U-B)$	Spectral class	$(B-V)$	$(U-B)$
O5	−0.32	−1.15	B9	−0.06	−0.19
O6	−0.32	−1.14	A0	0.00	0.00
O7	−0.32	−1.14	A5	0.15	0.10
O8	−0.31	−1.13	F0	0.27	0.10
O9	−0.31	−1.12	F5	0.45	0.07
B0	−0.30	−1.09	G0	0.65	0.30
B1	−0.26	−0.95	G5	0.84	0.52
B2	−0.24	−0.88	K0	1.03	0.87
B3	−0.20	−0.72	K5	1.45	1.65
B5	−0.16	−0.56	M0	1.57	1.8
B7	−0.12	−0.42	M5	1.80	2.1
B8	−0.09	−0.30			

Table 3-3. *(continued)*

Supergiants (luminosity class I)				
Spectral class	$(B - V)$	$(U - B)_{Ib}$	$(U - B)_{I}$	$(U - B)_{Ia}$
O5	−0.32		−1.16	
O6	−0.32		−1.15	
O7	−0.31		−1.14	
O8	−0.29		−1.13	
O9	−0.28		−1.12	
B0	−0.24	−1.05		−1.07
B1	−0.19	−0.96		−1.00
B2	−0.17	−0.91		−0.96
B3	−0.13	−0.82		−0.87
B5	−0.09	−0.72		−0.78
B7	−0.05	−0.62		−0.68
B8	−0.02	−0.53		−0.60
B9	0.00	−0.48		−0.56
A0	0.01		−0.35	
A5	0.07		...	
F0	0.21		0.22	
F5	0.40		0.30	
G0	0.70		0.60	
G5	1.07		0.83	
K0	1.37		1.35	
K5	1.65		1.7	
M0	1.9		1.7	
M5	2.1		1.8	

SOURCES: Adapted from (**A3**, 206), (**G3**, 79), (**S24**, 214)

can expect that its continuous energy distribution, and hence colors, will be closely correlated with its MK spectral type. This is indeed found to be the case, and, after allowance for interstellar reddening, one can associate distinctive colors with most types as shown in Table 3-3 for $(B - V)$ and $(U - B)$. [In the table, the data for supergiants represent all subclasses (Ia, Iab, Ib) lumped together unless otherwise indicated.]

The *UBV* color indices clearly show that early-type stars are rich in ultraviolet radiation and that late-type stars are distinctly red. Note that, for the hottest stars, both $(B - V)$ and $(U - B)$ approach limiting values. This is a result of the fact that, at high temperatures ($T \gtrsim 40{,}000$ K), the Planck function in the visible reduces to the Rayleigh–Jeans form $B_\nu(T_R) = (2kT_R\nu^2/hc^2)$, where T_R is a characteristic *radiation temperature* of the radiation field. It follows from equation (2-15) that, in this limit, the color indices will be essentially independent of T_R. The bulk of the radiation emitted by

Table 3-4. Infrared Colors of Main-Sequence Stars

Spectral class	Color index					
	$(V-R)$	$(V-I)$	$(V-J)$	$(V-K)$	$(V-L)$	$(V-N)$
A0	0.00	0.00	0.00	0.00	0.00	0.00
F0	0.30	0.47	0.55	0.74	0.8	0.8
G0	0.52	0.93	1.02	1.35	1.5	1.4
K0	0.74	1.4	1.5	2.0	2.5	
M0	1.1	2.2	2.3	3.5	4.3	
M5	. . .	2.8				

SOURCE: (**A3**, 208), by permission

late-type stars emanates in the infrared as is shown clearly in Table 3-4. Similarly, colors constructed from far-ultraviolet magnitudes (observed from space) and V show that the peak energy emission by early-type stars moves to progressively shorter wavelengths as the temperature of the stellar atmosphere rises.

The data in Table 3-3 can be used to construct *color-color* (or *two-color*) *diagrams*, such as those shown in Figure 3-7. As can be seen, it is possible to distinguish stars of different spectral and luminosity classes from one another, assuming that adequate allowance can be made for interstellar reddening. Similar diagrams using other photometric indices can sometimes be used to carry out a two-dimensional photometric "spectral classification"

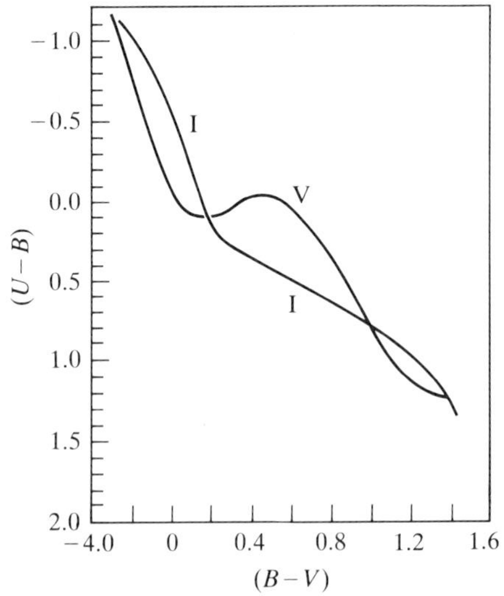

Figure 3-7. Two-color diagram for main-sequence stars (luminosity class V) and supergiants (luminosity class I).

Table 3-5. The Effective-Temperature and Bolometric-Correction Scales

Spectral class	Luminosity class V		Luminosity class III		Luminosity class I	
	T_{eff}	*B.C.*	T_{eff}	*B.C.*	T_{eff}	*B.C.*
O5	47,000	−4.0				
O7	38,000	−3.5				
O9	34,000	−3.2			30,000	−2.9
B0	30,500	−3.00			25,500	−2.6
B2	23,000	−2.30				
B3	18,500	−1.85				
B5	15,000	−1.40			13,500	−0.9
B7	13,000	−0.90				
B8	12,000	−0.70				
A0	9,500	−0.20				
A5	8,300	−0.10				
F0	7,300	−0.08			6,400	−0.2
F5	6,600	−0.01				
G0	5,900	−0.05	5,400	−0.1	5,400	−0.3
G5	5,600	−0.10	4,800	−0.3	4,700	−0.6
K0	5,100	−0.2	4,400	−0.5	4,000	−1.0
K5	4,200	−0.6	3,600	−1.1	3,400	−1.6
M0	3,700	−1.2	3,300	−1.5	2,800	−2.5
M5	3,000	−2.5	2,700:	−3:		
M8	2,500	−4:				

SOURCES: Adapted from (**A3**, 206), (**C13**), (**J2**)

within limited regions of the H-R diagram. This is valuable because we can then classify very faint stars.

As was discussed in §3-4, the spectrum of a star is essentially determined by its effective temperature and surface gravity. Furthermore, because a stellar spectrum can be characterized by giving its MK spectral type, it follows that, to each MK spectral type, there should correspond definite values of T_{eff}, *B.C.*, and the surface gravity g. In §3-4, we described various methods for determining T_{eff}; Table 3-5 assembles representative results of this work. The most reliable values given are for main-sequence *B–K* stars. The accuracy of the estimates deteriorates for O and M stars and for giants and supergiants. Table 3-5 shows that MK spectral classes correlate closely with T_{eff}. The MK luminosity classes are primarily sensitive to the surface gravity g. The relationships $L \propto R^2 T_{\text{eff}}^4$ and $g \propto \mathscr{M}/R^2$ imply $L \propto (\mathscr{M}/g)$ at fixed spectral type (fixed T_{eff}). Figure (3-8) shows how the luminosity of

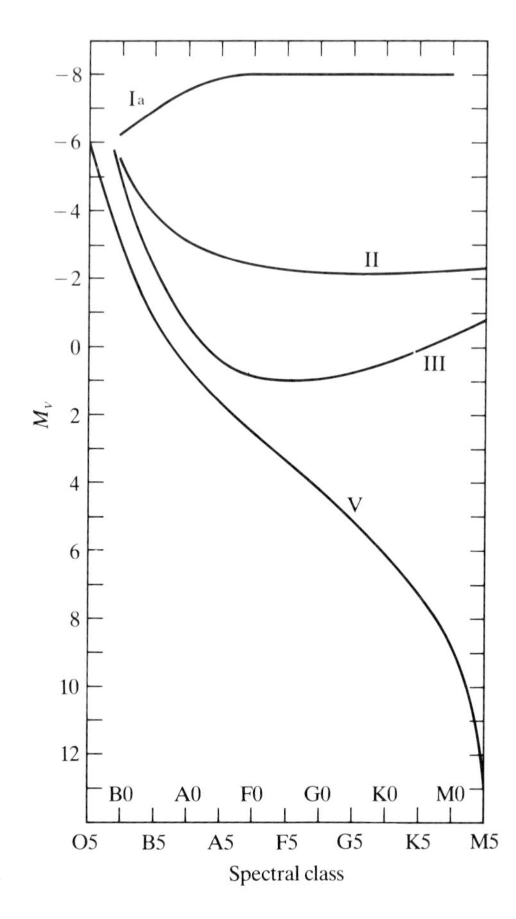

Figure 3-8. Absolute visual magnitude as a function of MK spectral type. Individual curves are labeled with luminosity classes.

a star depends on its MK spectral type. High-luminosity stars tend to have large masses and low gravities.

The Mass-Luminosity and Mass-Radius Relations

Naively, one would expect massive stars to be very luminous and stars with small masses to be faint. Broadly speaking, it is true that faint stars have small masses, but it turns out that luminous stars are not necessarily massive. As we shall see, the luminosity of a star tends to increase as it evolves, and this increase is very pronounced for low-mass stars. Thus, toward the end of its life, even quite a low-mass star can become a luminous giant. If one confines one's attention to main-sequence stars, however (and, in any given cluster, most stars will at any one time be on the main sequence), one finds

Table 3-6. Physical Properties of Main-Sequence Stars

$\log(\mathscr{M}/\mathscr{M}_\odot)$	Spectral class	$\log(L/L_\odot)$	M_{bol}	M_V	$\log(R/R_\odot)$
−1.0	M6	−2.9	12.1	15.5	−0.9
−0.8	M5	−2.5	10.9	13.9	−0.7
−0.6	M4	−2.0	9.7	12.2	−0.5
−0.4	M2	−1.5	8.4	10.2	−0.3
−0.2	K5	−0.8	6.6	7.5	−0.14
0.0	G2	0.0	4.7	4.8	0.00
0.2	F0	0.8	2.7	2.7	0.10
0.4	A2	1.6	0.7	1.1	0.32
0.6	B8	2.3	−1.1	−0.2	0.49
0.8	B5	3.0	−2.9	−1.1	0.58
1.0	B3	3.7	−4.6	−2.2	0.72
1.2	B0	4.4	−6.3	−3.4	0.86
1.4	O8	4.9	−7.6	−4.6	1.00
1.6	O5	5.4	−8.9	−5.6	1.15
1.8	O4	6.0	−10.2	−6.3	1.3

SOURCE: Adapted from (**A3**, 209), by permission

both empirically and theoretically that a *mass-luminosity relation* holds; $L \propto \mathscr{M}^{3.2}$, approximately. For main-sequence stars, there is also a *mass-radius relation*; $R \approx R_\odot(\mathscr{M}/\mathscr{M}_\odot)^{0.7}$. These results, which are summarized in Table 3-6, can be explained satisfactorily by the theory of stellar structure.

A very important consequence of the main-sequence mass-luminosity relation and the tendency of the luminosity of a star to increase as it evolves is that luminous stars must have lifetimes much shorter than the Sun's. The amount of fuel available for thermonuclear burning is approximately proportional to a star's mass, and the rate at which it is consumed is proportional to its luminosity. Hence the lifetime τ of a star is proportional to $\mathscr{M}/\bar{L}$, where $\bar{L}$ is the mean luminosity of a star averaged over its lifetime. Because stars leave the main sequence only when they are running short of fuel and then rapidly squander their slender remaining reserves, a star's mean luminosity $\bar{L}$ does not differ significantly from its main-sequence luminosity L_{ms}. Therefore, from the main-sequence mass-luminosity relation, we can estimate for the lifetime, $\tau \propto \mathscr{M}/L_{ms} \propto \mathscr{M}^{-2.2}$. It is thus clear that massive stars have short lifetimes. In fact, the theory of stellar structure predicts [see equation (3-44)] that an O star cannot live longer than about a million years.

It was essentially this quite simple calculation that led L. Spitzer to conclude (**S20**) in 1948 that star formation must be a continuous, ongoing process. The mere existence of very luminous stars indicates that not all star formation can have occurred in the distant past. To determine how the

Table 3-7. Correlation of Stellar Radius and Luminosity with MK Spectral Type

Spectral class	$\log(R/R_\odot)$			$\log(L/L_\odot)$		
	V	III	I	V	III	I
O5	1.25			5.7		
B0	0.87	1.2	1.3	4.3		5.4
B5	0.58	1.0	1.5	2.9		4.8
A0	0.40	0.8	1.6	1.9		4.3
A5	0.24		1.7	1.3		4.0
F0	0.13		1.8	0.8		3.9
F5	0.08	0.6	1.9	0.4		3.8
G0	0.02	0.8	2.0	0.1	1.5	3.8
G5	−.03	1.0	2.1	−0.1	1.7	3.8
K0	−.07	1.2	2.3	−0.4	1.9	3.9
K5	−.13	1.4	2.6	−0.8	2.3	4.2
M0	−.20		2.7	−1.2	2.6	4.5
M5	−.5			−2.1	3.0	

SOURCE: Adapted from (**A3**, 209), by permission

rate of star formation changes in time is an important task of the theory of *galactic evolution*. Conversely, any such theory will be constrained in significant ways by the numbers of stars with various masses observed to exist at present, essentially because there *is* a characteristic lifetime for each mass, which implies that a study of stars of decreasing mass gives a glimpse of our Galaxy at increasing mean stellar age and hence at increasing look-back times.

Each MK spectral type is associated with characteristic values of L and R. Table 3-7 gives typical values for these quantities, expressed in solar units. The great range of R values quoted in Table 3-7 for M stars follows from the huge range of luminosities displayed by these stars, which all have similar effective temperatures. Actually, the range in the radii of stars is even greater than this table would suggest. Some of these stars are 1000 solar radii in diameter, whereas a typical white dwarf has a radius of about $10^{-2}R_\odot$, comparable to that of the Earth. The mean densities of stars vary over a correspondingly gigantic range, from less than 10^{-6} of the solar mean density for the typical red giant to nearly 10^6 of solar for a white dwarf. The latter are, in fact, so dense that they are sustained against gravitational collapse by the zero-point energies of their electrons, just as an atom is sustained against electrostatic collapse by the residual motion of its electrons. One says that the material of white dwarfs is *degenerate*. During the life of a star, its radius, and hence its mean density, will change considerably. Its mass may also change as a result of mass loss or mass exchange, but this change is probably a less important phenomenon than the change in radius. In any case, we do not know how to model mass loss theoretically.

3-6. SYSTEMATICS OF STELLAR PROPERTIES: SPHEROIDAL-COMPONENT STARS

In addition to the metal-rich stars in the disk and spiral arms, our Galaxy contains a second major group of stars in its *spheroidal component.* These stars are represented in the solar neighborhood by the metal-poor *subdwarfs* and at larger distances by the stars in *globular clusters* and the *galactic bulge.* The physical properties of spheroidal-component stars are quite different from those of disk stars, and this fact is clearly manifested in their H-R and CM diagrams. As was mentioned in Chapter 1, these stars are all old and have a very different history from the disk stars. In all probability, they were formed before the disk itself existed.

The Subdwarfs

Among the F, G, and K stars observed in the neighborhood of the Sun are a few that fall distinctly below the main sequence in a CM diagram. Because they appear somewhat underluminous (in M_V) for their $(B - V)$ color, these stars are called subdwarfs. Compared to common field stars, the subdwarfs also have an *ultraviolet excess*; that is, they have brighter $(U - B)$ colors than normal stars having the same $(B - V)$.

When the subdwarfs are examined spectroscopically, it is found that their spectral lines are abnormally weak, and quantitative spectroscopic analyses show them to be extremely metal poor relative to the Sun, with abundances down by a factor of 10^{-2} or more. As soon as it was known that the subdwarfs are weak lined, it was recognized that this implied that they should appear to be too blue relative to stars having normal line strengths but otherwise identical properties (L and T_{eff}). In turn, it was realized that they might appear to be subdwarfs in a CM diagram, not because they are really underluminous but rather because they are anomalously blue and therefore displaced to the left of the normal main sequence.

Thus, to compare stars with markedly different compositions, we must be able to account for the effects of different metal abundances on photometric indices and to determine the position of a star of arbitrary metal abundance relative to some fiducial sequence (the Hyades is invariably used as the standard) in both CM diagrams and the theoretical H-R diagram. Furthermore, we must develop methods that allow photometric indices to be used directly to infer information about the metal content of stars. To accomplish these goals, we must evaluate the photometric effects of line-blanketing in stars with differing metal abundances. In the following discussion, we shall focus mainly on blanketing corrections in the *UBV* system because this system reaches the faintest stars and has been widely used, for example, in studies of globular clusters. Similar considerations apply to other systems as well.

Line-Blanketing The mere presence of dark spectral lines in any photometric band obviously reduces the energy emitted in that band. This direct effect of *line-blocking* is described by the *blocking coefficient*

$$\varepsilon_\lambda \equiv 1 - \left[\frac{\int_{\lambda-\Delta\lambda}^{\lambda+\Delta\lambda} F_\lambda \, d\lambda}{\int_{\lambda-\Delta\lambda}^{\lambda+\Delta\lambda} F_\lambda^c \, d\lambda} \right] \tag{3-36}$$

where F_λ denotes the observed flux at wavelength λ, F_λ^c is the continuum flux near λ, and $\Delta\lambda$ is a prechosen wavelength interval (typically 25 Å to 50 Å). Blocking coefficients have been measured by direct planimetry of tracings of high-dispersion spectra for a variety of stars **(M14)**, **(W7)**; some typical results are shown in Figure 3-9. The basic points to be noticed in the figure are that (1) the blocking coefficient increases rapidly from the visual toward the blue and ultraviolet (indeed, so strongly that, in the ultraviolet, it is difficult to find unblocked windows within which to set the continuum level) and (2) the later the spectral type of a star, the larger is the amount of line-blocking. Notice particularly in Figure 3-9*b* how much weaker the blocking is in the subdwarf HD 19445 than it is in the Sun. Trustworthy blocking coefficients for very late-type stars are almost impossible to determine empirically, because the lines simply swamp the continuum and there is no way to set the continuum-flux level reliably in much of the visible spectrum.

From the variation of ε_λ with wavelength, it is clear that, if we start with a star having zero metal abundance and therefore no spectrum lines (other than hydrogen lines) and gradually increase the metal abundance up to the Hyades value, the star should become fainter in $(B - V)$ and fainter yet in $(U - B)$. But these line-blocking effects produce only part of the total blanketing effect. An additional effect of increasing the metal abundance while holding a star's effective temperature constant is *backwarming* of the atmosphere. If we demand that the same total energy flux (which is fixed by thermonuclear processes in the star's interior) must ultimately escape from the star (that is, that T_{eff} remains constant), then the energy blocked in the lines must be redistributed in wavelength and escape from continuum windows between the lines. This redistribution raises the continuum level above the value it would have had in the absence of lines, and it simulates the effect of a higher effective temperature of the star. The net photometric changes produced by adding lines are determined by a competition between these two effects, blocking tending to increase each observed magnitude and

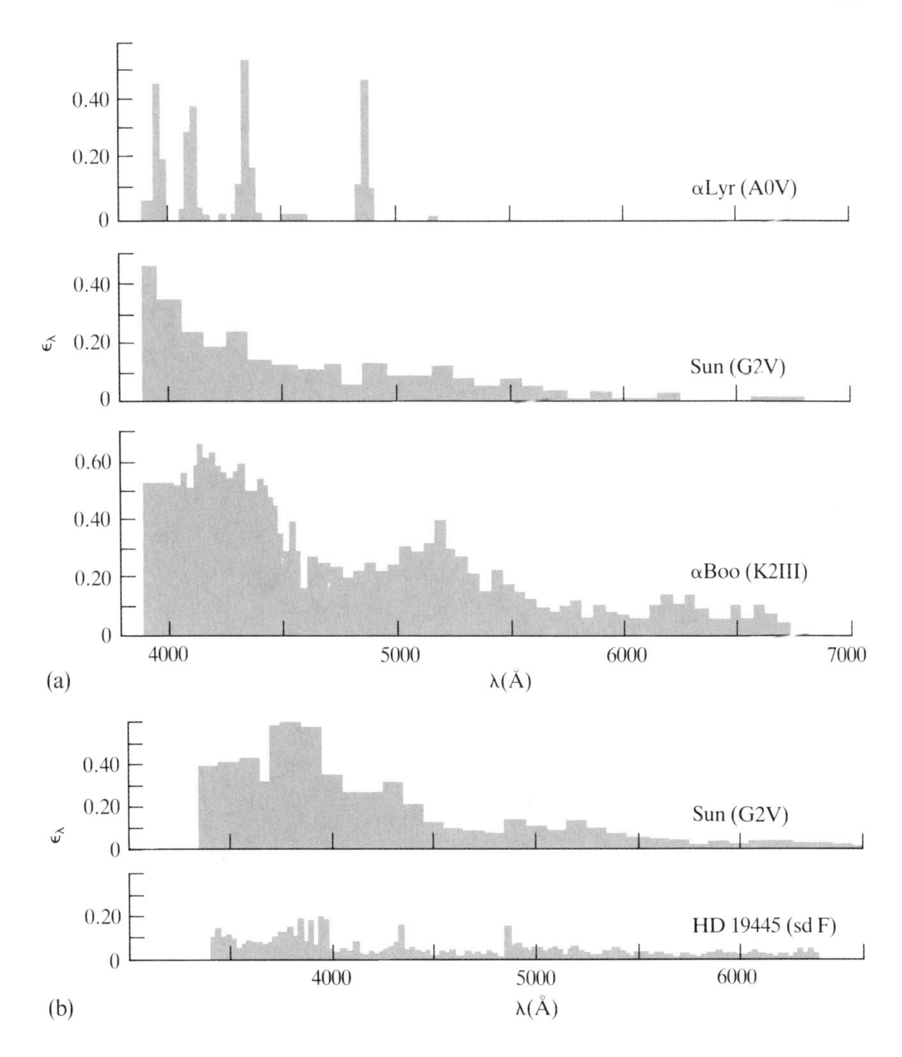

Figure 3-9. Observed line-blocking coefficients for (a) an early-type dwarf (α Lyr), the Sun, and a late-type giant (α Boo); and (b) the Sun and an extreme subdwarf, HD 19445. *Ordinate*: ε_λ as defined in equation (3-36); *abscissa*: wavelength in Ångstroms. [Adapted from (**M14**) and (**W7**) by permission. The latter is © 1962 by the University of Chicago.]

backwarming tending to decrease it. In practice, blocking dominates over backwarming in the U band, exceeds it slightly in the B band, and is less important in the V band; in the end, adding lines increases $(U - B)$ and $(B - V)$ and decreases V.

The final result in the two-color diagram is that the star moves along a *blanketing vector*, such as that shown in Figure 3-10. We see in the figure

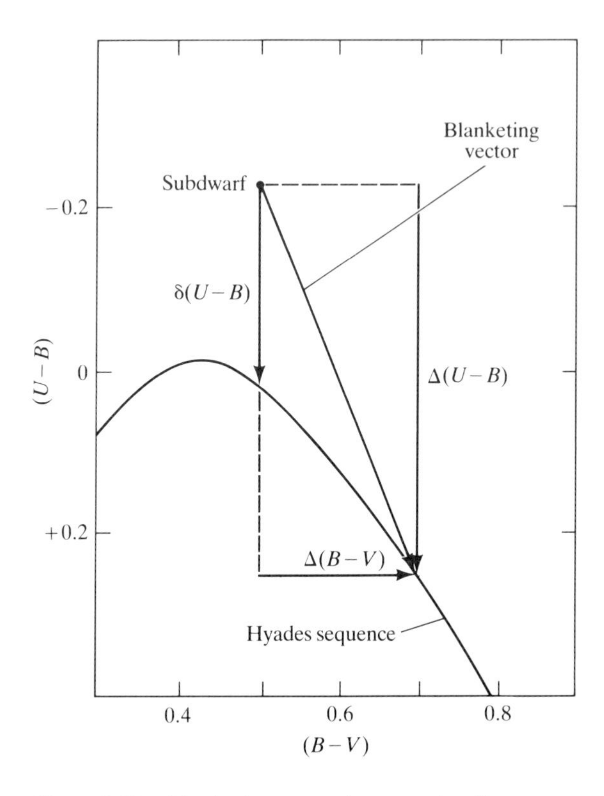

Figure 3-10. Blanketing vector in two-color diagram for a metal-deficient subdwarf. The subdwarf has an ultraviolet excess $\delta(U - B)$ compared to a Hyades star (which has near-solar metal abundance) of the same $(B - V)$. A Hyades star *of the same effective temperature* has colors that differ by amounts $\Delta(B - V)$ and $\Delta(U - B)$ from those of the subdwarf. The effect of adding metal lines to the subdwarf's atmosphere while holding $T_{\rm eff}$ constant is to move the observed colors along the blanketing vector as defined in (**S3**) and (**W7**).

that a metal-poor star that has an *ultraviolet excess* $\delta(U - B)$ relative to Hyades stars having the same $(B - V)$ will differ from Hyades stars of the same effective temperature by amounts $\Delta(B - V)$ and $\Delta(U - B)$ in $(B - V)$ and $(U - B)$, respectively. If we can determine the slope of blanketing vectors as a function of $(B - V)_{\rm Hyades}$, then we can clearly determine the *blanketing corrections* $\Delta(B - V)$ and $\Delta(U - B)$ for any star of known $\delta(U - B)$ and $(B - V)$. If, in addition, we can calculate ΔV, we are then in a position to correct the observed colors and brightnesses of stars for differential blanketing effects and thus map cluster CM diagrams onto one another, reduced to a common abundance.

Empirical blanketing vectors have been derived by Wildey et al. (**W7**). Blocking effects were calculated directly as

$$\Delta m_{\text{bk}} = 2.5 \log\left[\frac{\int_{\lambda_1}^{\lambda_2} F_\lambda S_\lambda \, d\lambda}{\int_{\lambda_1}^{\lambda_2} F_\lambda(1 - \varepsilon_\lambda) S_\lambda \, d\lambda}\right] \tag{3-37}$$

where S_λ is the photometer response on the interval (λ_1, λ_2). To compute the backwarming effect, the *integrated blocking coefficient*

$$\eta \equiv \frac{\int_0^\infty \varepsilon_\lambda F_\lambda \, d\lambda}{\int_0^\infty F_\lambda \, d\lambda} \tag{3-38}$$

is used to evaluate T_1, the (larger) effective temperature of an atmosphere that has the same continuum level as does the backwarmed line-blanketed atmosphere, whose actual effective temperature is T_2, from the relation

$$(1 - \eta)\sigma T_1^4 = \sigma T_2^4 = \int_0^\infty F_\lambda \, d\lambda \tag{3-39}$$

The magnitude change produced by the backwarming effect is then

$$\Delta m_{\text{bw}} = 2.5 \log\left[\frac{\int_{\lambda_1}^{\lambda_2} F_\lambda(T_1) S_\lambda \, d\lambda}{\int_{\lambda_1}^{\lambda_2} F_\lambda(T_2) S_\lambda \, d\lambda}\right] \tag{3-40}$$

To actually evaluate this expression, one must use fluxes from model atmospheres at the appropriate effective temperatures.

Applying equations (3-37) and (3-40), one can compute the net changes in ΔU, ΔB, and ΔV produced by the combined effects of blocking and backwarming. These results can then be used to construct tables (**W7**) or graphs (**B7**, 365) giving $\Delta(B - V) = f[\delta(U - B), (B - V)_{\text{obs}}]$, and similar relations for $\Delta(U - B)$ and ΔV. By custom, $\delta(U - B)$ is taken as increasingly positive for more metal-deficient stars; to map such stars onto the Hyades sequence, one will have $\Delta(U - B) > 0$, $\Delta(B - V) > 0$, and $\Delta V < 0$. For example, the Hyades are slightly metal rich compared to the Sun, and one finds $\Delta(U - B)_\odot \approx 0.07$, $\Delta(B - V)_\odot \approx 0.055$, $\Delta(U - B)_\odot \approx 0.13$, and $\Delta V_\odot \approx -0.026$.

Thanks to recent advances in model-atmosphere calculations, blanketing vectors can now also be determined theoretically [see (**B5**), (**G8**), (**N3**), (**P6**, 271), and (**P6**, 319)]. An example of such work is shown in Figure 3-11. Similar calculations can also be made for other photometric systems, such as the Strömgren *uvby* system. It is worth noting that theoretical model atmosphere analyses of subdwarf spectra are actually easier than for ordinary stars because the subdwarfs' line spectrum is so weak, and accurate effective

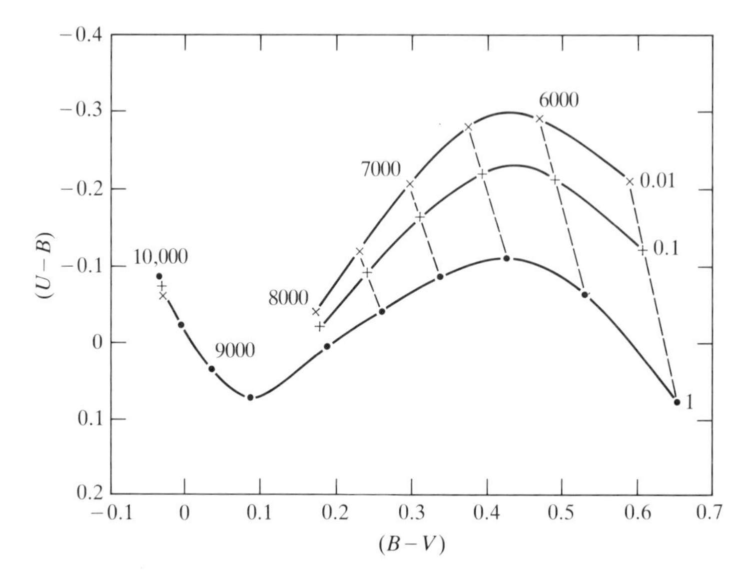

Figure 3-11. Theoretical $(U - B)$ and $(B - V)$ colors for model atmospheres having $\log g = 4$ and effective temperatures between 5500 K and 10,000 K. *Solid dots*: normal solar metal abundance. *Plus signs*: 0.1 times solar metal-abundance. x's: 0.01 times solar metal abundance. Dashed lines show the resulting theoretical blanketing vectors. [From (**P6**, 271).]

temperatures can be obtained for these stars by fitting their observed absolute energy distributions to theoretical fluxes, as was first done by Melbourne (**M11**). Certain photometric indices, for example, $(G-I)$ in the six-color system, are almost unaffected by blanketing effects in F–K stars because the blocking and backwarming effects almost exactly cancel (**C12**), (**D2**). For nearly solar-type stars, such indices can thus be used to make fairly reliable estimates of T_{eff} for stars of all metallicities.

Photometric Metal-Content Indicators By combining photometric data for subdwarfs with spectroscopically determined metal abundances, one can calibrate a photometric index, say, the ultraviolet excess $\delta(U - B)$, in terms of a star's metal abundance. This index can then be used to estimate metal abundances in other stars, a procedure that is highly advantageous because (1) many stars of interest that are too faint for high-dispersion spectroscopy (the practical limit is $m \approx 12$) can be measured with UBV photometry (practical limit $m \approx 20$), and (2) several days of laborious measurement and analysis are required to produce a spectroscopic abundance estimate even when spectra are available, whereas a photometric determination requires only a small fraction of this effort.

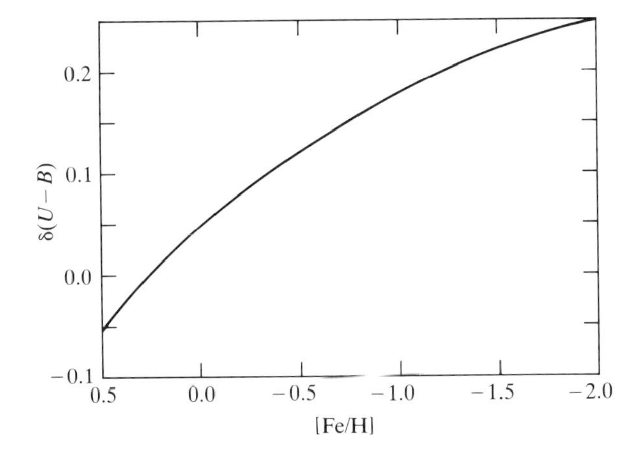

Figure 3-12. Correlation of the ultraviolet excess $\delta(U - B)$, measured with respect to the Hyades sequence, with logarithmic metal abundance [Fe/H] relative to the Sun.

Metal abundances arc usually expressed in terms of the stellar iron-to-hydrogen ratio compared to that of the Sun via the parameter

$$[\text{Fe/H}] \equiv \log[n(\text{Fe})/n(\text{H})]_* - \log[n(\text{Fe})/n(\text{H})]_\odot \qquad (3\text{-}41)$$

The correlation of $\delta(U - B)$ with [Fe/H] has been established by several investigations [see **(B9)**, **(E3)**, **(H10)**, **(P1)**, **(P9)**, **(W1)**, **(W2)**], and a representative curve is shown in Figure 3-12; it should be recalled that $\delta(U - B)$ is measured with respect to the Hyades, which has $[\text{Fe/H}]_{\text{Hyades}} \approx +0.025$. For modest values of $\delta(U - B)$, the relation

$$[\text{Fe/H}] - [\text{Fe/H}]_{\text{Hyades}} \approx -5\delta(U - B) \qquad (3\text{-}42)$$

can be used for $\delta(U - B)$ both greater than or less than zero (the latter applying to metal-rich stars).

Similarly, in the Strömgren system (see §2-6), if Δm_1 is the difference between the m_1 index of a star and that of a Hyades star of the same $(b - y)$, then one finds **(G8)**, **(N3)**

$$[\text{Fe/H}] - [\text{Fe/H}]_{\text{Hyades}} \approx -14\Delta m_1, \qquad [0.2 \leq (b - y) \leq 0.3] \quad (3\text{-}43\text{a})$$

and

$$[\text{Fe/H}] - [\text{Fe/H}]_{\text{Hyades}} \approx -12.5\Delta m_1, \qquad [0.3 < (b - y) < 0.4] \quad (3\text{-}43\text{b})$$

Calibrations also exist for other photometric systems [see, for example, **(P6)**].

Globular Clusters

Physical Characteristics Globular clusters are very rich, roughly spherically symmetric star clusters found at positions ranging from near the galactic center to remote regions in the halo. These systems typically contain hundreds of thousands of stars within a volume having a typical radius of about 20 to 50 pc. They have high star densities at their centers (10^2 to $\geq 10^4$ stars/pc^3) and are dynamically very stable and long lived. In fact, as we shall see in Chapter 5, in many ways they may be regarded as a type of extremely low-mass elliptical galaxy. In particular, their radial surface-brightness profiles are so similar to those of many elliptical galaxies that the same family of dynamical models has been widely used to describe both types of system. [We shall discuss these models, the King models (**K1**), in some detail in Chapters 5 and 14.] The radial star-density profiles of globular clusters show differing degrees of central concentration, which led H. Shapley and H. B. Sawyer (**S17**) to classify clusters into *concentration classes*, designated with Roman numerals I, . . . , XII. In this classification, class I clusters show the highest degree of concentration, and class XII clusters the lowest. The categories were defined such that clusters are found with nearly equal frequency in all twelve classes. Quantitative measurements of surface brightness and star density in globular clusters are compiled in (**I8**), (**K2**), and (**P4**).

The observed departures of globular clusters from circular symmetry are normally quite small, most systems having axial ratios in the range 0.9:1 to 1:1, although a few clusters show appreciable elongation (down to 0.6:1) in their outer regions.

Globular clusters are very luminous systems. This fact partly accounts for their usefulness in galactic-structure research, because it implies that they can be seen at large distances (**S16**). As we shall see in Chapter 5, globular clusters are also important because observations of these relatively nearby systems can help us understand our necessarily less-detailed observations of elliptical galaxies. The total apparent magnitude of a cluster follows from surface-brightness measurements (extrapolated to infinity, if necessary, by the use of a theoretical model; see Chapter 5 for details). If we can estimate the interstellar absorption and the cluster distance modulus (see §4-4), then, from the integrated apparent magnitude, we can derive an integrated absolute magnitude $(M_V)_0$. Typically, one finds $-5 > (M_V)_0 \gtrsim -10$. The distribution of globular-cluster absolute magnitudes shows a peak at $(M_V)_0 \approx -8.5$, with a half-width of about ± 1 mag [see (**B7**, 424) and (**O3**)]. The mass-to-luminosity ratio (in solar units) of globular-cluster stars is known to be of order unity (**I7**). Hence we find that a typical cluster mass is of the order of $2 \times 10^5 \mathscr{M}_\odot$.

Integrated $(B - V)$ and $(U - B)$ colors for globular clusters can be measured by the same techniques used to measure integrated magnitudes. After correction for reddening, intrinsic $(B - V)_0$ colors are typically found to lie on the range $0.4 \lesssim (B - V)_0 \lesssim 0.8$, with a well-defined peak near

$(B - V)_0 \approx 0.57$. These colors show immediately that, unlike the case for galactic clusters, the observed light from globular clusters is dominated by contributions from stars somewhat cooler than the Sun. The interpretation of the integrated colors is somewhat ambiguous. For example, a bluer color might indicate the presence of some relatively hot stars, or it might be produced by a deficiency of metals in the cluster stars.

The integrated spectral types of globular clusters cover a wide range from about F2 to G5 (**M8**), (**M17**), (**K3**), (**K6**). Characteristically, halo clusters have early-type spectra and show strong similarities to subdwarf spectra, which indicates a large metal deficiency in the cluster stars. Clusters with later-type spectra are typically found near the galactic center, and their spectra are compatible with nearly normal metal abundances. These conclusions are supported by examinations of the line strengths in the spectra of individual luminous giants in the clusters (**B4**), (**D3**). Thus cluster spectra provide key information about the chemical compositions of the clusters themselves and about the distribution of chemical elements within our Galaxy (a point to which we shall return in §4-4).

As Morgan has stressed, the interpretation of composite spectra is difficult even under favorable circumstances, and it can often be quite ambiguous [see (**M18**) for an illuminating discussion]. To focus clearly on metallic-line intensity, Morgan developed a classification scheme having eight *metallic-line groups*, denoted by Roman numerals I, . . . , VIII (the *Morgan class* of the cluster). Class I clusters are extremely weak lined and very metal poor, and class VIII clusters are the strongest lined and have essentially normal (solar) metal content.

Metal abundances for globular cluster stars can be estimated quantitatively by several methods. Abundances in the atmospheres of individual cluster giants (the brightest cluster stars) can be obtained from curve-of-growth analyses of their spectra or from photometric indices, such as $\delta(U - B)$, or indices in the DDO system (**H7**), (**H8**), (**H16**) and other specially devised systems (**C2**), when the latter have been calibrated using field stars of similar spectral type whose abundances have been determined spectroscopically. In clusters that are close enough for *UBV* photometry to reach the main sequence, the observed $\delta(U - B)$ of cluster main-sequence stars can be used to infer a metal abundance via the calibration of [Fe/H] versus $\delta(U - B)$ for field subdwarfs. Finally, a cluster metallicity can be inferred from the ΔS parameter (see §3-8) of RR Lyrae stars in the cluster (**B13**). In those cases where several different methods can be applied, the agreement among the results is usually good, which encourages confidence in their reliability. Metal abundances in globular clusters cover the entire range from $-2.2 \lesssim$ [Fe/H] $\lesssim 0.0$, that is, from being extremely metal deficient to having essentially solar abundances. Some representative results are given in Table 3-8 (note that the values for NGC 5139, NGC 6356, and NGC 6553 are based only on secondary indicators and are quite rough). From the data in the table, it is easy to see that [Fe/H] correlates closely with a cluster's

Morgan class and integrated spectral type, and moderately well with $(B - V)_0$.

Once fairly reliable metal abundances are known for several clusters, they can be used to calibrate secondary abundance indicators, such as the form and character of particular features in the CM diagram (see next subsection) or some suitably chosen integrated photometric index. Integrated indices are advantageous for statistical studies because they apply to the cluster as a whole, which is a hundred times brighter than the brightest individual stars. Whereas the indices $(B - V)$ or $(U - B)$ themselves are difficult to use in this connection because they are affected by interstellar reddening, the parameter Q [defined in equation (3-70)] is essentially reddening free and correlates closely with both a cluster's spectral type and its Morgan class (**V2**). Hence it provides an effective photometric metallicity indicator.

Very useful collections of data—including distances, positions, luminosities, spectrophotometric properties, and CM diagram characteristics—are given in (**A2**) and (**P5**) for more than 100 globular clusters.

Color-Magnitude Diagrams Since the pioneering work of Shapley (**S15**) in 1917, it has been known that the globular clusters have a very distinctive stellar content and that their CM diagrams contrast sharply with those of galactic clusters (except very old galactic clusters, such as M67 and NGC 188, whose CM diagrams show some similarities to those of the globular clusters). An observed CM diagram for M92 is shown in Figure 3-13, with principal sequences and other important features indicated schematically. This CM diagram, which is typical, shows the following characteristic features.

1. The upper end of the main sequence terminates at a late spectral type, and normally luminous blue stars (except a few "blue stragglers"; see below) are absent. As we shall see in §3-7, this immediately implies that all globular clusters are old.
2. A *subgiant branch* joins continuously onto the main sequence at the turnoff point; there is no Hertzsprung gap.
3. A *red-giant branch* (or *first*, or *ascending*, giant branch) extends upward and to the right. The brightest stars in globular clusters are always red giants with absolute magnitudes about $M_V \approx -2$ to -3 (much fainter than the extremely luminous blue stars and supergiants found in young galactic clusters).
4. A *horizontal branch* (*HB*) near $M_V \approx +0.5$, which may contain a few hot blue stars (as is the case for M92), extends redward until it rises in the *asymptotic giant branch* lying above the first giant branch.
5. Within the horizontal branch there is an *instability strip* containing RR Lyrae variables.
6. A few *blue stragglers* are occasionally found lying along an extension of the main sequence above its turnoff point.

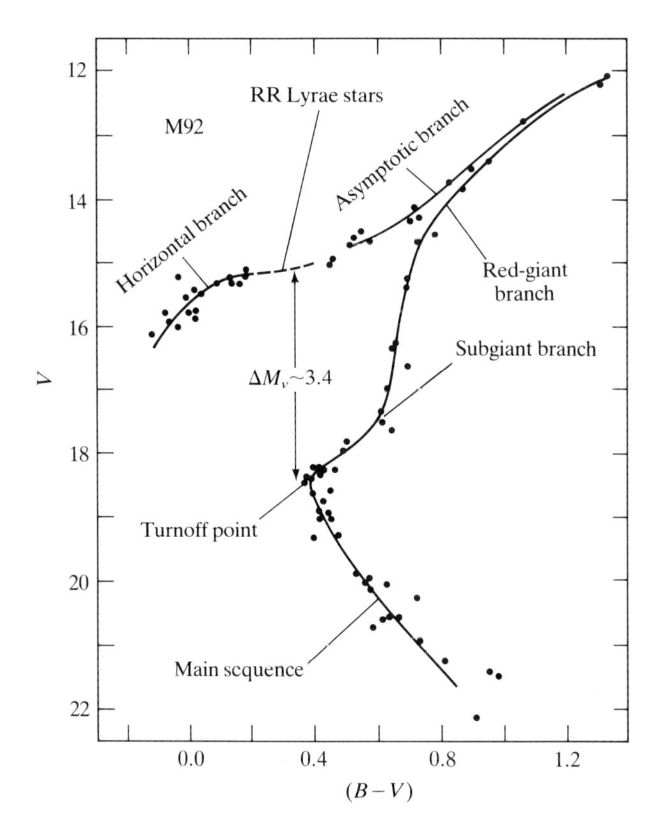

Figure 3-13. Observed color-magnitude diagram for the metal-poor globular cluster M92, with principal sequences and other important features indicated schematically. [Adapted from (**C4**, Chapter 2).]

Not every cluster CM diagram shows all of these features, and, as we shall see, there is considerable variation in the detailed morphology from cluster to cluster.

In the two-color diagram, globular-cluster stars (whether giants, HB stars, or main-sequence stars) show a clear ultraviolet excess; detailed data for four clusters are given by Sandage (**S2**). From an analysis of the two-color diagram, one can deduce both the amount of line-blanketing and interstellar reddening. As expected, the average ultraviolet excess $\langle\delta(U - B)\rangle$ correlates well with metal abundance: for example, $\langle\delta(U - B)\rangle_{\mathrm{M92}} = 0.24$ and $\langle\delta(U - B)\rangle_{\mathrm{M3}} = 0.17$.

In principle, the most fundamental method for determining globular-cluster distance moduli, and hence the absolute magnitude calibration of cluster stars, is based on *main-sequence fitting*. Because globular clusters lie at large distances, their main-sequence stars are found only at extremely faint apparent magnitudes (see Figure 3-13). The observational problems of obtaining reliable magnitudes and colors for these stars are severe, and they can be overcome only for nearby clusters by concentrated effort using the largest telescopes—a situation that should improve when observations can be made with the Space Telescope. The fitting procedure is complicated by the wide spread in globular-cluster metal abundances, and careful allowance must be made for the effects of line-blanketing on the measured color and of the metal abundance on the absolute magnitudes of main-sequence stars (see §3-7).

A procedure that avoids the direct use of blanketing corrections and is independent of the position of the Hyades main sequence (and hence of the Hyades distance modulus) was developed by Sandage (**S2**). This approach has been applied to the metal-poor and intermediate clusters M3, M13, M15, and M92 (**S2**) and the metal-rich clusters 47 Tuc and NGC 6838 (**H6**), (**H9**). Results for M92, M3, and 47 Tuc and the old, metal-rich galactic clusters M67 and NGC 188 are shown in Figure 3-14. These studies provide the most fundamental estimates of cluster-star absolute magnitudes available at the present time. In particular, they provide important (though conflicting; see §3-8) information about the absolute magnitudes of the RR Lyrae stars. Notice in Figure 3-14 the clear effect of metal abundance on the observed positions of cluster main sequences. The main sequences of M67 and NGC 188 have been fitted to the main sequence for nearby field disk stars with reliable trigonometric parallaxes, and they agree with one another because both clusters have essentially normal metal abundances. As a cluster's metal content decreases, its observed main sequence lies farther to the blue and farther below the disk-star sequence, as is illustrated nicely by the progression from 47 Tuc through M3 to M92.

Unfortunately, the main-sequence fitting procedure just described is vulnerable to significant observational errors. Sandage (**S2**) showed that absolute magnitudes of globular-cluster stars as obtained from main-sequence fitting are uncertain by at least ± 0.5 mag. These uncertainties necessarily introduce very significant uncertainties into the evolutionary analyses of globular-cluster CM diagrams.

Globular-cluster CM diagrams vary considerably from cluster to cluster [see the extensive data in (**A2**), (**P5**), and the catalogs listed in Table 2-6]. Their morphology depends systematically on metal abundance and at least one other variable. Sandage and Wallerstein (**S8**) first demonstrated that the quantity $\Delta V[(B - V) = 1.4]$, which is the height of the giant branch at $(B - V) = 1.4$ over the horizontal branch, correlates very closely with decreasing metal abundance. This behavior is consistent with the results of stellar evolution theory. For example, ΔV for such metal-poor clusters as M92 is about 3 mag, whereas for such intermediate clusters as M3 and M5, ΔV is about 2.6 mag, and, for such metal-rich clusters as 47 Tuc M71, NGC 6352, and NGC 6553, it is 2 mag or less [see Figure 3-14, Table 3-8, and (**S8**), (**A11**), (**H4**), (**H5**)]. These results imply that ΔV can be used as a convenient (but secondary) metallicity indicator and that older methods of distance determination, based on an assumed unique luminosity for stars

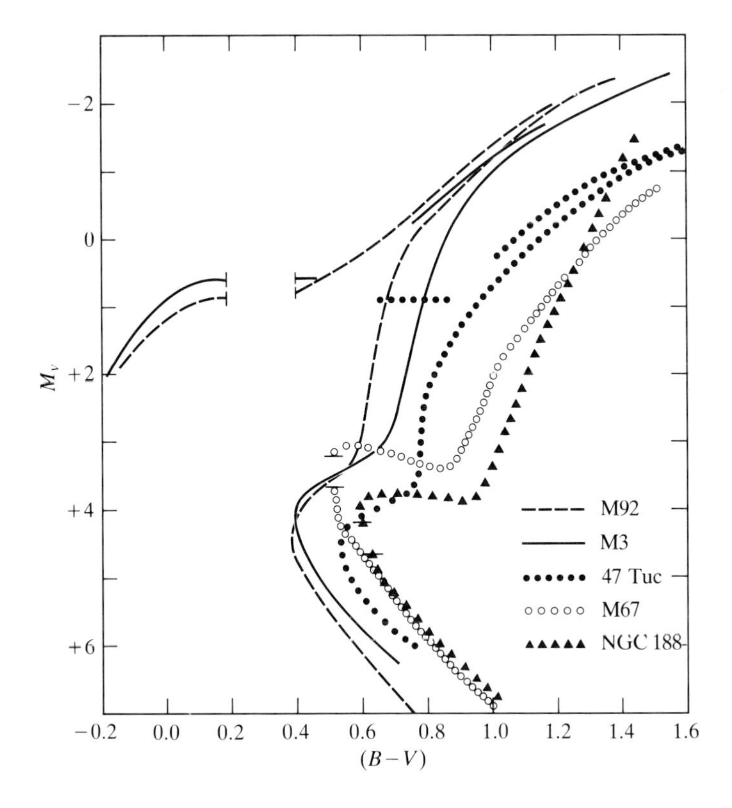

Figure 3-14. Observed color-magnitude diagrams for the globular clusters M92, M3, and 47 Tuc, and the old galactic clusters M67 and NGC 188. Absolute magnitudes have been set using the main-sequence fitting procedure of Sandage (**S2**). [Data from (**H6**), (**S2**), and (**S4**).]

on the tip of the giant branch of a cluster, are invalid unless they are calibrated for the effects of differences in cluster metal abundances. An alternative to ΔV is the *slope parameter* S, which is defined (**H3**) to be the slope of the line joining the intersection of the extension of the horizontal branch with the subgiant branch and a point on the giant branch 2.5 mag above the HB. This parameter is reddening-independent and independent of the color of the giant branch.

Initially, it appeared (**S8**) that horizontal-branch morphology also correlated closely with cluster metal abundance in the sense that metal-poor clusters (for example, M92) have most of their HB stars on the blue side of the RR Lyrae instability strip, intermediate clusters (for example, M3) have nearly equal numbers on both sides, and metal-rich clusters (for example, 47 Tuc) have most or all HB stars on the red side of the gap. These characteristics are denoted "B," "E," and "R" respectively in Table 3-8. Metal-rich

Table 3-8. Physical and Photometric Characteristics of Selected Globular Clusters

Name				Shapley concen.	log	Integrated spectral	Morgan				Horizontal branch	
NGC	Other	$(M_V)_0$	$(B-V)_0$	class	(r_t/r_c)	type	class	[Fe/H]	Q	ΔV	morphology	ΔS
6341	M92	−8.13	0.59	IV	1.78	F2	I	−2.2	−0.43	3.2	B	12
7089	M2	−8.96	0.64	II	1.61	F3	II	−1.4	−0.41	2.9	B	7.5
5139	ω Cen	−10.35	0.69	VIII	1.36	F7	II	(−1.4:)	−0.41	2.8	E	–
5272	M3	−8.74	0.69	VI	1.90	F7	II	−1.5	−0.41	2.6	E	8
6715	M54	−9.36	0.68	III	2.02	F7	III	−1.2	−0.35	2.6	E	5
6402	M14	−9.92	0.75	VIII	1.10	F8	IV	−1.1	−0.31	2.1	E	3.5
104	47 Tuc	−9.50	0.83	III	2.03	G3	–	−0.4	−0.26	2.0	R	–
6356	–	−8.90	0.84	II	1.19	G5	VI	(−0.2:)	−0.24	2.0	R	–
6838	M71	−5.84	0.84	–	0.78	G3	VI	−0.1	−0.23	2.0	R	−1
6352	–	−6.16	0.78	XI	–	G2	–	0.0	−0.24	1.6	R	–
6553	–	−7.92	0.82	XI	0.76	G4	VIII	(−0.2:)	−0.11	1.3	R	–

clusters with short, stubby, red-only horizontal branches often do not contain any RR Lyrae stars at all.

It was soon recognized, however, that there are clear violations of these results and that HB morphology does not correlate uniquely with metallicity. Hence at least one other parameter determines the distribution of stars in globular-cluster CM diagrams (**B1**), (**S9**), (**S14**), (**V2**), (**V3**). The physical parameters that actually produce the observed variations in HB morphology are as yet unidentified. Some authors have suggested that the "second parameter" is the cluster helium abundance, and a two-dimensional (Y,Z) classification scheme for clusters was even proposed (**H3**), but this hypothesis no longer seems tenable. Another possibility is that different clusters have significantly different ages (**S14**). Recent work [for example, (**H15**), (**H16**)] has also shown that there are large variations in the CNO abundances of the cluster red giants, not only from cluster to cluster, but even from star to star within a cluster! These abundances affect both the position of the giant branch and the position of stars on the HB. Thus the situation may actually be very complex, and the pronounced variations that we observe in the morphology of cluster CM diagrams may be the result of differences in several chemical-composition parameters and in the detailed structural and evolutionary histories of individual cluster stars (see §3-10).

3-7. STELLAR STRUCTURE AND EVOLUTION

In this section, we shall outline some of the major results obtained from the theory of stellar structure and evolution. We shall apply these results in §3-9 and §3-10 to interpret the evolutionary status of stars in clusters and thereby to develop some preliminary ideas about the evolution of our Galaxy as a whole. As it is our intent only to use the results as interpretive tools, we shall merely state them without detailed justification. There are several good monographs (**A5**, Chapter 11), (**C5**), (**C9**), (**C15**), (**T1**) and review articles (**I3**), (**I4**) where principles, methods, and detailed results are given.

Implications of the H-R Diagram and the Mass-Luminosity Relation for Stellar Structure and Evolution

The mere existence of such well-defined correlations among stellar properties as the mass-luminosity relation, and the fact that stars are found to lie along definite sequences in the H-R diagram, imply that the laws governing stellar structure must lead to highly systematic results. Furthermore, the fact that we can find stars of a given mass with strikingly different physical structures—for example, on the main sequence and among the supergiants or, again, on the main sequence and among the white dwarfs—suggests that any specific star must pass through a series of rather different structural configurations as it ages or *evolves*. Finally, it is clear that we find concentrations in the H-R diagram simply because those are phases in which stars spend a relatively large fraction of their lifetime and hence in which they have a high statistical probability of being found. In particular, it is clear that *the main sequence is a major phase of stellar evolution.*

The goal of the theory of stellar evolution is to explain these empirical facts. A basic theoretical picture of stellar evolution can be developed by assuming that stars are spherically symmetric and by writing a set of four nonlinear partial differential equations that describe the structure of a star as a function of radius from the star's center and as a function of time. One can then show (as was done by Russell, Vogt, Eddington, and others) that the structure of a nonevolving star is determined by its mass and its chemical composition. More generally, a star's structure is determined by its mass, its composition as a function of radius, and its age.

One of the key factors affecting the course of stellar evolution is the nature of the energy-production mechanism. Early in this century, it was believed that the energy radiated by a star was produced by the conversion of gravitational energy into thermal energy by a (relatively) slow contraction of a star. But, when it was discovered that, on the basis of geophysical evidence, the age of the earth is about 5×10^9 years and that therefore the Sun must be at least this old, it was realized that the time interval over which a gravitational energy-release mechanism could sustain the Sun's luminosity is two orders of magnitude too short to match the geophysical scale. The problem was solved when it was recognized that the required energies could be provided by *thermonuclear fusion* of light elements into heavier elements at the centers of stars. In particular, the main sequence can be identified with the episode of conversion of hydrogen (cosmically the most abundant element) into helium.

When the equations of stellar structure are solved, zero-age, chemically homogeneous stars of different masses are found to be arranged along the main sequence as observed; by changing the assumed chemical composition of the material, families of theoretical main sequences can be generated. While a star is on the main sequence, its luminosity is supported by the slow conversion of hydrogen into helium. As the hydrogen is progressively depleted, the star undergoes structural evolution. When an appreciable fraction of the original hydrogen in a star's core has been consumed, the star moves away from the main sequence in the first phase of a series of structural changes, which are driven by the changes in its chemical composition that are produced by successive episodes of thermonuclear burning. The structural changes in stars as they evolve are manifested in the H-R diagram. Hence, by analysis of H-R diagrams, we should be able to infer significant information about stellar evolution. But, if we examine the H-R diagram for a random sample—say, all stars within a given distance down to some limit of apparent brightness (see Figure 3-5)—it will include stars of all ages that are, in general, at markedly different stages in their evolution, all mixed together in hopeless confusion. The key advantages of cluster H-R diagrams for the development of an understanding of stellar evolution was recognized by Trumpler **(T3)**, **(T4)**, who, as early as 1925, stressed that *the stars in a cluster were most likely all formed from the same interstellar material at about the same time.* Hence clusters provide "snapshots" of the relative evolutionary behavior in the H-R diagrams of stars of different masses but equal age. We shall exploit this fact to good advantage in §3-9.

Evolution of Disk-Component Stars

Let us first consider the evolution of the common, metal-rich stars in the galactic disk, which range from objects forming now in extremely young clusters and associations in the spiral arms to very old stars, such as those in the old disk-cluster NGC 188.

Star Formation and Pre-Main-Sequence Evolution Although our Galaxy itself is of the order of 10^{10} years old, in some clusters and associations we find stars (for example, O stars) that cannot be much more than 10^6 years old. As mentioned before, this fact leads us to conclude that star formation continues to the present time. Clusters of young stars are invariably associated with dense nebulae and interstellar dust clouds in the galactic plane, and, as observations of external galaxies show, they are also associated with spiral arms, which are the sites of prominent nebulae and dust lanes. We therefore conclude that stars are formed by condensation from interstellar material.

The condensation process leading to star formation is activated by gravitational forces, by means of which a local density enhancement in the material tends to attract yet more material and hence grow in size, and it is resisted by disturbing forces, such as the shear produced by the differential rotation of material in the disk around the galactic center, thermal pressure in the gas, magnetic forces, internal turbulence in the clouds, and centrifugal forces produced by rotation of the collapsing material. In general terms, gravitational forces will dominate and overcome each resisting force if the density of the material rises above some critical level.

For the self-gravitation of the gas to exceed tidal-shear forces, the density must exceed a critical value that depends on the kinematic constants describing galactic rotation (the Oort constants; see §8-2). In the solar neighborhood, the critical density is about 2 to 3 atoms/cm^3 **(G4)**, which is larger than the observed gas density, but only by a small factor. The wavelength of the most unstable collapse-mode is of the order of 1 kpc; the amount of gas in a region of this size is of the order of a million solar masses. This mass is comparable to the observed masses of large interstellar clouds. The required initial density enhancement over ambient values could be produced by a very modest compression of the gas, perhaps as it passes through a spiral density wave (see §9-1).

As a cloud complex contracts, it fragments into smaller pieces. The basic instability next encountered is governed by the *Jeans criterion* (see Chapter 15), which shows that gravity forces will overpower internal pressure if the cloud's mass exceeds a critical value that depends on its temperature and density. For typical conditions in the interstellar medium, the critical mass is of the order of a few thousand solar masses, roughly comparable to the total mass of stars and interstellar material in young clusters and associations. As the material in such a unit contracts further and its density rises, smaller and smaller masses become unstable, reaching the range of reasonable stellar masses at densities comparable to those actually observed in the *interstellar molecular clouds* and the *dark globules* seen in some nebulae (see §9-3). These are, in fact, sites of star formation.

Once the material has fragmented to individual star-sized masses, *protostars*—initially having very large radii and low temperatures—evolve into stars. A crucial element of the theory was discovered by C. Hayashi, who showed that, for a star of a given mass and chemical composition, at each luminosity there exists a maximum radius (and hence minimum effective temperature) at which the star becomes wholly convective. The nearly vertical locus of this curve in the H-R diagram defines the *Hayashi limit* for a star of that mass (see Figure 3-15). To the right of this curve is a region in which a star (unless it is of very low luminosity and nearly wholly degenerate) must evolve dynamically on a very short time-scale; that is, it must collapse. The lower the stellar mass, the farther to the right the Hayashi track lies in the H-R diagram.

Suppose that we start with a nonluminous protostar essentially in free fall. As the protostar collapses, gravitational energy is released and must be radiated away. However, because the material is initially quite transparent, radiative energy loss is efficient, the luminosity rises rapidly, and the temperature remains quite low (10 K to 20 K). Eventually the material becomes opaque in a dense core; this opacity traps the radiation and allows the core to heat. As the core heats, pressure builds and arrests the collapse. The core continues to heat until its temperature rises to about 2000 K, at which point molecular hydrogen dissociates into atomic hydrogen. This dissociation provides an energy sink and triggers another free-fall collapse of the core, which continues to heat until the hydrogen not only dissociates but ionizes. At the onset of ionization, pressure balance can again be achieved, and core collapse halts. During these phases, both the luminosity and the temperature of the star continue to rise until, as shown schematically in Figure 3-15, the star finally penetrates to the left of its Hayashi-limit locus and emerges in the upper right-hand part of the H-R diagram.

Once a star has moved to the left of its Hayashi limit, it can achieve a state of hydrostatic equilibrium, and it then evolves on a much slower time-scale—essentially the *Kelvin-Helmholtz time-scale*—somewhat prolonged by

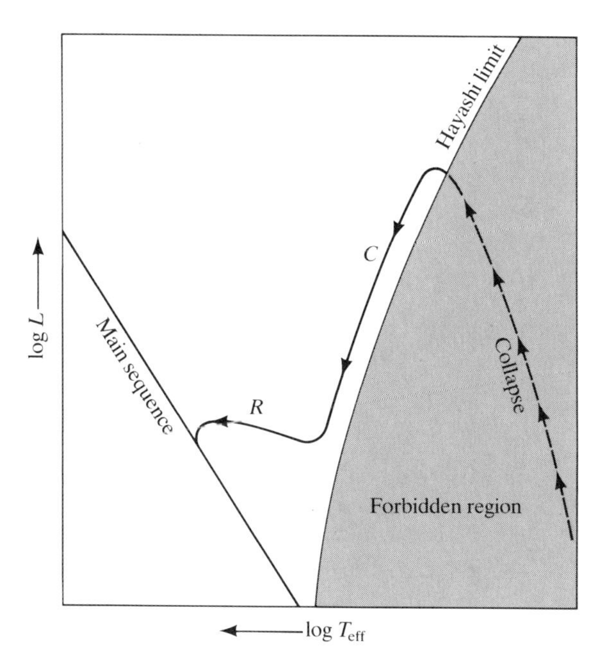

Figure 3-15. Schematic evolution track for pre-main-sequence stellar evolution. On the Hayashi track (labeled C) the star is convective throughout. On the portion of the track labeled R, the core is radiative (and, if the star is massive enough, the envelope can be radiative as well). Nuclear energy generation occurs when the star alights on the main sequence.

nuclear burning of such light isotopes as deuterium and lithium. From the point of maximum luminosity where it first penetrates out of the forbidden region, a star evolves downward in the H-R diagram along its *Hayashi track*. Initially, the star is convective throughout, but, in due course, it develops a radiative core, which grows in size. Eventually, the evolution track turns sharply to the left, as shown in Figure 3-16. Subsequent evolution occurs with the star contracting at nearly constant luminosity and ever-increasing effective temperature. Ultimately, the temperature in the core rises to the point where thermonuclear fusion of hydrogen into helium becomes possible, and the star settles onto the main sequence. As can be seen in Figure 3-16, the pre-main-sequence evolution of a star proceeds very rapidly.

From the point of view of the large-scale structure and evolution of our Galaxy, we are interested in overall rates of star formation per unit volume per unit time. This rate is usually described by the *total birthrate function* $\beta(t)$ (see also §4-2). The birthrate function is only poorly known because the physical situation is so complex. We do know that star formation occurs in

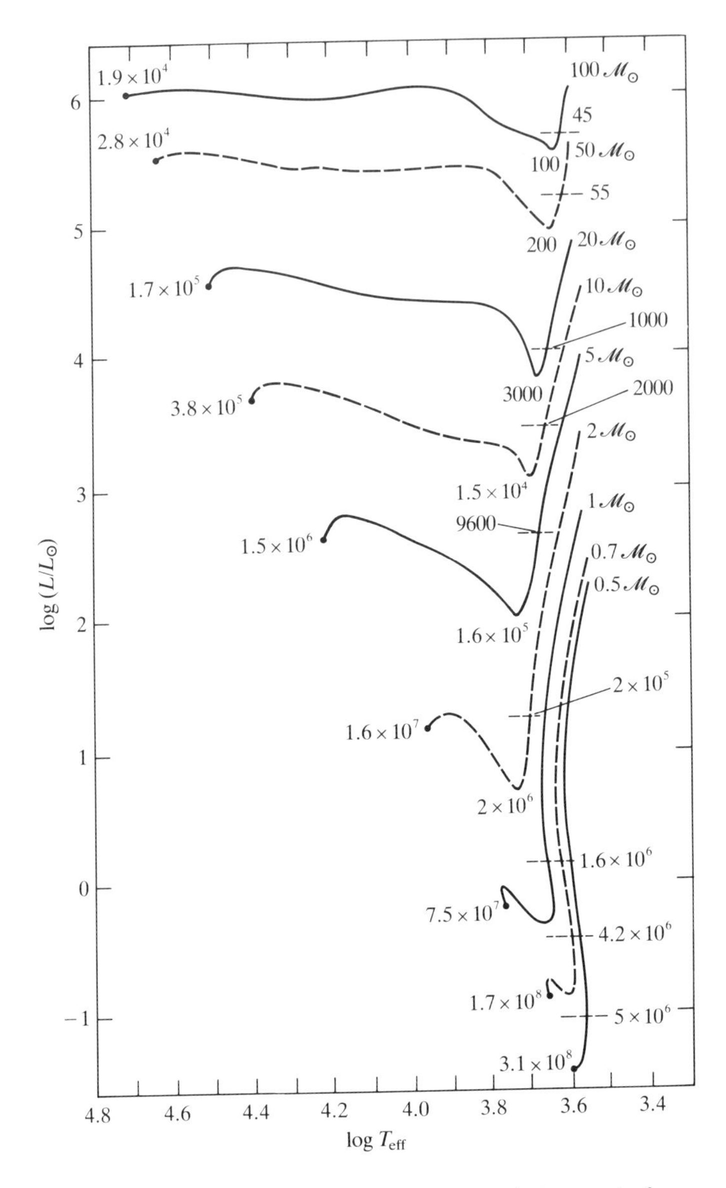

Figure 3-16. Calculated pre-main-sequence evolution tracks for stars with $0.5\ \mathcal{M}_\odot \leq \mathcal{M} \leq 100\ \mathcal{M}_\odot$. Dashed horizontal lines indicate the point where the star ceases to be convective throughout. Stellar ages in years are shown at two or three points along the track. The tracks terminate when the stars reach the ZAMS. [From (**E11**). Reproduced by permission of the National Research Council of Canada from the *Canadian Journal of Physics*, **45**, 1967, pp. 3429–3460.]

regions where the interstellar gas is compressed beyond its critical gravitational-collapse threshold. We also see star formation occurring in the spiral arms of some galaxies, which may indicate that it is a result of compression in a spiral density wave (which may contain shocks). Indeed, once some stars have formed in a region, they produce expanding H II regions that compress the surrounding gas and thus may trigger yet more star formation (that is, star formation is "contagious"). A commonly used approximation is the assumption that the total birthrate of stars is proportional to some power of the gas density; that is, $\beta(t) = k(t)\rho^{\alpha}$. We expect that $\alpha > 1$, so that star formation will be most rapid in regions of higher gas densities. Analyses of the distribution of stars in the solar neighborhood yield an empirical value of $\alpha \approx 2$ (**S10**), (**B2**, 165). Further broad discussion of problems of star formation and its significance for galactic evolution can be found in (**R1**).

The Main Sequence and Early Post-Main-Sequence Evolution When thermonuclear conversion of hydrogen into helium at the center of a star begins, the star enters its main-sequence hydrogen-burning phase, the longest stage of its existence (until it ends as one or another kind of degenerate object). The locus of points in the theoretical H-R diagram, where hydrogen burning first occurs (see Figure 3-16), defines the *zero age main sequence* (*ZAMS*). Initially, all main-sequence stars are chemically homogeneous, with composition parameters (X, Y, Z) as defined in §3-4. All have internal structures that are fixed by the mass of the star. Stars are gaseous throughout because high internal temperatures (10^7 K) guarantee that the matter is essentially completely ionized and behaves like a perfect gas, despite enormous central pressures. Theory shows that, with increasing mass, the central temperature T_c rises, whereas the central pressure p_c and density ρ_c decrease. These trends imply that high-mass and low-mass stars have substantially different physical structures, and it is convenient to divide the main sequence into *upper main-sequence* ($\mathscr{M} \gtrsim 1.5\mathscr{M}_{\odot}$) and *lower main-sequence* ($\mathscr{M} \lesssim 1.5\mathscr{M}_{\odot}$) domains.

A typical upper main-sequence star has a high-enough central temperature to permit the carbon-nitrogen cycle (CN cycle) to be the dominant source of energy production. The energy-generation rate of this mechanism rises so rapidly with increasing temperature that energy production is strongly concentrated in the innermost core. The energy flux rises so swiftly outward from the center of the star that it cannot be transported efficiently enough by radiation. Consequently, the star has a convective core, surrounded by a radiative envelope outside the region of energy generation. The size of the convective core (measured either as a fraction of the stellar radius or of the stellar mass that it contains) increases with increasing mass. For reference, a star with $\mathscr{M} = 10\mathscr{M}_{\odot}$ has $R \approx 4.5R_{\odot}$, $L = 7000L_{\odot}$, $T_{\rm eff} \approx 25{,}000$ K (B1V), $T_c \approx 3 \times 10^7$ K, $p_c \approx 3 \times 10^{16}$ dynes cm^{-2}, $\rho_c \approx 7$ g cm^{-3}, and $(r_{\rm conv}/R) \approx 0.2$ or $[\mathscr{M}(r_{\rm conv})/\mathscr{M}] \approx 0.3$, where the last two parameters define the outer edge of the convective core.

A typical lower main-sequence star derives its energy predominantly from the *proton-proton cycle* (p-p cycle). The temperature sensitivity of this cycle is much smaller than that of the CN cycle, and energy production is much less strongly centrally concentrated in the star; the core is thus radiative. The stellar mass at the transition point where the CN and p-p cycles contribute about equally to the total energy production is about $2\mathscr{M}_\odot$. Some convection in the core persists even when the CN cycle provides less than half the energy generation, and a fully radiative core is not achieved until $\mathscr{M}$ drops below about 1.1 or $1.2\mathscr{M}_\odot$. The effective temperatures of lower main-sequence stars are so low that hydrogen is neutral near the surface, and, as it ionizes in deeper layers, it produces a convection zone in the envelope outside the radiative core. For reference, a star with $\mathscr{M} = \mathscr{M}_\odot$ has $T_{\text{eff}} \approx 5900$ K, $T_c \approx 1.5 \times 10^7$ K, $p_c \approx 2 \times 10^{17}$ dynes cm^{-2}, $\rho_c \approx 100$ g cm^{-3}, $(r_{\text{conv}}/R) \approx 0.76$, and $[\mathscr{M}(r_{\text{conv}})/\mathscr{M}] \approx 0.99$, where the last two parameters define the inner edge of the convective envelope.

For stars of progressively lower mass, the convection zone penetrates deeper and deeper, and, at about $0.4\mathscr{M}_\odot$, the whole star is convective; that is, it is at its Hayashi limit. The theory of energy transport by superadiabatic convection is at present primitive, and, in astrophysical work, the problem is usually treated by the *mixing-length theory*. This theory is fundamentally only a phenomenological one containing an adjustable parameter: the ratio of the mixing length (the distance over which a typical convective eddy moves) to the pressure scale height. Changes in the assumed mixing length produce important changes in the computed radius of a model, but they scarcely affect its luminosity. Thus models of stars with convective envelopes have significant uncertainties in their effective temperatures (at least of the order of $\Delta \log T_{\text{eff}} \approx 0.03$), but they give fairly reliable luminosities. These facts must be borne in mind in fitting observed H-R diagrams to theoretical models. The problem just mentioned becomes most severe for giants and supergiants.

Because central temperatures drop and the central densities rise for lower-mass stars, the material in the core becomes partially degenerate at a sufficiently low mass. In the limit of complete nonrelativistic degeneracy, the pressure depends only on the density and not on the temperature: $p = K\rho^{5/3}$. This fact has the consequence of setting a lower limit $\mathscr{M}_{\text{cr}}$ on the mass of a star in which thermonuclear energy release can occur. Stars with $\mathscr{M} < \mathscr{M}_{\text{cr}} \approx 0.08\mathscr{M}_\odot$ can reach the main sequence in hydrostatic equilibrium without the central temperature having risen to the point at which nuclear reactions can occur. These stars thus never tap their nuclear energy source, but simply cool without radiating significant amounts of energy and spend essentially their entire existence as degenerate *black dwarfs*. The planet Jupiter $(\mathscr{M}/\mathscr{M}_\odot \approx 10^{-3})$ is an example of such an object. Other bodies of this type would be extremely hard to detect observationally, and, for all we know, our Galaxy could contain large numbers of them without our being able to discover them by the usual approach of detecting the energy they emit. On the other hand, it is possible, at least in principle, to sense the effects produced by the gravitational fields of such bodies.

Computations of chemically homogeneous models for a wide range of masses yield both a theoretical mass-luminosity relationship and a theoretical ZAMS. The agreement between these theoretical relations and the corresponding empirical ones is quite satisfactory. The theoretical mass-luminosity law shows that $L \propto \mathscr{M}^3$ for $\mathscr{M} \gtrsim 7\mathscr{M}_\odot$, $L \propto \mathscr{M}^{4.75}$ for $\mathscr{M} \approx \mathscr{M}_\odot$, and $L \propto \mathscr{M}^{1.9}$ for the fully convective stars on the lower main sequence ($\mathscr{M} \approx 0.4\mathscr{M}_\odot$).

The position of the ZAMS in the H-R diagram depends on the chemical composition of the models. This situation can be expressed most conveniently in terms of the luminosity difference between two points on neighboring ZAMSs at the same value of T_{eff} [and hence practically the same $(B - V)$ and $B.C.$—which also implies $\Delta M_V = \Delta M_{\text{bol}}$]. One finds that, for $T_{\text{eff}} =$ constant, $\Delta \log L \approx \Delta X + 5\,\Delta Z$ for $L \approx L_\odot$. The masses of stars on different ZAMSs at a given T_{eff} differ from one another.

The *main-sequence lifetime*, τ_{ms}, of a star is fixed by the length of time that its luminosity can be supported by thermonuclear conversion of hydrogen to helium. The fusion of four protons into one helium nucleus releases an energy of 26.7 MeV. Thus conversion of Δm grams of hydrogen releases $E = 0.007\,\Delta mc^2$ ergs. If a fraction α of the total mass of a star can be converted, then $\tau_{\text{ms}} = (0.007\alpha\mathscr{M}c^2/L)$. Detailed computations show that, when about one-tenth of the stellar mass has been converted, the star evolves rapidly away from the main sequence. Hence, to an order of magnitude, we find

$$\tau_{\text{ms}} \approx \frac{10^{10}(\mathscr{M}/\mathscr{M}_\odot)}{(L/L_\odot)} \text{ years} \tag{3-44}$$

More precise results are given in Table 3-9.

The conversion of hydrogen to helium in the core of a star results in a continuous depletion of hydrogen at the star's center. Eventually, the hydrogen is exhausted, and an inert (that is, unproductive of energy) helium core forms. The details of how this occurs depend strongly on the mass of a star, but the main result, true for all stars, is that the region of energy generation shifts to a hydrogen-burning shell surrounding a growing helium core. The fundamental significance of this situation is that the star now must be considered chemically inhomogeneous. It was realized by E. Öpik (**O4**) in 1938 that such models could explain why the red giants have structures that are so different from those of main-sequence stars [see also (**H20**), (**G1**)]. Subsequent calculations by Oke, Sandage, and Schwarzschild (**O2**), (**S5**) confirmed that, after a long period of slow evolution on the main sequence, stars become inhomogeneous and evolve very rapidly through the Hertzsprung gap (which explains the paucity of stars there) onto the giant branch. A satisfactory basic understanding of the observed morphology of H-R diagrams was thereby achieved. Let us now consider briefly some of the main results obtained from modern stellar-evolution calculations.

Table 3-9. Stellar-Evolution Times (Years)

Evolution-track interval	Mass ($\mathscr{M}_\odot$)							
	1.0	1.25	1.5	2.25	3	5	9	15
1–2	7×10^9	2.8×10^9	1.5×10^9	4.8×10^8	2.2×10^8	6.5×10^7	2.1×10^7	1.0×10^7
2–3	2×10^9	1.8×10^8	8.1×10^7	1.6×10^7	1.0×10^7	2.2×10^6	6.1×10^5	2.3×10^5
3–4	1.2×10^9	1.0×10^9	3.5×10^8	3.7×10^7	1.0×10^7	1.4×10^6	9.1×10^4	
4–5	1.6×10^8	1.5×10^8	1.0×10^8	1.3×10^7	4.5×10^6	7.5×10^5	1.5×10^5	7.5×10^4
5–6	$\geq 10^9$	$\geq 4 \times 10^8$	$\geq 2 \times 10^8$	3.8×10^7	4.2×10^6	4.9×10^5	6.6×10^4	
6–7	–	–	–	–	2.5×10^7	6.1×10^6	4.9×10^5	7.2×10^5
7–8	–	–	–	–		1.0×10^6	9.5×10^4	6.2×10^5
					4.1×10^7			
8–9	–	–	–	–		9.0×10^6	3.3×10^6	1.9×10^5
9–10	–	–	–	–	6.0×10^6	9.3×10^5	1.6×10^5	3.5×10^4

SOURCE: (**13**). Reproduced, with permission, from the *Annual Review of Astronomy and Astrophysics*, Volume 5. © 1967 by Annual Reviews, Inc.

Evolution of High-Mass Stars As an example of the evolutionary history of a moderately high-mass star, we shall summarize results from computations by I. Iben **(I1)**, **(I2)**, **(I3)**, **(I4)** for a star of $5\mathcal{M}_\odot$ with a composition $(X, Y, Z) = (0.71, 0.27, 0.02)$. The star's evolution track in the H-R diagram is shown in Figure 3-17, and the time intervals between the numbered points on the track are listed in Table 3-9. At point number 1, the star begins its

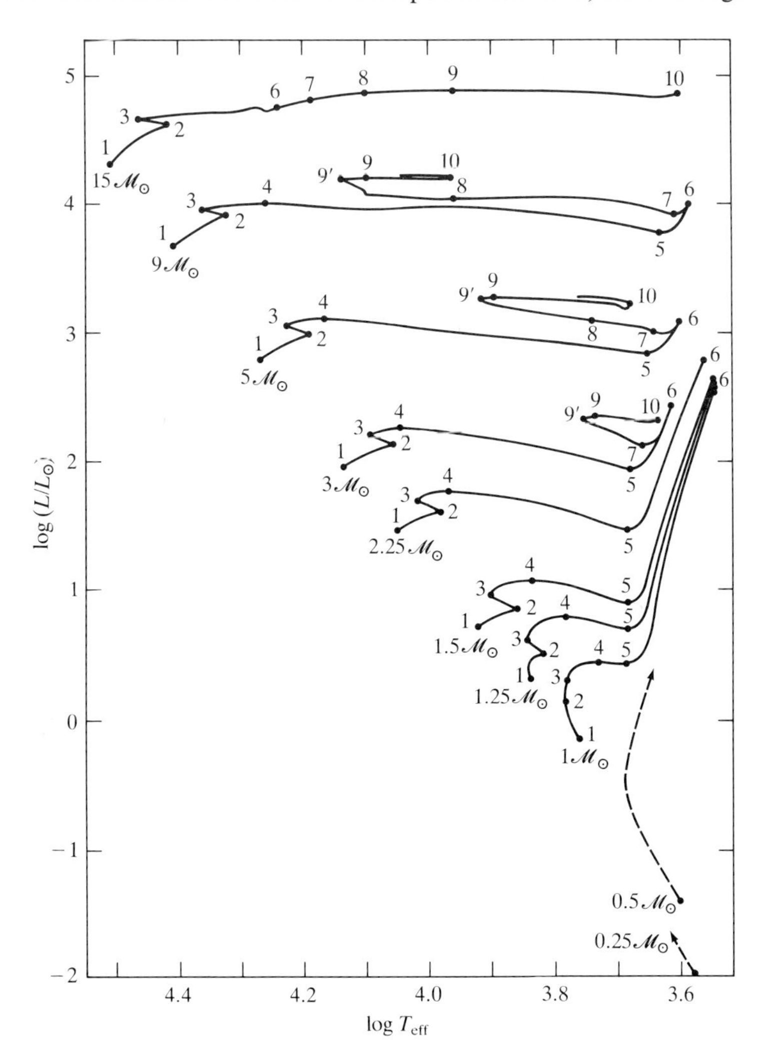

Figure 3-17. Post-main-sequence evolution tracks for stars with $0.25\ \mathcal{M}_\odot \leq \mathcal{M} \leq 15\ \mathcal{M}_\odot$. Ages at the labeled points are given in Table 3-9. For $\mathcal{M} \leq 2.25\ \mathcal{M}_\odot$, the tracks are terminated at the point of core helium ignition. For $\mathcal{M} \geq 3\mathcal{M}_\odot$, the tracks are terminated shortly before helium-core exhaustion. [Reproduced with permission from the *Annual Review of Astronomy and Astrophysics*, Volume 5. Copyright © 1967 by Annual Reviews, Inc.]

life on the main sequence, burning hydrogen in a convective core that contains about 20% of the stellar mass. Within this core, the material is well mixed, and thus depletion of hydrogen occurs throughout the entire convective core. Between points 1 and 2, hydrogen burning continues, and the star evolves upward and to the right in the H-R diagram; its luminosity rises by about 60%. During this time, the convective core contracts, shrinking until at point 2 it contains only about 8% of the star's mass. The contracting core leaves behind regions in which some helium enrichment has occurred but which are now outside the region of mixing, thus creating a zone with a chemical-composition gradient.

The processes of core contraction (both in physical size and mass fraction) and heating gradually accelerate until finally, at point number 2, where the core hydrogen content has dropped to 0.05, the whole star contracts and moves rapidly to the left in the H-R diagram to point 3, where hydrogen is exhausted in the core. An appreciable part of the luminosity between points 2 and 3 is provided by conversion of gravitational potential energy into heat in the envelope. When core exhaustion occurs, energy production shifts to a thick hydrogen-burning shell that was established just outside the convective core shortly before core exhaustion. Once energy generation in the core ceases, it becomes isothermal. The hydrogen-burning shell moves outward in mass fraction, adding more mass to the helium core. The star resumes its evolution upward and to the right until it arrives at point 4, where the isothermal core reaches the *Schönberg-Chandrasekhar limit* (**S11**), where it contains roughly 10% of the star's mass. At this limiting core mass (which applies to all stars with nondegenerate cores), pressure balance in the core can no longer be maintained, and the care undergoes a very rapid contraction.

Beyond this point, while the core is contracting rapidly, the shell achieves higher temperatures and densities, and the rate of nuclear energy generation rises almost explosively. At the same time, the shell becomes narrower, leading to a drop in the total rate of energy production, and the envelope expands, and in doing so consumes energy in order to drive the expansion. As a result, the stellar luminosity drops by almost a factor of two. The star thus evolves very rapidly to the right and somewhat downward in the H-R diagram. At about point number 5, the envelope becomes convective, and the convection zone extends rapidly downward, until, at its maximum extension, it contains more than 50% of the mass of the star. As the envelope convection zone deepens, it sweeps past zones in which nuclear burning has occurred, and it produces a depletion of some elements (for example, Li) in the envelope and an enrichment of others (for example, He^3). As the stellar envelope continues to cool, it reaches a critical temperature where further cooling leads to a decrease in the opacity of the envelope material, and energy from the interior escapes more efficiently, thus producing a rise in luminosity. At the same time, increasing hydrogen-burning shell temperatures offset shell narrowing, and the total rate of energy production rises.

The star then evolves very rapidly upward along the red-giant branch, nearly parallel to its Hayashi track.

Evolution up the giant branch terminates at point number 6, where the core temperature becomes high enough to ignite the *triple-alpha process*, in which three nuclei of He^4 are fused to form one C^{12} nucleus. The star now has a convective helium-burning core surrounded by an inert helium shell contained within a hydrogen-burning shell inside the inert hydrogen-rich envelope. The energy released in the core causes an expansion of the inner regions of the star, which leads to a drop in the temperature of the hydrogen-burning shell and thus to a temporary drop in the luminosity. Evolution continues at a moderate pace, and the star contracts slightly until it reaches point number 7, where the convection zone in the envelope rapidly retreats and vanishes. There is then a major structural readjustment of the envelope in a period of rapid contraction. This contraction breaks the era of core helium burning into two major episodes that are separated by a phase of rapid transit in the H-R diagram. Between points 8 and 9, evolution again proceeds at a modest pace, and the ratio of the luminosity produced in the helium-burning core to that in the hydrogen-burning shell rises from about 5% to 40%.

As the helium-burning core begins to become more important, the star again evolves to the right (between points 9 and 10) in a rapid phase of core contraction and modest envelope expansion reminiscent of initial post-main-sequence hydrogen burning, when the structure of the star also was dominated by a strong core energy source. The precise location of point 9 in the H-R diagram depends sensitively on the details of the previous evolutionary history of the star and on the input physics (opacities, reaction rates), and thus is rather uncertain at present (**I4**).

As the helium in the helium-burning convective core approaches exhaustion, the star undergoes an overall contraction (moves to the left of point number 10) reminiscent of the contraction that precedes hydrogen exhaustion near the main sequence (points 2 to 3). The helium in the core is soon exhausted, and a helium-burning shell develops around a contracting isothermal carbon core. The star continues to evolve to the left in the H-R diagram. During this period, the matter between the helium-burning shell and the surface expands, and the temperature in the hydrogen-burning shell drops. At a certain point, hydrogen burning ceases. The star then has only a helium-burning shell source; it reverses the direction of its evolution track and moves rapidly to the right until it again develops a convective envelope, and then rises up a *second red-giant branch*, a more luminous prolongation of the track between points 5 and 6. The convection zone in the envelope again sweeps downward and mixes nuclear-processed material into the surface layers, enriching their helium content substantially. The base of this zone reaches down almost to the helium-burning shell.

As the star rises along its new giant branch, the hydrogen shell reignites, and the material in the carbon core becomes relativistically degenerate. Our knowledge of the subsequent evolution of the star is at present uncertain, both because the structure of the star becomes inherently complex and because the detailed evolution depends on the rate of energy loss from the core by neutrino-emission via various mechanisms for which we have only (conflicting) theoretical estimates [see **(I4)** for a review].

Evolutionary tracks for several models of different masses are shown in Figure 3-17, and the time intervals between the labeled points on the curves are listed in Table 3-9. For $\mathscr{M} \leq 2.25\mathscr{M}_{\odot}$, the tracks end when helium burning in the core ignites, which terminates evolution up the first giant branch. For $3\mathscr{M}_{\odot} \leq \mathscr{M} \leq 9\mathscr{M}_{\odot}$, the tracks are carried to the onset of overall stellar contraction prior to helium-core exhaustion. For $15\mathscr{M}_{\odot}$, the tracks run only to the beginning of the red-giant phase.

From the fact that the evolution tracks of massive stars run nearly horizontally until a star reaches its red-giant branch we can see immediately how two very different kinds of stars (for example, a blue main-sequence star and a red supergiant) can be found to lie on essentially the same mass-luminosity relation. In contrast, stars on the nearly vertical giant branch cannot be expected to obey the same law. Furthermore, scrutiny of the evolution times in Table 3-9 quickly reveals why the Hertzsprung gap exists. The fact that the main sequence runs diagonally across the H-R diagram while the Hayashi tracks run vertically explains why the gap is wide at high luminosities and becomes narrower at lower luminosities. Finally, we see that stars spend their most advanced evolutionary phases as red giants and supergiants. It is during this time that substantial mixing of material processed by nuclear burning into the observable envelope can occur. This mixing probably explains the observed existence of very unusual chemical compositions in many late-type giants and supergiants (for example, C or S stars), even though our understanding of the details of the mixing process is extremely poor.

The late evolutionary stages of the $5\mathscr{M}_{\odot}$ star whose life history has just been sketched become complicated because of the onset of degeneracy in the carbon core left by helium burning. The situation is simpler for stars with masses $\mathscr{M} \gtrsim 10\mathscr{M}_{\odot}$, because carbon ignition occurs before the core becomes degenerate, and it is possible to follow the evolution of such stars rather clearly until their final catastrophic demise as *supernovae*. Massive stars pass smoothly through successive episodes of core burning of fuels having progressively higher atomic numbers. During each episode, a new fuel is ignited and produces a convective core. The fuel is ultimately exhausted in the core, which then becomes inert, and the central source is replaced by a shell source. After each core-exhaustion event, the core contracts, and both the central temperature and pressure rise until, finally, the temperature becomes high enough to produce ignition of the next burning process. In this way, elements of successively higher atomic weights are built up by the process of *nucleosynthesis*, in which H → He, He → C, C → O, O → Ne, Ne → Si, and so on, and the star develops an "onion-shell" distribution of layers of elements ranging from a hydrogen–helium envelope to a core containing the heaviest element yet created. In this structure, there can be several distinct nuclear shell sources.

The process of nucleosynthesis terminates with the creation of the *iron-peak elements*, which have the highest binding energies of all nuclei and hence cannot be transmuted to other elements with a net release of energy. At this point, the progression of *exoergic* (energy-releasing) thermonuclear reactions has run to completion. When this occurs, the core can continue to contract, reaching ever-higher temperatures and densities but with no further hope of delaying the contraction by means of nuclear energy release. This process cannot continue indefinitely, however, because the core material eventually passes through a phase transition (at temperatures near 5×10^9 K) in which the iron nuclei can photodisintegrate into α particles and free neutrons. This reaction is *endoergic*; that is, thermal energy is consumed in driving the disintegration process. It thus causes a drop in the core temperature and pressure, and therefore the core can no longer support the overlying layers but collapses violently in an implosion. At the same time, neutrons are released copiously and become available to transmute elements in the outer layers.

During this collapse, the *Fermi energy level* of the (degenerate) free electrons rises in step with the shrinkage of the volume within which the free electrons are confined. This rise of the Fermi level drives a further endothermic reaction in which the electrons are absorbed by protons to form neutrons. Eventually, the pressure generated by the by-now degenerate neutrons is able, in some cases, to halt the collapse. In this event, a *neutron star* about 20 km in diameter is formed, which may manifest itself via its spinning magnetic field as a *pulsar*. However, if the core mass is greater than about $2\mathscr{M}_\odot$ (the precise figure is still uncertain), relativistic effects enable gravity to overwhelm even the degeneracy pressure of the neutrons. At this point, the collapse becomes a catastrophic instability of spacetime in which the matter is caught up like a helpless picnicker who has inadvertently started a forest fire. The end point of this catastrophe is the formation of what we call a *black hole*. It is not yet known what happens to the star's envelope when the core collapses; one possibility is that a large part of it is ejected at great velocity as a supernova, but the physics of this process remains obscure.

Early Evolution of Low-Mass Stars As an example of the evolution of a low-mass star, let us consider the early evolutionary behavior of a star of 1 solar mass. An evolution track computed by Iben (**I1**), (**I2**), (**I3**) for a $1\mathscr{M}_\odot$ model is shown in Figure 3-17. The star starts on the main sequence, burning hydrogen in a radiative core by means of the proton-proton cycle. Because there is no mixing in the hydrogen-burning region, unlike the case for a massive star, there is no buildup of an extended core region within which there is a uniform hydrogen depletion. Rather, there is a smooth, monotonic decrease in hydrogen abundance from its envelope value at the outer edge of the energy-generation zone down to a minimum value at the star's center.

As hydrogen burning progresses, the central hydrogen content diminishes continuously, and the zone in which some depletion has occurred grows steadily in fractional mass. The time interval to the point of hydrogen exhaustion at the center is long, about 7×10^9 years for the Sun. As the Sun is about 4.5×10^9 years old, it has not yet reached the point of central hydrogen exhaustion, and current models indicate that its central hydrogen content has been reduced by about one-third, to $X \approx 0.48$.

Hydrogen exhaustion eventually occurs, but only in a very small central region, which soon becomes isothermal, forming a small core whose mass $\mathscr{M}_c$ is much less than the Schönberg-Chandrasekhar limit $\mathscr{M}_{SC}$. This core starts contracting slowly well before $\mathscr{M}_c$ approaches $\mathscr{M}_{SC}$. Finally, the core material becomes degenerate, and contraction in that region halts. The transition from hydrogen burning in the core to hydrogen burning in a thick shell thus occurs slowly and smoothly, and the star does not experience a rapid overall contraction leading to a jog to the left in the H-R diagram, such as is found at higher masses. The initial evolution of low-mass stars through most of the thick-shell phase is more nearly parallel to the main sequence than it is for high-mass stars, so they tend to linger near the main sequence while their luminosity rises by $\Delta M_{bol} \approx 1.5$ mag. Owing to the contribution of degeneracy to the core pressure, low-mass stars can actually build larger isothermal cores than can massive stars ($0.13\mathscr{M}_\odot$ for a $1\mathscr{M}_\odot$ star versus $0.1\mathscr{M}_\odot$ for a $10\mathscr{M}_\odot$ star) before shell narrowing occurs.

Eventually, the hydrogen-burning shell becomes narrow, the core mass exceeds $\mathscr{M}_{SC}$, and the star then evolves to the right in the H-R diagram. Although this phase is relatively rapid compared to the star's mainsequence lifetime (see Table 3-9), it is much slower than the corresponding episode for massive stars, and so there is a relatively high probability of actually finding stars in these evolutionary stages. As a result, the Hertzsprung gap vanishes in H-R diagrams in which the turnoff point is near $1\mathscr{M}_\odot$, and it is replaced by a well-populated subgiant branch. The star continues to evolve to the right at nearly constant luminosity until the evolution track approaches the Hayashi limit. The star—which has an isothermal helium core, hydrogen-burning shell, and a deep convective envelope—then evolves up the red-giant branch on a track that runs nearly parallel to its Hayashi track.

Evolution up the red-giant branch is terminated when helium ignition occurs in the core. Unlike the situation for massive stars, the isothermal core is degenerate and contains a mass that far exceeds the Schönberg-Chandrasekhar limit. As was first pointed out by L. Mestel (**M12**), the ignition of helium in such a degenerate core is explosive, leading to a *helium flash.* As the helium begins to burn, it releases a large amount of energy. However, because the electrons in the material are degenerate, the gas pressure is independent of the temperature. Therefore, in contrast to the case of a perfect gas, a release of energy, which raises the temperature of the nuclei, does not significantly raise the gas pressure. Hence this release cannot produce an expansion and cooling of the gas, which would slow the rate of energy release and thus permit an equilibrium to be attained. Rather, an increase in temperature swiftly increases the rate of energy release, which in turn raises the temperature still further. Hence there is a runaway that terminates only when either the core helium is exhausted or when the temperature of the gas rises sufficiently to remove the electron degeneracy, at which point the gas pressure *can* increase with increasing temperature, thus allowing the core to expand and cool.

Detailed model computations show that the mass of the hydrogen-exhausted core at the onset of the helium flash is almost independent of Z and relatively independent of the mass of the star outside the core. For $X \approx 0.6$ to 0.7, the core mass is about $0.4\mathscr{M}_{\odot}$. This is the critical mass at which thermonuclear ignition in a pure helium core is forced.

At the peak of the brief helium flash, the rate of energy release in the core corresponds to about $10^{11} L_{\odot}$! This enormous luminosity (comparable to the luminosity of our entire Galaxy) forces the temperature gradient in the material to rise above the adiabatic gradient, and the core becomes convective. Virtually none of the energy generated in the core during the helium flash penetrates beyond the outer convection zone because it is trapped by the opaque overlying layers, and the surface luminosity of the star is practically unchanged. During the helium flash, the inner convection zone reaches almost out to the hydrogen-burning shell, and there is even a possibility that it may penetrate all the way out to the convective envelope. In this case, the whole star would be convective, mixing would take place, and the star would start life over again on a helium-rich main sequence. However, current results suggest that this mixing does not take place, but rather that the energy released in the helium flash goes to lift degeneracy in the core, and evolution again proceeds on a nuclear-burning time scale in a star that has a helium-burning core and a hydrogen-burning shell and now resides on the horizontal branch. We suspend the discussion of the evolution of low-mass stars at this point and resume it later in our discussion of the horizontal-branch evolution of globular-cluster stars.

Finally, we note that the initial evolution of very low-mass stars, such as M dwarfs, moves them almost vertically in the H-R diagram along tracks that are essentially parallel to their Hayashi tracks. The rate of evolution is so slow, however, that, even during the entire lifetime of our Galaxy, such stars have barely moved from their initial positions on the zero-age main sequence.

Final Stages of Evolution To close our discussion of the evolution of disk-component stars, we summarize briefly a few important points about the final stages of stellar evolution. Detailed reviews (including numerous references) on white dwarfs are given in (**W3**) and (**O5**), on neutron stars in (**C1**) and (**R3**), and on black holes in (**M16**, Chapter 33).

Consider first the final state of a star like the Sun. As evolution progresses, a stage will be reached where all nuclear fuels that can be burned at the prevailing core temperatures are exhausted, and the material is strongly degenerate and hence able to support the overlying envelope, no matter how cold the core becomes. The star is then unable to contract further, and the central temperature will never rise above the ignition point of any reaction that could burn the remaining material in the core. Eventually, all energy-producing shells in the overlying layers burn themselves out, and essentially the entire star then becomes degenerate. Such a star will be extremely dense and will shine with a low luminosity that is fed by cooling of the material; that is, the star will be a white dwarf, such as the ones observed in the H-R diagrams of clusters and field stars.

The structural properties of white dwarfs are fixed by the characteristics of the equation of state for degenerate material. The key factor in the problem is whether or not the degeneracy is relativistic. In the former case, the pressure varies as $\rho^{4/3}$, and, in the latter case, as $\rho^{5/3}$. It is not difficult to show from what amounts to a dimensional analysis that, if the material is *non-relativistically degenerate*, the gravitational and pressure forces acting on a star will vary with different powers of the stellar radius. This result implies that, for a star of a given mass, it is always possible to find a radius at which an equilibrium between the two forces can be achieved and the pressure in the core can support the weight of the overlying layers. The larger the mass of the star, the smaller is its radius.

However, as the mass of the star increases, the material becomes *relativistically degenerate*. In this limit, the pressure and gravitational forces vary as the same power of the stellar radius, and an accommodation between the two forces cannot always be achieved. Rather, as was first shown by S. Chandresekhar, there is a critical mass—the *Chandrasekhar limiting mass* $\mathcal{M}_{\text{lim}}$—above which the gravitational force always prevails over pressure, and the star will collapse to zero radius (formally) and infinite central density. For a nonrotating, nonmagnetic star composed of any element other than hydrogen, $\mathcal{M}_{\text{lim}} = 1.44\mathcal{M}_{\odot}$. White dwarfs with $1\mathcal{M}_{\odot} < \mathcal{M} < \mathcal{M}_{\text{lim}}$ are relativistically degenerate through much of their inner regions, become nonrelativistically degenerate near the surface, and are surrounded by a thin skin of nondegenerate material. From stability considerations, it is found that, except in this thin outer (observable) skin, a white dwarf must be essentially devoid of hydrogen.

Once a star is a white dwarf, its radius is fixed by its mass, and it thenceforth remains constant. Thus, as the star cools, it slides down a *constant-radius line* in the H-R diagram on which $L \propto T_{\text{eff}}^4$ or $M_{\text{bol}} = -10 \log T_{\text{eff}} + c$. These constant-radius lines have a somewhat shallower slope than the theoretical main sequence. The observed positions of white dwarfs in the H-R diagram are in good agreement with the theory, and they imply that these objects typically have masses comparable to, but below, the predicted critical mass.

The existence of a limiting mass for white dwarfs naturally poses an evolutionary problem for stars whose masses initially are larger than $\mathcal{M}_{\text{lim}}$. What is the ultimate fate of such stars? One possibility is that, during its evolution, a star loses some of its mass. We do, in fact, observe stars to be losing mass in *stellar winds* and through the ejection of *planetary nebulae*. Dense winds that imply significant mass-loss rates are found in OB supergiants and in most red giants and supergiants. Also, planetary-nebula formation occurs during the post-red-giant evolution of stars. Together, these mechanisms provide a means by which stars having initial masses up to some characteristic limit, say $\mathcal{M}_w$, can lose enough mass to reach the white-dwarf regime with masses less than $\mathcal{M}_{\text{lim}}$.

But some stars will have initial masses greater than $\mathcal{M}_w$, and it is thought that the cores of these stars eventually implode violently, leading to the catastrophic ejection of the overlying envelope as spectacular supernovae. As we have just seen, protons will be transmuted to neutrons by inverse beta decay during the collapse of the core, and either a degenerate neutron star or a black hole will form, depending on whether the core mass is greater or less than some (rather uncertain) limiting mass. Neutron stars were first discussed theoretically in the 1930s, but it was only with the discovery of pulsars in the late 1960s that their existence was verified observationally, and their link to supernovae established. Unfortunately, it is harder to establish conclusively the presence of black holes in particular systems. At present, the best we can do is to identify any very massive ($\mathcal{M} > 3\mathcal{M}_\odot$, say) compact object as a potential black hole.

The actual value of the initial mass, $\mathcal{M}_w$, below which a star can shed enough material during its lifetime to become a white dwarf with a mass less than $\mathcal{M}_{\text{lim}}$, is hard to determine. Present evolutionary theory is too uncertain to yield $\mathcal{M}_w$ reliably, and it must therefore be estimated semiempirically. This estimate can be made by several methods: (1) comparing the formation rate for white dwarfs to the death rate for main-sequence stars of different masses; (2) equating $\mathcal{M}_w$ to the main-sequence turnoff point mass in young clusters (for example, the Hyades and Pleiades) in which white dwarfs are observed to be present; (3) using the observed numbers and ages (on their cooling tracks) of white dwarfs in clusters to determine the main-sequence positions of their precursors; and (4) assigning limits on $\mathcal{M}_w$ to obtain consistency with the observed rates of formation of planetary nebulae and supernovae. The conclusion that emerges from such analyses is that $\mathcal{M}_w \lesssim 5\mathcal{M}_\odot$. Stars initially less massive than this eject enough mass in stellar winds and planetary nebulae to end as white dwarfs, while more massive stars become supernovae and end as neutron stars or black holes.

Evolution of Spheroidal-Component Stars

Let us now consider the structure and evolution of the metal-poor spheroidal-component stars.

The Subdwarf Main Sequence and the Primordial Helium Abundance In addressing the problem of the evolution of spheroidal-component stars, we are immediately confronted with the question, "What is the helium content of these stars?" Knowledge of the helium content of spheroidal-component stars is important, not only because it determines their structure and evolution, but also because these objects are as old as our Galaxy itself. Therefore their chemical composition gives vital clues about the composition of the primordial material from which our Galaxy was formed.

On the theoretical side, there are two possibilities to be considered: (1) the primordial material was essentially pure hydrogen, and both helium and the heavy elements were manufactured by stellar nucleosynthesis processes, or (2) the primordial material had a significant helium content, and only the heavier elements were made in stars. Analysis of the first alternative shows that it is possible, but difficult, to explain present-day abundances by means of nucleosynthesis in stars alone, because element-building scenarios that produce enough helium tend to give metal abundances in conflict with observations. In contrast, in an initial hot big bang, as indicated by the existence of the cosmic microwave background, helium can easily be manufactured in sufficient quantity to explain even the helium abundance in disk material, $Y \approx 0.28$. In fact, a determination of the helium content of the oldest known stars—those in globular clusters and the halo—can place valuable constraints on possible models for the early evolution of the Universe.

Observationally, the helium content of spheroidal-component stars is, unfortunately, extremely difficult to determine. Most spheroidal-component stars are too cool to excite the spectrum of He I, hence we cannot obtain any spectroscopic information about their helium content. Helium lines *are* observed in blue HB stars, but the abundances derived are anomalously low—a factor of ten below that in disk stars. The present belief is that this situation results from a gravitational settling of helium, which depletes the helium abundance of the observable layers. In short, the spectroscopic evidence is inconclusive, and we must turn to other methods.

Analysis of the location of the subdwarf main sequence in the theoretical H-R diagram provides a powerful method of estimating the helium content of the spheroidal-component stars. The luminosity and effective temperature of a chemically homogeneous star of a given mass depends on Y and Z in a known way. For stars with normal (that is, solar) Y, a decrease in Z produces a main sequence lying below the normal zero-age main sequence. For stars with $X = 0.70$, a change in Z from 0.02 to 4×10^{-4} lowers the main sequence (measured at constant T_{eff}) by about 1 mag (**P3**). On the other hand, if Z is fixed and Y is decreased, then the main sequence rises above the normal ZAMS. If $Z = 0.02$, then lowering Y from 0.28 to 0.18 raises the main sequence (at constant T_{eff}) by about 0.4 mag. The effect of a low value of Z on the position of the main sequence can, in principle, be almost exactly compensated by lowering Y; stars that are both metal poor and He poor by just the right amounts would lie almost on the normal ZAMS (the mass-luminosity relation—which is unknown for spheroidal-component stars—would, however, be different in the two cases). Because we know Z spectroscopically, we can determine Y if we can fix the position of the subdwarf main sequence in the theoretical H-R diagram.

Unfortunately, neither M_{bol} nor T_{eff} is directly observable, and the determination of the relative positions of the subdwarfs and the Hyades main sequence in the (M_{bol}, log T_{eff}) plane poses difficulties. First, because subdwarfs are rare in the solar neighborhood, very few of them are close enough to the Sun to have reliably known parallaxes. Indeed, only five extreme subdwarfs [$\delta(U - B) \geq 0.16$] have trigonometric parallaxes $\geq 0''.035$ (**E8**), and only one or two more (that are cool enough to be of relevance) have distances that can be estimated from membership in moving groups. Second, at the time this problem was first approached, it was rather difficult to estimate reliable effective temperatures for subdwarfs.

In 1959, A. D. Code (**C12**) showed that the subdwarfs fall about 1 mag below the normal main sequence in a plot of M_V versus an index analogous to $(G - I)$. As the latter is known to be insensitive to line-blanketing, one concludes that the subdwarfs are indeed intrinsically 1 mag less luminous than normal stars of the same T_{eff}. Knowing that $Z_{sd} \ll Z_{\odot}$, one also infers that these stars must have nearly solar He content. At about the same time, O. J. Eggen and A. R. Sandage (**E8**), (**S3**) attempted to make a direct comparison of the subdwarfs and the Hyades using UBV photometry and line-blanketing corrections as described in §3-6. In contradiction to Code's results, they found that, after allowance was made for line-blanketing effects, the subdwarfs mapped almost exactly onto the Hyades sequence, and hence these stars at not subluminous. Their results would imply that the subdwarfs are severely helium deficient.

As was subsequently pointed out by R. Cayrel (**C3**), one is faced with a serious dilemma when using the deblanketing procedure for UBV photometry. On the one hand, the procedure works best for stars hotter than the Sun, because the blocking coefficients can be determined reliably only for solar and hotter stars. However, the spheroidal component is so old (see §3-10) that such stars are already appreciably evolved and hence could well be a magnitude (or more) more luminous than the true subdwarf main sequence. Therefore, one would like to analyze stars cooler than the Sun. But here the empirical deblanketing procedure becomes unreliable, for two reasons. First, the blocking coefficients, and hence the slope of the blanketing vectors, are ill determined. Second, the blanketing vectors become nearly parallel to the Hyades sequence in the two-color diagram, which implies that inferred blanketing corrections will be extremely sensitive, both to the exact value of the adopted slope of the blanketing vector and to errors in the measured colors.

Cayrel analyzed several cool subdwarfs with accurately known distances, particularly *Groombridge 1830*, and he derived effective temperatures for them from several different indices chosen specifically to be almost unaffected by line-blanketing. He showed convincingly that the subdwarfs do, in fact, lie about 0.75 ± 0.3 mag below the Hyades main sequence in the theoretical H-R diagram. He suggested that Eggen and Sandage had found their result because the stars they analyzed were already appreciably evolved. This suggestion was later confirmed by Eggen (**E6**), who then extended the photometry to much cooler subdwarfs and showed that, in a plot of M_V versus $(R - I)$, the subdwarfs lie 0.75 mag below the Hyades main sequence. Thus we know today that the subdwarfs are actually underluminous compared to metal-normal stars of the same effective temperature and hence that they must have a nearly solar helium abundance, as would be expected if helium is made primordially. If Y were exactly the same in the subdwarfs as it is in the disk, then they should be about 1 mag less luminous than the normal main sequence. From their actual position, Eggen concluded that $Y \approx 0.23$ in the subdwarfs. This figure is to be compared to $Y \approx 0.28$ for the disk. In *number* ratios, $N(\text{He})/N(\text{H}) \approx 0.075$ for subdwarfs and 0.10 in the disk.

In principle, another method of obtaining the He abundance of spheroidal-component stars is to determine their mass-luminosity relation, which depends sensitively on Y for a given value of Z. Unfortunately, masses are not known for even one spheroidal-component star. As noted by T. R. Dennis (**D1**), the most promising candidate for a mass determination is μ Cas, which has a reliable trigonometric parallax (and hence known luminosity), and for which an absolute orbit of the primary star is known. Dennis pointed out that, if one could measure the separation of the two components (even once), then one could obtain $(\mathscr{M}_1 + \mathscr{M}_2)$ from Kepler's law and $(\mathscr{M}_1, \mathscr{M}_2)$ from the scale of the orbit of the primary. This information would yield $\mathscr{M}_1$ and $\mathscr{M}_2$ individually. Attempts have been made to resolve this close binary, but the results are not decisive (**F1**), (**G2**), (**H13**). The measurements should be easily made with the Space Telescope.

Some additional evidence is provided by spectroscopic analyses of the emission-line spectra of two planetary nebulae, one found in the globular cluster M15 (**O1**), (**P2**), and the other in the halo in the direction of the galactic pole (**M15**). In both cases, one finds $Y \approx 0.3 \pm 0.05$, which again argues for a large value for the primordial helium abundance. These results are not in themselves conclusive, however, because there is the possibility that the helium content of the matter from which these nebulae were formed was enriched by mixing with material that underwent hydrogen burning in the star during its evolution.

There are yet other indications that the helium content in the spheroidal component lies in the range $Y \approx 0.25 \pm 0.03$. These indications are based on evolutionary interpretations of globular-cluster CM diagrams and the properties of cluster RR Lyrae stars, which we shall discuss in §3-8 and §3-10. In summary, the current evidence supports a relatively high value for the helium abundance of primordial material created in the cosmic big bang.

The Evolution of Globular-Cluster Stars Observations of globular clusters provide us with very complete CM diagrams that contain a wealth of information about the evolution of very old, metal-poor, solar-mass stars. Although all globular-cluster stars having masses above solar have long since evolved into white dwarfs (and hence have vanished below our present-day observational capabilities), the advanced evolutionary phases of approximately solar-mass stars are well represented.

Up through the tip of the ascending giant branch, which is terminated by the helium flash, the evolution of a globular-cluster star is qualitatively the same as for low-mass disk stars, as described previously. However, an important difference in their evolutionary tracks is that, as can be seen in Figure 3-14, the subgiant branch for metal-rich stars is flatter and more extended, and the giant branch lies farther to the right in the H-R diagram than is the case for metal-poor stars. Early work on giant-branch evolution by F. Hoyle and M. Schwarzschild (**H21**) showed that these observed features are well accounted for by theory, the differences arising from differences in the envelope structure produced by the very different opacity sources present in the two cases.

As was true for disk stars, mixing of nuclear-processed material into the outer envelope occurs on the giant branch. In particular, some helium enrichment of the envelope occurs ($\Delta Y \approx 0.02$). During the helium flash itself, there is a possibility that material is mixed from the convective helium-burning core into the hydrogen-burning shell by convective overshoot, but, at present, detailed calculations suggest that this does not in fact occur.

As a result of the helium flash, a star evolves very rapidly onto the *zero-age horizontal branch* (*ZAHB*), as shown schematically in Figure 3-18. At that point, electron degeneracy in the core has been lifted, and evolution is again proceeding on a nuclear-burning time-scale. The structure of the star is well approximated by a static model with a helium-burning core and a hydrogen-burning shell. The core mass is found to be nearly independent of cluster age, and typical values that give the correct luminosity for the HB are about $0.4\mathscr{M}_{\odot}$. The total mass of the zero-age horizontal branch star lies in the range of 0.6 to $0.8\mathscr{M}_{\odot}$. This mass is somewhat smaller than the masses of stars on the red-giant branch, from which one infers that mass-loss has occurred (as is in fact observed in giant-branch stars). The primary factors determining a star's position on the zero-age horizontal branch are its total mass, CNO abundance, and envelope helium abundance.

Detailed calculations during core helium-burning phases near the horizontal branch show that, even when the most favorable assumptions are made to prolong their evolution on the HB itself, the evolutionary tracks of stars of a given mass span an interval in log T_{eff} that is small compared to the observed widths of globular-cluster HBs. From this, one concludes that there must be a distribution of stellar mass among the stars that compose real horizontal branches in clusters. This spread in mass is presumably produced by variations in the total amount of mass-loss a star experiences during its stay on the ascending giant branch and as a result of the helium flash.

For all other parameters fixed, the location of a star on the ZAHB depends strongly on the metal abundance parameter Z. The smaller the value of Z,

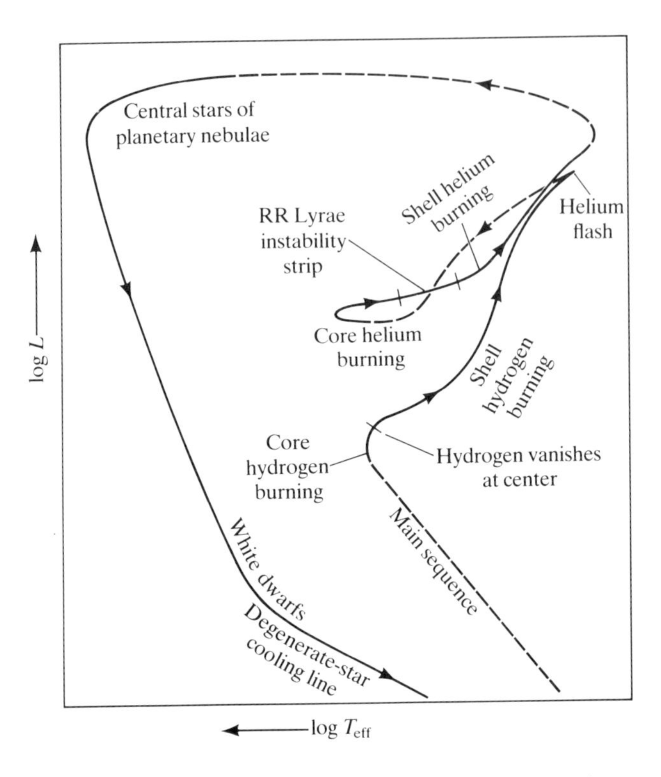

Figure 3-18. Schematic evolution track for a representative low-mass, globular-cluster star from the main sequence to its ultimate demise as a white dwarf. The major energy sources are indicated at several key phases. Dashed lines indicate episodes of very rapid evolution, during which details of the structure of the star are, at present, not too well known. Compare this figure with Figure 3-13.

the bluer is the color of a star on the HB. A match to observed globular-cluster HBs is achieved by models with $Z \approx 10^{-4}$, which is in just the right range. An interesting result is that, for values of Z appropriate to disk stars, the ZAHB moves so far to the right that it merges into the giant branch. This merger explains the existence of the *clump* found in the distribution of stars along the giant branches of metal-rich disk clusters at just the luminosity appropriate to the HB as predicted by theory.

The lifetime of a star during its core helium-burning stages near the HB is typically of the order of 5×10^7 to 10^8 years, and it is fixed mainly by the initial mass of the helium core. As helium becomes exhausted in the core, a star's evolution proceeds very rapidly on a contorted path above the HB in the H-R diagram. When helium burning in a thick shell is finally established, evolution again slows to a nuclear-burning time-scale, and the star evolves from blue to red and rises along the *asymptotic giant branch* (see Figure 3-18). As the evolution proceeds, the asymptotic giant branch approaches the first giant branch more and more closely. In the initial phases of the rise along the asymptotic branch, the helium-burning shell narrows, the rate of helium burning rises, and, finally, the hydrogen-burning shell is extinguished. But, as the tip of the curve is approached, the hydrogen-burning shell reignites, and both the hydrogen-burning and helium-burning shell approach one another, being separated by only a thin shell of almost pure helium.

At this point, the star becomes vulnerable to several strong instabilities. One of these is the Schwarzschild-Härm (**S12**) double-shell-source thermal instability that produces *relaxation oscillations* consisting of a succession of extremely rapid thermal pulses, each followed by a slower adjustment of the star back to a long interval of quiescent evolution. During each pulse, there is a thermal runaway in the helium burning shell that generates an extensive convection zone outside that shell, which may lead to mixing of the stellar material and could bring exotic products of nucleosynthesis into the envelope. It is not yet known whether mixing actually occurs, although current models suggest that it does not. On the other hand, stars in this part of the H-R diagram sometimes do show striking abundance anomalies.

As a star rises to the tip of the asymptotic giant branch, we observe that it again begins to lose mass in dense stellar winds. These winds may be driven by radiation pressure on grains that form in a star's atmosphere at very cool temperatures, or by energy input from the dissipation of shock waves that are generated by pulsational instabilities. Most red supergiants are in fact observed to be regular, semiregular, or irregular pulsating variables. As the stellar luminosity increases, ordinary pulsations give way to relaxation oscillations of the envelope and, finally, dynamical instability of the whole envelope. This instability culminates in the ejection of planetary nebulae as the envelope material passes escape velocity, leaving the core exposed. The star evolves very rapidly over to the upper left of the H-R diagram and becomes a small, very blue *planetary-nebula nucleus*. After a brief existence there, its nuclear shell sources burn out, and the star descends into the region of the white dwarfs and thence moves down a cooling line to a dark, cold oblivion.

The achievements of the modern theory of stellar evolution are indeed impressive, and we undoubtedly do have at least a qualitative understanding of the entire history of stars from their birth to their death. Nevertheless, we should mention some of the uncertainties that remain, and it will be prudent to bear these in mind when we apply the theory for interpretive purposes in §3-9 and §3-10. To begin, there are always uncertainties in some of the basic physical properties of stellar material (opacities, energy-generation rates, neutrino-production rates, and so on) that introduce uncertainties of various degrees of seriousness into the results. Likewise, considerable uncertainties are caused by the lack of an adequate theory of convection.

More fundamentally, several effects are normally neglected in stellar-evolution computations that may be of great importance. Thus allowance is not usually made for the effects of magnetic fields, rotation, mass exchange between binary partners, or the internal hydrodynamics of a star upon its evolution. For some stars in some evolutionary phases, these phenomena may dominate the course of evolution and lead to results quite different from those generally accepted at present. In addition, we have assumed that stellar evolution proceeds without major mass-loss, although we in fact observe copious mass-loss via stellar winds for virtually all giants and supergiants. Integrated over the lifetime of a star, such noncatastrophic mass-loss might have major evolutionary consequences. But, as we do not have a complete theory of stellar winds, we cannot account for them unless we simply adopt empirical mass-loss rates.

Finally, it is worth noting that, in a general way, the uncertainties in the models grow more severe as evolution proceeds to more advanced stages, with errors at each successive step tending to amplify and combine with others as the computation proceeds. Although the main-sequence and early post-main-sequence phases are probably fairly well understood, our understanding of later phases may at present be only qualitative.

3-8. PULSATING VARIABLE STARS

Basic Properties

Although the luminosity of most stars is essentially constant, some stars exhibit very regular light variations of moderate to large amplitude, which result from a *pulsation* of the star's envelope. The variable stars that are the best understood and most useful for galactic-structure research are the *Cepheid* and *RR Lyrae variables.* We shall confine our discussion to these stars exclusively.

Cepheids The Cepheid variables are supergiants of spectral classes F to K, which are found in a narrow instability strip in the H-R diagram (see Figure 3-19). All stars in this region of the diagram have envelopes that are pulsationally unstable. Typical Cepheid periods lie in the range $1^{d} \leq P \leq 50^{d}$. The Cepheids are usually divided into two subgroups, the *classical Cepheids* (or δ Cephei stars) and the *W Virginis stars*, each named after its prototype. These two categories can easily be discriminated from one another by differences in their light curves and their spectrophotometric properties.

The classical Cepheids are known to be high-mass stars in a phase of core helium burning. The intersection of the instability strip with evolutionary tracks of stars in their major helium-burning phase determines where concentrations of Cepheids of a given chemical composition and within a given mass range will be found in the H-R diagram. From their concentration to

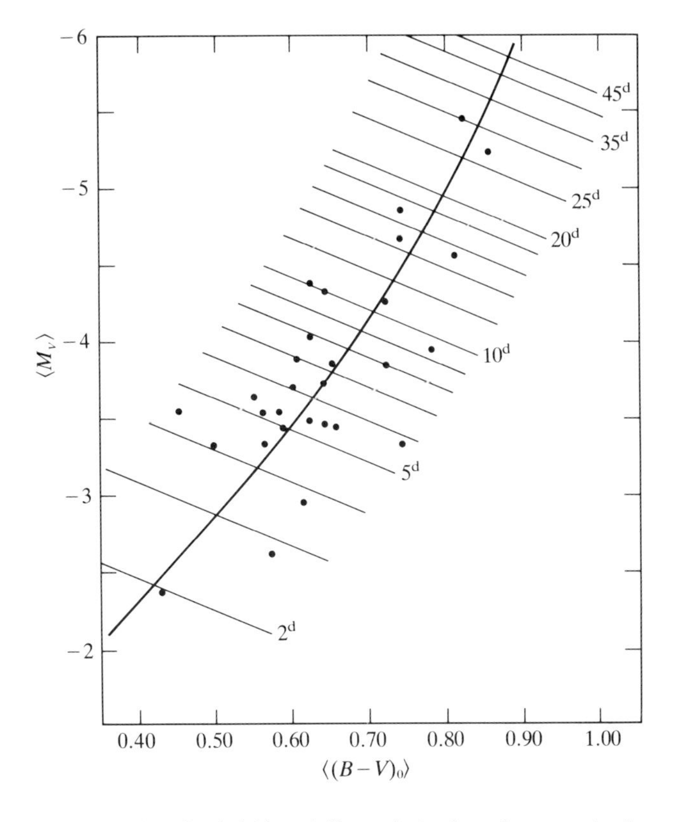

Figure 3-19. Cepheid instability strip in the color-magnitude diagram. Individual points represent the observed positions of galactic Cepheids. Lines of constant period are labeled with periods in days. [From (**S24**, 435), by permission. Copyright © 1963 by the University of Chicago.]

the galactic plane and the fact that they are found in galactic clusters, we know that classical Cepheids are disk-component stars. From observations of external galaxies, we find that they are associated with spiral arms. In contrast, the W Virginis stars are found at high galactic latitudes and in globular clusters, and they show a concentration toward the galactic center. These are spheroidal-component stars, and hence we infer that they are low-mass, metal-poor stars in a core helium-burning phase.

Cepheid light curves show a strict periodicity (in all properties), and δ Cephei stars typically have an abrupt rise (in about 20% of the period) followed by a slower decline (about 50% of the period). In W Vir stars, the rise times and fall times are more nearly equal, and the maximum is less pronounced. At maximum light (phase = 0.0), the star has its highest temperature, earliest spectral type, and greatest outward radial velocity; at minimum light (phase $\approx$ 0.75), it has its lowest temperature, latest spectral type, and greatest inward velocity. The star achieves its maximum radius on the descending part of the light curve (phase $\approx$ 0.4) and its minimum radius shortly after the next light-rise begins.

RR Lyrae Stars The RR Lyrae stars (or *cluster variables*, because they are found in globular clusters) are mostly metal-poor giants, roughly of spectral type A, that pulsate with periods $P < 1^{d}$. RR Lyrae stars with periods $\gtrsim 0^{d}.5$ and $\Delta S \gtrsim 5$ [see equation (3-45)] are commonly found in globular clusters, in the galactic halo, and concentrated toward the galactic center. Hence they are clearly spheroidal-component stars. RR Lyrae stars with periods $\lesssim 0^{d}.4$ and $\Delta S \lesssim 5$ are concentrated to the galactic plane and are identified as old disk-component stars. In globular clusters, the RR Lyrae stars always fall in a narrow color range on the horizontal branch, and hence they can be identified as low-mass, post-giant-branch, helium-core-burning stars that have envelopes with just the right properties to be pulsationally unstable.

RR Lyrae stars were classified into *Bailey types a, b*, and *c* according to the shapes of their light curves. Classes *a* and *b* have been amalgamated into one class, RR_{ab}. These stars have asymmetric (Cepheid-like) light curves with periods $P_{ab} \gtrsim 0^{d}.4$ and an average period $\langle P_{ab} \rangle \approx 0^{d}.55$. In general, these stars have large amplitudes, $0.5 \lesssim \Delta m \lesssim 1.5$ mag, and are in the *fundamental mode* of pulsation. The RR_c variables have symmetric (practically sinusoidal) light curves with smaller amplitudes, $\Delta m \lesssim 0.5$ mag, and shorter periods, $P_c \lesssim 0^{d}.4$ (average $\langle P_c \rangle \approx 0^{d}.3$). These stars are pulsating in the *first-overtone mode*. About 90% of the RR Lyrae stars are RR_{ab}, and 10% are RR_c. The former are spheroidal-component stars, and the latter are old disk stars (see also §4-5).

Because most RR Lyrae stars are metal poor, their metallic line spectral types (classified by the standard criteria) tend to be earlier than the spectral types appropriate to their hydrogen lines. G. Preston (**P10**) introduced the parameter

$$\Delta S \equiv 10[\text{spectral type (H lines)} - \text{spectral type (Ca II K lines)}] \quad (3\text{-}45)$$

where the spectral types are measured at minimum light in tenths of a spectral class, as usual. This parameter correlates closely with the metal content of a star; $\Delta S \approx 0$ for stars of normal (that is, disk) abundances and rises to $\Delta S \approx 10$ to 12 for the most extreme metal-poor halo stars. ΔS has been calibrated in terms of [Fe/H] by means of curve-of-growth analyses of field RR Lyrae stars (**B13**). A very tight correlation,

$$[\text{Fe/H}] \approx -0.16\Delta S - 0.23 \quad (3\text{-}46)$$

is observed. This relation can be applied to cluster variables to estimate metal abundances in globular clusters.

Significance for Galactic-Structure Research

Cepheids and RR Lyrae stars are of importance to galactic-structure research because (1) their very distinctive light variations allow them to be identified unambiguously and easily even in a crowded field of stars, and (2) their absolute magnitudes can be calibrated accurately, and subsequently the absolute magnitude of any particular star can be determined with good precision from its observed light variation.

In 1912, it was discovered by Henrietta Leavitt (**L4**) that the Cepheids obey a *period-luminosity relation* (*PL relation*). If the PL relation can be calibrated properly, then a simple observational determination of a Cepheid's period of variation yields its absolute magnitude and hence its distance from its distance modulus. The classical Cepheids and W Vir stars obey different PL laws because of their different masses and internal structures and because of the differences in the pulsation properties of their envelopes resulting from their very different metal abundances. Because the Cepheids are highly luminous, they can be seen to fairly large distances within our own Galaxy, and hence they can be used to trace parts of spiral arms. Furthermore, they can be identified as individuals in other galaxies, and hence they are used as important distance indicators in setting the extragalactic distance scale (see §5-3 for details).

The RR Lyrae stars, being confined to the horizontal branch, all have nearly the same absolute magnitude, which can be calibrated by a variety of independent methods. Although these stars are much fainter than Cepheids, they can nevertheless be seen all the way to the galactic center in one or two regions of low interstellar absorption (the so-called *Baade windows*). In fact, their observed numbers in these windows show a peak at a definite apparent magnitude and then a falloff. These observations can be used to estimate the Sun's distance from the galactic center (see §8-2). Moreover, because RR Lyrae stars are so readily identified, they can be sampled in a given field to a high level of completeness, and hence they can be used in stellar-density analyses of the spheroidal component of our Galaxy (see §4-4). Finally, the RR Lyrae stars are typically only 2 to 3 mag fainter than the brightest stars in globular clusters. Thus they are relatively easy to observe, in addition to being easily identified. Because they have well-defined absolute magnitudes, they provide good distance indicators for clusters, and they allow us to avoid having to observe down to the cluster main sequence (which is 5 mag fainter).

Physics of Cepheid and RR Lyrae Pulsations

The envelopes of both the Cepheids and RR Lyrae stars undergo self-excited radial pulsations. The physics of this process is fascinating, and by now it is fairly well understood through detailed hydrodynamical calculations. It would take us too far afield to describe these computations; excellent reviews can be found in (**C6**) and (**C14**). We shall sketch only some of the basic physical features of the problem.

The pulsation is driven by those zones in the envelope in which He^+ is ionizing to He^{++}. The basic process involved, the *κ mechanism*, was first described by Eddington, but it was erroneously thought by early investigators to apply to the hydrogen-ionization zone. Only when detailed models were constructed was the correct identification of the driving zone made. The instability is the result of the variation of the opacity of the material with temperature. As He^+ first begins to ionize, its opacity increases with increasing temperature; this situation is contrary to the normal one (in stellar interiors) in which the opacity decreases with increasing temperature. Because of this peculiarity, when the material is compressed and therefore heats, it becomes more opaque and hence traps radiation emanating from below. This trapping tends to cause yet more heating and raises the internal energy of the gas. Thus, as the material compresses, it builds up a reservoir of extra thermal energy.

Eventually, the material reaches maximum compression and starts to expand. During the expansion, the extra energy in the thermal reservoir is tapped and performs work that drives the material to higher velocities than it would have had if no radiant energy had been trapped earlier. As the material expands, it cools, and because it cools, it becomes more transparent. Therefore, during the expansion phase, energy escapes too freely, and the material cools so much that, when the envelope reaches maximum extension, the gas pressure is too low to support it (that is, the oscillation has "overshot"). The envelope then collapses in another, even stronger compression stroke.

Thus the He^+ ionization zone acts like any common heat engine, with the opacity acting as a valve that regulates the cycle. Once a star is in the instability strip, any small perturbation from an initially static configuration will produce a disturbance that will amplify into an ever-growing pulsation; thus the pulsation is self-excited. The final amplitude of the pulsation is set by nonlinear effects, which prevent the oscillation from growing indefinitely.

From a simple dimensional analysis, which in essence argues that the pulsation period should be of the order of the time required for a sound wave to move through the star, one can show that a relationship of the form

$$P\left(\frac{\bar{\rho}}{\bar{\rho}_\odot}\right)^{1/2} = Q \tag{3-47}$$

exists. Here, P is the period, $\bar{\rho}$ denotes the mean density, and Q is the *pulsation constant*, which will, in general, depend (weakly) on P and other stellar parameters such as the chemical composition. This *period-mean-density* relation, which is verified by detailed model calculations, is very important and has interesting implications in terms of the period-luminosity-color relation. For example, we see immediately that the mere fact that L increases as T_{eff} decreases in the instability strip (as shown in Figure 3-19) implies that

Cepheid radii increase with increasing L. Thus their mean densities decrease with increasing L, and, on the basis of equation (3-47), we expect the most luminous stars to have the largest periods. This is just what is observed.

Not only do modern theoretical pulsation calculations give an accurate description of the observed variations of light, color, radial velocity, and so on during a cycle, but also theoretical evolution tracks for stars in the appropriate evolutionary stages explain the observed distribution of Cepheids in the CM diagram of such clusters as NGC 1866 [see (**I3**)]. We thus appear to have a fairly sound understanding of Cepheids. As we shall see later, the situation for the RR Lyrae stars is not quite as satisfactory.

Period-Luminosity-Color Relation for Classical Cepheids

The original PL relation discovered by Henrietta Leavitt was a linear correlation of the apparent magnitudes of Cepheids in the Small Magellanic Cloud with log P. To use this relation for distance estimation, we must express it in terms of absolute magnitudes by determining the zero-point; this problem has an interesting history [see (**S25**, Chapter 15)].

No Cepheid is found close enough to the Sun to have a measurable trigonometric parallax, so for the calibration Shapley used statistical parallaxes (see §6-6) derived from their observed proper motions and radial velocities. A flaw in this work was that interstellar absorption was ignored, and hence the estimated Cepheid luminosities were too small. In reality, most galactic Cepheids are found at such large distances that their light is quite significantly dimmed by interstellar absorption. The problem was compounded when the classical Cepheids were assumed to have the same properties as the W Virginis stars in globular clusters. When this assumption was made, it was found that the RR Lyrae stars (using the then-available estimates of their absolute magnitudes) and Cepheids all fell along a single apparently well-defined period-luminosity law, and this seeming harmony lulled astronomers into a false sense of security about the accuracy of the results.

Hubble's use in 1929 of this PL relation to determine the distance to M31 provided a hint that there was a problem. At the estimated distance, M31 turned out to be about half the size of our Galaxy, and its globular clusters were about a factor of four fainter than ours. But, given the uncertainties at that time about the size of our Galaxy, the extragalactic distance scale, and even the nature of external galaxies, these discrepancies did not loom large.

The root problem was uncovered by Baade in 1952 when he discovered that 200-inch telescope photographs of M31 failed to reveal RR Lyrae variables in the globular clusters at the magnitude predicted by the then-adopted distance scale, but instead showed only the brightest cluster giants. Inasmuch as the absolute magnitudes of the RR Lyrae stars had been determined independently from statistical parallaxes and cluster CM diagrams, he realized that the error lay in the absolute magnitudes of the classical Cepheids, whose luminosities had been set about $1^{m}_{.}5$ too faint. Making this revision doubled the estimated distance to, and the size of, M31 (hence making it comparable to our Galaxy), and the revision brought its globular-cluster luminosities into line with those of our Galaxy. Furthermore, it doubled the estimated size of the Universe, because the extragalactic distance scale had been calibrated using Cepheids.

With the development of pulsation theory, we have come to recognize that the period-luminosity (-color) relation must depend on chemical composition, so that the calibration for the classical Cepheids and W Virginis stars must be different. The present calibration for the classical Cepheids rests securely on absolute magnitudes determined for Cepheids that are members of galactic clusters.

The classical Cepheids are also found to obey a rather precise *period-spectral type* or *period-color relation.* A. D. Code (**C11**) showed that, at maximum light, all Cepheids have spectral types in the range F5 Ib to F7 Ib. This fact means that they all have a well-defined intrinsic color at a convenient time in their cycle (that is, when they are easiest to observe), which facilitates corrections for interstellar reddening. From the intrinsic colors of Cepheids in galactic clusters, R. P. Kraft (**K5**) has established a period-mean-color relation that shows a very small scatter.

In observed PL relations for Cepheids, there is always much more scatter around the mean curve than can be accounted for by observational errors in the magnitudes, colors, or interstellar-absorption corrections. This scatter is intrinsic; we can understand why it should exist from the following simple argument. Given a period–mean-density relation of the form of equation (3-47), and remembering that $\bar{\rho} = \bar{\rho}(\mathcal{M}, R)$ and $R = R(L, T_{\text{eff}})$, it is clear that some relationship of the form

$$f_1(P, \mathcal{M}, L, T_{\text{eff}}) = Q \tag{3-48}$$

must exist. Furthermore, for a given chemical composition, we know that $L = L(\mathcal{M}, \text{age})$ and that $T_{\text{eff}} = T_{\text{eff}}[(B - V), L]$ and $B.C. = B.C.[(B - V), L]$. Equation (3-48) then implies a general relationship of the form

$$f_2[P, L, (B - V), \text{chemical composition, age}] = Q \tag{3-49}$$

where Q may also depend on composition, period, and age. In other words, stars of a given age and composition do not obey a period-luminosity relation but rather a *period-luminosity-color relation* (*PLC relation*).

Careful observational work over the past two decades has resulted in a well-established PLC relation for classical Cepheids in our Galaxy. The most complete analysis is that of A. R. Sandage and G. A. Tammann (**S6**). They combined Cepheid data from the Large and Small Magellanic Clouds,

M31, and NGC 6822 to derive the shape of a mean PL relation in the form of M_V (or M_B) versus log P. The zero-point was set using nine Cepheids in galactic clusters. For $3^d \leq P \leq 40^d$, their mean PL relation is well approximated by the expression

$$M_V(\text{max}) = -2.0 - 2.8 \log P \tag{3-50}$$

where a correction of -0.2 mag has been added to their numbers to allow for the revision of the Hyades distance modulus. Then, using the best available relations linking T_{eff} and *B.C.* to $(B - V)$, they rewrite equation (3-49) as

$$\log P + 0.239 M_{\langle V \rangle} - 0.602(\langle B \rangle - \langle V \rangle) = 0.838 + \log Q \tag{3-51}$$

where the angular brackets denote intensity averages. A prediction of this equation is that, at fixed P, Cepheids should scatter in absolute magnitude by an amount

$$\Delta M_V = 2.52 \Delta(B - V) \tag{3-52}$$

where $\Delta(B - V)$ is the departure of the observed (unreddened) color of the star from the color given for stars of the same period by the ridge line in the period-color relation:

$$[\langle B \rangle - \langle V \rangle]_{\text{mean}} = 0.264 \log P + 0.37 \tag{3-53}$$

Sandage and Tammann (**S7**) have also demonstrated the existence of a *period-luminosity-amplitude relation.* This relation permits a very accurate determination of M_V for any particular star from the light curve alone, and it has the advantage that colors are not required (except as needed to correct for interstellar absorption).

Theoretical pulsation calculations (**I5**), (**I6**) yield a relationship between the period P, mass $\mathscr{M}$, luminosity L (in solar units), and effective temperature T_{eff} of the form

$$\log P \approx 0.65 + 0.83(\log L - 3.25) - 0.63(\log \mathscr{M} - 0.7) \\ - 3.44(\log T_{\text{eff}} - 3.77) \tag{3-54}$$

Adopting estimates of L and T_{eff} for galactic Cepheids, one can derive pulsation-theory estimates $\mathscr{M}_{\text{puls}}$ of their masses from equation (3-54). These estimates may be compared with the masses $\mathscr{M}_{\text{evol}}$ deduced from fitting evolution tracks to the positions of Cepheids in the H-R diagram. Taking into account the revised Hyades distance modulus, one finds $\log(\mathscr{M}_{\text{evol}}/\mathscr{M}_{\text{puls}}) \approx 0.04$, which indicates good agreement from the two independent theories.

Finally, it is worth remembering that the PLC relation just discussed applies, strictly speaking, only to our Galaxy. At the present time, we do not know how large the variations in the PLC relation are from galaxy to galaxy.

Absolute Magnitudes of RR Lyrae Stars

The RR Lyrae stars are confined to the horizontal branch, and hence, in contrast to the Cepheids, they all have about the same absolute magnitude in a given cluster. The position of the horizontal branch in a cluster and the pulsation properties of stellar envelopes both depend on metallicity, hence M_V(RR) should vary from cluster to cluster. There should also be an intrinsic spread in the absolute magnitudes of field RR Lyrae stars induced by the spread in metal content among them.

There are essentially three independent methods of estimating M_V(RR): (1) from the statistical parallaxes (see §6-5) of field RR Lyrae stars; (2) by the calibration of globular-cluster CM diagrams by main-sequence fitting; and (3) by comparison of RR Lyrae stars with classical Cepheids in the Magellanic Clouds, using the known zero-point for the Cepheids. Each method has advantages and disadvantages; the statistical parallaxes suffer from any errors present in the proper motions and from the intrinsic spread produced by variations in metal content among the stars. The problems with main-sequence fitting of cluster CM diagrams were discussed in §3-6. The comparison of RR Lyrae stars and Cepheids in the LMC and SMC presumes that those stars in the clouds are in fact identical to their counterparts in our Galaxy. This may or may not be true.

Results from several studies are listed in Table 3-10. An overall average is $\langle M_V(\text{RR})\rangle \approx +0.6 \pm 0.2$, which is probably the best estimate that can be made from the data available at the present time.

The differences, from cluster to cluster, among the values of M_V(RR) obtained by main-sequence fitting are, unhappily, in disagreement with the predictions of theoretical pulsation models. Specifically, the theory predicts that the variables with the highest metallicity should be the faintest; only the variables in 47 Tuc obey this rule, while those in M3 and M92 are in clear violation of it. In the past, the theory has been given precedence over the main-sequence-fitting results, and M_V(RR) values for the various clusters have been force fitted to the predictions of the theory (**S2**). The required adjustments are alarmingly large. For M3, M_V(RR) must be increased by 0.2 mag, and for M92 it must be decreased by 0.45 mag, implying relative errors of the order of 0.65 mag! This is a very serious uncertainty indeed. If the cluster HBs are forced to have the "correct" relative relationship according to pulsation theory, then their main sequences all coincide in the CM diagram [see Figure 17 of (**S2**)], which is contrary to what would be expected if any reasonable allowance is made for line-blanketing effects. Furthermore, in view of the complexity of the HB resulting from its sensitivity

Table 3-10. Absolute Magnitudes of RR Lyrae Variables

Location of stars	Method	M_V(RR)	Reference
M3	Main-sequence fitting	0.60	(**S2**)
M13	Main-sequence fitting	0.05	(**S2**)
M15	Main-sequence fitting	0.98	(**S2**)
M92	Main-sequence fitting	0.91	(**S2**)
NGC 6838	Main-sequence fitting	0.35	(**H9**)
47 Tuc	Main-sequence fitting	0.90	(**H6**)
		$\langle M_V \rangle_{\text{cluster}} \approx 0.63 \pm 0.2$	
Galactic field	Statistical parallax	0.7 ± 0.2	(**V4**)
Galactic field	Statistical parallax	0.6 ± 0.2	(**W9**)
Galactic field	Statistical parallax	0.5 ± 0.2	(**H11**)
Galactic field	Statistical parallax	0.5 ± 0.4	(**H14**)
LMC, SMC	Comparison with Cepheids	0.6 ± 0.2	(**G5**)

to several diverse parameters (Y, Z, $\mathscr{M}_{\text{core}}$, $\mathscr{M}_{\text{star}}$, age), the validity of forcing a fit to the pulsation theory is somewhat questionable. In any event, the dilemma cannot be resolved with present data. Observations with the Space Telescope should help considerably by providing accurate cluster H-R diagrams down to the main sequence.

One of the most reliable predictions of the theory (**I6**) is the fundamental pulsation period as a function of mass and luminosity (in solar units) and effective temperature:

$$\log P_F \approx -0.34 + 0.825(\log L - 1.7) - 0.63(\log \mathscr{M} + 0.19) - 3.34(\log T_{\text{eff}} - 3.85) \tag{3-55}$$

The first overtone period is given by $\log P_H \approx \log P_F - 0.127$; these expressions are nearly independent of the stellar composition parameters. Using the period at the transition between the RR_{ab} and RR_c variables in a cluster to fix P_H, and adopting luminosities and effective temperatures determined observationally, one can estimate masses for the variables from equation (3-55). For M3, we find $0.55\mathscr{M}_\odot \lesssim \mathscr{M} \lesssim 0.75\mathscr{M}_\odot$, which is in the range predicted by evolutionary theory to the main sequence and giant branch.

The pulsation-theory results also give a fairly sensitive method of estimating the helium abundance in RR Lyrae stars, which is what one would expect considering that the pulsations are driven by the He^+ ionization zone. It is found that the properties of cluster variables support a relatively high helium content in the spheroidal component. For example, in M3 one finds (**I6**) $Y \approx 0.22$. We cannot, however, yet claim that these results are unequivocally accurate in view of the uncertainties just described in M_V(RR).

3-9. EVOLUTIONARY ANALYSIS OF SPIRAL-ARM AND DISK STARS

Ages of Galactic Clusters and Associations

Equipped with the results from the theory of stellar evolution, we can now make evolutionary analyses of the CM diagrams of galactic clusters and associations. Ultimately, cluster age-dating is based on the fact that the main-sequence lifetimes of stars diminish rapidly with increasing mass (see Table 3-9) and that, during its shell hydrogen-burning phase, a star evolves rapidly away from the main sequence to the right in the CM diagram. This rightward evolution is so rapid for massive stars that, in a typical cluster, there is essentially no chance of finding stars in those phases, and hence the observed main sequence simply terminates at some maximum luminosity (see Figure 3-6). For less-massive stars (say, around $1\mathscr{M}_{\odot}$), the rightward evolution is much slower and produces a subgiant branch. In either case, there will be a definite main-sequence *turnoff point*, which is a sensitive function of age. In principle, one can make age estimates using only the turnoff-point luminosity, but to do this would be to ignore the information available elsewhere in the CM diagram, which, of course, we cannot afford to do. Thus we need a more comprehensive procedure.

On the assumption that the stars in a cluster all formed at about the same time, the locus of stars in a cluster CM diagram clearly gives us information about the relative evolutionary status of stars of differing masses at a single instant after their arrival onto the ZAMS. Thus, to fit these data, we use sets of *isochrones*, which are constructed by connecting together the points representing the positions of model stars on their evolution tracks at a specified time after the onset of core hydrogen burning. These points are found by interpolation along each evolutionary track. Different sets of isochrones are obtained for different choices of the composition parameters (Y, Z) and of the ratio of the mixing length to scale height (ℓ/H) in convective layers in the envelope. A representative set of isochrones is shown in Figure 3-20, and an extensive collection is available in (**C7**).

To estimate a cluster's age, one attempts to fit its H-R diagram with a unique isochrone. In the isochrone-fitting process, one must first choose the appropriate chemical composition parameters. Spectroscopic estimates of Z can be made for most stars, but they can be made for Y only for upper main-sequence stars. Furthermore, it is necessary to map the theoretical isochrones onto observed CM diagrams, or vice versa, and thus we need to know an effective temperature scale and bolometric corrections.

On the upper main sequence, the isochrones are nearly vertical in the CM diagram, and isochrone fitting is essentially equivalent to a determination of the color of the bluest part of the main sequence. For these stars, mass loss and rotation effects are both significant, and the assumption of coevality suspect, so there can be important uncertainties in the results. On the lower

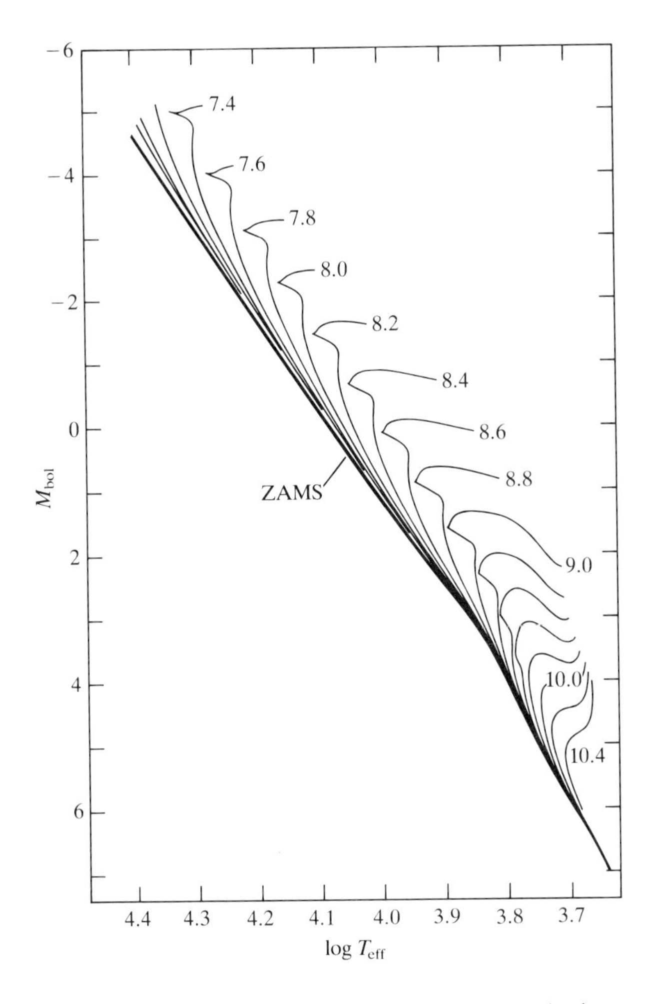

Figure 3-20. Representative theoretical isochrones in the H-R diagram for models with $(X, Y, Z) = (0.70, 0.27, 0.03)$, and mixing length to scale-height ratio $(\ell/H) = 1.5$. The curves are labeled with log (age in years). [From **(C4**, Chapter 17).]

main sequence, one fits not only the luminosity and color near the turnoff point, but also the shape of the isochrone as it curves into the subgiant branch. The comparison between theory and observation is more secure for these stars.

In some cluster CM diagrams, there is a distinctive *gap* in the main sequence below the turnoff point. Such gaps appear quite plainly for M67 and NGC 188 (see Figure 3-14) and several other clusters. These gaps are the result of rapid evolution in the H-R diagram during the period of overall stellar contraction at the point of hydrogen exhaustion in a convective core, which occurs between two much slower phases of evolution. This phenomenon happens only for a limited range of stellar masses, and gaps are therefore expected only in clusters with ages in the range of about 10^9 to 7×10^9 years. The luminosity at which the gap is found is a good age indicator, which, however, depends sensitively on composition.

In the older disk clusters, in which even low-mass stars have evolved appreciably, one gains additional leverage by fitting the subgiant and giant branches. Unfortunately, stars in these parts of the CM diagram have convective envelopes, and the models are therefore sensitive to shortcomings in the convection theory, and the predicted values of T_{eff} are not reliable. To make matters worse, an accurate conversion from $(B - V)$ to T_{eff} for red giants is also difficult. Thus, on the subgiant and giant branches, primary attention is given to fitting the luminosity.

As an example, if we assume that M67 and NGC 188 have about the same chemical composition, then the fact that NGC 188 has the less luminous turnoff and subgiant branch (see Figure 3-14) shows that it is unquestionably older than M67. The shape of the subgiant branch is also a useful age indicator because, as can be seen in Figure 3-17, the ratio of the turnoff-point luminosity to the minimum luminosity on the subgiant branch is closely correlated with the mass of stars at the turnoff point, and hence with the age of the cluster. Using this criterion, one again concludes that M67 is younger than NGC 188. Despite the uncertainties in the position of the theoretical giant branch and its mapping into the CM diagram, relative positions of observed giant branches yield important information about relative cluster ages quite directly, because we know that (for a given composition) the Hayashi tracks, and hence giant branches, of stars move progressively farther to the red as the mass decreases. On this basis, it is again obvious from Figure 3-14 that NGC 188 is older than M67. Of course, we cannot compare the giant branches of clusters with different compositions (for example, the globulars and NGC 188) in this way.

In analyzing cluster CM diagrams, a fairly wide variety of problems arise. First, age estimates depend sensitively on the luminosity of the features being fitted, and hence on the absolute magnitudes assigned to cluster stars. Any error in the distance scale has serious consequences. Thus the revision of the Hyades distance modulus upward by 0.2 mag increased the luminosities assigned to stars in all clusters whose CM diagrams had been fitted to the Hyades main sequence by $\Delta \log L \approx 0.08$. In old clusters, such as M67 and NGC 188, this reduces their estimated ages by about 30%.

Second, as mentioned previously, the fitting process is vulnerable to errors in assumed values of the abundance parameters Y and Z. Fortunately Z can usually be estimated spectroscopically, and recent work has shown that the age estimates for disk clusters are fairly insensitive to (reasonable) departures of Y from the standard disk-component value (**B2**, 193).

Third, almost all cluster CM diagrams show a larger spread in the main sequence, a *main-sequence band*, than is found in the Hyades. Observational errors and errors in interstellar-absorption corrections are inadequate to explain this spread in well-observed clusters; it is intrinsic. Part of the spread may result from a spread in the time of formation of individual stars in the cluster (see later discussions). Part may result from abundance variations from star to star. In some cases, stars may have suffered mass exchange in a close binary (a mechanism that may also explain the blue stragglers). Part of the spread may be the result of the operation of other factors, for example, rotation, magnetic fields, mass exchange, mass loss, or accretion. The width of the main-sequence band tends to make estimates of the age of a cluster uncertain. For instance, in the α Per cluster, estimates range from 2×10^7 to 7×10^7 years, and for the Pleiades, from 6×10^7 to 8×10^7 years. In some cases, these ranges can be narrowed by a star-by-star analysis.

Finally, special problems are encountered in the analysis of very young clusters and associations. Many of the stars contracting onto the main sequence are surrounded by dense circumstellar envelopes which contain absorbing particles. As a result, both their intrinsic luminosities and effective temperatures are difficult to estimate. Furthermore, there is clear evidence that star formation occurs more or less continuously over a finite period that amounts to a significant fraction of the entire age of a very young cluster. In a typical young cluster, we find stars on the ZAMS up to some maximum brightness, and, at lower masses, we find large numbers of stars still contracting onto the ZAMS, spread out to the right in the CM diagram. We can then make two different estimates of the cluster's age. On the one hand, we can determine a *nuclear age* t_n from the turnoff point at the bright end of the main sequence; t_n measures the time interval since the arrival of the brightest blue stars onto the ZAMS. On the other hand, we can determine a *contraction age* t_c from the luminosity of the faintest stars that are already on the ZAMS and have no counterparts of equal mass still contracting onto the main sequence. If the contraction of all protostars started simultaneously, we should find $t_c \approx t_n$; but generally we find $t_n < t_c$, which means that star formation has been proceeding continuously over a period of $t_c - t_n$. For example, in h and χ Persei, $t_n \approx 3 \times 10^6$ years and $t_c \approx 15 \times 10^6$ years, which implies star formation has occurred over some 10^7 years. A spread of this size obviously complicates the interpretation of CM diagrams of young clusters, but, if it is a representative number, it would be inconsequential for old clusters.

Bearing in mind the uncertainties and difficulties just discussed, let us now consider briefly some typical results. Evolution ages for a number of very young clusters and associations are given in (**S24**, 383), (**S23**). Some of the noteworthy ones are: Ori OB Id (Sword), (3–4) $\times 10^6$ years; Sco OB I, 4×10^6 years; Ori OB Ia, 10^7 years; h and χ Per, 1.5×10^7 years; Lac OB Ib, 2×10^7 years. Ages for young-to-intermediate clusters are: Pleiades, 6×10^7 years; Coma, 5×10^8 years; Hyades, 7×10^8 years. Ages for a number of intermediate-to-old clusters as determined by P. Demarque and R. D. McClure (**T2**, 199) and R. D. McClure and A. Twarog (**B2**, 193) are

Table 3-11. Ages and Metal Abundances for Disk Clusters

Cluster	Age (10^9 years)	$\delta(U - B)$	[Fe/H]
Hyades	0.7	0.0	0.20
NGC 2477	0.7	0.02	0.10
NGC 5822	0.9	0.03	−0.06
NGC 2360	1.3	0.04	−0.12
NGC 7789	1.6	0.08	−0.23
NGC 752	1.7	0.05	−0.18
NGC 3680	1.8	0.03	−0.06
M67	3.2	0.03	0.01
NGC 6819	3.5	0.06	−0.16
NGC 2243	3.9	0.11	−0.48
NGC 2506	4.0	0.13	−0.53
NGC 2420	4.0	0.10	−0.39
NGC 188	5.5	0.03	−0.06

SOURCE: (**B2**, 193)

listed in Table 3-11. Taken together, these results show clearly the enormous range in ages of the galactic clusters.

Perhaps the most important single result in Table 3-11 is the age of NGC 188, the oldest known disk cluster, which (assuming $Y \approx 0.3$) turns out to be $(5-6) \times 10^9$ years old. The isochrones and observed CM diagram for this cluster are shown in Figure 3-21. This estimate implies that NGC 188 is less than one-half as old as the globular clusters (see §3-10), and it implies that the galactic disk was formed much later than the spheroidal component (see also §4-5 and §6-2). This age for NGC 188 (and also for M67) is significantly smaller than was obtained in earlier work, for example, by Sandage and Eggen (**S4**), who found $(8-9) \times 10^9$ years for NGC 188 and $(5.5 \pm 0.5) \times 10^9$ years for M67. They concluded that the galactic disk is essentially as old as the spheroidal component. Virtually all of the reduction in the estimated age is a result of the increase in the Hyades distance modulus, which raises the luminosity assigned to stars in this cluster. At the higher luminosity, the observed main-sequence gap is in harmony with theoretical predictions, whereas previously the existence of a gap posed problems [see (**I3**, 613)].

Although NGC 188 is the oldest disk cluster, it does not necessarily contain the oldest disk stars. On statistical grounds, one might argue that the oldest disk cluster that we find might be only about half as old as the disk itself. On the other hand, the CM diagram for NGC 188 forms a clear lower envelope for late-type field giants and subgiants with reliably known distances in the solar neighborhood (**W8**). Hence it is likely as old as the disk unless all older disk stars are metal deficient by at least a factor of two relative to solar abundances [see (**B2**, 193) and (**T2**, 199) for further discus-

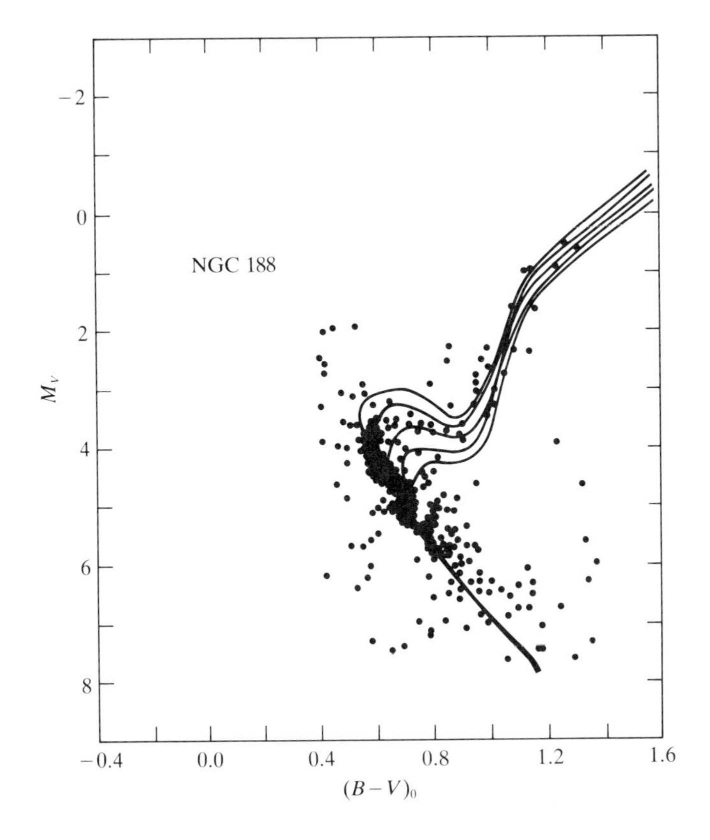

Figure 3-21. Observed CM diagram and theoretical isochrones for the old disk cluster NGC 188. The absolute-magnitude scale was determined by fitting to the Hyades main sequence [using the revised distance modulus $(m-M)=3.30$]. The isochrones are for models with (X, Y, Z) = (0.68, 0.30, 0.02) at ages (from top to bottom) of 4, 5, 7, 10, and 13 ($\times 10^9$) years. The main-sequence turnoff point indicates an age of (5–6) $\times$ 10^9 years. [From (**T2**, 199).]

sion of this point]. As we shall see here and in §4-5, there is evidence that the oldest disk stars do indeed have lower metal content, so the question is at present unresolved.

The question "How reliable are these ages (and those of the globular clusters)?" stimulates lively debate; probably it is unanswerable at present. There are many possible sources of error, both theoretical and observational, in the fitting procedure. These sources of error are usually seriously under-estimated when (formal) error estimates are assigned to published results [see (**A5**, Chapter 11) for an interesting discussion]. In practice, the changes that have been made in age estimates (for example, for NGC 188) as the fitting procedure has "improved" have been much larger than the error bounds quoted for any particular age determination. Whether the process has now converged and current estimates can be regarded as reliable is simply unknown.

Ages of Disk Field Stars

Under favorable circumstances, the ages of individual field stars can be determined from an analysis of their location in the theoretical H-R diagram, the (ΔM_{bol}, log T_{eff}) diagram, or the (log T_{eff}, log g) diagram, provided that their compositions are reasonably well known and that the evolutionary tracks from which the isochrones are constructed are not too complicated (**C4**, Chapter 17), (**P3**). Even at best, age estimates for individual stars are much more difficult to make (and are much less reliable) than are estimates of cluster ages, simply because much less information is available in the fitting procedure. However, despite the difficulties, such studies can yield key information and must be pursued.

Consider first the fitting of a single star in the H-R diagram or (ΔM_{bol}, log T_{eff}) diagram. (The latter is a plot of curves giving the difference in M_{bol} between points on an isochrone of specified age and the ZAMS at the same T_{eff}. These are sometimes called *curves of evolutionary deviation.*) To make the fit, we obviously require accurate values of T_{eff} and M_{bol}. The former can now be obtained fairly precisely from model-atmosphere analyses and the empirical scale. To obtain accurate luminosities, we need accurate distances, and this limitation restricts the sample to nearby stars having reliably known parallaxes. To avoid problems associated with uncertainties in stellar-model radii (and hence effective temperatures), late-type stars having convective envelopes must be excluded [see (**C4**, Chapter 11) for an illuminating discussion of this problem using the Sun as an example]. Thus, in practice, we can only hope to analyze F, G, and K dwarfs and subgiants, and even then, for the later-type stars, the results depend strongly on the assumptions made about the physics of convective transport.

If now we regard M_{bol} and T_{eff} as known, we can fit the data to isochrones, which will depend on two abundance parameters, say (Y, Z). We can determine Z from spectrophotometric analyses. Hence, for an assumed value of Y, we can fit the star's observed position and thus estimate its age and mass. It must be stressed that these results depend on the value of Y chosen, and they can be no more accurate than our estimate of Y is reliable.

An extensive study of this kind was made by Perrin et al. (**P3**) for about 140 nearby ($\pi'' \geq 0''.04$) F–K stars whose effective temperatures and metal abundances were derived from detailed spectroscopic analyses. Those authors deduced ages and masses assuming $X = 0.7$. They showed that, for $T_{eff} \lesssim 5500$ K, the observed width of the main-sequence band can be accounted for by the known variations of Z within the sample; for $T_{eff} \gtrsim$

6000 K, evolutionary effects dominate. They obtained masses that agree to within $\pm 10\%$ with "semi-astrometric" masses. (An error of ± 0.01 in Z would induce a $\pm 10\%$ error in their mass estimate.) They concluded that all the stars in their sample are old, having ages $\gtrsim 4 \times 10^9$ years.

Isochrone fitting in the H-R diagram can also be carried out for eclipsing spectroscopic binaries. Here one determines R and $\mathscr{M}$ from orbital and light-curve data. From a spectroscopic analysis, one can find T_{eff} and Z. Given T_{eff} and R, one can compute L. Thus no distance determination is needed, and the method can be applied to early-type stars, which are all too distant for parallax measurements. (But, it must be remembered that $L \propto T_{\text{eff}}^4$, so that very precise values of T_{eff} are needed.) For each star in the binary, one can then determine the age and one composition parameter (Y) because $\mathscr{M}$ is known. If the full set of data is available for both stars, one can demand that both stars fit the same isochrone (have equal ages). Because the masses are known, this requirement determines both Y and Z. In favorable cases [see, for example, (**A6**), (**A7**), (**A9**)], such analyses lead to a rather precise picture of the evolutionary status of the system, but in others they result in a conundrum that poses a challenge to the theory (**A8**).

Isochrones in the (log T_{eff}, log g) plane can be obtained from evolutionary tracks inasmuch as these explicitly give L and R as a function of time for models of known $\mathscr{M}$; a representative set (**C4**, Chapter 17) is shown in Figure 3-22. The advantages of this diagram are that it provides a fairly close connection between theory and observation because both T_{eff} and log g can be derived directly from a spectroscopic analysis and that it is not

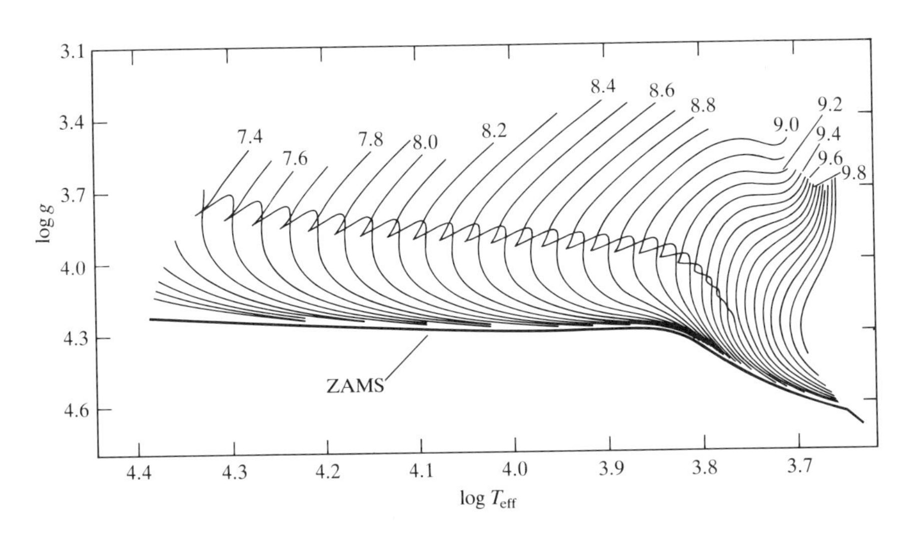

Figure 3-22. Representative theoretical isochrones in the (log T_{eff}, log g) diagram for models with $(X, Y, Z) = (0.70, 0.27, 0.03)$ and mixing length to scale-height ratio $(\ell/H) = 1.5$. The curves are labeled with log (age in years). [From **C4**, Chapter 17).]

necessary to know the distance to the star (or to calculate L from T_{eff}). It is found that the position of the ZAMS in this diagram for log $T_{eff} \gtrsim 3.85$ is almost independent of X and that decreasing Z by 0.01 increases log g for the ZAMS by about 0.06. Thus, in principle, the method is well suited to the analysis of individual stars, although in practice it is difficult to estimate log g accurately spectroscopically. For a single star, knowledge of T_{eff}, g, and Z yields the age and mass for an assumed value of Y. The derived age is reasonably independent of Y, but the derived mass is not. For an eclipsing spectroscopic binary, where masses are known in addition to T_{eff} and g, isochrone fitting of one component yields its age and helium content (if Z is known). If both components are used, and if we demand that they have the same age, we can determine both Y and Z. As was true for H-R diagram fitting, this method sometimes yields very convincing results (**C8**) and sometimes produces only a puzzle (**G7**). At least in some cases, the problems may be produced by mass loss and mass exchange between the binary members.

Time Variation of the Chemical Composition of Disk Stars

We have just seen how we can obtain estimates of the metal content and the age of disk stars, both in the field and in clusters. We are now in a position to pose the question "Is there evidence for a systematic variation of the metallicity of the material in the galactic disk with stellar age?" The answer to this question has been sought for at least two decades, but only now is it beginning to emerge clearly.

Without doubt, spectroscopic studies yield the most reliable abundances, but the sample of stars that has been studied in this way is not very large. If we wish to determine the average variation of the metal content of disk material with time, we must extend the analysis to as large a sample of stars as possible. This extension can be accomplished by using photometric metallicity estimates for field stars whose ages have been derived via the methods described earlier. The uncertainties in these estimates for individual stars are often comparable to the average variations being studied, and it is only because the present sample is large ($\sim 10^3$ stars) that definite, statistically significant trends begin to emerge. Several studies have been made, and a nice discussion by M. Mayor can be found in (**B2**, 213). We shall summarize the results by quoting estimates of the change $\Delta\langle[Fe/H]\rangle$ *per* 10^{10} *years.*

B. E. J. Pagel and B. Patchett (**P1**) used abundance estimates made by Eggen (**E5**), Powell (**P9**), and M. Mayor (**M9**) for various stellar age groups to construct Figure 3-23. Eggen's data for 160 F and G stars near the main sequence are divided into four age groups, and a mean value of [Fe/H] is assigned to each group using observed values of $\delta(U - B)$ in equation (3-42). Powell's $\delta(U - B)$ data for 95 stars with $\pi'' \geq 0''.04$ were divided into five groups and analyzed similarly. Mayor's Strömgren-photometry data, which

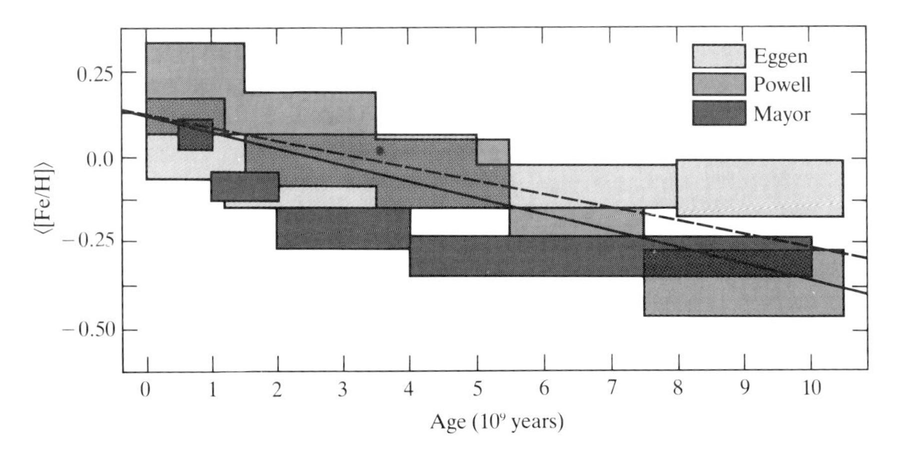

Figure 3-23. Variation of $\langle[\text{Fe/H}]\rangle$ with stellar age as derived from three photometric studies (**E5**), (**P9**), (**M9**) by Pagel and Patchett (**P1**). In the abscissa, each box spans the estimated age range of the group of stars that it represents, and, in the ordinate, it spans $\pm$ two standard deviations in [Fe/H] around its geometric mean. The *dashed line* shows $\Delta\langle[\text{Fe/H}]\rangle/\Delta(\text{age}) = -0.3/10^{10}$ years, and the *solid line* shows $\Delta\langle[\text{Fe/H}]\rangle/\Delta(\text{age}) = -0.5/10^{10}$ years. [Adapted from (**P1**), by permission.]

yield Δm_1 for 380 stars, are divided into four age groups and analyzed with equations (3-43). It is clear that, on the average, there is a decrease $\Delta\langle[\text{Fe/H}]\rangle$ of the order of 0.3 to 0.5. Using *uvby* data, R. E. S. Clegg and R. A. Bell (**C10**) estimate $\Delta\langle[\text{Fe/H}]\rangle \approx 0.8$, and Mayor (**M10**) estimates $\partial(Z/Z_{\odot})/\partial t \approx 0.6 \pm 0.3$, which is equivalent to $\Delta\langle[\text{Fe/H}]\rangle \approx 0.4$. From spectroscopic analyses, J. B. Hearnshaw (**H10**) estimates $\Delta\langle[\text{Fe/H}]\rangle \approx 0.8$, and Perrin et al. (**P3**) find 0.3. Less direct evidence comes from use of known correlations of [Fe/H] with kinematic properties, coupled with correlations of kinematic properties with age (see Chapter 7) to deduce an estimated variation of [Fe/H] with age. In this way, Mayor (**B2**, 213) obtains $\Delta\langle[\text{Fe/H}]\rangle \approx 0.7$ from data by K. A. Janes (**J1**), and 0.6 from data by M. Grenon (**B2**, 169).

Averaging over all estimates, we conclude that $\Delta\langle[\text{Fe/H}]\rangle/\Delta t \approx 0.5$ to 0.7 per 10^{10} years; *that is, the metallicity in the disk has increased by about a factor of three to five during its lifetime.* Lines with these slopes are shown in Figure 3-23, and it is seen that they are not inconsistent with the data.

Similar estimates can be made for galactic clusters, using the data in Table 3-11. Although the number of clusters for which ages and metallicities are known is relatively small, these data are inherently more reliable because the ages are determined by fits to whole CM diagrams, and the abundances are averages for several stars. On the other hand, it can be argued that the composition of the material in the oldest clusters could have been contaminated by metal-enriched material recycled from cluster stars themselves, perhaps at a very early epoch in the cluster's formation. Hence the composition may not be truly representative of the metallicity of the old disk material out of which they were formed. Using the results in Table 3-11, one finds that, with the exception of NGC 188 (which, though old, has an almost normal metal content), the data suggest a trend of $\Delta\langle[\mathrm{Fe/H}]\rangle \approx 0.5$ in 6×10^9 years or about 0.8 per 10^{10} years. But, as McClure and Twarog (**B2**, 193) emphasize, part of the trend reflects *spatial gradients*—a general decrease in [Fe/H] away from the galactic center and galactic plane (see Chapter 4)—and, when these are taken into account, the trend is weakened or even erased altogether. Further work on disk clusters in the solar neighborhood is needed to decide the issue.

All together, the bulk of the data indicate that the material from which the disk formed was already metal enriched to within a factor of two to five of present-day levels, in marked contrast to the material from which the spheroidal component formed (excluding the galactic bulge). The stars in the disk appear to have formed well after the formation of the spheroidal component, and, since the formation of the disk, its material appears to have undergone a progressive metal enrichment as a result of nucleosynthesis in stars. The results bear upon detailed theoretical models of the chemical evolution of our Galaxy, which we shall discuss in Chapter 19.

3-10. EVOLUTIONARY ANALYSIS OF SPHEROIDAL-COMPONENT STARS

Ages of Globular Clusters and Subdwarfs

When we compare theoretical isochrones with observed H-R diagrams of globular clusters, we find good basic agreement, and it would be reasonable to suppose that we should be able to assign cluster ages with fair confidence. But, in practice, a large number of difficulties arise, and current results are still subject to considerable uncertainties.

For instance, on the observational side, the absolute-magnitude scale in globular clusters is still uncertain. One can fix the scale, using the main-sequence fitting procedure described in §3-6 or by means of estimated absolute magnitudes of the cluster RR Lyrae variables (§3-8). Unfortunately, the two scales are not mutually consistent. Furthermore, on the theoretical side, we noted earlier that the radii of models with convective envelopes depend sensitively on convection theory. These difficulties can be surmounted partially by the use of the *relative* positions of features in the H-R diagram. For example, one can show that the difference in luminosity between the main-sequence turnoff point and the horizontal branch can be fitted only by theoretical isochrones in a limited age range. Next, we must admit that the value of the helium abundance in the spheroidal components is not accurately known, and we must regard it as one of the parameters to be derived. Naturally, this problem reduces the precision with which we can hope to determine ages. Furthermore, there are yet other abundance parameters (particularly CNO—see the following discussion) that can critically affect the structure of a star in certain phases, and these parameters are often unknown or poorly known. Finally, we must realize that many of the most prominent features that we wish to interpret represent stars in very advanced evolutionary phases, and we must remember that the models for these stars are subject to considerable uncertainty.

Faced with these difficulties, we must exploit the information available as fully as possible, and it is worthwhile to be aware of what kinds of information are provided by specific features in cluster H-R diagrams [see (**B2**, 133) for further details].

1. The location of the main sequence depends primarily on Y and Z. For-low-mass, metal-poor stars, the variation in luminosity produced by changes in (Y, Z) at fixed $T_{\text{eff}} \approx 6000$ K is

$$\Delta M_{\text{bol}} \approx 2.84 \Delta Y - 1.33 \Delta \log(Z + 0.001) \qquad (3\text{-}56)$$

2. The turnoff-point luminosity depends on the cluster age and on (Y, Z) through the position of the main sequence. It is insensitive to the treatment of convection, which affects mainly R, but not L, of a model.
3. The slope of the subgiant branch depends on stellar mass (hence age) and Y, being steeper for low-mass stars and large Y. It is almost independent of Z.
4. The luminosity of the subgiant branch depends fairly strongly on Y.
5. The location of the red-giant branch depends on mass (hence age) because Hayashi tracks move farther to the right for low-mass stars, and also upon (Y, Z) and (ℓ/H).
6. The slope of the giant branch depends mainly on Z (see §3-6).
7. The position of the horizontal branch depends on age, mass (after mass loss), Y, and Z. The distribution of stars along the HB depends on the variation in mass loss that occurs in pre-HB phases.
8. The separation in luminosity of the HB from the turnoff point depends primarily on cluster age.
9. Both evolutionary theory and pulsation theory yield information about RR Lyrae variables, and they often provide two independent estimates for the same parameter, thus affording a consistency check. Bounds can be placed on masses, luminosities, and Y for these stars. In particular, the blue edge of the instability strip depends sensitively on Y.

10. Comparison of observed and theoretical luminosity functions for stars at various positions along the cluster locus provides estimates of age and Y.
11. Finally, theoretical lifetimes of stars in different evolutionary stages can be compared with observed number ratios in those stages. In particular, the relative numbers of red giants and HB stars is sensitive to Y.

Armed with these tools, one can attempt to estimate globular-cluster ages and helium abundances. We consider the latter first. From the pulsation properties of the RR Lyrae stars, Sandage (**S2**) estimated $Y \approx 0.3$ for M3, M13, M15, and M92. More recent estimates, based on better theory, (**I4**) yield $Y \approx 0.22$ for M3 (see §3-8). Variations in the pulsation properties of cluster variables suggest variations of the order of $\Delta Y \approx 0.02$ from cluster to cluster. Using luminosity functions, several authors have estimated $Y \approx 0.2$, but the precision of these estimates is not high. Another approach is to compare observed and predicted properties of HB stars in the ($\log T_{\text{eff}}$, $\log g$) plane. This method is sensitive to Y and insensitive to uncertainties in the other model parameters (for example, core mass and CNO abundance). Present estimates give $Y \gtrsim 0.3 \pm 0.04$. Comparisons of the predicted number ratio of red giants to HB stars with the observed values give estimates of Y that range from 0.16 to 0.28. A serious problem with this approach is that it is difficult to disentangle asymptotic giant-branch stars from stars on the first giant branch observationally; hence the observed number ratio is relatively rough. Finally, we recall that the position of the subdwarfs in the H-R diagram yielded $Y \gtrsim 0.23$. Overall, it seems safe to conclude that, in the spheroidal component, $Y \gtrsim 0.2$, but it is not as high as in the disk. A reasonable estimate is $Y \approx 0.25 \pm 0.03$.

Let us now consider age estimates. The essential difficulty always encountered here arises in the choice of an absolute-magnitude scale for the cluster. Differences among models also introduce appreciable (but less serious) uncertainties. For example, using the absolute magnitudes for cluster variables that are obtained from a force fit to theoretical predictions about their dependence on Z (as described in §3-8), and adopting $Y = 0.3$, Sandage derived (**S2**) the following ages for four globular clusters: M3, $(11\text{–}13.5) \times 10^9$ years; M13, $(9.5\text{–}11.5) \times 10^9$ years; M15, $(9.5\text{–}11.5) \times 10^9$ years; M92, $(9.5\text{–}11.5) \times 10^9$ years. The age spread quoted is for two different sets of models. These results imply that all the globular clusters have essentially the same age.

In contrast, fixing $M_V(\text{RR}) = +0.6$, taking $Y = 0.2$, and using isochrones computed for the observed values of Z, P. Demarque and R. D. McClure (**T2**, 199) have recently derived the following results, which are essentially independent of the adopted value of Y: M92, $(14\text{–}16) \times 10^9$ years; M15, $(14\text{–}16) \times 10^9$ years; M13, $(13\text{–}15) \times 10^9$ years; M3, $(13\text{–}15) \times 10^9$ years; 47 Tuc, $(10\text{–}11) \times 10^9$ years. A typical fit (for M92) is shown in Figure 3-24.

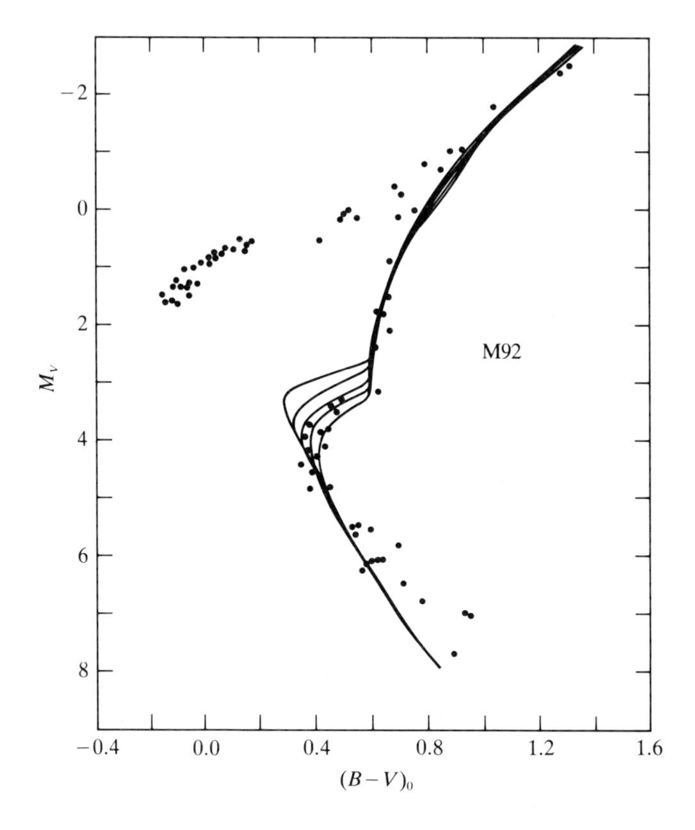

Figure 3-24. Observed CM diagram and theoretical isochrones for the globular cluster M92. The absolute-magnitude scale was fixed by setting $M_V(\text{RR}) = +0.6$. The isochrones are for models with $(X, Y, Z) = (0.80, 0.20, 1 \times 10^{-4})$ at ages (from top to bottom) of 10, 12, 14, 16, and 18 ($\times 10^9$) years. The turnoff point indicates an age of $(14\text{–}16) \times 10^9$ years. [From (**T2**, 199).]

These last results indicate that there is an appreciable spread in the ages of globular clusters. For 47 Tuc, F. D. A. Hartwick and J. E. Hesser (**H6**) estimate an age of 10.5×10^9 years if $Y \approx 0.3$, or 13.5×10^9 years if $Y \approx 0.2$ (substantially larger than the values just quoted). From the slope of the subgiant branch, they estimate $Y \approx 0.23 \pm 0.1$.

For an adopted value $Y = 0.25$, it appears that all of the preceding clusters can be fit to isochrones having $T = 12.5 \pm 1.5$ billion years. But, because neither of the absolute-magnitude scales in this discussion agree with that obtained from the basic main-sequence fitting technique (see §3-6), which would yield yet another set of cluster ages, we must admit there is still considerable uncertainty about cluster ages, despite the considerable effort that has been devoted to determining them. Certainly, the present evidence will not support the claim (**S2**) that we can limit the spread in cluster ages to $(\Delta T/T) \approx 0.014$ or even that we know ages to ± 1 or 2 billion years. Such claims are obviously a strong function of the assumptions used to derive ages, and their apparent precision is illusory. On the other hand, the conclusion (**C4**, Chapter 11) that all we can say is that cluster ages lie in the range of 8×10^9 to 18×10^9 years seems pessimistic. The situation will improve fundamentally only when we can observe, with very high accuracy, down to the main sequence in a large sample of globular-cluster CM diagrams. As mentioned in §3-8, the Space Telescope should enable us to do this.

Little work has been done on age determinations for field subdwarfs, mainly because their luminosities are relatively uncertain. Perrin et al. (**P3**) have analyzed nine extreme subdwarfs. Their work shows conclusively that the hotter subdwarfs (for example, HD 19445 and HD 140283) with $T_{\text{eff}} \approx 6000$ K are appreciably evolved. But their age estimates for these stars are alarmingly large ($\gtrsim 25 \times 10^9$ years if $Y \approx 0.3$), and they are in conflict with the expansion age of the Universe. Part of the problem may be that their chosen value for Y is too large. It would be extremely helpful to know the masses of these stars!

In addition to the difficulties just described concerning cluster age dating, there are yet other problems. For example, it has been known for some time that, in ω Cen, the color spread on the giant branch at a given luminosity is far too large to be explained by photometric errors or by allowable spreads in stellar mass or Y. The only remaining interpretation is that there must be a spread, from star to star in the cluster, in the abundances of one or more of the heavier elements (**I4**). This interpretation has been shown to be correct by direct spectrophotometric analyses (**B14**), (**D4**), (**F4**), (**M6**), (**N4**). Furthermore, narrow-band photometric analyses [see, for example, (**H15**), (**H16**), (**M1**), (**M5**)] of red giants in several other globular clusters have shown conclusively that there are major variations in the CN-band strengths from star to star within a cluster. These may result from large differences in the primeval (C/O) ratio among cluster stars, from abundance gradients within the cluster that result from successive waves of star formation, or, more likely, from a sporadic variation from star to star of the mixing of internally produced elements into the envelope. In any case, the basic conclusion that *stars at a given location in a globular-cluster CM diagram do not necessarily all have identical structural or chemical-evolutionary histories* seems inescapable. It is known that red-giant models are very sensitive to assumed CNO abundances. Thus the large variations observed in these abundances raise some very discomforting questions about our basic interpretive techniques for cluster CM diagrams.

In summary, it seems that, despite our best efforts, at present we have only a rudimentary understanding of the evolutionary status of globular clusters. Our ideas about them will be substantially revised by future work.

Time Variation of the Chemical Composition of Globular Clusters

Having discussed estimates of the age and metal content of globular clusters, we may now ask, "Are they correlated?" The answer to this question is of tremendous importance because, when coupled with information about the time variation of the metal content within the disk component (§3-9) and about spatial gradients of metallicity within our Galaxy (§4-3 and §4-4), it helps us to build a picture of the chemical evolution of our Galaxy.

It is interesting to note how much the accepted opinions on the question of the time variation of the metal content of galactic material have altered in the past two decades. For example, in 1962, Eggen, Lynden-Bell, and Sandage (**E7**) argued on kinematic grounds that, in its early history, our Galaxy must have undergone a very rapid collapse. They estimated the collapse time to be about 2×10^8 years. They suggested that, during this collapse, the spheroidal-component stars were formed, evolved, and progressively enriched the infalling material, which, at the end of the collapse, resulted in the galactic disk. In later work, Eggen and Sandage (**E9**) concluded that (1) all globular clusters have about the same age, but their metal content rose extremely rapidly during the rapid collapse; (2) the oldest galactic cluster, NGC 188, is as old as the globulars but has normal metallicity, which implies that the material was fully enriched during the rapid collapse; and (3) the metallicity of disk material has not been significantly increased since disk formation, although there are random variations of metallicity by a factor of two around the solar metal abundance.

On the basis of the information now available, the picture could look rather different.

1. As we have just seen, the most recent results may indicate a significant spread in globular-cluster ages. If we accept these ages, we would find a clear correlation of metallicity with age [see (**A10**), (**B2**, 193)], with [Fe/H] rising from about -2 for clusters 15×10^9 years old to about -0.4 for a cluster 11×10^9 years old. Furthermore, they would imply a slow halo collapse, extending over billions of years.
2. As we saw in §3-9, there is some evidence that the oldest objects in the disk could be as little as half the age of the globular clusters. This evidence seems to indicate that the disk formed well after the initial spheroidal-component collapse. If this is true, then, in the time interval between the initial halo collapse and the settling of material into the disk, there would be time for a substantial amount of metal-rich material to be "rained" into the disk from dying stars.
3. The material in the disk at its earliest epochs of star formation was already fairly metal-rich. But, in addition, there has been a progressive metallicity enrichment by about a factor of three during the lifetime of the disk. Furthermore, despite the uncertainties in

our present estimates of Y for the spheroidal component, it seems likely that its helium content ($Y \approx 0.22$–0.25) is in fact slightly lower than that of the disk component ($Y \approx 0.28$). It thus appears that at least some helium enrichment of the interstellar material in the disk has resulted from stellar nucleosynthesis. Over and above these time variations, we shall see in the next chapter that there are clear spatial variations of the metallicity within both the disk and the spheroidal component.

It would obviously be premature to assert that the picture we have just described is correct, but it does account for the available observations satisfactorily. Furthermore, we shall see in Chapter 19 that models of this type account nicely for important additional evidence concerning the chemical evolution of galaxies that is reviewed in later chapters. However, no model of the chemical evolution of the Galaxy can be more secure than the interpretation of the observations on which it is founded—and we have seen that the relative ages and metallicities of globular-cluster and field halo stars are still very insecurely determined. Therefore, until more conclusive observations are available, it would be wrong to commit oneself to any particular theory of galactic chemical evolution.

3-11. INTERSTELLAR ABSORPTION

The Interstellar Medium

Since the late nineteenth century, there has been an accumulation of evidence pointing to the existence of an *interstellar medium* (*ISM*), that is, material distributed throughout the space between stars. The ISM is composed of *gas* (molecules, atoms, ions) and *dust* (small grains of material made of ices of various kinds, graphite, silicates, and possibly metals). The gas reveals its presence through characteristic absorption lines, both optical and radio, seen in the spectra of stars and radio-continuum sources and through emission features, ranging from optical emission lines emitted by diffuse nebulae (H II regions in the vicinity of hot stars) to radio emission in numerous molecular lines, the H I 21-cm line, radio recombination lines, and the radio continuum. Dust appears prominently in dark absorbing nebulae and in reflection nebulae, which shine by light from a star near them reflected off its grains. The presence of a pervasive general absorbing medium in the galactic disk was demonstrated by Trumpler, who showed that it produced both a systematic dimming and reddening of the light from distant stars. Roughly speaking, there is an absorption of about 1 mag/kpc and a reddening in $(B - V)$ of about 0.3 mag kpc^{-1}.

A vast amount of both theoretical and observational study in the X-ray, UV, optical, IR, and radio spectral regions has been devoted to the ISM, and we now have a reasonably coherent picture of its composition and physical properties [see Spitzer's excellent monographs (**S21**), (**S22**) for a summary]. In broad terms, we know that the gaseous material in the ISM comprises at least five distinct components:

1. Very cold, dense molecular gas ($T \approx 20$ K, $n > 10^3$ cm^{-3}) that is distributed in giant molecular clouds. These clouds contain an appreciable fraction of the galactic interstellar medium, although they occupy only a very small fraction ($<1\%$) of interstellar space (**S13**).
2. Cold gas ($T \approx 100$ K, $n \approx 20$ cm^{-3}) that is predominantly in neutral atomic form and is distributed in relatively dense clouds, which occupy about 2% to 4% of the volume.
3. Hot neutral gas ($T \approx 6 \times 10^3$ K, $n \approx 0.3$ cm^{-3}) that envelops the cooler interstellar clouds and fills about 20% of space.
4. Hot ionized material ($T \approx 8 \times 10^3$ K, $n > 0.5$ cm^{-3}), exemplified by the material of ordinary H II regions that surround groups of hot stars. This component may occupy as much as 10% of interstellar space.
5. A very hot, low-density medium ($T \approx 10^6$ K, $n \approx 10^{-3}$ cm^{-3}), which is gas that has been heated by blast waves from supernovae and occupies perhaps 70% of the volume [see (**F3**), (**M3**)].

All these components, except the first (each cloud of which is probably probably gravitationally bound) and the fourth (which is expanding into the volume occupied by the other components) exist in approximate pressure equilibrium with one another. The gaseous material is composed of a wide variety of constituents, ranging from simple particles such as free electrons, protons, neutral H atoms, and H_2 molecules to fairly complicated organic molecules. The dust grains have also been studied extensively (**F2**), (**W6**), and questions concerning their formation and composition present many fascinating problems in solid-state physics.

Despite the great inherent interest of the problems encountered in analyzing the physical properties of the ISM, we shall not be able to discuss them in this book. Rather, we shall focus on the absorption and reddening effects produced by dust grains because these affect stellar-distance estimates and hence star-density analyses, which makes these effects of primary importance to galactic-structure research. We therefore concentrate exclusively on the question of correcting for absorption and reddening observationally.

The Absorption Law

Trumpler showed that interstellar reddening (measured in magnitudes) is approximately described by the law $r_\lambda = a + (b/\lambda)$. If the dust grains were large compared with a wavelength of light, then the absorption would be neutral, that is, independent of wavelength. If the grains were of molecular size, their reddening effects would follow a *Rayleigh scattering law*, $r_\lambda = c + (d/\lambda^4)$. The observed λ^{-1} variation implies that the particles fall between these two extremes and have a grain size of the order of 10^{-5} cm, roughly the size of a wavelength of light. The average mass density associated with the grains is about 10^{-26} g cm^{-3} in the galactic plane, about 1% of the average mass density of the gas.

Suppose the grains have a *number density* $n(x)$(cm^{-3}), where x measures the position along a line of sight from the observer, and an average *extinction cross section* per particle k_λ(cm^2) at wavelength λ. Then, a beam of light of incident intensity I_λ passing through a slab of thickness dx suffers an intensity drop dI_λ given by

$$dI_\lambda = -I_\lambda n(x) k_\lambda \, dx = -I_\lambda \, d\tau_\lambda \tag{3-57}$$

where $d\tau_\lambda \equiv n(x) k_\lambda \, dx$ is an increment of *optical depth* along the path. Integrating equation (3-57), we find that the intensity I_λ received by an observer from a star that emits an intensity I_λ^0 is

$$I_\lambda = I_\lambda^0 e^{-\tau\lambda} \tag{3-58}$$

where

$$\tau_\lambda = k_\lambda \int_0^x n(x') \, dx' = k_\lambda N(x) \tag{3-59}$$

is the total optical depth between the observer and the star at wavelength λ. $N(x)$ in equation (3-59) is the *column density* of absorbing grains along the line of sight. The optical depth is clearly just a measure of the total absorptivity of the material along the line of sight.

If we express the absorption suffered by the starlight in magnitude units, we find

$$A_\lambda \equiv -2.5 \log_{10}(I_\lambda/I_\lambda^0) = -2.5(\log_{10} e)\ln(e^{-\tau_\lambda}) = 1.086\tau_\lambda \tag{3-60}$$

That is, the absorption in magnitudes is numerically almost equal to the optical depth along the line of sight—a handy result to remember. Inasmuch as $N(x)$ is a fixed number for a given star, it is obvious from equations (3-60) and (3-59) that the variation of A_λ with λ reflects faithfully the variation of k_λ with λ. The absorption is largest in the ultraviolet and smallest in the infrared. As we shall see, it is because A_λ depends on λ that we can detect it at all and hence correct for it. A corollary is that, if there is a strictly neutral absorption component present, we could be completely unaware of it.

As a first crude model of the distribution of absorbing material, let us suppose that it is in a homogeneous, finite layer of equal thickness above and below the galactic plane; a representative thickness is ± 150 pc. Suppose that, at some specified wavelength, the optical thickness of half of this layer, measured in the direction perpendicular to the galactic plane, is τ_1 (see Figure 3-25). Then, light from a distant source—for example, an external galaxy—at galactic latitude b would have to pass through an optical thickness

$$\tau(b) = \tau_1 \csc b \qquad (3\text{-}61)$$

and hence its apparent magnitude as a function of latitude would increase by an amount

$$\Delta m(b) = 1.086\tau_1 \csc b = \Delta m(90^\circ) + 1.086\tau_1(\csc b - 1) \qquad (3\text{-}62)$$

Hubble realized that, with this model, one could explain the observed variation of the number of external galaxies ("extragalactic nebulae") counted per square degree down to some limiting apparent magnitude, and, in particular, he could explain why there is a *zone of avoidance* near the galactic plane [because clearly $\Delta m(0) \to \infty$ in this model]. If galaxies are uniformly distributed in space, and if there is no interstellar absorption, then $N_0(m)$, the number that can be counted down to a limiting apparent magnitude m, will be given by the relation [see equation (4-8)]

$$\log N_0(m) = 0.6m + C \qquad (3\text{-}63)$$

But galaxies that would have appeared at magnitude m_0 in the absence of absorption will appear only at magnitude $m(b) = m_0 + \Delta m(b)$ when absorption is present. Thus the number we can actually count down to a given magnitude m will necessarily be smaller than the true number. If $N(m, b)$ is the observed number at latitude b down to magnitude m, including absorption, then, from equations (3-62) and (3-63), we have

$$\log N(m, b) = \log N_0(m) - 0.6\Delta m(b) \qquad (3\text{-}64a)$$

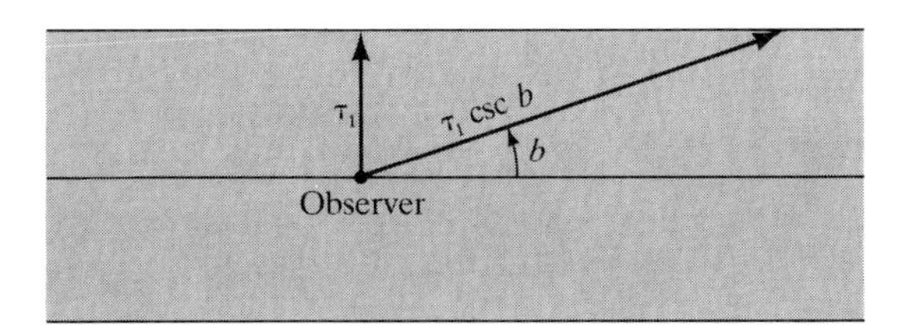

Figure 3-25. Optical depth in a homogeneous absorbing layer in the galactic disk having an optical half thickness τ_1 measured perpendicular to the galactic plane.

or

$$\log N(m, b) = \log N(m, 90^\circ) - 0.6[\Delta m(b) - \Delta m(90^\circ)]$$
$$= \log N(m, 90^\circ) - 0.6[1.086\tau_1(\csc b - 1)] \qquad (3\text{-}64b)$$

or, yet more simply

$$\log N(m, b) = A - B \csc b \qquad (3\text{-}65)$$

The coefficient B in equation (3-65) can be determined empirically from observed galaxy counts; knowing B we can immediately compute τ_1. In the V band, one finds $(\tau_1)_V \approx 0.36$ mag, and in the B band, $(\tau_1)_B \approx 0.47$ mag.

Given a value for τ_1, equation (3-62) can then be used to derive rough statistical corrections for the absorption suffered by the light from objects well outside the galactic plane—for example, globular clusters and external galaxies—when no other estimates are feasible. However, it is well known, from studies such as those by Shane and Wirtanen at Lick Observatory, that the absorption in the disk is extremely patchy and that the corrections predicted by equation (3-62) can be very rough indeed. They should therefore be applied with caution. For example, the region near the north galactic pole is apparently relatively "cloudless," with a total absorption certainly less than 0.3 mag, and possibly even unobservably small. Furthermore, the estimates provided by equation (3-62) are worthless for objects inside the layer, because then we must know what fraction of the layer the line of sight to the object traverses. We must therefore confront the problem of determining directly the absorption for any particular object that we observe.

Interstellar Reddening

One of the enigmatic problems of stellar spectrophotometry in the 1920s was the observation that some stars having O- and B-type spectra have red colors characteristic of much later spectral types. This problem posed the puzzling question "How does a star that appears to have a high atmospheric temperature, as judged by the degree of excitation and ionization of the material, radiate a continuous spectrum that is appropriate to a much cooler temperature?" Once the existence of interstellar absorption was recognized, the mystery evaporated and the answer emerged clearly: It doesn't! It radiates the same continuous spectrum as any other star of its type. But the light we receive is reddened by selective absorption processes in the interstellar medium, which dim the light more efficiently at short wavelengths than at long wavelengths. The process is very similar to the one that reddens the sun and moon when they are seen near the horizon on hazy, dusty, or polluted days. And therein we find the key to a diagnostic procedure for the determination of the amount of interstellar absorption: Granting that stars with identical spectra all have a unique intrinsic color, then from any particular star's observed color we derive the amount of reddening the light has suffered, and from that information, the amount of absorption itself.

Suppose that, at some wavelength λ_i, the amount of interstellar absorption is $A(\lambda_i)$ mag. Then the observed magnitudes m_i and m_j at two different wavelengths λ_i and λ_j are related to the intrinsic magnitudes m_i^0, m_j^0 by the expressions

$$m_i = m_i^0 + A(\lambda_i) \tag{3-66a}$$

$$m_j = m_j^0 + A(\lambda_j) \tag{3-66b}$$

Thus the *observed color index* $C_{ij} \equiv m_i - m_j$ is related to the *intrinsic color index* $C_{ij}^0 \equiv m_i^0 - m_j^0$ by

$$C_{ij} = C_{ij}^0 + [A(\lambda_i) - A(\lambda_j)] \equiv C_{ij}^0 + E_{ij} \tag{3-67}$$

where E_{ij} is called the *color excess.* In the UBV system, the customary notation for color excesses is

$$E(B - V) \equiv (B - V) - (B - V)_0 \tag{3-68a}$$

$$E(U - B) \equiv (U - B) - (U - B)_0 \tag{3-68b}$$

where the unadorned color indices denote observed values, and those with subscript zero denote intrinsic values. Because $A(\lambda_i) > A(\lambda_j)$ for $\lambda_i < \lambda_j$ (at least in the visible spectrum—see Figure 3-28), E_{ij} is positive; that is, the colors defined in the usual way become redder in the presence of interstellar absorption.

Interstellar reddening effects, measured by color excesses, can be determined directly from observation. The classic studies in the UBV system are those by Hiltner, Johnson, and Morgan **(J5)**, **(H18)**. These authors analyzed a large sample of O–B stars, which are ideal for determining reddening effects because (1) they are highly luminous and hence can be seen to large distances, over which the reddening can accumulate to large values; (2) they are intrinsically blue and hence strongly susceptible to reddening; and (3) they have distinctive spectra that can be classified with high precision. Accurate MK spectral types were determined for all stars in the sample, and their UBV colors were measured. Figure 3-26 shows how stars of identical MK type are distributed along a *reddening line* in the two-color diagram. The bluest stars on this line are taken to be unreddened, and these define the intrinsic two-color sequence given in Table 3-3 and Figure 3-7. More reddened stars move down the line to larger values of $(U - B)$ and $(B - V)$. This procedure can be applied to a group of stars of any spectral type (so

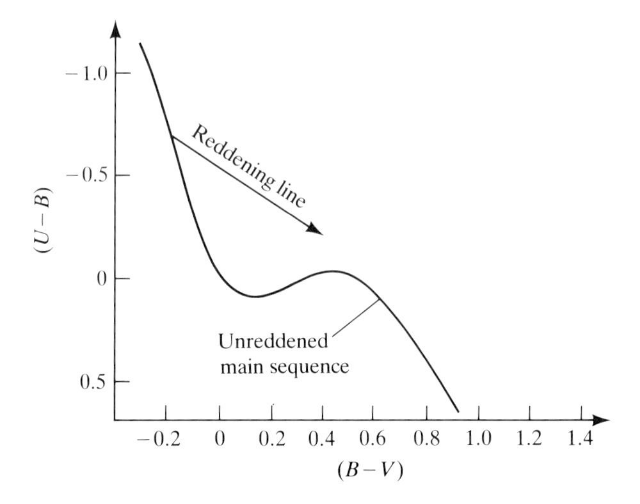

Figure 3-26. Effects of interstellar reddening in the *UBV* system two-color diagram.

long as it is known in advance) to determine both the intrinsic color and the reddening line for the type.

Of course, the reddening lines for stars of differing MK types originate at different points in the two-color diagram, but they are all found to have practically identical slopes. From a careful analysis of the data for stars of types O through B9, the slope of the reddening line is found to be given by

$$\frac{E(U - B)}{E(B - V)} = 0.72 + 0.05E(B - V) \tag{3-69}$$

The second term on the right-hand side of equation (3-69) is small and can usually be neglected. It arises mainly because the changes in shape of the stellar energy distribution produced by reddening change the effective wavelengths of the filters.

Thus far, we have assumed that the spectral types of the stars under study are known, so that we can determine color excesses by comparing the observed colors of each star with those of an unreddened star of the same (known) type. However, in many cases, we do not have spectral types for the stars we wish to study. Nevertheless, given the slope of the reddening line, it is possible to define a photometric parameter that depends only on the spectral type of a star and is independent of the amount of reddening. In the *UBV* system, this parameter is

$$Q \equiv (U - B) - \frac{E(U - B)}{E(B - V)}(B - V) \tag{3-70}$$

or

$$Q = (U - B) - 0.72(B - V) \tag{3-71}$$

If we substitute equations (3-68) into (3-70), we can verify immediately that Q is indeed reddening independent, because then

$$\begin{aligned} Q &= (U - B)_0 + E(U - B) - \frac{E(U - B)}{E(B - V)}[(B - V)_0 + E(B - V)] \\ &= (U - B)_0 - \frac{E(U - B)}{E(B - V)}(B - V)_0 \equiv Q_0 \end{aligned} \tag{3-72}$$

Values of Q for stars of spectral types O through B9 are listed in Table 3-12. The importance of this parameter is that, for early-type stars, it uniquely determines a star's intrinsic color from photometric data alone without the need for spectra. Furthermore, Q can be measured easily for stars that are much too faint for spectral classification. From the data given in Tables 3-3 and 3-12, one finds that, to a good approximation,

$$(B - V)_0 = 0.332Q \tag{3-73}$$

To summarize, we can determine $(B - V)_0$ for early-type stars from Q. Once $(B - V)_0$ and $(U - B)_0$ are known, we find $E(B - V)$ from $(B - V)$, and, as we shall see later, the visual absorption A_V (or A_λ at any other λ) follows from $E(B - V)$.

For spectral types later than A0, Q is not a unique function of spectral class [see (**S24**, Chapter 13) for the full dependence of Q on spectral type], and hence it is no longer useful in estimating the amount of reddening present.

Table 3-12. Q Versus Spectral Class for Early-Type Stars

Spectral type	Q	Spectral type	Q
O5	−0.93	B3	−0.57
O6	−0.93	B5	−0.44
O8	−0.93	B6	−0.37
O9	−0.90	B7	−0.32
B0	−0.90	B8	−0.27
B0.5	−0.85	B9	−0.13
B1	−0.78	A0	0.00
B2	−0.70		

SOURCE: (**J5**), by permission. Copyright © 1953 by the University of Chicago.

This failure occurs because the reddening line happens to have almost the same slope as the unreddened main-sequence curve for late-type stars, shown in Figure 3-26, which means that it is nearly impossible to tell (from *UBV* data alone) whether a given star has been reddened or is unreddened but of a later type. Nevertheless, even for late-type stars, Q is still reddening free, and, as mentioned in §3-6, it provides a useful metallicity indicator for globular clusters.

Finally, it is worth noting that reddening-free indices can be defined in other photometric systems as well. For example, in the Stromgren *uvby* system, we can define

$$[c_1] \equiv c_1 - 0.2(b - y) \tag{3-74a}$$

$$[m_1] \equiv m_1 + 0.18(b - y) \tag{3-74b}$$

$$[u - b] \equiv (u - b) - 1.84(b - y) = [c_1] + 2[m_1] \tag{3-74c}$$

from which intrinsic stellar properties can be inferred in certain ranges of spectral types.

Let us now turn to the primary problem at hand, namely, that of determining the total absorption, in magnitudes, in some photometric band—say A_V in the visual band—directly from observed data. From equations (3-67), (3-60), and (3-59), we see that

$$R_V \equiv \frac{A_V}{E(B - V)} = \frac{\tau_V}{(\tau_B - \tau_V)} = \frac{k_V}{(k_B - k_V)} \tag{3-75}$$

This result tells us that the absorption A_V is proportional to the color excess $E(B - V)$ or, indeed, any other color excess, and that the value of the constant of proportionality R_V is fixed by the wavelength dependence of the extinction coefficient. Our goal is to determine R_V. Once we know R_V, then an observational determination of $E(B - V)$ for any star immediately yields A_V. An assumption we shall make for the present is that the physical properties (hence k_λ) of the interstellar material are everywhere the same in our Galaxy, so that a unique value of R_V actually exists. We can check this assumption later. There are basically two approaches to determining R_V.

Extinction-Curve Method Ideally, we wish to determine not just A_V but the absorption A_λ, as a function of λ over the entire spectrum, that is, the complete interstellar extinction curve. To do this, we compare, at several wavelengths, the energy distribution of a reddened star with that of another star that is assumed to be unreddened and of identical spectral type and hence is assumed to have the same intrinsic energy distribution. The energy distributions can be compared using either spectrum-scanner data or filter-photometry measurements. For expository convenience, we shall assume that the measurements have been made using the 10-color Johnson system described in Table 2-5.

From a comparison of the colors of the chosen pair of stars, we derive a set of color excesses for example, $E(U - V)$, $E(B - V)$, $E(I - V)$, $E(J - V)$, ..., $E(N - V)$—all relative to a single reference waveband, V. For any particular star, the observed color excesses will be proportional to the number of absorbing grains along the line of sight [see equations (3-67), (3-60), and (3-59)]. This number will naturally vary from star to star. We must therefore reduce the observations to some standard optical depth. It is customary to do this by normalizing $E(B - V)$ to unity and scaling the color excesses in all the other bands accordingly. For each star in our sample, we then plot the normalized values of $E(X - V)$, where X stands for any one of the measured photometric bands, against $1/\lambda$, the reciprocal of the corresponding effective wavelength of that band. Typically we obtain a curve like the one shown in Figure 3-27.

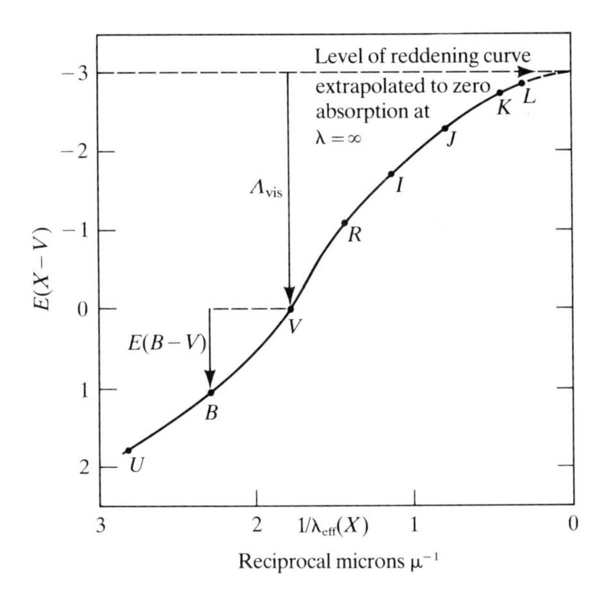

Figure 3-27. The interstellar absorption curve in the visible and infrared. The data are for the Perseus region as measured by H. L. Johnson (**J3**). The observed color excesses are scaled so that $E(B - V) = 1$ mag. The resulting curve is extrapolated to infinite wavelength ($1/\lambda = 0$), where the absorption is presumed to be zero. The absorption at any wavelength λ, per unit $(B - V)$ color excess, can be read directly from the curve. For example, in the visible, $A_V = 3$ mag. Thus, in this region of the sky, $R_V = A_V/E(B - V) = 3$. [Adapted from (**J3**). by permission. Copyright © 1965 by the University of Chicago.]

We must now set the zero-point of the absorption scale. To do this, we argue that, as $\lambda \to \infty$ (so that $1/\lambda \to 0$), we expect $A_\lambda \to 0$ because the small interstellar grains are physically unable to absorb radiation whose wavelength is much larger than grain size. Of course, in practice, we cannot observe to $1/\lambda = 0$, and to estimate the zero-point of the scale, we must extrapolate the available data. To minimize the uncertainties in the extrapolation, it is crucial to push the photometry as far into the infrared as possible. An example of the procedure is illustrated in Figure 3-27.

Having fixed the zero-point, we see that the absorption A_λ [per unit $E(B - V)$] at any λ is just the vertical distance in magnitudes from the level of zero absorption to the appropriate point on the curve. Thus, for the star illustrated in Figure 3-27, we see that, in the V band, $A_V = 3$. Hence for this absorption curve, we find $R_V = A_V/E(B - V) = 3.0$.

One of the classic studies of the interstellar absorption curve was made by A. E. Whitford (**W5**), who used spectrum-scanner data extending to 2μ. From observations in several regions of the sky, he established an average

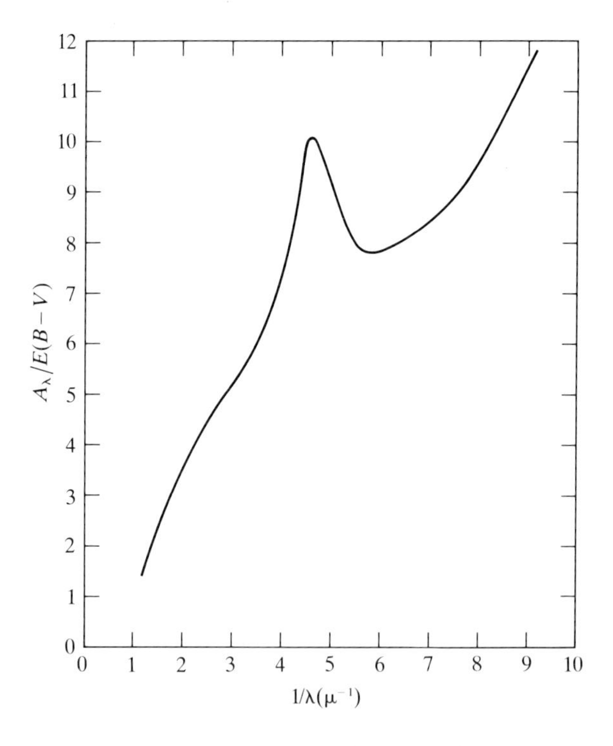

Figure 3-28. The mean interstellar absorption curve in the visual and ultraviolet [normalized to $E(B - V) = 1$] as determined from OAO-2 observations. [From (**B8**), by permission. Copyright © 1972 by the University of Chicago.]

absorption curve quite similar to that shown in Figure 3-27. In later work, H. L. Johnson (**J3**) extended the coverage to 10μ using broad-band photometry, which made possible a much more reliable extrapolation of the data to $1/\lambda = 0$. Yet more recently, it has been possible to determine the interstellar absorption curve in the ultraviolet using measurements made from space observatories. A mean curve obtained from OAO-2 data is shown in Figure 3-28. Notice the very high absorption peak near $\lambda 2150$ Å. This peak has been variously attributed to absorption by graphite or silicates or small oxide grains, according to different views about the composition of the grains [see, for example, (**B8**), (**D5**), (**F2**, 89)].

Cluster Method When we examine stars in a cluster, we know that they are all at essentially the same distance from us, and we expect $V - M_V$ to be the same for every cluster star except for differences produced by variations in the amount of interstellar absorption along the line of sight to each star. That is, we should find

$$V - M_V = C + A_V = C + R_V E(B - V) \tag{3-76}$$

where C is a constant for the whole cluster, depending only on the cluster's distance, and A_V is the total absorption suffered by the light from each cluster star. Variations in A_V can result from changes in the amount of absorption along the line of sight both between the observer and the cluster and within the cluster itself.

To find R_V, we determine the MK spectral types and UBV magnitudes of several cluster stars. The MK types yield M_V and $(B - V)_0$ directly from the calibrations given in Tables 3-2 and 3-3. We then compute $(V - M_V)$ and $E(B - V)$ for each cluster star, and we plot $(V - M_V)$ versus $E(B - V)$. By making a least-squares fit, we determine the slope of the straight line that best fits the data. As can be seen from equation (3-76), this slope is R_V. For most clusters, R_V is found to be near the "standard" value of 3.2.

Variations in R_V Most regions of the sky yield values of R_V close to the standard value of 3.2. The data are usually summarized by saying that $R_V \approx 3.2 \pm 0.2$. A few special regions have long been known to have anomalous values of R_V. For example, in the Trapezium in the Orion nebula, we find $R_V \approx 6$ (**S18**). At one point, work by Johnson (**J3**) seemed to indicate that there were numerous regions where the absorption curve, determined by 10-color photometry, differed significantly from the standard Whitford curve and yielded much larger values of R_V. Johnson found that highly reddened stars sometimes showed an abrupt rise in brightness in the far infrared, and he attributed this to additional absorption at wavelengths somewhat shorter than the longest wavelength bands observed. Recent work has shown that this interpretation was wrong and that, in most of these cases, the stars actually emit excess far-infrared radiation, which arises in extensive cool *circumstellar shells*. These shells are composed of grains that are heated to moderate temperatures by energy absorbed from stellar radiation at short wavelengths. They then reradiate copiously at much longer wavelengths, near the peak of the Planck curve appropriate to the low temperature of the material in the shell.

In summary, the value $R_V = 3.2$ can be considered as representative; but, for the most precise work, one should ideally attempt to determine R_V directly for each specific field under study.

REFERENCES

(A1) Abell, G. O. 1964. *Exploration of the Universe*. (New York: Holt, Rinehart & Winston).

(A2) Alcaino, G. 1977. *Pub. Astron. Soc. Pacific*. **89**:491.

(A3) Allen, C. W. 1973. *Astrophysical Quantities*, 3rd ed. (London: Athlone Press).

(A4) Aller, L. H. 1963. *The Atmospheres of the Sun and Stars*. 2nd ed. (New York: Ronald Press).

(A5) Aller, L. H. and McLaughlin, D. B. (eds.). 1965. *Stellar Structure*. (Chicago: University of Chicago Press)

(A6) Andersen, J. 1975. *Astron. and Astrophys.* **44**:355.

(A7) Andersen, J. 1975. *Astron. and Astrophys.* **44**:445.

(A8) Andersen, J. 1975. *Astron. and Astrophys.* **45**:203.

(A9) Andersen, J., Gjerløff, H., and Imbert, M. 1975. *Astron. and Astrophys.* **44**:349

(A10) Arimoto, N. and Saio, H. 1978. *Pub. Astron. Soc. Japan* **30**:483.

(A11) Arp, H. C. and Hartwick, F. D. A. 1971. *Astrophys. J.* **167**:499.

(B1) Baade, W. and Swope, H. H. 1961. *Astron. J.* **66**:300.

(B2) Basinska-Grzesik, E. and Mayor, M. (eds.). 1977. *Chemical and Dynamical Evolution of Our Galaxy*. I.A.U. Colloquium No. 45. (Geneva: Observatoire de Geneve).

(B3) Batten, A. H. 1973. *Binary and Multiple Systems of Stars*. (Oxford: Pergamon Press).

(B4) Baum, W. A. 1952. *Astron. J.* **57**:222.

(B5) Bell, R. A. 1971. *Mon. Not. Roy. Astron. Soc.* **154**:343.

(B6) Bertiau, F. C. 1958. *Astrophys. J.* **128**:533.

(B7) Blaauw, A. and Schmidt, M. (eds.). 1965. *Galactic Structure*. (Chicago: University of Chicago Press).

(B8) Bless, R. C. and Savage, B. D. 1972. *Astrophys. J.* **171**:293.

(B9) Branch, D. and Alexander, J. B. 1973. *Mon. Not. Roy. Astron. Soc.* **161**:409.

(B10) Brown, R. H. 1968. *Ann. Rev. Astron. and Astrophys.* **6**:13.

(B11) Brown, R. H., Davis, J., and Allen, L. R. 1974. *Mon. Not. Roy. Astron. Soc.* **167**:121.

(B12) Brown, R. H., Davis, J., Lake, R. J. W., and Thompson, R. J. 1974. *Mon. Not. Roy. Astron. Soc.* **167**:475.

(B13) Butler, D. 1975. *Astrophys. J.* **200**:68.

(B14) Butler, D., Dickens, R. J., and Epps, E. 1978. *Astrophys. J.* **225**:148.

(C1) Cameron, A. G. W. 1970. *Ann. Rev. Astron. and Astrophys.* **8**:179.

(C2) Canterna, R. 1975. *Astrophys. J. Letters.* **200**:L63.

(C3) Cayrel, R. 1968. *Astrophys. J.* **151**:997.

(C4) Cayrel de Strobel, G. and Delplace, A. M. 1972. *L'Age des Etoiles*, I.A.U. Colloquium No. 17. (Meudon: Observatoire de Paris).

(C5) Chandrasekhar, S. 1939. *An Introduction to the Study of Stellar Structure.* (Chicago: University of Chicago Press).

(C6) Christy, R. F. 1966. *Ann. Rev. Astron. and Astrophys.* **4**:353.

(C7) Ciardullo, R. B. and Demarque, P. 1977. *Trans. Yale Univ. Obs.* Vol. **33**.

(C8) Clausen, J. V., Gyldenkerne, K., and Grønbech, B. 1976. *Astron. and Astrophys.* **46**:205.

(C9) Clayton, D. D. 1968. *Principles of Stellar Evolution and Nucleosynthesis.* (New York: McGraw-Hill).

(C10) Clegg, R. E. S. and Bell, R. A. 1973. *Mon. Not. Roy. Astron. Soc.* **163**:13.

(C11) Code, A. D. 1947. *Astrophys. J.* **106**:309.

(C12) Code, A. D. 1959. *Astrophys. J.* **130**:473.

(C13) Code, A. D., Davis, J., Bless, R. C., and Brown, R. H. 1976. *Astrophys. J.* **203**:417.

(C14) Cox, J. P. 1974. *Reports Prog. Phys.* **37**:563.

(C15) Cox, J. P. and Giuli, R. T. 1968. *Principles of Stellar Structure* (2 vols.). (New York: Gordon and Breach).

(D1) Dennis, T. R. 1965. *Pub. Astron. Soc. Pacific.* **77**:283.

(D2) Dennis, T. R. 1968. *Astrophys. J. Letters.* **151**:L47.

(D3) Deutsch, A. J. 1955. In *Principes Fondamenteux de Classification Stellaire.* (Paris: Centre National de Recerche Scientifique). p. 25.

(D4) Dickens, R. J. and Bell, R. A. 1976. *Astrophys. J.* **207**:506.

(D5) Duley, W. W. 1976. *Astrophys. Space Sci.* **45**:253.

(D6) Dunham, D. W., Evans, D. S., and Sandmann, W. H. 1974. *Astron. J.* **79**:483.

(E1) Eggen, O. J. 1958. *Mon. Not. Roy. Astron. Soc.* **118**:65.

(E2) Eggen, O. J. 1960. *Mon. Not Roy. Astron. Soc.* **120**:563.

(E3) Eggen, O. J. 1964. *Astron. J.* **69**:570.

(E4) Eggen, O. J. 1967. *Ann. Rev. Astron. and Astrophys.* **5**:105.

(E5) Eggen, O. J. 1970. *Vistas in Astron.* **12**:367.

(E6) Eggen, O. J. 1973. *Astrophys. J.* **182**:821.

(E7) Eggen, O. J., Lynden-Bell, D., and Sandage, A. 1962. *Astrophys. J.* **136**:748.

(E8) Eggen, O. J. and Sandage, A. R. 1962. *Astrophys. J.* **136**:735.

(E9) Eggen, O. J. and Sandage, A. R. 1969. *Astrophys. J.* **158**:669.

(E10) Evans, D. S. 1970. *Astron. J.* **75**:589.

(E11) Ezer, D. and Cameron, A. G. W. 1967. *Can. J. Phys.* **45**:3429.

(F1) Feibelman, W. A. 1976. *Astrophys. J.* **209**:497.

(F2) Field, G. B. and Cameron, A. G. W. (eds.). 1975. *The Dusty Universe.* (New York: Watson Academic).

(F3) Field, G. B., Goldsmith, D. W., and Habing, H. J. 1969. *Astrophys. J. Letters.* **155**:L149.

(F4) Freeman, K. C. and Rodgers, A. W. 1975. *Astrophys. J. Letters.* **201**:L71.

(G1) Gamow, G. 1944. *Phys. Rev.* **65**:20.

(G2) Gezari, D. Y., Labyrie, A , and Stachnik, R. V. 1972. *Astrophys. J. Letters.* **173**:L1.

(G3) Golay, M. 1974. *Introduction to Astronomical Photometry.* (Dordrecht: Reidel).

(G4) Goldreich, P. and Lynden-Bell, D. 1965. *Mon. Not. Roy. Astron. Soc.* **130**:97.

(G5) Graham, J. A. 1975. *Pub. Astron. Soc. Pacific.* **87**:641.

(G6) Gray, D. 1976. *The Observation and Analysis of Stellar Photospheres.* (New York: Wiley).

(G7) Grønbech, B., Gyldenkerne, K., and Jørgensen, H. E. 1977. *Astron. and Astrophys.* **55**:401.

(G8) Gustafsson, B. and Nissen, P. E. 1972. *Astron. and Astrophys.* **19**:261.

(G9) Gyldenkerne, K. and West, R. M. (eds.). 1970. *Mass Loss and Evolution in Close Binaries.* I.A.U. Colloquium No. 6. (Copenhagen: University of Copenhagen Observatory).

(H1) Hanson, R. B. 1975. *Astron. J.* **80**:379.

(H2) Hanson, R. B. 1977. *Bull. Amer. Astron. Soc.* **9**:585.

(H3) Hartwick, F. D. A. 1968. *Astrophys J.* **154**:475.

(H4) Hartwick, F. D. A. 1975. *Pub. Astron. Soc. Pacific.* **87**:77.

(H5) Hartwick, F. D. A. and Hesser, J. E. 1972. *Astrophys. J.* **175**:77.

(H6) Hartwick, F. D. A. and Hesser, J. E. 1974. *Astrophys. J. Letters.* **194**:L129.

(H7) Hartwick, F. D. A. and McClure, R. D. 1972. *Astrophys. J. Letters.* **176**:L57.

(H8) Hartwick, F. D. A. and McClure, R. D. 1974. *Astrophys. J.* **193**:321.

(H9) Hartwick, F. D. A. and Vanden Berg, D. A. 1973. *Pub. Astron. Soc. Pacific.* **85**:355.

(H10) Hearnshaw, J. B. 1972. *Memoirs Roy. Astron. Soc.* **77**:55.

(H11) Heck, A. 1973. *Astron. and Astrophys.* **24**:313.

(H12) Heck, A. 1978. *Vistas in Astron.* **22**:221.

(H13) Hegyi, D. and Curott, D. 1970. *Phys. Rev. Letters.* **24**:415.

(H14) Hemenway, M. K. 1975. *Astron. J.* **80**:199.

(H15) Hesser, J. E., Hartwick, F. D. A., and McClure, R. D. 1976. *Astrophys. J. Letters.* **207**:L113.

(H16) Hesser, J. E., Hartwick, F. D. A., and McClure, R. D. 1977. *Astrophys. J. Supp.* **33**:471.

(H17) Hiltner, W. A. (ed.). 1962. *Astronomical Techniques.* (Chicago: University of Chicago Press).

(H18) Hiltner, W. A. and Johnson, H. L. 1956. *Astrophys. J.* **124**:367.

(H19) Hodge, P. W. and Wallerstein, G. 1966. *Pub. Astron. Soc. Pacific.* **78**:411.

(H20) Hoyle, F. and Lyttleton, R. A. 1942. *Mon. Not. Roy. Astron. Soc.* **102**:218.

(H21) Hoyle, F. and Schwarzschild, M. 1955. *Astrophys. J. Supp.* **2**:1.

(I1) Iben, I. 1964. *Astrophys. J.* **140**:1631.

(I2) Iben, I. 1966. *Astrophys. J.* **143**:483.

(I3) Iben, I. 1967. *Ann. Rev. Astron. and Astrophys.* **5**:571.

(I4) Iben, I. 1974. *Ann. Rev. Astron. and Astrophys.* **12**:215.

(I5) Iben, I. and Tuggle, R. S. 1972. *Astrophys. J.* **173**:135.

(I6) Iben, I. and Tuggle, R. S. 1972. *Astrophys. J.* **178**:441.

(I7) Illingworth, G. 1976. *Astrophys. J.* **204**:73.

(I8) Illingworth, G. and Illingworth, W. 1976. *Astrophys. J. Supp.* **30**:227.

(J1) Janes, K. A. 1975. *Astrophys. J. Supp.* **29**:161.

(J2) Johnson, H. L. 1965. *Astrophys. J.* **141**:170.

(J3) Johnson, H. L. 1965. *Astrophys. J.* **141**:923.

(J4) Johnson, H. L. and Iriarte, B. 1958 *Lowell Obs. Bull.* **4**:47.

(J5) Johnson, H. L. and Morgan, W. W. 1953. *Astrophys. J.* **117**:313.

(K1) King, I. R. 1966. *Astron. J.* **71**:64.

(K2) King, I. R., Hedemann, E., Hodge, S. M., and White, R. E. 1968. *Astron. J.* **73**:456.

(K3) Kinman, T. D. 1959. *Mon. Not. Roy. Astron. Soc.* **119**:538.

(K4) Kopal, Z. 1959. *Close Binary Systems.* (London: Chapman and Hall).

(K5) Kraft, R. P. 1961. *Astrophys. J.* **134**:616.

(**K6**) Kron, G. E. and Mayall, N. U. 1960. *Astron. J.* **65**:581.

(**K7**) Kuiper, G. P. 1938. *Astrophys. J.* **88**:429.

(**L1**) Labeyrie, A. 1970. *Astron. and Astrophys.* **6**:85.

(**L2**) Labeyrie, A. 1978. *Ann. Rev. Astron. and Astrophys.* **16**:77.

(**L3**) Lacy, C. H. 1977. *Astrophys. J. Supp.* **34**:479.

(**L4**) Leavitt, H. S. 1912. In *Harvard Coll. Obs. Circ. No. 173*, p. 1.

(**M1**) McClure, R. D. and Norris, J. 1974. *Astrophys. J.* **193**:139.

(**M2**) McGraw, J. T. and Angel, J. R. P. 1974. *Astron. J.* **79**:485.

(**M3**) McKee, C. F. and Ostriker, J. P. 1977. *Astrophys. J.* **218**:148.

(**M4**) McNally, D. 1974. *Positional Astronomy.* (London: Muller Educational).

(**M5**) Mallia, E. A. 1977. *Astron. and Astrophys.* **60**:195.

(**M6**) Manduca, A. and Bell, R. A. 1978. *Astrophys. J.* **225**:908.

(**M7**) Matthews, T. A. and Sandage, A. R. 1963. *Astrophys. J.* **138**:30.

(**M8**) Mayall, N. U. 1946. *Astrophys. J.* **104**:290.

(**M9**) Mayor, M. 1974. *Astron. and Astrophys.* **32**:321.

(**M10**) Mayor, M. 1976. *Astron. and Astrophys.* **48**:301.

(**M11**) Melbourne, W. G. 1960. *Astrophys. J.* **132**:101.

(**M12**) Mestel, L. 1952. *Mon. Not. Roy. Astron. Soc.* **112**:598.

(**M13**) Mihalas, D. 1978. *Stellar Atmospheres.* 2nd ed. (San Francisco: W. H. Freeman).

(**M14**) Milford, N. 1950. *Ann. d'. Astrophys.* **13**:251.

(**M15**) Miller, J. S. 1969. *Astrophys. J.* **157**:1215.

(**M16**) Misner, C., Thorne, K., and Wheeler, J. 1973. *Gravitation.* (San Francisco: W. H. Freeman).

(**M17**) Morgan, W. W. 1956. *Pub. Astron. Soc. Pacific.* **68**:509.

(**M18**) Morgan, W. W. 1959. *Astron. J.* **64**:432.

(**N1**) Nather, R. E. and Evans, D. S. 1970. *Astron. J.* **75**:575.

(**N2**) Nather, R. E. and McCants, M. M. 1970. *Astron. J.* **75**:963.

(**N3**) Nissen, P. E. 1970. *Astron. and Astrophys.* **6**:138.

(**N4**) Norris, J. and Bessell, M. S. 1975. *Astrophys. J. Letters* **201**:L75.

(**O1**) O'Dell, C. R., Peimbert, M., and Kinman, T. D. 1964. *Astrophys. J.* **140**:119.

(**O2**) Oke, J. B. and Schwarzschild, M. 1952. *Astrophys. J.* **116**:317.

(**O3**) Oort, J. H. 1977. *Astrophys. J. Letters.* **218**:L97.

(**O4**) Öpik, E. 1938. *Pub. Tartu Obs.* **30**:No. 4.

(**O5**) Ostriker, J. P. 1971. *Ann. Rev. Astron. and Astrophys.* **9**:353.

(P1) Pagel, B. E. J. and Patchett, B. 1975. *Mon. Not. Roy. Astron. Soc.* **172**:13.

(P2) Peimbert, M. 1972. In *Les Nebuleuses Planetaires. Mem. Soc. Roy. des Sci. de Liege.* Sixth Series. **5**:307.

(P3) Perrin, M. N., Hejelsen, P. M., Cayrel de Strobel, G., and Cayrel, R. 1977. *Astron. and Astrophys.* **54**:779.

(P4) Peterson, C. J. and King, I. R. 1975. *Astron. J.* **80**:427.

(P5) Philip, A. G. D., Cullen, N. F., and White, R. F. 1976. *UBV Color-Magnitude Diagrams of Galactic Globular Clusters.* Dudley Observatory Report No. 11. (Albany: Dudley Observatory).

(P6) Philip, A. G. D. and Hayes, D. S. (eds.). 1974. *Multicolor Photometry and the Theoretical H-R Diagram.* Dudley Observatory Report No. 9. (Albany: Dudley Observatory).

(P7) Popper, D. M. 1959. *Astrophys. J.* **129**:647.

(P8) Popper, D. M. 1967. *Ann. Rev. Astron. and Astrophys.* **5**:85.

(P9) Powell, A. L. T. 1970. *Mon. Not. Roy. Astron. Soc.* **148**:477.

(P10) Preston, G. 1959. *Astrophys. J.* **130**:507.

(R1) Reddish, V. C. 1978. *Stellar Formation.* (Oxford: Pergamon).

(R2) Ridgeway, S. T., Wells, D. C., and Carbon, D. F. 1974. *Astron. J.* **79**:1079.

(R3) Ruderman, M. 1972. *Ann. Rev. Astron. and Astrophys.* **10**:427.

(S1) Sandage, A. R. 1957. *Astrophys. J.* **125**:435.

(S2) Sandage, A. R. 1970. *Astrophys. J.* **162**:841.

(S3) Sandage, A. R. and Eggen, O. J. 1959. *Mon. Not. Roy. Astron. Soc.* **119**:278.

(S4) Sandage, A. R. and Eggen, O. J. 1969. *Astrophys. J.* **158**:685.

(S5) Sandage, A. R. and Schwarzschild, M. 1952. *Astrophys. J.* **116**:463.

(S6) Sandage, A. R. and Tammann, G. A. 1968. *Astrophys. J.* **151**:531.

(S7) Sandage, A. R. and Tammann, G. A. 1971. *Astrophys. J.* **167**:293.

(S8) Sandage, A. R. and Wallerstein, G. 1960. *Astrophys. J.* **131**:598.

(S9) Sandage, A. R. and Wildey, R. L. 1967. *Astrophys. J.* **150**:469.

(S10) Schmidt, M. 1959. *Astrophys. J.* **129**:243.

(S11) Schönberg, M. and Chandrasekhar, S. 1942. *Astrophys. J.* **96**:161.

(S12) Schwarzschild, M. and Härm, R. 1965. *Astrophys. J.* **142**:855.

(S13) Scoville, N. Z. and Solomon, P. M. 1975. *Astrophys. J. Letters.* **199**:L105.

(S14) Searle, L. and Zinn, R. 1978. *Astrophys. J.* **225**:357.

(S15) Shapley, H. 1917. *Astrophys. J.* **45**:118.

(S16) Shapley, H. 1930. *Star Clusters.* (Cambridge: Harvard University Press).

(S17) Shapley, H. and Sawyer, H. B. 1927. *Harvard Obs. Bull. No. 849.*

(S18) Sharpless, S. 1952. *Astrophys. J.* **116**:251.

(S19) Smart, W. M. 1977. *Textbook on Spherical Astronomy*. (Cambridge: Cambridge University Press).

(S20) Spitzer, L. 1948. *Phys. Today*. **1**:6.

(S21) Spitzer, L. 1968. *Diffuse Matter in Space*. (New York: Interscience).

(S22) Spitzer, L. 1978. *Physical Processes in the Interstellar Medium*. (New York: Wiley).

(S23) Stothers, R. 1972. *Astrophys. J.* **175**:431.

(S24) Strand, K. A. (ed.). 1963. *Basic Astronomical Data*. (Chicago: University of Chicago Press).

(S25) Struve, O. and Zebergs, V. 1962. *Astronomy of the 20th Century*. (New York: Macmillan).

(T1) Tayler, R. J. 1972. *The Stars: Their Structure and Evolution*. (London: Wykeham).

(T2) Tinsley, B. and Larson, R. B. (eds.). 1977. *Proc. Conf. on the Evolution of Galaxies and Stellar Populations*. (New Haven: Yale University Observatory).

(T3) Trumpler, R. J. 1925. *Pub. Astron. Soc. Pacific*. **37**:307.

(T4) Trumpler, R. J. 1930. *Lick Obs. Bull.* **14**:154.

(V1) van Altena, W. F. 1974. *Astron. J.* **86**:217.

(V2) van den Bergh, S. 1967. *Astron. J.* **72**:70.

(V3) van den Bergh, S. 1967. *Pub. Astron. Soc. Pacific*. **79**:460.

(V4) van Herk, G. 1965. *Bull. Astron. Inst. Netherlands*. **18**:71.

(W1) Wallerstein, G. 1962. *Astrophys. J. Supp.* **6**:407.

(W2) Wallerstein, G. and Helfer, H. L. 1966. *Astron. J.* **71**:350.

(W3) Weidemann, V. 1968. *Ann. Rev. Astron. and Astrophys.* **6**:351.

(W4) White, N. M. 1974. *Astron. J.* **79**:1076.

(W5) Whitford, A. E. 1958. *Astron. J.* **63**:201.

(W6) Wickramasinghe, N. C. and Morgan, D. J. 1976. *Solid State Astrophysics*. (Dordrecht: Reidel).

(W7) Wildey, R. L., Burbidge, E. M., Sandage, A. R., and Burbidge, G. R. 1962. *Astrophys. J.* **135**:94.

(W8) Wilson, O. C. 1976. *Astrophys. J.* **205**:823.

(W9) Woolley, R. and Savage, A. 1971. *Roy. Obs. Bull. No. 170*.

4

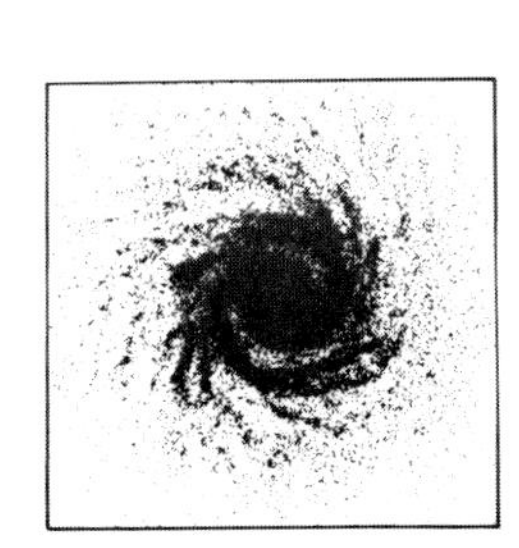

The Space Distribution of Stars and the Chemical Elements in Our Galaxy

Having considered a qualitative description of our Galaxy, and having discussed properties of individual stars, we may now begin assembling a quantitative picture of our Galaxy's structure and dynamics. As a first goal, we shall attempt to determine something about the distribution of stars in space and the nature of the major structural features of our Galaxy. As we shall see, it is possible to obtain fairly complete statistical information about the *space densities* of stars of various types in the solar neighborhood, which we can assume is a typical region of the galactic disk. From these data, we can construct a reasonably detailed picture of the structure of the galactic disk and a model of the spheroidal component, which, though less detailed, permits us to estimate the radial star-density gradient in the halo, star densities in the galactic bulge, and also the Sun's distance from the galactic center. By coupling information about the distribution of stars with information about their composition, we can begin to develop a rough picture of the nature and size of chemical-composition gradients, both in the disk and in the spheroidal component. Finally, we can integrate this information into a picture that describes the *stellar populations* within our Galaxy, that is, one that shows how our Galaxy is made up of stars of different ages, compositions, and kinematic properties (see Chapters 6 and 7). From this picture, we get important hints about the evolutionary history of our Galaxy (see Chapter 19).

How are we to proceed? Naively, we might ask whether we can determine distances for individual stars directly, and thus map them one by one. The answer is obviously no. Direct distance-measurement techniques are restricted to stars in a very small volume of space around the Sun, and, even if they were not, there are too many stars to analyze in this way. We must therefore proceed indirectly. One approach is to use *star counts*—that is, the numbers of stars that can be seen in an area on the sky at successively fainter apparent magnitudes—to find space densities. Alternatively, we can use particular objects that are characteristic of a specific structural feature—for example, a spiral arm—as *tracers* to delineate the outlines of that feature.

Limits on these procedures are set by the presence of interstellar absorbing material and the apparent magnitude which can be reached by existing telescopes. As we shall see, we cannot derive much useful information about space densities of most stellar types in the galactic disk beyond about 1 kpc from the Sun. (This rises to about 2.5 kpc for B stars, which are more luminous.) Perpendicular to the disk, where there is much less absorption, we can derive information about intrinsically luminous stars, such as A dwarfs or G and K giants, out to distances of about 2 kpc. For intrinsically faint stars, such as K dwarfs, we are limited to only 200 pc or so. In the spheroidal component, we can derive useful information from RR Lyrae stars even at distances comparable to the size of our Galaxy because these stars are reasonably bright, can be found in large numbers outside the obscuring dust layer, and show distinctive light variations that permit their identification with high reliability even at faint apparent magnitudes. And, of course, the very brightest objects—the globular clusters—can be seen at immense distances, and, as described in Chapter 1, outline the large-scale structure of our whole Galaxy.

4-1. THE APPARENT DISTRIBUTION OF STARS

Observations

When we look at the sky on a clear night, we see a fairly uniform distribution of the brighter stars over the entire sky and a rather faint, diffuse band of light—the Milky Way—which runs a full 360° in a great circle around the sky and which, when examined with even a small telescope, is found to be composed of innumerable faint stars. We mentioned in Chapter 1 that the basic significance of these observations (as was first realized by Herschel) is that our Galaxy is a disklike system, much broader than it is thick, and that the Sun is immersed in the disk near the central plane. Nearby (and hence apparently bright) stars are thus seen in all directions, but distant (and hence apparently faint) stars will be seen only in a narrow strip of the sky that contains those lines of sight that are confined to the disk. One can also easily see with the naked eye a dark rift through the Milky Way. Such a feature is seen as well in other edge-on galaxies (see Figures 1-6 and 1-7), and it is attributed to the presence of a layer of obscuring interstellar dust in the galactic plane. The dust layer has a characteristic thickness somewhat smaller than that of the disk of stars.

We can describe the distribution of stars on the sky quantitatively by star counts. Let $A(m, b)$ be the number of stars at apparent magnitude m, per unit magnitude interval, per square degree, at galactic latitude b. In practice, one determines $A(m, b)$ by counting the number of stars on the magnitude range from $m - \frac{1}{2}$ to $m + \frac{1}{2}$; given a smooth run of $A(m, b)$, for purposes of analysis we can consider $A(m, b)\,dm$ to be the number of stars per square degree with apparent magnitudes in the range $(m, m + dm)$. Star-count data for our Galaxy in the galactic plane ($b = 0°$, averaged over longitude) and toward the north galactic pole ($b = 90°$) are shown in columns 2 and 4 of Table 4-1. The magnitude scale is the photographic apparent magnitude m_{pg}. The last two columns give the light received (in the photographic band) per square degree from stars of magnitude m, in units of an equivalent number of tenth-magnitude stars per square degree; that is,

$$\ell(m, b) \equiv A(m, b) 10^{0.4(10-m)} \tag{4-1}$$

An inspection of Table 4-1 immediately shows the following:

1. From the ratio $A(m, 0°)/A(m, 90°)$, we can see that the bright stars ($m \lesssim 9$) are fairly uniformly distributed between the galactic pole and plane. There are, of course, more of these stars in the plane than near the pole, but the ratio is nearly constant as a function of apparent magnitude. In contrast, there are enormously many more faint stars in the galactic plane than near the pole. The data merely give quantitative expression to the qualitative impression one obtains from a direct visual inspection of the sky.
2. The columns $\ell(m, 0°)$ and $\ell(m, 90°)$ show that, whereas the major contribution to the light we observe in the direction of the galactic poles comes from brighter stars, the diffuse glow in the galactic plane is produced mainly by light from the much more numerous faint stars (maximum light coming from stars with $m \approx 14$), each of which is much too faint to be seen individually by the naked eye.

From differential star counts $A(m, b)$, we can determine an *integrated star count* $N(m, b)$, the cumulative number of stars, per square degree, having apparent magnitudes less than or equal to m. Thus

$$\frac{dN(m, b)}{dm} \equiv A(m, b) \tag{4-2}$$

or

$$N(m, b) = \int_{-\infty}^{m} A(m', b)\, dm' \tag{4-3}$$

In practice, we compute $N(m, b)$ from real data as

$$N(m, b) = \sum_{m'=-2}^{m} A(m', b) \tag{4-4}$$

Table 4-1. Star Counts in the Galactic Plane and North Galactic Pole

m_{pg}	log $A(m, 0°)$	Δ log $A(m, 0°)$	log $A(m, 90°)$	Δ log $A(m, 90°)$	$A(m, 0°)/A(m, 90°)$	$\ell(m, 0°)$	$\ell(m, 90°)$
4	0.2–2	0.52	0.68–3	0.50	3.3	4.0	1.2
5	0.72–2	0.47	0.18–2	0.46	3.5	5.2	1.5
6	0.19–1	0.44	0.64–2	0.44	3.5	6.1	1.7
7	0.63–1	0.47	0.08–1	0.42	3.6	6.7	1.9
8	0.10	0.48	0.50–1	0.42	4.0	7.9	2.0
9	0.58	0.46	0.92–1	0.36	4.6	9.6	2.1
10	1.04	0.46	0.28	0.35	5.8	11.0	1.9
11	1.50	0.44	0.63	0.35	7.4	12.6	1.7
12	1.94	0.41	0.98	0.31	8.7	13.8	1.5
13	2.35	0.41	1.29	0.26	11.5	14.1	1.2
14	2.76	0.35	1.57	0.23	15.5	14.4	0.9
15	3.15	0.31	1.80	0.26	22.4	14.1	0.6
16	3.46	0.38	2.06	0.22	25.0	11.5	0.5
17	3.84	0.36	2.28	0.22	36.4	11.0	0.3
18	4.2	0.3	2.50	0.2	50	10.0	0.2
19	4.5	0.2	2.7	0.1	63	7.9	0.1
20	4.7		2.8	0.1	80	5.0	0.06
21	4.9		2.9		100	3.1	0.03

SOURCE: (**A1**, 245), by permission

Table 4-2. Integrated Star Counts in the Galactic Plane, at the North Galactic Pole, and Summed over the Sky

m_V	log $N(m, 0°)$	log $N(m, 90°)$	$N_{tot}(m)$
4	−1.55	−2.20	5.2×10^2
5	−1.08	−1.69	1.6×10^3
6	−0.60	−1.20	4.8×10^3
7	−0.16	−0.74	1.4×10^4
8	0.29	−0.30	4.1×10^4
9	0.78	0.14	1.2×10^5
10	1.25	0.55	3.3×10^5
11	1.73	0.96	9.0×10^5
12	2.18	1.33	2.4×10^6
13	2.60	1.69	6.1×10^6
14	3.02	2.01	1.5×10^7
15	3.42	2.27	3.6×10^7
16	3.78	2.54	8.2×10^7
17	4.13	2.78	1.8×10^8
18	4.50	3.02	3.7×10^8
19	4.8	3.2	6.5×10^8
20	5.0	3.4	1×10^9
21	5.3	3.5	2×10^9

SOURCE: (**A1**, 244), by permission

in which case $N(m, b)$ gives, strictly speaking, the total number down to apparent magnitude $m + \frac{1}{2}$. Integrated star counts in the galactic plane and at the north galactic pole are given in columns 2 and 3 of Table 4-2; the fourth column gives the total number $N_{tot}(m)$ summed over the entire sky. In the table, the magnitude scale is visual apparent magnitude m_v (or V). The total amount of starlight received from the whole sky is equivalent to about 4.6×10^6 tenth-magnitude stars in V, or 2.3×10^6 tenth-magnitude stars in m_{pg}.

The concentration of stars toward the galactic plane is strikingly different for stars of different spectral classes, as is illustrated in Figure 4-1. These data were drawn from counts of stars in the Henry Draper Catalog in several small zones on the sky, and thus they apply to stars with $m_{pg} \lesssim 8.5$. As can be seen in the figure, O and B stars show an extremely strong concentration to the plane. They also show a marked tendency to clump into clusters and associations. Likewise, open clusters and classical Cepheids are very strongly concentrated to the plane. All of these objects are very young, and they are generally closely associated with spiral arms; as we mentioned earlier, the O and B stars have such short lifetimes and small random velocities that they must be found today very close to their places of formation (which are the clusters and associations).

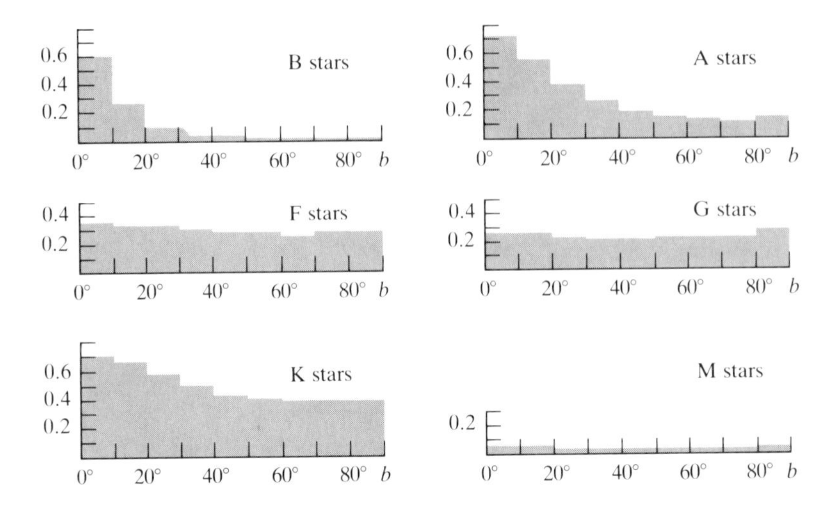

Figure 4-1. Apparent distribution of stars of various types as a function of galactic latitude. Ordinate gives average number of stars per square degree. [From (**N2**).]

The A stars show a fairly strong concentration to the plane, but they are less concentrated than the O–B stars; the A stars are, on the average, much older than the O–B stars, and they have existed a sufficient time for their random velocities to disperse them over a fairly large volume of space in the disk. The average A star with $m_{\rm pg} \lesssim 8.5$ will also be nearer than the average O–B star of similar apparent brightness, and hence, even if both groups had the same distribution relative to the plane, the A stars would *appear* to be more widely spread at a given level of apparent brightness. The concentration to the plane of T Tauri stars, novae, and planetary nebulae is similar to that of the A stars.

The F stars and late-type variables show only a slight concentration to the plane, and stars of types G and M show essentially no concentration at all. The latter two groups are composed mainly of stars that are, in fact, contained in the disk, but they are intrinsically faint. Consequently, down to $m_{\rm pg} \approx 8.5$, we sample only nearby stars within a small volume surrounding the Sun, and, because the Sun is itself immersed in the disk, the distribution of these late-type dwarfs appears isotropic. The RR Lyrae stars and globular clusters also show no concentration to the plane (although they do show a concentration toward the galactic center). These objects are intrinsically luminous and we know, therefore, that they are distributed at large within the spheroidal component and are not confined to the disk. The distribution of K stars shown in Figure 4-1 appears anomalous at first sight (more strongly concentrated than G stars, which are of earlier type), but this appearance is a result of the fact that most of the K stars in the sample are K giants, which are actually evolved massive stars that were themselves strongly concentrated in the plane. We shall see, in the material which follows, that these distinctive space distributions are correlated with stellar kinematic properties, ages, and compositions. Taken together, these correlations delineate important stellar-population characteristics.

Another point of interest is that, while we find the apparently faint ($m > 6$) and hence more distant B stars fairly uniformly distributed above and below the galactic plane, the brightest B stars lie along a great circle, *Gould's belt*, inclined at about 16° to the galactic equator. This distribution is clearly a local feature, and it probably arises because these stars all belong to a common spiral arm.

Galactic nebulae, both luminous and dark, are also strongly concentrated to the plane. As we mentioned in Chapter 1, this heavy concentration of absorbing material to the plane gives rise to the zone of avoidance, within which almost no external galaxies are observed. Numerous dark nebulae can be seen in even the most casual examination of the *Palomar Sky Survey* prints, or the *Atlas of Selected Regions of the Milky Way* (**B2**). A comprehensive map of the major absorbing clouds is shown in (**L4**). One can readily see on this map that the whole region around the galactic center (except for one or two small windows) is extremely heavily obscured by dark nebulae, and the galactic center itself is virtually invisible except at X-ray, far-infrared, and radio wavelengths.

Predictions for a Uniform Galaxy

Let us now consider what star counts we might expect to observe in our Galaxy. The simplest conceivable model is obtained if we assume that (1) stars are strictly uniformly distributed in space with density D stars per cubic parsec, (2) our Galaxy is infinite in extent, and (3) there is no interstellar absorption. While none of these assumptions is correct, the model provides basic orientation. Suppose that, in some area of the sky, we examine a star field subtending a solid angle ω. [Note that one square degree subtends $(\pi/180)^2 = 1/3283$ steradians (sr), and the entire sky contains 4π sr $= 41{,}253$ square degrees.] Then the volume element contained in the distance interval $(r, r + dr)$ is $\omega r^2\, dr$ (see Figure 4-2), so that the total number of stars out to

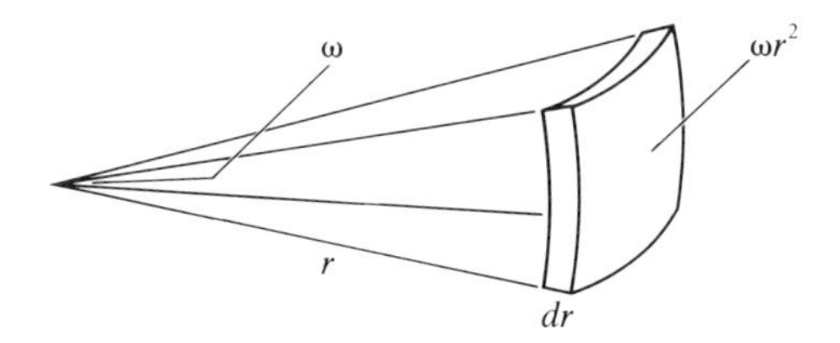

Figure 4-2. The volume element on the distance interval $(r, r + dr)$ within a field subtending solid angle ω is $\omega r^2 dr$.

distance r is

$$N(r) = \omega D \int_0^r (r')^2 \, dr' = \tfrac{1}{3}\omega D r^3 \tag{4-5}$$

where D is the (uniform) star density. If all the stars have an absolute magnitude M, then, from equation (2-13),

$$r = 10^{[0.2(m-M)+1]} \text{ (pc)} \tag{4-6}$$

Combining equations (4-5) and (4-6), we have

$$N(m) = 10^{(0.6m-c)} \tag{4-7}$$

or

$$\log N(m) = 0.6m + C \tag{4-8}$$

where C is a constant that depends on D, ω, and M.

In deriving equation (4-8), we have assumed that all stars have the same absolute magnitude M, but this assumption is not really necessary—it merely simplifies the derivation. Indeed, inasmuch as equation (4-7), with an appropriate value of C, is true for each group of stars at any specified absolute magnitude, it is clear that an equation of the same form (that is, $N \propto 10^{0.6m}$) can be written for a superposition of several distinct groups of stars, each with a different absolute magnitude and space density. It follows that, in passing to the limit of infinitely many groups, equation (4-8) will be valid for an arbitrary distribution of stars over absolute magnitude, provided that this distribution is everywhere the same and that the three assumptions just stated still hold. Only the constant in the equation is changed when stars with a spread of absolute magnitudes are present.

According to equation (4-8), in each unit magnitude interval, $\log N$ should change by 0.6; that is, the total number of stars should go up by very nearly a factor of four as we count to each successive apparent magnitude. Before we compare with actual data, let us first ask whether this is a plausible result. In fact it is not, because it leads to an absurd conclusion. Let us write

$$N(m) = C_1 10^{0.6m} = C_1 e^{0.6\mu m} \tag{4-9}$$

where $\mu \equiv \ln 10 = 2.303$. Then

$$A(m) = dN(m)/dm = 0.6\mu C_1 e^{0.6\mu m} \equiv C_2 10^{0.6m} \tag{4-10}$$

Now, if the light we receive from a star with $m = 0$ is ℓ_0, then the light we receive from each star of magnitude m is

$$\ell(m) = \ell_0 10^{-0.4m} \tag{4-11}$$

Hence the light received from all stars of magnitude m is

$$\mathscr{L}(m) = \ell(m)A(m) = \ell_0 10^{-0.4m} C_2 10^{0.6m} \equiv C_3 10^{0.2m} \tag{4-12}$$

The total light received from all stars with magnitudes $m' \leq m$ is obtained by integrating equation (4-12), which yields

$$\mathscr{L}_{\text{tot}}(m) = \int_{-\infty}^{m} \mathscr{L}(m')\,dm' = C_3 \int_{-\infty}^{m} 10^{0.2m'}\,dm' = K 10^{0.2m} \tag{4-13}$$

where $K = C_3/(0.2\mu)$. This shows that, as m grows larger, $\mathscr{L}_{\text{tot}}$ diverges exponentially!

Thus for an infinite, homogeneous, transparent Galaxy, the sky should be a blazing bright surface, which is contrary to observation. This contradiction, commonly called *Olbers' paradox*, shows that one of our assumptions is wrong. In particular, for a discussion of the structure of our Galaxy, we can obviously abandon the assumption of infinite extent. Then, provided that the star density is finite (as it must be), the total number of stars (and hence $\mathscr{L}_{\text{tot}}$) in the system will be bounded. Precisely the same problem arises again in discussions of the whole Universe, where now galaxies are the luminous bodies rather than stars. In this context, Olbers' paradox is not so easily evaded; it is resolved only by taking into account the effects of the large-scale expansion of the Universe.

The Kapteyn Universe

Let us now ask to what extent the magnitude dependence of the observed star counts in our Galaxy resembles equation (4-8). We immediately see from Table 4-1 that the counts do not agree with the predictions of this simple model. Specifically, the changes in log $A(m)$ for $\Delta m = 1$ are always less than the theoretical value of 0.6. The observed differences Δ log $A(m)$ are fairly constant for the brighter stars, but they become progressively smaller for fainter stars; the number of faint stars grows much more slowly than predicted by equation (4-8).

One interpretation of these results is that our Galaxy is finite, or, more generally, that the space density of stars decreases with increasing distance from the Sun. In either case, we will observe fewer stars than predicted by equation (4-8). This explanation is essentially the one offered by Kapteyn, who (as described in Chapter 1) undertook a study of star counts early in this century, using the techniques to be described in §4-2. From this study, he derived a model for the space density of stars in our Galaxy. Kapteyn did not allow for the effects of interstellar absorption, the existence of which had not yet been recognized. From his analysis, he concluded that the star density fell to half its value in the solar neighborhood at a distance of about 150 pc in the directions of the galactic poles and about 800 pc in directions lying in the galactic plane. (Corresponding figures for a decrease to 1% were about 1700 pc and 8500 pc, respectively.) Kapteyn's work thus suggested a rather small ellipsoidal Galaxy with the Sun near its center, a model commonly referred to as the Kapteyn Universe. We know today that this model for our Galaxy is fundamentally incorrect, the flaw in its derivation being the omission of absorption effects. We also described in Chapter 1 how Shapley's observations of globular clusters led to a much more realistic model, which has been confirmed by all subsequent work.

An alternative explanation of the slow growth of $N(m)$, then, is that there is interstellar absorption of starlight. Suppose that there is an absorption of $a(r)$ magnitudes at distance r. If m is the apparent magnitude that a star would have if no absorption were present, then the apparent magnitude at which we will actually observe it is $m + a(r)$. If we were to ignore the effects of absorption, we would conclude that the star is more distant than it really is. Thus, if r is the *true distance* to the star, given by $\log r = 0.2(m - M) + 1$, then ρ, the *apparent distance* to the star, which is larger than r because of absorption, is $\log \rho = 0.2(m - M) + 1 + 0.2a(r)$, so that

$$\log \rho = \log r + 0.2a(r) \tag{4-14}$$

or

$$\rho = 10^{0.2a(r)}r \tag{4-15}$$

For example, if the light from stars in some field were dimmed by an absorption of 1.5 mag that we did not know about, then we would overestimate their distances by a factor of two. The volume assigned to a group of stars scales as the cube of their average distance, so we would underestimate their average space density by about a factor of eight, thus producing a very strong apparent (but actually spurious) drop in density away from the Sun.

We know today that both effects just mentioned are important. On the one hand, in the directions of the galactic poles, the amount of interstellar absorption is relatively small, and the main factor affecting star counts is a steep decrease in the actual star density with increasing distance away from the galactic plane. In contrast, within the galactic disk, the star density in the solar neighborhood is roughly uniform (increasing gradually toward the galactic center), and the main effect seen in the counts is strong absorption by the interstellar medium. A fair estimate of the absorption in the disk is about 1 mag kpc^{-1}, which implies a factor-of-fifteen reduction in the apparent star density over a distance of 2 kpc, a result not too different from the Kapteyn model. Thus it is clear that, if we wish to determine the space density of stars in the galactic disk, we shall have to formulate methods that account fully both for density variations as a function of position along the line of sight and for the effects of interstellar absorption. Let us now turn to this problem.

4-2. STAR-COUNT ANALYSIS

Formulation

Our ultimate objective is to determine the space density $\nu(r, \ell, b, M, S)$ (stars pc^{-3}) of stars of absolute magnitude M and spectral type S at distance r from the Sun, in the direction whose galactic coordinates are (ℓ, b). By S, we shall mean here the full MK spectral type, that is, both spectral class and luminosity class. Thus G V and G III stars, for example, would each be denoted by a different S in this analysis. The basic data from which we hope to deduce $\nu(r, \ell, b, M, S)$ are the observed star counts $A(m, \ell, b, S)$ of stars of spectral type S in direction (ℓ, b). For brevity, we shall henceforth drop the variables ℓ and b in our notation, with the understanding that any particular analysis will be carried out for a definite field with a specific ℓ and b.

It is usually assumed that $\nu(r, M, S)$ can be represented as the product of two factors: a function $D_S(r)$ representing the density of stars of spectral type S at distance r in units of the corresponding stellar density near the Sun, and a distribution function $\Phi(M, S)$ giving the actual number of stars having absolute magnitude M and spectral type S per cubic parsec in the solar neighborhood. We can then write

$$\nu(r, M, S)\,dM\,dV = \Phi(M, S)\,dM\,D_S(r)\,dV \tag{4-16}$$

where dV is a volume element and dM is an increment of absolute magnitude. In the astronomical literature, $D_S(r)$ is called the *relative density function*, and $\Phi(M, S)$, the *luminosity function*.

We now inquire how the star counts $A(m, S)$ in a given field on the sky are related to the space density of stars. If a star has an absolute magnitude M, then it will have an apparent magnitude m if it is at a distance r such that

$$m = 5 \log r + a(r) + M - 5 \tag{4-17}$$

where $a(r)$ is the absorption in magnitudes along the line of sight. Now, in a field subtending a solid angle ω, stars on the distance interval $(r, r + dr)$ occupy a volume $dV = \omega r^2\,dr$. Hence, from this volume we obtain a contribution to the star count at apparent magnitude m of

$$dA(m, S) = \Phi[m + 5 - 5 \log r - a(r), S]D_S(r)\omega r^2\,dr \tag{4-18}$$

from stars at the appropriate absolute magnitude $M(m, r)$ as determined from equation (4-17). Stars at some other distance, say, r', and of a different absolute magnitude, M', can also contribute to $A(m, S)$ if the combination m, r', M' satisfies equation (4-17). Thus the total number of stars $A(m, S)$ is obtained by summing over all distances or, equivalently, over stars at all absolute magnitudes distributed along the line of sight in such a way as always to have the chosen apparent magnitude m. That is,

$$A(m, S) = \omega \int_0^\infty \Phi[m + 5 - 5 \log r - a(r), S] D_S(r) r^2 \, dr \qquad (4\text{-}19)$$

The analysis as formulated thus far has presumed that we actually have star-count data for stars of a specified MK spectral type S (or a narrow range of types). This means, of course, that somehow we must have selected just these stars within the field on the basis of some spectrophotometric criterion. As this assumption clearly implies that we have examined every single star in the field, however casually, it is obvious that the practical data-acquisition and data-analysis problems are formidable. Often these data simply do not exist, and many studies of star counts, particularly the early ones such as Kapteyn's, have been based on *general star counts* giving $A(m)$, the total number of stars at apparent magnitude m, irrespective of spectral type S. While the approach of lumping all spectral types together seems much simpler than a separate treatment of each type because it demands fewer data, we shall see that general star-count analyses yield much less information, have very restricted validity, and are subject to great inaccuracies.

For general star counts, we can write an equation of the same form as equation (4-19), namely

$$A(m) = \omega \int_0^\infty \Phi[m + 5 - 5 \log r - a(r)] D(r) r^2 \, dr \qquad (4\text{-}20)$$

Here $\Phi(M)$ now denotes the *general luminosity function*

$$\Phi(M) \equiv \sum_{\substack{\text{all spectral} \\ \text{types } S}} \Phi(M, S) \qquad (4\text{-}21)$$

and the density function $D(r)$ appearing in equation (4-20) is to be interpreted as the total number of stars of all types per unit volume at distance r in units of the total number of all types per unit volume in the solar neighborhood. But this interpretation will be strictly correct only if the relative proportions of stars of different spectral classes per unit volume are the same everywhere along the line of sight, so that the luminosity function $\Phi(M)$ determined in the solar neighborhood is, in fact, valid throughout the volume of integration. That this is necessary can be seen by summing equation (4-19) over all S, which yields equation (4-20) only if

$$\Phi(M, r) = \sum_S \frac{D_S(r)}{D(r)} \Phi(M, S) \equiv \Phi(M) \tag{4-22}$$

This equation can be true only if the fraction $[D_S(r)/D(r)]$ is everywhere unity for each S, that is, if the mixture of stars of different types is everywhere the same as in the solar neighborhood. (Recall that both D_S and D are *relative* densities and hence are identically unity near the Sun.) If the mixture of stars of different spectral types changes from position to position, then the assumptions on which general star-count analyses are based become invalid.

Equations (4-19) and (4-20) are *integral equations* for $D_S(r)$ and $D(r)$, respectively. In principle, they can be solved provided that we know (1) $A(m, S)$ or $A(m)$ from observation, (2) $a(r)$, either from observation or an assumed model, and (3) $\Phi(M, S)$ or $\Phi(M)$. Later in this section, we shall discuss how we handle absorption effects, how we determine the luminosity functions, and how we solve the integral equations. But before we consider these questions in detail, we want to mention a few general points.

First, although the integrations in equations (4-19) and (4-20) formally extend to infinity, in reality they are truncated at some maximum distance by the fact that there is a physical upper limit to stellar luminosities. Thus, from Table 3-2, we recall that there are no stars to which we assign absolute magnitudes $M_V \lesssim -9$, and, indeed, stars with $M_V \leq -6$ are extremely rare (see Table 4-4). Thus, in practice, there will be essentially no contribution to the integrals from distances greater than those determined by the relation

$$\log r_{\max} \approx 0.2(m + 6) + 1 - 0.2a(r_{\max}) \tag{4-23}$$

Most star-count analyses have typically been carried down only to about $m \approx 16$, so that a representative upper limit on $\log r_{\max}$ is $5.4 - 0.2a(r_{\max})$. As mentioned earlier, a reasonable estimate for $a(r)$ along lines of sight lying in the disk is about 1 mag/kpc. Therefore, even the very brightest stars that exist will not contribute to the observed star counts from distances beyond about 7.5 kpc. For a more realistic example, say A stars with $M = 1$, the limit is $\log r_{\max} \approx 4 - 0.2a(r_{\max})$, which leads to $r_{\max} \approx 2.3$ kpc for the same assumed absorption. For fainter stars, the limiting distances become even smaller. Thus star-count analyses will yield useful information only in a rather restricted volume of space around the Sun, unless they are pushed to very faint apparent magnitudes in fields of very low absorption. On the other hand, observations made from satellites in astutely chosen X-ray or infrared bands in which interstellar absorption is not important may possibly allow the classical method of star-count analysis to be applied on a Galaxy-wide basis for objects that emit at those wavelengths.

Second, one should realize that there are important differences in the amount and reliability of the information we can obtain from general star counts and from counts for a specific spectral type. Broadly speaking, the general counts are inferior, even though the data for them are much easier to acquire, for the following reasons:

1. Both equations (4-19) and (4-20) are essentially *convolution integrals* of the density function against the luminosity function. In these integrals, the effective width of the luminosity function—that is, the range of absolute magnitudes for which it contains large numbers of stars—determines the range of distances that can contribute to the observed number of stars at some specified apparent magnitude. For the sake of argument, suppose that Φ is constant on some range $(M, M + \Delta M)$ and identically zero outside this range. If stars of absolute magnitude M contribute to $A(m)$ for some m when they are at distance r_2, then stars with absolute magnitude $M + \Delta M$ can also contribute to $A(m)$ when they are at some distance $r_1 < r_2$, and stars within the whole distance range (r_1, r_2) will affect $A(m)$. Clearly, the larger the value of ΔM, the larger the distance spread $\Delta r \equiv r_2 - r_1$ within which stars contribute to the observed counts. The larger Δr becomes, the more severely information about $D(r)$ is smeared out, and the more difficult it is to recover, because a fluctuation in, say, $D(r')$ can be masked by a suitable opposite fluctuation in $D(r'')$, where both r' and r'' are on the range (r_1, r_2). Given that the observational data $A(m)$ inevitably contain errors, a large width ΔM in Φ generally implies that the density function is poorly determined or may even be indeterminate. As we shall see, the general luminosity function has an enormous width in ΔM (about 10 mag). Hence analyses of general star counts suffer from a severe lack of resolving power, and $D(r)$ is often very poorly known. In contrast, stars of a single MK type have well-defined absolute magnitudes (see Table 3-2), which scatter over a very narrow range, typically less than 1 mag. Thus, if we choose stars of a single type S, only a very narrow range Δr can contribute to $A(m, S)$, and $D_S(r)$ is correspondingly much better determined.
2. As we have already mentioned, when we group all spectral types together in a general star-count analysis, we are tacitly assuming that the general luminosity function $\Phi(M)$ is the same at every point within our range of observation, as it is within the solar neighborhood. This assumption means that the relative proportions of various spectral types in the sample must remain the same within the entire volume analyzed; it simply ignores changes in these proportions that result from stellar-revolutionary effects (for example, O–B stars are for the most part confined to spiral arms because

their lifetimes are so short), and it ignores changes in the relative numbers of stars from different age groups or population groups that result from spatial gradients or differences in kinematical properties. These problems are particularly severe along lines of sight out of the galactic plane, for, as we shall see in §4-3, stars of different types have markedly different scale heights above and below the galactic plane (already apparent in Figure 4-1). In contrast, for stars of a specific spectral type S, $\Phi(M, S)$ is essentially a description of the distribution of stars around a well-defined mean value of M. Insofar as stars of a definite MK type have well-defined physical properties and a definite evolutionary status, there is no reason to suppose that the form of this distribution or its associated mean magnitude will change significantly, even over large distances in our Galaxy. Hence the use of a particular $\Phi(M, S)$ is probably valid within a very large volume of space.

From the points just discussed, it should be clear why most recent investigations have focused on determining space densities for samples of stars having a narrow range of spectral type. The disadvantage, of course, comes in having to identify those particular stars among all stars found within a field. Either one must be able to screen a vast number of stars photometrically or spectroscopically with a very efficient technique, or one must be able to choose the desired stars in some other way (for example, by means of a distinctive light variation, such as that used to pick out RR Lyrae variables).

Elimination of Absorption Effects

Before we can find the true space density of stars, we must eliminate absorption effects. There are basically two ways in which this can be done. First, for some spectral types (particularly early types), it is possible to determine the amount of absorption directly for each star observed. For example, we could measure the reddening-free parameter Q [see equation (3-71)] for all the stars in a field, and, from these data, find intrinsic colors, color excesses, and hence absorptions for each star. If we then merely subtract the measured absorptions from the observed apparent magnitudes of the stars before we construct the counts $A(m, S)$, we can solve equation (4-19) directly for $D_S(r)$, using mathematical techniques to be discussed later. This approach, while straightforward in principle, requires that we make the appropriate photometric measurements for every star in the sample, which may not be practicable.

If we cannot eliminate absorption effects directly before analyzing the data, then we can take a second approach and derive a *fictitious density distribution* by initially ignoring absorption and working with apparent stellar distances rather than true distances. If we can later determine $a(r)$ by some independent means, we can ultimately convert the fictitious density function to the *true density function*. Thus, let ρ be the apparent distance of a star, and r its true distance; ρ and r are related by equation (4-14) or (4-15). For $A(m, S)$, we then write

$$\begin{aligned} A(m, S) &= \omega \int_0^\infty D_S(r)\Phi[m + 5 - 5 \log r - a(r), S]r^2\, dr \\ &\equiv \omega \int_0^\infty \Delta_S(\rho)\Phi(m + 5 - 5 \log \rho, S)\rho^2\, d\rho \end{aligned} \tag{4-24}$$

where $\Delta_S(\rho)$ is the fictitious density function, which will be the same as the true density function $D_S(r)$ only if there is no absorption. Clearly, these two functions are related by the expression

$$D_S(r) = \left(\frac{\rho^2}{r^2}\frac{d\rho}{dr}\right)\Delta_S(\rho) \tag{4-25}$$

or, from equation (4-15),

$$D_S(r) = \left[\left(1 + 0.2\mu r \frac{da}{dr}\right)10^{0.6a(r)}\right]\Delta_S[\rho(r)] \tag{4-26}$$

where again $\mu = \ln 10$. Thus, if we can find $\Delta_S(\rho)$ and determine $a(r)$ by one means or another, then we can calculate the true density $D_S(r)$, corrected for absorption, from $\Delta_S(\rho)$ via equation (4-26).

In the approach just described, we determine $a(r)$ from direct measurements of the absorption suffered by a selected sample of stars in the field, chosen such that their individual absorptions are readily determinable and their distribution along the line of sight defines the run of $a(r)$ with the desired degree of precision. These stars are employed merely as tracers, and they need not be of the same spectral type as the group under analysis. Moreover, we do not need to measure absorptions for all stars under analysis, but only for a sufficient number to find $a(r)$. We then assume that this run of $a(r)$ is valid for all stars in the field, which is reasonable unless the absorption in the field is patchy (a matter that can be investigated observationally), and then we can use it in equation (4-26) to convert Δ_S to D_S.

If $a(r)$ has not been determined directly from observation for some particular field, one can attempt to correct for absorption effects by using some assumed run of $a(r)$. A common choice is $a(r) = kr$, where k is the extinction in magnitudes per parsec or per kiloparsec. In this simple case, equation (4-26) reduces to

$$D_S(r) = (1 + 0.2\mu kr)10^{0.6kr}\Delta_S[\rho(r)] \tag{4-27}$$

Table 4-3. Illustration of Effects of Interstellar Absorption on Distance and Space-Density Estimates*

r(pc)	$a(r)$(mag)	ρ(pc)	$\Delta(\rho)/D(r)$
0	0.00	0	1.00
10	0.01	10	0.98
100	0.10	105	0.83
200	0.20	219	0.69
500	0.50	630	0.41
800	0.80	1,155	0.24
1000	1.00	1,580	0.17
4000	4.00	25,200	0.0014

* $a(r) = kr$, $k = 1$ mag kpc^{-1}

The choice of k is naturally rather problematical, and, although one might be guided in setting its value if direct estimates of $a(r)$ are available for adjacent fields, the final results are actually extremely sensitive to the run of $a(r)$, and one cannot expect to obtain results of high accuracy in this way. At best, the use of an assumed form for $a(r)$ merely provides an illustration of the effects of absorption.

To see how important absorption effects can be, consider Table 4-3, which was constructed assuming $k = 1$ mag kpc^{-1} $= 0.001$ mag pc^{-1}. It is obvious from the table that, beyond a few hundred parsecs, the correction factors required to convert the fictitious density function to the true density function become enormous, and thus even small errors in $a(r)$ will translate to huge errors in $D_S(r)$. This fact demonstrates the need for determining $a(r)$ with as much precision as possible.

Finally, before leaving the subject of the effects of absorption on star counts, it is worth mentioning that, in a plot of $A(m)$ versus m, one occasionally sees a rather distinctive break in the curve at some particular m, where the slope suddenly decreases in some magnitude interval and then rises again with the same slope as before, as illustrated schematically in Figure 4-3. Such plots are called *Wolf diagrams* in honor of M. Wolf, who used them in pioneering studies of the properites of interstellar absorption clouds. He realized that the observed effect is caused by the presence of a discrete interstellar cloud along the line of sight. The near edge of the cloud is located at the distance corresponding to apparent magnitude m_1, and the far edge is located at the distance corresponding to apparent magnitude m_2 (the absolute magnitude of the stars presumably being known). The total absorption in the cloud is Δm, as shown in the figure. These diagrams were of importance historically in helping to define (roughly) the physical properties of dark absorption clouds.

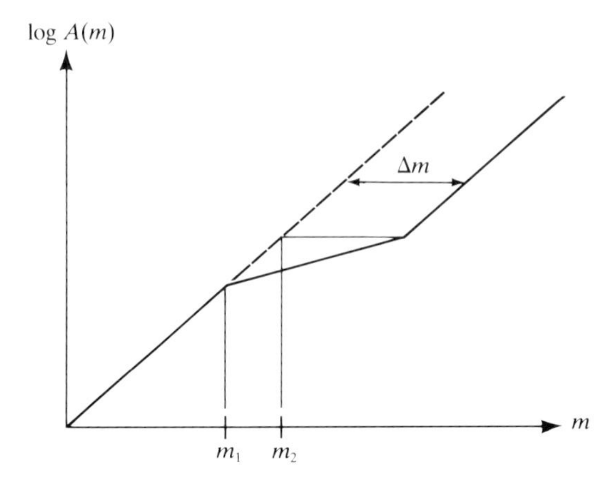

Figure 4-3. Schematic Wolf diagram. *Ordinate*: star count $A(m)$; *abscissa*: apparent magnitude m; *solid curve*: observed counts; *dashed curve*: counts that would have been obtained in absence of absorbing cloud. The spatial extent of the cloud can be determined from m_1 and m_2, and the total absorption in the cloud is Δm.

The Luminosity Function

Let us now turn to the problem of determining the luminosity functions $\Phi(M)$ and $\Phi(M, S)$. These functions can be derived in several different ways. We shall consider here only the most direct approach, which utilizes trigonometric parallax data. Our discussion will rather closely parallel that of Trumpler and Weaver in their book *Statistical Astronomy* (**T4**, Chapter 4.2), which should be consulted for further details and for discussion of other methods.

In the determination of $\Phi(M)$, a number of difficulties are encountered because of the *incompleteness* of the observations and because of the operation of *selection effects*. As problems of this sort often arise in the analysis of galactic data, we shall examine these difficulties in modest detail as an example of the elaborate chain of reductions and corrections that must sometimes be applied to extract the desired information from incomplete data. The methodology has been rather highly developed in this case, and it thus provides a good model (unfortunately often disregarded) of how one should proceed in determining true distributions for the more recently discovered exotic objects, such as pulsars, X-ray sources, radio galaxies, and quasars. For these objects, the observational data are necessarily woefully incomplete. Nevertheless, careful attention to the details of how the observed sample is selected allow statistical corrections to be applied, at least to some extent.

The General Luminosity Function, $\Phi(M)$ Consider the stars within a sphere of radius 50 pc centered on the Sun. This volume is large enough to contain a statistically useful sample of stars while still small enough that interstellar absorption can be neglected within it. Inside this sphere, stellar parallaxes are all 0''.02 or more, and they can therefore be measured trigonometrically with reasonable accuracy. To obtain $\Phi(M)$ for this sample of stars, we would in principle (1) select all stars with parallaxes $\geq$ 0''.02 from parallax catalogs, (2) calculate M from m and π for each star, (3) add the total number of stars in each interval of M, and finally (4) divide by the volume of the 50-pc sphere.

While in theory one might think that this procedure would give the number of stars per cubic parsec in the solar neighborhood at successive values of M, in practice it is worthless, and the actual procedure is much more complicated, for several reasons. First, and most important, the parallax data are very incomplete. Parallaxes have thus far been measured for only a tiny fraction of the stars within 50 pc of the Sun. For example, suppose we want to analyze all stars with absolute magnitudes $M \leq 15$ within 10 pc of the Sun. In principle, we want to choose, from all stars with apparent magnitudes $m \leq 15$, just those stars with parallaxes $\geq$ 0''.10. But there are roughly 4×10^7 stars with $m \leq 15$ on the sky, and parallaxes are known for only a few thousand stars! Moreover, the task of extending the body of presently available data by orders of magnitude is obviously hopeless, given that even a single parallax determination requires an extensive series of observations over a period of years and then a large effort to measure and reduce the observations. A practical way around this impasse is to make use of proper-motion data, exploiting the fact that, in general, stars that are close to the Sun tend to have large values of μ. Such stars are relatively easily found from proper-motion surveys, and an extensive body of data exists (see Table 2-6). Of course, there will still be stars close to the Sun whose motions relative to it happen to lie nearly along the line of sight, so that their proper motions are small. These stars will be missing from the proper-motion surveys, but, as we shall see, an allowance can be made for them by means of an appropriate statistical correction.

Second, the measured parallaxes contain random errors of the order of 0''.01. Some stars whose true parallaxes are really 0''.02 or larger will have measured parallaxes less than 0''.02, and hence they will inadvertantly be excluded from the sample. Likewise, some stars that have true parallaxes less than 0''.02 will have measured parallaxes of 0''.02 or more, and hence they will spuriously be included in the sample. As the number of stars rises rapidly with decreasing parallax (increasing distance) because larger volumes are surveyed, the net effect of these errors causes us to include more stars in the 50-pc sphere than really belong there. Again, a statistical correction can be made for this effect.

Finally, high-luminosity stars are very rare within 50 pc of the Sun. In fact, from this volume of space, essentially no information can be derived about $\Phi(M)$ for $M \lesssim -2$. For the intrinsically brightest stars, we must survey

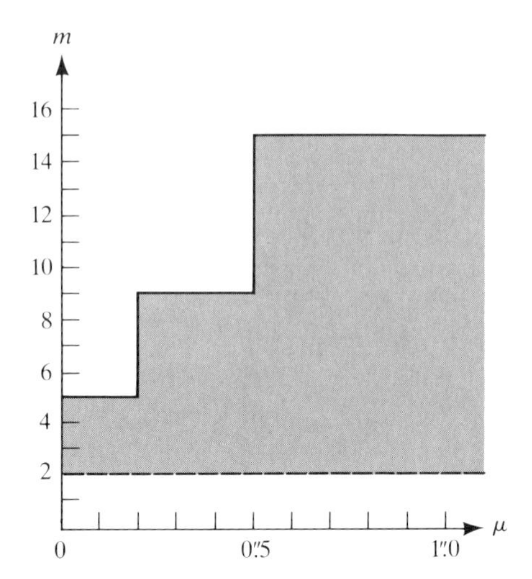

Figure 4-4. Region in the (m, μ) plane where the observed sample of stars is complete. At very bright magnitudes, there are too few stars to permit an adequate statistical discussion. [From (**T4**, 373), by permission.]

through larger volumes and carry out the analysis for one spectral class at a time.

With the foregoing remarks in mind, let us now consider the actual procedure by which $\Phi(M)$ has been determined. We begin by counting the number of stars for a specified set of values of m and μ in proper-motion catalogs. We restrict the region in the (m, μ) plane over which counts are made (see Figure 4-4) by fixing, for each m, a minimum value μ_0 chosen to be large enough to ensure that, at the apparent magnitude specified, most stars with $\mu \geq \mu_0$ have in fact been discovered in the proper-motion surveys. Representative choices for μ_0 in various ranges of m are as follows: $(m, \mu_0) = (5, 0)$; (5–9, 0″.2); (9–15, 0″.5). Now, for all stars in the proper-motion catalogs that have a specified m and μ, only a fraction will have measured parallaxes. To allow for this fact, we define the quantity $K_1(m, \mu)$ as

$$K_1(m, \mu) \equiv \frac{\text{number of stars with chosen } (m, \mu)}{\text{number of stars with chosen } (m, \mu) \text{ and measured } \pi} \tag{4-28}$$

Those stars referred to in the denominator of this ratio will, of course, have some distribution over parallax. If we can assume that all stars in the (m, μ) group have the same distribution, then $K_1(m, \mu)$ is a valid estimate of the factor by which any star with some measured parallax within some (m, μ) group should be multiplied to give the total number that really have that value of the parallax in that (m, μ) group.

If we now turn to the parallax catalogs, we can find the number of stars with a specified m and *observed parallax* π'. Each such star has a known proper motion μ, and, to correct for incompleteness of the parallax data, we count it as $K_1(m, \mu)$ stars using the factors just derived. By summing over all values of μ at a set of fixed values of (m, π'), we can then construct numerically the distribution function $G_0(m, \pi' \mid \mu \geq \mu_0)$, which gives the number of stars of apparent magnitude m and parallax π' for all proper motions down to the limit μ_0 specified earlier.

We must now consider the effects of errors in the measured parallaxes. If the errors of measurement are random, then it is reasonable to compute the probability of measuring a parallax π' for a star of *true parallax* π from the Gaussian distribution

$$p(\pi', \pi) = \left(\frac{0.56}{\sigma}\right) e^{-(\pi' - \pi)^2/\sigma^2} \tag{4-29}$$

where σ is $\sqrt{2}$ times the standard deviation of the measuring errors. If the true distribution of stars over parallax is $f(\pi)$, then the observed distribution $f_{\text{obs}}(\pi')$ will be given by

$$f_{\text{obs}}(\pi') = \left(\frac{0.56}{\sigma}\right) \int_{-\infty}^{\infty} f(\pi) e^{-(\pi - \pi')^2/\sigma^2} \, d\pi \tag{4-30}$$

We wish to determine $f(\pi)$ from the observed function $f_{\text{obs}}(\pi')$; to do this we must deconvolve equation (4-30). If σ is small, we can perform the deconvolution approximately in the following way. Suppose we expand $f(\pi)$ in a power series around $\pi = \pi'$; that is, we write

$$f(\pi) \approx f(\pi') + (\pi - \pi')f'(\pi') + \tfrac{1}{2}(\pi - \pi')^2 f''(\pi') + \cdots \tag{4-31}$$

Substituting this expression into equation (4-30), we find that the dominant terms that survive the integration are

$$f_{\text{obs}}(\pi') = f(\pi') + \tfrac{1}{4}\sigma^2 f''(\pi') + 0(\sigma^4)$$

If σ is small, then we can neglect the higher-order terms, and because we expect the correction term to be small (it is proportional to σ^2) we can approximate $f''(\pi')$—the second derivative of the true distribution with respect to π, evaluated at π'—by $f''_{\text{obs}}(\pi')$—the second derivative of the observed distribution function, which can be estimated numerically by differencing the data. Hence we can write

$$f(\pi) \approx f_{\text{obs}}(\pi) - \tfrac{1}{4}\sigma^2 f''_{\text{obs}}(\pi) \tag{4-33}$$

In practice this result is useful only if σ, which can be estimated directly from the parallax observations, is small and if the data are smooth enough to survive two differencing operations and still yield a reasonably good estimate of f''.

We make use of equation (4-33) in the following way. If we simply disregard proper motions for the moment, then, at a specified value of m, the function $G_0(m, \pi' | \mu > \mu_0)$ gives the number of stars as a function of π'. We can identify this distribution with $f_{obs}(\pi')$ and apply equation (4-33) to obtain the true distribution $f(\pi)$. By carrying this procedure through at each successive apparent magnitude m, we obtain the corrected distribution function, $G(m, \pi | \mu > \mu_0)$.

Next, we must allow for the fact that some stars within the 50-pc sphere have proper motions below our completeness limit, that is, $\mu < \mu_0$. These stars are missing from the proper-motion catalogs, but they must, of course, be included in the determination of $\Phi(M)$. Given a kinematic model for the distribution of stellar random velocities, both in direction and in speed, we can calculate the distribution of proper motions that we will observe for a group of stars at some specified distance. Hence it is possible to evaluate a correction factor $K_2(m, \pi | \mu_0)$, such that the function

$$F(m, \pi) \equiv K_2(m, \pi | \mu_0) G(m, \pi | \mu \geq \mu_0) \tag{4-34}$$

is the total number of stars of apparent magnitude m and parallax π within the 50-pc sphere, irrespective of proper motion. The value obtained for $K_2(m, \pi | \mu_0)$ depends mainly on the direction and magnitude of the solar motion (see §6-3) with respect to the stars under study and on their random-velocity dispersion. The dependence of K_2 upon m is very weak. The calculation of $K_2(m, \pi | \mu_0)$ can be based on an *ellipsoidal random-velocity distribution* (see §7-1), which is a superposition of three independent Gaussian distributions having different dispersions along three orthogonal axes, oriented in a particular way with respect to the galactic center. The details of the computation are a bit too lengthy to discuss here, but they are described briefly in (**T4**, 375–377). Typical values for K_2, computed for an isotropic random-velocity distribution and averaged over the sky, are given in Table 4-4 for a representative range of π. When K_2 is large, say 2 or greater, it is obvious that the majority of the stars in the corresponding (m, π) group are missing from the catalogs, and hence the statistical sample is too small to yield reliable results. Such groups are therefore excluded from the analysis.

Table 4-4. Correction Factor $K_2(m, \pi | \mu_0)$

π	$K_2(m, \pi \| 0''.2)$	$K_2(m, \pi \| 0''.5)$
0''.02	2.7	830
0''.05	1.18	2.7
0''.10	1.04	1.28
0''.20	1.01	1.06

SOURCE: (**T4**, 377), by permission

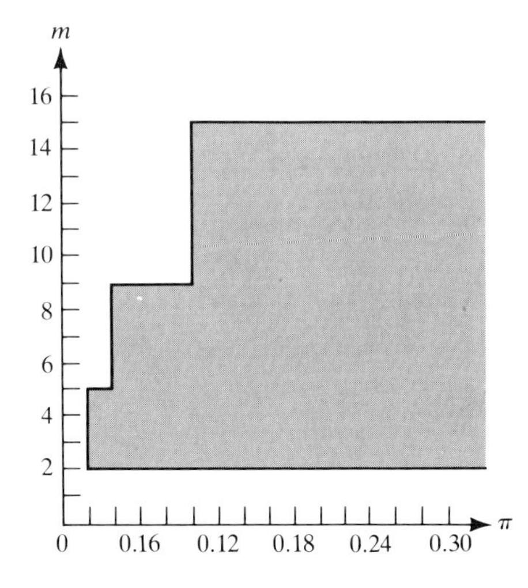

Figure 4-5. Region in the (m, π) plane where the function $G(m, \pi)$ can be determined reliably. [From (**T4**, 378), by permission.]

The region of the (m, π) plane within which $G(m, \pi)$ can be determined reliably is shown in Figure 4-5.

The final step in deriving $\Phi(M)$ is to convert $G(m, \pi)$, the distribution over apparent magnitude m, into a distribution $H(M, \pi)$ over absolute magnitude M by use of the relation

$$M = m + 5 + 5 \log \pi \tag{4-35}$$

This conversion can be done reliably within the shaded region shown in Figure 4-6. We then choose a specific value of M and sum the tabulated values

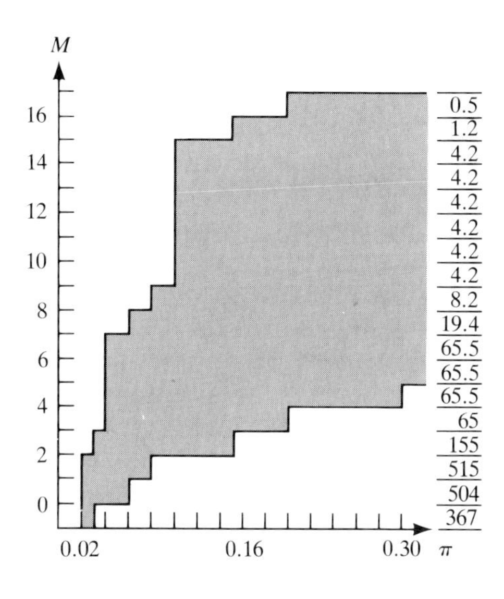

Figure 4-6. Limits over which the data are sufficiently reliable for the transformation of $G(m, \pi)$ to $H(M, \pi)$. The numbers at the right-hand edge of the diagram are the volumes (in units of 10^3 pc^3) employed in the magnitude ranges indicated, as described in the text. [From (**T4**, 379), by permission.]

of $H(M, \pi)$ over the range of π for which data exist. Finally, we divide this sum by the volume corresponding to that range of π, which yields an estimate of $\Phi(M)$, the number of stars of absolute magnitude M per cubic parsec in the solar neighborhood. Repeating this procedure for successive values of M yields the full run of $\Phi(M)$.

The results for the general luminosity function as obtained by van Rhijn, McCuskey, and, most recently, Luyten (**L3**) are presented in Table 4-5. This luminosity function is often referred to as the *van Rhijn luminosity function* in

Table 4-5. General Luminosity Function

	$\log \Phi(m) + 10$		$\Phi(M)$ per 10^4 pc^3		$(\mathscr{L}/\mathscr{L}_\odot)$ per 10^4 pc^3		
M	M_{pg}	M_V	M_{pg}	M_V	Photo	Vis	$(\mathscr{M}/\mathscr{M}_\odot)$ per 10^4 pc^3
<-6					3	1	
-6	2.4	2.1	0.0002	0.0001	6	3	0.005
-5	3.1	2.8	0.0012	0.0006	13	6	0.02
-4	3.63	3.46	0.0043	0.0029	17	11	0.06
-3	4.21	4.10	0.016	0.013	26	20	0.17
-2	4.77	4.72	0.06	0.05	37	33	0.5
-1	5.31	5.40	0.20	0.25	51	63	1.6
0	5.87	6.05	1	1	74	112	4
1	6.36	6.54	2	3	91	138	10
2	6.70	6.80	5	6	79	100	12
3	6.98	7.06	10	12	60	72	18
4	7.19	7.28	15	19	39	48	23
5	7.34	7.53	22	34	22	34	37
6	7.47	7.63	30	42	12	17	38
7	7.53	7.55	34	35	5	6	26
8	7.61	7.62	41	42	3	3	26
9	7.70	7.73	59	54	1	1	29
10	7.81	7.89	65	78	1	1	34
11	7.90	7.99	80	98			35
12	7.97	8.03	93	107			34
13	8.01	8.07	102	117			28
14	8.06	8.11	115	129			23
15	8.10	8.10	126	125			20
16	8.08	8.08	120	120			15
17	8.03	8.03	107	107			9
18	7.95	7.92	89	83			6
19	7.8	7.7	63	50			4
20	7.6	7.5	40	30			2
21	7.3	7.1	20	13			1
22	6.9	6.7	8	5			1
Total			1247	1310	540	669	437

SOURCE: (**A1**, 248), by permission

recognition of van Rhijn's pioneering efforts in its determination. The first column of the table gives the absolute magnitude of the group considered. The second and third columns give logarithms of $\Phi(M)$, the number of stars of absolute magnitude M per cubic parsec, for two different magnitude scales, M_{pg} and M_V, respectively. The fourth and fifth columns give $\Phi(M)$ directly. The sixth and seventh columns give the light emitted by each absolute-magnitude group per unit volume in the photographic and visual bands, respectively, in solar units. The last column gives the mass density per unit volume in solar units. A scale factor of 10^4 has been incorporated into columns 4 through 8 for convenience, so that the numbers given there are appropriate to a volume of 10^4 pc^3.

The basic significance of the results in Table 4-5 is that they begin to provide the outlines of a picture of the properties of the stars found in a typical volume element of the galactic disk. Details of the picture will begin to emerge when we break the data down according to spectral types and stellar-population groups. Inspection of Table 4-5 reveals several noteworthy features.

First, we see that *most stars in the solar neighborhood are intrinsically faint.* The maximum in $\Phi(M)$ occurs near $M_{pg} \approx 15.5$. These stars are typically K and M dwarfs. This fact alone shows the extreme value of being able to make a detailed analysis of the local neighborhood, for it is obvious that we shall never be able to analyze such faint stars at remote locations in our Galaxy (not to mention other galaxies!), and yet these are by far the most common stellar constituent of galaxian disks.

Second, we see that *essentially all the light emitted per cubic parsec is contributed by intrinsically luminous stars* despite the fact that they are extremely rare. The peak emission comes from stars with $M_V \approx +1$. These stars are typically A dwarfs and K and M giants. It is worth stressing just how strong this bias toward high luminosities really is. For example, stars at $M_V \approx -3$, typically B stars, emit as much light per unit volume in the V band as do the stars at $M_V \approx +6$, mostly G and K dwarfs, even though the latter are 4000 times as numerous and contain 225 times as much mass! The average light output per unit volume is about 0.067 solar unit pc^{-3} in the visual band and about 0.054 solar units pc^{-3} in the photographic band, which implies that the average color of the light emitted from the disk is somewhat redder than solar radiation. The redder color results from the fact that a large fraction of the most luminous stars in the disk are red giants, whose spectrum would in fact dominate in the integrated spectrum of the disk that would be measured by an external observer.

Third, we see that *most of the stellar mass density in the solar neighborhood is contributed by the vast number of low-luminosity stars.* This situation has the unpleasant implication that the dynamics of our Galaxy are dominated by stars that are at best inconspicuous, while those very luminous objects that we can so readily observe in our own Galaxy and in other galaxies contribute but little to the gravitational forces in the system. Therefore, except insofar as the few high-luminosity stars can be used as tracers of the entire population, it is dangerous to assume that the luminosity density in galaxies is proportional to the mass density. The only safe procedure is to infer the mass distribution by studying the gravitational field of the whole system through dynamics of the population of bright stars. We shall often have occasion to return to this point in the following chapters. The total mass density from the stars in Table 4-5 is about $0.044\mathscr{M}_\odot$ pc^{-3}. However, these data omit most of the white dwarfs, which are numerous but intrinsically very faint ($M_V \approx +13$ to $+16$) and yet have masses of the order of 0.5 to $1\mathscr{M}_\odot$ per star. When the white dwarfs are taken into account, one finds (**W1**) that they contribute approximately an additional $0.02\mathscr{M}_\odot$ pc^{-3} (see Tables 4-7, 4-9, and 4-10). Hence the total stellar mass density is about $0.065\mathscr{M}_\odot$ pc^{-3}. This number gives a feeling for the relative emptiness of space in the galactic disk in the solar neighborhood.

Fourth, we see that the average *mass-to-light ratio*, that is, the mass per unit volume divided by the light emission per unit volume (in solar units), is about $\langle \mathscr{M}/\mathscr{L} \rangle \approx 0.97\mathscr{M}_\odot/\mathscr{L}_\odot$ in the visible band and about 1.2 in the photographic band. In round numbers, the light emitted by the diverse mixture of stars found in the solar neighborhood averages out to about one solar luminosity per solar mass. These numbers are useful, for, insofar as they are typical for a galaxian disk, they can be used to convert observed light distributions in other galaxies into rough first estimates of stellar mass distributions, thereby providing some input into dynamical analyses of these objects. On the other hand, note that even for the solar neighborhood the value for $\langle \mathscr{M}/\mathscr{L} \rangle_{\text{observed}}$ just quoted is surely a lower limit, because many kinds of objects (for example, faint companions in multiple stars, dead white dwarfs and neutron stars, stars lost in dense interstellar—particularly molecular—clouds, and so on) would evade discovery. Cumulatively, these objects could add appreciably to the mass density while contributing nothing to the emitted light.

We have mentioned that the actual stellar composition of a sample volume in the galactic disk is strikingly different from that which would be inferred from examining a sample of apparently bright stars, because the latter contains a disproportionately high number of intrinsically luminous stars. These stars can be seen to large distances and hence through a larger volume than faint stars. This disparity is shown clearly in Table 4-6, which gives $\phi(M_V|V \leq V_0)$, the expected distribution of 100 stars having $M_V \leq 5$ in samples that are complete down to an apparent magnitude V_0 (for two values of V_0), and $\phi(M)$, the general luminosity function normalized to 100 stars with $M_V \leq 5$.

If any one point emerges from the preceding discussion, it is that the behavior of $\Phi(M)$ at faint magnitudes is of great importance to galactic research. For this reason, a tremendous effort has been lavished on its determination. A perennial question has been whether the decrease of $\Phi(M)$ for $M_{pg} \geq 15.5$ shown in Table 4-5 is real or whether it is a spurious result produced by incompleteness of the observational data. For example, a few years ago, the

Table 4-6. Absolute-Magnitude Distribution of Apparently Bright Stars Contrasted to General Luminosity Function

M_V	$\phi(M \mid V \leq V_0)$		$\phi(M)$
	$V_0 = 2.5$*	$V_0 = 6$	
<-6	6	0	0
-6	7	1	0
-5	9	1	0
-4	10	3	0
-3	10	7	0
-2	9	10	0
-1	12	14	0.5
0	17	18	1.5
1	12	21	4
2	5	15	8
3	2	6	16
4	1	3	25
5	0	1	45

* These data are smoothed.

claim was made that the standard general luminosity function grossly underestimates the density of M dwarfs and that these stars actually occur in such large numbers as to almost double the local mass density attributable to stars. For reasons that we will discuss in the companion volume to this book, this claim produced considerable interest among astronomers studying the dynamics of our Galaxy, but, in the end, it was shown to be false, having been the result of a systematic error in the photometry of the faint red stars being studied.

The practical difficulties faced in the determination of $\Phi(M)$ for $M > 15$ are formidable. First, observations to very faint apparent magnitudes, $m \approx 18$ to 20, are required. Second, an extensive search for faint stars with large proper motions must be made to identify, among the multitudes of stars in this apparent-magnitude range, those stars that are likely to be close to the Sun. Finally, accurate distances must be determined for a sizeable fraction of these stars. The most extensive body of data for this type of analysis has been accumulated with the Palomar 48-inch Schmidt telescope by Luyten, who has unselfishly devoted a lifetime of effort to determining $\Phi(M)$ at the faintest magnitude limits now accessible. An alternate approach has been followed by van de Kamp, who has attempted to make a complete census of all the stars within a few parsecs of the Sun. The difficulty here is that this census can be done in only a very small volume of space, and the resulting sample is small and totally omits many stellar types. Nevertheless, it provides important information about the number of stars in binaries and multiples and about the occurrence of white dwarfs. For example, out to about 10 pc, one finds a sample of 8 white dwarfs, 2 subgiants, and 245 main-sequence stars down to $M_V = 18$. This sample includes no O–B stars and only 4 A stars. It contains no giants or supergiants, and at least 50% of the stars are members of binaries or multiple systems. Similarly, one can use Gliese's catalog of stars within 20 pc to estimate $\Phi(M)$. The results (**W2**) for $M_V \lesssim 13$ are in excellent agreement with Luyten's results; for stars fainter than this limit, the data in Gliese's catalog become seriously incomplete and yield only lower bounds on $\Phi(M)$.

At the present time, the evidence seems categorically to support the existence of a maximum in $\Phi(M)$ near $M_{pg} \approx 15.5$, with a subsequent decline at fainter luminosities. Nevertheless, we may not yet have the complete picture. In particular, there could be huge numbers of dark objects (for example, degenerate stars, black holes, or merely substellar masses) that are undetectable by current radiation-measuring techniques, but which nevertheless contain enough material to contribute significantly to the total mass density and to influence galactic dynamics. As this issue is, at present, only speculative, we shall leave it for future research and turn to another matter.

The Luminosity Function for a Specified Spectral Type, $\Phi(M, S)$ The van Rhijn luminosity function lumps together all spectral types, and, at a given M, several radically different kinds of stars may contribute to $\Phi(M)$. As mentioned earlier, the use of $\Phi(M)$ in general star-count analyses leads to a loss of resolution in the solution for stellar space densities. Moreover, $\Phi(M)$ does not apply universally throughout our Galaxy, particularly at points away from the galactic plane. Most important, by lumping spectral types together indiscriminately, we obliterate information of great astrophysical interest. It is essential, therefore, to study the luminosity function for stars of specified spectral types.

To determine $\Phi(M, S)$, we proceed essentially as we did before for $\Phi(M)$, but for one spectral type S at a time. The same problems arise as before, one of the more troublesome being the scarcity of intrinsically bright stars. Results from several studies of $\Phi(M, S)$ are summarized in Table 4-7, which gives numbers of stars per 10^6 pc^3. These data can be used to construct a *Hess diagram*, which is, in effect, a relief map of the conventional H-R diagram, in which one plots contours of equal stellar space density in the $(M_V$, spectral type) plane [see (**T4**, 393) for an example].

Detailed studies of individual groups of stars of specified MK type show that, to a very good approximation, the data can be represented by a distribution function of the form

$$\Phi(M, S) = \frac{\Phi_0}{(2\pi)^{1/2}\sigma} e^{-(M - M_0)^2/2\sigma^2} \qquad (4\text{-}36)$$

Table 4-7. Luminosity Function by Spectral Type (Stars/10^6 pc^3)

M_V	O	B	A	F	G	K	M
−7	2×10^{-4}	5×10^{-4}	3×10^{-4}	3×10^{-4}	3×10^{-5}		
−6	5×10^{-4}	2.5×10^{-3}	10^{-3}	10^{-3}	10^{-4}	4×10^{-4}	4×10^{-4}
−5	0.001	0.025	0.01	0.006	0.008	0.004	0.010
−4	0.003	0.16	0.01	0.016	0.025	0.012	0.012
−3	0.01	0.50	0.05	0.08	0.08	0.1	0.06
−2	0.01	2.5	0.08	0.2	0.3	0.6	0.4
−1	0.01	12.5	1	1.6	1	2.5	3
0	0.001	20	20	2	8	25	10
1	–	30	100	30	30	120	10
2	–	20	200	160	50	110	–
3	–	10	80	700	150	100	–
4	–	–	30	1200	700	100	–
5	–	–	–	600	2000	300	–
6	–	–	–	200	1500	1500	10
7	–	–	–	100	800	3000	100
8	–	–	–	10	400	2500	1000
9	–	–	–	–	200	1500	3000
10	–	10	–	–	–	400	8000
11	–	100	30	10	–	200	9000
12	–	200	400	100	–	100	10^4
13	–	400	600	300	100	400	10^4
14	–	800	1000	1000	600	800	10^4
15	–	1500	2000	1000	1500	1200	8000
16	–	3000	5000	3000	3000	–	6000

SOURCE: (**A1**, 247), by permission

Table 4-8. Parameters M_0 and σ in $\Phi(M, S)$

Spectral type	M_0	σ
B0–B1	−4.0	1.0
B2–B2	−2.0	1.1
B5	−1.0	0.9
B8–A0	+0.5	1.0
A2–A5	+1.5	1.0
dF0–dF5	+3.0	1.0
dF8–dG2	+4.5	1.1
dG5	+5.5	1.1
dG8–dK3	+6.0	1.2
dK	+7.5	1.3
gG	+2.0	0.9
gK	+2.4	0.8

SOURCE: (**A1**, 200), (**B7**, 23), (**M5**)

That is, the stars scatter around a mean magnitude M_0 in a Gaussian distribution with dispersion σ. This is an extremely useful result, for, as we shall see, it makes possible a clever analytical method for solving equation (4-19) for the stellar space-density function. The coefficient Φ_0 in equation (4-36) is just the number of stars per pc^3 for the chosen spectral type (or group of types), as given in Table 4-7. The value of M_0 is the mean absolute magnitude assigned to the type or types within a group by the fundamental calibration procedures described in §3-5 (see Table 3-2). The dispersion σ represents (1) the intrinsic cosmic dispersion associated with individual MK types (recall that the classification procedure segregates stars into discrete categories, which necessarily must embrace a spread of stellar properties), (2) the additional spread that may result if a range of types are grouped together, and (3) the scatter caused by errors in the actual spectral-classification procedure used in selecting stars for the star counts (usually done at much lower dispersion than fundamental MK classification work). Typical values of M_0 and σ for representative stellar groups are listed in Table 4-8. Note that the stars listed as G and K giants are a mixture of giants and subgiants.

Given the information about $\Phi(M, S)$ contained in Table 4-7, we can now partition the data in the direction orthogonal to that chosen in constructing $\Phi(M)$, that is, we can group stars according to spectral class and luminosity class only, irrespective of their absolute magnitudes M. We then obtain the values listed in Table 4-9 for the total numbers of stars (down to $M_V = +16$) per 10^6 pc^3 in each spectral group. This table shows that the total space density of white dwarfs in the solar neighborhood is about 30% that of main-sequence stars, while the total density of giants and supergiants is about 1% that of the main sequence. The local mass density of these stellar types is

Table 4-9. Number Density of Stars by Spectral Type; log (number/10^6 pc^3)

Luminosity group	Spectral class							Totals
	O	B	A	F	G	K	M	
Giants and supergiants				1.7	2.2	2.6	1.4	2.8
Main sequence	−1.6	2.0	2.7	3.4	3.8	4.0	4.8	4.9
White dwarfs		3.8	4.0	3.7	3.7	3.4		4.4

SOURCE: Adapted from (**A1**, 247), (**A1**, 249), by permission

Table 4-10. Mass Densities ($\mathscr{M}_\odot/10^3$ pc^3) of Various Objects in the Solar Neighborhood

Object	Mass density	Object	Mass density
O–B	0.9	White dwarfs	20
A	1	Cepheids	0.001
F	3	Long-period variables	0.001
dG	4	Planetary nebulae	5×10^{-6}
dK	9	Galactic clusters	0.04
dM	25		
gG	0.8	Subdwarfs	0.15
gK	0.1	RR Lyrae variables	10^{-6}
gM	0.01	Globular clusters	0.001

SOURCE: Adapted from (**A1**, 247), (**A1**, 251), (**O4**), and (**S5**), by permission

given in Table 4-10, which again shows how the dynamical effects arising from stars in the disk are dominated by the most inconspicuous stars—the K and M dwarfs and the white dwarfs. For completeness, the table also lists average mass densities from other distinctive disk objects. The last three entries give local densities of members of the spheroidal component (that from globular clusters being averaged over a spherical shell of the appropriate radius). Of these objects, only the subdwarfs have an appreciable local mass density, about 3% of the total. As we shall see in §4-4, the situation is entirely different near the galactic center, where the mass density of the spheroidal component dominates that of the disk.

The Initial Luminosity Function and Initial Mass Function The general luminosity function $\Phi(M)$ describes the stellar composition of a unit volume in the solar neighborhood at the present time. The distribution of stars over luminosity, or mass, implied by $\Phi(M)$ is not, however, an accurate representation of the relative frequencies with which these stars are formed. This is because the main-sequence lifetime $\tau_{MS}(\mathcal{M})$ of stars decreases rapidly with increasing mass. Therefore, below some critical mass $\mathcal{M}_0$, somewhat smaller than a solar mass $\mathcal{M}_\odot$, stars will persist on the main sequence for periods longer than τ_G, the age of our Galaxy, whereas stars with $\mathcal{M} > \mathcal{M}_0$ will evolve away from the main sequence and subsequently "die" quietly as white dwarfs or explosively as supernovae, in a period $\tau_{MS}(\mathcal{M}) < \tau_G$. Thus the faint end of $\Phi(M)$ contains stars that have been created throughout the entire lifetime of our Galaxy (or at least the lifetime of the disk), while, for brighter and brighter stars, $\Phi(M)$ comprises the stars formed over an ever-decreasing fraction of that period. Clearly, this effect leads to a great preponderance of low-mass, faint stars, and it gives a very distorted picture of the direct results of star formation.

To describe the rate of star creation as a function of luminosity, E. E. Salpeter introduced (**S1**) the *initial luminosity function* $\Psi(M_V)$, which gives the relative numbers of stars formed, per unit magnitude interval per unit volume, as a function of M_V. An equivalent description is given by the *initial mass function* $\xi(\mathcal{M})$, which gives the numbers of stars formed per unit mass interval in a unit volume as a function of mass $\mathcal{M}$. Although these functions have no direct relevance to star-count analyses, they are of considerable theoretical importance, and it is convenient to discuss them here.

One can attempt to derive the initial luminosity function Ψ in two ways: (1) By determining the relative numbers of main-sequence stars as a function of M_V in extremely young clusters, one finds Ψ for stars up to the turnoff-point luminosity. By studying several clusters and using matching techniques similar to those employed in the derivation of the ZAMS, and by normalizing Ψ to Φ at the faint end, one can estimate Ψ all the way up to the intrinsically most-luminous stars. (2) An alternative approach is to use theoretical main-sequence lifetimes and write

$$\Psi(M_V) = \Phi(M_V) \qquad \text{for} \qquad \tau_{MS} \geq \tau_G \tag{4-37a}$$

and

$$\Psi(M_V) = \frac{\tau_G}{\tau_{MS}}\,\Phi(M_V) \qquad \text{for} \qquad \tau_{MS} < \tau_G \tag{4-37b}$$

Equations (4-37a and b) tacitly assume that $\Psi(M_V)$ has been independent of time. If the form of $\Psi(M_V)$ has remained unchanged, but the rate of birth of stars has been decreasing with (suppose) an exponential rate $\exp(-\alpha t)$, then the factor (τ_G/τ_{MS}) in equation (4-37b) should be replaced by $(\tau_G\alpha)\exp(\tau_G\alpha)/[\exp(\tau_{MS}\alpha) - 1]$. Then, by requiring that the two methods of deriving $\Psi(M_V)$ just described should agree, we derive some information about α, which indicates that star formation has been a slowly declining function of time: $\frac{1}{2} < \tau_G\alpha < 2$.

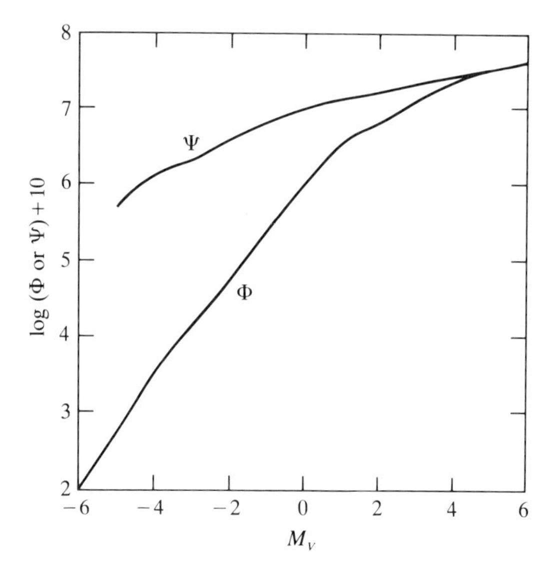

Figure 4-7. Comparison of the initial luminosity function $\Psi(M_V)$ with the general luminosity function $\Phi(M_V)$. *Ordinate* gives ($\log \Phi + 10$) or ($\log \Psi + 10$), that is, the logarithm of the number of stars in a volume of 10^{10} pc^3; *abscissa* gives absolute visual magnitude M_V.

Several different studies have been made to determine the initial luminosity function [see, for example, **(M3)**, **(S1)**, **(S2)**, **(S3)**, **(S4)**]. Both $\Psi(M_V)$ and $\Phi(M_V)$ are plotted in Figure 4-7, where we can see that the formation of a luminous star is not in reality a very rare occurrence. Even B stars are created about 10% as frequently as solar-type stars.

The initial luminosity function $\Psi(M_V)$ yields the initial mass function $\xi(\mathscr{M})$ via the known mass-luminosity relation for main-sequence stars (see Table 3-6). Salpeter found **(S1)** that, to a fair approximation, the results could be represented by a simple power law,

$$\xi(\mathscr{M}) = C(\mathscr{M}/\mathscr{M}_\odot)^{-2.35} \tag{4-38}$$

More recent work has shown that $\xi(\mathscr{M})$ is significantly flatter than this for low-mass stars [exponent ≈ -1.25 for $0.4 \lesssim (\mathscr{M}/\mathscr{M}_\odot) \lesssim 1$] and significantly steeper for high-mass stars [exponent ≈ -3.2 for $3 \lesssim (\mathscr{M}/\mathscr{M}_\odot) \lesssim 20$]. Certain important statistical calculations will contain significant errors if a constant exponent is used for all masses.

If $\xi(\mathscr{M})$ is regarded as known and is assumed to have the same form at all times, then one can write the *stellar birthrate function* in the galactic disk, that is, the number of stars born per unit volume in the mass range ($\mathscr{M}$, $\mathscr{M} + d\mathscr{M}$) and time interval ($t$, $t + dt$) as

$$b(\mathscr{M}, t)\, d\mathscr{M}\, dt = \xi(\mathscr{M})\beta(t)\, d\mathscr{M}\, dt \tag{4-39}$$

where $\beta(t)$ is the *total birthrate function* (see §3-7) and $\xi(\mathscr{M})$ is presumed to be normalized over all masses. If this description is accurate, then one can compute $\nu(\mathcal{M})$, the present space density of stars of mass $\mathcal{M}$ (at time τ_G), as

$$\nu(\mathcal{M}) = \int_0^{\tau_G} b(\mathcal{M}, t)\,dt = \xi(\mathcal{M}) \int_0^{\tau_G} \beta(t)\,dt \equiv \xi(\mathcal{M})\langle\beta\rangle\tau_G \tag{4-40}$$

for stars whose main-sequence lifetimes exceed τ_G, and as

$$\nu(\mathcal{M}) \approx b(\mathcal{M}, \tau_G)\tau_{\rm MS}(\mathcal{M}) = \xi(\mathcal{M})\beta(\tau_G)\tau_{\rm MS}(\mathcal{M}) \tag{4-41}$$

for short-lived stars. By fitting the predicted results to presently observed conditions in the galactic disk, one can develop an understanding of the time variation of star formation in the disk, and with it the history of the chemical composition of the disk, topics to which we shall return in Chapter 19.

There is no strong reason to suppose that $\xi(\mathcal{M})$ is a universal function, valid at all positions and times in our Galaxy. Indeed, considering that the physics of star formation ought to depend on the physical properties of the material (for example, its composition) and on local conditions (shear and turbulence in the primordial gas, magnetic fields, and so on), one would expect it not to be. It may thus be necessary to admit functions of the form $\xi(\mathcal{M}, t)$ or even $\xi(\mathcal{M}, \mathbf{r}, t)$, where t and $\mathbf{r}$ denote age and position in our Galaxy of the material from which the stars form.

The Luminosity Function for Spheroidal-Component Stars The spheroidal-component stars have a markedly different chemical composition from disk stars, and they are, on the average, much older. We therefore expect them to have a distinctive luminosity function. Again, although this function is not of direct relevance to space-density analyses, it is convenient to discuss it here. The representatives of the spheroidal component in the solar neighborhood, namely, the subdwarfs, are very rare, and only a small sample have reliable parallaxes. It is therefore quite difficult to use these objects to deduce a luminosity function.

The first attempts to derive a luminosity function for spheroidal-component stars used star counts in globular clusters that had been observed down to the main sequence (for example, M3, M13, M15, M92, and 47 Tuc) and for which reasonably accurate absolute-magnitude scales had been established. These data gave relative numbers of stars down to about $M_V \approx +5$. On the (reasonable) assumption that, at fainter magnitudes, $\Phi(M_V)$ for the globular-cluster stars has about the same shape as the general luminosity function for disk stars, it has been customary to normalize the two to the same value at $M_V = +5$ and to extrapolate the globular-cluster function to fainter magnitudes by using the disk function. The results of this procedure are given in Table 4-11 and are plotted in Figure 4-8. At some future time, observations of globular clusters with the Space Telescope should directly yield $\Phi(M_V)$ down to much fainter absolute magnitudes.

Table 4-11. Luminosity Function for Globular-Cluster Stars*

M_V	$\log \Phi(M_V) + 10$	M_V	$\log \Phi(M_V) + 10$
−3	4.0	4	7.3
−2	5.4	5	7.5
−1	5.7	6	7.6
0	6.2	7	Probably the same as the general luminosity function for disk stars (Table 4-5)
1	6.0	8	
2	6.3	9	
3	6.8	10	

* Arbitrarily normalized to disk luminosity function at $M_V = 5$.
SOURCE: (**A1**, 249), by permission

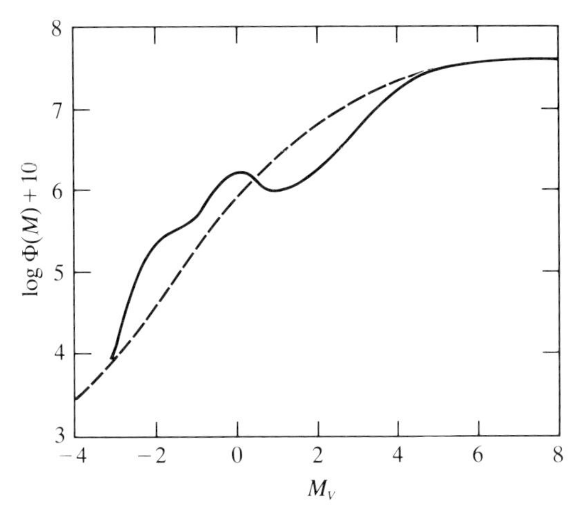

Figure 4-8. Comparison of the general luminosity function for disk stars (*dashed curve*) with that for globular-cluster stars (*solid curve*).

In comparison to the general luminosity function, the globular-cluster function (normalized as just described) shows a deficiency of stars on the range $1 \lesssim M_V \lesssim 4$. This situation occurs because, in globular clusters, such stars are subgiants and giants passing through evolutionary phases with lifetimes that are short compared to the main-sequence lifetimes of disk stars on the same absolute-magnitude range. The bump at $M_V \approx 0$ is produced by the concentration of globular-cluster stars on the horizontal branch. The relative excess of stars on the range $-3 \lesssim M_V \lesssim 0$ results from the well-developed giant branches in globular clusters and from the rapid decrease in the main-sequence lifetime of massive disk stars, which produces a rapid drop in the general luminosity function $\Phi(M_V)$ for luminous stars. [Notice that the initial luminosity function $\Psi(M_V)$ lies well above the globular-cluster luminosity function $\Phi(M_V)$ in this absolute-magnitude range.] Finally, the globular-cluster luminosity function terminates at $M_V \approx -3$, because these are the brightest stars in globular clusters, while the luminosity function for the disk, of course, persists to brighter magnitudes.

Only recently have enough data accumulated to permit one even to attempt a direct estimate of the space density and luminosity function for spheroidal-component stars in the immediate solar neighborhood. An analysis by Schmidt (**S5**) of some 125 stars, constituting a complete sample of subdwarfs with $M_{pg} \lesssim 16$ and $\mu \gtrsim 1''.3$ year, yielded a total number density of about 2×10^{-4} stars pc^{-3}, a mass density of about $1.5 \times 10^{-4} \mathscr{M}_\odot$ pc^{-3}, and a luminosity density of about $4 \times 10^{-5} \mathscr{L}_\odot$ pc^{-3}. Thus the average mass-to-luminosity ratio for halo stars in the solar neighborhood is roughly $\langle \mathscr{M}/\mathscr{L} \rangle \approx 5\mathscr{M}_\odot/\mathscr{L}_\odot$. It should be stressed that this analysis is extraordinarily difficult, and great skill and care must be exercised in applying appropriate statistical corrections to the very limited body of data now available. The situation should improve as continued observational work yields proper motions and parallaxes of a larger group of halo stars.

Mathematical Solution for the Stellar Density Function

Let us now consider how one can actually solve the integral equations (4-19) or (4-20) for $D_S(r)$ of $D(r)$. More precisely, we shall focus on the solution of equation (4-24) for the fictitious density function $\Delta_S(\rho)$, with the understanding that we will later convert $\Delta_S(\rho)$ into $D_S(r)$ using a known run of the absorption with distance, $a(r)$. Thus, suppose that, in a field subtending a solid angle ω, we are given a set of observed star counts $A(m, S)$, the number of stars of spectral type S in the apparent magnitude range $(m - \frac{1}{2}, m + \frac{1}{2})$, along with a known luminosity function $\Phi(M, S)$. We then wish to solve the equation

$$A(m, S) = \omega \int_0^\infty \Delta_S(\rho)\Phi(m + 5 - 5 \log \rho, S)\rho^2 \, d\rho \qquad (4\text{-}24)$$

for $\Delta_S(\rho)$. There are two commonly used methods of effecting this solution. We shall discuss each of these in some detail as illustrations of the practical difficulties involved in obtaining physically acceptable solutions to integral equations of the type of equation (4-24) from noisy data. We shall encounter similar difficulties on many occasions in subsequent chapters, and it is important to understand how one treats problems of this type, and why.

***Method of* (m, log π)** Of the two methods that we shall describe, the method of (m, log π) is the more general and flexible because it makes few restrictive assumptions and can be used with an arbitrary luminosity function. We begin by replacing the integral by a summation over discrete spherical shells. This may be done in several ways, the simplest being to represent the product of Δ times Φ as a step function having a constant value within each shell. Then

$$A(m, S) = \sum_{k=1}^{\infty} \Delta_S(\rho_k)\Phi(m + 5 - 5\log\rho_k, S)\delta V_k \tag{4-41}$$

where δV_k equals the volume of shell number k. The shells are customarily chosen [see, for example, (**B7**, 26–37)] so that their midpoints lie at distances

$$\log \rho_k = -\log \pi_k = \tfrac{2}{10}k \qquad (k = 1, 2, \ldots) \tag{4-42}$$

All stars in the kth shell are, in effect, assumed to lie at distance ρ_k. The spacing $\Delta \log \rho_k = 0.2$ between successive shell centers is a convenient choice, which causes the brightnesses of identical stars in neighboring shells to differ by exactly 1 mag. The edges of the kth shell are chosen to be at $\rho_{k\pm 1/2}$, where

$$\log \rho_{k+1/2} = \tfrac{2}{10}(k + \tfrac{1}{2}) \qquad (k = 1, 2, \ldots) \tag{4-43}$$

The first shell is taken to be a complete sphere having $\rho_{1/2} \equiv 0$. The volume of the kth shell within a field subtending a solid angle ω (in steradians) is

$$\delta V_k = \tfrac{1}{3}\omega(\rho_{k+1/2}^3 - \rho_{k-1/2}^3) \tag{4-44}$$

Typical values of ρ_k, and of δV_k for a $1^\circ \times 1^\circ$ field, are listed in Table 4-12. It is evident that the volume of each successive shell increases by about a factor of four over its predecessor [$\Delta(\log \delta V_k) \equiv 0.6$].

Table 4-12. Shell Centers, Edges, and Volumes in Method of $(m, \log \pi)$

Shell number k	Shell center $\log \rho_k$	Shell edges		Shell volume δV_k (pc³) for $1^\circ \times 1^\circ$ field
		$\log \rho_{k-1/2}$	$\log \rho_{k+1/2}$	
1	0.2	–	0.3	8.12×10^{-4}
2	0.4	0.3	0.5	2.40×10^{-3}
3	0.6	0.5	0.7	9.56×10^{-3}
4	0.8	0.7	0.9	3.80×10^{-2}
5	1.0	0.9	1.1	0.15
6	1.2	1.1	1.3	0.60
7	1.4	1.3	1.5	2.40
8	1.6	1.5	1.7	9.56

Once we have chosen volume elements δV_k and have evaluated $\Phi(m + 5 - 5 \log \rho_k, S)$, equation (4-41) for several values of m constitutes a set of linear equations for the unknowns, $\Delta(\rho_k)$. These equations can, in principle, be solved directly by standard methods. Such an approach usually fails, however, because the data $A(m, S)$ normally contain substantial random errors, and the system of equations is so poorly conditioned (especially for general star counts) that it is unstable, so that one obtains nonsensical results [for example, negative values of $\Delta_S(\rho)$ or wild fluctuations in Δ_S from shell to shell]. To overcome these problems, one can use specially developed least-squares techniques, or one can even turn to a trial-and-error hand-calculation procedure, which allows one to monitor the solution and find one that fits the data to within the errors and is physically plausible.

As an example, suppose we wish to analyze a given body of general star-count data. We choose a set of shells as described, and, using the general luminosity function given in Table 4-5, we form a table of the product $\Phi(m + 5 - 5 \log \rho_k)\delta V_k$, in which the rows are evaluated at constant $\log \rho_k$, and successive columns at constant m, in 1-mag steps over the relevant range. An example for a $1^\circ \times 1^\circ$ field is shown in Table 4-13 [see also (**B7**, 26–30)]. The last column in the table gives the logarithm of the volume elements δV_k in each shell, except for the first entry, which is for the volume of a complete sphere of radius $\rho_1 = 12.6$ pc. Each entry in this table corresponds to a definite (integer) value of the absolute magnitude M, and it gives the number of stars at this absolute magnitude in the volume element δV_k if $\Delta(\rho_k) \equiv 1$. Note that lines of constant M run diagonally across the table from upper left to lower right. Successive rows correspond to stars that are 1 mag brighter in absolute magnitude than in the preceding row.

To determine $\Delta(\rho)$, we begin by making some initial estimate of this function. For example, as a first guess, we might assume that $\Delta(\rho) \equiv 1$ for all values of ρ, that is, that the stellar density is everywhere the same as in the solar neighborhood. Then, for each successive column, we multiply the appropriate entries in Table 4-13 by the assumed $\Delta(\rho)$ and sum down that column to obtain, at a definite m, a computed value for $A(m)$. Ideally, this computed value should agree with the observed value. In general, we will find that the computed and observed values do not agree. We therefore go back and adjust $\Delta(\rho_k)$ at each ρ_k in such a way as to yield computed values of $A(m)$ that more nearly equal the observed values. This adjustment process can be repeated again and again until we obtain agreement of the computed and observed values of $A(m)$ to within the accuracy of the observational data. The final step is to use the absorption function $a(r)$ (determined observationally or merely estimated) to convert ρ into r by equation (4-15), and $\Delta(\rho)$ into $D(r)$ by equation (4-26). Improvements and extensions of this basic technique are described in (**T4**, 454–481).

To illustrate the method, suppose we analyze the observed values of $A(m, 0^\circ)$ given in Table 4-1. Following the procedure just outlined, we obtain the fictitious density distribution $\Delta(\rho)$ given in Table 4-15. This

Table 4-13. Table of $\Phi(m + 5 - 5 \log \rho_k)\delta V_k$ for Method of $(m, \log \pi)$

$\log \rho_k =$ $-\log \pi_k$	m												$\log \delta V_k$
	8	9	10	11	12	13	14	15	16	17	18	19	
−1.0	0	0	0										−0.69
−1.2	0	0	0										−0.22
−1.4	0.01	0.01	0.01	0.01	0.01	0.02	0.02	0.02	0.03	0.03			0.38
−1.6	0.02	0.03	0.03	0.03	0.04	0.06	0.08	0.09	0.10	0.12	0.13		0.98
−1.8	0.05	0.08	0.10	0.12	0.13	0.16	0.25	0.30	0.34	0.42	0.49	0.54	1.58
−2.0	0.14	0.21	0.32	0.40	0.46	0.51	0.63	1.00	1.18	1.35	1.66	1.95	2.18
−2.2	0.26	0.58	0.85	1.26	1.58	1.82	2.04	2.52	3.98	4.72	5.38	6.60	2.78
−2.4	0.58	1.05	2.29	3.40	5.00	6.30	7.25	8.14	10.0	15.8	18.8	21.4	3.28
−2.6	0.66	2.29	4.18	9.13	13.5	19.9	25.1	28.9	32.4	39.8	63.0	74.8	3.98
−2.8	0.60	2.63	9.14	16.6	36.3	53.8	79.4	100	115	129	158	251	4.58
−3.0	0.60	2.40	10.5	36.4	66.0	144	214	316	398	458	513	630	5.18
−3.2	0.76	2.40	9.55	41.7	145	262	575	853	1260	1580	1820	2040	5.78
−3.4	0.96	3.02	9.55	38.0	166	575	1040	2290	3400	5000	6300	7250	6.38
−3.6	0.60	3.80	12.0	38.0	151	660	2290	4170	9130	1.35 + 4	1.99 + 4	2.51 + 4	6.98
−3.8	0.30	2.40	15.1	47.8	151	604	2630	9140	1.66 + 4	3.63 + 4	5.38 + 4	7.94 + 4	7.58
−4.0	0.15	1.20	9.55	60.4	190	604	2400	1.05 + 4	3.64 + 4	6.60 + 4	1.45 + 5	2.14 + 5	8.18
−4.2	0	0.60	4.78	37.9	240	760	2400	9550	4.17 + 4	1.45 + 5	2.62 + 5	5.75 + 5	8.78
−4.4	0	0	2.4	15.0	151	955	3020	9550	3.80 + 4	1.66 + 5	5.75 + 5	1.05 + 6	9.38
−4.6	0	0	0	9.6	60	600	3800	1.20 + 4	3.80 + 4	1.51 + 5	6.66 + 5	2.29 + 6	9.98
−4.8	0	0	0	0	38	240	2400	1.52 + 4	4.80 + 4	1.51 + 5	6.04 + 5	2.66 + 6	10.58

NOTE: $1.35 + 4 \equiv 1.35 \times 10^4$, etc. Tabular entries are based on van Rhijn Luminosity Function.
SOURCE: (**B7**, 23), by permission. Copyright © 1938 by the University of Chicago.

Table 4-14. Example of Observed and Computed Star Counts

	Observed		Computed	
m	$A(m)$	$\log A(m)$	$A(m)$	$\log A(m)$
8	1.26	0.10	1.26	0.10
9	3.80	0.58	3.85	0.59
10	11.0	1.04	10.7	1.03
11	31.6	1.50	31.6	1.50
12	87	1.94	87	1.94
13	224	2.35	229	2.36
14	575	2.76	572	2.75
15	1410	3.15	1368	3.14
16	2885	3.46	3160	3.50
17	6920	3.84	7032	3.85
18	15850	4.2	15170	4.18

Table 4-15. Example of Stellar Density Distribution Derived from Star Counts

$\log \rho$	$\Delta(\rho)$*	$\log r$†	r(pc)	$D(r)$*
1.0	1.00	1.00	10	1.00
1.2	0.98	1.19	15.5	1.00
1.4	0.96	1.38	24	1.00
1.6	0.91	1.57	37	0.98
1.8	0.81	1.76	57.5	0.90
2.0	0.69	1.96	91	0.82
2.2	0.55	2.15	141	0.72
2.4	0.43	2.34	218	0.64
2.6	0.33	2.52	331	0.60
2.8	0.25	2.70	502	0.61
3.0	0.19	2.85	708	0.67
3.2	0.14	3.00	1000	0.81
3.4	0.10	3.12	1320	0.99
3.6	0.06	3.24	1740	1.19
3.8	0.031	3.34	2190	1.33
4.0	0.018	3.44	2760	1.85
4.2	0.010	3.52	3330	2.51
4.4	0.005	3.60	4000	3.56
4.6	0.0025	3.65	4460	3.61

* In units of the density in the solar neighborhood
† Obtained on the assumption that $a(r) = kr$, with $k = 1$ mag kpc^{-1}

density distribution yields the computed counts shown in Table 4-14, which are seen to be in satisfactory agreement with the observed counts. (Note that to obtain the computed values for $m \geq 16$, the contributions of Δ out to $\log \rho \approx 5.2$, extrapolated linearly in a plot of $\log \Delta$ versus $\log \rho$, must be taken into account.) To convert $\Delta(\rho)$ to $D(r)$ in this example, we have assumed that $a(r) = kr$, with $k = 1$ mag/kpc. From these results, we sce that, when absorption is ncglected, there is a sharp apparent decrease of the density with distance. Indeed, our results strongly resemble the Kapteyn Universe. When absorption is accounted for, however, we obtain a radically different solution. The density now shows a modest decrease out to about 500 pc and then increases. The density increase at large distances shown in Table 4-15 is not real but is an artifact produced by an overcorrection for interstellar absorption by our ad hoc absorption law. Had we chosen a smaller value for k in $a(r)$, we would have found a deeper initial drop in $D(r)$, with a final rise back to about unity. This ambiguity shows that it is imperative to determine the interstellar absorption in a field with the highest accuracy possible if one is to obtain results that are at all physically meaningful from star-count analyses.

Malmquist's Method If one obtains a solution to equation (4-24) by the method of $(m, \log \pi)$ just discussed, one suppresses the tendency of the noise in the data to generate physically absurd predictions for $\Delta(\rho)$ by hand fitting, through trial and error, a run of stellar density $\Delta(\rho)$ that is both plausible and compatible with the data. What one is really doing here is picking, out of a range of mathematically possible solutions, that one which conforms with one's a priori conviction that the run of stellar density must be a smooth function of distance. But, it is not necessary to do this by hand. One can equally well pick out the desired physically acceptable solution by constraining some part of the solution to be of a certain functional form. Malmquist (**M1**), (**M2**) developed a technique of this second type for solving equation (4-24). For the sake of definiteness, we shall assume that the effects of interstellar absorption have been eliminated from the star counts prior to their analysis (for example, by the use of multicolor photometry). If one analyzes uncorrected data, the calculation proceeds analogously in terms of the pseudodistances ρ introduced earlier.

Malmquist started by assuming that the luminosity function $\Phi(M, S)$ is of Gaussian form, with known amplitude Φ_0, mean M_0, and dispersion σ. As we saw [equation (4-36) and Tables 4-7 and 4-8], these assumptions are valid if $\Phi(M, S)$ describes the distribution of stars of any one MK type (or a narrow range of types). Malmquist then sought the distribution $\Phi_m(M, S)$ of the absolute magnitudes of the stars that are observed to have a certain apparent magnitude m. If one can determine this distribution from the data, one is close to a solution to $D_S(r)$. Indeed, any scatter in the M values of stars of fixed m must arise because the stars are distributed through a shell of a certain thickness in r, so, if one knows the distribution with respect to M, one can readily deduce that in r.

Now, any reasonable distribution can be written as a Gaussian plus an (infinite) series of functions that describe the deviation of the given distribution from the normal distribution. Malmquist expanded $\Phi_m(M, S)$ in this way, writing

$$\Phi_m(M, S) = \frac{N(m)}{(2\pi)^{1/2}\sigma_m} \exp\left\{-\frac{[M - \overline{M(m)}]^2}{2\sigma_m^2}\right\} + A_3\Phi^{\text{III}}(M) + A_4\Phi^{\text{IV}}(M) + \cdots, \tag{4-45}$$

where $\overline{M(m)}$ is the mean absolute magnitude of the $N(m)$ stars observed to have apparent magnitude m, σ_m is the corresponding dispersion of absolute magnitudes, and the symbols $\Phi^{\text{III}}(M)$, etc., stand for certain polynomials (Hermite polynomials) in $y \equiv [M - \overline{M(m)}]/\sigma_m$ multiplied by $\exp(-\frac{1}{2}y^2)$. The precise specification of the functions Φ^{III}, Φ^{IV}, and so on does not matter, however. All one needs to know is that there exist certain functions that allow expansion of $\Phi_m(M, S)$ in the form specified by equation (4-45), and they are such that

$$\int_{-\infty}^{\infty} \Phi^{\text{III}}(M)\, dM = \int_{-\infty}^{\infty} M\Phi^{\text{III}}(M)\, dM = \int_{-\infty}^{\infty} M^2\Phi^{\text{III}}\, dM = 0 \tag{4-46}$$

and that the constants A_n can be determined from the nth moments $\int M^n\Phi_m(M, S)\, dM$ of the distribution $\Phi_m(M, S)$.

Malmquist's next step was the evaluation of these moments from the data. Let $q(m, r)$ be the number of stars at distance r that appear to be of magnitude m in a field subtending a solid angle ω. Then $\overline{M(m)}$ is nothing but

$$\overline{M(m)} = \frac{\int_0^{\infty} M(m, r)q(m, r)\, dr}{\int_0^{\infty} q(m, r)\, dr} \tag{4-47}$$

where $M(m, r) = m + 5 - 5 \log r$ as usual. Now, from equation (4-18), we see that

$$q(m, r)\, dr = \Phi[M(m, r), S]D_S(r)\omega r^2\, dr = dA(m, S) \tag{4-48}$$

and thus that

$$A(m, S) = \int_0^{\infty} q(m, r)\, dr \tag{4-49}$$

Therefore,

$$\overline{M(m)} = \frac{\omega}{A(m, S)} \int_0^\infty M(m, r)\Phi[M(m, r), S]D_S(r)r^2\, dr$$
$$= \frac{\omega}{A(m, S)} \frac{\Phi_0}{(2\pi)^{1/2}\sigma} \int_0^\infty (m + 5 - 5 \log r)$$
$$\times \exp\left[-\frac{(m + 5 - 5 \log r - M_0)^2}{2\sigma^2}\right] D_S(r)r^2\, dr \qquad (4\text{-}50)$$

where we have made use of the explicit form of $\Phi(M, S)$ given by equation (4-36). Similarly,

$$A(m, S) = \frac{\omega\Phi_0}{(2\pi)^{1/2}\sigma} \int_0^\infty \exp\left[-\frac{(m + 5 - 5 \log r - M_0)^2}{2\sigma^2}\right] D_S(r)r^2\, dr \quad (4\text{-}51)$$

and, by differentiation with respect to m, we obtain

$$\frac{dA(m, S)}{dm} = -\left(\frac{1}{\sigma^2}\right) \frac{\omega\Phi_0}{(2\pi)^{1/2}\sigma} \int_0^\infty (m + 5 - 5 \log r - M_0)$$
$$\times \exp\left[-\frac{(m + 5 - 5 \log r - M_0)^2}{2\sigma^2}\right] D_S(r)r^2\, dr \quad (4\text{-}52)$$

Comparing equation (4-52) with (4-50), we easily see that

$$-\sigma^2 \frac{dA(m, S)}{dm} = A(m, S)\overline{M(m)} - A(m, S)M_0 \qquad (4\text{-}53)$$

and hence that

$$\overline{M(m)} = M_0 - \frac{\sigma^2}{A(m, S)} \frac{dA(m, S)}{dm} = M_0 - \mu\sigma^2 \frac{d \log A(m, S)}{dm} \qquad (4\text{-}54)$$

Differentiating (4-52) again with respect to m and dividing by $A(m, S)$, one obtains, in analogous fashion, the dispersion σ_m of the stars observed to have apparent magnitude m:

$$\sigma_m^2 = \sigma^2 \left[1 + \mu\sigma^2 \frac{d^2 \log A(m, S)}{dm^2}\right] \qquad (4\text{-}55)$$

Continuing to differentiate (4-52) with respect to m, one obtains, successively, the third, fourth, and so on moments of $\Phi_m(M, S)$, and thus (as Malmquist showed) the initially unknown coefficients A_n of the expansion specified by equation (4-45). For example, one has

$$A_3 = -\frac{\mu\sigma^3 N(m)}{6(2\pi)^{1/2}} \frac{d^3 \log A(m, S)}{dm^3} \tag{4-56}$$

Before proceeding further, it is worthwhile to notice an important consequence of equation (4-54). If, as will normally be the case, more stars are observed at faint magnitudes than at bright ones, so that $(dA/dm) > 0$, then equation (4-54) predicts $\overline{M(m)} < M_0$—that is, under these circumstances, the stars one sees at a given apparent magnitude are, on the average, more luminous than the average for all the stars in a given volume. This effect, called the *Malmquist bias*, arises because, when one selects stars of fixed apparent magnitude, the volume element containing the more distant, intrinsically luminous stars is larger than that occupied by the nearer, fainter objects. The Malmquist bias also plays an important role in connection with counts of radio galaxies, quasars, and other objects that have been used as cosmological probes.

Let us now return to the problem of solving (4-24) for $D_S(r)$. Malmquist's technique for preventing the noise in the data generating a physically unacceptable solution for $\Phi_m(M, S)$, and hence for $D_S(r)$, was simple. He set all the A_n in equation (4-45) to zero. That is, he considered that the expansion (4-45) neatly divided the $\Phi_m(M, S)$ that one would derive from a complete analysis of the data into signal, in the form of the first (Gaussian) term, and noise (everything else). Equations (4-56) and the analogous equations coupling the other A_n to higher derivatives of the data indicate that it is indeed unlikely that any useful information is contained in the part of equation (4-45) neglected by Malmquist.

The procedure for determining $D_S(r)$ from $A(m, S)$ is now fairly straightforward. First, from the observed counts $A(m, S)$, using equations (4-54) and (4-55) we compute $\overline{M(m)}$ and σ_m at each value of m (which we assume are separated by a constant increment Δm) for which we have data. The practical difficulty encountered here is the evaluation of physically meaningful estimates for (dA/dm) and (d^2A/dm^2) from data that inevitably contain appreciable random errors. Assuming that satisfactory values of $\overline{M(m)}$ and σ_m can, in fact, be derived, then, from each m at which we know $A(m, S)$, we compute $\Phi_m(M, S)$ over that range of a predetermined set of M values within which Φ_m is significantly different from zero. For convenience, we choose $\Delta M = \Delta m$. The size of these steps should be smaller than the dispersion σ_m to ensure that the Gaussian distribution is sampled adequately. This operation produces a table in which each entry gives the number of stars of absolute magnitude M that have apparent magnitude m.

We next compute $\log r = 0.2(m - M) + 1$ for each entry in the table. Because we chose $\Delta m = \Delta M$ for the magnitude differences between neighboring entries in our table, only a discrete set, $\log r_k$, of values of $\log r$ will arise in this way. These values will be constant on diagonal lines $m - M =$ constant through our table. Next, add the numbers along one of these lines of constant r to find the total number of stars at each r_k, irrespective of m or M. We consider these stars to be distributed in a shell whose midpoint radius is r_k and whose boundary surfaces are at $r_{k+1/2} = (r_{k+1}r_k)^{1/2}$. Finally, dividing by the volume δV_k of this shell (equation 4-44), we obtain $D_S(r_k)$, the density of stars of all magnitudes at r_k [see (**B7**, 37) for further details]. Malmquist's method and variants based on it have often been used to analyze the space distributions of well-defined classes of objects, such as RR Lyrae stars (see §4-4).

4-3. THE DISTRIBUTION OF STARS AND THE CHEMICAL ELEMENTS WITHIN THE DISK

Space Distribution of Stars in the Galactic Plane near the Sun

We saw in §4-2 that, in principle, we can map the space distribution of stars in the galactic disk by a suitable analysis of star-count data. We can thereby begin to build a picture of the structure of our Galaxy in the solar neighborhood. This goal has been pursued industriously in numerous studies over the past fifty years, but, although some interesting results have emerged, in some ways the return on the effort has been disappointingly small. In particular, the early work, using general star counts, was hampered by a lack of resolving power induced by the vast range of intrinsic brightnesses of the stars that were lumped together in the general luminosity function. It also suffered from uncertainties in this luminosity function and questions about its applicability at different positions within our Galaxy, and it was seriously impaired by inadequate precision in the determination of the amount and distribution of interstellar absorption.

Despite these difficulties, a few apparently trustworthy results emerged [see (**B6**, Chapter 1)]. Average relative density distributions have been determined for three to six fields centered around each of a dozen lines of sight in the galactic plane, separated by $\Delta\ell \approx 20°$ to $30°$ in the range $20° \lesssim \ell \lesssim 220°$. From these results, one finds evidence for an elongated high-density region with its maximum 300 to 500 pc from the Sun in the direction of the galactic anticenter. This feature is probably associated with the local spiral arm (the Orion-Cygnus arm—see Figure 4-11). In longitudes toward the galactic center, there is a rapid decrease in relative density to 0.2 at 600 pc, followed by a rise to about 1.0 at 2 to 3 kpc from the Sun. These results are much less reliable than those for the anticenter regions because of the difficulty of making adequate allowance for absorption effects, yet they are compatible with the existence of an inner arm (the Sagittarius arm—see Figure 4-11) and the location of the Sun near a relative density maximum. A high-density region is found in the range $210° \lesssim \ell \lesssim 215°$, with $D(r) \approx 1$

out to 2.5 kpc from the Sun. This apparent concentration may be associated with the fact that these lines of sight will be nearly tangent to the Orion arm and hence will pass through regions of high density along considerable path lengths. In the longitude range $170^\circ \lesssim \ell \lesssim 200^\circ$, there is a marked deficiency, relative to the immediate solar neighborhood, of stars at distances beyond 1 kpc. Here one probably has looked through the local arm into the interarm region beyond. Results for individual regions indicate large local density fluctuations within 1 kpc of the Sun. These fluctuations may be partly real, but most probably they reflect irregularities in the absorbing material. Such irregularities remain a key factor limiting the reliability of stellar density maps, and unfortunately the reality of many individual features remains questionable.

Somewhat more reliable information can be deduced from counts of individual spectral types. The most comprehensive studies of this kind are those of S. W. McCuskey, discussed thoroughly in (**M5**) and (**B6**, Chapter 1). We shall summarize just a few results of this work. A map of the space density of B5 stars is shown in Figure 4-9. The prominent concentrations running diagonally from upper left to lower right are probably associated with the local arm. The feature at 1.5 to 2 kpc in the direction $\ell = 130^\circ$ is probably associated with the Perseus arm. The early A stars (B8–A0) show a very strong concentration within 500 pc in the directions $\ell = 85^\circ$ to 135°. This feature is in the local arm (see Figure 4-10). In addition, between 1 and 2 kpc from the Sun in the directions $165^\circ \lesssim \ell \lesssim 215^\circ$, they show a

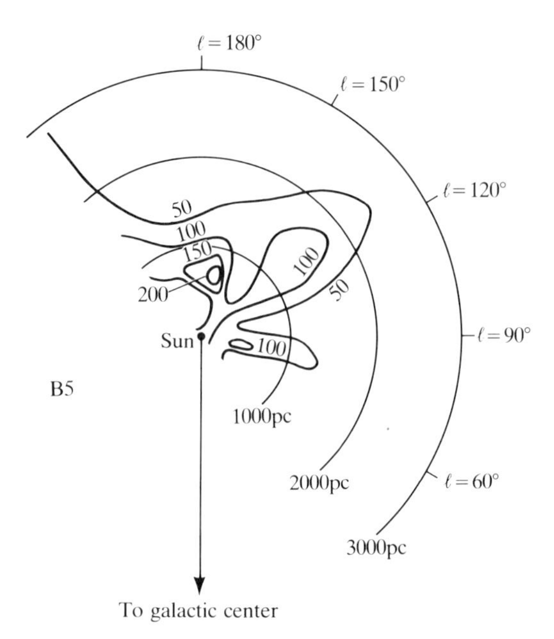

Figure 4-9. Distribution of B5 stars in the galactic plane in the solar neighborhood. Curves give contours of equal space density in numbers of stars per 10^7 pc^3. [Adapted from (**B6**, Chapter 1), by permission. Copyright © 1965 by the University of Chicago.]

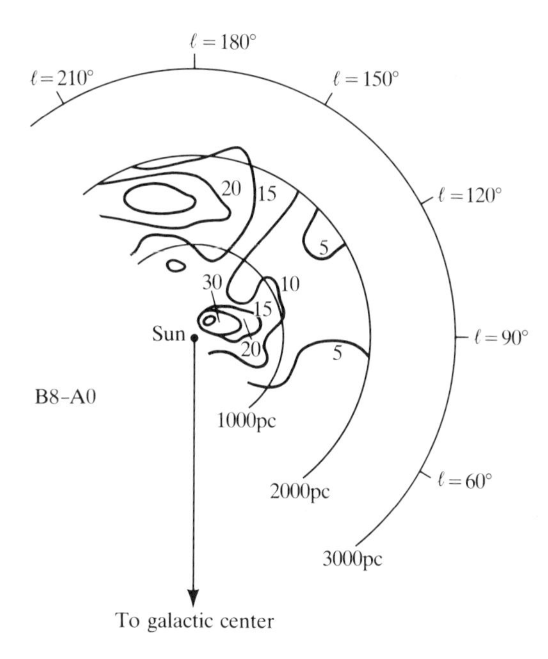

Figure 4-10. Distribution of B8–A0 stars in the galactic plane in the solar neighborhood. Curves give contours of equal space density in numbers of stars per 10^5 pc^3. [Adapted from (**B6**, Chapter 1), by permission. Copyright © 1965 by the University of Chicago.]

very strong concentration with about four to five times the density in the immediate solar neighborhood. This feature is probably associated with the Perseus arm.

The space distribution of the middle A stars (A2–A5) qualitatively resembles that for the B5 stars. The G giants (gF8–gK3) show a density peak on the range $90° \lesssim \ell \lesssim 145°$ within 500 pc, which is similar to that shown by the early A stars. This resemblance is not surprising because many of these stars are probably evolved A stars. Except for this one concentration, the relative space density of G giants is quite uniform within 700 pc of the Sun.

Our knowledge of the space distribution of late-type dwarfs is limited by their low intrinsic luminosity. In the region within 500 pc of the Sun that has been investigated, the space distribution of these stars appears to be fairly uniform. No significant concentrations can be identified, but there are fluctuations of the order of a factor of two up and down from region to region.

We remind the reader that, when assessing the results described here and shown in Figures 4-9 and 4-10, one must bear in mind that many of the details seen in these maps may be more the result of interstellar absorption effects than real variations in the space density. Unfortunately, it does not appear that this situation can readily be improved.

What can be said about the large-scale variation of stellar density within the disk? Unhappily, we have almost no direct observational data with which to answer this question, because, as we have repeatedly emphasized, the location of the Sun (immersed in a dust-filled disk) is extremely unfavorable for observations of the disk itself, and attempts to infer the variation of stellar space density in the disk at distances beyond 1 or 2 kpc are invariably thwarted. We know, however, from observations of external galaxies (see Chapter 5), that the density in galaxian disks rises exponentially toward their centers, with an *e* folding length of about 3 to 5 kpc. We have every reason to suppose that the disk of our own Galaxy has a similar density rise toward the center, despite our lack of direct proof that it does.

Spiral Structure Inferred from Optical Data

From both its large-scale disk structure and from the kinds of astrophysical objects we observe within the galactic disk (for example, young open clusters and O associations, Cepheids, and H II regions) we believe that our Galaxy is a spiral of an intermediate type—about Sb or Sbc (see §5-1 for a discussion of galaxian classification schemes). This conclusion is strongly supported by wide-angle infrared photographs of our Galaxy, which show a central nuclear bulge, an equatorial belt of obscuring material, and a general appearance strikingly similar to that of Sb spirals seen edge-on (see Figures 1-6 and 1-7). We expect, therefore, that our Galaxy should have spiral arms. If we were to attempt to delineate spiral structure from density analyses of common field stars, we would probably fail for at least three reasons: (1) As we have seen, the analyses suffer severe limitations of accuracy and penetrating power because of interstellar absorption. (2) There is no dynamical reason to suppose that spiral arms represent major density concentrations. Current work suggests that the potential well associated with arms is relatively weak. (3) Most of the field stars are so old they they would have diffused randomly to rather large distances away from their original sites of formation, and they would no longer show a clear spiral pattern, even if one had originally existed. Typical field stars have random velocities of about 20 to 30 km s^{-1} (see §7-1), which means that stars, say, 3×10^7 years old have already diffused through a volume 0.5 to 1 kpc in diameter, and hence any initial pattern would have been largely obliterated. At best, these stars would respond only weakly to a spiral potential fluctuation, and they would therefore show only a modest concentration into a broad spiral pattern [see §5-2 and (**S6**)].

Tracers To isolate the galactic spiral arms, we thus take a different approach and analyze the distribution, relative to the Sun, of carefully selected objects that serve as *spiral-arm tracers*. The ideal tracers should satisfy the following criteria:

1. They should be known, from observations of external galaxies, to be closely associated with spiral arms.

2. They should be young, so that, during their lifetime, they cannot have moved far from their original birthplace.
3. They should be luminous, so that they can be seen at large distances, thus allowing us to trace the arms coherently over substantial arc lengths.
4. They should be objects whose intrinsic brightnesses are well defined and easily determined and for which interstellar absorption and reddening effects can be estimated precisely.

Several types of objects satisfy these criteria; among them are H II regions, O associations, young galactic clusters, Cepheids, and certain types of supergiants. The close association of H II regions with spiral arms was demonstrated by Baade and Mayall (**B1**), and, for the other objects, by several observers (among them Hubble and Baade) in many studies. H II regions are easy to find and identify, and, because their sources of excitation are O stars, they satisfy the criteria of youth and high luminosity. Their distances can be estimated from knowledge of the spectral types, reddenings, and apparent magnitudes of the stars associated with them. Young clusters satisfy all the criteria we have listed. Very precise distances can be determined for these objects, because their stars lie very nearly on the ZAMS and hence will have very little scatter in absolute magnitude and have accurately known colors. Furthermore, averages can be formed from many stars within a single cluster, which tends to reduce the effects of errors in photometry, in reddening corrections, and in the estimated absolute magnitude of each individual star. Similar remarks apply to O associations, which are extremely young. A problem with clusters, and particularly with associations, is that, when they are at large distances, they tend to be hidden by the foreground field and become hard to identify. This difficulty is somewhat ameliorated by their very distinctive stellar content. Classical Cepheids are known to be fairly good spiral-arm tracers, and, because they obey well-established period-color-luminosity and period-amplitude relations (see §3-8), their distances can be estimated quite accurately. Supergiants of types A–M are commonly observed in spiral arms in external galaxies, and they are often found in associations that appear as large clumps along the arms. However, they are also found singly in the field. These stars cannot be expected to trace the arms as sharply as H II regions or O associations, because they are appreciably older than those objects (being evolved early-type main-sequence stars) and hence have probably diffused a significant distance away from their places of formation. Moreover, the absolute-magnitude calibration for these stars is not terribly precise, and it is often extremely difficult to determine the amount of interstellar reddening they have suffered. Together, these difficulties induce uncertainties of the order of $\pm 30\%$ in their estimated distances.

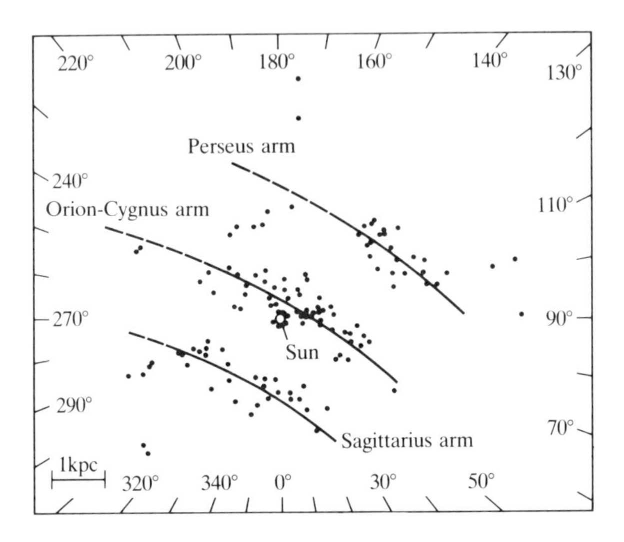

Figure 4-11. Spiral arms in the solar neighborhood as inferred from optical evidence. Individual points are at positions of young star clusters and H II regions. Three arms are clearly indicated, separated from one another by about 1.5 kpc: the outer Perseus arm, the local Orion-Cygnus arm, and the inner Sagittarius arm. The Sun is near the inner edge of the Orion-Cygnus arm. [From (**B5**, 205), by permission.]

Results The spiral arms in our Galaxy were traced optically for the first time by W. W. Morgan, S. Sharpless, and D. E. Osterbrock (**M8**) through a study of galactic H II regions. A spiral pattern was established definitely, and segments of two arms were delineated (the local Orion–Cygnus arm and the Perseus arm; see Figure 4-11). Shortly afterward, Morgan, Whitford, and Code (**M9**) determined positions for about thirty O associations, thus tracing the local and Perseus arms more clearly and establishing the existence of the Sagittarius arm. Since that time, a large number of studies of spiral-arm tracers have been made [see, for example, (**B6**, Chapter 7) for a summary]. A recent map by W. Becker and R. Fenkart (**B5**, 205) of the positions of H II regions and young galactic clusters is shown in Figure 4-11. All three arms show quite unambiguously. The arms are separated by about 1.5 kpc, are inclined by about 25° to a radius vector from the galactic center, and trail with respect to the direction of galactic rotation (clockwise in the diagram). The significance of this map is strikingly demonstrated by comparing it to a map of the positions of old clusters, constructed using the same techniques (**B5**, 205). The old clusters scatter at random in the map and show no hint whatsoever of spiral structure.

Plots of the positions of supergiants on a map such as Figure 4-11 show that these stars do not, in themselves, define spiral arms clearly (the picture is blurred by their uncertain distances). However, they do indeed follow the spiral pattern established by the other objects. Some associations (for example, h and χ Persei) pass through a brief evolutionary phase during which the most massive members become M supergiants. Reasonably good distances can be determined for four such groups of M-supergiants, and, when plotted on a map of the spiral arms, they are found to be clearly in the local and the Perseus arms (**B6**, Chapter 7). Studies of the distribution of classical Cepheids (**B6**, Chapter 8) have shown that they outline the Perseus and the Orion-Cygnus arms.

In this section, we have considered only the optical evidence for spiral arms. Information can also be obtained from H I 21-cm line observations, as will be discussed in §9-1.

Space Distribution Perpendicular to the Plane

Analyses of the space densities of common stars perpendicular to the galactic plane can be made in a limited region around the position of the Sun. The results of this work are of great interest, for, as we shall discuss in Chapter 14, they can be used in dynamical discussions to infer information about the gravitational potential of our Galaxy in the direction perpendicular to the plane. Furthermore, as we shall see in §7-1, these data yield information about the time variation of the random velocities of disk stars. In general terms, density analyses in the z direction (that is, perpendicular to the plane) have been relatively more successful than analyses in the plane, partly because density gradients in this direction are large and hence easy to detect, and partly because interstellar absorption is generally considerably smaller away from the plane than in it.

It has been recognized since the earliest work on this problem by Lindblad (**L2**) more than fifty years ago that the most striking feature of the distribution of stars perpendicular to the plane is that the density gradient is markedly different for different spectral types, the earlier types showing a much stronger concentration to the plane. This fact vitiates general star-count analyses in the z direction, for it implies that the general luminosity function $\Phi(M)$ depends strongly on z and must be substantially different from what it is in the plane at even a few hundred parsecs above the plane. Analyses of individual spectral types naturally do not suffer from this defect because, insofar as the physical properties (in particular the absolute magnitudes) of stars of a given type are the same at all distances from the plane, the function $\Phi(M, S)$, which merely gives a distribution around the mean magnitude for the group, remains valid.

The mathematical techniques used to determine $D_S(z)$ are the same as those already described for the determination of $D_S(r)$ in the plane. Several studies have been made, and they are summarized by T. Elvius in (**B6**, Chapter 3). We shall quote merely a few representative results here. Because they are reasonably luminous, A dwarfs and G and K giants can be studied out to about 2 kpc from the plane. In contrast, K dwarfs can be reached only out to about 200 to 300 pc. For A and F stars, it is usually assumed (and has been verified observationally) that essentially all the stars in the sample are dwarfs. For G and K stars, however, the separation of dwarfs from giants is a long-standing problem, and spectroscopic classification is required to segregate the two groups. In addition, in samples of later-type stars, the relative proportions of disk stars and spheroidal-component stars—which have substantially different physical properties and space distributions—change as one proceeds farther from the plane, the latter group becoming more numerous.

Some typical results for the run of $D_S(z)$ with z are shown in Figure 4-12. The strong concentration of A stars to the plane relative to later types is quite apparent. As can be seen from the figure, the data can be at least roughly represented by the standard barometric equation

$$\log D_S(z) = \log D_S(0) - (|z|/\beta_S)/\mu \tag{4-57a}$$

or

$$D_S(z) = D_S(0)e^{-(|z|/\beta_S)} \tag{4-57b}$$

where $\mu = \ln 10$, and β_S is the *scale height*, or *mean z distance*, from the plane of stars of spectral type S. This description is quite convenient, as it

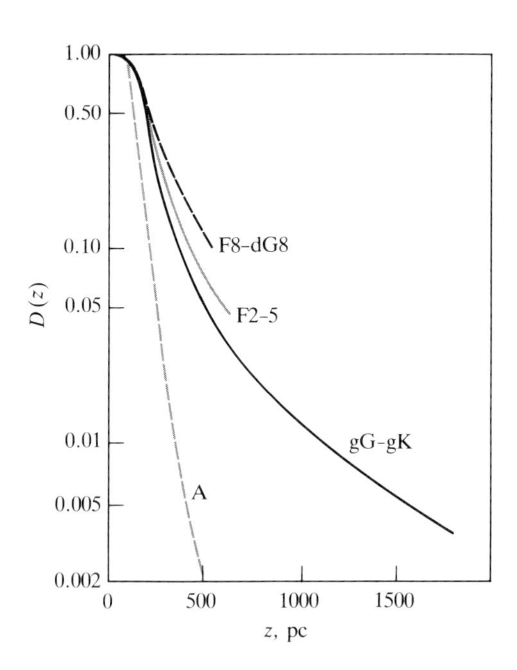

Figure 4-12. The relative density distribution perpendicular to the galactic plane of stars of various spectral classes. [From (**B6**, Chapter 3), by permission. Copyright © 1965 by the University of Chicago.]

gives a succinct and easily comprehended parameterization of the space distribution. It is, however, not exact. Near the plane $z = 0$, a parabolic or Gaussian fit to $D_S(z)$ is better. Moreover, equation (4-57) implies a discontinuity in the density gradient at $z = 0$, whereas physical intuition suggests that $(dD_S/dz) \to 0$ as $z \to 0$.

The product $D_S(0)\beta_S$ has an important physical significance. It is, in fact, equal to half the total *surface density* Σ_S of the stellar type S

$$\Sigma_S = \int_{-\infty}^{\infty} D_Z(z)\,dz = 2D_S(0)\beta_S \tag{4-58}$$

The surface density Σ_S is actually a much more fundamental quantity than a volume density like $D_S(z)$ or $D_S(0)$, because, during the course of time, stars diffuse away from their initial birthplaces near $z = 0$ into a more extended distribution. By comparing relative values of $\Sigma_S/\Sigma_{S'}$ for different types of stars, we compensate for this diffusion and obtain realistic estimates of the relative total numbers of these stars born in the disk over the age of our Galaxy. Furthermore, it is surface densities that we measure directly for other galaxies. Representative values of Σ_S for various classes of disk objects in our Galaxy are shown in Table 4-16. Results are not given for spheroidal-component objects, as equations (4-57) and (4-58) do not provide a quantitatively realistic representation of their distribution. We have already noted the increase in stellar numbers in the solar neighborhood with diminishing luminosity and increasing age. We see now that the preponderance of low-mass stars is even greater if we look at total surface densities Σ_S because of the increase of both β_S and $D_S(0)$ with decreasing mass.

The values of the scale heights β_S for various types of stars and other objects given in Table 4-16 show a number of interesting features. First, notice that the scale height of the youngest stars (O–B stars) is smaller than that of the interstellar gas whence they form. This fact can be understood if we suppose that the total birthrate function (see §3-7) for stars varies as a power $\alpha > 1$ of the density, that is, $\beta = k\rho^{\alpha}$, for the densest regions of the gas would then produce stars most efficiently. Consequently, the stars would have a smaller scale height. Turning the argument around, one can use the observed scale heights to estimate the exponent α; typical results (**B3**, 165) are $1.5 \lesssim \alpha \lesssim 2$. Second, we see that, for disk stars, β_S generally increases with increasing mean age of the stellar group in question. Now, the mean height away from the plane to which stars can penetrate increases as their average z velocity (or kinetic energy in the z direction) increases. The data thus suggest that, for stars in the galactic disk, some mechanism produces a secular "heating" of the "gas" of stars during their lifetimes. We shall return to this question in §7-1 and in the next volume. Third, in the last four lines of the table, we give scale heights for spheroidal-component objects. Obviously, these are distributed throughout the halo, even at enormous distances above and below the plane (see §4-4). Finally, knowing $D_S(z)$ for

Table 4-16. Scale Heights β_S in the Direction Perpendicular to the Galactic Plane and Surface Density Σ_S for Various Objects

Object	β_S(pc)	$\Sigma_S \left(\frac{\text{stars}}{\text{pc}^2}\right)$	$\Sigma'_S \left(\frac{\mathscr{M}_\odot}{\text{pc}^2}\right)$
O stars	50	1.5×10^{-6}	10^{-4}
Classical Cepheids	50	7.5×10^{-6}	5×10^{-5}
B stars	60	6×10^{-3}	6×10^{-2}
Galactic clusters	80	–	–
Interstellar dust and gas	120	–	
A stars	120	6×10^{-2}	0.1
F stars	190	0.6	0.6
Planetary nebulae	260	–	–
gK stars	270	1.2×10^{-3}	3×10^{-2}
Novae	300	–	–
dG stars	340	2	2
dK stars	350	3.5	2.5
dM stars	350	20	9
gG stars	400	6×10^{-2}	1.6×10^{-1}
White dwarfs	500	12.5	10
Long-period variables (M5–M8)	700		
RR Lyrae variables (P < 0ᵈ.5)	900		
Long-period variables (M0–M4)	1000		
RR Lyrae variables ($P >$ 0ᵈ.5)	2000		
W Virginis variables (spheroidal-component Cepheids)	2000		
Subdwarfs	2000		
Globular clusters	3000		

SOURCE: Adapted from (**A1**, 247), (**A1**, 249), and (**A1**, 251), by permission

various spectral types, one can reconstruct a general luminosity function $\Phi(M, z)$ as a function of z. Typically, one finds [see (**A1**, 251), (**B6**, 54)] that use of the general luminosity function for the plane would provide a reasonable estimate of $\Phi(M, z)$ for stars with $M_V \gtrsim 2$ up to perhaps 100 pc away from the plane, but it would greatly overestimate the number of intrinsically very luminous stars. At 200 pc, the cutoff is already $M_V \gtrsim 4$, and, at 500 pc, the cutoff is at $M_V \gtrsim 8$. Indeed, at 500 pc, the issue is problematical anyway because the stellar-population characteristics (see §4-5) of the stars have changed to include a substantial admixture of spheroidal-component stars.

Chemical-Composition Gradients in the Disk

In addition to variations in the space density and the distribution over spectral type of stars as a function of R and z (the radial distance from the galactic center in the galactic plane and the distance away from the plane), a number of studies have shown that there are gradients in R and z of the chemical composition of disk stars. Recent reviews of this work have been given by M. Mayor (**B3**, 213) and M. Peimbert (**B3**, 149). The existence and nature of these gradients are matters of great interest, for they provide important constraints on possible evolutionary histories of our Galaxy.

Results from several studies of abundances in disk stars are given in Table 4-17. As was true for the problem of estimating the variation of the metallicity of disk stars with age, the accuracy of these (mainly photometric) determinations is often quite poor for individual stars, and significant trends emerge only because a large sample of stars can be analyzed.

The results definitely show a decrease in average metallicity with increasing radial distance (in the plane) from the galactic center. Although the derivatives given are, of course, purely local values, they suggest (excluding young objects) a decrease in average metal abundance by about a factor of two to three over 10 kpc. The average metallicity of stars in the nuclear bulge is thus probably about a factor of two above solar. Observations of integrated starlight in other spiral galaxies, for example, M31 and M81, also show strong

Table 4-17. Radial Composition Gradients in the Galactic Disk in the Vicinity of the Sun

Stars		
Group	$d[\text{Fe/H}]/dR$ (kpc^{-1})	References
G and K dwarfs and giants	−0.05	(**C3**, Chapter 55)
Nearby stars earlier than K5	−0.07	(**B3**, 169)
K giants	−0.03	(**B3**, 173), (**C3**, Chapter 28)
dF and gG–dK, (all stars)	−0.04	(**M4**)
dF and gG–gK, (intermediate-age)	−0.02	(**M4**)
dF and gG–gK, (young)	−0.08	(**M4**)
Cepheids	−0.08*	(**B3**, 149)

Nebulae				
Objects	$d[\text{He/H}]/dR$	$d[\text{O/H}]/dR$	$d[\text{N/H}]/dR$	References
H II Regions	−0.2	−0.13	−0.23	(**B3**, 149)
H II Regions		−0.05	−0.10	(**H2**)
Planetary nebulae	−0.02	−0.06	−0.18	(**B3**, 149)

* Value given is for $d[\text{O/H}]/dR$, not $d[\text{Fe/H}]\ dR$

increases in CN-band strengths in their disks at decreasing radial distances from their centers [see (**V1**) for references]. The most straightforward interpretation of these observations is an increase in average stellar metal content toward the galaxian nucleus.

The two samples of young stars in Table 4-17 show a much swifter rate of change than do the older stars. Furthermore, for the sample of K giants studied by K. A. Janes (**B3**, 173), (**J1**), it was found that the gradient is steeper in the outer part of the galactic disk. He estimates $d[\mathrm{Fe/H}]/dR \approx -0.05$ for $R \approx 10$ kpc and $d[\mathrm{Fe/H}]/dR \approx -0.1$ for $R \gtrsim 10$ kpc.

The stellar results we have quoted are supported by analyses of the spectra of emission nebulae. We defer full discussion of nebular abundance determinations to §5-4, because the majority of studies of this type have been concerned with external galaxies [see (**P1**), (**S7**), (**S10**), (**V1**)]. Here we note only the most important results obtained for galactic H II regions. The abundances of oxygen and nitrogen relative to hydrogen have been estimated for about two dozen galactic H II regions covering galactocentric distances in the range 7 kpc $< R <$ 14 kpc. Typically, the nebulae show approximately solar abundances of oxygen and nitrogen, although the abundances of both these elements tend to be higher in emission regions observed in the direction to the galactic center than in the anticenter direction. In Table 4-17, we collect the best available estimates of the gradients in [He/H], [O/H] and [N/H] from galactic H II regions. Notice two points: (1) The gradients in [O/H] and [N/H], derived by different workers in contemporary studies, are significantly different in magnitude, although the relative sizes and senses of the [O/H] and [N/H] gradients do agree. (2) Both the studies quoted conclude that the gradient in [N/H] is about twice as great as that in [O/H]. The suggestion that the abundance of nitrogen increases toward the galactic center faster than does that of oxygen has received a good deal of attention, because it would seem to confirm the classical theory of nucleosynthesis, according to which the production of nitrogen occurs only in stars that already contain carbon and oxygen. We shall see in Chapter 5, however, that recent studies of H II regions in other galaxies have tended not to confirm the early suggestion of especially steep gradients in [N/H]. Furthermore, determinations of element abundances in galactic H II regions are severely hampered by interstellar absorption, which restricts the range of radii (and therefore abundances) at which emission regions can be observed, and renders uncertain the determination of nebular temperatures from temperature-sensitive line ratios, by selectively absorbing the shorter wavelength line (see §5-4). Therefore, the inference that the gradient in [N/H] is steeper than that in [O/H] should be treated with caution.

Observations of the stellar and nebular content of other galaxies are of great importance because, unlike observations within our Galaxy, which are confined by interstellar absorption to relatively nearby objects, they can sample the whole disk. Thus they show directly that the radial abundance gradient is, in fact, a large-scale phenomenon. All the evidence available from such studies confirms that metallicity gradients are a general property of galaxian disks.

Composition gradients perpendicular to the plane have been studied by A. Blaauw (**M10**, 51) and M. Grenon (**M10**, 55) by determining directly the variation with z of photometric composition indicators, such as the Strömgren-system index Δm_1. A weakness of this method is that it requires very accurate photometry and an accurate absolute-magnitude calibration. Martinet and Grenon (**B3**, 289) have synthesized the z gradient from data for stars in the galactic plane by considering a mixture of four subgroups of differing metallicities and average z velocities and calculating the variation in the properties of this mixture as a function of height, using an assumed force law in the z direction. All the results show a swift drop in metallicity with height, $d[\mathrm{Fe/H}]/dz \approx -0.5$ to $-0.6\ \mathrm{kpc}^{-1}$.

The question that immediately occurs is "What are the origin and the significance of these observed composition gradients?" We shall offer only a couple of brief comments on this question here and defer fuller discussion until a later chapter. One possibility is that the gradients in the disk were established before much star formation occurred there, and they have been but little modified since. In this case, the spatial variation of the composition of disk material would have to reflect mainly the manner in which enrichment by recycled material, shed by evolved halo stars, occurred, and the effects of enrichment by nucleosynthesis in the disk stars themselves would have to be negligible. Alternatively, the gradients may reflect more frequent or more efficient recycling of material by disk stars in the inner regions of the disk than in the outer regions, perhaps a result of more rapid rates of star formation in regions of higher surface density. One could then argue that the gradient should progressively steepen in time (at least until the gas in the disk is exhausted), which would seem to be in harmony with the steeper gradient found for young stars and nebulae. We shall return to these matters in Chapter 19, where we discuss the chemical history of our Galaxy in more detail and try to incorporate the observations of the present state of our Galaxy into a coherent picture of its formation and dynamical evolution.

4-4. THE DISTRIBUTION OF STARS AND THE CHEMICAL ELEMENTS IN THE SPHEROIDAL COMPONENT

Having discussed the distribution of stars and of the chemical elements within the galactic disk, we now turn to the same questions for the spheroidal component. As we shall now deal with a system whose constituents are rare near the Sun (being represented in the solar neighborhood by subdwarfs and RR Lyrae stars), and as we are interested in its properties on the largest observable scales, our approach must necessarily be rather different, and the information we shall be able to derive is much less detailed.

Globular Clusters

The most prominent members of the spheroidal component are the globular clusters, which are found throughout the halo and in the vicinity of the galactic center. These objects make ideal tracers for the spheroidal component because they are luminous and easy to detect, the properties of their stars are sufficiently well understood that they can be used to obtain fairly reliable distance estimates, and the sample of known clusters is relatively complete. It is likely that the space density of globular clusters is representative of the density within the spheroidal component as a whole, and, insofar as this is true, they provide a uniquely powerful tool for studying its structure. As a result of recent comprehensive studies [see, for example, (**H1**), (**O3**), (**W3**), and the summary of older work in (**B6**, Chapter 19)], there have been major improvements in the quantity and quality of the data. In particular, dozens of new color-magnitude diagrams have been constructed and homogeneous, precise measurements of foreground interstellar reddening and absorption have been made.

The apparent distribution of globular clusters on the sky is totally different from that of any constituent of the disk population. They are found predominantly at high galactic latitudes, and they show a distinct deficit near the galactic plane, where they are hidden by interstellar absorption. In addition, they are concentrated toward the direction of the galactic center (more than 50% lie in the longitude range $-15^\circ \leq \ell \leq 15^\circ$). As was early recognized by Shapley, this apparent distribution reflects a true distribution that is basically spherically symmetric about, and strongly concentrated toward, the galactic center. The group of all known globular clusters is small enough that, to determine their distribution in space, we can employ a direct cluster-by-cluster mapping technique based on measurements of individual cluster directions and distances from the Sun, combined with an estimate of the Sun's distance from the galactic center.

Distance Indicators The most fundamental distance indicators for globular clusters rest on estimates of the intrinsic brightness of some class of cluster stars or of some specific feature in their CM diagrams. In the final analysis, these estimates are based on cluster main-sequence fitting, which was described in §3-6.

The most important single quantity of this type is $\langle M_V \rangle_{RR}$, the average absolute magnitude of RR Lyrae stars (which is almost identical with the average absolute magnitude $\langle M_V \rangle_{HB}$ of the horizontal branch). As we saw in §3-8, the current best estimate is $\langle M_V \rangle_{RR} \approx 0.6 \pm 0.2$. The question whether (and if so, how) this result depends on the metallicity of the stars is still not satisfactorily resolved, and the best we can do at present is to adopt the mean value. But it should be noted that even an uncertainty of only ± 0.2 mag in $\langle M_V \rangle_{RR}$ induces an uncertainty of $\pm 10\%$ in cluster distances, and thus $\pm 30\%$ in space densities. Indeed, the intrinsic spread in $\langle M_V \rangle_{RR}$ produced by differences in cluster metallicities (and in other evolutionary parameters) could be even larger (see the discussion of main-sequence fitting in §3-6).

A second primary distance indicator is obtained by calibrating M_{25}, the mean absolute magnitude of the 25 brightest cluster stars. Here it is necessary to account for two effects: (1) The mean magnitude M_{25} depends on the total cluster population (and hence luminosity), simply because the more cluster members there are, the higher is the probability of finding very rare stars of the highest luminosity in the sample. (2) The mean magnitude also depends on the metallicity of the cluster stars because, as we saw in §3-6, the height of the tip of the giant branch above the horizontal branch is a strong function of metal content. Details of methods that compensate for these effects and allow M_{25} to be obtained directly from observable data are described in **(H1)** and **(W3)**.

Secondary distance indicators that make use of *integrated* properties of clusters can be devised. They have the advantage that they can be measured in many instances where information about features in the CM diagram is unavailable. These secondary indicators are calibrated using clusters in which both the secondary indicators and a primary indicator can be measured.

Of the 110 clusters for which comprehensive data are available, primary indicators can be used to derive distance moduli for about 75 clusters, and secondary indicators can be used for the remainder. When combined with reddening and absorption measurements, the distance moduli yield positions of the clusters relative to the Sun.

The Sun's Distance from the Galactic Center From the positions of clusters with respect to the Sun, one can estimate the Sun's distance from the galactic center on the hypothesis that the clusters are distributed symmetrically about the center. Let (x, y, z) measure a cluster's position relative to the Sun in a system with axes aligned along the Sun-center line ($\ell = 0°$, $b = 0°$) toward the direction of galactic rotation ($\ell = 90°$, $b = 0°$) and toward the north galactic pole ($b = 90°$), respectively. Forming means for 106 clusters within 40 kpc of the center, W. E. Harris **(H1)** finds

$$\langle x \rangle = 7.3 \pm 0.6 \text{ kpc} \tag{4-59a}$$

$$\langle y \rangle = 0.4 \pm 0.6 \text{ kpc} \tag{4-59b}$$

$$\langle z \rangle = 0.3 \pm 0.6 \text{ kpc} \tag{4-59c}$$

From the essentially zero values for $\langle y \rangle$ and $\langle z \rangle$, we see that the clusters are indeed distributed symmetrically around the center, and, from $\langle x \rangle$, we obtain a lower bound on R_0, the Sun's distance from the center.

The value of $\langle x \rangle$ quoted sets only a lower bound on R_0 because of selection effects. Globular clusters lying beyond the center (particularly those near the galactic plane) suffer heavier interstellar obscuration and tend to be lost from the sample, which is therefore biased toward too small an average distance $\langle x \rangle$. To overcome this effect, one can consider samples of clusters that lie beyond some specified distance away from the plane, that is, that have $z \geq z_{min}$, and calculate the centroid $\langle x \rangle$ for various values of z_{min} (**B6**, Chapter 19). One expects to find an asymptotic value for $\langle x \rangle$ for large z_{min}, because then absorption effects become negligible, and the sample should be unbiased. But, in practice, for very large z_{min}, the sample contains too few clusters to yield a statistically reliable result, and a compromise must be struck at some intermediate value. Harris finds a fairly clear upper envelope in a plot of $\langle x \rangle$ versus z_{min} for $z_{min} \gtrsim 2.5$ kpc, and, from this, he estimates $R_0 \approx 8.6 \pm 1.5$ kpc.

L. Woltjer (**W3**) has noted that, if one segregates halo (metal-poor) and disk (metal-rich) clusters into two groups, the latter yield a value for $\langle x \rangle$ some 2.6 kpc larger than the former. This situation is precisely the opposite of what is expected, because the disk clusters, which on the average lie much closer to the plane than the halo clusters, should suffer more obscuration, and hence that sample should be more strongly biased toward clusters that lie closer to the Sun. As Woltjer points out, this discrepancy could indicate that the mean absolute magnitude of RR Lyrae stars is, in fact, larger for the more metal-rich clusters, perhaps around $\langle M_V \rangle_{RR} \approx +1.0$. This conclusion is only tentative at present, and it requires confirmation. In any case, the value of R_0 derived by Harris should be but little affected, even if this turns out to be true, because the value of z_{min} at which R_0 is estimated is already so large that the sample excludes most metal-rich clusters, and $\langle M_V \rangle_{RR}$ is calibrated with metal-poor stars near the Sun.

Completeness Having a reliable value for R_0, one can estimate the completeness of the known sample of clusters. Although some earlier investigators [see, for example, (**B6**, Chapter 19)] considered the presently known sample to be essentially complete, it is readily apparent that severe selection effects still exist in the data. For example, a plot of cluster positions in the (x, z) plane shows that, for $x \gtrsim 9$ kpc (that is, beyond the galactic center), the number of clusters with $|z| \lesssim 2$ kpc drops sharply. Indeed, a wedge-shaped gap centered on the galactic plane appears in the observed distribution. It is obvious that many clusters near the galactic center have suffered such heavy obscuration that they have not been identified and still await discovery.

At present, about 130 globular clusters are known (a number larger than the total estimated to exist in early analyses of the completeness question!). A reasonable lower bound on the total number can be obtained by doubling the number of clusters known to lie on the Sun's side of the galactic center $(x \leq R_0)$, as this sample suffers the least absorption and hence is known to the highest degree of completeness. W. E. Harris (**H1**) estimates that there are 85 ± 15 clusters with $x \leq R_0$ (the "error" allows for the uncertainty in R_0). Thus, by symmetry arguments alone, one concludes that there should be at least 170 globular clusters in our Galaxy. Allowing for a 20% incompleteness (a mere guess) for low-latitude clusters on the Sun's side of the galactic center, Harris concludes that a reasonable estimate for the total is about 200 ± 30 clusters, which implies that the present sample is about 65% to 75% complete. From the discussion presented later on the number of clusters very near to the center, it appears that the larger value (that is, 230) derived by Harris for the total number is likely to be the more realistic.

Space Distribution The space distribution of the system of globular clusters is characterized by an approximately spherical symmetry about the galactic center. This situation is most readily seen in a plot of cluster positions projected onto the (y, z) plane, as shown in Figure 4-13. There we see that the clusters are distributed at large throughout a halo of at least a 20-kpc radius and that they show a strong concentration toward the galactic center.

The question whether the space distribution of globular clusters is truly spherical or has a significant ellipticity is an important one, for it has dynamical implications. The conventional picture built on several older studies (**B6**, 401), (**K2**), (**K3**), (**O1**, 303) is that the halo clusters form an essentially

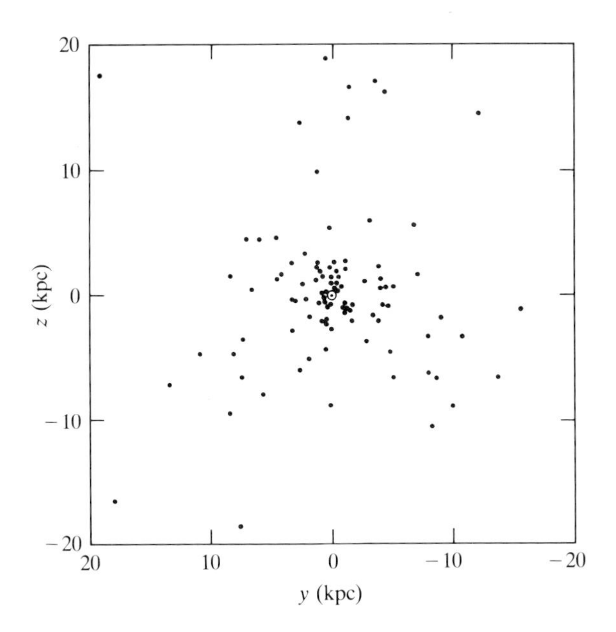

Figure 4-13. Distribution of globular clusters projected on the (y, z) plane, that is, the plane through the galactic center perpendicular to the Sun-center line. This projected distribution is plainly consistent with a spherically symmetric true space distribution around the center. [From (**H1**), by permission.]

unflattened system, and the metal-rich (G type) disk clusters and the innermost F type clusters are concentrated toward the galactic plane in a subsystem showing a significant degree of flattening. However, recent work (**H1**), (**W3**) has shown that this result is spurious and is an artifact of random errors in estimated cluster distances. These errors tend to induce a large spread in cluster positions along the line of sight, which in turn produces an illusory ellipticity. Woltjer has stressed that, taken at face value, the apparent distribution of "disk" clusters would imply that the subsystem is cigar-shaped, with its major axis directed toward the Sun. Such an arrangement is so improbably fortuitous that it can be immediately rejected.

Actually, it is much better to avoid projections onto the (x, y) or (x, z) planes and study only the (y, z) plane, in which errors in the distances do not alter the ratios (y/z) of the coordinates of individual clusters. Recently, S. van den Bergh (**V2**) has studied the way absorption affects the apparent distribution of globular clusters in this plane. One hundred "globular clusters" were distributed in spheroidal distributions through the halo of a model galaxy according to the radial density profile $\rho \sim R^{-3}$ $[0.2 < R(\text{kpc}) < 20]$, which is found to approximate that of the globular clusters (see below). Van den Bergh then assembled the y and z coordinates that would be assigned to these clusters when viewed from the solar neighborhood through a 240-pc thick disk of absorbing material having an extinction $A_V = 1.5$ mag kpc^{-1}. A cluster was assumed to become invisible when its apparent distance modulus $m - M \approx 20$. Using this model, van den Bergh concluded that the (centrally

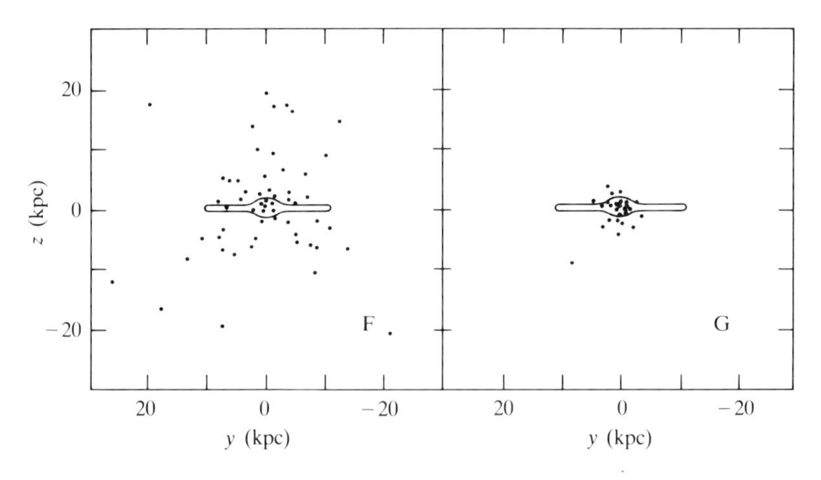

Figure 4-14. Distribution of F type (metal-poor) and G type (metal-rich) globular clusters projected onto the (y, z) plane. It is clear that the space distribution is basically spherically symmetric about the galactic center for both groups, with the metal-poor group distributed throughout the entire halo, and the metal-rich group strongly concentrated toward the center. [From (**H1**), by permission.]

concentrated) G type clusters can appear to be as nearly circularly symmetrical in the (y, z) plane as they actually do (see Figure 4-13 and 4-14), only if in reality they form a flattened system (axial ratio about 1:2). By contrast, the F type clusters, which also appear to be circularly distributed in the (y, z) plane, really are spherically distributed. The counts of these more outlying clusters are only slightly affected by absorption, so their apparent distribution is a true reflection of their actual distribution. Thus, according to this model, the system of globular clusters is rather flattened very close to the galactic center and becomes essentially spherical beyond 2–3 kpc. The approximate spherical symmetry of the distribution of more outlying globular clusters may indicate that the equipotential surfaces of our Galaxy are more spherical than would be expected from the distribution of the highly luminous material that dominates the measured light, because we might expect the globular clusters (as test particles in the general galactic gravitational field) to adopt at least as flattened a configuration as the galactic potential. We shall return to this question in later chapters.

The average space density of globular clusters (clusters kpc^{-3}) as a function of galactocentric distance can be obtained directly, by counting the number of clusters contained in successive spherical shells and by dividing by the volume of each shell. To minimize incompleteness effects, the sample can be limited to clusters on the Sun's side of the center. The results obtained by Harris (**H1**) are shown in Figure 4-15; R measures the radial distance from the center. The points plotted for the innermost shells ($R \lesssim 3$ kpc) are only lower limits because of incompleteness of the sample near the galactic center. The basic conclusion that emerges is that, over the entire range $R \geq 3$ kpc, the space density is well approximated by a power-law distribution of the form $\nu(R) \propto R^{-\alpha}$, with $\alpha \approx 3.5 \pm 0.25$. There is a suggestion that the slope may be shallower for the inner regions, say, $\nu \propto R^{-3}$ for $R < 10$ kpc, and steeper outside, say, $\nu \propto R^{-4}$ for $R > 10$ kpc, but the observational errors are too large for one to be sure that this result is significant. It seems best to use a single power law for $R \gtrsim 3$ kpc.

The result obtained for the exponent α is in harmony with results for halo stars (RR Lyrae variables—see below) and density estimates made via a dynamical analysis of the observed *asymmetric drift* of clusters relative to the Sun (see §6-4 and Chapter 14). It also agrees with the light distributions observed in the spheroidal components of other nearby edge-on disk galaxies (see §5-2). On the basis of this agreement, it seems reasonable to suppose that the globular clusters are truly representative spheroidal-component tracers whose space distribution describes that of the spheroidal component as a whole.

A luminosity can be assigned to each cluster once its distance is known, and, by adopting an assumed average mass-to-luminosity ratio of 1.5 in solar units, one can estimate that the average cluster's mass is about 2×10^5 and that the total mass of the halo cluster system is about $4 \times 10^7 \mathscr{M}_\odot$. A detailed mass-density distribution $\rho(R)$, as a function of R for the globular-cluster

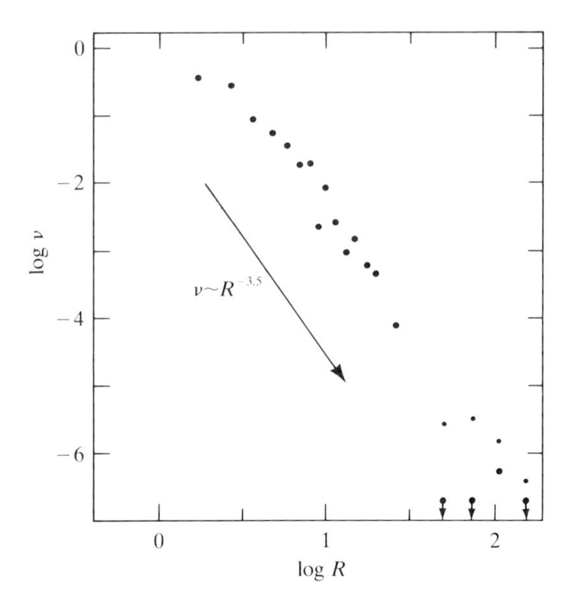

Figure 4-15. Average space density $\nu(R)$ (clusters kpc^{-3}) of globular clusters in spherical shells around the galactic center as a function of log R (kpc). For the outermost four shells, the small dots correspond to the densities obtained if the low-concentration Palomar clusters and the dwarf ellipticals are included, while the large dots correspond to the densities found if these systems are omitted. The distribution is well approximated by a power law of the form $\nu \propto R^{-3.5}$ over the entire range $R \gtrsim 3$ kpc. [From (**H1**), by permission.]

system, is tabulated by Harris (**H1**). The average mass density of globular clusters at the Sun's distance from the center is about $1.4 \times 10^{-6} \mathscr{M}_{\odot}\ \text{pc}^{-3}$, which is miniscule compared to the density of the disk. As can be seen in Table 4-10, the mass density of extreme subdwarfs, which have kinematic and spectroscopic properties identical to the globular-cluster stars, is orders of magnitude larger than the average mass density resulting from the globular clusters. The physical similarity of these two groups of stars, coupled with the implausibility of forming individual halo stars singly in the very low-density regions in which they are found, has led several workers to speculate that most of the stars now found in the general field of the spheroidal component were actually born in globular clusters that have subsequently "dissolved" [see (**F1**) for references]. If this speculation is true, then we now observe only a small fraction of the original population of globular clusters, an inference of importance for theories of galactic evolution and of the formation and dynamics of globular clusters.

Number Density at the Galactic Center Almost no direct observational information is available about the number of globular clusters that may exist within 1 kpc of the galactic center. Such clusters would be extremely heavily obscured, and they would be most readily detectable (if at all) by far-infrared observations. Until such data become available, one can do little more than attempt a theoretical extrapolation of the existing data by use of a dynamical model; this approach has recently been taken by Oort (**O3**). The results obtained depend on the random-velocity dispersions assigned to the globular-cluster system. Oort makes estimates for two values of $\langle v^2 \rangle^{1/2}$, namely 134 and 147 km s^{-1}, the latter being a plausible upper bound on the true dispersion and yielding the more conservative estimate of the number of clusters. Normalizing the predictions of the model to the observed space densities for $R \geq 2.5$ kpc, he finds that initially there must have been of the order of 110 clusters within 1.25 kpc of the center. Of these, roughly 20% were massive systems with integrated absolute magnitudes $M_V \lesssim -9.5$, and the remainder were less massive, fainter systems.

The numbers quoted are markedly changed over the lifetime of our Galaxy ($\sim 10^{10}$ years) by the effects of *dynamical friction*, which tends selectively to drag the more massive clusters into the center, where they coalesce to form a small nucleus. On the basis of theoretical work by Tremaine et al. (**T2**), (**T3**), Oort estimates that roughly 5 clusters with $M_V \leq -9.5$ and 50 clusters with $M_V \geq -9.5$ survive today. Assigning these clusters mean masses of $1.8 \times 10^6 \mathscr{M}_\odot$ and $7 \times 10^5 \mathscr{M}_\odot$, respectively, he concludes that the nuclear globular clusters presently account for about $4.5 \times 10^7 \mathscr{M}_\odot$ and that the mass of the nuclear core itself is about $2 \times 10^7 \mathscr{M}_\odot$. The fact that the total mass in the clusters near the center is probably equal to that of the clusters in the entire halo again emphasizes the strong central concentration of this system.

Chemical-Composition Gradients We saw in §3-6 how the metallicity of cluster stars can be inferred from spectroscopic and photometric data, and we have noted that metal abundances in clusters range from near solar to less than 10^{-2} times solar. If one combines information about cluster metal abundances with knowledge of their positions with respect to the galactic center, one finds evidence for a correlation within the inner halo ($R \lesssim 10$ kpc) between metallicity and galactocentric distance (**H1**), (**M6**), (**M7**), (**S8**). The basic result is that, within the inner halo, the most metal-rich clusters are found closest to the galactic center, and clusters of progressively lower metallicity tend to fill larger and larger spherical volumes around the center. For example, Harris (**H1**) divides his sample of globular clusters into three metallicity groups, F^- (most metal poor), F^+ (intermediate), and G (most metal rich), and he finds that no G clusters are found beyond about 7 kpc from the center, no F^+ clusters beyond about 20 kpc, and only F^- clusters beyond 20 kpc. Thus, the closer to the center one observes, the greater is the range in abundances among the clusters found there. Indeed, near the center, one finds clusters of low, intermediate, and high metallicity occurring with roughly equal frequencies.

It is generally believed at present that these results imply a composition gradient within the halo. However, it is possible (as suggested by some authors) that there is no gradient and that the central location of metal-rich clusters is a result of their being the tail in a distribution function of abundances, which is seen only when the density of all clusters is high (that is, at the galactic center). Further research is required to decide the issue. Because metal-poor clusters fill the whole volume of the halo and bulge, while the most metal-rich clusters are confined to a central region, the metallicity gradient (assuming that there is one) of inner-halo material is best measured not by average abundances but by the upper envelope of the metal-abundance distribution as a function of galactocentric distance.

The situation in the outer halo ($R > 10$ kpc) is rather different. Here, at $30 < R(\text{kpc}) < 100$, one finds, besides ordinary globular clusters, the low-concentration *Palomar systems* and the *dwarf-spheroidal* systems (for example, the Draco system; see §5-1). These systems show no clear evidence for a metallicity gradient (**C1**), (**C2**), (**C4**), (**S8**), (**Z1**), but rather they seem to have an essentially random distribution of metallicities in the range $-2.5 \lesssim [\text{Fe/H}] \lesssim -1.5$. At face value, the observations indicate that the metal abundance of objects in the very outer halo ($R > 40$ kpc) is always above a floor value of around $[\text{Fe/H}] \approx -2.2 \pm 0.3$, but it is not clear at the present time whether this floor value is real or is the result of the spectrophotometric metallicity indicators—for example, $\delta(U - B)$—becoming very insensitive to [Fe/H] once the opacity of the stellar atmospheric material is no longer determined by the number of electrons liberated by the ionization of the metals. Further observational and theoretical work will be required to provide definitive results.

To explain the results just described, one can offer the following scenario. During the collapse of the halo, clusters with low metallicities were apparently formed at large distances from the galactic center, and now they fill the entire halo as a result of their orbital motion (on highly eccentric plunging orbits). Clusters of progressively higher metal contents were formed from progressively more metal-enriched material closer and closer to the center, and hence they are now confined to smaller volumes around it. The assumptions made here, of course, are that the apparent metallicity gradient in the halo is real and that present-day cluster metallicities do still accurately reflect the metal content of the material out of which they were formed. Presumably, the progressive enrichment of the inner-halo material is the result of the recycling, by evolved halo and cluster stars, of nuclear-processed material into the ambient medium. Furthermore, the floor value of [Fe/H] ~ -2.3 suggests that even the primoidal protogalactic material had been enriched by nucleosynthesis processes from some primeval stellar population.

Halo Stars

The distribution of material in the spheroidal component can also be studied by determining the space densities of individual halo and nuclear-bulge stars. The large-scale structure of the halo is best revealed by RR Lyrae variables. A rough picture of the three-dimensional structure of the spheroidal component in the vicinity of the Sun's position in our Galaxy can be derived from analyses of the z distribution of various kinds of stars.

The Large-Scale Distribution of RR Lyrae Variables The RR Lyrae variables can be observed to distances beyond the galactic center in certain fields of high transparency near the center, and they have therefore been intensively studied, first by Baade, later by Kinman and others (**K4**), (**K5**), (**L1**), and most recently and most completely by Oort and Plaut (**O4**) using data from the Palomar-Groningen Survey with the 48-inch Schmidt telescope. The limiting apparent magnitude in the survey is $M_{pg} \approx 19.5$–20. The variables were identified by blink-comparator searches in ten plate pairs for each of six fields. The stars selected were RR_{ab} variables with amplitudes $\Delta m \gtrsim 0.5$ mag, which can be discovered with a high degree of completeness. Oort and Plaut estimate average completeness fractions, in the magnitude ranges indicated, to be as follows: ($m < 15, 0.9$), ($15 \leq m \leq 16, 0.8$), ($16 \leq m \leq 17$, 0.75), ($17 \leq m \leq 18, 0.65$), ($18 \leq m \leq 19, 0.45$), and ($m > 19, 0.25$). To reach the galactic center, one must observe stars in the range $16 \lesssim m \lesssim 16.5$, for which the completeness fractions are quite high. Corrections for interstellar absorption were derived directly for each star from its observed colors.

The distances to individual stars were assigned by adopting $\langle M_V \rangle = +0.6$, in agreement with the results cited in §3-8. Plots of the observed numbers of stars in interval $\Delta m = 0.25$ mag show sharp maxima along each line of sight. These maxima are presumed to be associated with a density maximum at the galactic center, and hence they can be used to determine R_0.

To analyze the data, Oort and Plaut adopted a model-fitting procedure that assumes (1) the equidensity surfaces are oblate spheroids centered on the galactic center with major axis a, minor axis c, and equatorial plane coincident with the plane of our Galaxy; (2) the space density varies as $v = v_{1.5}$ $(1.5/a)^{\alpha}$, where $v_{1.5}$ is the density at 1.5 kpc, and α is a constant; and (3) the average random error (Gaussian-distributed) of a stellar magnitude corrected for absorption is ε. Their model is specified by the set of independent parameters (c/a), α, ε, $v_{1.5}$, and R_0. They proceeded by first plotting $\log n$, the logarithm of the observed number of stars in the field, versus $\log r$, the logarithm of the distance from the Sun in kiloparsecs. For each model, specified by a choice of (c/a), α, and ε, they then computed a theoretical $\log n$ curve by (1) calculating the product of the number density (using the assumed density law) and the volume element associated with the field as a function of (R/R_0) along the line of sight, (2) convolving this distribution with a Gaussian error function for the assumed value of ε, and (3) multiplying the resulting numbers by the appropriate incompleteness functions. In this procedure, the shape of the computed curve depends only on (c/a), α, and ε. Hence, by fitting the theoretical curve to the observed curve by horizontal and vertical shifts (on a logarithmic scale), they found absolute values for R_0 and $v_{1.5}$.

Even this relatively simple model has a fairly large number of free parameters. But, from detailed analysis, rather close limits can be set on each of them. The parameter α is found unambiguously from the asymmetry of the curves around their peak. The parameters (c/a) and ε cannot be determined independently, but one can find a lower limit for (c/a) by using the minimum possible value for ε (conservatively estimated to be ± 0.25 mag). Overall, Oort and Plaut find that the best representation of the data is given by $\alpha = -3.0$, and, in the best fields, they find that lower bounds on (c/a) lie between 0.8 and 1.0, that is, the distribution is nearly spherical (a conclusion made firmer if a larger value of ε is admitted).

Thus, to a good approximation, the space distribution of RR Lyrae variables is given by $v(R) = v_{1.5}(1.5/R)^3$, where R is in kilopersecs, and $v_{1.5}$ is estimated to be about 2.6×10^{-7} stars pc^3. This density variation is fairly well established for $R \lesssim 5$ kpc, and it is best determined on the range $1 \lesssim R \lesssim 3$ kpc. The exponent α agrees well with the results obtained for the globular clusters, and it also agree with the results from the analysis of the asymmetric drift of subdwarfs and RR Lyrae variables in the solar neighborhood. The weighted mean value for R_0 is found to be $R_0 = 8.7 \pm 0.6$ kpc, which Oort and Plaut consider to be one of the most reliable estimates of this parameter.

General Density Distribution in the Halo near the Sun Some additional information about the distribution of stars in the spheroidal component near the Sun can be obtained from star-count analyses at intermediate and high galactic latitudes. Only a few studies of this kind have been made; an important example is the early work of Oort (**O2**) using general star counts (comprising mainly K giants).

Oort attempted to deduce the large-scale variations in stellar density in regions away from the galactic plane by comparing the observed star counts in high-latitude fields with the numbers that would be expected (on the basis of counts toward the galactic pole) if the surfaces of equal density were exactly parallel to the galactic plane. He concluded that most of the interstellar absorption occurs close to the plane and can be treated as a foreground effect. (This conclusion is somewhat open to question, however.) Then, from the fact that, in a field of angular size ω, the contribution to $A(m)$—the number of stars with apparent magnitude m—from stars in the range $(r, r + dr)$ scales as $n\omega r^2\, dr$, and from the fact that, on a surface with z constant, the distance r_b of a point at latitude b is $r_b = r_{90} \csc b$, it follows that, if the equal-density surfaces were parallel to the galactic plane, then at latitude b and apparent magnitude m_b we would expect to observe star counts

$$A(m_b, b) = \csc^3 b\, A(m_{90}, 90^\circ) \tag{4-60}$$

where

$$\begin{aligned} m_{90} &= m_b - 5\log(r_b/r_{90}) - \Delta a \\ &= m_b + 5\log(\sin b) - \Delta a \end{aligned} \tag{4-61}$$

and Δa is the excess absorption at latitude b relative to the pole.

From an analysis of the deviations of the observed star counts from those predicted by equations (4-60) and (4-61), Oort was able to deduce the deficiency (or excess) of the space density of stars, relative to the hypothesis $D(z) \equiv$ constant, as a function of (x, y) around the direction to the galactic pole. He concluded that the isodensity surfaces were inclined to the galactic plane by about 10°, rising toward the galactic center, as sketched in Figure 4-16. He also concluded that the radial density gradient was $(\partial \ln \nu/\partial \ln R) \approx -3.2$, which is in good (perhaps fortuitous) agreement with the recent results

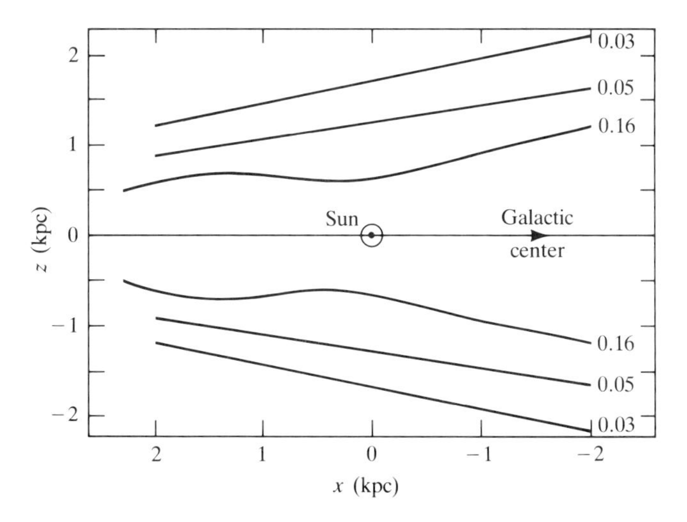

Figure 4-16. Density distribution of all stellar types in the vicinity of the Sun projected on the (x, z) plane perpendicular to the galactic plane along the Sun-center line. Curves give isodensity contours in units of the density near the Sun. [Adapted from (**O2**).]

for globular clusters and RR Lyrae stars discussed earlier. Qualitatively similar density contours have been obtained in recent work by W. Becker (**B4**), whose data extend to much larger distances and show an increasing tilt at higher z's (that is, an approach to a more nearly spherical distribution). Oort also plotted contours close to the plane, but we have omitted these because they are very uncertain owing to the very rough treatment of interstellar absorption in his work. Other studies of this general character are discussed in some detail by Elvius (**B6**, Chapter 3).

Taken at face value, Oort's results imply a rather surprisingly flat distribution for the halo if the stars in his sample were all true halo objects. In fact, this is not the case. Within ± 1 kpc of the plane, one finds mainly old disk stars with fairly high metallicities. These are not true halo objects at all, but merely stars that can penetrate to relatively large distances from the plane because they have substantial random velocities in the z direction. These disk stars strongly bias the sample studied by Oort. As we shall see in Chapter 7, stars of progressively decreasing metallicity show a progressive increase in their velocity dispersion $\langle w^2 \rangle^{1/2}$ perpendicular to the plane, and, if one confines attention to true halo objects $[\delta(U - B) \gtrsim 0.2]$, one finds that these stars must be fairly uniformly distributed through distances up to ± 6 kpc around the plane. Hence they really do have an unflattened density distribution.

Mass Densities of Halo Stars; Mass of the Inner Halo The space densities just derived for the globular clusters and RR Lyrae stars can be used to make rough mass estimates for the inner halo. The basic hypothesis is that, because the form of the space-density distribution is the same for both classes of object, either one provides a valid indicator of the form of the mass distribution for the halo as a whole. In particular, we hypothesize that the ratio $\mu \equiv \rho(\text{halo})/\rho(\text{globular clusters})$ is constant throughout the halo and equals $\mathscr{M}(\text{halo})/\mathscr{M}(\text{globulars})$. We have seen that the system of globular clusters is thought to contain about $4.5 \times 10^7 \mathscr{M}_\odot$ within about 1 kpc of the center and about $4 \times 10^7 \mathscr{M}_\odot$ outside this central region and within about 20 kpc, for a total mass of about $9 \times 10^7 \mathscr{M}_\odot$ within 20 kpc.

Woltjer (**W3**) has estimated that, within 9 kpc of the center ($R \lesssim R_0$), the total number of RR Lyrae variables in globular clusters is about 200 and, from the Oort–Plaut data, that the general field contains about 2×10^4 RR Lyrae variables. Assuming that the mass-distribution function (or luminosity function) is the same for field stars and cluster stars, one then finds $\mu \approx 100$, and hence that $\mathscr{M}(\text{halo}) \sim 9 \times 10^9 \mathscr{M}_\odot$ within 20 kpc of the center. Alternatively, one can try to estimate the ratio μ by using observed values of $\rho_\odot(\text{halo})$ near the Sun, knowing (**H1**) that $\rho_\odot(\text{globular clusters}) \approx 1.4 \times 10^{-6} \mathscr{M}_\odot\ \text{pc}^{-3}$. To fix $\rho_\odot(\text{halo})$, we could use the observed density of 10^{-9} RR Lyrae stars per pc^3 in the solar neighborhood (**O4**) and the luminosity function for spheroidal-component stars (Table 4-11) to estimate $\rho_\odot(\text{halo}) \approx 3 \times 10^{-5} \mathscr{M}_\odot\ \text{pc}^{-3}$, and hence $\mu \approx 20$. Alternatively, the space density of subdwarfs in the immediate solar neighborhood implies $\rho_{\odot}(\text{halo}) \approx 1.5 \times 10^{-4} \mathscr{M}_{\odot}\ \text{pc}^{-3}$ (see Table 4-10), and hence $\mu \approx 100$. Thus $\mathscr{M}(\text{halo})$ within 20 kpc of the center appears to lie in the range of 2×10^9 to $9 \times 10^9 \mathscr{M}_{\odot}$; this result must be regarded as fairly uncertain, however. In the companion volume, we shall make more definite estimates of $\mathscr{M}(\text{halo})$ by combining observational data with dynamical considerations. In conclusion, it is worth emphasizing that the halo mass estimates just made apply only to the easily detected luminous material in the halo. The mass of the hypothetical dark population mentioned in Chapter 1 remains undetermined by the evidence we have considered thus far.

The Nuclear Bulge

Although the analyses just described tell us something about the density of stars in the central regions of our Galaxy, they tell us little about their physical nature. Such information is difficult to obtain because the galactic nucleus is extremely heavily obscured (see §9-4). It can be observed clearly only at X-ray, infrared, and radio wavelengths, and it can be partially observed in the visible spectrum through the low-absorption "Baade windows" mentioned earlier.

Just about the only information we have about the stellar content in the nuclear region of our Galaxy comes from observations of the integrated spectra of stars in the Great Star Cloud in Sagittarius by W. W. Morgan (**M7**). He used the same spectrographic techniques as he had employed in his work on the classification of globular clusters and on the integrated spectra of galaxies. His primary conclusion was that, in the blue-violet (that is, photographic) region of the spectrum, the principal contributors to the light received from the galactic nuclear bulge are probably K giants, and the integrated spectrum strongly resembles that of the nuclear bulge of M31. These results are of enormous importance, for they indicate that stars in the innermost regions of the spheroidal component are metal rich (that is, have solar or above solar metal content).

Morgan further showed that, as one examines regions with increasing amounts of interstellar absorption (within which the line of sight will penetrate less and less closely to the center), one sees a progressively larger contribution from stars of earlier types; this is direct evidence for a change in the characteristic stellar population as one proceeds from the bulge to the inner disk. The same effect is observed in spectra of M31, where one can see unambiguously that the stars in the inner disk differ systematically from those in the nuclear bulge itself.

Further evidence about the nature of the stars in the galactic nuclear bulge is given by the observation that, although the density of RR Lyrae stars in the central regions of our Galaxy is extremely high, these stars are nevertheless outnumbered two-to-one (**N1**) by ordinary M giants! Moreover, the color-magnitude diagram of nuclear-bulge stars (**V1**) resembles that of the old metal-rich cluster NGC 188, and it is markedly different from those of typical halo globular clusters. This effect is also seen in integrated colors; for example, the unreddened values of $[(U - B), (B - V)]$ for M3, the nuclear bulge, and NGC 188 are (0.54, 0.0), (0.77, 0.33), and (0.80, 0.37), respectively. The inescapable conclusion is that the stars in the nuclear bulge of our Galaxy (and of M31) are old and metal rich. Their properties seem entirely consistent with an inward-rising, radial metallicity gradient of the spheroidal component, as is apparently exhibited by the globular clusters.

4-5. STELLAR POPULATIONS

Historical Perspective

In Chapter 1, we described the introduction of the concept of *stellar populations* by Baade. Baade discriminated two population groups: *Population I*, characterized by the spiral-arm stars, and *Population II*, characterized by stars found in the smooth spheroidal components of spirals (halo plus bulge) and throughout elliptical galaxies. He emphasized the fundamentally different nature of the H-R diagrams for these two groups, Population I having an H-R diagram resembling those of open clusters in our Galaxy, and Population II having one resembling those of globular clusters (in particular, halo clusters). It was soon recognized that the concept of stellar populations provided a unifying framework for describing the stellar content in a galaxy in terms of a number of distinctive subgroups, each with a characteristic space distribution, age, chemical composition, set of kinematic properties, and distribution in the H-R diagram. In the next two decades after Baade's discovery, there was an intensive development and elaboration of a scheme of stellar populations in our Galaxy, including subdivisions of Baade's original types, and there were numerous attempts to apply these ideas to other galaxies.

By the middle 1960s, a rather definite conventional picture had emerged. While it is not our intention to discuss this picture in great detail—particularly in the face of excellent review articles written, for example, by Blaauw (**B6**, Chapter 20), Gratton (**S9**, 13), Oort (**B6**, Chapter 21), (**S9**, 85), and the whole book *Stellar Populations* (**O1**) summarizing the famous Vatican conference on the subject—we must nevertheless sketch the broad outlines of the ideas that were developed in order to indicate the conceptual background against which later developments are to be viewed. The reader is urged to consult the references just cited for details of the rationale and implications of this scheme, for we cannot do them justice in our brief discussion. We must also mention that we shall have to draw on a certain amount of kinematical evidence in this section, even though we shall not discuss that material until Chapters 6 and 7; it will pay the reader to reread this section after reading those chapters.

As a result of the vigorous research initiated by Baade's introduction of the idea of stellar populations, a number of important facts emerged. First, from the theory of stellar evolution, it was realized that the globular clusters—the paradigm of Population II—are all old, while Population I objects are young. Second, both evolutionary theory and direct spectroscopic evidence indicated that Population I is metal rich and that Population II stars in the solar neighborhood are metal poor. Third, the two populations were recognized to have markedly different kinematic properties. Population I stars move dominantly on nearly circular orbits around the galactic center with a small random-velocity dispersion around the exact circular velocity; in contrast, Population II objects (globular clusters and the extreme subdwarfs) were found to have a low (perhaps zero) systemic rotation and a very large random-velocity dispersion, which implies that they move on highly eccentric orbits through large ranges of galactocentric radial distance. Fourth, it was found that the space distribution of Population I is strongly flattened, and the space distribution of Population II is essentially unflattened.

Putting these various lines of evidence together, and considering what seemed to be a reasonable rough outline of the dynamical evolution of our Galaxy, astronomers developed a scenario that went something like this: The Galaxy may be supposed to have been formed by the condensation and collapse of the primordial material contained initially in a huge volume of space. During this collapse, a sequence of distinctive populations was established, whose properties bear a one-to-one relationship to the dynamical state of the system at definite epochs. The sequence of populations is thus conceived to be fundamentally one-dimensional with all relevant properties being a function of a single parameter (say, stellar age). The first stars that formed were widely distributed in an unflattened system, of low rotation velocity, out of material with very low metal content. Successive groups of stars, each corresponding to a definite population subdivision, formed a sequence of systems that were progressively more flattened (concentrated to the galactic plane), contained stars that moved on more nearly circular orbits, and were composed of progressively more metal-rich material. It was thought that this sequence progressed from the distended *Halo Population II* through a more flattened *Intermediate Population II* and culminated in a great burst of star formation that produced the *Disk Population*, whose identification with Population I or II varied from author to author. After the formation of the disk, it was supposed that star formation proceeded at a slower rate; the spiral-arm stars now forming were designated *Extreme Population I*, while somewhat older (but still young compared to globular clusters) metal-rich stars in the general field, such as the Sun, were called *Older Population I*.

From among the arguments advanced in support of this scheme, we single out the following two for discussion: (1) The metal-rich globular clusters were thought to be selectively concentrated toward the galactic plane compared to metal-poor halo clusters. (2) In the solar neighborhood, it was found, from kinematic evidence, that groups of stars having progressively larger velocity dispersions perpendicular to the galactic plane, and which can therefore move to larger distances away from the plane, show progressively lower metal contents as inferred from spectrophotometric data. This progression seems, at first sight, to merge continuously into the halo stars exemplified in the solar neighborhood by the most extreme metal poor subdwarfs. As we shall see later, both of these bits of evidence are false leads.

As research progressed, it became evident that the basic, one-dimensional scheme just described is inadequate in that it leads to contradictions and serious ambiguities. We shall not attempt to discuss in detail all the problems that arose, as excellent critiques focusing on different problem areas can be found in review articles by King (**K1**), (**T1**, 1), Spinrad (**S11**), Unsöld (**U1**), van den Bergh (**V1**), and others; most of these make very good reading. Let us just choose two problems as examples. First, as time went by, stellar metal content became the primary population discriminant used. This choice was probably a result of its being well correlated with easily measured photometric indicators, such as $\delta(U - B)$ and Δm_1. In particular, "Population II" came to mean, for all practical purposes, metal-poor stars. Astronomers who adopted this view were then confronted with a logical cramp when it was shown that stars in the nuclear bulge in our own Galaxy and others (for example, M31), and the stars in the central regions of elliptical galaxies—all of which are Population II by every other criterion—are metal rich! Second, detailed analyses of stars in the solar neighborhood showed that the picture of a one-to-one correlation of stellar metal content and kinematic properties simply does not hold. Specific examples were identified of groups of stars that all have practically identical values of $\delta(U - B)$ and yet fall neatly into two distinct kinematic subgroups, one clearly having "halo population" characteristics (high z velocities, eccentric orbits) and the other clearly having "disk population" characteristics (low z velocities, circular orbits). Beyond these two specific problems, many precepts of the scheme that were self-consistent in the solar neighborhood gave rise to contradictions when applied to other parts of our Galaxy. In fact, so many other contradictions, dilemmas, counterexamples, and inconsistencies have arisen that it has become clear that the picture just outlined is fundamentally oversimplified, and a new picture is required.

Structural and Evolutionary Framework

In what follows, we shall start afresh and attempt to build a simple descriptive system of stellar populations which will provide a framework for further discussion. We stress that this system is not comprehensive and that we have no illusions that it will be "definitive." Yet, it is compatible with, and coordinates well, the observed facts as we understand them at present. From a purely structural point of view, it is clear that our Galaxy consists of two major components: the spheroidal component and the disk. In developing a scheme of stellar populations, it is fruitful to focus on each of these distinctive entities more or less separately, an approach that is supported by contemporary views that the dynamical histories of the two are quite different. We shall omit any discussion, necessarily speculative, of the hypothetical "dark population" here, and we shall refrain from (probably) artificial distinctions between galactic "nucleus," "bulge," and "halo," because there seems to be a very smooth progression of physical properties within the spheroidal component.

In Chapter 1, we sketched briefly an evolutionary picture upon which a discussion of stellar populations can be based, and, as we shall consider it in detail in the next volume, we shall not discuss it here beyond the following remarks. To begin, it is now thought likely that the entire spheroidal component was formed in an essentially radial collapse that produced a roughly spherical system (or sequence of subsystems) with a radial composition gradient of increasing metallicity toward the center. The composition gradient was presumably established through progressive enrichment of the infalling material as a result of the recycling of nuclear-processed material from terminally evolved massive stars, formed slightly earlier in the collapse of the halo. (The evolutionary lifetimes of massive stars are short compared with the freefall time of the halo.) An essential point of difference between this scheme and the old one is that we now realize that the metal-rich globular clusters are not concentrated to the galactic plane; rather, they are concentrated toward the center (see discussion in §4-4). Therefore, the spheroidal component probably does not comprise a set of subsystems that become progressively more flattened and merge continuously into the disk in the galactic plane. It is more likely a sequence of relatively unflattened systems, each of which occupies a characteristic volume (out to, say, some $R = R_{\max}$) and cuts through that part of the disk that lies within its volume (that is, $R \leq R_{\max}$). Thus, at any point within the disk, we should find a mixture of disk stars plus that range of subgroups of spheroidal-component stars whose containment volumes extend far enough from the center.

In the disk, the situation is rather different. Current estimates of cluster ages (see §3-9 and §3-10) suggest that, well after the spheroidal component collapsed, the disk formed out of material that was already metal enriched to within about a factor of three to five of solar, and it has since that time been progressively enriched with metals at a rate of perhaps a factor of three to five per 10^{10} years (see §3-9). An essential ingredient in our present picture of the disk is the realization that stars in the disk have had their random velocities significantly increased over the lifetime of the disk (possibly by encounters with interstellar clouds) and that this increase has substantially altered their kinematic properties and space distributions. In particular, recent work (see §7-1) has shown that the increase in stellar random velocities during the lifetime of the disk is sufficiently large to allow an unambiguous identification of many previously enigmatic groups as very old disk stars that have been widely dispersed, both in the plane and away from it. In effect, the "gas" of disk stars grows "hotter" as it grows older, and the "molecules" (stars) diffuse around within the disk on increasingly eccentric orbits and into a thicker layer above and below the galactic plane on relatively steeply inclined orbits.

The progressive "heating" of disk stars as a function of time has the important implication that at least part of the correlation of decreasing [Fe/H] with increasing z_{max} (the maximum distance a star can move away from the plane, which is determined by the size of its z velocity) must be produced by the progressive increase of the disk thickness for older disk stars, which are less metal-rich than currently forming disk stars. Hence at least part of the evidence advanced in the old picture for a progressive flattening of the collapsing halo system (as seen locally) can be explained instead by precisely the opposite effect—namely, a progressive rediffusion outward of older and older stars away from a disk whose metal content is progressively increasing with time.

There is, in addition, another effect that leads to a systematic overestimate of the degree of flattening of the space distribution of halo stars observed in the solar neighborhood. Suppose we examine a sample of extreme subdwarfs in the galactic plane in the immediate vicinity of the Sun. (Recall that the data are restricted to a volume whose radius is only 25 pc.) Assume that these stars are, in reality, distributed throughout a spherical volume $R \leq R_{max}$ on highly eccentric (in-and-out) orbits. If we choose a particular radius vector, then stars whose instantaneous orbital major axes lie along that vector, and for which $R_{max} \geq R_0$, will be confined to a rather narrow cigar-shaped volume around it (that is, a volume much longer in the radial direction than in the tangential directions), because the orbits are so eccentric and their orbital planes are oriented isotropically around this chosen axis. Now, in actuality, the whole sphere $R \leq R_{max}$ is filled with these cigar-shaped volumes surrounding randomly oriented radius vectors. But, if we can sample only one such volume (specifically, that one whose axis passes through the Sun's position in the plane) and no other because we can neither identify a complete sample of the relevant stars nor measure their space motions beyond 25 pc, then the sample is biased. From an analysis of the orbits of the stars in this biased sample, we would be led to the false conclusion that all such stars were distributed in a narrow range of distances (at most two "cigar-widths") above and below the galactic plane. We would simply miss all those stars whose "cigar-volumes" do not happen to include the Sun, that is, all orbits whose major axes are inclined so steeply to the plane that the orbit cannot pass through the galactic plane at $R = R_0$. Put another way, a study of the kinematics of the sample of halo stars that happen to pass through the Sun's neighborhood in the disk provides, at best, a lower bound on the width of the distribution of those stars away from the disk.

If we have tacitly assumed that these stars are confined to a flattened system, then, by analyzing a biased sample, we can easily "verify" our assumption, despite the fact that it is false!

The failure to realize the importance of the two effects just discussed no doubt contributed to the development of the older (probably incorrect) picture of a progressive flattening of a collapsing, metal-enriching halo, which ultimately settles smoothly and continuously into the present disk.

Archetype Population Groups

Let us now delineate population groups compatible with the broad structural and evolutionary framework we have chosen and describe their properties in moderate detail. For convenience, we shall discuss the spheroidal-component and disk populations separately. It should be borne in mind that the stars found in the solar neighborhood consist of a complex mixture of subpopulations of both of these larger groups.

Spheroidal Component The situation for the spheroidal component (SC)—that is, the galactic halo, bulge, and nucleus—is relatively simple, although it differs significantly from Baade's original conception of it. The space distribution of these stars is nearly spherical and is strongly centrally concentrated, with the number densities of all objects rising as $\nu \propto R^{-3}$ for R in the range 100 pc $\lesssim R \lesssim$ 30 kpc. Globular clusters and RR Lyrae stars are typical constituents of the SC that can be identified throughout its whole volume. In the solar neighborhood, the SC is also represented by the extreme subdwarfs, which of course exist everywhere within it, but can be detected only locally. All of these objects are old, 10×10^9 to 15×10^9 years in age. The masses of SC stars are about $0.8\mathscr{M}_\odot$ or less, more-massive stars having already evolved to the white-dwarf stage. The SC contains little dust or interstellar gas, except in the galactic nucleus (see §9-4).

Within the SC there is a strong radial gradient of metallicity, in the sense of increasing metallicity toward the galactic center. In the outermost regions of the SC, there is a scatter in the values of [Fe/H] between a floor value of about -3 and an upper envelope near -1.5. This upper envelope begins to rise near $R = 10$ kpc, increasing to [Fe/H] ≈ 0.0 or above at the galactic center (**S8**). The existence in the outermost halo of a floor value for [Fe/H] that is not characteristic of primordial cosmic material is still somewhat uncertain, but, if true, it suggests that, either the protogalaxy formed out of material that had somehow already been nuclear-processed, or that some "population zero" (now vanished from sight) processed the protogalactic material at the earliest instant of our Galaxy's collapse.

For the purposes of our discussion, it is sometimes convenient to divide the SC into a metal-poor halo, a metal-rich bulge, and, to interpolate between, one or more (at present rather vaguely defined) *intermediate spheroidal-component* systems. These divisions are somewhat arbitrary, and they may well be replaced at a later time or subdivided in some other way. Each of these metallicity groups fill almost the whole volume from the center out to some maximum distance, because SC objects move on plunging orbits.

As we have already mentioned, at any given position in the SC, there is a large dispersion in metallicity (about a factor of ten) around the mean, being largest near the center and becoming smaller in the outer regions, as successive groups of less and less metal-rich clusters are left behind within their limiting volumes. In the solar neighborhood, the dispersion of metallicity is quite large. Indeed, it is large enough that the range of metal abundances of genuine halo objects overlaps that of the oldest, most metal-poor disk stars, thereby introducing considerable confusion into attempts to untangle the two groups. Almost all the information we have concerning the metallicity gradient in the SC comes from observations of the globular clusters. A study of, say, the run of $\langle \Delta S \rangle$ for RR Lyrae stars as a function of galactocentric distance might be quite rewarding in helping to confirm or to modify this picture.

Everywhere within the SC, the brightest stars are red giants, and the H-R diagram for stars in a sample volume will qualitatively resemble that of a globular cluster. In the outer regions, a representative CM diagram should resemble that of the Draco system or of M92 (which differ significantly from one another—see §3-6) and, toward the center, should change smoothly through a sequence such as M3, M71, and 47 Tuc. In the nuclear bulge, the CM diagram should probably look like that of NGC 188 or that of a group of the oldest disk stars in the solar neighborhood, and it would not resemble that of a metal-poor globular cluster or of halo stars near the Sun.

Kinematically, the SC is a system having a low rotation velocity around the center and a large random-velocity dispersion. If we adopt the ratio of the energy in random motions to the total kinetic energy as a measure of a dimensionless "temperature" of a system, we would characterize the SC as being quite "hot." Both SC stars and globular clusters move on highly eccentric orbits around the galactic center. In practice, we can determine these orbits only for those halo subdwarfs and RR Lyrae stars found in the solar neighborhood for which the requisite measurements can be made. The high orbital eccentricities are consistent with a fairly rapid radial halo collapse.

Disk Component As was described in Chapter 3, current age estimates for disk and halo clusters suggest that the disk may be significantly younger than the spheroidal component. It is likely that the disk was formed from material of high metallicity, perhaps pre-enriched by the ejecta of evolved SC stars. Star formation in the disk apparently was delayed until the motion of the gas had already been circularized by inelastic cloud-cloud collisions, so that even the oldest disk stars were born on initially nearly circular orbits.

As time progressed, the metal content of the disk rose, at a rate of perhaps a factor of three to five in 10^{10} years. Also, the random velocity of stars relative to the circular velocity increased progressively as a result of encounters between stars and massive interstellar clouds or passing spiral arms.

The stars at any given position in the disk—for example, in the solar neighborhood—are composed of a broad mixture of disk populations of widely varying ages and properties, plus stars from the interpenetrating spheroidal component within which it is immersed. An observer at the Sun's position is therefore presented with a rather complicated tangle, which historically was not immediately understood. Actually, the apparent difficulties encountered in understanding the stellar populations near the Sun may be a result of the very fact that it is possible to measure stellar characteristics in this local sample fairly completely. Only our ignorance of the details of the situation in other parts of our Galaxy may shield us from having to unravel equally complex circumstances existing elsewhere.

To penetrate this thicket, let us first consider the situation in the solar neighborhood in detail and then examine the implications of the picture we develop there for other positions in the disk. We shall suppose that we can examine groups of stars of progressively greater age at a given location, and we shall ask what the physical and kinematic properties of each of these groups is likely to be. To facilitate the discussion, we have summarized in Table 4-18 the expected properties of some representative population groups, whose ages range from very young to very old. We stress from the outset that this is a risky business, for, while the identification of certain objects (for example, the spiral-arm population) is unequivocal, in other cases it may be no more than a guess. The table is intended only as a guide.

The numerical quantities listed in the lower rows of the table apply only to the solar neighborhood; the values given are likely to be different elsewhere in our Galaxy. For each group of objects, the numbers listed are: an age category (in 10^9 years); the total root-mean-square (rms) random-velocity dispersion $\langle\sigma^2\rangle^{1/2}$ (in kilometers per second); the random-velocity dispersion $\langle u^2\rangle^{1/2}$ (in kilometers per second) along the Sun-galactic center line; the average lag $\langle v\rangle$ (in kilometers per second) of the group behind the local circular velocity; the random-velocity dispersion $\langle w^2\rangle^{1/2}$ (in kilometers per second) perpendicular to the galactic plane; the scale height $\langle|z|\rangle$ (in parsecs) of the space distribution of the objects perpendicular to the plane; an orbital eccentricity parameter e (defined more precisely in Chapter 7) describing the departure of the group's orbits from circularity; a characteristic range of metallicity relative to solar; and a probable morphology of their H-R diagram. Some of the representative objects listed are found rarely, or not at all, in the solar neighborhood. In this event, the objects are assigned to the subgroup to which we guess they might belong at the appropriate positions in the disk (usually on the basis of their kinematics or space distribution). Finally, note that not every numerical value applies to every object listed in the same column of the table. For example, the quoted

Table 4-18. Archetype Disk-Population Groups

Property	Spiral-arm population	Disk population		
		Young	Intermediate	Oldest
Representative objects	Interstellar gas and dust Open clusters and stellar associations O–B stars Supergiants Classical Cepheids T Tauri stars *Some* stars of type A and later	A stars F stars A–K giants Me dwarfs *Some* G, K, and M dwarfs and white dwarfs	Sun Most G dwarfs *Some* K and M dwarfs and white dwarfs *Some* subgiants and red giants Planetary nebulae	*Some* K and M dwarfs and white dwarfs *Some* subgiants and red giants Moderately metal-poor (weak-lined) stars Long-period variables RR Lyraes, $\Delta S < 5$, $P < 0^d.5$
Age (10^9 years)	$\lesssim 0.1$	~ 1	~ 5	$\lesssim 10$ (?)
$\langle \sigma^2 \rangle^{1/2}$ (km s^{-1})	15	25	50	80
$\langle u^2 \rangle^{1/2}$ (km s^{-1})	10	20	40	60
$\langle v \rangle$ (km s^{-1})	≈ 0	-10	-25	-50
$\langle w^2 \rangle^{1/2}$ (km s^{-1})	10	15	25	40
$\langle \lvert z \rvert \rangle$ (pc)	120	200	400	700
e	0	$\lesssim 0.1$	$\lesssim 0.2$	$\lesssim 0.4$
$(Z/Z_\odot)$	1 to 2	1 to 2	0.5 to 1	0.2 or 0.3 to 0.5
H-R diagram	h and χ Persei	Hyades	M67	NGC 188

metallicity is meant to be the initial content of the group and is not necessarily appropriate to an evolved object such as a nova, planetary nebula, or, of course, a white dwarf.

The very youngest stars in the solar neighborhood belong to the *spiral-arm population.* These stars are concentrated into clumps (clusters and associations) in the galactic plane in a dust- and gas-filled environment, from which they were recently formed when the interstellar material was compressed and star formation was triggered, and they have solar or above-solar metallicity. The spiral-arm population includes stars of all masses, and it has few or no white dwarfs. Its brightest members are the blue supergiants and upper-main-sequence stars, and their H-R diagram will be like that of NGC 2362 or h and χ Persei. The presence of these intensely luminous stars makes this population so conspicuous, even though it is, metaphorically speaking, nothing more than the froth of the spiral wave, and it contributes a negligible amount to the local mass density. Spiral-arm stars move on essentially circular orbits around the galactic center ($e \approx 0$) and have small random velocities characteristic of the random-velocity dispersion within interstellar clouds. These stars are more closely confined to the galactic plane than typical interstellar clouds of neutral hydrogen, and they correlate more closely with the distribution of massive molecular clouds, in which they probably form.

After a few hundred million years, loose associations and clusters are torn apart by tidal shear and by the evaporation of individual cluster members, and their constituent stars are dispersed into the general background field. (Of course, tightly bound clusters will survive, and some are still seen today.) Thus, if we were to examine a sample of stars, say, a billion years old, we would find both stars that were initially formed in the field and former cluster members fairly uniformly spread out into the disk. We shall call these stars the *young disk population.* From this sample, the massive blue stars would be missing, having evolved to white dwarfs and neutron stars, so the group's mean stellar mass would be lower than for the spiral-arm population. The H-R diagram for this population is like that of the Hyades, and, again, these stars have solar or above-solar metal abundances. Essentially all A and F stars belong to the young disk population, as do a large fraction of the normal A–G giants, which are evolved upper-main-sequence stars. *Some* of the G, K, and M dwarfs and white dwarfs in the field belong to this age group, but most are probably much older (assuming a roughly uniform rate of star formation), so, in practice, the problem of how to distinguish this particular subset arises. For some stars, this can be done spectroscopically; for example, the emission-line M dwarfs (dMe) are readily identifiable and are known (from their presence in young clusters and associations) to be young stars. The kinematics of young disk stars begin to show the gradual heating effect of encounters with interstellar clouds. The random-velocity dispersions are larger in all directions, and the stars begin to lag, on the average, behind the circular orbital speed. They thus diffuse to larger distances away from the galactic plane, and their orbits have an appreciable eccentricity, reflecting the inward-outward undulation of their paths about a perfectly circular orbit, induced by their nonzero radial velocities with respect to the center.

If we now choose a significantly older sample, which we shall call the *intermediate disk population*, we will find that the trends just outlined have progressed. Thus, in a group, say, 5×10^9 years old, the A and F stars have vanished, and they are accounted for in an increased number of white dwarfs. The mean stellar mass of the sample would be yet smaller than for young disk stars. The H-R diagram for these stars would be like that of M67, and their metallicities probably range from perhaps a factor of two below solar up to solar. Most field G dwarfs (for example, the Sun) are in this group (outnumbering those in the groups considered earlier by virtue of the longer time interval spanned by their epochs of formation). Some of the field K and M dwarfs and white dwarfs belong to this population, but, again, the practical problem of identifying which of these stars belongs to this population arises. (One can invoke kinematic criteria, but this assumes a correlation exists.) The intermediate disk population must include large numbers of subgiants and red giants on a well-developed giant branch. As we look toward the galactic center (in longitude), we see increasing numbers of planetary nebulae and novae, and, on the basis of their z scale height and kinematics, it is reasonable to suppose that they belong to the intermediate disk population; but clearly this identification is not firm. However, if these objects do belong to the intermediate disk population, then we would, in fact, expect to pick them out preferentially toward the center, because: (1) the rising star density in the inner disk implies that more of them should be created per unit volume there; (2) it is likely that star formation started earlier in the inner disk (higher gas densities) and progressed more rapidly there, and hence that the ratio of old to young stars is higher in the inner disk; and (3) we thus would expect large numbers of low-mass stars to be dying in these regions and to be producing planetaries as the last spectacular phase that a star passes through before it sinks to the white-dwarf domain. Finally, both novae and planetaries are prominent because they are luminous and distinctive, and thus they attract attention. The fact that they are far enough from the plane to avoid being lost in strong interstellar absorption also helps. Kinematically, the intermediate disk population is yet "hotter" than the young disk population, and it shows a greater average lag behind the circular velocity. A typical member has an appreciably eccentric orbit, and thus it can move over a greater range of galactocentric distances within the disk. Furthermore, these stars will be found in a thicker layer above and below the galactic plane.

The *oldest disk population* stars may be as old as 10^{10} years, but they are probably somewhat younger. In a sample of these stars, we would find only late G, K, and M dwarfs on the main sequence, large numbers of white dwarfs, and a good representation of subgiants and red giants. The latter would be cooler and redder than those of the intermediate disk population, owing to their lower average mass, which implies that their Hayashi tracks lie farther to the right in the H-R diagram. Overall, their H-R diagram might resemble that of NGC 188, and we would expect average metallicities to be perhaps a factor of three to five lower than solar. Kinematically, this would be the "hottest" group in the disk, with a large random-velocity dispersion, and lag behind circular velocity. The orbits of a fair fraction of these stars will be quite eccentric, with some stars that pass through the solar neighborhood being able to move through a zone of radial distances on the range from about 5 kpc to 13 kpc. They have a scale height of 700 pc or more above and below the galactic plane. In longitudes toward the galactic center, we find concentrations of apparently faint long-period, late-type variables (the apparently bright ones are scattered at random on the sky) and of relatively metal-rich ($\Delta S < 5$), short-period RR Lyrae stars. On the basis of their kinematics and their space distribution, especially their vertical scale height, we identify these as probable members of the oldest disk population. Again, we find these stars readily because they are prominent (bright and easy to identify) even though they are numerically only a minor constituent of the population, and their apparent concentration toward the center again most likely reflects higher star densities and greater average stellar ages in the inner disk. Of course *some* of these stars are probably intermediate disk-population members; but, because we must judge the kinematics and z dispersion for the group as a whole, we naturally tend to assign them to the oldest category into which they fit. Just as was the case for G, K, and M dwarfs, we could segregate subgroups by age only if we had a suitable spectroscopic criterion that would permit us to do so.

How would this picture change if we moved to some other position in the disk, say, toward the galactic center? To within at least 3 kpc of the center (where the gas density in the disk suddenly drops to nearly zero and the dynamics of the gas changes radically—see §9-1 and §9-4), we would expect to find the same kinds of population groups having qualitatively the same relative properties. Overall, the star density would rise, and the ratio of gas to star density might fall somewhat. Stars on the average would be older, and hence we would see relatively more highly evolved objects. The stars would have larger random-velocity dispersions, and the average metallicity, group-by-group, would rise by perhaps a factor of two to three above the corresponding values in the solar neighborhood. The H-R diagrams would again range from h and χ Persei through the Hyades to NGC 188, with minor modifications to allow for changes in metal content, and the physical properties of the innermost disk stars probably merge essentially continuously into those of the bulge-component stars.

Some parts of the picture we have presented are accurate and firm; some parts are merely plausible; some are probably speculative at best, and mere wishful thinking at worst. Only future research will reveal which parts fall into which categories. But, for the moment, this picture fits the data we have, and it will provide a useful reference system around which we can build further discussion in this book.

The Dark Population For completeness, we mention again that there is dynamical evidence, to be discussed in later chapters, for the existence of a massive, widely dispersed component in our Galaxy. Because we have no evidence for radiation from these hypothetical objects, we have called them the dark population. Possibly this component comprises degenerate stars, such as black dwarfs (cooled white dwarfs), neutron stars, and black holes, or perhaps it contains a wide range of substellar masses, anywhere from, say, baseball-sized objects to Jupiter-sized ones. Perhaps these objects, if they are indeed stellar, were responsible for the contamination of the primordial protogalactic material with the heavy elements that we observe today, to set a floor on [Fe/H] in the outermost regions of our Galaxy.

Implications for Other Galaxies An advantage of the structure-based categorization of stellar populations that we have developed is that it can also be applied to other galaxies. Thus, what we have called the spiral-arm population is precisely the population seen in the spiral arms of other galaxies and throughout irregular systems (except those that happen to be rather metal poor). The disk population is the same as that seen in S0 systems. The spheroidal component would, if seen by itself, appear to be a perfect example of a normal, if not very luminous, elliptical galaxy. Even the details of the light and color distribution, and the relation between color and luminosity, are the same as those found for ellipticals. We shall exploit this correspondence in Chapter 5, and we will draw there, from classes of data that are more readily obtainable in external galaxies than in ours, plausible inferences that help fill in the picture for our own Galaxy.

REFERENCES

(**A1**) Allen, C. W. 1973. *Astrophysical Quantities*, 3rd. ed. (London: Athlone Press).

(**B1**) Baade, W. and Mayall, N. U. 1951. *Problems of Cosmical Aerodynamics*. (Dayton: Central Air Documents Office), p. 165.

(**B2**) Barnard, E. E. 1927. *A Photographic Atlas of Selected Regions of the Milky Way*. (Washington, D. C.: Carnegie Institution).

(**B3**) Basinska-Grzesik, E. and Mayor, M. (eds.). 1977. *Chemical and Dynamical Evolution of Our Galaxy*. I.A.U. Colloquium No. 45. (Geneva: Observatoire de Genève).

(**B4**) Becker, W. 1970. *Astron. and Astrophys.* **9**:204.

(**B5**) Becker, W. and Contopoulos, G. (eds.). 1970. *The Spiral Structure of Our Galaxy*. I.A.U. Symposium No. 38. (Dordrecht: D. Reidel).

(B6) Blaauw, A. and Schmidt, M. (eds.). 1965. *Galactic Structure*. (Chicago: University of Chicago Press).

(B7) Bok, B. J. 1937. *The Distribution of Stars in Space*. (Chicago: University of Chicago Press).

(C1) Canterna, R. 1975. *Astrophys. J. Letters*. **200**:L63.

(C2) Canterna, R. and Schommer, R. A. 1978. *Astrophys. J. Letters*. **219**:L119.

(C3) Cayrel de Strobel, G. and Delplace, A. M. (eds.). 1972. *L'Age des Etoiles*. I.A.U. Colloquium No. 17. (Meudon: Observatoire de Paris).

(C4) Cowley, A. P., Hartwick, F. D. A., and Sargent, W. L. W. 1959. *Astrophys. J.* **220**:453.

(F1) Fall, S. M. and Rees, M. J. 1977. *Mon. Not. Roy. Astron. Soc.* **181**:37P.

(H1) Harris, W. E. 1976. *Astron. J.* **81**:1095.

(H2) Hawley, S. A. 1977. *Astrophys. J.* **224**:417.

(J1) Janes, K. A. 1979. *Astrophys. J. Supp.* **39**:135.

(K1) King, I. R. 1971. *Pub. Astron. Soc. Pacific*. **83**:377.

(K2) Kinman, T. D. 1959. *Mon. Not. Roy. Astron. Soc.* **119**:499.

(K3) Kinman, T. D. 1959. *Mon. Not. Roy. Astron. Soc.* **119**:538.

(K4) Kinman, T. D. 1965. *Astrophys. J. Supp.* **11**:199.

(K5) Kinman, T. D., Wirtanen, C. A., and Janes, K. A. 1965. *Astrophys. J. Supp.* **11**:223.

(L1) Lafler, J. and Kinman, T. D. 1965. *Astrophys. J. Supp.* **11**:216.

(L2) Lindblad, B. 1926. *Medd. Uppsala Obs.* No. 14.

(L3) Luyten, W. J. 1968. *Mon. Not. Roy. Astron. Soc.* **139**:221.

(L4) Lynds, B. T. 1962. *Astrophys. J. Supp.* **7**:1.

(M1) Malmquist, K. G. 1924. *Medd. Lund Astron. Obs., Ser. II.* No. 32, p. 64.

(M2) Malmquist, K. G. 1936. *Stockholms Obs. Medd.* No. 26.

(M3) Mathis, J. S. 1959. *Astrophys. J.* **129**:259.

(M4) Mayor, M. 1976. *Astron. and Astrophys.* **48**:301.

(M5) McCuskey, S. W. 1956. *Astrophys. J.* **123**:458.

(M6) Morgan, W. W. 1956. *Pub. Astron. Soc. Pacific*. **68**:509.

(M7) Morgan, W. W. 1959. *Astron. J.* **64**:432.

(M8) Morgan, W. W., Sharpless, S., and Osterbrock, D. E. 1952. *Astron. J.* **57**:3.

(M9) Morgan, W. W., Whitford, A. E., and Code, A. D. 1953. *Astrophys. J.* **118**:318.

(M10) Müller, E. A. (ed.). 1977. *Highlights of Astronomy*. XVIth General Assembly of the I.A.U. Vol. **4**, part 2 (Dordrecht: Reidel).

(N1) Nassau, J. J. and Blanco, V. M. 1958. *Astrophys. J.* **128**:46.

(**N2**) Nort, H. 1950. *Bull. Astron. Inst. Netherlands.* **11**:181.

(**O1**) O'Connell, D. J. K. (ed.). 1958. *Stellar Populations.* (New York: Interscience).

(**O2**) Oort, J. H. 1938. *Bull. Astron. Inst. Netherlands* **8**:233.

(**O3**) Oort, J. H. 1977. *Astrophys. J. Letters.* **218**:L97.

(**O4**) Oort, J. H. and Plaut, L. 1975. *Astron. and Astrophys.* **41**:71.

(**P1**) Peimbert, M. 1975. *Ann. Rev. Astron. and Astrophys.* **13**:113.

(**S1**) Salpeter, E. E. 1955. *Astrophys. J.* **121**:161.

(**S2**) Salpeter, E. E. 1959. *Astrophys. J.* **129**:608.

(**S3**) Sandage, A. R. 1957. *Astrophys. J.* **125**:422.

(**S4**) Schmidt, M. 1959. *Astrophys. J.* **129**:243.

(**S5**) Schmidt, M. 1975. *Astrophys. J.* **202**:22.

(**S6**) Schweizer, F. 1976. *Astrophys. J. Supp.* **31**:313.

(**S7**) Searle, L. 1971. *Astrophys. J.* **168**:327.

(**S8**) Searle, L. and Zinn, R. 1978. *Astrophys. J.* **225**:357.

(**S9**) Setti, G. (ed.). 1975. *Structure and Evolution of Galaxies.* (Dordrecht: Reidel).

(**S10**) Shields, G. A. and Searle, L. 1978. *Astrophys. J.* **222**:821.

(**S11**) Spinrad, H. 1966. *Pub. Astron. Soc. Pacific.* **78**:367.

(**T1**) Tinsley, B. and Larson, R. B. (eds.). 1977. *Proc. Conf. on Evolution of Galaxies and Stellar Populations.* (New Haven: Yale University).

(**T2**) Tremaine, S. D. 1976. *Astrophys. J.* **203**:345.

(**T3**) Tremaine, S. D., Ostriker, J. P., and Spitzer, L. 1975. *Astrophys. J.* **196**:407.

(**T4**) Trumpler, R. J. and Weaver, H. F. 1953. *Statistical Astronomy.* (Berkeley: University of California Press).

(**U1**) Unsöld, A. 1969. *Science* **163**:1015.

(**V1**) van den Bergh, S. 1975. *Ann. Rev. Astron. and Astrophys.* **13**:217.

(**V2**) van den Bergh, S. 1979. *Astron. J.* **84**:317.

(**W1**) Weidemann, V. 1967. *Z. für Astrophys.* **67**:286.

(**W2**) Wielen, R. 1974. In *Highlights of Astronomy.* XVth General Assembly of the I.A.U., G. Contopoulos (ed.). Vol. **3**. (Dordrecht: Reidel).

(**W3**) Woltjer, L. 1975. *Astron. and Astrophys.* **42**:109.

(**Z1**) Zinn, R. 1978. *Astrophys. J.* **225**:790.

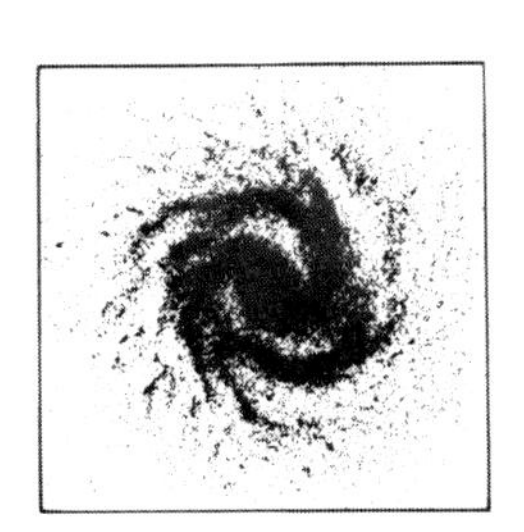

5 Large-Scale Structure and Stellar Content of Galaxies

In the preceding chapters, we were concerned primarily with the structure of our own Galaxy. By piecing together information from star counts and from observations of bright distant objects like globular clusters and Cepheid variables, we concluded that the Galaxy is made up of three components: (1) the young disk population (Baade's Population I), characterized by young, massive stars, dust, and gas; (2) the old disk population (intermediate population), characterized by relatively metal-rich older stars, which are distributed into a somewhat thicker disk structure than that formed by the extreme Population I objects; and (3) a nearly spherically distributed population (Baade's Population II), characterized by very old, low-mass stars concentrated into a central nucleus and bulge, and distributed at large throughout the halo. In Chapters 6 and 7, we shall return to our discussion of our Galaxy, showing that these different components are also characterized by different kinematic properties, but now let us turn briefly to the study of external galaxies, which also prove to be, in varying degrees, made up of the three components just mentioned.

The types of observational data available for external galaxies are quite different from those available for our own Galaxy. Because we enjoy a bird's-eye view of other galaxies, the interpretation of the data is often much more straightforward than the elaborate forensic exercises which sometimes characterized the considerations of the preceding chapter. On the other hand, because of the great distances to other galaxies, the quality of the data is generally much poorer than that available for our Galaxy, and the spatial resolution is much lower. These data thus yield the broad picture easily, but they lack the detail available from observations of our own Galaxy. Therefore, it is essential to develop our understanding of our Galaxy and of other galaxies in parallel, complementing our understanding now from this source and now from that.

In this chapter, we begin our study of external galaxies by discussing their shapes and something of what can be learned about their composition from the spectrum of the light they emit. In §5-1, we describe the various attempts that have been made to classify galaxies into types by their appearance and spectral properties. In §5-2, we discuss the brightness distributions that seem to characterize the various morphological classes defined in the previous section. In particular, we present empirical laws which approximately describe the radial brightness distributions of the various components of galaxies, and we explore what information can be gleaned about the three-dimensional forms of galaxies from their observed surface distributions of light. In §5-3, we describe briefly the methods used to estimate total luminosities of galaxies, and we summarize integrated luminosities and colors for different morphological types. In §5-4, we discuss the spectral properties of the light emitted by galaxies of various types, and we point out some implications of a number of systematic correlations that have been found to exist between the gross morphological properties of galaxies and their spectra. Finally, in §5-5, we review some of the more important results to emerge from this discussion, and we inquire whether the objects we have studied are, in fact, genuinely typical of the galaxian content of the Universe, or whether observational difficulties and selection effects may have distorted our picture of what a typical extragalactic object is.

5-1. MORPHOLOGY AND CLASSIFICATION OF GALAXIES

Photographs of galaxies show them to have a wide range of appearances. When confronted with a complex set of objects like this, a natural first step is to attempt to group them into classes of similar objects. If one wishes the class to which a galaxy is to belong to reflect something of that galaxy's intrinsic structure, rather than merely details of the observational data used to make the classification, then it is essential that, as far as possible, all galaxies be classified on the basis of similar data—that is, that the data employed must be homogeneous. Furthermore, one should take care that the criteria used to assign a galaxy to a class permit a unique classification in each case. In particular, it is a bad idea to employ more than one criterion when establishing a classification scheme, lest the different criteria lead one to place a given galaxy into more than one class. Finally, a flexible and unambiguous notation for expressing the results of the classification procedure must be developed. Unfortunately, most of the classification schemes described in this section violate one or more of these criteria, a circumstance which has introduced considerable confusion into the area. The reader should be constantly on the watch for such ambiguities and inconsistencies arising from the classification procedure.

We suggest that the reader approach the classification schemes we describe here with a measure of skepticism. One has constantly to bear in mind (1) that,

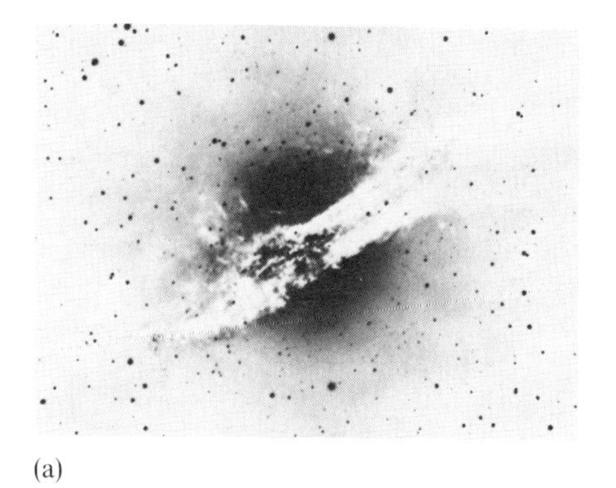

(a)

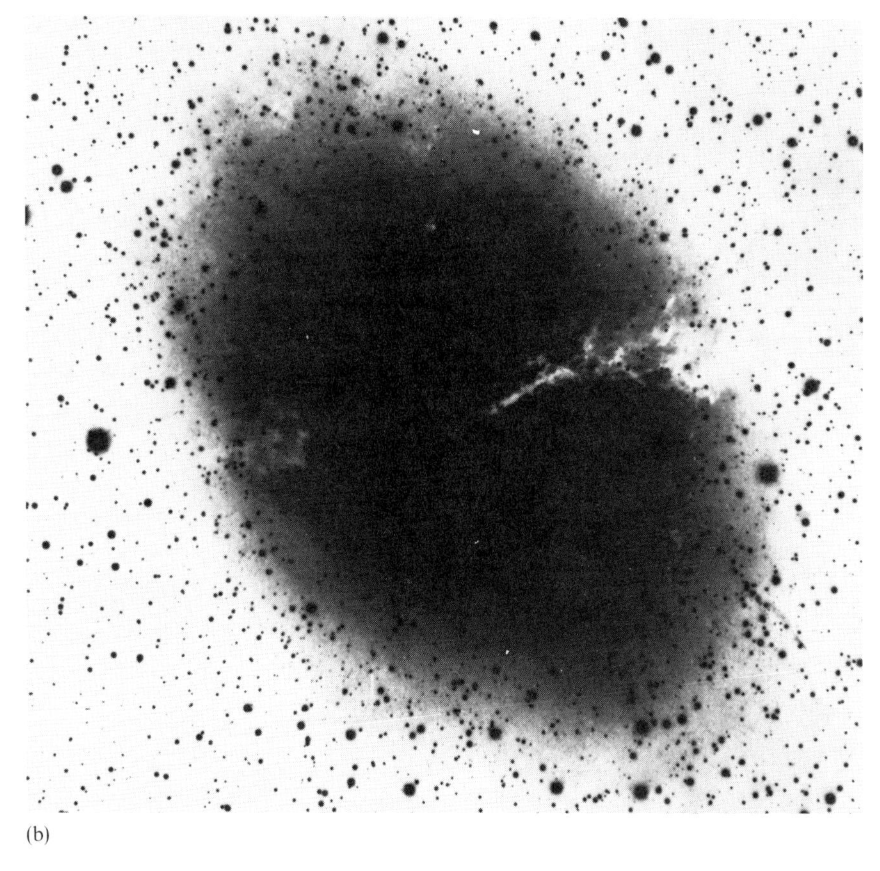

(b)

Figure 5-1. An illustration of how different the appearance of a galaxy can be on plates of different exposures. Both (a) and (b) are reproductions of plates of the peculiar galaxy NGC 5128, which has been identified with the radio source Centaurus A. The scale and orientation of these reproductions are identical, as one may verify by matching bright stars in the surrounding field, but (a) is from the Palomar Sky Survey (Palomar Observatory, California Institute of Technology), while (b) is a deep IIIa-J plate taken with the 4-m CTIO telescope in Chile. [From (**B1**) by permission. Copyright © 1978 by the American Astronomical Society.]

in the absence of a full understanding of the true nature of galaxies, the criteria on which any given classification scheme is based are to certain extent arbitrary and can as easily give rise to false dichotomies as they can provide useful insights and (2) that galaxies are not hard-edged objects of the type that surround us in our daily lives. Indeed, galaxies fall far outside the range of our ordinary experience as regards both their self-illumination and their ghostlike insubstantiality. It is not always easy to bear these points in mind when one sees the apparently rather crisp image of a galaxy in a book or journal. Figure 5-1 illustrates just how misleading such pictures can be. Here one sees two photographs of the same galaxy on the same scale and in the same orientation. Yet, because of the different exposure times of the two plates, the appearance of the object is so different that one may feel a degree of disbelief that both illustrations are of the same system. Similar differences arise between plates taken through different color filters. A feature that is conspicuous in blue light may totally vanish at longer wavelengths. Therefore, one has to try to avoid being misled by superficial appearances that are little more than optical illusions, and one must seek quantitative confirmation of qualitative impressions wherever possible.

The Hubble System

The most widely used classification scheme is that introduced by Hubble (**H4**) in his 1936 book, *The Realm of the Nebulae*, and subsequently modified as his collection of large-scale plates of giant galaxies grew. Reproductions

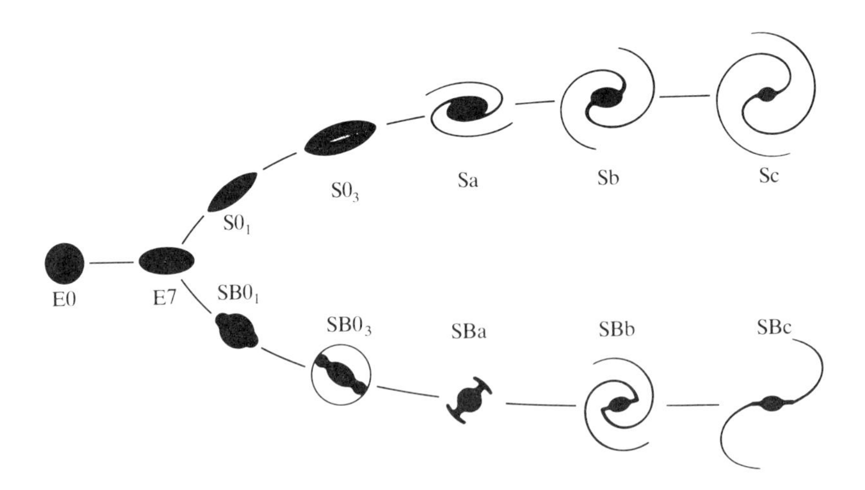

Figure 5-2. The Hubble tuning-fork diagram. The diagram shown here differs from Hubble's original in that it shows various stages of lenticular galaxies interposed between the ellipticals and the spirals.

of many of these plates were published posthumously in 1961 by Sandage (**S1**) as *The Hubble Atlas of Galaxies*, alongside an introduction by Sandage which is generally regarded as the definitive exposition of Hubble's scheme. Figure 5-2 illustrates the scheme in its final form.

On the left of Hubble's *tuning-fork diagram* are placed those galaxies which appear smooth and structureless. These galaxies vary in shape from round to fairly highly elliptical in form, and they are referred to as "early-type" galaxies. A galaxy of this type is designated En, the letter E standing for *elliptical*, and the number n describing the apparent axial ratio (b/a) by the formula $n = 10[1 - (b/a)]$. Thus, a galaxy which appears round on the sky is designated E0, and one whose major axis is twice as long as its minor axis is an E5 galaxy. Figure 5-3 shows examples of each type of elliptical galaxy, ranging from the round E0 galaxy NGC 4636 through the highly elongated (E6) system NGC 3377.

The Hubble system, as conceived by Hubble and further developed by Sandage, is defined in terms of type-examples which are almost exclusively *giant* galaxies ($M_{pg} \leq -19$). Thus the galaxies shown in Figure 5-3 are all giants. However, the most numerous type of galaxy in the Local Group (the collection of 25 to 30 galaxies of which our Galaxy, the Andromeda Nebula M31, and M33 are the most conspicuous members; see Table 5-3) is either *dwarf elliptical* or *dwarf spheroidal*. As will be explained later, the distinction between dwarf elliptical and dwarf spheroidal galaxies is only one of degree, and both types are designated dE because they are dwarfs and have roughly elliptical contours of equal star density. M32, the compact companion of the Andromeda galaxy (M31), which appears conspicuously in, for example, the *Hubble Atlas* photograph of M31 (**S1**), is a good example of a dwarf elliptical galaxy. The surface brightness of M32 is perfectly normal for an elliptical galaxy, so that it differs from a giant elliptical galaxy only in linear size and absolute magnitude (see Tables 5-3 and 5-4). Dwarf spheroidal galaxies, on the other hand, are very low surface brightness objects. The Sculptor system, discovered in 1937 by Shapley (**S8**) on plates from the Boyden Observatory in South Africa, was the first dwarf spheroidal to be found. Several other dwarf spheroidals have been discovered since that time, although, at present, we are able to detect these objects only if they lie within the Local Group. On plates taken with even the largest ground-based telescopes, dwarf spheroidals appear as mere clusterings of faint (but intrinsically moderately luminous) stars. No smooth background of light produced by less luminous stars is visible. We can expect to discover several thousand more dwarf spheroidals with the Space Telescope, whose great resolving power will enable us to detect distant dwarf spheroidals whose brightest stars are, at present, too faint to be visible.

Globular clusters, which were discussed in Chapter 3, are not normally accorded the dignity of being termed elliptical galaxies, being considered mere appendages of more massive systems like our Galaxy. But, as we shall see in §5-2, from a structural point of view, it is useful to regard them as a

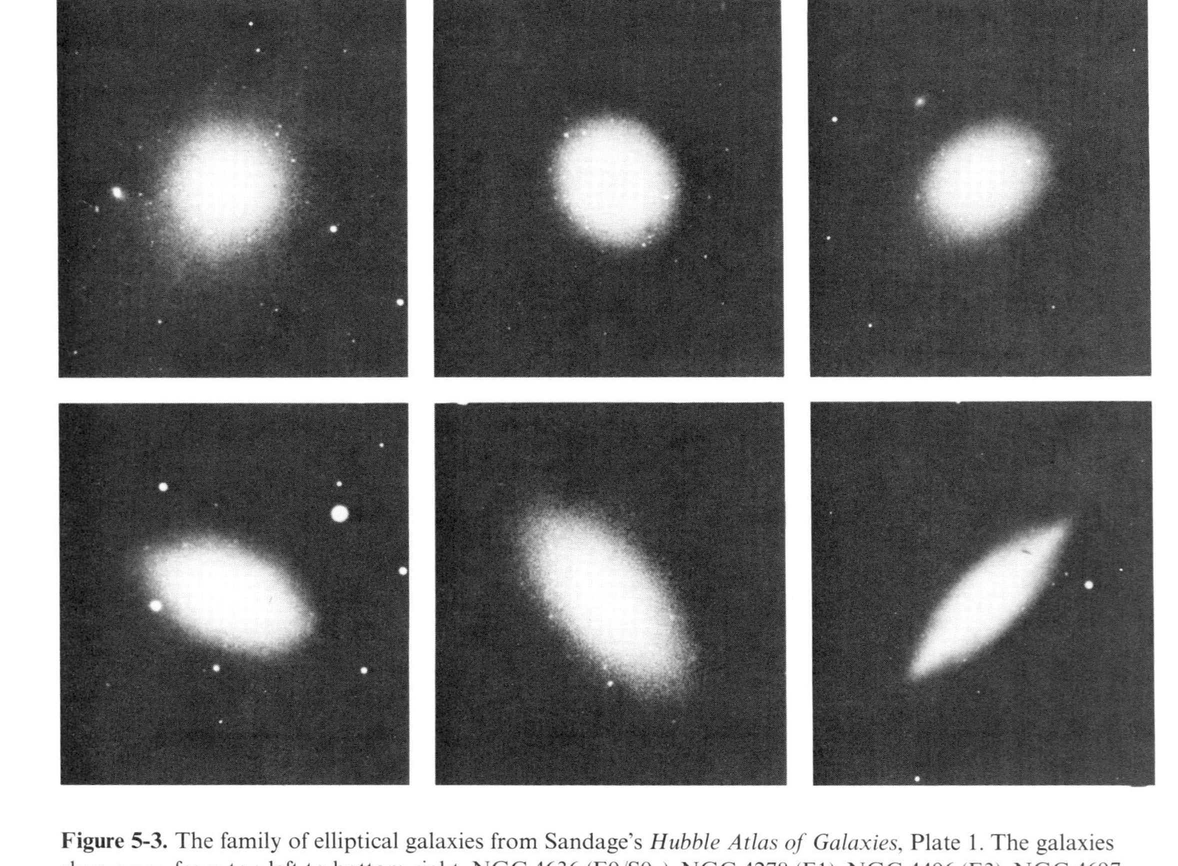

Figure 5-3. The family of elliptical galaxies from Sandage's *Hubble Atlas of Galaxies*, Plate 1. The galaxies shown are, from top left to bottom right, NGC 4636 (E0/$S0_1$), NGC 4278 (E1), NGC 4406 (E3), NGC 4697 (E5), NGC 3377 (E6), and NGC 3115 (E7/$S0_1$). [From (**S1**), by permission.]

species of extremely low-mass, elliptical galaxy. Indeed, one finds that the stellar content of globular clusters is frequently quite similar to what one might expect to find in a very low-mass elliptical galaxy (see §5-4), and their density profiles prove to be intermediate in type between those of giant ellipticals and of dwarf spheroidals. Many of the globular clusters that are located far out in the galactic halo, for example, the *Palomar clusters* (**A2**), probably provide a link between the classical clusters, which tend to be tightly bound to giant galaxies, and the more massive but less centrally concentrated dwarf spheroidals. If one extends Hubble's original elliptical class in this way, one finds that elliptical galaxies span an enormous range (at least six powers of ten) of intrinsic brightness (see §5-3).

After the elliptical galaxies, Hubble's diagram bifurcates into two branches—the *normal* and the *barred* galaxies. The word "normal" should not mislead one into supposing these galaxies are more common than the barred species; the two types actually occur with similar frequency. The reader should also beware of imagining that the division of galaxies into normal and barred type is clear and unambiguous. Most galaxies show some of the characteristics of barred galaxies, and it is only the more extreme examples that are classified as barred systems. Furthermore, barlike characteristics may be lost on small-scale plates, so that misclassification in this respect is quite common.

In the middle of Hubble's tuning-form diagram, at the junction of the elliptical and the *spiral galaxies*, comes a class of galaxies known as *lenticulars*. These galaxies are designated as type S0 or type SB0 according to whether or not they are barred. The S0 galaxies are characterized by a smooth central brightness condensation (the bulge or spheroidal component) similar to an elliptical galaxy, surrounded by a large region of less steeply declining brightness. This latter component, which is generally rather structureless (although it may sometimes contain some dust), is believed to be intrinsically rather flat. Indeed, when systems of this type are seen edge-on, one sees clearly that they are very flat in their outermost parts. Between the inner, elliptical-like bulge region and the outer (*envelope*) parts, many lenticular galaxies possess smooth subsidiary features called *lenses*. The family of S0 galaxies is subdivided into three classes, $S0_1$, $S0_2$, and $S0_3$ galaxies, according to the strength of dust absorption within their disks. The $S0_1$ galaxies do not show any signs of absorption by dust (see Figure 5-4), whereas the $S0_3$ galaxies have a complete dark band of dust absorption running within their disks around their elliptical-like components. The strength of dust absorption in $S0_2$ galaxies is intermediate between these two extremes. The *barred lenticular* galaxies are also divided into three types, $SB0_1$, $SB0_2$, and $SB0_3$, but now the division is made according to the prominence of the bar rather than the presence of dust-absorption lanes. In $SB0_1$ galaxies, the bar shows only as two broad regions of slightly enhanced brightness on either side of the central bulge. In $SB0_3$ galaxies, the bar is narrow and well defined and

Figure 5-4. A normal $S0_1$ galaxy, NGC 1201. [Mount Wilson and Las Campanas Observatories, Carnegie Institution of Washington, by permission.]

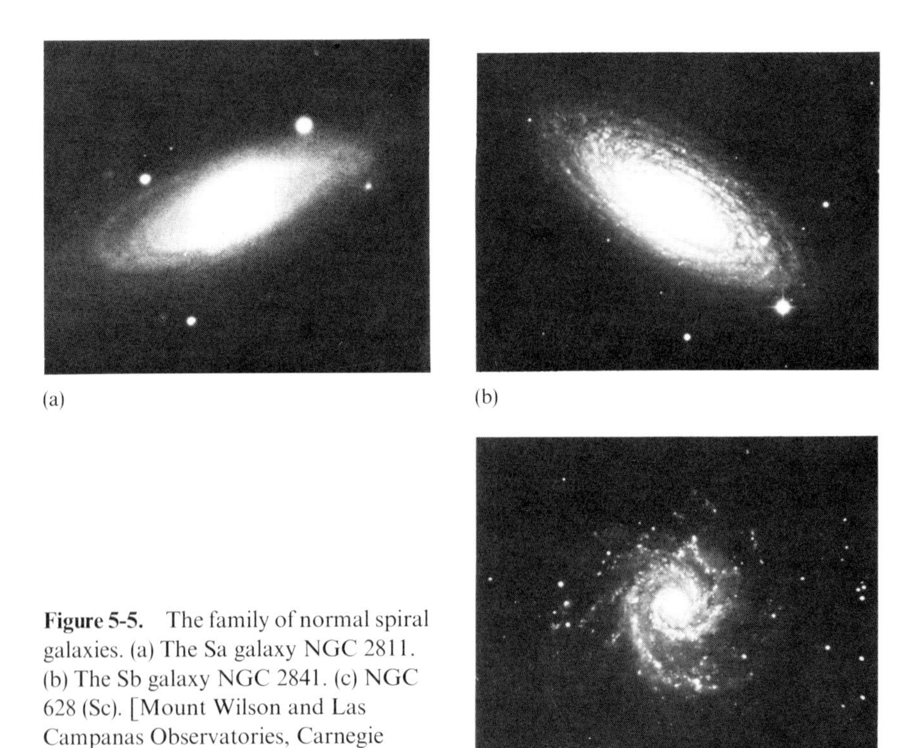

Figure 5-5. The family of normal spiral galaxies. (a) The Sa galaxy NGC 2811. (b) The Sb galaxy NGC 2841. (c) NGC 628 (Sc). [Mount Wilson and Las Campanas Observatories, Carnegie Institution of Washington, by permission.]

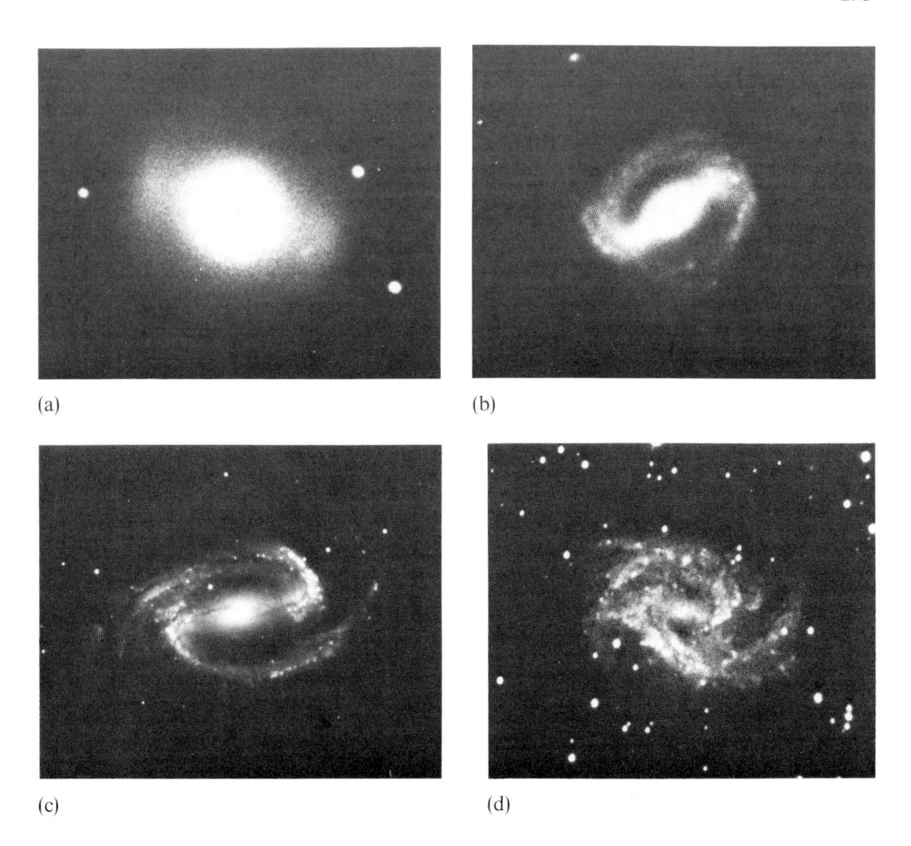
(a) (b) (c) (d)

Figure 5-6. The family of barred spirals. (a) NGC 2859 (SB0). (b) NGC 175 (SBab). (c) NGC 1300 (SBb). (d) NGC 2525 (SBc). [Mount Wilson and Las Campanas Observatories, Carnegie Institution of Washington, by permission.]

extends completely across the lens. $SB0_2$ galaxies have bars of intermediate prominence.

In the Hubble sequence, the lenticulars are followed by the spiral galaxies. A normal spiral galaxy comprises a central brightness condensation, which resembles an elliptical, located at the center of a thin disk containing more or less conspicuous spirals of enhanced luminosity, the *spiral arms.* A barred spiral has, interior to the spiral arms, a *bar*, often containing dark lanes believed to be produced by absorption of light by dust. The spiral arms of barred spirals generally emanate from the ends of the bar.

Within each class of spiral, normal, or barred, a sequence of subtypes is identified by division according to a combination of three criteria (see Figures 5-5 and 5-6): (1) the relative importance of the central luminous bulge and the outlying disk in producing the overall light distribution of the galaxy; (2) the tightness with which the spiral arms are wound; and (3) the degree to which the spiral arms are resolved into stars and individual

Table 5-1. Galaxy Classification Schemes

System	Principal criteria	Symbols employed	Examples of classifications	References
Hubble-Sandage	barrishness; openness of arms/disk-bulge ratio; degree of resolution of arms into stars	E, S0, S, SB, Ir a, b, c	M87 = E1 M31 = Sb M101 = Sc LMC = Irr I	(**S1**)
de Vaucouleurs	barrishness; openness of arms/disk-bulge ratio; ring or *s* shapes	E, S0, S, SA, SB, I a, b, c, d, m, (r), (s)	M87 = E1P M31 = SA(s)b M101 = SAB(rs)cd LMC = SB(s)c	(**D2**)
Yerkes	central condensation of light; barrishness/smoothness	k, g, f, a E, R, D, S, B, I	M87 = kE1 M31 = kS5 M101 = fS1 LMC = afI2	(**M2**), (**M8**)
Revised DDO	young-star richness of disk; barrishness; central concentration of light; quality and length of arms	E, S0, A, S, Ir B a, b, c I, . . . , V	M87 = E1 M31 = Sb I–II M101 = Sc I LMC = Ir III–IV	(**V5**)

emission nebulae (*H II regions*). Properly speaking, this superposition of criteria is highly unsatisfactory, but it turns out that there is sufficient correlation among these three parameters—in the sense that galaxies with conspicuous central bulges tend to have tightly wound spiral arms that are not highly resolved into stars—that experienced observers independently classifying objects usually agree rather closely in their class assignments. Early-type spirals (that is, those more to the left in the Hubble diagram; see Figure 5-2) are those having conspicuous bulges and tightly wound, smooth arms. These spirals are designated Sa or SBa according to whether they are barred (SBa) or not (Sa). Late-type spirals, having small central brightness condensations (bulges) and loosely wound, highly resolved arms, are designated Sc and SBc. In between are Sb and SBb galaxies. Intermediate stages of central condensation and tightness of winding of the arms are designated Sab, Sbc, and so on.

Later workers have generally considered Hubble's original scheme satisfactory in regard to the ellipticals, but many have said that Hubble's classification of the spirals is incomplete and that his treatment of irregular galaxies—of which more will be said presently—was quite inadequate. A few of the more important attempts at reclassification of types later than E (that is, to the right of the ellipticals in Hubble's diagram) are described later in this chapter.

De Vaucouleurs' Extension of the Hubble System

G. de Vaucouleurs, in an article in the *Handbuch der Physik* (**D2**), has argued that Hubble's division of the spirals of a given degree of condensation and tightness of winding of the arms into just two types, the normal and the barred spirals, does not do justice to the great variety of morphologies of real galaxies. De Vaucouleurs suggests that, at each stage of the Hubble sequence of spiral galaxies, one must distinguish a two-dimensional continuum of types rather than just the ordinary and barred spirals recognized by Hubble. De Vaucouleurs' revised classification diagram is shown in Figure 5-7. This diagram differs from that of Hubble in that (1) several extra stages E^+, $S0^-$, $S0^+$, Sd, Sm, and Im have been introduced, (2) normal spirals are now called *ordinary* and designated SA instead of just S (a symbol now reserved for an ambiguous spiral), and (3) additional symbols r and s have been introduced to indicate whether a spiral or lenticular galaxy has a *ring-shaped* or an *s-shaped* structure. Galaxies designated E^+ are objects showing some lenticularlike structure. The classes SAa, SAb, and SAc correspond to Hubble's classes Sa, Sb and Sc; the Sd class overlaps Hubble's Sc class to some extent, but it also contains more extreme objects which were classified as Type I Irregulars (Irr I) in Hubble's scheme. The Sm and Im classes contain the remaining galaxies of Hubble's Irr I class. For example, the Large Magellanic Cloud is classified SB(s)m. Only very irregular and loose objects like IC 1613 (IBm) are placed in the Im stage by de Vaucouleurs.

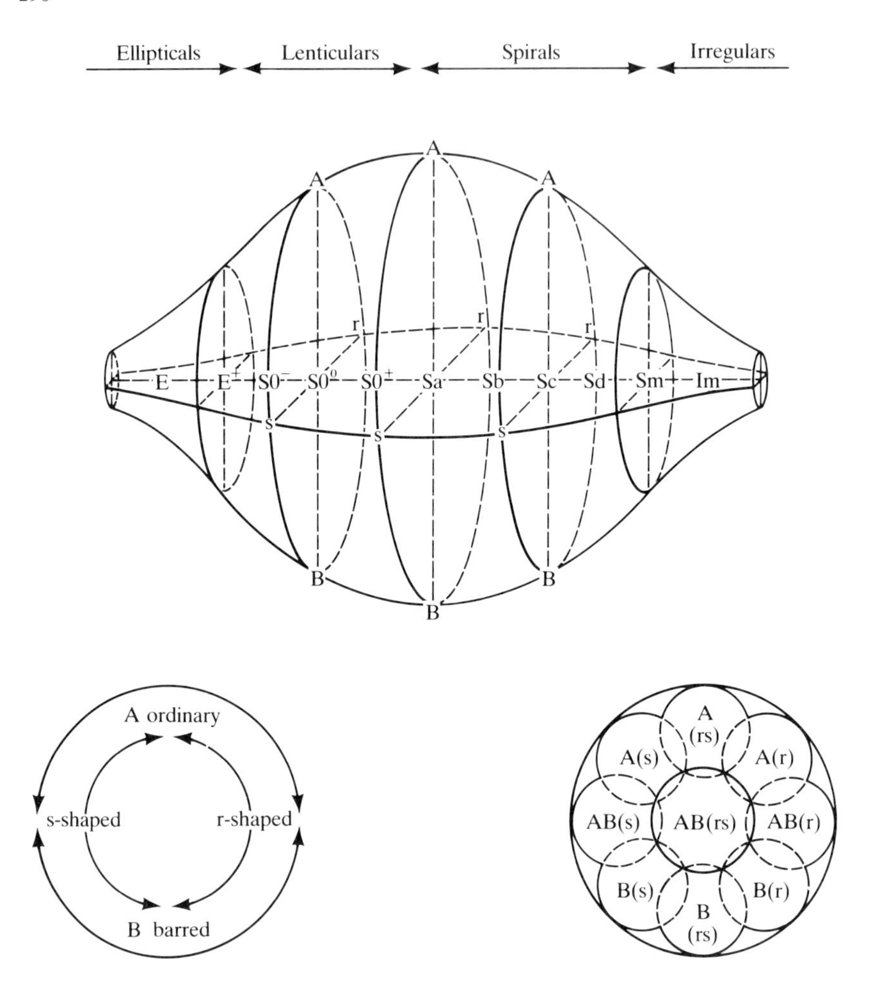

Figure 5-7. De Vaucouleurs' extension of the Hubble system. Hubble's two-dimensional scheme has been made three-dimensional to include explicit reference to rings and s-shaped objects, and, in addition, the sequence has been extended to Sd and Im galaxies. [From (**D2**), by permission.]

Table 5-2. The Correspondence Between the Hubble and de Vaucouleurs Systems and the Hubble stage *t*

Hubble	E	E–S0	S0	S0/a	Sa	Sa–b	Sb	Sb–c	Sc	Sc–Irr	Irr I
de Vaucouleurs	E	L^-	L	S0/a	Sa	Sab	Sb	Sbc	Scd	Sdm	Im
t	−5	−3	−2	0	1	2	3	4	6	8	10

The Yerkes System

W. W. Morgan (**M6**) has pointed out that Hubble's use of three criteria (the relative prominence of bulge and disk and the openness and degree of resolution of the spiral arms) in his classification scheme created an Sa class which includes objects that exhibit a wide range of nuclear concentration of luminosity, and it led to a pronounced overlap between the classes Sa and Sb in this characteristic. As we have emphasized, the use of possibly conflicting classification criteria in this way is in itself undesirable. The situation is made all the more unfortunate by the fact that there appears to be a strong correlation between the *nuclear concentration* of a galaxy's light and its *integrated spectrum*. One finds, in fact, that, in the violet region of the spectrum, giant elliptical galaxies have integrated nuclear spectra resembling those of giant K stars, whereas galaxies of Hubble's Sc type, which have only small central bulges, have integrated spectra resembling those of A stars. Galaxies with intermediate central light concentration have more composite spectra, which most nearly resemble those of F and G main-sequence stars. We return to a more detailed discussion of the spectra of galaxies in §5-4. We note these facts here merely to provide motivation for Morgan's classification of galaxies according to *the degree of central concentration of their light* only; the observed correlation of central concentration and integrated spectral type suggests that there exists also a correlation between the kinds of stars populating a galaxy and the central condensation of its light, and hence that fundamental significance is to be attached to the latter parameter.

Morgan defines a fundamental parameter designating the *population group* or *concentration class* of a galaxy. This parameter runs from k to a such that, among normal galaxies (that is, not active galaxies), those of type k have the highest degree of central concentration of their light, and those of type a have the smallest central bulges and the most diffuse light distributions. Galaxies of type k include giant ellipticals, such as those found in the Virgo cluster, and spiral systems like M31 in which the major share of the luminosity of the main body comes from a bright amorphous central region. Galaxies of type a include irregular systems of the Magellanic Cloud type and spirals having an insignificant central condensation of luminosity. Galaxies of type gk, g, fg, f, and af have various intermediate light concentrations. Two subsidiary parameters follow the fundamental parameter. An upper case letter S, B, E, I, R, or D indicates the *form-family* of a galaxy; S indicates spiral; B, barred spiral; E, elliptical; I, irregular; R, galaxies having rotational symmetry without pronounced spiral or elliptical structure; and D, a galaxy having an elliptical-like nucleus set in an extended envelope. The last parameter is a number in the range from 1 to 7, which indicates the apparent shape of the galaxy (1 = circular, 7 = highly elliptical, with $a/b = 10$). Each galaxy type is defined, ultimately, by standards listed in Morgan's papers (**M4**), (**M5**), (**M6**).

A few minor points should be mentioned here about Morgan's classification system (or the *Yerkes system*, as it is generally known, after the observatory where it was developed). First, it is basically a linear system which suppresses mention of the morphological complexities that motivate de Vaucouleurs' elaborate extension of Hubble's system. The ultimate value of the Yerkes system depends on the extent to which it has isolated the morphological characteristic which is really most fundamental for developing an understanding of the structure of galaxies. In contrast, schemes of the de Vaucouleurs type endeavor merely to describe the appearance of galaxies without any particular theoretical or interpretive overtones. Second, note that there is an unfortunate clash in the nomenclature relative to the Hubble system. Early-type galaxies (in the sense of Hubble) have late-type spectra (dominated by K giants), whereas Hubble's late-type galaxies have early-type spectra. One simply has to learn to live with these inappropriate but conventional designations inherited from obsolete evolutionary notions. Finally, the new form-family designations R and D have undergone some changes in definition since their introduction in 1958 (**M4**). The Yerkes system R family [which was designated D until the presently accepted definitions of the form-families were introduced in (**M8**)] closely corresponds to the Hubble-de Vaucouleurs S0 class, although there is also some overlap with their E and Sa classes. The significance of the family of D objects, as it is presently interpreted, is not immediately grasped in a casual inspection of plates. The characteristic morphology of this family of elliptical-like objects becomes fully apparent only when one traces their radial brightness profiles using the methods we describe later. One then finds that they deviate from elliptical galaxies in the sense that, far from their centers, their brightness profiles decline much more slowly than those of ordinary giant ellipticals. These objects thus appear to be surrounded by unusually extensive (and presumably massive) outer envelopes. Certain galaxies of this family are denoted cD (the c stands for supergiant and is not a concentration class designation), and they are of considerable importance for our understanding of the evolution of clusters and groups of galaxies (see Chapter 20).

The DDO System

S. van den Bergh (**V1**), (**V2**), (**V5**) has developed a scheme for the classification of galaxies which incorporates many features of both the classical Hubble system and the Yerkes system. The framework of the system is illustrated in Figure 5-8. The elliptical galaxies are arranged by apparent elongation, as in the Hubble sequence, and they are designated by the same En notation. Galaxies possessing disks are then arranged in three parallel sequences, those of the lenticular galaxies (designated S0), the *anemic spirals* (designated A), and the *gas-rich spirals* (designated S). In this scheme, a galaxy is classified as lenticular on exactly the same criteria as in the Hubble-Sandage system;

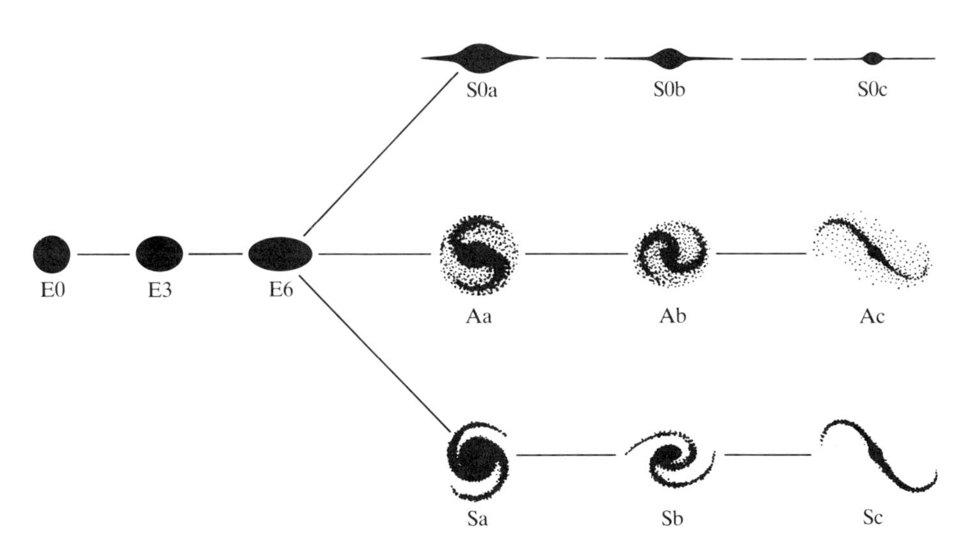

Figure 5-8. Van den Bergh's trident diagram. The lenticular galaxies have been arranged parallel to the spirals rather than before them, and a new class of galaxies, the "anemic" spirals, has been interposed between the spirals and the lenticulars. One may imagine a sequence of barred galaxies, corresponding to each species of disk galaxy illustrated, to be arranged in a separate trident below the trident shown. [From (**V5**) by permission. Copyright © 1976 by the American Astronomical Society.]

a lenticular galaxy is a disk galaxy that lacks spiral arms. Most of the galaxies that are classified as gas rich by van den Bergh would be classified as normal spirals in the Hubble-Sandage system. This class includes most of the well-known spiral galaxies (M31, M33, M81, M101, and so on). Anemic spirals are disk galaxies that are intermediate in type between lenticular galaxies and gas-rich spirals in that, although they have spiral arms, these are unusually diffuse features. Also, their surface brightnesses tend to be lower than those of gas-rich spiral galaxies.

Within each sequence, the position of a galaxy is (as in the Yerkes system) determined entirely by the degree of central concentration of its light, and it is denoted by the letters a, b, and c as in the somewhat different Hubble ordering of the spirals. The presence of a bar in a disk galaxy is indicated by the letter B as follows: SBa or S0Bb. The Magellanic irregulars form a natural extension of the gas-rich spirals and are designated Ir. This framework is motivated by the fact that there is a degree of continuity from lenticular galaxies (which generally show dust absorption lanes, though little, if any, gas, and certainly no H II regions), through anemic spirals such as one finds in the Virgo cluster of galaxies (which again possess dust but usually have less gas for their luminosity than spirals in smaller groupings and are correspondingly poor in H II emission regions), to gas-rich spirals like M101 or M51, having many H II regions.

In addition to assigning galaxies positions in the framework shown in Figure 5-8, van den Bergh subdivides spirals and anemic spirals of type Sb or Ab and later into *luminosity classes* designated by Roman numerals I through V. The criterion employed in this subdivision is the quality and length of a galaxy's spiral arm. Galaxies having well-developed arms that can be traced across a substantial part of the image of the galaxy are assigned to luminosity class I, and galaxies with the least well-developed arms are put into luminosity class V. Together with intermediate classes III–IV and so on, there are nine subdivisions in all.

The combination of a position in the framework of Figure 5-8 and (where appropriate) a luminosity class assignment is known as a *revised DDO type*, after the David Dunlap Observatory where this system was developed. Thus for example, M33 has revised DDO type Sc II–III, M31 is of type Sb I–II, and M101, the pinwheel galaxy, is classified Sc I.

The luminosity classes derive their name from a correlation that is found to exist between the luminosity class to which a galaxy is assigned and the actual luminosity of the galaxy in question. Thus, although luminosity assignments are made on entirely morphological grounds, it is found empirically that class I galaxies tend to be very luminous and class V galaxies are usually faint, with all intermediate degrees between these two extremes. This correlation is interesting in itself, and it is of key importance in determinations of the cosmological distance scale, which we shall discuss in §5-3.

The DDO system comes close to employing separately the two of the three classification criteria (disk-to-bulge ratio and openness of the spiral arms) which Hubble combined in his system. Thus the position of a galaxy from left to right in Figure 5-8 depends only on the degree of central concentration of its light, while its luminosity class is entirely dependent on the degree of development of its spiral arms. We have emphasized that, in practice, the rather confused criteria on which the Hubble system is based lead to unique classifications of individual galaxies because there is a high degree of correlation between a galaxy's disk-to-bulge ratio and the tightness with which its arms are wound. It is interesting to ask whether van den Bergh's slightly different criteria are correlated. One finds that there is, in fact, some correlation in that galaxies of type Sb and Ab always fall in luminosity classes I to III, whereas galaxies of type Ir are usually of luminosity class III and below. There are Sc galaxies in each of the luminosity classes I–IV.

Classes of Special Galaxies

The Hubble classification system and its modifications and derivatives refer principally to normal galaxies. However, one goal of any classification scheme is to discover which galaxies are abnormal and therefore merit further investigation. Hubble labeled all abnormal galaxies as Type II Irregulars (Irr II galaxies). However, since Hubble's day, many classes of unusual galaxies have been identified, and we here enumerate a few of the more important types. Many unusual galaxies are considered peculiar because they have abnormally active nuclei. This is, for example, true of all galaxies belonging to the first two classes we discuss next. Other galaxies may simply have odd distributions of brightness or dust. In fact, the range of the peculiar-galaxy phenomenon is so wide, and the physical processes involved are so little understood, that classification of these objects can be very difficult. Thus the definitions of types given here are to be regarded as provisional and subject to change in the light of new understanding of the essential similarities and differences characterizing peculiar extragalactic objects. For a discussion of the classification criteria that should be applied to these systems, see (**M7**).

Seyfert Galaxies Seyfert galaxies are systems, usually spirals, whose nuclei show strong, fairly broad emission lines of high excitation. There are two types of Seyfert galaxy. In the spectra of *Type I Seyferts*, the permitted lines (especially the Balmer lines) are made up of cores, which are about as broad as the forbidden lines, together with very broad wings. The wings of the permitted lines of Type I Seyferts correspond to Doppler broadening (if that is the right model) by velocities of 1000–5000 km s^{-1}. *Type II Seyferts* differ from Type I Seyferts in that the permitted lines in their spectra have no broad wings. In the spectra of these galaxies, both the permitted and the forbidden lines appear to be broadened by Doppler velocities of the order of 500 km s^{-1}. About one-third of the Seyferts are of Type II. Seyferts of both types are usually strong, variable X-ray sources.

N Galaxies Morgan (**M7**) has defined N galaxies as "systems having small, brilliant nuclei superposed on a considerably fainter background." Thus the definition of an N galaxy is a morphological one, and an N galaxy may, for example, be spectroscopically classified as a Seyfert.

Quasars and BL Lac Objects Extremely distant galaxies whose nuclei are abnormally luminous may be visible only as point sources of light arising in or near their nuclei. A great many objects that are believed to be of this type are known. These objects are classified as *quasi-stellar objects* (*QSOs* or *quasars*, for short) or as *BL Lac objects* (after the prototype, BL Lacertae), depending on whether their spectra show emission lines (QSO) or not (BL Lac objects). The emission-line spectra of quasars resemble (after allowance is made for their often large redshifts) the spectra of Type I Seyferts very closely, which suggests that Seyfert galaxies are nearby, very low-luminosity quasars whose associated galaxies are clearly visible. By contrast, the spectra of BL Lac objects do not show emission lines. Essentially all objects of these classes are X-ray sources, and some, but not all, are also powerful radio sources [often referred to as quasi-stellar sources (QSS)].

The optical and X-ray emission from these objects appear to arise in a very small volume (≤ 1 pc in radius), whereas the radio emission, in some cases, comes from a volume only slightly larger than this, and, in other cases, it comes from a region which is several megaparsecs on a side. At the present time, we understand very little about the processes which lead some galactic nuclei to radiate so spectacularly at all wavelengths, and we shall not be concerned with these phenomena in this book. Accounts of quasars, BL Lac objects, and similar objects will be found in (**B2**) and (**U1**).

Zwicky Compacts and Markarian Galaxies A number of lists of nonstellar objects that are singled out because they exhibit some unusual property have been published (**M1**), (**Z1**). These lists are usually very inhomogeneous, containing objects of quite different fundamental properties. Here we shall merely mention the selection criteria of two of the more important lists and remark that much detailed work will be necessary before the objects on the lists can be satisfactorily classified according to their underlying physical nature.

A *Zwicky compact* is a galaxy that can just be distinguished from a star on a plate taken with the Palomar 1.2-meter Schmidt telescope and has a diameter of 2 to 5 seconds of arc. A *Markarian galaxy* is a galaxy with a strong ultraviolet excess as observed with an objective prism camera at a dispersion of about 1800 Å/mm near Hγ.

Interacting and Post-eruptive Galaxies H. C. Arp, B. A. Vorontsov-Velyaminov, F. Zwicky, and others (**A7**), (**V9**), (**Z1**) have published catalogs of odd-shaped galaxies. These systems often involve strange jets, tails, or ringlike structures. A. Toomre, J. Toomre (**T2**), and others have shown that many of these features can be understood in terms of colliding galaxies. It is possible that many galaxies have, at one time or another, suffered a catastrophic collision of this type. In Chapter 17, we shall discuss the ways in which the structure of a galaxy can be permanently altered by such collisions.

5-2. QUANTITATIVE MORPHOLOGY: SURFACE BRIGHTNESS DISTRIBUTIONS

In this section, we describe how one may begin to put the study of the morphology of galaxies on a quantitative basis. We shall concentrate principally on the radial behavior of the surface brightness of galaxies, because it turns out that the different components of galaxies have characteristic radial brightness profiles. Indeed, one may usefully sharpen one's classification criteria by *defining* components in terms of these characteristic radial brightness profiles. Also, an understanding of the constitution and dynamics of galaxies clearly requires, at the outset, a knowledge of their brightness distributions, because these data give clues about the mass distribution within the galaxies. Unfortunately, the determination of brightness distributions proves to be extremely difficult, and much less work has been done in this area than one would ideally wish. Recent technical advances have begun to make this problem much more tractable, however, so that many high quality data should become available over the next few years. In our discussion, we shall first outline the nature of the photometric problems to be solved, and then describe some of the results now available. In the course of this discussion, we hope to indicate ways in which future work in this area could make important contributions to our understanding of the structure and dynamics of galaxies.

The Night Sky

The determination of the brightness profiles of systems other than globular clusters and dwarf ellipticals (whose profiles may be derived merely by counting stars in annuli or strips) is difficult, because galaxies are very faint by comparison with the brightness of the night sky. To a first approximation, the surface brightness of an extended object is independent of its distance from us. As one removes the object to a distance r from the observer, the amount of light received falls off as $1/r^2$, but the solid angle subtended by the object, and over which its light is distributed, falls off in the same proportion, so that the amount of light per square second of arc remains constant. Thus the surface brightness at a given point on the surface of the galaxy is a well-defined, distance-independent quantity. As we shall see later, the peak brightnesses achieved by giant galaxies are about 17 mag per square second of arc in the blue (often written μ_B), but at least half of the light of a typical galaxy comes from outlying regions of surface brightness less than $22\mu_B$, the approximate brightness of the night sky. The brightness distributions of galaxies are commonly followed down to below $26\mu_B$, that is, to less than 3% of the night sky brightness, and they are occasionally followed down to less than 0.5% of the sky.

The brightness of the moonless night sky is made up of four contributions **(A4)**, **(D11)**:

1. *Air glow* produced by photochemical processes in the upper atmosphere. This component has a very irregular spectrum, fluctuates in magnitude by about 20% from hour to hour, and increases from latitude 20° to latitude 70° by about a factor of two. At many observatories, this component is augmented by mercury and sodium line radiation from the street lamps of nearby cities.
2. *Zodiacal light*, which is sunlight scattered off particulate matter in the solar system.
3. *Faint and unresolved stars* in our Galaxy.

4. *Diffuse extragalactic light*, coming from distant, faint, unresolved galaxies.

The relative proportions of these various components, and the total intensity they produce when added together, vary from one observing site to another. They also vary with galactic and ecliptic longitude and latitude. Generally, zodiacal light is the greatest source of brightness, followed by air glow and diffuse galactic light. Diffuse extragalactic light usually makes the smallest contribution to the night sky. Together, these components shed more light onto the surface of the Earth than all the resolved stars, nebulae, and galaxies put together.

The night sky is rather red. It has a color index near $(B - V) = 0.7$, similar to that of a fairly red galaxy. It is important to note that the sky seen from a space vehicle will be nearly as bright as that seen from Earth. The Earth's atmosphere contributes only about 30% of the ground-based night sky brightness. Thus, even with the Space Telescope, accurate photometric measurements of the outer regions of galaxies will be difficult and delicate, although it should prove easier to correct Space Telescope observations for night sky brightness because the rapidly varying component will have been eliminated.

In order to trace the brightness profile of a galaxy, one has to subtract the contribution of the sky background from the flux measured in some region of the sky, and it is apparent from the numbers just given that this will, in general, be much greater than the contribution of the galaxy itself. Any experimental procedure of this kind, in which one obtains the desired quantity as the small difference of two large numbers, is hazardous, and the situation for the problem at hand is rendered worse yet because the light detector most widely used in astronomy—the photographic plate—does not readily lend itself to signal subtraction of this sort. Other types of detectors better suited to signal subtraction are available—for example, photoelectric cells, television cameras, or solid-state light-sensitive devices—but, as we shall now explain, these are unlikely to replace photographic plates as photometric detectors in the near future.

Photoelectric cells have been widely used for the photometry of both stars and galaxies since the 1930s. As was described in Chapter 2, the signal current produced by the photometer is directly proportional to the amount of light entering the device. Therefore, measurement of the current produced by an area of the blank sky, when subtracted from all the other readings, yields directly the currents produced by the galaxy alone. These signals are converted to actual surface brightnesses by observing, with the same equipment, a photometric standard star of known magnitude. Extremely low brightness levels may be measured by switching the field of the photometer back and forth between the region of interest and a blank sky field at regular intervals (*chopping*). In this way, one can correct rather precisely for the effects of instrumental drift and of fluctuations in the strength of the air-glow contribution to the night sky. Photoelectric photometry of an extended object is, however, a painfully slow business, because each point on the object's surface requires a separate measurement at the telescope. Thus, to carry out such a program for more than a few objects is, unfortunately, prohibitively costly in telescope time.

Television systems and solid-state detectors are recent technical developments that may eventually replace photographic plates as the preferred detector for photometric work. However, at the present time, these advanced systems suffer from several severe disadvantages by comparison with photographic plates. These disadvantages include a small number of independent picture elements (*pixels*) per frame, image distortion, spatially variable light sensitivity, unreliability, and expense. Thus it seems likely that the majority of the photometric work carried out on galaxies over the next few years will continue to be performed by a judicious combination of photoelectric and photographic measurements.

Photographic Photometry

In order to understand the quality and limitations of this kind of work, it is necessary to gain some insight into the workings of the photographic process. The blackening of a photographic emulsion on exposure to light is described by the *characteristic curve* of the emulsion, which is a plot of the photographic *density* D against the logarithm of *exposure* E (which equals the product It of the intensity of light I and the exposure time t). The density D is defined by

$$D = -\log T \tag{5-1}$$

where T is the *transmission* of the plate, giving the fraction of normally incident light that passes through the developed plate. A typical characteristic curve has four distinct parts, customarily called (1) the *toe*, (2) the *linear portion*, (3) the *shoulder*, and (4) the *region of solarization*. All photographic emulsions have a threshold; that is, a finite exposure is required to produce any blackening at all. On the toe of the curve, the density rises very slowly with increasing exposure. Eventually, the curve rises into the linear portion where D is nearly linearly proportional to $\log E$. On this part of the curve, the slope, or *contrast*, usually denoted γ, is maximum. A large value of γ ensures that a small increment in the exposure on some part of the plate, resulting, say, from the addition of a faint galaxian halo to the ubiquitous night sky, will produce a substantial blackening on that part of the plate. An astute use of the variation of the slope of the characteristic curve with varying exposure can often be used to enhance the appearance of the images of galaxies like those in the *Hubble Atlas of Galaxies* (**S1**). Finally, the plate approaches *saturation*, and, on the shoulder at the top of the characteristic, curve D rises only very slowly with $\log E$. In the region of solarization, exposure to further light actually bleaches the plate, thus reducing the photographic density recorded. The contrasts of surface brightness within galaxies and the *dynamic response range* of photographic plates are such that, if the exposure time is short enough that the central regions of a typical galaxy are still below the shoulder of the characteristic curve, the outer portions fail to register because the exposure there is below the threshold. Conversely, if the exposure time is sufficiently long that the night sky (and hence also the outer parts of the galaxy) is beginning to register, much of the central structure will be *burned out*, that is, be overexposed. Thus successful photographic photometry of a whole galaxy requires a series of plates of different exposures.

A program of photographic photometry therefore ideally involves the following steps. (1) A series of plates is obtained whose exposure times are such that every part of the galaxy under investigation lies on the linear portion of the characteristic curve of at least one plate. (2) The plates are scanned by a densitometer, a machine which measures the transmission at each point on the plate, takes the logarithm, and stores the result in digital form. (3) The density measures thus obtained are converted to intensity measures by comparing them with the densities produced by exposure of a portion of the plate to a series of spots of illumination having accurately known relative brightnesses. (4) The contribution of the sky to the measured brightness is estimated and subtracted from the brightness ascribed to each point on the galaxian image. This is the crucial step in determining the brightness distribution of a galaxy far from its center. Unfortunately, there is no general agreement as how best to estimate the sky's contribution. We shall return to this point when discussing the brightness profiles of elliptical galaxies. (5) Finally, the absolute scale in magnitudes per square second of arc is set by comparison with a photoelectric measurement of some portion of the same field of view.

It is clear that the procedure we have described is time consuming and prone to error. The body of reliable data for galaxian brightness distributions is not large, because, until the advent of high-speed plate-scanning machines and the computers required to produce and process the vast amount of data generated in photographic work, and until the advent of linear devices of the television or solid-state variety, photometry of galaxies was painfully slow, and the results often contained serious errors [see, for example, (**D6**)]. Nevertheless, let us now turn to a discussion of the results that are available, bearing in mind that they will often be more limited and less reliable than we should like.

The Radial Brightness Distributions of Globular Clusters and Elliptical Galaxies

Globular Clusters The brightness profiles of globular clusters are relatively easily obtained by combining photoelectric photometry of the inner regions with star counts in the more sparsely populated parts. Moreover, their profiles are fairly well understood. Figure 5-9 shows the radial star-density

The results shown in Figure 5-12 demonstrate that the brightness distribution of NGC 3379 is well fitted by the relation proposed by de Vaucouleurs in 1948 (**D1**), namely

$$\begin{aligned}\Sigma(r) &= \Sigma_e 10^{\{-3.33[(r/r_e)^{1/4}-1]\}} \\ &= \Sigma_e \exp\{-7.67[(r/r_e)^{1/4} - 1]\}.\end{aligned} \tag{5-2}$$

This formula is generally known as *the de Vaucouleurs* $r^{1/4}$ *law*. The length scale r_e is known as the *effective radius*, and the numerical factor 3.33 in equation (5-2) is chosen such that, if one were to extrapolate this formula to all radii, both large and small (and it must be remembered that its validity has been established only for a restricted range of radii), then one-half of the total light of the system would be emitted interior to r_e. Thus

$$\begin{aligned}2\int_0^{r_e} \Sigma(r) 2\pi r\, dr &= \int_0^{\infty} \Sigma(r) 2\pi r\, dr \\ &= \frac{8!\exp(7.67)}{(7.67)^8}(\pi r_e^2 \Sigma_e) = 7.22\pi r_e^2 \Sigma_e\end{aligned} \tag{5-3}$$

The parameter Σ_e is clearly the surface brightness at $r = r_e$, and the central brightness of the galaxy is $10^{3.33}\Sigma_e \simeq 2000\Sigma_e$ according to the de Vaucouleurs law.

The de Vaucouleurs surface brightness distribution, which fits the observed data for many giant ellipticals extremely well, is quite different from the King models just discussed. This fact is clearly shown by Figure 5-13, in which King's curves are plotted on the scales $\mu \equiv -2.5 \log \Sigma$ versus $r^{1/4}$. On these scales, the de Vaucouleurs law is a straight line. None of the curves run straight over 10 mag like the data of Figure 5-12, hence none of these curves can fit the observations for NGC 3379 as well as the de Vaucouleurs law does. Conversely, the data for NGC 4472 shown in Figure 5-11 are not as well fitted over their entire range by the de Vaucouleurs law as they are by the King models. It is thus clear that no one empirical or theoretical law will describe all elliptical galaxies.

Other formulae have been used at one time or another to fit galaxian brightness profiles [see (**D6**) for references]. Aside from the relations just discussed, the most important of the other empirical laws is the earliest and simplest, namely

$$\Sigma(r) = \frac{\Sigma_0}{(1 + r/r_0)^2} \tag{5-4}$$

which was introduced by Reynolds in 1913 (**R1**) and popularized by Hubble (**H3**). It is usually called the *Hubble law*. The quantity r_0, sometimes called the *structural length* of the model, is the radius at which the surface brightness

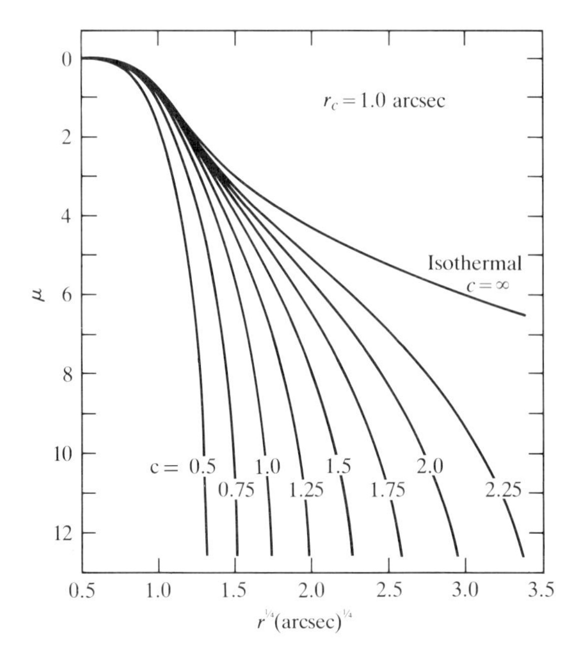

Figure 5-13. King model brightness profiles plotted on scales for which the de Vaucouleurs law is a straight line. Giant ellipticals are generally best fit by curves having $2 < c < 2.35$. [From (**K4**), by permission. Copyright © 1977 by the American Astronomical Society.]

falls to a quarter of its central value. It is thus quite distinct from the effective radius r_e of the de Vaucouleurs law, where the brightness is less than 10^{-3} of its central value. It is, however, similar to the core radius r_c of the King models. By comparing the best fit of the King, Hubble, and de Vaucouleurs models for various objects (**K4**), one typically finds $r_c \simeq r_0 \simeq r_e/11$. It is interesting to note that the Hubble law, which enjoys the merit of great simplicity, usually fits the observed brightness profiles at least as well as the King or de Vaucouleurs laws (**K4**).

It is perhaps worth stressing that the King curves just described have a definite theoretical basis, whereas the de Vaucouleurs and Hubble laws are both purely empirical. The fact that the King curves do not fit all galaxies satisfactorily indicates that the theoretical assumptions on which they rest are not sufficiently general to provide an adequate description of the structure and dynamics of many galaxies. The ultimate significance of a fit (or lack of fit) of any particular model to the data will emerge only when we have a much more complete understanding of the dynamics of galaxies (see Chapter 14).

Brightness Profiles Compared; Night-Sky Subtraction and Seeing Corrections
We have seen that some galaxies are well fitted by one of the profiles we have discussed, and others by another. What are the important differences among the model profiles? These differences prove to be in their behaviors at small and very large radii. We discuss each of these regions in turn.

Figure 5-14 shows that the model profiles can be ranked in the order de Vaucouleurs, Hubble, King according to their gradients at $r = 0$ in a linear-linear plot. The de Vaucouleurs law has an infinite gradient at $r = 0$, the Hubble law has a finite gradient, and the King model has zero gradient. If galaxies consisted of nothing but stars moving in the field of their mutual attraction, the star density gradient at their centers (and hence the gradient of their projected light distributions at $r = 0$) would have to be zero, for (as we shall see in Chapter 14) thc star-density gradient at any point is a measure of the gravitational force field there, and this force can change suddenly only where there is an exceedingly large mass concentration. If strong point sources of light, or of gravitational attraction, were located at the centers of galaxies, then nonzero gradients in the light densities across galaxian centers of the type implied by the Hubble and de Vaucouleurs profiles could occur. Are these gradients found in reality?

To answer this question, one has to eliminate from the observed images of galaxies the effects of seeing created by the Earth's atmosphere and by

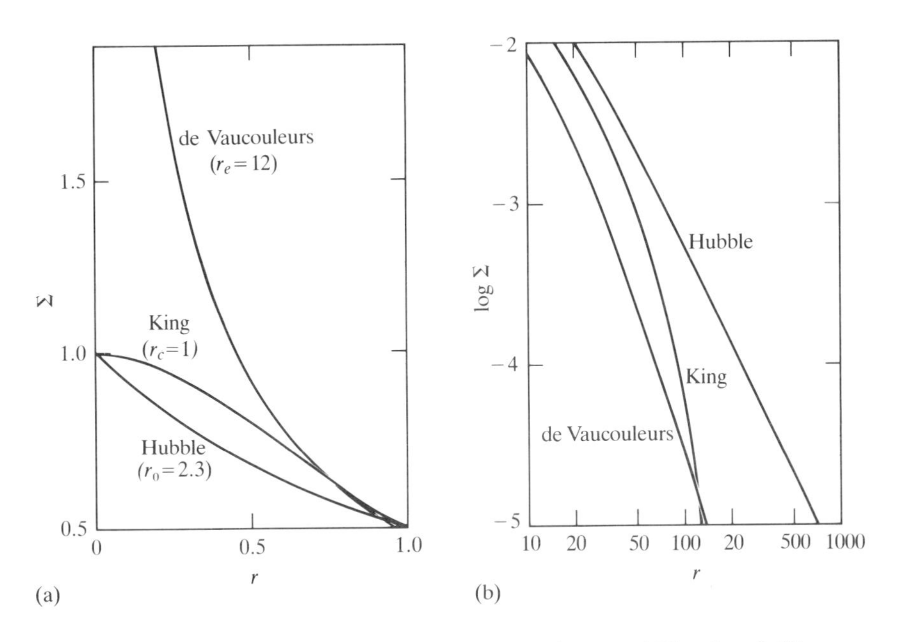

Figure 5-14. Comparison of the Hubble, de Vaucouleurs, and King ($c = 2.25$) profiles. (a) The linear-linear plot shows differences among these profiles at small r. (b) The log-log plot shows differences at large r.

scattering within the telescope (see Chapter 2), which blur the images of point sources into disks of light 1 or 2 arcsec across. Thus, even if the de Vaucouleurs law were to hold right to the center of a galaxy, the brightness gradient there becoming arbitrarily large, the profile registered on the photographic plate would appear to have a zero gradient at the origin on account of these blurring effects. In principle, if one knows the *point spread function* of the observations—that is, the function $f(r)$ that specifies the fraction $f(r)\delta A$ of the light reaching the plate from a point source that falls in the small element of area δA located a distance r from the center of the source's image—then one may correct for seeing to obtain, from the observed light distribution $\Sigma_{\text{app}}(r)$, the true light distribution $\Sigma_t(r)$, for one has

$$\Sigma_{\text{app}}(r) = \int \Sigma_t(r') f(|\mathbf{r} - \mathbf{r}'|)\, dA' \tag{5-5}$$

Figure 5-15 shows the radial brightness profiles $\Sigma_{\text{app}}(r)$ that are obtained by convolving a circularly symmetric King profile $\Sigma_t(r)$ (shown as the dotted curve in the figure) with two point-spread functions $f(r)$. The underlying

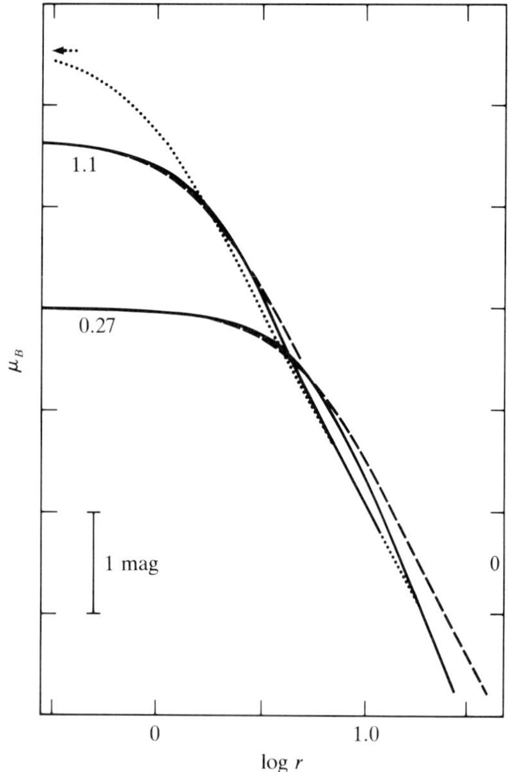

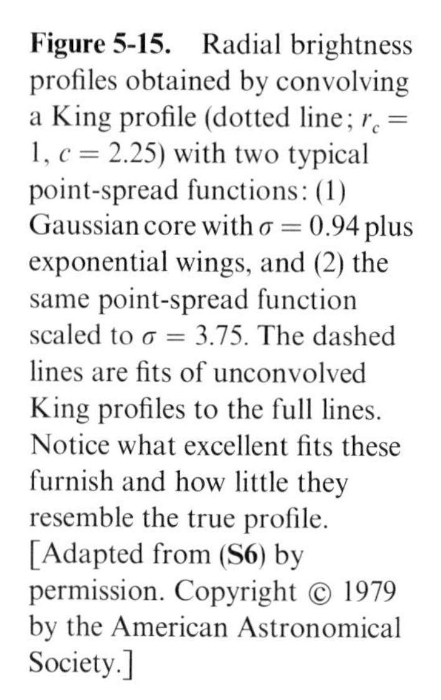
Figure 5-15. Radial brightness profiles obtained by convolving a King profile (dotted line; $r_c = 1$, $c = 2.25$) with two typical point-spread functions: (1) Gaussian core with $\sigma = 0.94$ plus exponential wings, and (2) the same point-spread function scaled to $\sigma = 3.75$. The dashed lines are fits of unconvolved King profiles to the full lines. Notice what excellent fits these furnish and how little they resemble the true profile. [Adapted from (**S6**) by permission. Copyright © 1979 by the American Astronomical Society.]

King profile has unit core radius and $c = 2.25$, but very similar results would be obtained for King models with any value of c greater than about 1.5, because the central profiles of these models do not depend sensitively on c. The point-spread functions $\mathring{f}$ are made up of central Gaussian cores surrounded by exponential wings. This type of profile is found to characterize stellar images on photographic plates. The predicted observational profiles $\Sigma(r)$ are given by the solid curves in Figure 5-15 and labeled by the corresponding ratio (r_c/σ) of the intrinsic core radius to the dispersion of the Gaussian core of the seeing disk.

Understandably enough, one finds that, for values of (r_c/σ) larger than about 4, $\Sigma_{\text{app}}(r)$ differs from $\Sigma_t(r)$ only at very small radii. As (r_c/σ) declines through 1, the peak of Σ_{app} drops more and more below that of Σ_t. However, even for small values of (r_c/σ), the true and observed profiles remain similar at large radii, with the result that $\Sigma_{\text{app}}(r)$ may be well fit by an unblurred King profile for all values of (r_c/σ). Thus, in Figure 5-15, the dashed line is a least-squares fit of a King profile to the Σ_{app} curve for $(r_c/\sigma) = 0.27$. But, it is important to note that the core radius $r_{c,\text{app}} = 5.9$ of the best-fitting King profile is, under these circumstances, very much greater than the true core radius, $r_c = 1$, of the underlying profile. Indeed, it is even greater than the radius, $\sigma = 3.7$, of the seeing disk with which the underlying profile has been convolved. Therefore, if one observes a galaxy in excellent seeing, say $\sigma = 0.75''$, and fits the observed profile with an unconvolved King model to obtain an apparent core radius of, say, $r_{c,\text{app}} = 1.2''$, one cannot assume that one has resolved the core just because $r_{c,\text{app}} > \sigma$. On the contrary, it is entirely possible that the true core radius is as small as $0.2''$. In Figure 5-16, we plot against resolution (r_c/σ) the ratios $(r_{c,\text{app}}/r_c)$ and $(r_{c,\text{app}}/\sigma)$ of the observed core radius, obtained as described earlier, to the true radius and

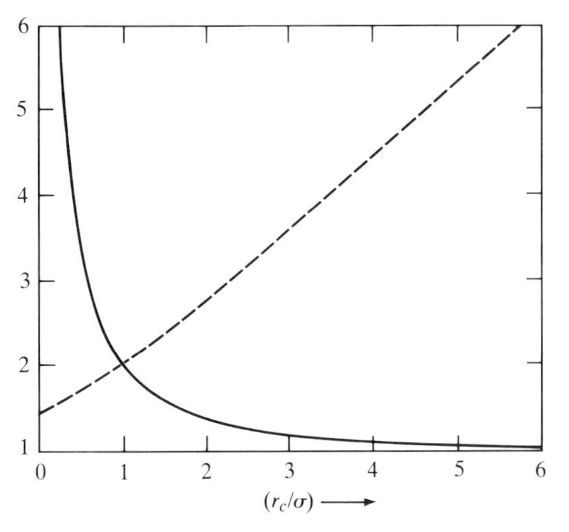

Figure 5-16. The variation of the apparent core radius $r_{c,\text{app}}$ of a seeing-convolved King model as a function of the ratio of the true core radius r_c to the radius σ of the seeing disk. The full curve shows $(r_{c,\text{app}}/r_c)$, and the dashed curve shows $(r_{c,\text{app}}/\sigma)$. Notice that $r_{c,\text{app}}$ is significantly greater than r_c even for large values of (r_c/σ). [From data published in (**S5**).]

to the radius of the seeing disk. One sees that, even for quite large values of (r_c/σ), $(r_{c,\mathrm{app}}/r_c)$ can differ significantly from unity. Thus, for $(r_c/\sigma) = 4$, the apparent core radius is 12% greater than the true core radius. For values of (r_c/σ) less than 1, the ratio $(r_{c,\mathrm{app}}/r_c)$ rises extremely steeply, so that, for these values of (r_c/σ), even small errors in the measured value of $r_{c,\mathrm{app}}$ will lead to a large error in the inferred value of the true core radius.

Figure 5-17 shows the effects of convolving the $r^{1/4}$ law with Gaussian point-spread functions. The solid curves show the seeing-convolved profiles. They are labeled by value of the ratio (r_e/σ) of the effective radius r_e to the dispersion of the Gaussian point-spread function, and have been plotted against the same axes as those of Figure 5-15. The underlying profile is again represented by a dotted curve. Notice how closely the seeing-convolved curves in Figure 5-17 corresponding to values of $(r_e/\sigma) \approx 60$ resemble the seeing-convolved curves of Figure 5-15. This resemblance indicates that, for certain ranges of the parameters, it will be very difficult to distinguish, by ground-based photometry of their cores, between two galaxies, one of which

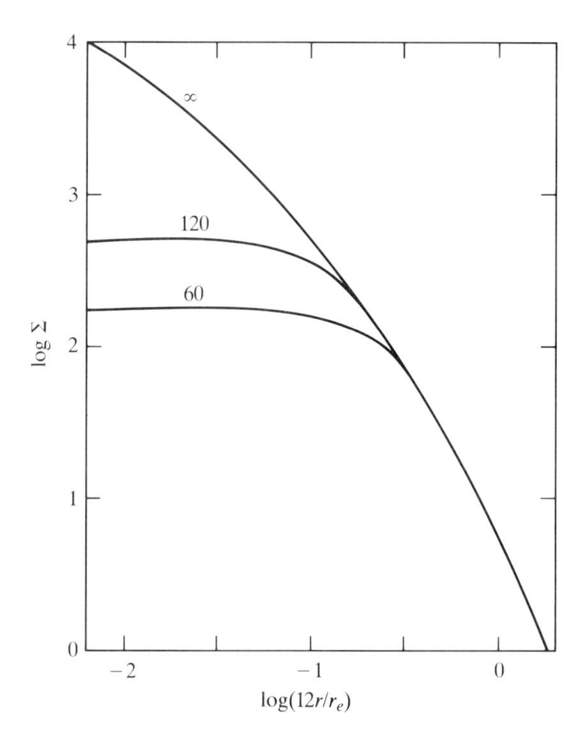

Figure 5-17. Seeing-convolved $r^{1/4}$ law profiles. The curves are labeled by the appropriate values of the ratio (r_e/σ) of the effective radius to the seeing-disk radius. Notice how closely some of these curves resemble those of Figure 5-15.

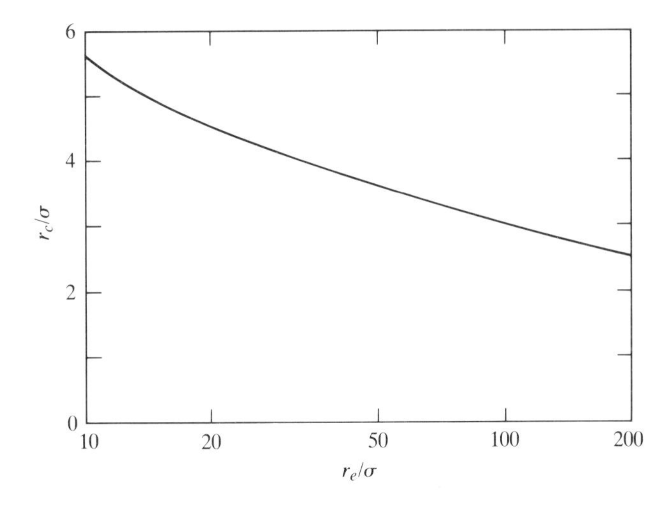

Figure 5-18. The variation of the apparent core radius of seeing-convolved $r^{1/4}$ law profiles with the resolution parameter (r_e/σ).

has in reality a pure King profile, and the other of which exactly obeys the $r^{1/4}$ law. By analogy with a King profile, one can define the core radius $r_{c,\mathrm{app}}$ of one of the seeing-convolved profiles of Figure 5-17 as the radius at which the apparent brightness falls to one-half of its central value. In Figure 5-18, we plot the ratio $(r_{c,\mathrm{app}}/\sigma)$ obtained in this way against (r_e/σ), which is a measure of how well resolved the underlying profile is. Notice that, for essentially all values of (r_e/σ), $r_{c,\mathrm{app}}$ is at least twice as large as σ. Thus we see that, independently of whether the true profile of a galaxy is of the King type but has a small true core radius, or follows the $r^{1/4}$ law at all radii, the apparent core radius which one derives from the observed profile will be at least twice as large as the radius σ of the seeing disk. Therefore, to be sure that the galaxy does have a real core and does not follow the $r^{1/4}$ law all the way into the center, one requires $r_{c,\mathrm{app}} \geq 3\sigma$ at the very least.

Consider now the situation as regards King's observations of fifteen nearby giant ellipticals that we showed in Figure 5-10. In Table 5-3, we list, after the NGC number of each galaxy, the seeing-disk radius σ on the best of King's plates, the best-fitting effective radius r_e, the apparent core radius $r_{c,\mathrm{app}}$ of the observed profile, the ratios (r_e/σ) and $(r_{c,\mathrm{app}}/\sigma)$, and two predicted core radii. The first of these radii, $r_{c,\sigma}$, is the apparent core radius that corresponds to the given value of (r_e/σ) according to the relation shown in Figure 5-18. This then is the apparent core radius one would observe in the given seeing if the galaxy obeyed the $r^{1/4}$ law all the way to its center. The last radius given is the maximum core radius that is, according to Figure 5-16, compatible with the given seeing and apparent core radius. Comparison of

Table 5-3. Apparent and Seeing-Corrected Radii for Selected Elliptical Galaxies

NGC	σ	r_e	$r_{c,\text{app}}$	r_e/σ	$r_{c,\text{app}}/\sigma$	$r_{c,\sigma}$	$r_{c,K}$
4452	1.41	24.4	3.51	17	2.49	3.05	2.29
4347	1.33	41.3	4.12	31	3.10	3.31	3.22
2300	0.67	24.6	2.14	37	3.19	1.74	1.69
4589	0.74	36.6	1.62	49	2.19	2.05	0.90
4621	0.61	34.7	1.38	57	2.26	1.75	0.80
4261	0.69	40.8	2.73	59	3.95	2.00	2.33
4649	0.69	44.1	5.37	64	7.78	2.04	5.37
5846	0.66	44.7	2.89	68	4.38	1.99	2.56
7626	0.46	38.5	1.43	84	3.11	1.46	1.10
4365	1.00	86.3	5.01	86	5.01	3.20	4.55
4406	0.62	59.1	2.04	95	3.29	2.03	1.62
4697	0.62	66.6	2.45	107	3.25	2.09	2.08
4486	0.59	68.4	7.94	116	13.46	2.03	7.94
4472	0.85	117	5.50	138	6.47	3.06	5.29
4636	0.58	103	3.16	178	5.45	2.24	2.95

SOURCE: **(S6)**, by permission.

columns 4 and 7 of the table shows that only four of the fifteen galaxies have measured core radii that are more than 50% greater than the apparent core radii predicted by the $r^{1/4}$ law. This observation has led F. Schweizer **(S6)** to suggest that the majority of these galaxies do, in fact, follow the $r^{1/4}$ law right into their centers. However, Table 5-3 shows that the observed central profiles of these galaxies are also compatible with the assumption that their true profiles are of the King type. Comparison of columns 4 and 8 shows furthermore that, except in the cases of NGC 4589 and NGC 4621, if the true profiles are of the King type, then the true core radii $r_{c,K}$ are reasonably well determined by the observations.

The Space Telescope will resolve this question in a few years' time, but, as we shall see in later chapters, the exact way in which the brightness profiles of ellipticals behave near their centers is of considerable dynamical importance. Therefore, it is worthwhile to consider what further information we can immediately bring to bear on the problem.

A strong argument in favor of the hypothesis that ellipticals obey the $r^{1/4}$ law even at small radii is that the observed central surface brightnesses of very nearby galaxies tend to be very much higher than those of the more distant ones observed by King. Both M32 and M31 (which is an Sb galaxy whose central light is dominated by an elliptical-like spheroidal component) have observed central surface brightnesses that are higher than any values determined by King. This effect may be because M32 and M31 are systems of lower luminosity than those observed by King, but one may equally well suspect that the effect arises because the very nearby galaxies have more highly resolved $r^{1/4}$ law profiles than do the more distant galaxies. A second argument in favor of centrally peaked true brightness profiles is that, not only does M31 have *more* brightness at its center than is envisaged by even the $r^{1/4}$ law (**L1**), but de Vaucouleurs and Capaccioli (**D6**) have concluded that the same is true of the well-observed giant elliptical NGC 3379 we discussed earlier. The brightness near the center of M31 has been securely determined by E. S. Light, R. E. Danielson, and M. Schwarzschild (**L1**) from observations made with the balloon-borne telescope, whereas the conclusion of de Vaucouleurs and Capaccioli (**D6**) concerning NGC 3379 rests on an iteratively obtained inversion of equation (5-5) with ground-based photoelectric measurements.

The de Vaucouleurs, Hubble, and King profiles differ markedly in their behavior at large as well as at small radii. Figure 5-14 shows that, in order of compactness, they rank: King, de Vaucouleurs, Hubble. Indeed, according to the King model, the luminosity of a galaxy should vanish at some finite radius r_t. The brightness predicted by the de Vaucouleurs model never vanishes, but it does tend toward zero sufficiently rapidly that the total light contained within an arbitrarily large circle is finite. In contrast, according to the Hubble profile, the light contained within a circle on the sky of radius r diverges logarithmically: $\int(1 + r)^{-2} r\,dr \sim \log r$ for large r. Which of these laws is in closer agreement with observation?

It was emphasized earlier in this section that the determination of the brightness profile of a galaxy requires an accurate knowledge of the night sky background on which the image of the galaxy is superposed. Consider now the way in which a derived galaxian profile can be distorted by small errors in the adopted night sky background. Figure 5-19 shows the profile of a galaxy, whose true profile does cut off at some radius r_t, as one would observe it for three different background brightness levels. Figure 5-19a shows the profile one obtains with too low a background level. The profile declines almost as a power law [a straight line in the $(\mu, \log r)$ plane] before flattening off at some brightness level μ_e, which gives the error in the assumed background level. Figure 5-19b shows the galaxian profile one obtains with a slightly higher but still underestimated sky level. A tidal cutoff has begun to emerge from the background, although the steep decline in the profile associated with the cutoff soon gives way to a flat tail at brightness level μ_e', which is the new error in the assumed sky level. Finally, Figure 5-19c shows the shape of the profile one obtains with the correct background; here, at some finite radius, the brightness goes to zero and $\mu \to -\infty$. But note that one would also obtain a profile of exactly the same form for too large an assumed background level. Indeed, whenever one overestimates the sky level, the brightness attributed to the galaxy will necessarily go to zero at some radius at which the galaxy actually contributes a finite amount of light, thus causing μ_{gal} to plunge toward $-\infty$ at that radius, independently of the true behavior of the galaxy's brightness distribution far from its center. Thus a profile like

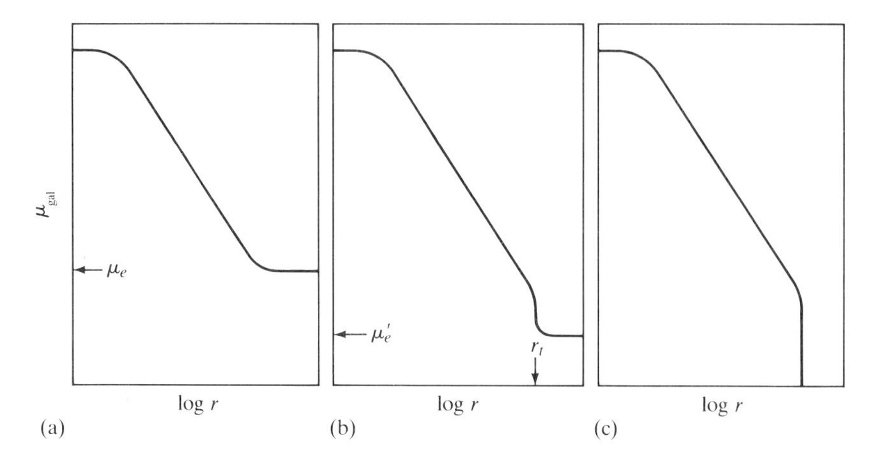

Figure 5-19. The effects of errors in the subtracted sky background on the profile of a galaxy with a truncated brightness distribution. (a) The background has been underestimated so that the night sky contributes to the "galaxy" brightness at large radii. (b) The background is higher than in (a) but still too low. (c) The correct background level.

that of Figure 5-19c cannot by itself be taken as evidence that the galaxy in question has no luminous material beyond r_t, for even a de Vaucouleurs or Hubble-type brightness distribution would appear to possess a cutoff if the galaxian brightness profile had been obtained using an overestimated sky level. The correct choice of the sky background level is therefore of critical importance.

Three methods of night-sky estimation have been used in recent photometric work: (1) photoelectric measurement of one or more nearby blank fields [see, for example, (**D6**)]; (2) polynomial fitting of the brightness distribution surrounding the galaxy under observation (**J3**); and (3) direct estimation from the brightness profiles of galaxies [see, for example, (**C2**)]. At the present time, it is not possible to come down in favor of one of these methods rather than another. The first method is, in principle, simple, but it requires many prolonged photoelectric observations for good accuracy. Reliance on measurements of only a few "blank fields," which may be contaminated by faint images, vignetting, and so forth, could be dangerous. The polynomial fitting procedure is attractive because it exploits brightness measurements at a large number of different blank-sky points around the galaxy's image. One must, however, take great care to eliminate the images of stars from the data before fitting the polynomial. One must also ensure that the order adopted for the polynomial is not so high as to cause large oscillations in the brightness interpolated across the galaxian image. The third method of sky-background estimation involves plotting out galaxian brightness profiles for progressively increasing background brightness levels. One then estimates the true sky-brightness level to be that value at which the final shoulder of Figures 5-19a and 5-19b disappears. According to this view, no galaxy has been demonstrated to have a brightness cutoff of the sort envisaged by the King models unless a plot of the type shown in Figure 5-19b has been obtained, for only a figure of that type can demonstrate that an observed cutoff in the galaxy profile is not caused by overestimation of the night-sky level.

Typical Results for Giant Ellipticals The radial brightness profile of an elliptical galaxy is conveniently characterized by (1) the approximate slope α of the straightest portion of a plot of the surface brightness μ against log radius, and (2) whether or not the slope of this curve increases markedly at large radii. Slopes in the range $\alpha = 2.1$ to $\alpha = 1.65$ are commonest (the Hubble law predicts a slope $\alpha = 2$). The brightness profiles of some galaxies show, at radii in excess of 30 kpc, the marked steepening predicted by the de Vaucouleurs and King models, though not by the Hubble law. NGC 6158 is an example of such a galaxy (Figure 5-20). However, the presence or absence of this cutoff is sensitive to the sky-background estimation procedure adopted. Other galaxies show no sign of having a cutoff in their brightness profiles down to $\mu_B = 28$ mag per square arcsecond (about 0.4% of the sky background). Galaxies not showing cutoffs may be divided into two types: (1) those whose profiles disappear into the noise in the observations as continuous straight lines, and (2) those whose profiles actually become less steep as the background noise level is approached. NGC 6166 is an example of a

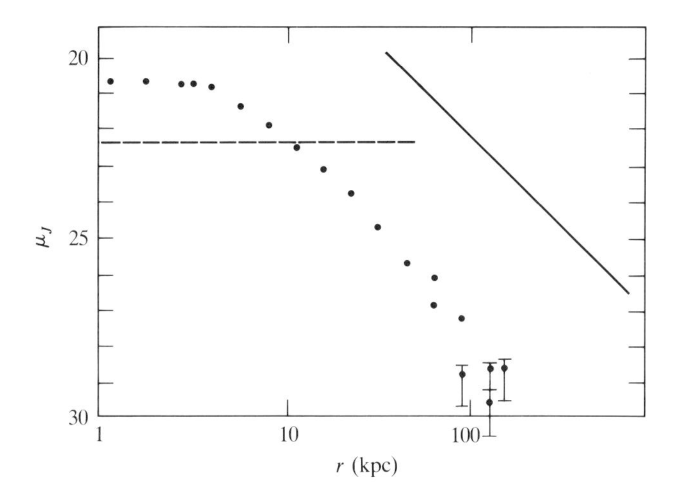

Figure 5-20. The IIIa-J brightness profile of NGC 6158 seems to turn down at 80 to 90 kpc. The dashed line shows the level of the sky brightness. The full line here and in Figure 5-21 shows the slope of the relationship $\Sigma(r) \sim r^{-2}$. [From (**C1**).]

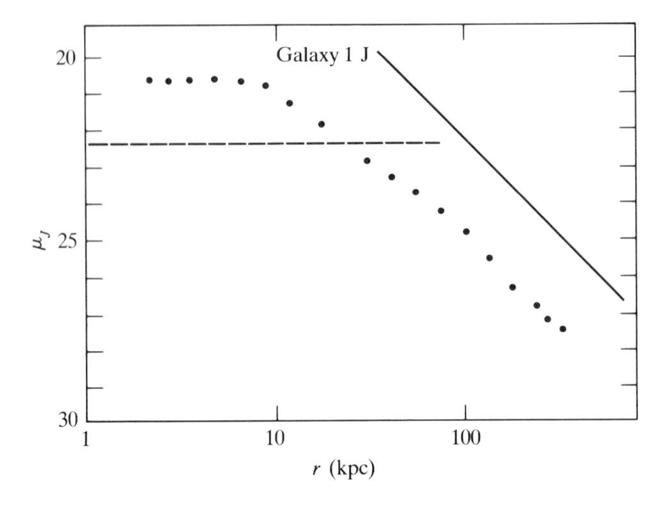

Figure 5-21. The IIIa-J brightness profile of NGC 6166 has a shallower slope than that of NGC 6158 and shows no sign sign of steepening to beyond 120 kpc. [From (**C1**).]

galaxy of the first type (Figure 5-21), whereas the giant galaxy in the rich cluster of galaxies Abell 1413 is a galaxy of the second type (**O3**). Galaxies of the second type, which are always giants and almost always the brightest galaxies in compact clusters of galaxies, are called cD galaxies (see §5-1).

The linear size of giant elliptical galaxies is very great. Obviously one can define only lower limits on the radii of galaxies whose profiles show no cutoffs, and, when cutoffs are detected, they are generally found to occur only at very great radii. For example, few of the galaxies in the Coma cluster of galaxies, which are bright enough to be NGC galaxies, have cutoff radii smaller than 50 kpc (for a Hubble constant of 50 km s^{-1} Mpc^{-1}). Some cD galaxies are detectable out to beyond 1 Mpc (**O3**). By contrast, it will be remembered that the Sun is less than 10 kpc from the galactic center.

The Radial Brightness Distributions of Disk Galaxies

Although elliptical galaxies are the simplest systems observed, we have seen that, on closer inspection, even these display considerable variety and individuality of form. Now we turn to a discussion of the brightness distributions of galaxies which possess some sort of highly flattened disk structure. The complexity of these systems can be very great, for, in addition to the disk and elliptical (or bulge or "spheroidal") components, they usually have yet other structures—for example, spiral arms or bars or lenses—and these complicate their brightness distributions considerably. These other structures tend to contribute to the brightness distribution least at positions very close to and very far from the center. Very near the center, the total light is dominated by the bulge component, whereas far from the center the disk contribution is the more important. With this fact in mind, we start our investigation of the profiles of disk galaxies by concentrating on the behavior of their brightness profiles at small and large radii, hoping thereby to isolate the contributions of the two most important components.

The Disk and Bulge Components Figure 5-22 shows surface brightness plotted against radius for six fairly face-on spirals. Notice that, in this plot of log Σ versus r, the outermost points tend to lie on straight lines; that is, at large radii, one has

$$\Sigma = \Sigma_s \exp(-r/r_s) \tag{5-6}$$

This brightness distribution is known to fit the brightness profiles of the outer regions of a large class of disk galaxies (**D2**) and has, in fact, come to *define* the normal disk component of all flat galaxies. Deviations from this profile are generally ascribed to the existence of other components.

This state of affairs suggests that we might be able to express the brightness profiles of many disk galaxies as the sums of an elliptical-like component obeying one of the laws discussed in the last section and a disk component

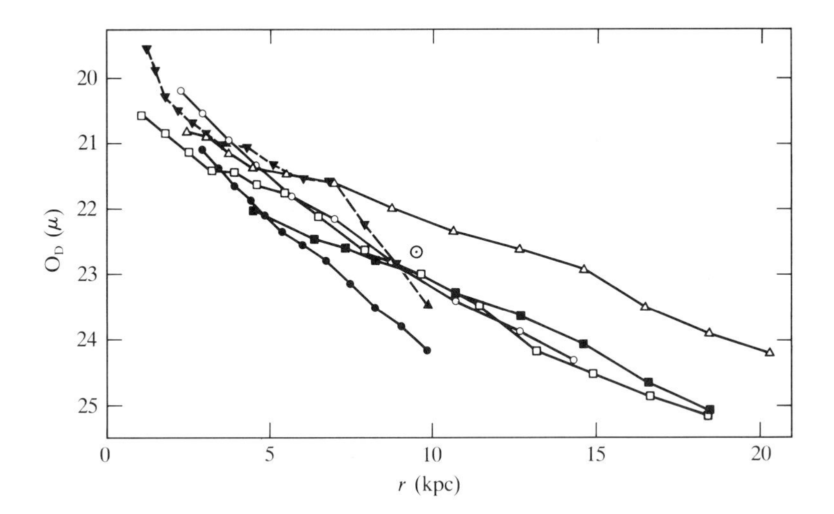

Figure 5-22. Orange surface brightness versus radius for six spiral galaxies [NGC 3031 (●), NGC 4255 (○), NGC 4321 (△), NGC 5194 (▼), NGC 5364 (■), and NGC 5457 (□)]. The solar symbol ⊙ indicates the estimated surface brightness of our Galaxy near the Sun. Notice that the outer parts of the profiles tend to be fairly straight in accord with equation (5-6). [From (**S5**) by permission. Copyright © 1976 by the American Astronomical Society.]

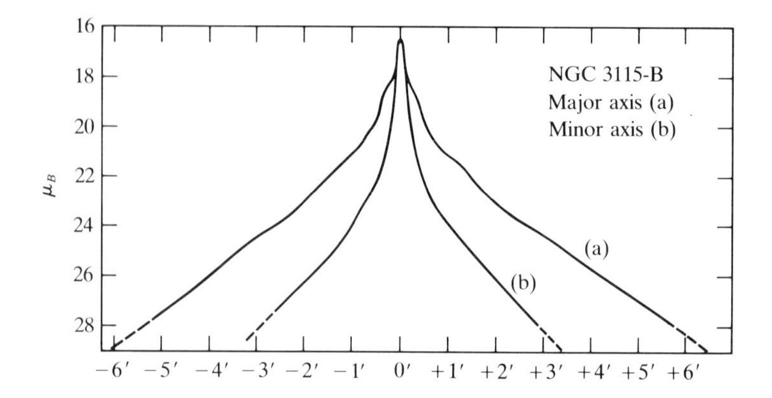

Figure 5-23. The major-axis (a) and minor-axis (b) brightness profiles of the S0 galaxy NGC 3115 versus radius in arcminutes. [From (**T3**).]

having an exponential brightness distribution of the form of equation (5-6). The brightness profiles of the nearby, edge-on S0 galaxy NGC 3115 shown in Figure 5-23 offer encouragement to this expectation. The profile along the minor axis of an edge-on galaxy like NGC 3115 should be dominated (except at very small radii) by the elliptical-like component, and in fact we find that the minor axis profile of NGC 3115 is well fitted by the de Vaucouleurs $r^{1/4}$ law. Subtracting the brightness predicted by this elliptical-like distribution from the major axis profile, one obtains a crude estimate of the disk's contribution to the galaxy's light, which is approximately exponential in form. Such an estimate is only rough, however, because the elliptical-like component is not expected to be quite spherical. Rather, it is reasonable to suppose that the effective radius r_e of the profile of the bulge component measured in the plane of the disk will be greater than that measured perpendicular to the disk, because the bulge will, in general, be flattened.

To obtain accurate estimates of the disk and bulge-component parameters, one must develop a general procedure for fitting the major axis profiles of disk galaxies of any orientation to the line of sight with the sum of the profiles of elliptical and disk components. This one may do iteratively by least-squares fitting, say, an $r^{1/4}$ profile to the inner part of the profile, then fitting at large r an exponential component to the residuals between the true profile and the chosen $r^{1/4}$ law, then fitting at small r a new $r^{1/4}$ profile to the residuals between the true profile and the fitted disk, and so on, iteratively, until convergence is achieved. Alternatively, one can fit both components to the observed profile over the two extreme ranges of radius simultaneously. J. Kormendy (**K3**) has shown that one obtains similar parameters Σ_s, r_s (for the disk) and Σ_e and r_e (for the bulge) describing the two components by either of the methods just described, and that such models fit the observations fairly well outside a range of intermediate r values. A decomposition of the brightness profile of V Zw 257 is shown in Figure 5-24.

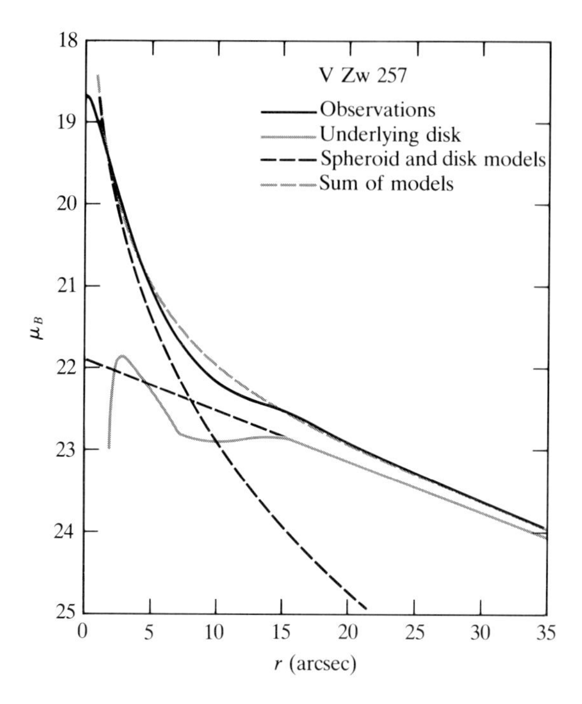

Figure 5-24. The brightness profile of V Zw 257 decomposed into disk and bulge components. Notice that the disk-component ordinate intercept μ_{B_0} is smaller than that one would obtain by simply fitting a straight line to the outer data points.

When one has in this way isolated the contributions of the disk and the bulge components to the overall galaxian brightness distribution, one may evaluate the ratio D/B of the luminosities of these two components. Thus, integrating the brightness distribution of equation (5-6) over the entire galaxian surface, and employing equation (5-3) for the bulge luminosity, one finds

$$\frac{D}{B} = 0.28\left(\frac{r_s}{r_e}\right)^2 \frac{\Sigma_s}{\Sigma_e} \tag{5-7}$$

In Table 5-4, we list some typical values of (D/B) for S0 and spiral galaxies. Notice that the S0 galaxies in this sample all have significantly smaller values of the disk-to-bulge ratio than those of many spiral galaxies. It is possible that biases in the selection of the galaxies chosen for photometric study may artificially increase the mean (D/B) of the quoted spirals, but it is not clear that this is the case. Therefore, it seems that lenticular galaxies generally have more prominent bulges than do typical spirals.

Table 5-4. Disk and Bulge Parameters for Spiral and Lenticular Galaxies

NGC	Type	$-M_B$	r_s/kpc	B_s/μ_B	r_e/kpc	B_e/μ_B	D/B
3384	S0	20.4	3.9	22.13	1.2	20.51	0.7
4270	S0	20.3	3.1	21.36	2.7	22.49	1.0
4281	S0	21.2	4.4	20.93	1.1	19.71	1.6
4459	S0	20.6	3.7	22.19	3.7	22.19	0.2
4526	S0	21.3	6.9	22.10	4.8	22.41	0.7
4570	S0	20.2	2.5	20.78	0.4	18.98	1.8
224 (M31)	Sb	19.8	4.5	21.4	3.5	17.5	3.2
598 (M33)	Sc	17.8	1.4	21.4	<0.5	–	>12

SOURCE: Data published in (**B4**) and (**F7**)

An important point to notice about the decomposition into disk and bulge components shown in Figure 5-24 is that the slope of the derived disk component is smaller than the slope of the straight line that is tangent to the entire galaxy profile at large radii. This difference in slope arises because the contribution of the bulge component to the brightness of the galaxy is nonnegligible even at large r. In fact, if the brightness of the bulge continues to obey the $r^{1/4}$ law at very large radii, there will be a radius beyond which the bulge's light once again dominates that of the disk; clearly, the exponential decrease of disk brightness predicted by equation (5-6) will ultimately lead to a lower brightness at large r than the less steeply falling $r^{1/4}$ law. The bulge contribution drops below that of the disk at intermediate radii only because the scale length of the bulge is smaller than that of the disk. It is therefore dangerous to ignore the brightness of the bulge and fit a straight line to the large-r behavior of the composite profile. Such a fit to the composite profile will inevitably be steeper than the line of the true disk contribution and thus yield a smaller scale length r_s (from the steeper slope) and a higher intercept Σ_s than the true values.

The last point we made is important because K. C. Freeman (**F7**) has noted a remarkable fact. If one does fit straight lines to the outer parts of observed disk-galaxy profiles, one finds that the intercept values Σ_s obtained in this way for different galaxies show a remarkably small scatter. Freeman obtained $\Sigma_s = 21.65\mu_B$ with dispersion $0.3\mu_B$ for twenty-seven galaxies, values which have recently been confirmed by Schweizer from an independent sample of six galaxies (**S5**). According to this result, the absolute magnitudes of the disks of these galaxies are entirely determined by their scale lengths. If one galaxy has a disk which is four times as luminous as that of another, it follows that it has twice the scale length of the other. Freeman and Schwiezer do indeed observe a large scatter of scale lengths r_s: for a Hubble constant, $H_0 = 50$ km s^{-1} Mpc^{-1}, they find $2.6 < r_s < 5.7$ kpc.

Two questions about this intriguing result arise immediately. First, is it real or a selection effect? Detailed photometry is available for very few galaxies, and these certainly have not been selected at random, so one must be cautious in drawing conclusions from statistical analysis of the photometric data. We shall soon see that there are reasons to believe that one may tend to pick precisely those galaxies whose disks happen to satisfy Freeman's criterion. Second, even if Freeman's result were to hold for an unbiased sample of *compound* profiles, would it hold for the disk components alone? We have seen that the bulge contribution to the compound profiles can appreciably affect the intercepts Σ_s derived from straight lines fitted to the compound profiles, and Kormendy (**K3**) has argued that this effect increases the Σ_s values of galaxies with underluminous disks just enough to bring their effective Σ_s's back into Freeman's range.

In practice, disk-galaxy profiles tend to be more complex than our simple disk-plus-bulge model would allow. Three lines of evidence suggest that, either the bulges of disk galaxies differ from ellipticals, or that many, perhaps all, disk galaxies possess a third major component in addition to the disk and the bulge.

1. The minor axis profile of the well-studied, near edge-on Sb galaxy NGC 4565 does not closely resemble that of an elliptical galaxy except at very large radii (**K6**).
2. Even those edge-on disk galaxies whose minor axis profiles *are* well fitted by the $r^{1/4}$ law tend to have profiles along cuts parallel to the minor axis which differ systematically from the profile predicted by the simple disk-plus-bulge model (**B4**). In these cases, the discrepancy between the predicted and the observed profiles, which increases as the major axis intercept of the cut increases, gives rise to a characteristic "box-shaped" appearance of the fainter isophotes.
3. Recent measurements of rotation in S0 galaxies indicate that their bulges are rotating rather rapidly (**K7**). We shall see, in Chapter 8, that this situation contrasts with the low rotation speeds typically observed in giant ellipticals.

In summary, the relation between the bulges of disk galaxies and ellipticals is very uncertain at the present time. Bulges certainly do resemble ellipticals in many respects and may even be indistinguishable from them, but we must emphasize that this possibility is not the only one and that elucidation of this point is an important task for the future.

Brightness Profiles in the Transition Region The profiles of disk galaxies at intermediate radii can be quite complex and appear to obey no general rules. Freeman has divided the profiles into types I and II (**F7**). Typical profiles

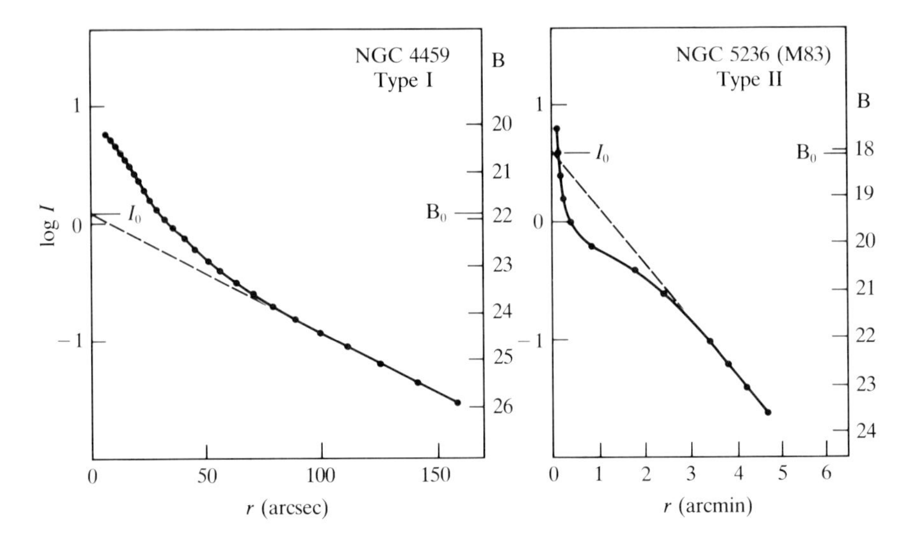

Figure 5-25. Freeman's definitions of type I and type II luminosity profiles of disk galaxies. [From (**F7**) by permission. Copyright © 1970 by the University of Chicago.]

of these classes are shown in Figure 5-25. Galaxies with type I profiles have, in addition to the bulge and disk, one or more components whose light makes important contributions to the overall profile just outside the region dominated by the bulge. Little is known about these components. Sometimes bar and ring structures of various types are discernible on direct plates, but we do not know what types of dynamical entities these might be. Type II profiles are thought to arise because the exponential disks of galaxies showing this type of profile have holes in their centers. Indeed, excellent fits to the profiles of such galaxies may be obtained with the sum of an $r^{1/4}$ profile and a modified exponential disk having a brightness distribution

$$\Sigma = \Sigma_0 \exp[-\alpha r + (\beta/r)^3] \tag{5-8}$$

A disk with a profile given by equation (5-8) has little luminosity interior to $r = \beta$ (**K3**). One cannot, on the other hand, satisfactorily model the brightness distribution of a type II galaxy with the sum of an $r^{1/4}$ bulge and a simple disk; such models always show excessive luminosity at intermediate radii. We shall see, in Chapter 9, that the gas distributions of many spiral galaxies also have central holes. This behavior may result from the disturbance of an originally simple disk by the gravitational field of a barlike bulge component (**D8**), or it may arise because the material that was originally where the hole now is has had time, since the formation of the galaxy, to fall into the galaxian nucleus.

Azimuthal Brightness Distributions

Having considered the radial distribution of brightness in galaxies, we now turn to the variation of brightness with angle around the nucleus, that is, the azimuthal distribution of brightness. Even less is known about azimuthal light distributions than about radial profiles, although the physical significance of this information is just as great.

Clearly a crucial factor determining the angular distribution of a galaxy's light on the sky is the inclination of the galaxy's three-dimensional form to the line of sight. For example, the image on the sky of a very flat galaxy can be either round or highly elongated, depending on whether the galaxy is seen face-on or edge-on. Other factors affecting the azimuthal distribution of light, like bumps produced by spiral arms, are intrinsic to the galaxy. We shall see later that the separation of the observed distribution into intrinsic and extrinsic components can be difficult. But first we must consider how one even obtains the apparent brightness variations. Let us commence once again with a discussion of the elliptical galaxies.

Elliptical Galaxies The brightness distributions of elliptical galaxies are smooth, so that their *apparent isophotes* (contours of equal brightness on the sky) are a series of nested ellipselike figures. Three methods of obtaining the shapes of these figures have been used in recent work. The most economical method involves tracing the brightness of a galaxy along a series of lines passing through its center. For example, one might trace along the approximate major and minor axes and two intermediate axes. Let the position angle (conventionally measured from north through east) between the ith such line and the hour circle through the galaxian nucleus be ϕ_i. Then, from the brightness profiles along these lines, one can find the radius r_i at which the appropriately smoothed brightness along the ith line falls to a certain level μ. The parameters of the ellipse most closely approximating the shape of the contour of a specified brightness level μ may then be obtained by least-squares solution of the equations

$$\frac{1}{r_i^2} = C - A\cos 2\phi_i - B\sin 2\phi_i \tag{5-9}$$

for all position angles ϕ_i. Equation (5-9) is the polar equation of an ellipse having axial ratio $q = \{[1 - (A^2 + B^2)^{1/2}/C]/[1 + (A^2 + B^2)^{1/2}/C]\}^{1/2}$, semimajor axis length $a = [C - (A^2 + B^2)^{1/2}]^{-1/2}$, and principal axis position angle $\phi = \frac{1}{2}\arctan(B/A)$.

Two conceptually straightforward methods of determining the shapes of isophotes start by measuring the brightness of the galaxian image over a wide area of the plate. This procedure yields a grid of brightness values. One may interpolate in some way among the points of the grid to obtain an estimate of the brightness at any given point, and then construct contours that trace out lines of equal brightness through the resulting distribution. The contours one obtains in this way will usually be jagged and irregular because of the noise in the observations, but they can nonetheless be fitted with ellipses to obtain an estimate of their shapes. An alternative to this contouring procedure is to guess an approximate center (x_0, y_0), semimajor axis length a, axial ratio q_0, and position angle ϕ_0 of a certain contour. One may then fit the observed brightness Σ_{obs} in a strip around the guessed ellipse to the functional form

$$\Sigma_m(x, y) = \Sigma_1 a^{2\beta}\{[(x - x_1)\cos\phi_1 - (y - y_1)\sin\phi_1]^2 + q_1^{-2}[(x - x_1)\sin\phi_1 + (y - y_1)\cos\phi_1]^2\}^{-\beta} \tag{5-10}$$

One varies $\Sigma_1, x_1, y_1, q_1, \phi_1$, and β in the fit to obtain new estimates of the parameters of the ellipse that most nearly fits the isophote with semimajor axis a. If these differ appreciably from the values originally chosen, one can repeat the process until the sequence of parameter sets obtained at each stage converges (**W3**).

Procedures of the type just described, which analyze brightness measurements over the entire of the galaxian image, have the advantage that they enable one to analyze the distribution for deviations of the isophotes from perfect ellipses. The best way of seeking such deviations is to express the residuals between the observed and the model brightnesses along an isophotal line as a Fourier series in $(\phi - \phi_p)$, where ϕ_p is the position angle of the major axis of the model isophote, and ϕ is the analogously defined angle of a general point on the isophote. That is, one writes

$$R(\phi) \equiv \Sigma_{\text{obs}}(\phi) - \Sigma_m(\phi) = A_1 \sin(\phi - \phi_p) + A_2 \sin 2(\phi - \phi_p) + A_3 \sin 3(\phi - \phi_p) + \cdots + B_0 + B_1 \cos(\phi - \phi_p) + B_2 \cos 2(\phi - \phi_p) + \cdots \tag{5-11}$$

The numbers A_1, A_2, B_0, B_1, and B_2 are not very interesting and should be small if the ellipse-fitting procedure has been carried out properly. The higher coefficients A_3, B_3, and so on tell one whether the true isophotes show a tendency to be "egg-shaped" or "box like" or whatever. From the theory of Fourier series, one has

$$A_n = (1/\pi)\int_0^{2\pi} R(\phi)\sin n\phi\, d\phi \qquad n = 1, \ldots \tag{5-12}$$

$$B_n = (1/\pi)\int_0^{2\pi} R(\phi)\cos n\phi\, d\phi \qquad n = 1, \ldots \tag{5-13}$$

If the true isophotes were rather box-shaped, so that Σ_{obs} at $\phi - \phi_p = 45^\circ$, 135°, and so on were greater than Σ_m, one would obtain $B_4 > 0$. Pointed isophotes yield $B_4 < 0$. Egg-shaped isophotes give A_3 or $B_3 \neq 0$. Remarkably enough, it is found that all deviations of the isophotes of elliptical galaxies from true ellipses are no larger than can be accounted for by the noise in the data [for example, (**C2**), (**W3**)]. It is not known why ellipses should be such good approximations to the shapes of isophotes of elliptical galaxies.

A quantity of some dynamical interest is the variation of the ellipticity $\varepsilon \equiv 1 - (b/a) = 1 - q$ of the isophotes with radius. Near the center, one has to correct the directly observed values for the effects of seeing, which tend to smear all intrinsic features into circular disks, and thereby systematically reduce the measured ε values. This correction can be made by tabulating true and seeing-smoothed ellipticities for different ratios σ/r of the diameter, σ, of the characteristic seeing disk to the mean radius r of an isophote. For this purpose, it is adequate to use a Hubble or de Vaucouleurs model of the radial brightness variation. After one has made this correction to the observed ε values, one finds that ε sometimes increases with radius (that is, image becomes more elongated), sometimes stays roughly constant, sometimes decreases with radius, and sometimes peaks or shows a minimum as a function of radius. The cause and significance of these different behaviors are not yet understood.

The behavior of the major axis position angle as a function of radius is also of great interest, for it, like the ellipticity, often shows a significant variation with radius. This implies that the elliptical isophotes are frequently *twisted* with respect to one another on the sky. Figure 5-26 shows ε and ϕ for NGC 6173, a galaxy studied by D. Carter (**C1**). This object shows both a complex ε variation and twist of the major axis. Between $r = 10''$ and $100''$, the major axis twists by about $20°$.

The importance of this twist phenomenon is that it strongly suggests that the galaxies in question do not possess an axis of rotational symmetry but are, rather, triaxial bodies. Indeed, if an elliptical galaxy has surfaces of equal luminosity density (*isodensity* surfaces) which are figures of rotation, and if these figures are so aligned that they share a common symmetry axis and center, then, no matter how the ellipticities of the isodensity surfaces may vary with radius, the galaxy will appear to all observers to have isophotal contours that are aligned, ellipselike curves, and no twist will be detected. In fact, it is not hard to convince oneself that one principal axis of the image of such a rotationally symmetric galaxy will be the line in which the equatorial plane of the system cuts the plane of the sky (the *line of nodes*). The other principal axis, which will be the apparent minor axis of the image if the galaxy is oblate spheroidal (saucer shaped) and the major axis in the prolate spheroidal case, will be perpendicular to the line of nodes. One could model a galaxy whose isophotes twist by imagining that the isodensity surfaces are rotationally symmetric figures that are tilted with respect to one another. However, such a model is unattractive dynamically and seems much less plausible than a model in which the isodensity surfaces are triaxial (that is, not figures of rotation) but do share common principal axes. Models of this

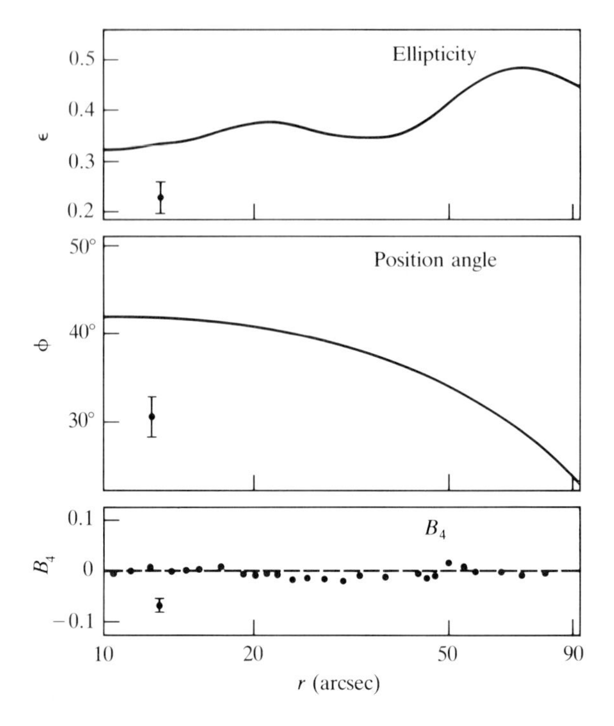

Figure 5-26. Ellipticity, position angle, and B_4 coefficient for NGC 6173 as functions of radius. Notice the twist of the major axis position angle through more than 20° and the negligible deviations of the isophotes from pure ellipses as indicated by the parameter B_4 [see equations (5-12) and (5-13)]. [From (**C1**).]

type are known to be dynamically acceptable (see Chapter 14) and have, in general, twisted isophotes, as we now show.

Figure 5-27 shows a prolate ellipsoid inside an oblate ellipsoid. Now, although the coordinate axes in the figure are the principal axes of both the inner and the outer ellipsoids, it is evident that the apparent major axis of the outer ellipsoid makes an angle of 50° with the apparent major axis of the inner body. This skew alignment of the apparent major axes arises from the fact that the axial ratios of the two bodies are different. The inner body has $a{:}b{:}c = 1{:}2{:}1$, whereas the outer body has $a{:}b{:}c = 1{:}1{:}0.5$. Imagine that these two bodies are two of the isodensity surfaces of an elliptical galaxy, and that the isodensity surfaces between these two have axial ratios such as 1:1.8:0.9, 1:1.1:0.6, and so on, which form a continuous sequence between the ratios of the two surfaces drawn in Figure 5-27. Then, if the isodensity surfaces do have axial ratios that vary with increasing mean radius, the isophotes will be seen (from the given observing position) to twist contin-

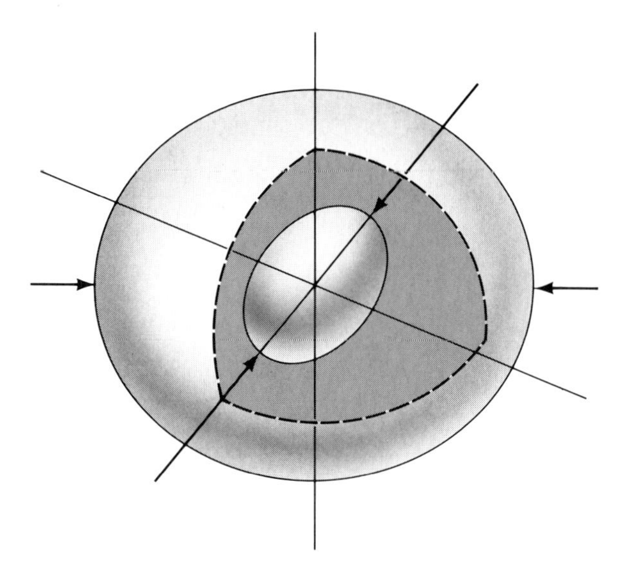

Figure 5-27. Isophotal twist as a consequence of triaxiality. Two concentric, coaxial ellipsoids are shown. The dashed lines mark the intersections of the ellipsoids with the coordinate planes, while the solid lines show their outlines to the observer. The arrows mark the directions of their apparent principal axes.

uously through a substantial angle. In this way, we can obtain a dynamically acceptable model of the isophotal twist phenomenon.

Three points are worth noting in connection with the model just described. (1) If a galaxy happens to be aligned in space such that the line of sight is one of the principal axes, then its isophotes will not twist. (2) Although this model interprets the twist phenomenon as being driven by changes with radius in the true axial ratios (such as we know must occur in most galaxies because we usually observe some radial variation of apparent axial ratio), it does not follow that, in any particular galaxy, an isophotal twist must be accompanied by a change in the apparent axial ratio. To specify a triaxial model, one has to determine the two true axial ratios, but fixing the apparent axial ratio at some definite value still leaves one degree of freedom, which allows the position angle of the observed isophotes to vary. (3) Isophote twist is not confined to elliptical galaxies. If one inspects the picture of the Sb galaxy M31 in the *Hubble Atlas* (**S1**), one can see that the major axis of the bulge twists by about 10° near the center. Stark (**S1**) has fitted Lindblad's (**L2**) observations of this phenomenon with detailed triaxial models.

Distribution of True Ellipticities of Elliptical Galaxies The observed distribution of the ellipticities of elliptical galaxies reflects the true distribution of ellipticities as modified by projection effects. For example, it is easy to see that any ellipsoidal galaxy will look round when viewed from some particular direction, so the existence of galaxies whose images are circular does not prove that any genuinely spherical galaxies exist. We must, therefore, ask what one can learn about the distribution of *true* ellipticities of galaxies from the observed distribution of their *apparent* ellipticities.

In the analysis, it is useful to make two simplifying approximations. We assume that all elliptical galaxies are spheroidal (that is, either oblate or prolate figures of revolution) and that any particular galaxy can be characterized by a unique ellipticity. We have seen that, strictly speaking, both of these approximations are incorrect; elliptical galaxies are, in general, more complex bodies than simple spheroids, and they usually show variable ellipticity. However, by supposing that one is dealing with the ellipticity near a particular contour level—for example, the $25\mu_B$ contour level—one may justify the use of just one ellipticity rather than the complete function $\varepsilon(r)$ to characterize the appearance of a galaxy. (In practice, overall ellipticities are assigned to galaxies by inspection of their images on plates, and, if similar plate material is used for all galaxies in the sample, this procedure will tend to yield the ellipticity of the galaxies at a certain fairly uniform contour level.) Furthermore, by assuming that galaxies are spheroids (oblate or prolate), one simplifies the problem to the point where a given distribution of observed ellipticities uniquely determines the distribution of true ellipticities. If one allowed triaxial bodies, many distributions of true ellipticities would be possible. In addition, one learns about the two extremes which lie at either end of the spectrum of triaxial ellipsoids, that is, about those objects in which the intermediate axis has the same length as the longest axis (oblate spheroids) or as the shortest axis (prolate spheroids). Thus, distributions obtained using these two simplifying hypotheses can reasonably be expected to span qualitatively the entire spectrum of possible distributions.

Consider then a spheroidal galaxy whose isodensity surfaces are described by the equation

$$(ux)^2 + (uy)^2 + z^2 = a^2 \tag{5-14}$$

If $u > 1$, the galaxy will be prolate, and if $u < 1$, it will be oblate. In either case, a is half the length of the symmetry axis of the isodensity surface labeled by a, and the x and y semiaxes are each of length (a/u). Consider the appearance of the galaxy to an observer whose line of sight makes an angle θ with the z axis (Figure 5-28 shows the geometry of this situation). Without loss of generality, we may take the y axis to be the line of nodes, so that one apparent semiaxis has length (a/u) and the other principal semiaxis has length A, as shown in Figure 5-28. In the notation of the figure,

$$(ux_0)^2 + z_0^2 = a^2 \tag{5-15}$$

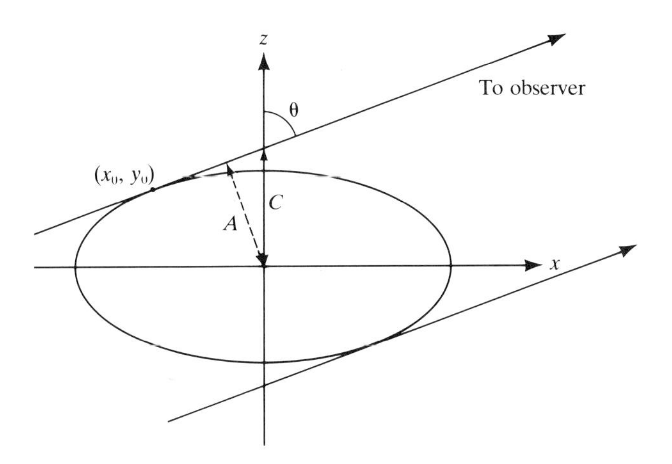

Figure 5-28. Evaluating the apparent axial ratio of a rotationally symmetric galaxy.

and, by differentiation of this equation, we find for the slope s of the tangent at (x_0, z_0)

$$s = \cot\theta = \frac{-u^2 x_0}{z_0} \tag{5-16}$$

so

$$\cot^2\theta = \frac{u^4 x_0^2}{z_0^2} = u^2\left(\frac{a^2}{z_0^2} - 1\right) \tag{5-17}$$

which implies

$$\frac{a^2}{z_0^2} = \frac{\cot^2\theta}{u^2} + 1 \tag{5-18}$$

But

$$\begin{aligned} C = \frac{A}{\sin\theta} &= z_0 - x_0 s \\ &= z_0 + \frac{u^2 x_0^2}{z_0} = \frac{a^2}{z_0} \end{aligned} \tag{5-19}$$

Thus, using equation (5-18), we obtain

$$A^2 = a^2\left(\frac{a^2}{z_0^2}\right)\sin^2\theta = a^2\left(\frac{\cos^2\theta}{u^2} + \sin^2\theta\right) \tag{5-20}$$

In the oblate case, the apparent axial ratio (when defined to be less than 1) is $q = uA/a$, and $q = a/uA$ in the prolate case. Therefore, we have

$$u^2 \sin^2 \theta + \cos^2 \theta \begin{cases} = q^2 \text{ (oblate)} \\ = \dfrac{1}{q^2} \text{ (prolate)} \end{cases} \tag{5-21}$$

Let us now use equation (5-21) to determine the distribution of apparent ellipticities that will arise from a large number of spheroidal galaxies whose symmetry axes are randomly oriented about the line of sight. If the symmetry axes of the galaxies are randomly oriented with respect to the line of sight, then, of the $n(u)\,du$ galaxies under discussion, a fraction $\sin\theta\, d\theta$ will have their symmetry axes directed at angle θ to the line of sight. By equation (5-21), q is a function $q(\theta)$ of θ. To have q between q and $q + dq$, the symmetry axis of the galaxy has to be oriented between θ and $\theta + d\theta$, where $d\theta = dq/|dq/d\theta|$. Therefore, galaxies with u values in the range $(u, u + du)$ contribute

$$\frac{n(u)\,du \sin\theta\, dq}{|dq/d\theta|} \tag{5-22}$$

to the observed $f(q)\,dq$ galaxies with axial ratios in the range $(q, q + dq)$. Summing the contributions from all u values and dividing both sides by dq, we have

$$f(q) = \int n(u) \left(\frac{\sin\theta}{|dq/d\theta|} \right) du \tag{5-23}$$

From equation (5-21), one has for *oblate* geometry,

$$\left| \frac{dq}{d\theta} \right| = \frac{1}{q} [(1 - q^2)(q^2 - u^2)]^{1/2} \tag{5-24a}$$

$$\sin\theta = \left[\frac{(1 - q^2)}{(1 - u^2)} \right]^{1/2} \tag{5-24b}$$

and for *prolate* geometry,

$$\left| \frac{dq}{d\theta} \right| = q[(1 - q^2)(q^2u^2 - 1)]^{1/2} \tag{5-25a}$$

$$\sin\theta = \frac{1}{q} \left[\frac{(1 - q^2)}{(u^2 - 1)} \right]^{1/2} \tag{5-25b}$$

For prolate geometry, equation (5-23) therefore becomes

$$f(q) = \frac{1}{q^2} \int_{1/q}^{\infty} \frac{n(u)\,du}{[(u^2 - 1)(u^2q^2 - 1)]^{1/2}} \tag{5-26}$$

In the prolate case, if the *true axial ratio* β is considered to be a quantity less than 1, then one has $\beta = 1/u$. Therefore, if we define $N(\beta)$ to be the number density of galaxies with true axial ratios near β, we have $N(\beta)\,d\beta = n(u)\,du$, and

$$f(q) = \frac{1}{q^2} \int_0^q \frac{N(\beta)\beta^2\,d\beta}{[(1 - \beta^2)(q^2 - \beta^2)]^{1/2}} \tag{5-27}$$

In the oblate case, $\beta = u$ and equations (5-23) and (5-24) lead immediately to

$$f(q) = q \int_0^q \frac{N(\beta)\,d\beta}{[(1 - \beta^2)(q^2 - \beta^2)]^{1/2}} \tag{5-28}$$

Equations (5-27) and (5-28) are integral equations relating the known (observed) frequency of galaxies of apparent axial ratio q to the unknown frequency $N(\beta)$ of galaxies of true axial ratio β.

In principle, one may solve these integral equations for $N(\beta)$ in terms of $f(q)$. The only difference between the two equations is the extra factor of (β^2/q^3) in the prolate case, which reduces the contribution of highly aspherical ($\beta \ll 1$) galaxies to the observed number of apparently round galaxies. However, the available data are neither sufficiently plentiful nor adequately noisefree to allow one to solve these equations directly. Any attempt to do so is likely to lead to physically meaningless distributions of true ellipticity. The reasonable way to proceed is to attempt to choose a true distribution which, on projection, is compatible with the observations. Figure 5-29 shows the observed distribution of ellipticities from (**D7**) and two inferred distributions of true axial ratios, derived by the same technique under the assumption that all the galaxies are prolate or oblate spheroids, respectively. Both of the "true" distributions produce "apparent" distributions that fit the observed statistics to within the expected random fluctuations. Notice that the distributions are rather similar, both peaking near $q = 0.62$ ($\varepsilon = 0.38$), and that both require the existence of some genuinely spherical galaxies. The additional factor (β^2/q^3) in equation (5-27) results in the distribution of prolate spheroids including a greater fraction of nearly spherical galaxies than are required under the oblate hypothesis. The distributions of true ellipticities just derived must be regarded as purely empirical results, inasmuch as we do not know what mechanisms are responsible for establishing a distribution of true axial ratios of this type.

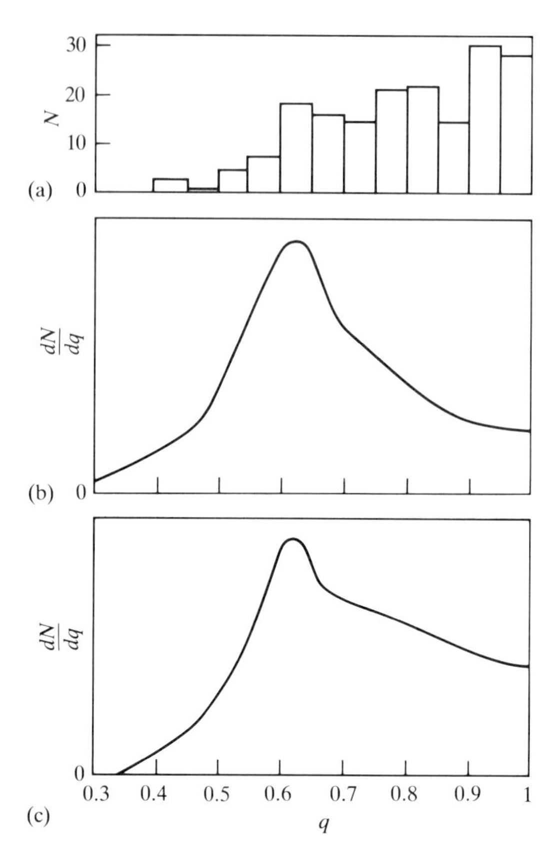

Figure 5-29. Observed apparent ellipticity distribution and inferred true ellipticity distributions for elliptical galaxies. Shown are (a) the frequency of observed ellipticities from the Second Reference Catalog (**D7**) along with derived true ellipticities when the galaxies are assumed to be all (b) *oblate* spheriods or (c) *prolate* spheroids.

Disk Galaxies Let us now consider what can be learned from the azimuthal brightness variations of disk galaxies. We shall discuss three basic questions. (1) Are disk galaxies very flat, and, if so, are they circular? (2) How does scattering by dust within disk galaxies affect their light distributions? (3) How do spiral arms and bars affect the light distributions of disk galaxies? Let us consider each of these in turn.

1. It turns out that disk galaxies are neither exactly flat nor quite round, but one has to study the distribution of neutral hydrogen in these systems to demonstrate this fact. We shall therefore delay full discussion of this point to Chapter 8, confining ourselves here to a statistical exercise for S0 and spiral galaxies similar to that conducted earlier for the ellipticals. We assume that disk galaxies are oblate spheroids, and, using equation (5-28), we seek a distribution of true axial ratios which is compatible with the observed distribution. Figure 5-30 shows the diagrams equivalent to Figure 5-27 for ellipticals. Notice that the distributions of apparent axial ratios of these galaxies show all axial ratios to be nearly equally common, and that

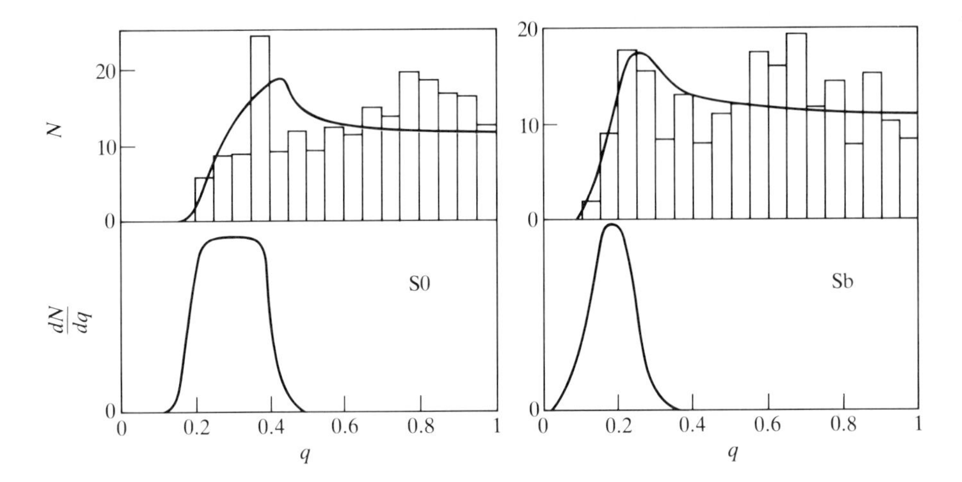

Figure 5-30. The apparent and true axial ratio distributions of disk galaxies. The histograms in the upper half of the figure are compiled from the Second Reference Catalogue (**D7**); the full curves show the fits furnished by equation (5-28) using the true distributions shown in the lower half of the figure.

this leads one to conclude that the distributions of true axial ratios are sharply peaked around some small value of the axial ratio. [It is, in fact, easy to see from equation (5-28) that, if $N(\beta)$ is constant from $\beta = 0$ up to some maximum value $\beta \ll 1$, then $f(q)$ is independent of q for $q > \beta_0$.] This result lends quantitative support to the subjective judgment that disk galaxies are intrinsically quite thin. It does not, however, establish that they are circular any more than our analysis of the distribution of ellipticals proved that they are spheroidal.

2. Dust can both scatter light from its original path and absorb it. Scattering by dust tends to make disk galaxies fainter when seen edge-on, because then more light is scattered out of the line of sight than is scattered into it, and therefore a galaxy looks brighter when viewed face-on. Blue light is more strongly absorbed and scattered than red light, and there is a tendency for dust to scatter light through only a small angle (that is, forward scattering predominates). Figure 5-31 shows how these effects lead to color and brightness asymmetries between the near and far sides of disk galaxies. The light received by the observer from the part of the bulge on the near side of the galaxy's major axis is more strongly absorbed than the light from the far side, because much of the former has to pass right through the absorbing dust within the disk. The absorption is so strong in type $S0_3$ galaxies that it produces a crescent of blackness on their near sides. This differential absorption also makes the light from the near side of a galaxy appear redder than that from the far side. Scattering by dust within the disk has precisely the opposite effect. Light emitted by the upper half of the bulge in Figure 5-31

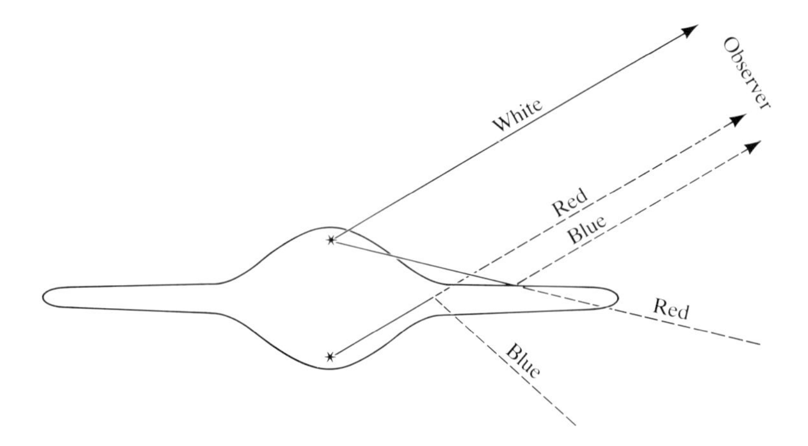

Figure 5-31. Effects of scattering and absorption of light by dust on the images of disk galaxies. Light from the top of the bulge reaches the observer without obstruction by dust in the disk. Light from the lower portion of the bulge is partially absorbed by the disk. Some light is forward scattered by the disk into the path to the observer.

is strongly forward scattered off the dust in the disk and thus contributes extra blue light to the galaxian image on the near side of the major axis, but little light is scattered toward the observer by the far side of the disk. In consequence the near side of the image tends to be bluer and brighter than the far side. The relative importance of these two competing effects varies with the angle of inclination i of the galaxy's disk to the plane of the sky. At small inclinations, absorption dominates over forward scattering, and the near side appears dim and red. At intermediate inclination ($80° > i > 60°$), forward scattering can make the near side bluer and brighter than the far side. At very large inclinations, light from the near side is very heavily absorbed, and forward scattering cannot make up for the light lost by absorption (which can be so heavy that it produces dark lanes across the image of the galaxy). Extensive discussions of this problem will be found in (**E1**) and (**V7**).

3. Very little is known about the brightness variations characterizing bars and spiral structure. The pioneering work in this area is that of F. Schweizer (**S4**), who has published three-color photographic photometry of six giant spirals. The brightness distributions of spirals, unlike those of ellipticals, are patchy and irregular, so that the use of carefully chosen averaging procedures is a prerequisite of any discussion of the systematics of their light distributions. Schweizer analyzed his data by dividing the image of each galaxy into about a dozen annuli, whose elliptical shapes were determined by the requirement that they be circular in the plane of the galaxy. These annuli were then subdivided by angle into a series of bins, each bin

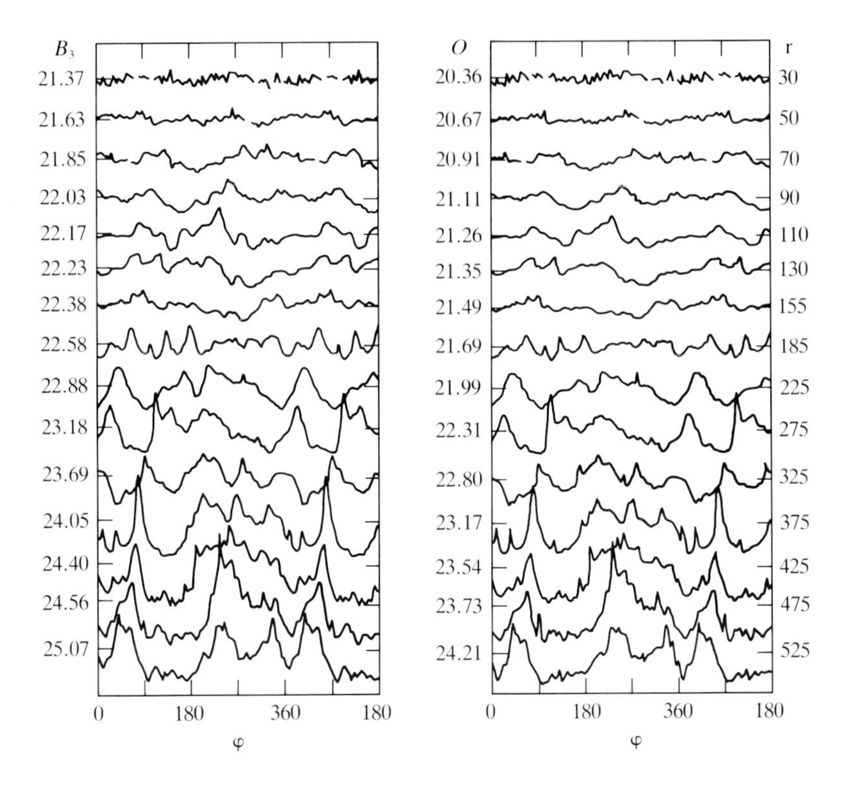

Figure 5-32. Spiral structure in M101. Shown are blue (left) and orange azimuthal profiles at 15 different radii, $r = 30$ arcsec (top) to $r = 550$ arcsec (bottom). Notice how the spiral structure grows in strength with increasing radius (top to bottom) and how similar the B and O profiles are. [Adapted from (S5) by permission. Copyright © 1976 by the American Astronomical Society.]

being figured like a small piece of a spiral arm, so that azimuthal brightness profiles could be constructed for each annulus by plotting the average brightness in a bin against the angle ϕ at which the bin crosses a certain circle. Figure 5-32 shows a series of such profiles for the giant Sc galaxy M101. In this diagram, radius increases from top to bottom, and the profiles have been continued periodically beyond $\phi = 360°$ to assist the inspection of the pattern near $\phi = 0°$. Notice that there is little evidence of two-armed structure, that the strength of the arms increases with increasing galactocentric radius (that is, toward the bottom in Figure 5-32), and that the blue (B_3) and orange (O) profiles are remarkably similar.

Schweizer endeavored to isolate the "arm" components of these profiles by drawing, at each radius, lines across the azimuthal profiles at the mean brightness of the two deepest troughs separated by more than 90°. He defined the brightness levels of these lines to be those of the underlying smooth disk.

From these data, one obtains the near-exponential variations shown in Figure 5-22 and discussed earlier in this section. With this definition of arm strength, Schweizer finds that, in the O passband, the arms typically contribute about 17% of the total brightness at 3 kpc, rising to 50% of the brightness at 15 kpc. The arms are generally 20% stronger in the B passband and 60% stronger in the U passband than they are in the O band.

The underlying disks prove to be very uniform in color and rather red. For the disks in his sample, $(B - V) \approx 0.75 \pm 0.05$, which is redder than many old galactic clusters, although it is a little over 0.1 mag bluer than the central parts of giant elliptical galaxies. As we shall soon see, the arms are in parts bluer than the disk, so the growth in strength of the arms relative to the disk with increasing radius causes the overall colors of the galaxy to become slightly bluer at large radii.

Figure 5-33 shows ultraviolet (U), blue, and orange azimuthal profiles for the giant Sb galaxy M81 at $r = 475''$, together with those in two of the

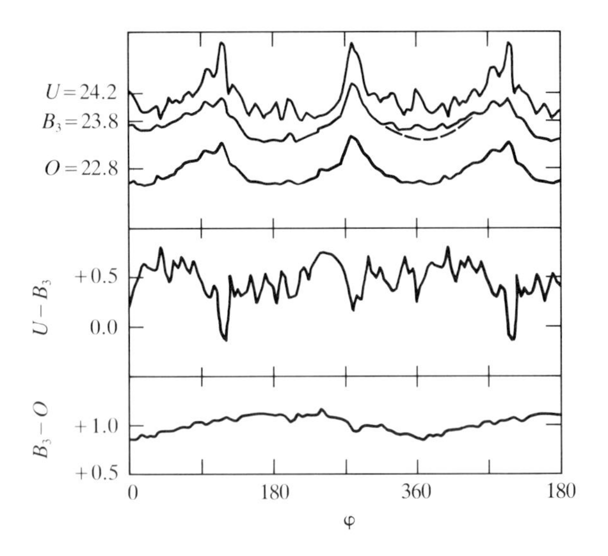

Figure 5-33. Spiral structure in M81. U, B, and O azimuthal profiles at $r = 475$ arcsec. The angle ϕ is measured from the apparent major axis in the direction of rotation. The profile has been continued periodically beyond $\phi = 360°$. Notice how the U profile consists of narrow peaks superposed on a broader wave pattern which resembles the B and O profiles. Much of the small fluctuation in the $(B_3 - O)$ color can be accounted for by an instrumental malfunction. [From (**S5**) by permission. Copyright © 1976 by the American Astronomical Society.]

associated color indices. Two points should be noted about this diagram. (1) The two spiral arms which feature so clearly in the top panel appear in the U band to be made up of a fairly symmetrical broad wave, on top of which is superimposed a pair of narrower peaks. In the B and O bands, these narrower peaks are much less conspicuous. (2) The bottom panel of Figure 5-33 shows that the $(B - O)$ color averages $\approx 0^{\mathrm{m}}.8$ and changes by no more than $\pm 0^{\mathrm{m}}.2$ across an arm. The narrowness of the arms in the ultraviolet probably reflects the narrowness of the spiral structure outlined by OB associations, which tend to occur in chains. These chains are particularly conspicuous on plates taken through Hα filters because of strong emission in the hydrogen Balmer lines by the excited gas within the constituent OB associations. The uniformity and numerical magnitude of the color of the broad components in Figure 5-33 is both rather surprising and of some importance. For if, as is often assumed to be the case, spiral arms result mainly from concentrations of newly formed stars, we would expect the arms (in the absence of enhanced reddening by dust) to be bluer than the interarm regions, because they would be abnormally rich in blue, upper main-sequence stars. The rather modest color changes actually observed by Schweizer do not support this expectation, and they suggest instead that either (1) the light from the young stars in the arms is unusually heavily reddened (by $A_V \approx 0^{\mathrm{m}}.5$), or (2) that old disk stars (of relatively late spectral types) contribute about 40% of the increased brightness found near the spiral arms. Unfortunately, it is very difficult to choose between these two alternatives because, in the broad-band colors, the reddening effects of dust are essentially identical with the effects of dilution of the light of young stars with that of old ones. From the point of view of theories of spiral structure, the resolution of this question is clearly a matter of the greatest importance. In the first case, we could consider spiral structure to be an entirely gas-dynamic phenomenon, which leaves the bulk of the mass in the disk quite unmoved. In the second case, one would have no alternative but to regard spiral structure as the result of some sort of gravity wave propagating through the stellar disk.

The azimuthal structures of the bars of barred spirals can be extremely complex and are not at all well understood. Two salient features require mention. The first is the presence, in many cases, of dust lanes running almost radially from the ends of the bar. These lanes tend not to run exactly toward the center, but rather to keep toward the side of the bar which leads in the rotation of the whole galaxy. An excellent example of this phenomenon can be seen in the photograph of NGC 1300 in the *Hubble Atlas* (**S1**). In galaxies later than the SBb galaxy NGC 1300, the dust lanes run along lines further removed from the principal axis of the bar, and through the less-well-illuminated disk. In very late-type specimens, the absorption lanes may be seen to curve in toward the nucleus as they reach the bar's minor axis, thus partially obscuring the center of the galaxy. In this event, a contour map of the galaxy's brightness will appear double peaked. It is known that,

Figure 5-34. The Small Magallenic Cloud photographed from Mount Stromlo Observatory in Australia. Notice the egg-shaped bar which is clearly displaced from the center of the surrounding luminosity. [Courtesy of G. de Vaucouleurs.]

in at least some cases, these dust lanes are produced by the streaming of gas *along* the length of the bar (**D8**). The second phenomenon deserving mention is that barred galaxies are commonly lopsided. This phenomenon is most clearly seen in the Magellanic-type galaxies, which usually have rather egg-shaped bars that are often not situated at the center of the surrounding light distribution (see Figure 5-34), but the same phenomenon may also be detected in giant SB systems—see, for example, the photograph of NGC 4548 in the *Hubble Atlas.*

The barred spirals are thus significantly more complex systems than ordinary spirals, which are themselves very much harder to understand than the ellipticals and unbarred lenticulars. De Vaucouleurs and Freeman (**D8**) and Kormendy (**K5**) have discussed the morphological complexities of these systems and Crane (**C5**) and Okamura (**O4**) have presented the photographic photometry of three specimens. However, much more work will be necessary before it is possible to give a coherent account of the light distributions of SB galaxies.

5-3. MAGNITUDES AND DISTANCES OF GALAXIES

Integrated Magnitudes and Apparent Diameters

Up to now, we have been primarily concerned with the detailed distribution of surface brightness within galaxies. As we have emphasized, photometric data of the quality necessary for the determination of such detailed brightness distributions are available for comparatively few galaxies. For many thousand other galaxies, only a few simple measurements of brightness and angular scale are available. It is therefore important to consider the definitions of some useful, conveniently measured quantities, whose values have been determined for a large sample of galaxies. These simple quantities also serve to summarize the large-scale properties of well-observed galaxies, and, therefore, tend to be very widely quoted. Let us first discuss estimates of total brightness.

Integrated Magnitudes Three types of integrated magnitude are to be found in the literature, metric magnitudes, isophotal magnitudes, and total magnitudes. Photoelectric measurements lead to estimates of the light contained within a certain aperture; these are called *metric magnitudes.* Because the metric absolute magnitudes of the brightest galaxies in rich clusters of galaxies show a remarkably small dispersion, photoelectric metric magnitudes play an important role in attempts to detect the deceleration of the expansion of the Universe. Photographic photometry, by contrast, leads naturally to *isophotal magnitudes*, that is, estimates of the light contained within a certain isophotal contour. For example, Holmberg (**H2**) has obtained magnitudes interior to the $26.5\mu_{pg}$ contour for several hundred objects. *Total magnitudes* are those estimated from metric or isophotal magnitudes by some sort of extrapolation procedure. In principle, these magnitudes represent the total amount of light emitted by a galaxy, but it is essential to remember that the brightness profiles of many galaxies drop so gradually that extrapolation can be both dangerous and strongly model dependent. Nonetheless, the most widely quoted magnitudes, those of the *Reference Catalogue* of de Vaucouleurs and his collaborators (**D7**), are total magnitudes in this sense. The introduction to this catalog describes the extrapolation procedures adopted and gives extensive bibliographies for the sources of galaxian magnitudes.

Galaxian Diameters Two simple measures of a galaxy's linear size are available for a wider sample of galaxies than those for which the model-specific quantities, such as r_c, r_s, and so on, defined in §5-2, have been determined. The *Holmberg radius* is the length of the semimajor axis of the 26.5 mag (pg) arcsec^{-2} contour of surface brightness (**H2**). The *standard diameter* D_0 of the *Second Reference Catalogue* (**D7**) is the diameter that one estimates the 25 mag arcsec^{-2} contour would have if the galaxy were seen face-on and were unobscured by dust. Half the standard diameter is often referred to as the *de Vaucouleurs radius*. Clearly, once the distance to a galaxy is known, its Holmberg radius or standard diameter can be expressed either in angular measure (usually in arcminutes) or as a true length (kpc).

In Tables 5-5 and 5-6, we list diameters and magnitudes for two samples of galaxies. Table 5-5 lists all galaxies known to be members of the Local

Table 5-5. The Local Group of Galaxies

System	Type	$-M_V$	$D(0)$(kpc)
Our Galaxy	Sb–Sbc	20(?)	23.8(?)
NGC 224 = M31	Sb	21.1	27.5
NGC 598 = M33	Sc	18.9	11.7
LMC	Irr I	18.5	9.3
SMC	Irr I	16.8	4.5
NGC 205	E5	16.4	2.8
NGC 221 = M32	E2	16.4	1.4
NGC 6822	Irr I	15.7	1.4
NGC 185	dE2	15.2	2.2
NGC 147	dE4	14.9	2.2
IC 1613	Irr I	14.8	2.6
Fornax	dE3 (dSph)	13.6	0.9
Sculptor	dE3 (dSph)	11.7	–
Leo I	dE3 (dSph)	11.0	0.7
Leo II	dE0 (dSph)	9.4	0.9
And I	dE0 (dSph)	11.0	0.5
And II	dE0 (dSph)	11	0.7
And III	dE2 (dSph)	11	0.9
LGS3	Irr I	9.0	0.5
Ursa Minor	dE6 (dSph)	8.8	0.6
Draco	dE3 (dSph)	8.6	0.7

SOURCE: Data published in (**A4**), (**D7**), (**T2**), (**V3**)

Table 5-6. Selected Luminous Galaxies

Galaxy	Type	D_0(kpc)	m_B	M_B	$(B - V)$	Remarks
NGC 3031 = M81	Sab	20.8	7.75	−19.95	0.93	
NGC 4486 = M87	E1	40.0	9.57	−21.8	0.94	Dominant in Virgo cluster
NGC 5457 = M101	Sc	29.8	8.2	−20.3	0.46	Pinwheel galaxy
NGC 4594 = M104	S0/a	12.7	9.27	−21.1	0.97	Sombrero galaxy
NGC 1275	S0p	61.7	12.35	−22.6	0.76	Brightest in Perseus cluster
NGC 3115	S0	15.2	10.05	−19.4	0.95	
NGC 3379	E1	16.0	10.2	−20.3	0.94	
NGC 4565	Sb	58.7	10.3	−21.1	0.64	
NGC 4889	E	90.5	12.45	−23.15	1.05	Brightest in Coma cluster
A 1153 + 23	cD	220.	15.3	−24.3		Brightest in Abell 1413

SOURCE: Data published in (**A4**), (**A9**), and (**D7**)

Group; this sample of galaxies is important because it goes to much fainter absolute magnitudes and much lower surface brightnesses than any other listing. Most bright galaxies are members of small groups like the Local Group (which contains only two or three moderately luminous systems), so the systems listed in Table 5-5 may be fairly typical of the galaxian content of the Universe. Notice how numerous the elliptical dwarfs are. In Table 5-6, we list by contrast some of the best-known giant galaxies. These systems are no more typical of the generality of galaxies than rajahs are of the people of India. They are, however, some of the most extensively studied galaxies, both because they are bright and large and therefore easily observed, and because they seem to be intrinsically more fascinating than lesser objects. Furthermore, the masses of these systems are so great that objects of their kind may contain most of the galaxian mass in the Universe, and they therefore enjoy an importance for our understanding of the Universe that is quite out of proportion to their numbers. The dichotomy in the properties of these two samples is closely analogous to that discussed in §4-2 in the properties of stars in the sample that determines the general luminosity function in the solar neighborhood, and those in the sample of apparently bright stars.

The Extragalactic Distance Scale

To pass from an apparent magnitude of whatever type to the corresponding absolute magnitude, one needs to know, for each galaxy observed: (1) the *distance*, (2) the interstellar *absorption* within our own Galaxy, (3) the *internal galaxian absorption*, and (4) the *K correction*, which describes the dimming of light from the galaxy caused by its recession from us. In general, considerable uncertainties attach to all these quantities, especially the first, which requires a knowledge of the *extragalactic distance scale* and, therefore, of the *Hubble constant*, H, which is used to relate the observed recession velocities v and distances r of typical galaxies through the equation $v = Hr$.

Here we shall not say anything about the effects of absorption beyond what was said earlier in this chapter and in Chapter 3, nor shall we give an account of K dimming, which is a comparatively small effect for nearby galaxies and is discussed in (**W1**). However, it is appropriate at this point to outline how the extragalactic distance scale is determined, for this determination poses one of the most important and difficult problems of extragalactic astronomy. At present, after fifty years of endeavor, the distance scale, and therefore H, is still uncertain by about a factor of two. Valuable reviews of the problem will be found in (**V4**), (**S2**), and (**D5**).

In outline, the procedure for determining the extragalactic distance scale is the following. The fundamental geometric considerations described in

Chapters 2 and 3 are employed to determine the characteristic absolute magnitudes for bright galactic objects. Probably the most important of these are the Cepheid variables, which are very luminous and whose distinctive light variations permit easy and certain identification even at enormous distances. In §3-8, we described the period-luminosity-color (PLC) relation obeyed by these stars, which is such that all Cepheids of a particular period and color have the same luminosity. It is worth noting here that one can establish the *existence* of such a relation by observing Cepheid variables in, for example, M31, without knowing the distance to M31, because all stars in M31 may be assumed to lie at the same distance. However, the *calibration* of the PLC relation has to be performed from observations of the very few (at present only fifteen) galactic Cepheid variables that are located in open clusters or associations whose main sequences can be fitted to the ZAMS of the Hyades. The determination of the distance to the Hyades was discussed in Chapter 3. If the PLC relation of the galactic Cepheids were universal, then observations of the Cepheid variables in nearby galaxies would enable us to determine the distances to galaxies nearer than about 4 Mpc. Unfortunately, the universality of the PLC relation is very doubtful, because we know that Cepheids in different galaxies have different mean colors. For example, the Cepheids in the Small Magellanic Cloud are much bluer than their galactic counterparts. Therefore, distances determined from observations of Cepheids in nearby galaxies, and the assumption that the PLC relation is universal, may be systematically in error. However, if one does assume a universal PLC relation, one may use Cepheids as what is called a *primary* distance indicator, that is, as an indicator whose calibration is directly based on fundamentally determined galactic distances, like that to the Hyades.

Other primary indicators are the novae (whose absolute magnitudes can be related to the rate of decline of their outbursts) and RR Lyrae variables (whose absolute magnitude calibration was discussed in §3-8). Novae can be seen to slightly greater distances than Cepheid variables, but they suffer from the disadvantage that they are harder to detect initially. RR Lyrae stars are fainter than Cepheid variables and cannot be seen much beyond the Magellanic Clouds, but, as was explained in Chapter 3, they occur in globular clusters and thus serve to calibrate clusters as secondary distance indicators (see later discussions).

When the distances to nearby galaxies have been established with the aid of primary distance indicators, the next step is to establish a connection between these galaxies and more distant ones. Recent determinations of the extragalactic distance scale exploit either (1) the fact that the peak rotation velocity of a spiral galaxy is correlated with its absolute magnitude, or (2) the correlation between the absolute magnitude and the DDO luminosity classifications of late-type spirals that we discussed in §5-1.

R. B. Tully and J. R. Fisher (**T5**) showed that more luminous spiral galaxies tend to have larger rotation velocities, and they suggested that one might be able to determine the distance to a spiral galaxy by comparing its apparent magnitude with the absolute magnitude one would infer from a knowledge of its rotation velocity. Unfortunately, optical magnitudes are not well suited to this work because they are subject to large and uncertain corrections for internal scattering and absorption by dust for exactly those fairly edge-on galaxies whose rotation velocities can be most securely determined. However, Aaronson et al. (**A1**) have shown that the infrared magnitudes m_H of edge-on galaxies are tightly correlated with rotation velocity and require negligible corrections for extinction by dust. In fact, observations of galaxies in the Virgo and Ursa Major clusters of galaxies (see Chapter 20) show that the scatter in the infrared absolute magnitudes of galaxies having the same peak rotation velocity is only ± 0.3 mag.

The DDO luminosity classes introduced by van den Bergh and discussed in §5-1 provide an alternative tool for determining the absolute magnitudes of distant galaxies, because they enable one to determine the absolute magnitudes of galaxies at different distances on purely morphological grounds. Indeed, it seems that the brightest Sc systems (those of luminosity class I) may be used as powerful standard candles which are visible to great distances. Unfortunately, the sample of galaxies reached by the Cepheid calibration is at present just too small to furnish a completely satisfactory calibration of the van den Bergh luminosity classes. The nearest Sc I galaxy, M101, is about a factor of two too distant to be directly calibrated from its Cepheids, and the available sample of galaxies in lower luminosity classes is too small to be satisfactorily calibrated. Several expedients have been proposed to overcome this difficulty. These include the study of the diameters of brightest H II regions, the luminosities of brightest blue or red variable stars, the luminosities of brightest nonvariable blue stars, and the luminosity functions of globular-cluster populations. However, with the possible exception of the globular clusters, none of these approaches is entirely satisfactory, because so far it has proved impossible to identify a class of standard luminous objects whose properties are constant from galaxy to galaxy. For example, a correlation is found to hold between the diameter of the brightest H II regions in a galaxy and the galaxy's absolute magnitude. Therefore H II regions cannot be used as simple secondary indicators in the same way that the Cepheids have been used as primary indicators (**S2**), (**D5**) until the effects of correlation have been removed by a preliminary normalizing procedure. Systematic errors in this procedure may incorrectly bias the final results.

Figure 5-35 shows the calibration of the DDO luminosity classes determined by Sandage and Tammann (**S2**) using a variety of the techniques just discussed. They find that the rms scatter in absolute magnitude at fixed luminosity class is about 0.5 mag. A random error of 0.5 mag in the absolute magnitude of a distance indicator leads to only a 25% uncertainty in the distance to the object. However, systematic effects, such as the likely non-universality of the PLC relation which was discussed earlier, lead to extra-

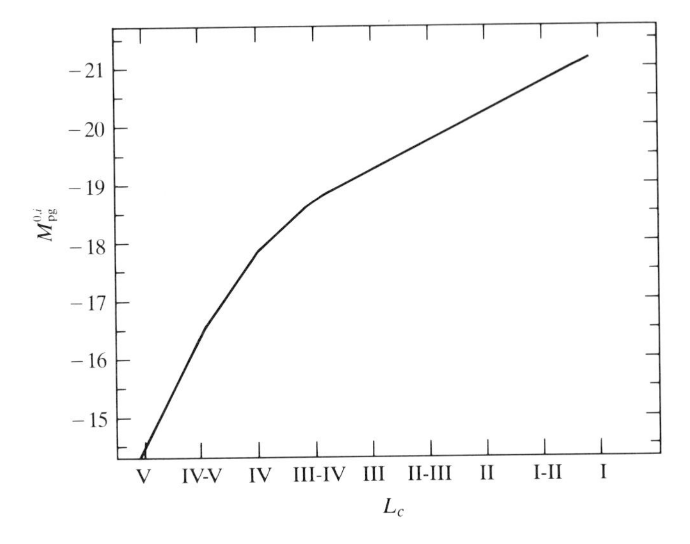

Figure 5-35. Late-type galaxies with well-developed spiral structure tend to be brighter. Here the absolute magnitudes (corrected for internal absorption) of 48 spirals are plotted against DDO luminosity class. The absolute magnitudes have been obtained by a variety of methods. [Adapted from (**S2**) by permission. Copyright © 1975 by the American Astronomical Society.]

galactic distances being uncertain by perhaps as much as a factor of two. When the Space Telescope becomes available, a considerably more accurate determination of the extragalactic distance scale will be possible, because it will allow us (1) to increase greatly the sample of galaxies whose distances can be directly determined from Cepheids, and (2) to identify globular clusters for a larger sample of galaxies and hence increase the utility of globular clusters as possible standard candles.

The Galaxy Luminosity Function

A quantity of great interest for extragalactic research is the function $\phi(L)$ that specifies the number

$$\delta N = \phi(L)\delta L\delta V \tag{5-29}$$

of galaxies that are located in volume of δV and have luminosities in the range

L to $L + \delta L$. The function ϕ is called the *luminosity function* of galaxies. A related quantity is the *luminosity density* $n(L)$ of a rich cluster of galaxies, which is defined such that $\delta N = n(L)\,\delta L$ is the number of galaxies in the cluster having luminosity near L. Notice that ϕ is a number of galaxies per unit luminosity within a certain specified volume.

The general luminosity function ϕ may be determined by a procedure which is analogous to that discussed in §4-2 for the determination of the stellar luminosity function $\Phi(M)$. J. E. Felten (**F5**) has recently reviewed the available determinations of ϕ, most of which are based on galaxy catalogs and on distances derived from redshifts. The common conclusion of these studies is that the number density of galaxies per unit luminosity falls steadily with increasing luminosity. The steepness of the decline increases sharply at a characterististic luminosity L^* that corresponds (for a Hubble constant, $H_0 = 50$ km s^{-1} Mpc^{-1}) to absolute magnitude $M_V = -21.5$.

The luminosity densities $n(L)$ of several rich clusters of galaxies have been derived by Abell (**A3**), Dressler (**D10**), Oemler (**O2**), and others. These workers find that, for most clusters, the shape of the density function $n(L)$ closely resembles the shape of the luminosity function $\phi(L)$. Thus $n(L)$ declines steadily with increasing L, and the steepness of the decline increases near the same characteristic luminosity L^* as the general luminosity function. Below L^*, the slope of the luminosity density is generally around -1.5 ($\delta N \sim L^{-1.5}\,\delta L$), and, above L^*, it becomes about -2.75.

Schechter (**S4**) has shown that both ϕ and n may be well fitted by the formula

$$\phi(L) = (N_0/L^*)(L/L^*)^{\alpha} \exp(-L/L^*) \tag{5-30}$$

where for $H_0 = 50$ km s^{-1} Mpc^{-1}

$$L^* = 3.4 \times 10^{10} L_0 \tag{5-31}$$

which corresponds to

$$M_V^* \approx -21.5 \qquad \text{and} \qquad M_B^* \approx -20.6 \tag{5-32}$$

The data set analyzed by Schechter yielded a best-fit value of $\alpha = -1.25$, but subsequent studies [(**A3**), (**G1**)] have shown that a better value of α is probably

$$\alpha = -1.5 \tag{5-33}$$

In the ($\log \phi$, $\log L$) plane, the function ϕ given by equation (5-30) declines with increasing L along a straight line of slope -1.5 when $L \ll L^*$, and then much more steeply, so that there should be very few galaxies with luminosities exceeding $3L^*$. Actually, galaxies (often of type cD) that have $L \approx 10L^*$ are observed, so equation (5-30) does not accurately represent the luminosity function at very large luminosities.

According to equation (5-30), the total number of galaxies in some volume whose luminosities exceed L is given by

$$N(>L) = \int_L^\infty \phi(L')\,dL' = N_0 \int_{x=L/L^*}^\infty x^\alpha e^{-x}\,dx$$
$$= N_0\Gamma(1 + \alpha, L/L^*) \tag{5-34}$$

In equation (5-34), Γ is the incomplete gamma function. If $\alpha \leq -1$, the total number of galaxies predicted by equation (5-34) diverges as $L \to 0$. However, the total luminosity of these galaxies is finite provided $\alpha > -2$;

$$L_{\text{tot}} = \int_0^\infty \phi(L) L\,dL$$
$$= N_0 L^* \Gamma(2 + \alpha)$$
$$\approx 1.77 N_0 L^* \text{ for } \alpha = -1.5 \tag{5-35}$$

In Table 5-7 we tabulate, from equation (5-30) with $\alpha = -1.5$, $N(>L)$, the relative number of galaxies with luminosities exceeding $x = L/L^*$, and $\mathscr{L}$ the fraction of the total light they produce.

Table 5-7. Numbers of Galaxies and the Integrated Luminosity Predicted by Equation (5-30).

L/L^*	3.0	2.5	2.0	1.5	1.0	0.5	0.25	0.1	0.05	0.01	10^{-4}
$M - M^*$	−1.2	−1.0	−0.75	−0.44	0	0.75	1.5	2.5	3.25	5	10
N	0.038	0.078	0.17	0.39	1	3.3	7.9	19	33	93	1100
$\mathscr{L}$	0.014	0.025	0.046	0.083	0.16	0.32	0.48	0.65	0.75	0.89	0.989

A magnitude-limited sample of galaxies, which are distributed homogeneously in space and are distributed in luminosity according to equation (5-30), is dominated by galaxies that have L near L^*. The volume $V(L)$ through which a galaxy of luminosity L has apparent magnitude above a fixed threshold magnitude limit is proportional to $L^{3/2}$. Therefore, the number δN of galaxies in a magnitude-limited sample that have luminosity between L and $L + \delta L$ is

$$\delta N = N_0 (L/L_*)^{\alpha + 3/2} \exp(-L/L^*)\,\delta L/L^* \tag{5-36a}$$

If we adopt $\alpha = -1.5$, we have

$$\delta N = N_0 \exp(-L/L^*)\,\delta L/L^* \qquad (5\text{-}36b)$$

and thus that half of the galaxies in the sample will have luminosities greater than $L_{1/2}$, where $\exp(-L_{1/2}/L_*) = \frac{1}{2}$, that is,

$$L_{1/2} \approx 0.7L_* \qquad (5\text{-}37)$$

By contrast, less than 5% of the sample will have luminosities exceeding $3L^*$.

Table 5-5 shows that all the galaxies of the Local Group are fainter than L^*, although M31 has luminosity very nearly equal to $L_{1/2}$. The majority of galaxies in the Local Group are, in fact, so faint that M31 accounts for 60% of the total luminosity, and M31 and our Galaxy account for more than 80% of the light emitted by the twenty-one galaxies listed in Table 5-5.

We have already noted that the luminosity distributions of rich clusters of galaxies resemble the general galaxian luminosity function. This similarity is important, because luminosity distributions of clusters are more easily determined than the general luminosity function because all the galaxies of a cluster may be assumed to be at the same distance from the Sun. However, there is clear evidence that there are significant differences between the luminosity functions of cluster and noncluster galaxies, particularly at the bright end. Thus van den Bergh and McClure (**V6**) have shown that the distribution of luminosities of elliptical galaxies extends to brighter absolute magnitudes than does that of disk galaxies. On the other hand, bright ellipticals are known to occur chiefly in rich clusters. From these facts, it follows that the luminosity function of cluster galaxies must extend to brighter absolute magnitudes than the general luminosity function. What is worse, it is uncertain whether the concept of a cluster luminosity function is appropriate; that is, one may not be able to account for the contents of different clusters of galaxies by supposing that these have been picked at random from a common pool of galaxies. It is possible that the galaxies of some clusters are systematically brighter than those of other clusters (**D10**). Furthermore, if the brightest galaxy in a particular cluster is exceptionally luminous, there is some evidence that the next brightest galaxy will be rather fainter than average, as if the luminosity of the brightest galaxy has been enhanced at the expense of the second brightest galaxy (**T3**). This behavior is incompatible with the existence of a common luminosity function for cluster galaxies.

Notwithstanding the uncertainties just mentioned, the concept of a universal galaxy luminosity function plays an important role in extragalactic studies. For example, the mean mass density in the Universe (a crucial quantity for cosmological theories) may be obtained by integrating $L\phi(L)$ over all values of L and multiplying by an estimate of the amount of matter required to produce a given luminosity (the mean mass-to-light ratio of matter in the Universe).

5-4. SPECTROPHOTOMETRIC PROPERTIES OF GALAXIAN LIGHT

Having studied the forms of galaxies and the quantitative distribution of light over the surfaces of galaxies, one might ask what variations there are in the *quality* (that is, spectrophotometric properties) of the emitted radiation. As we shall soon see, the quality of the light emitted by a galaxy gives important clues to the stellar composition of the system. Here we shall concentrate on the correlations that exist between the quality of the emitted light and gross morphological properties such as Hubble type and absolute magnitude.

There are basically four types of information available concerning the spectral distribution of light emitted by galaxies: (a) integrated broadband colors, for example, $(U - B)$, $(B - V)$, $(I - R)$, $(H - J)$, and so on; (b) line- and band-strength indices; (c) integrated MK spectral types; and (d) equivalent widths of nebular emission lines. Integrated colors (for definitions, see Chapter 2) constitute the simplest and most readily obtainable measures of the spectral quality of a galaxy's light. Near the nucleus of a galaxy, they are easy to measure by either photoelectric or photographic observations through filters having different passbands. For outlying regions of low surface brightness, however, even broadband colors are extremely difficult to obtain. This difficulty arises because colors are derived by differencing two surface brightnesses, each of which is, for a low-surface-brightness region, the small difference between the observed brightness and an estimated sky background. Thus small photometric errors can seriously affect estimates of galaxian colors in low-brightness regions.

A wide variety of line- and band-strength indices have been defined by choosing different sets of relatively narrow-band filters. These filters typically have half widths ≈ 100 Å, and they are strategically placed in wavelength so that one of a pair of filters passes light at the frequency of a strong spectral feature, while the other measures the brightness in an adjacent continuum band. The difference in the brightness observed in the two filters (expressed in magnitudes) provides a measure of the strength of the spectral feature in question. One of the first systems of this type to be introduced was the DDO six-color system of McClure and van den Bergh (**M2**). A twelve-color system has been defined by Wood (**W4**), (**W5**). Features that can be monitored in this way include CN absorption near 4200 Å, and magnesium absorption near 5200 Å.

Morgan and Mayall (**M9**) used low-dispersion spectra to demonstrate the correlation, mentioned in §5-2, between the central concentration of a galaxy's light and the spectral type of its emission. Low-dispersion spectra can also be used to derive line-strength indices.

The nuclei of early-type galaxies, and broader areas in the disks of late-type systems, have spectra showing emission lines which originate in very tenuous hot gas. These lines tend to be narrower than the absorption lines characteristic of stellar spectra. Thus higher dispersion spectra (resolution $\lesssim$ 4 Å) capable of resolving fine-structure doublets are valuable and not too difficult to obtain with modern equipment.

Colors and Line-Strength Indices

Figure 5-36 shows that the broadband colors of early-type galaxies are redder than those of late-type systems. In fact, after correction for galactic and internal extinction, the mean $(U - B)$ and $(B - V)$ colors of galaxies of the same Hubble type are found to be tightly correlated with Hubble stage; the dispersion of observed colors around the mean values of Figure 5-36 are only of the order of the combined errors in the color measure-

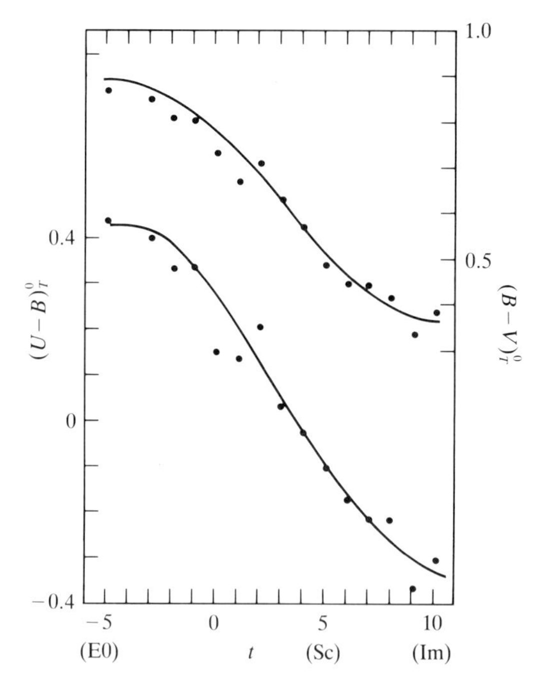

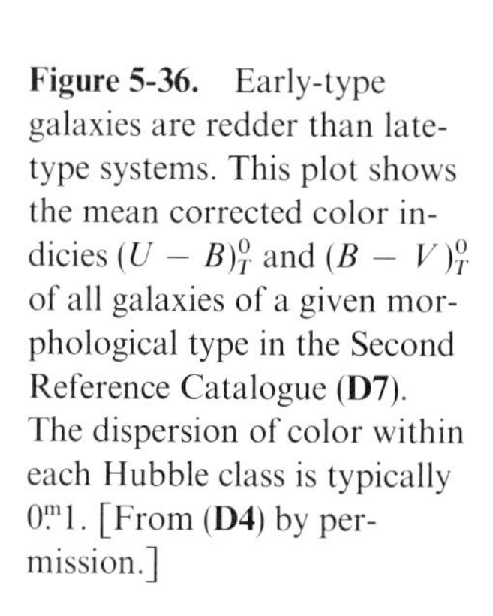

Figure 5-36. Early-type galaxies are redder than late-type systems. This plot shows the mean corrected color indicies $(U - B)_T^0$ and $(B - V)_T^0$ of all galaxies of a given morphological type in the Second Reference Catalogue (**D7**). The dispersion of color within each Hubble class is typically $0^{\mathrm{m}}.1$. [From (**D4**) by permission.]

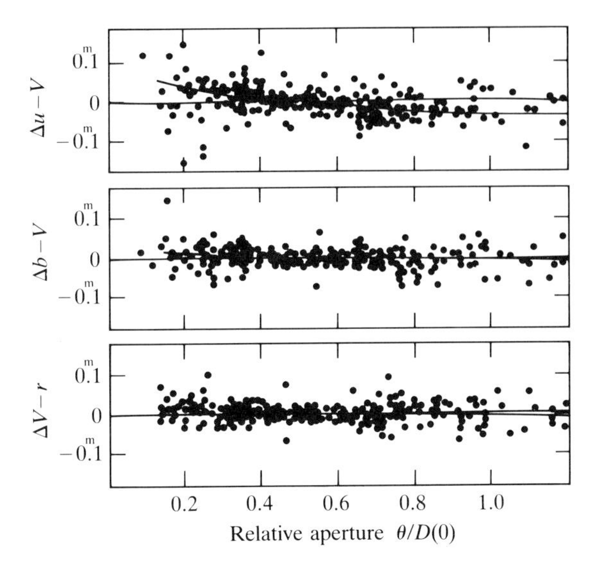

Figure 5-37. Color aperture effect for elliptical and S0 galaxies. Each point represents a photoelectric color measured through an aperture of angular size θ. $D(0)$ is the angular diameter of the $V = 25^m_.0$ isophote. The sets of points for each galaxy have been shifted vertically until the least-squares, linearly interpolated color at $\theta/D(0) = 0.5$ is 0. [From **(S3)** by permission. Copyright © 1978 by the American Astronomical Society.]

ments and in the Hubble classification. This correlation arises because the later a galaxy's Hubble type, the richer it is in young blue stars. However, Visvanathan and Sandage **(V8)**, **(S3)** have shown that E and S0 galaxies of similar absolute magnitude have similar colors. Thus the dependence of color on Hubble type is confined to galaxies later than type S0.

Figure 5-37 shows that galaxies tend to be redder near their centers than they are slightly further out; the larger the photometer aperture, the bluer the measured light. This *color-aperture effect* has been discussed by many workers over the years [see, for example, **(D3)**, **(D4)**]. Recently, it has become possible to extend our knowledge of the colors of galaxies from these central regions out to galactocentric radii up to 10 kpc. Figure 5-38 shows how the color of the S0 galaxy NGC 3115 varies along its major and minor axes. As we have already explained, these data are sensitive to the assumed sky level; the effect of a 1% change in the sky background is shown in the figure. Because colors are very hard to obtain far from the centers of galaxies, it is

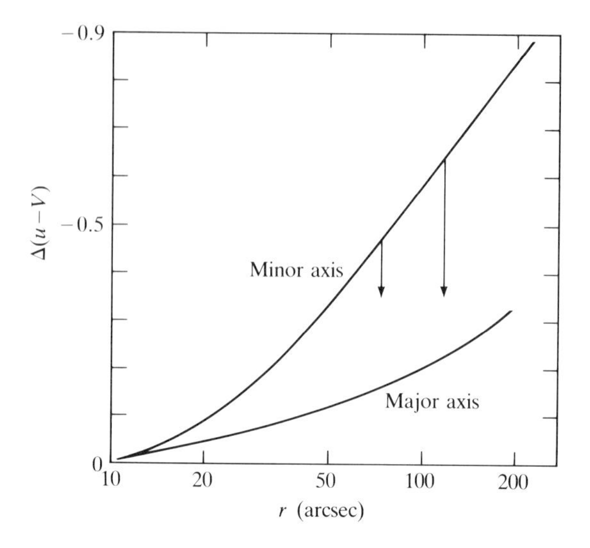

Figure 5-38. Major-axis and minor-axis color profiles for the S0 galaxy NGC 3115. The effect of a 1% error in the sky brightness is indicated by the arrows. [From data published in (**S13**).]

not known how general is the continuation out into galaxian halos of the color gradients which usually exist near the nuclei.

The colors of ellipticals and S0 galaxies are correlated with their absolute magnitudes. Figure 5-39 shows plots of three color indices, $(u - V)$, $(b - V)$, and $(V - r)$, versus apparent magnitude V_{26} for galaxies in the Virgo cluster. There is a clear correlation of color and brightness in the sense that the brighter galaxies are redder; this is known as the *color-magnitude effect*. The effect is particularly marked for $(u - V)$, the bluest of the color indices, for reasons that are made apparent in Figure 5-40, which shows the spectral energy distributions of five of the Virgo cluster galaxies plotted in Figure 5-39. The curves of Figure 5-40, which are labeled by the apparent magnitudes of the observed galaxies, show that the reddening of bright galaxies occurs mainly in the violet region shortward of 4250 Å, where stellar spectra are richest in metal absorption lines and the spectral energy distributions are rather steep. In fact, the u, b, V, r color system used for the observations shown in Figure 5-39 was specially developed to enhance this effect. The u filter has a bandwidth of 250 Å centered on 3550 Å, the b and r filters have widths of 300 Å centered on 4522 Å and 6738 Å, and the V filter is the standard Johnson and Morgan (**J2**) filter which is 800 Å wide and centered on 5500 Å. [Conversions between these colors and the standard *UBVR* colors are given in (**S3**).] Visvanathan and Sandage have demonstrated that the same color-magnitude effect is shown by clusters other than Virgo, and

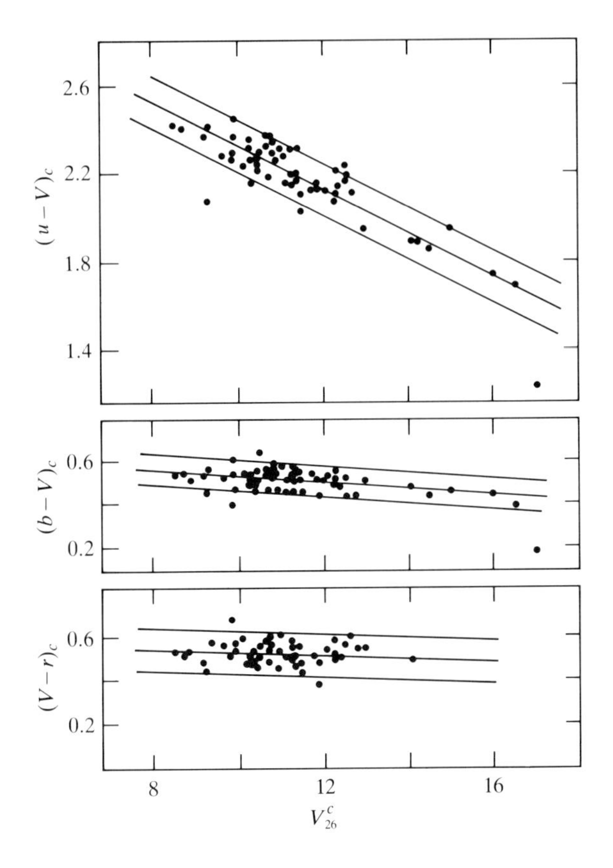

Figure 5-39. The color-magnitude effect for Virgo cluster galaxies. The plots show the dependence of mean colors $(u - V)$ and $(b - V)$ corrected for color aperture effect on apparent visual magnitude V_{26} for galaxies which are all at approximately the same distance. The bands of the u, b, and V colors are indicated in Figure 5-40. [Adapted from (**V8**), by permission. Copyright © 1977 by the American Astronomical Society.]

even by field galaxies. They also find that the color-magnitude effects shown by elliptical and lenticular galaxies are identical. Thus there appears to be some factor that governs the development of gas-poor galaxies, which is independent of environment or shape, and which causes the brighter galaxies to become redder. We shall see later that the color-magnitude relation probably arises from a dependence of mean metallicity on galaxian magnitude. In particular, Visvanathan and Sandage were able to rule out reddening by dust as the agent responsible for the correlation, because S0 galaxies of all

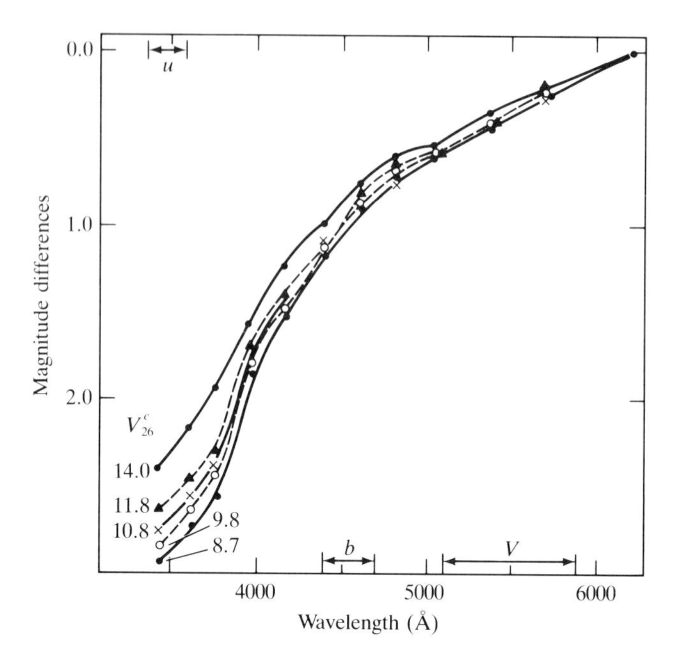

Figure 5-40. Spectral energy distributions of five Virgo cluster galaxies. The curves show the brightnesses in a series of bands relative to the brightness around λ6213 for galaxies of apparent magnitude $V_{26} = 8^m.7, 9^m.8, 10^m.8, 11^m.8$, and $14^m.0$. The color-magnitude effect is strongest in the violet region of the spectrum because the curves are steepest there. The bands *u*, *b*, and *V* used for the observations shown in Figure 5-37 and 5-39 are also indicated. [Adapted from (**V8**) by permission. Copyright © 1977 by the American Astronomical Society.]

orientations show the same color-magnitude relation (which indicates that internal absorption is negligible). They also showed that the $(u - V)$ and $(b - V)$ magnitude changes are not related to one another in the way one would expect if the changes were caused by normal dust reddening.

Intermediate-band colors of elliptical galaxies show color-magnitude relations that are important for the link they provide between studies of the integrated light of elliptical galaxies and the investigations of the metallicity of globular clusters discussed in §3-6. For example, Figure 5-41 is a plot for thirty-three elliptical galaxies of a reddening-free, intermediate-band color index, denoted $(\mathrm{CN} + \mathrm{Mg})_0$ because it is made up of indices that monitor CN and Mg absorption, against absolute magnitude derived from an assumed Hubble constant of 50 km s^{-1} Mpc^{-1} and the mean velocities of the groups to which the galaxies belong. If one ignores the three labeled points in this diagram, there is clearly a band of points running from the low-luminosity systems at bottom left all the way up to the supergiants at top

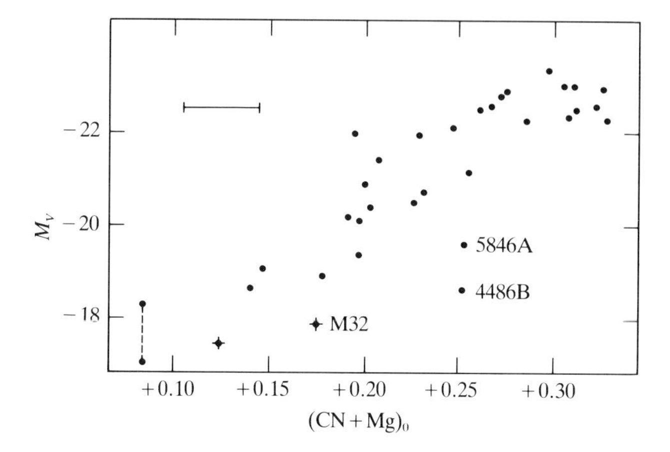

Figure 5-41. Bright galaxies tend to have spectra with stronger absorption lines. The combination of line strength indices denoted $(CN + Mg)_0$ is almost reddening free, and, therefore, it furnishes an accurate estimate of the intrinsic line strength of galaxian spectra. Apparent magnitudes have been converted to absolute magnitudes by assuming a Hubble constant of 50 km s^{-1} Mpc^{-1}. [Adapted from (**F3**) by permission. Copyright © 1973 by the American Astronomical Society.]

right. The three labeled points belong to galaxies which we believe, on several lines of evidence, to have been tidally stripped, so that their present absolute magnitudes are appreciably fainter than they were originally (**F2**). Thus it is probable that these objects have migrated downward in the figure from the band occupied by the other galaxies plotted, which therefore represents the natural color-magnitude relation for this color index.

In §3-6, we saw that the colors of stars are correlated with their metallicity. In particular, several different lines of evidence indicate that the color indices of globular clusters (which show many similarities with elliptical galaxies) are related to the metallicity of the cluster stars. Therefore it is natural to seek an understanding of the color-magnitude and color-aperture effects shown by elliptical galaxies in terms of variations in the metallicity of the stars in these systems. Figure 5-42 suggests that the same processes are responsible for causing color variations between globular clusters as those that cause variations within and between elliptical galaxies. The figure is a typical two-color plot, that of the intermediate-band index G_0, which monitors the strength of the G band in the integrated spectrum, against the Balmer discontinuity index $(35 - 55)_0 = M(\lambda 3450) - M(\lambda 5545)$, and one sees that globular clusters (marked with triangles) fall on the same sequence as elliptical galaxies (dots).

In the light of this figure, it seems reasonable to seek to establish the metallicities of elliptical galaxies by extrapolating the relationships discussed

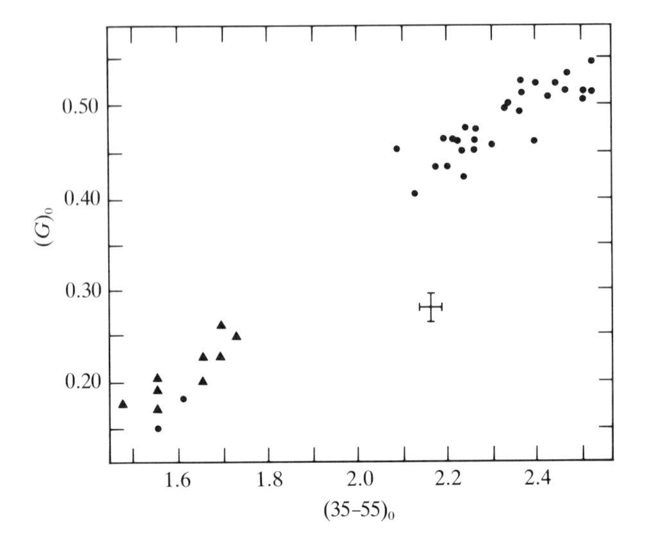

Figure 5-42. A two-color plot for elliptical galaxies (dots) and globular clusters (triangles). The color (35–55) monitors the Balmer discontinuity, while G monitors the G band. This diagram suggests that studies of globular clusters (which are nearby) may help us understand the colors of distant ellipticals. [Adapted from (**F3**) by permission. Copyright © 1973 by the American Astronomical Society.]

in §3-6 between line strength color indices and metal abundance [Fe/H] to the generally rather stronger line strengths characteristic of elliptical galaxies. Figure 5-43 shows that a straight-line extrapolation of the relationship between Faber's magnesium line-strength index Mg_0 and [Fe/H] that is established by the metal-poor globular clusters, would attribute metal abundances of up to one hundred times the solar value to giant elliptical galaxies (which typically have $M_B \approx -22$). Actually, one sees in Figure 5-43 that the slope of the [Fe/H] versus Mg_0 relationship may begin to steepen near $[\text{Fe/H}] = -0.5$. An alternative to extrapolation of the empirical relationship between Mg_0 and [Fe/H] is to synthesize artificially the integrated spectrum of an old, metal-rich stellar population (**F1**), (**M10**). In this technique, one uses the results of the theory of stellar evolution to determine the contributions of the various stellar types to the total light, and the theory of stellar atmospheres to relate the spectra of the stars involved to an assumed metal abundance. Unfortunately, both of these theories, especially the first, are subject to considerable uncertainty, with the result that the metal abundances one derives from the integrated spectra may be badly in error. However, such investigations confirm that the slope of the Mg_0 versus [Fe/H] relationship is very steep for $[\text{Fe/H}] > -0.5$. The dashed line in

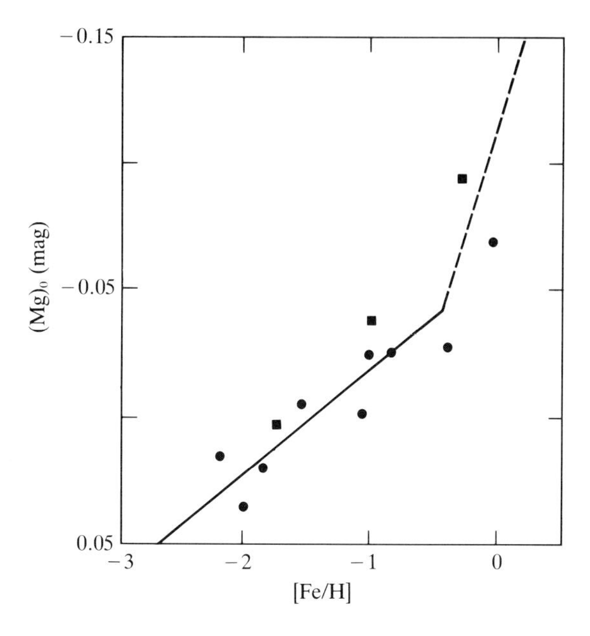

Figure 5-43. The relationship between Faber's magnesium line strength index $(Mg)_0$ and [Fe/H]. The dots represent globular clusters whose metallicities have been determined from spectra of individual stars. The solid line is a fit to the data for all clusters with $[Fe/H] < -0.5$. The dashed line represents a theoretical relationship based on the synthesis calculations of Mould (**M10**). [From (**B5**) by permission. Copyright © 1979 by the American Astronomical Society.]

Figure 5-43 has the slope determined by Mould (**M10**). If one accepts the validity of the theoretical line in Figure 5-43, one concludes that [Fe/H] varies in elliptical galaxies between 0.2 and -0.2 for absolute magnitude in the range $M_V = -22$ to $M_V = -17$. The analogous interpretation of the colour-aperture effect discussed earlier (which is also associated with line-strength variations) is that the metallicity of the stellar material in giant ellipticals and lenticulars varies by a factor of about four within 10 kpc of the centers (**F4**).

It must be emphasized, however, that the interpretation of the color-aperture and color-magnitude effects in terms of variations in [Fe/H] is not completely secure. There are two principal reasons why one should regard the interpretation just described with some caution. (1) Globular clusters do not form a simple one-parameter family. In particular, one of the most massive and metal-rich clusters, ω Cen, has a perplexingly abnormal CM diagram and is rich in stars that show strong abundance anomalies, with the result that one probably cannot characterize this cluster by a single metal-richness parameter like [Fe/H]. If the as yet unknown CM diagrams of elliptical galaxies are more like that of the massive cluster ω Cen than those of more normal clusters, then an attempt to deduce their metallicities by calibrating color indices against [Fe/H] for normal clusters will most likely lead to misleading results. (2) The line enhancements seen in elliptical galaxies are strikingly similar to the peculiar enhancements seen in certain galactic stars, called *super-metal-rich* (SMR) stars (**O1**). It is possible that the extra strong absorption lines of SMR stars are not due to a generally enhanced metal abundance but instead to enhancement of a specific element or elements, probably nitrogen (**P3**). Therefore, the color-aperture effect could conceivably be a measure of a rising nitrogen abundance toward galaxian centers rather than of a gradient in [Fe/H].

In defense of the conventional interpretation of color variations in early-type galaxies, two points should be made. The great majority of globular clusters *can* be characterized by a single metallicity parameter. Thus standard relative abundances of carbon to iron, or magnesium to carbon, are common to most clusters, and the great majority of stars within any one cluster have the same [Fe/H] (**C4**). Second, the abundances derived for giant ellipticals are very similar to those one would have predicted on the basis of the observed metallicity and estimated absolute magnitude of the galactic bulge component, by treating the latter as an isolated spheroidal system. Also, the color gradients derived for external galaxies would be natural counterparts of the similar gradient definitely detected in the galactic bulge component (see Chapter 4).

Integrated Spectra

In Chapter 4, we saw that most of the light emitted by stars in the solar neighborhood comes from a comparatively small number of very bright stars, mostly A dwarfs and G and K giants. This suggests that the spectrum of the integrated light of the solar neighborhood should have characteristics of an A type spectrum in the violet region, where the light of the A dwarfs may be expected to dominate, together with K type characteristics at longer wavelengths. It is interesting to ask whether effects of this type are observable in the integrated spectra of external galaxies.

W. W. Morgan and N. U. Mayall (**M9**) undertook just such an investigation of low-to-medium dispersion spectra of the central regions of nearby galaxies. As was mentioned in §5-2, they found that the nature of a galaxy's spectrum is closely correlated with the degree of central concentration of the galaxy's light. The least centrally condensed galaxies—for example NGC 672 (SBc), NGC 3889 (Sc), and NGC 4449 (Irr)—have spectral types in the region $\lambda\lambda$3850–4100 of class A, causing them to be designated by Morgan and Mayall as *A-systems*. The spectral type of these systems is around F8 at λ4340, indicating that they are composite—that is, that important contributions to their light come from more than one class of star. Systems of slightly higher central concentration tend to have spectra of type F0 to F2 in the region λλ3850–4100, while, at λ4340, their types are near F8; Morgan and Mayall call these *AF-systems*. Galaxies with spectra of this type include NGC 925, NGC 3556, and NGC 4244, all of Hubble type Sc. The well-known galaxies M33 (Sc) and M51 (Sc) have spectra of slightly later type than those just described and are designated *F-systems*. Galaxies with still later violet-region spectral types tend to have central concentrations of light that contribute a larger fraction of the galaxy's total luminosity. Morgan and Mayall distinguish two classes, the *FG-systems*, having only moderately late spectra, and the *K-systems*, a populous group of galaxies of a wide range of Hubble types, whose spectra in the blue-violet region are dominated by K type giant stars. Examples of K-systems include the Sb galaxies M31 and M81, the S0 galaxy NGC 3115, and the El giant, M87. The major part of the luminosity of a K-system originates in a brilliant, amorphous, central region.

The variations of spectral type with wavelength region observed by Morgan and Mayall in their pioneering work suggest that it might be possible to determine the detailed stellar content of the nuclear regions of nearby galaxies by discovering what mix of stars would reproduce the observed spectral features over the entire wavelength region studied. Morgan made a first attempt at such an analysis (**M5**). This work was followed by H. Spinrad and B. J. Taylor (**S12**), S. M. Faber (**F1**), T. B. Williams (**W2**), C. Pritchet (**P4**), and others. As we have indicated in our discussions of the integrated spectra of globular clusters (§3-6) and of the galactic nucleus (§4-4), the physical interpretation of integrated spectra is a very difficult task. Indeed, these workers have found that a wide variety of different mixes of stars can generally all produce composite spectra that fit the observations to within the errors. If, like Morgan, one seeks only a broad characterization of the nature of the systems, as being, for example, young star rich or poor, then one can obtain reasonably unambiguous results from low-dispersion spectra. If, however, one attempts to obtain a more detailed understanding of the stellar composition of the system, then *even excellent high-dispersion spectra do not suffice to determine a unique model* without the prior imposition of strong constraints on the solution, for example, the specification of the kind of H-R diagram which one requires. Unfortunately, there is no universally agreed way of constraining H-R diagrams, and therefore no accord as to the true stellar population of, for example, the nucleus of M31.

Spectra of Emission Regions

The spectra of H II regions, which are widely distributed over the surfaces of late-type galaxies, contain emission lines that are bright and narrow and hence easily measured. The theoretical interpretation of these spectra is, however, in some ways more difficult than the interpretation of the spectra of stars, because the conditions in the hot tenuous gas whence the emission lines arise are much more chaotic than those in the line-forming regions of stellar atmospheres. The spectra of H II regions do, nevertheless, show remarkable regularities which indicate that simple models of their formation may have much to teach us about the structure of late-type galaxies.

In 1942, L. H. Aller (**A6**) noted that the forbidden oxygen lines [O III] emitted by H II regions in M33 tend to be strong in the nebulosities more distant than 20′ from the nucleus, and they tend to be weak or absent in regions close to the nucleus. In 1971, L. Searle (**S7**) discussed the strengths of H II region emission lines in six Sc galaxies, including M33, M51, and M101. He showed that different lines of sight through the same H II region produce similar spectra, enabling one to characterize an entire H II region with a single spectral type, and H II regions situated at similar distances from the nuclei tend to have similar spectra, but the spectra of regions located at different galactocentric distances are generally quite unlike one another. In other words, the spectrum coming from an H II region depends systematically on the region's galactocentric distance and on little else. The sense of this dependence is that, in the spectra of H II regions which are located far from the center, the lines emitted by doubly ionized oxygen O^{++} are strong, whereas the spectra of regions at smaller galactocentric distances have relatively strong lines from singly ionized nitrogen N^+. Searle defines the *excitation* of an H II region to be the logarithm of the ratio of the sum of the intensities of a prominent pair of forbidden lines of doubly ionized oxygen to the intensity of the second Balmer line of hydrogen, that is, as log ([O III] $\lambda\lambda$4959, 5007/Hβ). Searle found that the excitations of the H II regions in his Sc galaxies increase outward. He also found that, to good accuracy, the spectra of these regions form a one-parameter family characterized by their excitation. Figures 5-44 and 5-45 from Searle's paper illustrate these results. Figure 5-44 furnishes evidence that H II regions do, in fact, form a one-parameter sequence; the ratio of the intensities of the forbidden doublet of singly ionized oxygen [O II] $\lambda\lambda$3726, 3729 to that of singly ionized nitrogen [N II] $\lambda\lambda$5648, 6584 rises in step with the excitation, log ([O III]/Hβ), allowing for scatter which is no larger than expected from observational errors. Figure 5-45 shows how the excitation of the H II regions of the giant Sc galaxy M101 varies with radius from -1 at the center to $+1$ at one Holmberg radius R_0. In other galaxies, Searle found a similar variation of excitation with radius, although the variation was over a smaller absolute range. Other surveys have confirmed the existence of excitation and [O II]/[N II] gradients across the faces of many late-type spirals [for example, (**S11**)]. Some galaxies, particularly those like the Magellanic Clouds with prominent bars, show small or undetectable gradients (**P1**).

The variation of the spectral type of H II regions with radius just described is conventionally interpreted in terms of a radial gradient of the metal

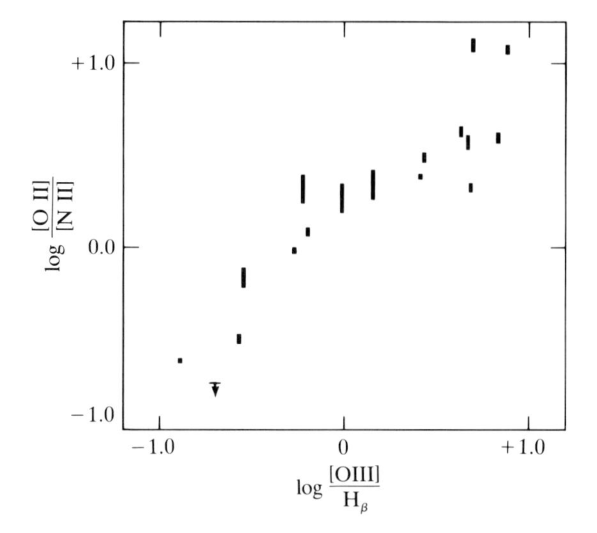

Figure 5-44. Nitrogen lines tend to be weak in high-excitation nebulae. The points are derived from the spectra of emission nebulae in six late-type galaxies. [From (**S7**) by permission. Copyright © 1971 by the University of Chicago.]

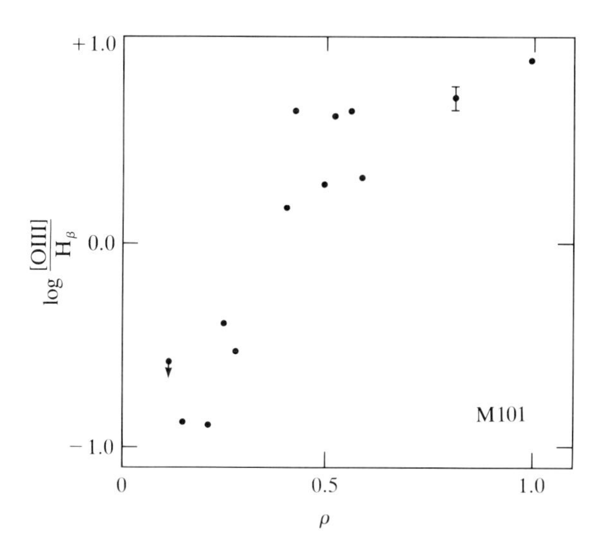

Figure 5-45. The excitation is an increasing function of radius in the Sc galaxy M101. The quantity ρ is a normalized radius $\rho = r/r_{25}$, r_{25} being the radius at which the mean surface brightness falls to $\mu_B = 25\overset{m}{.}0 \text{ arcsec}^{-2}$. [From (**S7**) by permission. Copyright © 1971 by the University of Chicago.]

content of the material in late-type galaxies. That is, the excitation is believed to be low in the inner H II regions because the overall metallicity is high there. A full account of the models of H II regions on which this conclusion rests is beyond the scope of this book. The interested reader should consult D. E. Osterbrock's book (**O5**). In very rough outline, the reasoning is as follows.

The spectra of emission regions depend sensitively on the metallicity of their material, because even very small numbers of heavy element ions (relative to hydrogen) can radically increase the cooling rate of the interstellar gas when heated to a given temperature by photoionization of H and He atoms. The actual temperature of the nebular gas (and with it the strengths of the various lines emitted) is determined by the requirement that the rate of heating by photoionization be balanced by radiative cooling by line emission. Thus metal-rich nebulae will, in general, be cooler (because they have higher densities of "refrigerant," heavy-metal atoms) than metal-poor ones. Furthermore, cooler gas will emit, for example, less optical [O III] radiation, because relatively fast-moving electrons are required to excite O^{++} ions into the level from which optical [O III] emission occurs. Thus we have the paradoxical situation that nebulae with large oxygen abundances have spectra weak in the prominent oxygen lines [O III] $\lambda\lambda$4959, 5007.

One may exploit the dependence of line strengths on metallicity just described to determine metallicities in two rather different ways. Much the better technique is to measure the relative intensities of a temperature-sensitive pair of lines like [S II] $\lambda\lambda$4070, 6720 or [O III] $\lambda\lambda$4363, 5007. From the ratio of the strengths of the members of such pairs, one can accurately estimate the electron temperature in the cloud and, from this, deduce reliable values for the abundances of ions whose line intensities have been observed. This last step involves equating the number of collisional excitations of each excited state to the number of radiative decays (which is, in principle, known).

A second and less satisfactory technique that has been widely used to obtain abundances from emission spectra involves modeling the energy and ionization balance of the nebula. The parameters of the model are adjusted until its predicted spectrum agrees with the observed one. This technique is the more widely used because observations of both members of a temperature-sensitive line pair are often unavailable. The results obtained by this second method are, however, substantially less certain than those derived on the basis of a reliable electron temperature measurement, because the models are sensitive to such poorly known parameters as the spectral type of the exciting star (or stars) and the dust content of the nebular gas.

In his original paper (S7), Searle used the second technique to conclude that his observations of normal Sc galaxies indicated that a modest increase in the oxygen-to-hydrogen abundance ratio toward the galaxian centers is accompanied by a substantial rise in the nitrogen-to-oxygen abundance-ratio; that is, he concluded that the nitrogen abundance relative to hydrogen rises more steeply toward galaxian centers than does the oxygen abundance. Subsequent analyses, using either more refined model-fitting techniques or direct electron temperature measurements, have confirmed Searle's inference that the oxygen-to-hydrogen ratio (O/H) rises toward the centers of galaxies, but they have not confirmed that the (N/O) ratio rises in step **(S9)**, **(S11)**, **(P2)**. Typically, the ratio (O/H) is found to increase by a factor of ten from a galactocentric radius equal to half the standard diameter D_0 (defined in §5-3) to the center. The nitrogen abundance (N/H) may rise slightly faster than this toward the center, but any increase in (N/O) is at most marginal and certainly not as large as Searle originally concluded. As we saw in Chapter 4, observations of H II regions in our Galaxy yield results compatible with this picture of the abundance gradients in the disks of normal spirals. The results are summarized in Figure 5-46, which also shows that the (O/H) ratio in three normal spirals declines nearly exponentially with increasing radius, (O/H) $\backsim \exp(-r/a)$; that the mean oxygen abundance in these three galaxies is similar to that in the solar neighborhood; and that the ratio (N/O), while fairly constant within any one galaxy, varies quite considerably from galaxy to galaxy.

The results just summarized are not particularly easy to understand from a theoretical point of view. First, simple models of metal enrichment of the disk population by supernovae cannot account for oxygen abundance distributions which, like the exponential distribution, have slopes that steepen considerably toward the centers **(S10)**. To account for the (O/H) distributions being of this form, one has either to adopt a more complex enrichment model or to imagine that most of the heavy elements now in disk material were actually produced in the halo component **(O6)**. Second, the conventional theory of the origin of the chemical elements predicts that galaxies should show strong (N/O) gradients, because nitrogen is considered to be a *secondary element*, that is—one that requires for its production the presence, in the synthesizing stars, of primary elements (like oxygen) that are directly synthesized from hydrogen and helium **(T1)**. Thus, if nitrogen were a secondary element and were primarily synthesized in galaxian disks, then an (N/O) gradient should be set up parallel to the (O/H) gradient actually observed, because, while the first generation of stars would have produced no nitrogen, each successive generation would have produced more and more nitrogen as the oxygen abundance in the newly formed stars increased. Therefore, the failure to detect appreciable (N/O) gradients in normal spirals either indicates that conventional nucleosynthesis theory is wrong or points again to a halo origin for most of the heavy elements now in disk material.

Galaxies with bars seem to show only small abundance gradients. Thus one finds that the giant SBc galaxy NGC 1365 has a very much shallower gradient in (O/H) than that typical of normal Sc galaxies **(P2)**, that the Large Magellanic Cloud has at most a slight gradient, and that the Small Magellanic Cloud has no gradient at all **(P1)**. It is likely that the strong, radial streaming

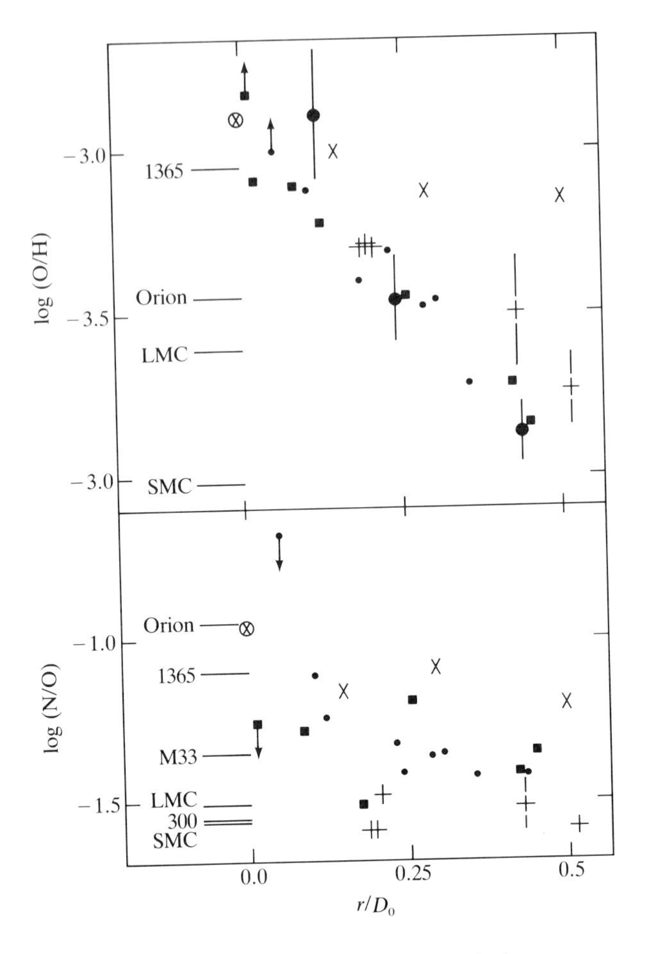

Figure 5-46. The abundances of oxygen and nitrogen decline with increasing radius in the H II regions of nearby spiral galaxies. Each point represents a determination of the abundances in the material of an emission region in either M33 (■), M101 (●), NGC 300 (+), or NGC 1365 (×) plotted against galactocentric distance (in units of the standard diameter D_0 of the galaxy in question). Notice that, in all three galaxies, the oxygen abundance declines linearly in this log-linear plot, and that the (N/O) values prevalent in the three galaxies differ appreciably. Values typical of the Orion nebula and of the Small and Large Magellanic Clouds are shown for comparison. [Adapted from (**P2**) by permission, kindly updated by Professor B. E. J. Pagel.]

motions observed in many barred galaxies (§5-1) are responsible for diminishing abundance gradients within these systems by mixing metal-rich material from the center with metal-poor material from the outlying regions.

5-5. SELECTION EFFECTS: A CAUTIONARY TALE

In this chapter, we discussed the light distributions of the more commonly observed types of galaxies. We saw that most well-studied systems may be characterized as either elliptical or disk galaxies. In either case, the light distribution $\Sigma(r)$ of the dominant component may be approximated by the formula

$$\log_{10}\left[\frac{\Sigma(r)}{\Sigma(0)}\right] = -\left(\frac{r}{r_s}\right)^{1/\beta} \tag{5-38}$$

where $\beta = 4$ for ellipticals and 1 for disk systems. The extrapolated central brightnesses $\Sigma(0)$ of photometrically studied disk systems cluster around $-2.5 \log[\Sigma(0)] = 21.5\mu$, while ellipticals generally have rather higher central surface brightnesses. In this section, we wish to warn that these seemingly well-established conclusions may actually tell us less about the objects that really populate the Universe than about the nature and limitations of astronomical observations.

Figure 5-47 indicates why one might be concerned that we may really know very little about the typical extragalactic object. In this figure, the absolute magnitudes of a number of galaxies and star clusters are plotted against the radii at which their surface brightnesses fall to 26.5μ. This radius (effectively the Holmberg radius defined in §5-3) has been converted to a linear dimension by using a Hubble constant $H = 50 \text{ km s}^{-1} \text{ Mpc}^{-1}$. In Figure 5-47, compact objects of high surface brightness are at top left, and diffuse objects are at bottom right. Also shown are two full lines and one dashed line. The lower full line is the locus at which the surface brightness of a uniformly illuminated object would be less than the level ($\approx 25.5\mu$) at which normal photographic techniques can readily detect the object against the background of the night sky. It is clear that objects whose representative points lie below this line will normally go undetected. Exceptions are the dwarf spheroidal galaxies like Draco and Fornax, which are near enough that their brightest stars are individually visible even though the integrated light of the unresolved stars is lost in the noise produced by the night sky.

The meaning of the upper two lines in Figure 5-47 is more subtle. To stand out from the myriads of faint distant galaxies and of stars near the Sun, and thus become the subject of detailed study, an object has to fulfill two requirements: it must be brighter than some limiting apparent magnitude, say, $m < 15$ mag, and it must be definitely nonstellar, say, $\theta > 15$ arcsec. The volume around the Sun, within which a galaxy may lie and still fulfill both

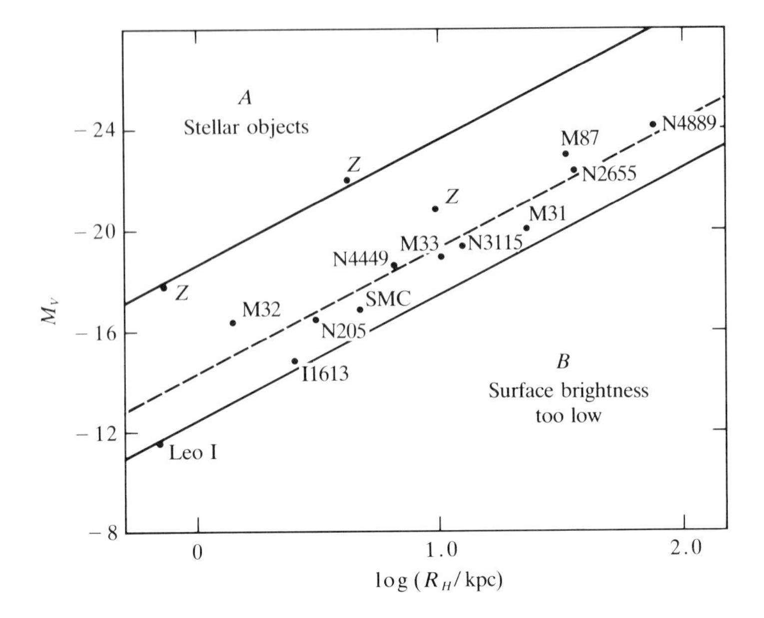

Figure 5-47. Observed galaxies might not be typical of all extragalactic objects. Here we plot the absolute magnitude of a number of galaxies against their Holmberg radii (both based on a Hubble constant $H = 50$ km s^{-1} Mpc^{-1}). Most objects whose representative points fell in region A would be indistinguishable from stars, while objects whose representative points fall in region B will not be visible against the brightness of the night sky. Notice that the swath of detectibility is well populated. The points marked Z are compact galaxies discovered by Zwicky (**Z2**). [After (**A7**).]

of these conditions, is determined in the lower part of Figure 5-47 by the magnitude limit, and it is determined in the upper part of the diagram by the requirement that the object appear nonstellar. Thus, at any absolute magnitude, there is a limiting true radius R_0 such that all objects whose radii are greater than R_0 will be selected on their apparent magnitudes alone, and they will therefore be detectable through the same volume around the Sun independent of their size. On the other hand, objects of the same absolute magnitude but $R < R_0$ will be recognized as galaxies only if they are very close to the Sun, because these objects will be selected by apparent size rather than by apparent magnitude. In Figure 5-47, the dashed line indicates, for each absolute magnitude, the linear size R_0 at which the effective selection criterion of a sample chosen by $m < 15$ and $\theta > 15$ arcsec switches from a magnitude limit (objects with $R > R_0$) to apparent diameter criterion (objects with $R < R_0$). One expects the numbers of objects known to be galaxies to fall off steeply to the left of the dashed line, because in this region the volume V around the Sun, within which galaxies of true radius R may be observed, falls off as $V \sim R^3$. At any absolute magnitude, the volume through which one can detect galaxies whose radii are given by the upper full line in the figure is less than 1/150 of the volume associated with galaxies of the same absolute magnitude whose representative points lie to the right of the dashed line.

What is disturbing about Figure 5-47 is that most of the region between the two full lines is fairly heavily populated by objects we recognize as galaxies. This situation suggests that there are probably objects in the Universe whose representative points lie outside this swath, but that they have not yet been recognized, either because they are not sufficiently well distinguished from stars, or because they are too faint to be seen at all against the sky background. We may, in fact, be aware of only a narrow "band" in the complete "spectrum" of galaxian morphology, just as early astronomers were wholly unaware of X-ray, ultraviolet, or radio emission from celestial objects.

One may investigate this effect more quantitatively as follows (**D9**). Suppose one asks: What will be the linear diameter r_{ap} at which we lose the image of an elliptical or disk galaxy of total luminosity L_t in the night-sky noise? Let the central surface brightness of the system be $\Sigma(0)$. Then the scale length r_s of the object may be determined from the condition, obtained by integrating equation (5-38) with respect to radius, that the total luminosity be L_t.

$$L_t = \int 2\pi r \Sigma(r)\, dr = \frac{(2\beta)!}{(\ln 10)^{2\beta}} \pi \Sigma(0) r_s^2 \tag{5-39}$$

With r_s thus determined, one has a relationship between r_{ap} and $\Sigma(0)$

$$r_{ap} = [L_t/\Sigma(r_{ap})]^{1/2}[\pi(2\beta)!]^{-1/2}(0.4 \ln 10)^{\beta}\, 10^{-0.2\Delta S}\, (\Delta S)^{\beta} \tag{5-40}$$

where $\Delta S = 2.5 \log_{10}[\Sigma(0)/\Sigma(r_{ap})]$.

Now ask how r_{ap} varies as a function of the central surface brightness $\Sigma(0) = 10^{-0.4S(0)}$. Figure 5-48 shows this dependence for elliptical and disk galaxies. Notice how r_{ap} peaks strongly at characteristic values of ΔS, which differ by $6^{m}\!.5\mu$ for elliptical and disk galaxies. These values of ΔS are such that, if one were to identify the characteristic central surface brightness derived in this way for disk galaxies with the characteristic brightness $21.6\mu_B$ derived observationally by Freeman (**F7**) and Schweizer (**S5**), then one would find $\Sigma(r_{ap}) = 23.8\mu_B$, a not unreasonable value for the effective limiting brightness of the conspicuous parts of the images of galaxies. Therefore, if typical photographs of spiral galaxies were such that the readily identifiable parts of galaxian images extended to surface brightness $\Sigma(r_{ap}) \approx 24\mu$, then the galaxies for which detailed surface photometry now exists would have central surface brightness that make their images as large as possible,

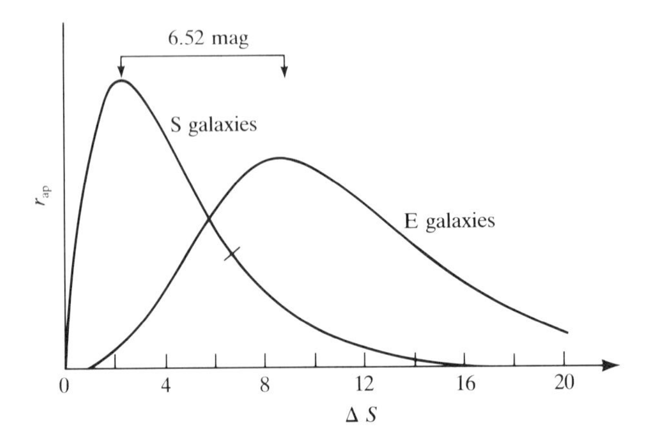

Figure 5-48. The dependence of $r_{\rm ap}$ on surface brightness contrast ΔS implied by equation (5-31) for elliptical and disk galaxies. [From (**D9**) by permission.]

compatible with fixed total luminosity. Put another way, if these systems had any other central brightnesses than the ones they do, then their images would be smaller and hence less conspicuous on photographic plates. Is it possible that those galaxies that have been selected for photometric study were chosen precisely because they are large and striking objects on photographic plates?

Figure 5-48 shows that the apparent radii of elliptical galaxies would be maximized if their central surface brightnesses were 8.7μ brighter than the limiting surface brightness $\Sigma(r_{\rm ap})$. Therefore, if selection effects of the type we are considering are also important for ellipticals, one might expect the central surface brightnesses of photometrically studied ellipticals to lie near $-2.5 \log[\Sigma(0)] = 15.3\mu$. In fact, data collected by R. A. Fish (**F6**) for twenty five bright ellipticals suggest $-2.5 \log[\Sigma(0)] \approx 14.8 \pm 0.9\mu$, in remarkably good agreement with our prediction. Thus we see, in two ways, that there *are* worrisome indications that the surface brightnesses and characteristic radii of the galaxies we observe and consider to be "typical" are strongly biased by observational considerations. Indeed, there is strong evidence that the sample of galaxies for which we have detailed photometry is strongly biased against low-surface-brightness objects.

The case for the existence of an observational bias against the selection of high-surface-brightness compact galaxies for photometric study is more controversial. It has been argued (**A5**) that, if high-surface-brightness compact objects really existed in significant numbers, then at least a few nearby specimens (whose apparent diameters would be appreciable despite their small linear size) would have been picked out as interesting nonstellar objects. However, if there really are no luminous compact galaxies, the ability of the argument just described to derive the approximately 6.5 mag difference between the characteristic central surface brightnesses of ellipticals and of disk galaxies has to be considered fortuitous. Therefore, it seems wise to keep an open mind as to how seriously the sample of well-studied galaxies is biased by observational considerations.

Biases of the kind just discussed may affect our understanding of the Universe in two important ways. (1) We may be unaware of the existence of numerous compact galaxies or diffuse, faint galaxies, which together might contribute significantly to the mass density and dynamics of the Universe, and which might feature prominently in the world views of observers whose skies are either darker or less hazy than our own. (2) We may seriously misjudge the nature and size of "normal" galaxies. It has often been assumed that galaxies do not extend much beyond their optical images, and "total" magnitudes, "total" masses, and the like, derived on the basis of this assumption, appear frequently in the literature. Yet, we have seen in our discussion of the photometry of galaxies that the light distributions of many ellipticals show no signs of converging to definite total magnitudes. And, in Chapter 8, we will see that the rotation curves of spiral galaxies generally do not lead to definite total masses. In fact, the idea that galaxies are bodies of limited, and readily determined, size and mass is quite likely an illusion produced by the nature of astronomical observations. This illusion is very dangerous. It makes us unwilling to recognize the existence of large amounts of relatively inconspicuous material in the Universe, and it inclines us to consider common those types of object that are merely striking in their appearance, and which, therefore, have been widely studied, even though these are not, in fact, the objects that actually occur most frequently.

REFERENCES

(A1) Aaronson, M., Huchra, J., and Mould, J. 1979. *Astrophys. J.* **229**:1.

(A2) Abell, G. O. 1955. *Pub. Astron. Soc. Pacific.* **67**:258.

(A3) Abell, G. O. 1977. *Astrophys. J.* **213**:327.

(A4) Allen, C. W. 1973. *Astrophysical Quantities*, 3rd ed. (London: Athlone Press).

(A5) Allen, R. J. and Shu, F. H. 1979. *Astrophys. J.* **227**:67.

(A6) Aller, L. H. 1942. *Astrophys. J.* **95**:52.

(A7) Arp, H. C. 1966. *Atlas of Peculiar Galaxies.* (Pasadena: California Institute of Technology).

(A8) Arp, H. C. 1965. *Astrophys. J.* **142**:402.

(A9) Austin, T. B. and Peach, J. V. 1974. *Mon. Not. Roy. Astron. Soc.* **168**:259.

(B1) Bertola, F. and Galletta, G. 1978. *Astrophys. J. Letters.* **226**:L115.

(B2) Burbidge, G. and Burbidge, E. M. 1967. *Quasi-Stellar Objects.* (San Francisco: W. H. Freeman).

(B3) Burstein, D. 1979. *Astrophys. J.* **234**:435.

(B4) Burstein, D. 1979. *Astrophys. J. Supp.* **41**:435.

(B5) Burstein, D. 1979. *Astrophys. J.* **232**:74.

(C1) Carter, D. 1977. Ph. D. Thesis, Cambridge University.

(C2) Carter, D. 1978. *Mon. Not. Roy. Astron. Soc.* **182**:797.

(C3) Carter, D. 1979. *Mon. Not. Roy. Astron. Soc.* **186**:897.

(C4) Cohen, J. G. 1978. *Astrophys. J.* **223**:487.

(C5) Crane, P. 1975. *Astrophys. J.* **197**:317.

(D1) de Vaucouleurs, G. 1948. *Ann. d'Astrophys.* **11**:247.

(D2) de Vaucouleurs, G. 1959. In *Handbuch der Physik*, vol. **53**, *Astrophysics IV: Stellar Systems*, S. Flügge (ed.). (Berlin: Springer-Verlag), p. 275.

(D3) de Vaucouleurs, G. 1961. *Astrophys. J. Supp.* **5**:223.

(D4) de Vaucouleurs, G. 1977. In *The Evolution of Galaxies and Stellar Populations.* Tinsley, B. M. and Larson, R. B. (eds.). (New Haven: Yale University Observatory).

(D5) de Vaucouleurs, G. 1978. *Astrophys. J.* **224**:14.

(D6) de Vaucouleurs, G. and Capaccioli, M. 1979. *Astrophys. J. Supp.* **40**:699.

(D7) de Vaucouleurs, G., de Vaucouleurs, A., and Corwin, H. G. 1976. *Second Reference Catalogue of Bright Galaxies.* (Austin: University of Texas Press).

(D8) de Vaucouleurs, G. and Freeman, K. C. 1970. *Vistas in Astron.* **14**:163.

(D9) Disney, M. J. 1976. *Nature.* **263**:573.

(D10) Dressler, A. 1978. *Astrophys. J.* **223**:765.

(D11) Dube, R. R., Wickes, W. C., and Wilkinson, D. T. 1972. *Astrophys. J. Letters.* **215**:L51.

(E1) Elvius, A. 1956. *Stockholm Obs. Ann.* **18**:No. 9.

(F1) Faber, S. M. 1972. *Astron. and Astrophys.* **20**:361.

(F2) Faber, S. M. 1973. *Astrophys. J.* **179**:423.

(F3) Faber, S. M. 1973. *Astrophys. J.* **179**:731.

(F4) Faber, S. M. 1977. In *The Evolution of Galaxies and Stellar Populations.* Tinsley, B. M. and Larson, R. B. (eds.). (New Haven: Yale University Observatory).

(F5) Felten, J. E. 1978. *Astron. J.* **82**:861.

(F6) Fish, R. A. 1964. *Astrophys. J.* **139**:284.

(F7) Freeman, K. C. 1970. *Astrophys. J.* **160**:811.

(G1) Godwin, J. C. and Peach, J. V. 1977. *Mon. Not. Roy. Astron. Soc.* **181**:323.

(H1) Hodge, P. W. and Michie, R. W. 1969. *Astron. J.* **74**:587.

(H2) Holmberg, E. 1958. *Medd. Lunds Astron. Obs.* Series 2, No. 136.

(H3) Hubble, E. P. 1930. *Astrophys. J.* **71**:231.

(H4) Hubble, E. P. 1936. *The Realm of the Nebulae.* (New Haven: Yale University Press).

(I1) Illingworth, G. and Illingworth, W. 1976. *Astrophys. J. Supp.* **30**:227.

(J1) Johnson, H. L. 1966. *Ann. Rev. Astron. and Astrophys.* **4**:193.

(J2) Johnson, H. L. and Morgan, W. W. 1953. *Astrophys. J.* **117**:313.

(J3) Jones, W. B., Obitts, D. L., Gollet, R. M., and de Vaucouleurs, G. 1967. *Pub. Dept. Astron. Univ. of Texas.*.**8**. (Austin: University of Texas).

(K1) King, I. R. 1966. *Astron. J.* **71**:64.

(K2) King, I. R. 1978. *Astrophys. J.* **222**:1.

(K3) Kormendy, J. 1977. *Astrophys. J.* **217**:406.

(K4) Kormendy, J. 1977. *Astrophys. J.* **218**:333.

(K5) Kormendy, J. 1979. *Astrophys. J.* **227**:714.

(K6) Kormendy, J. and Bruzual, G. 1978. *Astrophys. J. Letters.* **223**:L63.

(K7) Kormendy, J. and Illingworth, G. 1979. In *Photometry, Kinematics, and Dynamics of Galaxies.* Evans, D. S. (ed.). (Austin: University of Texas).

(L1) Light, E. S., Danielson, R. E., and Schwarzschild, M. 1974. *Astrophys. J.* **194**:257.

(L2) Lindblad, B. 1956. *Stockholm Obs. Ann.* **18**:No. 19.

(M1) Markarian, B. E., Lipovetskij, V. A., and Stepanyan, D. A. 1978. *Astrophys.* **13**:215 (and references cited therein).

(M2) McClure, R. D. and van den Bergh, S. 1968. *Astron. J.* **73**:313.

(M3) Miller, R. H. and Prendergast, K. H. 1962. *Astrophys. J.* **136**:713.

(M4) Morgan, W. W. 1958. *Pub. Astron. Soc. Pacific.* **70**:364.

(M5) Morgan, W. W. 1959. *Pub. Astron. Soc. Pacific.* **71**:92.

(M6) Morgan, W. W. 1970. In *Spiral Structure of Our Galaxy.* Becker, W. and Contopoulos, G. (eds.). (Dordrecht: Reidel).

(M7) Morgan, W. W. 1972. In *External Galaxies and Quasi-Stellar Objects.* Evans, D. S. (ed.). (Dordrecht: Reidel).

(M8) Morgan, W. W., Kayser, S., and White, R. A. 1975. *Astrophys. J.* **199**:545.

(M9) Morgan, W. W. and Mayall, N. U. 1957. *Pub. Astron. Soc. Pacific.* **69**:291.

(M10) Mould, J. 1978. *Astrophys. J.* **220**:434.

(O1) O'Connell, R. W. 1976. *Astrophys. J.* **206**:370.

(O2) Oemler, A. 1974. *Astrophys. J.* **194**:1.

(O3) Oemler, A. 1976. *Astrophys. J.* **209**:693.

(O4) Okamura, S. 1978. *Pub. Astron. Soc. Japan.* **30**:91.

(O5) Osterbrock, D. E. 1974. *Astrophysics of Gaseous Nebulae*. (San Francisco: W. H. Freeman).

(O6) Ostriker, J. P. and Thuan, T. X. 1975. *Astrophys. J.* **202**:353.

(P1) Pagel, B. E. J., Edmunds, M. G., Fosbury, R. A. E., and Webster, B. L. 1978. *Mon. Not. Roy. Astron. Soc.* **184**, 569.

(P2) Pagel, B. E. J., Edmunds, M. G., Blackwell, D. E., Chun, M. S., and Smith, G. 1979. *Mon. Not. Roy. Astron. Soc.* **189**:95.

(P3) Peterson, R. 1976. *Astrophys. J. Supp.* **30**:61.

(P4) Pritchet, C. 1977. *Astrophys. J. Supp.* **35**:397.

(R1) Reynolds, J. H. 1913. *Mon. Not. Roy. Astron. Soc.* **74**:132.

(S1) Sandage, A. 1961. *The Hubble Atlas of Galaxies*. (Washington: Carnegie Institution).

(S2) Sandage, A. and Tammann, G. 1975. *Astrophys. J.* **197**:265.

(S3) Sandage, A. and Visvanathan, N. 1978. *Astrophys. J.* **223**:707

(S4) Schechter, P. 1976. *Astrophys. J.* **203**:297.

(S5) Schweizer, F. 1976. *Astrophys. J. Supp.* **31**:313.

(S6) Schweizer, F. 1979. *Astrophys. J.* **233**:23.

(S7) Searle, L. 1971. *Astrophys. J.* **168**:333.

(S8) Shapley, H. 1938. *Nature*. **142**:715.

(S9) Shields, G. A. 1974. *Astrophys. J.* **193**:335.

(S10) Shields, G. A. and Searle, L. 1978. *Astrophys. J.* **222**:821.

(S11) Smith, H. E. 1975. *Astrophys. J.* **199**:591.

(S12) Spinrad, H. and Taylor, B. J. 1971. *Astrophys. J. Supp.* **22**:445.

(S13) Stark, A. 1977. *Astrophys. J.* **213**:368.

(S14) Strom, K. M., Strom, S. E., Jensen, E. G., Moller, J., and Thompson, L. A. 1977. *Astrophys. J.* **212**:335.

(T1) Tayler, R. J. 1972. *The Origin of the Chemical Elements*. (London: Wykeham).

(T2) Toomre, A. and Toomre, J. 1972. *Astrophys. J.* **178**:623.

(T3) Tremaine, S. D. and Richstone, D. O. 1977. *Astrophys. J.* **212**:311.

(T4) Tsikoudi, V. 1977. Pub. Dept. Astron. Univ. of Texas. **10**. (Austin: University of Texas).

(T5) Tully, R. B. and Fisher, J. R. 1977. *Astron. and Astrophys.* **54**:661.

(U1) Ulfbeck, O. (ed.). 1978. *Quasars and Active Nuclei of Galaxies*. (Stockholm: Roy. Swedish Acad. Sci.).

(V1) van den Bergh, S. 1960. *Astrophys. J.* **131**:215.

(V2) van den Bergh, S. 1960. *Astrophys. J.* **131**:558.

(V3) van den Bergh, S. 1972. *Astrophys. J. Letters*. **171**:L31.

(V4) van den Bergh, S. 1975. In *Galaxies and the Universe*. Sandage, A., Sandage, M., and Kristian, J. (eds.). (Chicago: University of Chicago Press).

(V5) van den Bergh, S. 1976. *Astrophys. J.* **206**:883.

(V6) van den Bergh, S. and McClure, R. D. 1979. *Astrophys. J.* **231**:671

(V7) van Houten, C. J. 1961. *Bull. Astron. Inst. Netherlands.* **16**:1.

(V8) Visvanathan, N. and Sandage, A. 1977. *Astrophys. J.* **216**:214.

(V9) Vorontsov-Velyaminov, B. A. and Arhipova, V. P. 1968. *Morphological Catalogue of Galaxies.* (Moscow: Moscow University Press).

(W1) Whitford, A. E. 1975. In *Galaxies and the Universe*. Sandage, A., Sandage, M., and Kristian, J. (eds.). (Chicago: University of Chicago Press).

(W2) Williams, T. B. 1976. *Astrophys. J.* **209**:716.

(W3) Williams, T. B. and Schwarzschild, M. 1979. *Astrophys. J.* **227**:156.

(W4) Wood, D. B. 1965, *Astrophys. J.* **145**:36.

(W5) Wood, D. B. 1969. *Astron. J.* **74**:177.

(Z1) Zwicky, F. 1971. *Catalogue of Selected Compact Galaxies and Post-Eruptive Galaxies.* (Guemlingen: Zwicky).

(Z2) Zwicky, F. 1964. *Astrophys. J.* **140**:1467.

6 Stellar Kinematics: The Solar Motion

In the preceding chapters, we studied the physical properties of stars and their distribution in both our Galaxy and external galaxies. We found that our Galaxy is composed of stars with a wide range of physical properties, ages, and chemical compositions. From analyses of star counts, we were able to deduce the relative frequencies of occurrence of various kinds of stars per unit volume of space. We could also determine approximate contours of the space densities of stars in the galactic disk and galactic halo in the vicinity of the Sun. By identifying those high-luminosity objects that are strongly associated with spiral arms and by using them as tracers, we were able to outline the pattern of spiral arms in our part of our Galaxy. A similar procedure, based on observations of high-luminosity halo objects like globular clusters and RR Lyrae variables, enabled us to study the space density of the halo and bulge components through most of our Galaxy. At the end of Chapter 4, we showed how the information obtained in this way could be integrated into a partial picture of stellar populations. Finally, in the preceding chapter, we complemented this work with a study of the brightness distributions and the spectra of external galaxies. We saw there that the components, like the disk and the bulge components, which we had already identified from observations in our Galaxy, could also be found in the brightness distributions of external galaxies. Similarly, many of the stellar population characteristics that emerged from our study of our Galaxy—for example, the metallicities characteristic of the various populations in our Galaxy—were found to have counterparts in other galaxies, as seen in color and line strength variations within these systems.

A number of lines of evidence thus suggest that galaxies are composed of several populations of stars that differ from one another both in their chemical compositions and in their spatial distributions. Now, if the different stellar population groups within galaxies have different spatial distributions, one might suspect on dynamical grounds that the kinematic properties of the various groups would differ from one another in a distinctive way. Thus one can expect that a star belonging to the halo population will normally pass through the galactic plane with a large component of velocity perpendicular to the galactic plane as it moves on its highly inclined orbit, whereas most of the stars of the disk population should have only small motions in the z direction. The location of the Sun in the galactic plane allows us to seek possible effects of this type by using radial velocities and proper motions to map the motions of the sample of stars (which contains both disk and halo stars) that are in some small volume around the Sun.

The next two chapters will be concerned with the attainment of this goal. In this chapter, we shall concentrate on two major problems, the *motion of the Sun* relative to the average motion of stars in the solar neighborhood and the *rotation velocity* around the galactic center of a point moving on a circular path at the Sun's distance from the center. Then, in Chapter 7, we shall examine the distribution of stellar *peculiar velocities* (or *residual velocities*) relative to the circular velocity, for spiral-arm, disk, and spheroidal-component stars. We shall find that the different population groups that we tentatively identified in Chapter 4 do indeed show very different kinematic properties. This information provides important clues about the formation and early dynamical history of our Galaxy. A discussion of the differential rotation of galaxian disks and the relationship of local kinematic properties to the global run of galactic rotation will be deferred until Chapter 8.

6-1. STANDARDS OF REST

The basic observational data from which we determine the motion of a star are its *radial velocity* v_R and its *tangential velocity* v_T relative to the Sun. The *speed* or *space motion* of a star relative to the Sun is

$$v_S \equiv (v_R^2 + v_T^2)^{1/2} \tag{6-1}$$

Although a radial velocity can be measured for any star that is bright enough to yield a suitable spectrogram, we must measure both the star's proper motion and its distance to obtain the tangential velocity. For a given v_T, the corresponding proper motion decreases inversely with the distance, and it eventually drops to a value that cannot be measured reliably. When a reliable proper motion is available for a star, it can be converted into a tangential velocity v_T only if the star's parallax is known. As we saw in Chapter 2, reliable direct-parallax measurements are limited to distances of 50 pc or less. Although much larger distances can be determined spectroscopically for stars of those spectral types that have had accurate absolute magnitudes and intrinsic colors associated with them, this technique fails for stars of unusual types whose intrinsic properties are unknown. Indeed, as we shall see in §6-6, in such cases we can turn the procedure around and use proper-motion data to estimate distances and hence absolute magnitudes. Therefore, in practice, most of our detailed knowledge of stellar kinematics is limited to a rather small volume of space around the Sun, and our knowledge for relatively uncommon stars, which typically are seen only at fairly large distances, is restricted to a narrow selection of objects of intrinsically high luminosity and becomes more and more sketchy with increasing distance.

The Fundamental Standard of Rest

From the point of view of galactic dynamics, the most useful description of the velocity of a star is in terms of the components (Π, Θ, Z) shown in Figure 6-1. We choose the galactic center (or, more precisely, the center of mass of our Galaxy) as the origin, which defines the *fundamental standard of rest*. We let Π be the component of velocity directed radially with respect to the center, taken to be positive for outward motion. We choose Z to be the velocity component perpendicular to the galactic plane, measured positive in the direction of the north galactic pole. We choose Θ to be the tangential component, normal to the other two, measured positive in the direction of galactic rotation. In terms of galactic coordinates, Π is positive in the direction ($\ell = 180^\circ$, $b = 0^\circ$), Θ is positive in the direction ($\ell = 90^\circ$, $b = 0^\circ$), and Z is positive in the direction $b = +90^\circ$.

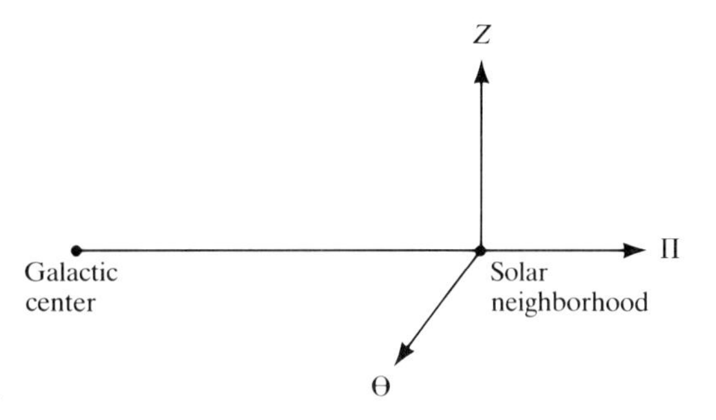

Figure 6-1. Definition of the velocity components (Π, Θ, Z). Π is directed along the radius vector from the galactic center and is positive outward. Θ is in the galactic plane, is perpendicular to Π, and is positive in the direction of galactic rotation. Z is perpendicular to the galactic plane and is positive northward.

The Local Standard of Rest and Stellar Peculiar Velocities

Suppose now that the mass distribution in our Galaxy is essentially axisymmetric, so that the gravitational forces at any position in the galactic plane are radial toward the galactic center, and that it is in a steady state, that is, that the forces are independent of time (or at least vary only extremely slowly during the time required for a full rotation of our Galaxy). Then, at each position in the galactic plane, there will be a particular vector velocity with components $(\Pi, \Theta, Z)_{\text{LSR}} = (0, \Theta_0, 0)$ such that a star moving with this velocity follows a circular orbit around the galactic center. We imagine a point moving with just this velocity and define it to be the *local standard of rest* (LSR); we call the velocity Θ_0 the *circular velocity*. Clearly, the circular velocity will be different at different positions in the plane inasmuch as the gravitational forces will depend on position in our Galaxy, that is, $\Theta_0 = \Theta_0(R)$. Note also that, even for two points at the same distance R, the vectors describing the motions of the two local standards of rest will point in different directions.

The motion of any individual star relative to the fundamental standard of rest is plainly the vector sum of the velocity of the LSR relative to the fundamental standard of rest and the *peculiar velocity* of the star relative to its LSR. We shall denote the (Π, Θ, Z) components of the peculiar velocity of a star as (u, v, w); that is,

$$u \equiv \Pi - \Pi_{\text{LSR}} = \Pi \tag{6-2a}$$

$$v \equiv \Theta - \Theta_{\text{LSR}} = \Theta - \Theta_0 \tag{6-2b}$$

and

$$w \equiv Z - Z_{\text{LSR}} = Z \tag{6-2c}$$

Our main tasks in this chapter are to determine the peculiar velocity of the Sun relative to its LSR (treated in §6-2 to §6-4) and the circular velocity of that LSR around the galactic center (discussed in §6-5).

The velocity of any star relative to the Sun consists of three contributions: (1) the peculiar velocity of the star relative to the LSR at its position (P_*) in space; (2) the peculiar velocity of the Sun relative to the LSR at its position ($P_\odot$); and (3) the differential velocity of the local standard of rest at P_* relative to that at $P_\odot$. The third contribution just mentioned arises from *differential galactic rotation*, which we shall analyze in Chapter 8. Henceforth, in Chapters 6 and 7, we shall assume for expository ease that the distance between the Sun and star under consideration is sufficiently small that the effects of differential galactic rotation can be neglected. With this assumption, the observed velocity of a star with respect to the Sun can be resolved into components (U_*, V_*, W_*), where

$$U_* = u_* - u_\odot = \Pi_* - \Pi_\odot \tag{6-3a}$$

$$V_* = v_* - v_\odot = \Theta_* - \Theta_\odot \tag{6-3b}$$

and

$$W_* = w_* - w_\odot = Z_* - Z_\odot \tag{6-3c}$$

We stress that equations (6-3) apply only in the immediate vicinity of the Sun ($d \lesssim 100$ pc).

The Centroid of Velocities and the Solar Motion

We now ask how we can infer the values of $(u_\odot, v_\odot, w_\odot) = (\Pi_\odot, \Theta_\odot - \Theta_0, Z_\odot)$, components of the Sun's peculiar velocity relative to its LSR. In practice, we must proceed in two steps. First, we must determine the *solar motion* relative to some specified group of stars. Suppose we have chosen N stars, say, according to their spectrophotometric properties, for example, MK spectral type or ultraviolet excess. Then, the mean peculiar velocity of the group is given by the *centroid of velocities*:

$$\langle u_* \rangle \equiv \frac{1}{N} \sum_{i=1}^{N} u_{*i} \tag{6-4a}$$

$$\langle v_* \rangle \equiv \frac{1}{N} \sum_{i=1}^{N} v_{*i} \tag{6-4b}$$

and

$$\langle w_* \rangle \equiv \frac{1}{N} \sum_{i=1}^{N} w_{*i} \tag{6-4c}$$

Now, suppose that our Galaxy is axisymmetric and in a steady state. We expect that, at any position in the disk, on the average there will be as many stars moving radially outward as inward, and as many stars moving upward with respect to the galactic plane as downward. Thus, for a large enough group of stars, we can set

$$\langle u_* \rangle \equiv 0 \tag{6-5a}$$

and

$$\langle w_* \rangle \equiv 0 \tag{6-5b}$$

We cannot, however, set $\langle v_* \rangle$ to zero, for, as we shall show in §6-4, the group will systematically lag behind the LSR, and $\langle v_* \rangle$ will average to a negative value, which is small for spiral-arm and disk stars and large for spheroidal-component stars.

In view of equations (6-5), it is clear that if we average equations (6-3) over a sufficiently large group of stars, we have

$$u_\odot = -\langle U_* \rangle \tag{6-6a}$$

$$v_\odot = -\langle V_* \rangle + \langle v_* \rangle \tag{6-6b}$$

and

$$w_\odot = -\langle W_* \rangle \tag{6-6c}$$

which shows that in the radial and z directions the average observed velocity of the group is simply a reflection of the peculiar velocity of the Sun. We cannot determine $v_\odot$ until we know $\langle v_* \rangle$, which requires an application of results from dynamical theory. In order to keep purely observational results separate from those that depend on theory, we defer the question of the determination of $\langle v_* \rangle$, and first calculate the reflection of the v component of the sun's motion from the average observed velocity of the group. That is, we define

$$v'_\odot \equiv -\langle V_* \rangle \tag{6-7}$$

and we then define the solar motion to be the velocity vector $(u_\odot, v'_\odot, w_\odot)$. It is apparent that, so defined, the solar motion can be determined directly from observations alone.

It is convenient to describe the solar motion in terms of a speed $S_\odot$,

$$S_\odot \equiv (u_\odot^2 + v_\odot'^2 + w_\odot^2)^{1/2} \tag{6-8}$$

and a direction, which is specified by the coordinates of the *apex of the solar motion*, the point on the sky toward which the Sun is moving. The position of the apex may be given in terms of either (α_A, δ_A) in the equatorial system, or (ℓ_A, b_A) in the galactic system.

Ultimately, having found the solar motion for all groups of interest, we finally determine $\langle v_* \rangle$, the difference between the motions of the LSR and the velocity centroid of each group. This determination allows us to convert the solar motion $(u_\odot, v'_\odot, w_\odot)$ to the Sun's peculiar velocity $(u_\odot, v_\odot, w_\odot)$. And, of course, once we know the peculiar velocity of the Sun, then from equations (6-3) we can determine the peculiar velocity (u_*, v_*, w_*) for any star from its observed velocity (U_*, V_*, W_*) relative to the Sun.

6-2. QUALITATIVE ANALYSIS OF THE SOLAR MOTION

Useful insight into the effects of a solar motion can be obtained from an extremely simple and instructive analysis presented by Russell, Dugan, and Stewart (**R2**, 658 ff).

Anticipated Observational Effects

Consider the effects of a solar motion with respect to the velocity centroid of a group of stars. As can be seen in Figure 6-2, the solar motion causes stars to appear to stream past the Sun, which gives rise to systematic effects in both the observed radial velocities and the proper motions of the stars. For example, the stars located near the *apex* of the motion (the direction in which the Sun is moving) will appear to be approaching systematically, and, on the average, this group will have negative radial velocities. Those in the opposite direction (the *antapex*) will appear to recede systematically, and, on the average, this group will have positive radial velocities. Stars located in directions perpendicular to the solar motion show no systematic change in radial velocity, and they will have a random distribution of velocities that average to zero relative to the Sun. Similarly, the proper motions of stars in the direction of the apex will appear to diverge away from the apex in the plane of the sky; in the directions normal to the solar motion, they will appear to stream across the sky; and, in the direction opposite to the motion, they appear to converge on the antapex.

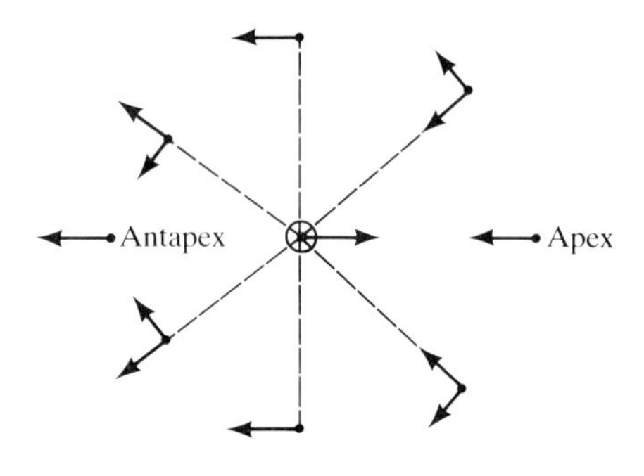

Figure 6-2. The observable effects of solar motion. The Sun (*center*) moves to the right relative to the stars, taken here to be fixed. As seen by an observer moving with the Sun, the stars have the apparent velocities shown, both along and across the line of sight. We see that stars in the direction of the apex appear to approach along the line of sight and to diverge from the apex on the plane of the sky. Stars in the opposite direction appear to recede and to converge upon the antapex.

The Direction of the Solar Motion

The systematic effects just described can be used to deduce the solar motion. We shall confine attention here to proper motions only, and hence we shall

Table 6-1. Proper Motion in Right Ascension for Stars in Right Ascension Zones $\alpha_i \pm 1^h_{\cdot}5$

α_i	0^h	3^h	6^h	9^h	12^h	15^h	18^h	21^h
Number of stars with $\mu_\alpha > 0$	132	152	107	39	46	47	101	143
Number of stars with $\mu_\alpha < 0$	68	48	93	161	154	153	99	57
Excess of + over −	64	104	14	−122	−108	−106	2	86

find only the direction of the solar motion, but not its speed. Suppose we examine the proper motions given in catalogs for groups of stars located in uniform intervals of right ascension centered at $\alpha = 0^h$, 3^h, 6^h, and so on, with, say, 200 stars in each group. Consider first the proper motions in right ascension μ_α. If we count the number of stars in each group with positive or negative μ_α, we find the results given in Table 6-1. A plot of the excess of the number of stars with positive μ_α over those with negative μ_α is shown in Figure 6-3. We see that the divergent point, and hence the apex, lies approximately in the direction of $\alpha = 18^h$ because, at this point, stars to the west ($\alpha < 18^h$) show a net westward movement, while those to the east ($\alpha > 18^h$) show a net eastward movement.

Next, let us consider two lunes on the sky containing stars with right ascensions in the ranges $\alpha = 6^h \pm 30^m$ and $\alpha = 18^h \pm 30^m$. We now divide these lunes into declination zones of 30° each and count the number of stars with positive and negative proper motions in declination μ_δ to obtain the results given in Table 6-2. Figure 6-4 is a plot of the excess of the number of stars with positive μ_δ over those with negative μ_δ. We see that the declination of the apex is approximately $\delta = +30°$. The direction $(\alpha, \delta) = (18\text{h}, +30°)$, which corresponds to $(\ell, b) = (56°, +23°)$, lies in the constellation of Hercules. As we shall see in the next sections, this rough analysis gives remarkably good values for the position of the apex.

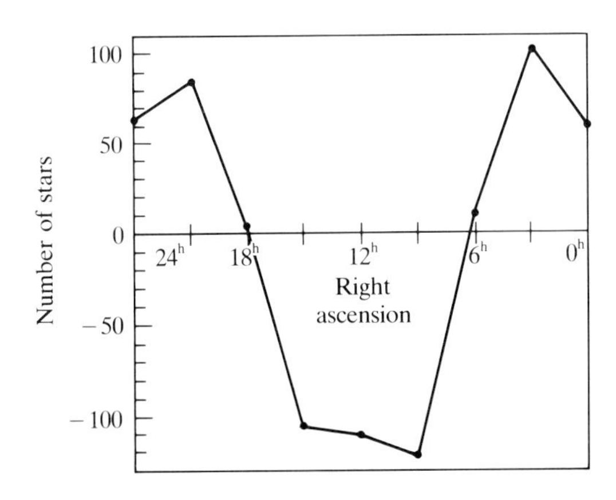

Figure 6-3. The excess number of stars with positive μ_α over stars with negative μ_α (see Table 6-1).

Table 6-2. Proper Motions in Declination for Stars in Declination Zones $\delta_i \pm 15^\circ$

Declination zones centered on $\alpha = 6^h$				Declination zones centered on $\alpha = 18^h$			
δ_i	Number of stars with $\mu_\delta > 0$	Number of stars with $\mu_\delta < 0$	Excess of + over −	δ_i	Number of stars with $\mu_\delta > 0$	Number of stars with $\mu_\delta < 0$	Excess of + over −
$+75^\circ$	3	10	−7	$+75^\circ$	10	7	3
$+45^\circ$	9	35	−26	$+45^\circ$	25	13	12
$+15^\circ$	22	72	−50	$+15^\circ$	22	38	−16
-15°	29	40	−11	-15°	8	59	−51
-45°	31	32	−1	-45°	6	46	−40
-75°	16	3	13	-75°	4	9	−5

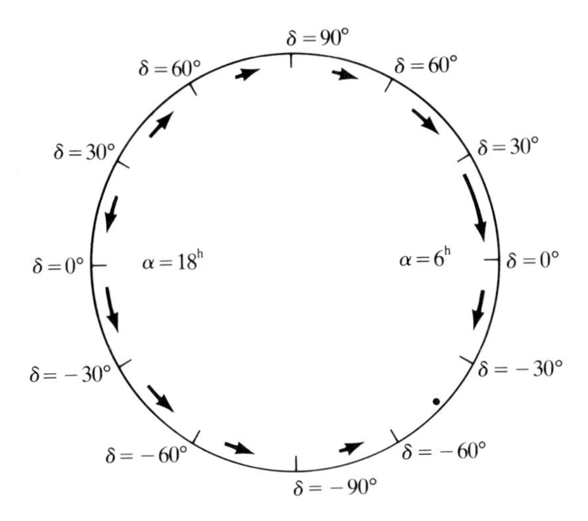

Figure 6-4. The arrows indicate the net motion in declination of stars in various declination zones (see Table 6-2).

6-3. RIGOROUS ANALYSIS OF THE SOLAR MOTION

Let us now consider more rigorous mathematical analyses of the observational data, which yield precise values both for the position of the apex and for the speed of solar motion. Three somewhat different methods can be employed, using radial-velocity data alone, proper-motion data alone, or space motions (both radial and transverse components). The first two methods can be used for stars whose distances are not known. The former yields both the direction and speed of the solar motion, while the latter yields only the direction. To use the third method, one must know the distances of the stars and one obtains both the direction and the speed of the solar motion. We remind the reader that we still assume that the analysis is being done for stars that are so close that the differential motions of the local standards of rest for the star and the Sun can be neglected.

In carrying out these analyses, it is essential to restrict attention to as homogeneous a group of stars as possible, for example, by considering only one spectral type at a time. As we shall see in §6-4, we obtain distinctly different solutions for the solar motion from stars of different spectral types. The reason for this is that differing types contain individual stars of more or less widely differing ages, and their velocity distributions will reflect a variety of different dynamical histories. For example, the youngest stars (spiral-arm stars or the youngest disk stars) will have velocities that may still reflect local conditions at the time of star formation, for example, in the spiral-arm pattern or within a star cluster or association. The kinematic properties of the earliest spectral types, which are all young stars, should reflect these conditions fairly faithfully. In contrast, older disk stars have velocities that reflect the accumulated effects of encounters with other stars and with interstellar clouds of dust and gas. Thus, samples of stars of later spectral types, which in general contain both old and young stars, will be a mixture of stars that have quite different kinematic properties. Indeed, within almost any group of stars chosen according to spectral type only, there will be a certain degree of kinematic heterogeneity, either because the group contains stars of widely differing ages or because it is composed of subgroups formed under rather different initial conditions. We shall return to these points in Chapter 7.

Basic Equations

We now derive the basic equations required for the analysis. Let d be the distance from the Sun to the star, and, because the observational data for proper motions are always given in the equatorial system, let us resolve positions and velocities into equatorial rectangular components. Then (see Figure 6-5),

$$x = d \cos \delta \cos \alpha \tag{6-9a}$$

$$y = d \cos \delta \sin \alpha \tag{6-9b}$$

$$z = d \sin \delta \tag{6-9c}$$

Differentiating equations (6-9) with respect to time, we have the equatorial components of the stellar velocity with respect to the Sun.

$$\dot{x} = X_* - X_\odot = \dot{d} \cos \delta \cos \alpha - \dot{\delta} d \sin \delta \cos \alpha - \dot{\alpha} d \cos \delta \sin \alpha \tag{6-10a}$$

$$\dot{y} = Y_* - Y_\odot = \dot{d} \cos \delta \sin \alpha - \dot{\delta} d \sin \delta \sin \alpha + \dot{\alpha} d \cos \delta \cos \alpha \tag{6-10b}$$

$$\dot{z} = Z_* - Z_\odot = \dot{d} \sin \delta + \dot{\delta} d \cos \delta \tag{6-10c}$$

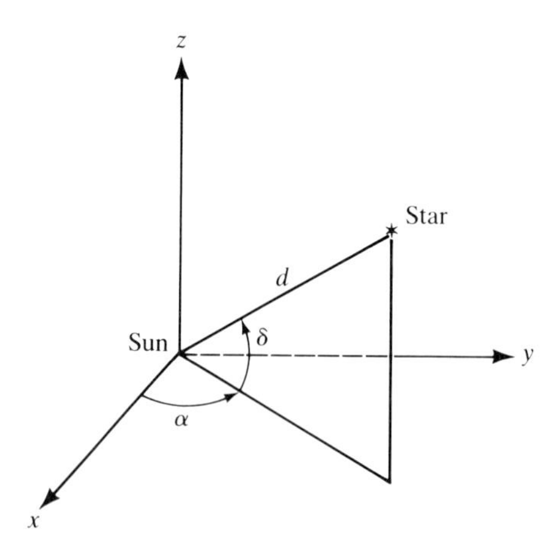

Figure 6-5. Equatorial rectangular coordinate system.

where (X_*, Y_*, Z_*) and $(X_\odot, Y_\odot, Z_\odot)$ are the velocity components of the star and the Sun, respectively, along the (x, y, z) axes in the equatorial system. The quantity $\dot{d}$ stands for the radial velocity, and $\dot{\delta}d$ and $\dot{\alpha}d \cos \delta$ are transverse velocities. These will be expressed in kilometers per second if $\dot{\alpha}$ and $\dot{\delta}$ are expressed in radians per second and d is expressed in kilometers. (We shall introduce more practical units shortly.)

To facilitate the analysis, we must solve equations (6-10) for $\dot{d}$, $\dot{\alpha}d \cos \delta$, and $\dot{\delta}d$, so that we can write equations that involve only radial velocities or only proper motions. The solution is easy to write because the matrix of coefficients in equations (6-10) is orthogonal, so that its inverse equals its transpose. We thus find

$$\dot{x} \cos \delta \cos \alpha + \dot{y} \cos \delta \sin \alpha + \dot{z} \sin \delta = \dot{d} \tag{6-11a}$$

$$-\dot{x} \sin \delta \cos \alpha - \dot{y} \sin \delta \sin \alpha + \dot{z} \cos \delta = \dot{\delta}d \tag{6-11b}$$

$$-\dot{x} \sin \alpha + \dot{y} \cos \alpha = \dot{\alpha}d \cos \delta \tag{6-11c}$$

Solution from Radial Velocities

Let us now examine how equation (6-11a) can be used to derive the solar motion from radial velocity data alone. Suppose we have chosen N stars uniformly distributed over the sky. For each star, we can write an equation of the form of equation (6-11a). Thus, for the ith star,

$$\dot{x}_i \cos \delta_i \cos \alpha_i + \dot{y}_i \cos \delta_i \sin \alpha_i + \dot{z}_i \sin \delta_i = \dot{d}_i \, (i = 1, \ldots, N) \tag{6-12}$$

By definition, the solar motion is determined relative to the centroid of observed stellar velocities. Therefore, we demand that the observed stellar space velocities average to zero, that is, $\langle X_* \rangle = \langle Y_* \rangle = \langle Z_* \rangle = 0$, where $\langle X_* \rangle \equiv (\sum_{i=1}^{N} X_{*i})/N$. Now, even though the coefficients $\cos \delta_i \cos \alpha_i$, $\cos \delta_i \sin \alpha_i$, and $\sin \delta_i$ are, strictly speaking, different from star to star, we can imagine that the sky is divided into small regions within which these coefficients are practically constant. Then, provided that we have chosen a large enough sample to assure that each region contains several stars, it is clear that the averages $\langle X_{*i} \rangle$ and so on can be set to zero region by region. That is, with the assumptions of a zero centroid velocity and a statistically satisfactory sample, we can, in effect, simply drop all terms involving X_{*i}, Y_{*i}, and Z_{*i} out of the left-hand side of equation (6-12). Equations (6-12) are thus effectively of the form

$$a_i X_\odot + b_i Y_\odot + c_i Z_\odot = k_i \qquad (i = 1, \ldots, N) \tag{6-13}$$

The system of equations (6-13) contains N equations in three unknowns. The system is therefore *overdetermined*. We can make an optimum solution for $(X_\odot, Y_\odot, Z_\odot)$ by the *method of least squares*. From the outset, we recognize that it is not possible to make a single choice for $(X_\odot, Y_\odot, Z_\odot)$ that will solve every one of equations (6-13) exactly. (This is true even for perfect data because the stellar velocity components can be set to zero only on the average.) For any specific choice of the solution $(X_\odot, Y_\odot, Z_\odot)$, each equation will yield a *residual* r_i,

$$r_i \equiv k_i - a_i X_\odot - b_i Y_\odot - c_i Z_\odot \qquad (i = 1, \ldots, N) \quad \text{(6-14)}$$

Thus we choose the particular solution that minimizes the sum of the squares of the residuals $R \equiv \sum_{i=1}^{N} r_i^2$. To do this, we demand that $(\partial R/\partial X_\odot) = (\partial R/\partial Y_\odot) = (\partial R/\partial Z_\odot) = 0$. It is easy to show that these requirements produce an equation of the form

$$X_\odot \sum_{i=1}^{N} a_i^2 + Y_\odot \sum_{i=1}^{N} a_i b_i + Z_\odot \sum_{i=1}^{N} a_i c_i = \sum_{i=1}^{N} a_i k_i \qquad \text{(6-15)}$$

and two others in which the common factor a_i inside the sums is replaced, in turn, by b_i and c_i. These equations are the result of taking moments against the coefficients of the unknowns, that is, by multiplying each equation through by its coefficient of $X_\odot$ (namely a_i) and summing over all equations, and so on.

Hence, the final system of equations to be solved, which we obtain by applying the method of least squares to equation (6-12), is

$$X_\odot \sum_{i=1}^{N} \cos^2\delta_i \cos^2\alpha_i + Y_\odot \sum_{i=1}^{N} \cos^2\delta_i \sin\alpha_i \cos\alpha_i + Z_\odot \sum_{i=1}^{N} \sin\delta_i \cos\delta_i \cos\alpha_i$$
$$= -\sum_{i=1}^{N} \dot{d}_i \cos\delta_i \cos\alpha_i \qquad \text{(6-16a)}$$

$$X_\odot \sum_{i=1}^{N} \cos^2\delta_i \sin\alpha_i \cos\alpha_i + Y_\odot \sum_{i=1}^{N} \cos^2\delta_i \sin^2\alpha_i + Z_\odot \sum_{i=1}^{N} \sin\delta_i \cos\delta_i \sin\alpha_i$$
$$= -\sum_{i=1}^{N} \dot{d}_i \cos\delta_i \sin\alpha_i \qquad \text{(6-16b)}$$

and

$$X_\odot \sum_{i=1}^{N} \sin\delta_i \cos\delta_i \cos\alpha_i + Y_\odot \sum_{i=1}^{N} \sin\delta_i \cos\delta_i \sin\alpha_i + Z_\odot \sum_{i=1}^{N} \sin^2\delta_i$$
$$= -\sum_{i=1}^{N} \dot{d}_i \sin\delta_i \qquad \text{(6-16c)}$$

This set of three linear equations can be solved straightaway for $(X_\odot, Y_\odot, Z_\odot)$ because $\dot{d}_i$, α_i, and δ_i are presumed known for each star.

Having obtained its components, we can calculate the speed of the solar motion as

$$S_\odot = (X_\odot^2 + Y_\odot^2 + Z_\odot^2)^{1/2} \tag{6-17}$$

and the position of the apex in equatorial coordinates follows from

$$\tan \alpha_A = \frac{Y_\odot}{X_\odot} \tag{6-18a}$$

$$\tan \delta_A = \frac{Z_\odot}{(X_\odot^2 + Y_\odot^2)^{1/2}} \tag{6-18b}$$

Equation (6-18b) determines δ_A uniquely in the range $-90^\circ \leq \delta_A \leq 90^\circ$. To resolve the choice of quadrant for α_A, one notes that $\cos \delta_A$ is greater than zero. Hence, the signs of $X_\odot$ and $Y_\odot$ determine the signs of $\cos \alpha_A$ and $\sin \alpha_A$, respectively. Finally, (ℓ_A, b_A), the position of the apex in galactic coordinates, is easily obtained from (α_A, δ_A) by standard formulae of spherical trigonometry. We can then easily compute

$$u_\odot = -S_\odot \cos \ell_A \cos b_A \tag{6-19a}$$

$$v'_\odot = S_\odot \sin \ell_A \cos b_A \tag{6-19b}$$

$$w_\odot = S_\odot \sin b_A \tag{6-19c}$$

Solution from Proper Motions

Consider now equations (6-11b) and (6-11c). We first express $\dot{\alpha} d \cos \delta$ and $\dot{\delta} d$ in terms of practical observational units. Both $\dot{\delta} d$ and $\dot{\alpha} d \cos \delta$ are to be velocities in kilometers per second. Thus if we express d in parsecs as $(1/\pi'')$, where π'' is the parallax in seconds of arc, and use proper motions μ''_δ and $\mu''_\alpha \cos \delta$ in seconds of arc per year, then, from equation (3-2), we see that $\dot{\delta} d = (4.74\mu''_\delta/\pi'')\,(\text{km s}^{-1})$ and $\dot{\alpha} d \cos \delta = (4.74\mu''_\alpha \cos \delta/\pi'')\,(\text{km s}^{-1})$. Therefore, equations (6-11b) and (6-11c) can be rewritten as

$$-\dot{x}_i \sin \delta_i \cos \alpha_i - \dot{y}_i \sin \delta_i \sin \alpha_i + \dot{z}_i \cos \delta_i = \frac{4.74\mu''_{\delta i}}{\pi''_i} \tag{6-20a}$$

and

$$-\dot{x}_i \sin \alpha_i + \dot{y}_i \cos \alpha_i = \frac{4.74\mu''_{\alpha i} \cos \delta_i}{\pi''_i} \tag{6-20b}$$

Suppose we choose N stars in a region of the sky that is sufficiently small that all stars in it have almost the same α_i and δ_i. Then, by the same line of reasoning as was used before, we argue that the averages $\langle X_{*i}\rangle$, $\langle Y_{*i}\rangle$, and $\langle Z_{*i}\rangle$ are zero within the region. Hence, when we average equations (6-20a) and (6-20b) over all stars in the region, we simply drop the terms in X_*, Y_*, and Z_*, and obtain

$$X_\odot \sin\bar{\delta}\cos\bar{\alpha} + Y_\odot \sin\bar{\delta}\sin\bar{\alpha} - Z_\odot \cos\bar{\delta} = \frac{4.74}{N}\sum_{i=1}^{N}\left(\frac{\mu''_{\delta i}}{\pi''_i}\right) \tag{6-21a}$$

and

$$X_\odot \sin\bar{\alpha} - Y_\odot \cos\bar{\alpha} = \frac{4.74}{N}\sum_{i=1}^{N}\left(\frac{\mu''_{\alpha i}\cos\bar{\delta}}{\pi''_i}\right) \tag{6-21b}$$

where $\bar{\alpha}$ and $\bar{\delta}$ are mean positions within the small region.

Generally speaking, the parallax π''_i for any individual star will be unknown. If, however, we assume that the stars all lie at some mean distance d_0 corresponding to a mean parallax π''_0, then we can replace π''_i by π''_0. In practice, to assure the validity of this approximation, we would initially choose stars of a definite spectral type (hence of a definite absolute magnitude) that all lie within a reasonably narrow range of apparent magnitudes. Then, writing

$$\bar{\mu}_\alpha \equiv \frac{\sum_{i=1}^{N}\mu_{\alpha i}}{N} \quad \text{and} \quad \bar{\mu}_\delta \equiv \frac{\sum_{i=1}^{N}\mu_{\delta i}}{N}$$

we have

$$X_\odot \sin\bar{\delta}\cos\bar{\alpha} + Y_\odot \sin\bar{\delta}\sin\bar{\alpha} - Z_\odot \cos\bar{\delta} = \frac{4.74\bar{\mu}_\delta}{\pi_0} \equiv K\bar{\mu}_\delta \tag{6-22a}$$

and

$$X_\odot \sin\bar{\alpha} - Y_\odot \cos\bar{\alpha} = \frac{4.74\bar{\mu}_\alpha\cos\bar{\delta}}{\pi_0} \equiv K\bar{\mu}_\alpha\cos\bar{\delta} \tag{6-22b}$$

If we choose R distinct regions on the sky, then we shall have R sets of equations (6-22), which can again be solve by the least-squares method by taking moments against the coefficients of the unknowns, as discussed previously. We thus obtain

$$X_\odot \sum_{j=1}^{R}\sin^2\bar{\delta}_j\cos^2\bar{\alpha}_j + Y_\odot \sum_{j=1}^{R}\sin^2\bar{\delta}_j\sin\bar{\alpha}_j\cos\bar{\alpha}_j - Z_\odot \sum_{j=1}^{R}\sin\bar{\delta}_j\cos\bar{\delta}_j\cos\bar{\alpha}_j$$
$$= K\sum_{j=1}^{R}\bar{\mu}_{\delta j}\sin\bar{\delta}_j\cos\bar{\alpha}_j \tag{6-23a}$$

$$X_{\odot} \sum_{j=1}^{R} \sin^2 \bar{\delta}_j \sin \bar{\alpha}_j \cos \bar{\alpha}_j + Y_{\odot} \sum_{j=1}^{R} \sin^2 \bar{\delta}_j \sin^2 \bar{\alpha}_j - Z_{\odot} \sum_{j=1}^{R} \sin \bar{\delta}_j \cos \bar{\delta}_j \sin \bar{\alpha}_j$$

$$= K \sum_{j=1}^{R} \bar{\mu}_{\delta j} \sin \bar{\delta}_j \sin \bar{\alpha}_j \qquad (6\text{-}23b)$$

$$X_{\odot} \sum_{j=1}^{R} \sin \bar{\delta}_j \cos \bar{\delta}_j \cos \bar{\alpha}_j + Y_{\odot} \sum_{j=1}^{R} \sin \bar{\delta}_j \cos \bar{\delta}_j \sin \bar{\alpha}_j - Z_{\odot} \sum_{j=1}^{R} \cos^2 \bar{\delta}_j$$

$$= K \sum_{j=1}^{R} \bar{\mu}_{\delta j} \cos \bar{\delta}_j \qquad (6\text{-}23c)$$

$$X_{\odot} \sum_{j=1}^{R} \sin^2 \bar{\alpha}_j - Y_{\odot} \sum_{j=1}^{R} \sin \bar{\alpha}_j \cos \bar{\alpha}_j = K \sum_{j=1}^{R} \bar{\mu}_{\alpha j} \sin \bar{\alpha}_j \cos \bar{\delta}_j \qquad (6\text{-}24a)$$

and

$$X_{\odot} \sum_{j=1}^{R} \sin \bar{\alpha}_j \cos \bar{\alpha}_j - Y_{\odot} \sum_{j=1}^{R} \cos^2 \bar{\alpha}_j = K \sum_{j=1}^{R} \bar{\mu}_{\alpha j} \cos \bar{\alpha}_j \cos \bar{\delta}_j \qquad (6\text{-}24b)$$

Finally, in order to use the proper-motion information as fully as possible, we add equations (6-24a) and (6-24b) to (6-23a) and (6-23b) to obtain three equations of the form

$$a_i X_{\odot} + b_i Y_{\odot} + c_i Z_{\odot} = K g_i \qquad (i = 1, 2, 3) \qquad (6\text{-}25)$$

which can be solved in a straightforward manner for $(X_{\odot}/K)$, $(Y_{\odot}/K)$, and $(Z_{\odot}/K)$.

We can then calculate the position of the apex from equations (6-18). Notice that we obtain only the direction of the solar motion, but not its magnitude. (That information was foregone when we assumed that the distances to the stars under analysis were unknown.)

Solution from Space Motions

If we know the distances to the stars under study, whether from direct parallax measurements or from estimates based on absolute magnitudes inferred from spectral-type information, we can convert their observed proper motions to transverse velocities. Combining the transverse components with radial velocities, we can compute space motions of the stars relative to the Sun. Velocity components in the equatorial system can be found directly from equations (6-10), and these components can then be expressed in the galactic coordinate system by a simple rotation of axes, which yields (U_*, V_*, W_*) as defined in equations (6-3). The solar motion in galactic coordinates

Table 6-3. Solar Motion with Respect to Disk and Spheroidal-Component Objects

Type of star or object	Components and speed (km s^{-1}) $u_\odot$	$v'_\odot$	$w_\odot$	$S_\odot$	Position of apex (°) ℓ_A	b_A
	Spiral-arm and disk objects					
Dwarfs						
B0	−9.6	14.5	6.7	19	57	21
A0	−7.3	13.7	7.2	17	62	25
A5	−8.5	7.8	7.4	14	43	33
F0	−11.1	10.8	7.2	17	44	25
F5	−10.1	12.3	6.2	17	51	23
G0	−14.5	21.1	6.4	26	56	15
G5	−8.1	22.1	4.3	24	70	11
K0	−10.8	14.9	7.4	20	55	23
K5	−9.5	22.4	5.8	25	68	14
M0	−6.1	14.6	6.9	17	68	25
M5	−9.8	19.3	8.6	23	64	23
Giants						
A	−13.4	11.6	10.3	20	41	30
F	−19.7	18.5	9.5	29	43	19
G	−7.2	11.1	6.9	15	57	28
K0	−10.6	18.6	6.5	22	60	17
K3	−9.0	17.6	6.4	21	63	18
M	−4.5	18.3	6.2	20	76	18
Supergiants						
Classical Cepheids	−8.6	12.0	7.6	17	54	27
O–B5	−9.0	13.4	3.7	17	56	13
F–M	−7.9	11.7	6.5	16	56	25
Other						
Interstellar Ca II	−11.4	14.4	8.2	20	52	24
Galactic (open) clusters	−7.0	19.5	8.2	22	70	22
Carbon stars	−10.7	31.8	3.5	34	71	6
Subgiants	−8.0	28.0	8.0	36	75	15
Planetary nebulae	−8	29	8	31	74	14
White dwarfs	−6	37	8	38	81	12
Late-type variables $P > 300^d$	−12	26	8	30	65	16
Late-type variables, $P < 300^d$	−27	42	11	50	58	12
RR Lyrae, $P < 0^d.45$	−47	31	4	56	32	4
	Spheroidal-component objects					
Subdwarfs	+8	150	3	150	93	1
RR Lyrae, $P > 0^d.45$	0	225	25	225	90	6
Globular clusters	−10	182	6	182	87	2

SOURCE: Adapted from (**B1**, Chapter 4), by permission. Copyright © 1965 by the University of Chicago.

then follows directly from equations (6-6a), (6-6c), and (6-7). A basic advantage of this method is that it makes full use of both radial velocity and proper-motion data, and hence it is less vulnerable to any systematic errors that may exist in one set of data or the other alone. The disadvantage of this approach, of course, is that it requires that accurate distances be available for the stars being studied.

6-4. SUMMARY OF RESULTS

A large number of determinations of the solar motion have been made relative to a wide variety of stellar types. The results are summarized in Table 6-3, which gives the components and total speed of the solar motion and the position of the apex with respect to each group analyzed. We immediately notice that the results obtained from spiral-arm and disk objects are quite different from those obtained from spheroidal-component objects, which characteristically yield large values of $v'_\odot$. We shall see in §7-2 that these large values of the solar motion relative to spheroidal-component objects arise because, on average, the latter rotate around the galactic center more slowly than do disk stars like the Sun.

The weighted mean values of $u_\odot$ and $w_\odot$ from the spiral-arm and disk stars, including the carbon stars, subgiants, planetary nebulae, and white dwarfs, is (**B1**, Chapter 4)

$$\langle u_\odot \rangle = -9.2 \pm 0.3 \text{ km s}^{-1} \tag{6-26a}$$

$$\langle w_\odot \rangle = 6.9 \pm 0.2 \text{ km s}^{-1} \tag{6-26b}$$

These results show that the peculiar velocity of the Sun is such that it moves inward toward the galactic center and "upward" (that is, toward the north galactic pole) out of the plane relative to the LSR. A similar average of $v'_\odot$ for all spectral types is not physically meaningful, because different types of stars have different systematic lags relative to the LSR, which depend on the average age and random velocity of stars of each type (see §7-1). But, even before we correct for this effect, we can see that $v'_\odot > 0$ for all stellar types, and we conclude that the Sun must surely lead the LSR in the direction of galactic rotation [see equations (6-31)].

As a different solar motion is obtained from stars of each spectral type, it is obvious that to talk about "the" solar motion without specifying the group of stars relative to which it has been measured lacks precision. Therefore, for ease of discussion, it is convenient to adopt a definite reference value. Two commonly used reference values are described next. As we mentioned earlier, these should be regarded as no more than approximate indicators of the Sun's peculiar velocity relative to the LSR.

Standard Solar Motion

The *standard solar motion* is defined to be the solar motion relative to the *stars most commonly listed in general catalogs of radial velocity and proper motion.* These stars are typically of spectral types A through G, including dwarfs, giants, and supergiants. Taking a weighted average of results for individual types, one obtains (**B1**, Chapter 4)

$$u_{\text{std}} = -10.4 \text{ km s}^{-1} \tag{6-27a}$$

$$v'_{\text{std}} = 14.8 \text{ km s}^{-1} \tag{6-27b}$$

$$w_{\text{std}} = 7.3 \text{ km s}^{-1} \tag{6-27c}$$

These values imply $S = 19.5 \text{ km s}^{-1}$ in the direction $\ell_{\text{std}} = 56°$, $b_{\text{std}} = 23°$. Although the standard solar motion is characteristic of the most commonly observed brighter stars, it does not have an obvious physical interpretation, and it is not necessarily the best representation of the Sun's peculiar velocity.

Basic Solar Motion

The *basic solar motion* is defined as the solar motion determined from *the most commonly measured velocities for stars in the solar neighborhood.* In practice, a sample of 400 A stars and 400 K giants within 100 pc give almost identical results, which are also in agreement with those obtained from a large sample of M dwarfs. One finds (**B1**, Chapter 4)

Type	u_{basic}	v'_{basic}	w_{basic}
A	−9.4	9.9	5.6
gK	−9.3	10.7	6.7
dM	−8	11.3	6

from which we adopt

$$u_{\text{basic}} = -9 \text{ km s}^{-1} \tag{6-28a}$$

$$v'_{\text{basic}} = 11 \text{ km s}^{-1} \tag{6-28b}$$

$$w_{\text{basic}} = 6 \text{ km s}^{-1} \tag{6-28c}$$

These components correspond to $S_{\text{basic}} = 15.4 \text{ km s}^{-1}$ in the direction $\ell_{\text{basic}} = 51°$, $b_{\text{basic}} = 23°$. As we shall see later, the basic solar motion agrees with the Sun's peculiar velocity relative to the LSR to within the errors of determination of the basic motion.

Reduction to the LSR

Let us now turn to the problem of determining the Sun's peculiar velocity with respect to the LSR. In terms of galactic dynamics, this description of the Sun's motion is the most interesting one physically. We first must estimate $\langle v_* \rangle$ for different types of stars. This is the quantity needed to reduce $v'_\odot$ to $v_\odot$ via equations (6-6b) and (6-7). As we shall show in detail in Chapter 14, dynamical theory leads to the relation

$$\langle v \rangle = \langle \Theta \rangle - \Theta_0 = \frac{\langle \Pi^2 \rangle}{2R_0(A - B)} \left[\frac{\partial \ln \nu}{\partial \ln R} + \frac{\partial \ln \langle \Pi^2 \rangle}{\partial \ln R} + \left(1 - \frac{\langle \theta^2 \rangle}{\langle \Pi^2 \rangle}\right) + \left(1 - \frac{\langle Z^2 \rangle}{\langle \Pi^2 \rangle}\right) \right] \tag{6-29}$$

where R is the distance from the galactic center (R_0 being the Sun's distance); ν is the number of stars per unit volume; $\langle \Pi^2 \rangle = \langle u^2 \rangle$, $\langle \theta^2 \rangle = \langle v^2 \rangle$, and $\langle Z^2 \rangle = \langle w^2 \rangle$ are dispersions of the velocity distribution; and A and B are constants characterizing galactic rotation. To see the physical significance of this result, let us focus on the term in $(\partial \ln \nu / \partial \ln R)$. Suppose stars in the solar neighborhood consist of a mixture of three groups (all of which have $Z \equiv 0$) as follows. (1) Stars moving on strictly circular orbits with $R \equiv R_0$ and $\Theta(R_0) \equiv \Theta_0$. These stars move *with* the LSR. (2) Stars moving on elliptical orbits with $R \leq R_0$. When these stars are at $R = R_{max} = R_0$, they will have $\Theta(R_0) < \Theta_0$; that is, these stars *lag* the LSR. (3) Stars moving on elliptical orbits with $R \geq R_0$. When these stars are at $R = R_{min} = R_0$, they will have $\Theta(R_0) > \Theta_0$; that is, these stars *lead* the LSR (see Figure 6-6).

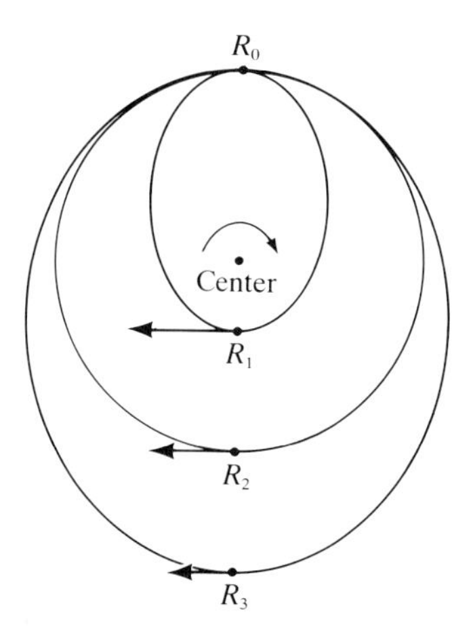

Figure 6-6. Stars starting at R_2 are assumed to move on circular orbits with $R \equiv R_0$ and $\Theta \equiv \Theta_0 = \Theta_{circular}(R_0)$. Stars starting at R_1 can reach $R_{max} = R_0$ if $\Theta(R_1) > \Theta_{circular}(R_1)$. When these stars arrive at R_0, they will have $\Theta(R_{max}) < \Theta_0$ and hence will lag the LSR. Stars starting at R_3 can reach $R_{min} = R_0$ if $\Theta(R_3) < \Theta_{circular}(R_3)$. When these stars arrive at R_0, they will have $\Theta(R_{min}) > \Theta_0$ and hence will lead the LSR. Because the density of stars at R_1 is greater than that at R_3, on the average the net effect is for stars at R_0 to lag the LSR.

Because the number density of stars in our Galaxy is greater for $R < R_0$ than $R > R_0$, we expect that, on the average, there will be a *net lag*, and so, for the group as a whole, $\langle v \rangle$ will be negative [Notice that, under the assumptions stated, $(\partial \ln v / \partial \ln R)$ is negative, hence equation (6-29) also predicts $\langle v \rangle < 0$.] The proportionality of the lag to $\langle \Pi^2 \rangle$ can be rationalized by noting that this quantity gives a measure of the stars' ability to migrate radially in our Galaxy (recall that $\langle \Pi \rangle$ is always zero for steady axisymmetric models). Clearly, stars with $\langle \Pi^2 \rangle \equiv 0$ are all on circular orbits, and hence can show no drift, while those with large $\langle \Pi^2 \rangle$ can come from positions with R quite different from R_0, and hence can lag or lead by large amounts.

For our present purposes, it suffices to note that equation (6-29) is equivalent to a relation of the form

$$v'_\odot = v_\odot - \langle v \rangle = v_\odot + c\langle \Pi^2 \rangle \tag{6-30}$$

a result first found empirically in 1924 by G. Strömberg (**S4**), (**S5**). If we plot $v'_\odot$ from Table 6-3 versus $\langle \Pi^2 \rangle$ from Table 7-1, we obtain the results shown in Figure 6-7. A linear fit to the points, extrapolated to $\langle \Pi^2 \rangle = 0$, gives $v_\odot = 12 \text{ km s}^{-1}$. Then, adopting the mean u and w components as given in equations (6-26), we have

$$u_\odot = -9 \text{ km s}^{-1} \tag{6-31a}$$

$$v_\odot = 12 \text{ km s}^{-1} \tag{6-31b}$$

$$w_\odot = 7 \text{ km s}^{-1} \tag{6-31c}$$

for the components of the peculiar velocity of the Sun relative to the LSR. These results imply a speed of 16.5 km s^{-1} in the direction $\ell = 53°$, $b = 25°$.

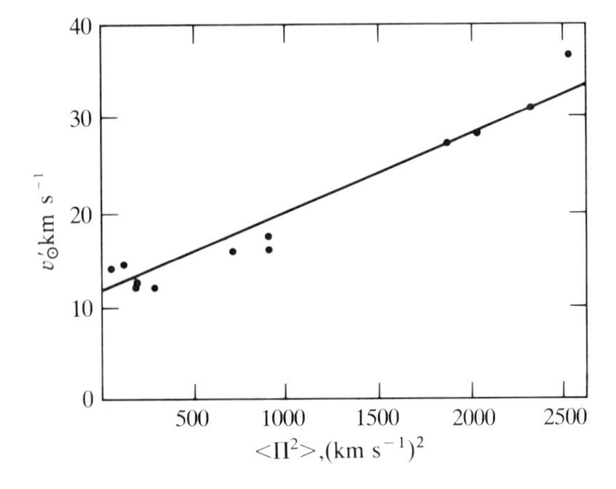

Figure 6-7. Correlation of $v'_\odot$ with $\langle \Pi^2 \rangle$. We see that the observed value of $v'_\odot$ is a linear function of $\langle \Pi^2 \rangle$ as predicted by equation (6-30). Extrapolating to $\langle \Pi^2 \rangle = 0$, we estimate $v_\odot = 12 \text{ km s}^{-1}$. [From (**B1**, Chapter 4), by permission. Copyright © 1965 by the University of Chicago.]

[For further details, see (**B1**, Chapter 4)]. It is interesting to note that the Sun's peculiar velocity agrees fairly well with the basic solar motion. If this agreement is not merely fortuitous, it may reflect a low value of the gradient $(\partial \ln v/\partial \ln R)$ in the solar neighborhood for the spectral types used in the determination of the basic solar motion, for this would imply that these stars would not lag the LSR by very much.

It is worth noting one final implication of equations (6-29) and (6-30). As we shall see in §7-1, the velocity dispersion $\langle u^2 \rangle$ (or $\langle \Pi^2 \rangle$) is an increasing function of age for disk stars. We therefore should expect old stars to lag the Sun systematically more than young stars. That this really is the case can be seen by comparing $v'_\odot$ in Table 6-3 for O and B stars and Cepheids to $v'_\odot$ for manifestly evolved stars such as the planetary nebulae and white dwarfs. The latter are seen to lag the former by about 25 km s^{-1}. The effect is much less clear in the data listed in Table 6-3 for G, K, and M dwarfs. The picture is blurred because, as mentioned earlier, these stars can live so long that a sample chosen by spectral class only lumps together stars of all ages (newly born to the oldest surviving disk objects) indiscriminately. However, we shall see in §7-1 that, if we choose subgroups within each spectral class according to some criterion that provides age discrimination (for example, position in the H-R diagram, or a spectroscopic criterion such as dMe stars versus nonemission dM stars), then we see the increase in $\langle u^2 \rangle$ and in $v'_\odot$ with time quite clearly.

6-5. MOTION OF THE LSR AROUND THE GALACTIC CENTER

Thus far in this chapter, we have focused attention on the determination of the Sun's peculiar velocity relative to the LSR. Let us now attempt to estimate Θ_0, the *circular velocity of the LSR* around the galactic center. To find Θ_0, we must first measure the Sun's velocity with respect to a group of objects that can be taken to be at rest relative to the fundamental standard of rest (that is, the center of our Galaxy). Then, because we know the Sun's peculiar velocity, it is trivial to correct this observed value for the Sun's motion with respect to the LSR, and hence to derive the motion of the LSR itself. (In practice, this correction is smaller than the probable error in the current estimates of the values of Θ_0, and we shall therefore ignore it henceforth in our discussion of Θ_0.)

In this section, we shall concentrate on the most direct observational determinations of Θ_0 that are possible at the present time, and we shall defer consideration of the constraints imposed by the local values of the differential rotation constants (Oort's A and B constants) until §8-2, and by the complete rotation curve $\Theta(R)$ for our Galaxy until §8-3. There are basically three different approaches that have been made to the problem of measuring Θ_0, each of which has its own advantages and difficulties. Two of these methods are based on the observed kinematic properties of objects within our Galaxy, and one makes use of data from external galaxies.

Globular Clusters and RR Lyrae Stars

The galactic halo is a very large, roughly spherical system around the galactic center. Because this system is not strongly flattened, it is reasonable to suppose that it does not rotate rapidly about the galactic center. If this assumption is true, halo objects such as globular clusters far from the galactic plane and metal-poor RR Lyrae stars [that is, those with large ΔS (see §3-8)] may be expected to move such that, at any point, their orbits are distributed nearly isotropically about the radius vector to the galactic center. The measured solar motion with respect to halo objects will therefore be a reflection of the rotation speed Θ_0 of the LSR. These considerations motivated the use of the halo globular clusters as an approximately nonrotating reference frame in some of the earliest attempts to infer Θ_0.

In general, we must allow for the possibility that the halo actually does rotate with nonzero velocity. When we determine the tangential component of the solar motion (in galactic coordinates) with respect to halo objects, what we really measure is

$$\langle -v_{\text{halo}} \rangle = \langle -V_{\text{halo}} \rangle - v_{\odot} = \Theta_0 - \langle \Theta_{\text{halo}} \rangle \qquad (6\text{-}32)$$

If we knew a priori that $\langle \Theta_{\text{halo}} \rangle = 0$, then, from the measured value of $\langle -v_{\text{halo}} \rangle$, the desired quantity Θ_0 would follow directly. Unfortunately, however, at the present time there is no way we can directly estimate $\langle \Theta_{\text{halo}} \rangle$. If, on the other hand, we assume that the sense of any possible rotation of the halo is the same as that of the disk—that is, if we assume $\langle \Theta_{\text{halo}} \rangle \geq 0$—then equation (6-32) enables us to obtain a lower bound on Θ_0.

In 1946, N. U. Mayall (**M1**) applied this method to the observed radial velocities for fifty globular clusters, and he found that the solar motion relative to these objects corresponds to a speed of 200 ± 25 km s^{-1} in the direction $\ell = 87°$, $b = 0°$ (that is, almost perpendicular to the direction to the galactic center). In 1959, T. D. Kinman (**K2**) rederived the solar motion with respect to the globular clusters from more extensive data (seventy clusters), and he found a speed (for all clusters) of 167 ± 30 km s^{-1}, again along a direction essentially perpendicular to the Sun center line. Kinman's data can be subdivided into clusters with $R < 9$ kpc, for which $\langle -V \rangle_{\text{g.c.}} = 141 \pm 40$ km s^{-1}, and those with $R > 9$ kpc, for which $\langle -V \rangle_{\text{g.c.}} = 182 \pm 50$ km s^{-1} (the value quoted in Table 6-3). Because the latter group is expected, on general principles, to have the smaller rotation speed around the galactic center, we conclude from these measurements that Θ_0 is at least 180 km s^{-1}.

As we shall see later, the best estimate of Θ_0 on the basis of all available evidence gives $\Theta_0 \approx 250$ km s^{-1}. From this value for Θ_0, we conclude that the system of globular clusters is indeed rotating, and, for the complete group measured by Kinman, $\langle \Theta \rangle_{\text{g.c.}} \approx 80$ km s^{-1}. These data also suggest a differential rotation of the system, with $\langle \Theta \rangle_{\text{g.c.}} \approx 110$ km s^{-1} for clusters with $R < 9$ kpc, and $\langle \Theta \rangle_{\text{g.c.}} \approx 70$ km s^{-1} for clusters with $R > 9$ kpc.

Another estimate of this type can be made using the radial velocities of halo RR Lyrae stars. From the radial velocities of a total of seventy-nine RR Lyrae variables with a mean $\Delta S \approx 7$, R. Woolley and A. Savage (**W1**) obtained $\langle V \rangle_{RR} = 225 \pm 25$ km s^{-1}. They found no significant difference between two subgroups of these stars having mean $\Delta S \approx 6$ and $\Delta S \approx 9$, respectively. These data indicate that Θ_0 must be at least 225 km s^{-1}. This conclusion is strengthened further by a study of the relative fractions of these stars that will be moving on direct and on retrograde orbits around the galactic center for different values of Θ_0. For each star, the rotational velocity Θ_* around the galactic center is given by $\Theta_* = V_* + v_\odot + \Theta_0$, where V_* is the observed tangential velocity of the star with respect to the Sun. It is obvious that the smaller the assumed value of Θ_0, the larger the number of stars that will be thought to have retrograde orbits (that is, $\Theta_* < 0$). Woolley and Savage find that, if $\Theta_0 = 250$ km s^{-1}, then about 48% of the sample move on retrograde orbits, whereas if $\Theta_0 = 200$ km s^{-1}, about 59% move on retrograde orbits. Because it seems extremely unlikely that more than 50% of the RR Lyrae stars should be counterrotating, the results of Woolley and Savage suggest that Θ_0 cannot be much less than 250 km s^{-1}.

Escape Speed of High-Velocity Stars

The kinematic properties of the halo-population stars observed to have the largest space velocities with respect to the Sun (that is, the extreme high-velocity stars) can be used to make an alternative indirect estimate of both the circular velocity Θ_0 and the *escape speed* S_{esc} at which stars in the solar neighborhood would have enough energy to escape completely from the gravitational field of our Galaxy. The basic idea underlying this approach was suggested by Oort, who noted that we observe (see §7-2) no stars leading the LSR in the direction of galactic rotation with a tangential peculiar velocity component greater than some critical value v_{crit}. He suggested that the reason we do not find such stars could conceivably be that they are moving at the escape speed and hence are lost from our Galaxy. That is, the observed value of v_{crit} might be such that

$$(\Theta_0 + v_{crit})^2 = S_{esc}^2 \tag{6-33}$$

The basic point is that stars moving with zero velocity relative to the Sun already have a tangential velocity component Θ_0 with respect to the galactic center and are on circular orbits of radius R_0. If we add even a small increment to the tangential component in the same direction, the star acquires a rather large amount of kinetic energy, and its energy very rapidly approaches the escape limit. Therefore, our a priori expectation is that, if we observe a cutoff in the stellar velocity distribution, it should occur first for stars moving in the same direction as the direction of galactic rotation.

More generally, any star with a peculiar velocity (u_*, v_*, w_*) with respect to the LSR will escape provided that the peculiar velocity satisfies the inequality

$$u_*^2 + (\Theta_0 + v_*)^2 + w_*^2 \geq S_{\text{esc}}^2 \tag{6-34}$$

Clearly, stars with large observed values of $|u_*|$ or $|w_*|$, or fairly large positive or exceedingly large negative values of v_*, are the best candidates for escape. If we equate the right-hand side of equation (6-34) with the left-hand side of equation (6-33) and solve for Θ_0, we find

$$\Theta_0 \leq \frac{u_*^2 + v_*^2 + w_*^2 - v_{\text{crit}}^2}{2(v_{\text{crit}} - v_*)} \tag{6-35}$$

Now, if we assume that we know v_{crit} and choose a sample of high-velocity stars with the largest observed peculiar velocities (u_*, v_*, w_*) according to the precepts just mentioned above, and if we hypothesize that these stars are moving at or near to the escape speed, then we can obtain an estimate of Θ_0.

This procedure was carried out by W. Fricke (**F1**) using $v_{\text{crit}} = 63$ km s^{-1}, the velocity at which Oort had found the distribution of stellar peculiar velocities to become markedly asymmetric (see §7-2). Using data for the most extreme high-velocity stars in the then available sample, he obtained $\Theta_0 = 275 \pm 25$ km s^{-1}. The pair of values, $v_{\text{crit}} = 63$ km s^{-1} and $\Theta_0 = 275$ km s^{-1}, implies by equation (6-33) that $S_{\text{esc}} = 338$ km s^{-1}. This value of S_{esc} is clearly an underestimate because, in the catalogs of high-velocity stars published by Eggen (**E1**) and Sandage (**S1**), one can find at least ten stars (conservatively chosen) that have speeds relative to the galactic center in the range from 380 to 400 km s^{-1}. It thus appears that the escape speed must be at least 400 km s^{-1}. S. Isobe (**I1**) has recently analyzed the best available data by the least-squares method to obtain new estimates of both Θ_0 and v_{crit}. He obtains $\Theta_0 = 275 \pm 20$ km s^{-1} in agreement with Fricke's value, and $v_{\text{crit}} \approx 110$ km s^{-1}, which implies $S_{\text{esc}} \approx 385$ km s^{-1}.

Independent considerations, which we shall discuss in Chapter 21, indicate, however, that the galactic escape velocity S_{esc} is very unlikely to be as small as 400 km s^{-1}. If the escape velocity really is larger than 400 km s^{-1}, then the apparent cutoff in the number density of rapidly moving stars with prograde motions noted by Oort cannot be ascribed to limitations on their kinetic energy imposed by a finite velocity of escape from our Galaxy. Rather, we must recognize that there is, for some reason, a deficiency of stars having very high angular momentum. If this deficiency of high angular-momentum halo stars applied equally to stars on both retrograde orbits and prograde orbits, then Oort's analysis would still yield a useful estimate of Θ_0, although the value of S_{esc} derived in this way would not be an estimate of the galactic escape velocity. However, even under these circumstances, application of Oort's method is likely to lead to an overestimate of Θ_0 because, from any given set of data, one is likely to underestimate v_{crit}, and equation (6-35) shows that this underestimate will tend to increase the derived value of Θ_0.

Local Group of Galaxies

In §5-3, we saw that our Galaxy is a member of a group of twenty to thirty galaxies known as the Local Group (LG). The galaxies of the LG are thought to be bound together by their mutual gravitational attraction. Several distinct subgroups of galaxies can be identified within the LG, each subgroup being centered on a giant galaxy. The subgroups centered on our Galaxy and on M31 contribute more than 90% of the luminosity, and, therefore, presumably of the mass, of the whole LG.

From measurements of the radial velocities of LG galaxies relative to the Sun, we can infer $\mathbf{v}_0$, the solar motion with respect to the velocity centroid of the LG. We shall assume that the velocity centroid coincides with the motion of the center of mass of the group. The observationally derived solar motion consists of three distinct components: (a) the velocity of revolution of the LSR around the galactic center, $\mathbf{v}_{rot} = \Theta_0\hat{\boldsymbol{\theta}}$, where $\hat{\boldsymbol{\theta}}$ denotes the unit vector in the plane of our Galaxy pointing in the direction of galactic rotation; (b) the (known) peculiar velocity of the Sun with respect to the LSR, $\mathbf{v}_\odot$; and (c) $\mathbf{v}_G$, the velocity of the center of mass of our Galaxy relative to the center of mass of the LG. That is,

$$\mathbf{v}_0 = \Theta_0\hat{\boldsymbol{\theta}} + \mathbf{v}_\odot + \mathbf{v}_G \qquad (6\text{-}36)$$

In principle, it is straightforward to determine $\mathbf{v}_0$, and several such determinations have been made (see Table 6-4). In practice, many difficulties are encountered, including the necessity of deciding whether a particular galaxy is, or is not, a member of the LG, and whether, if it *is* a member, its observed radial velocity should be used in the determination of $\mathbf{v}_0$. For example, it is clear that, although they do belong to the LG, the satellites of M31 and our

Table 6-4. Solar Motion Relative to the Velocity Centroid of the Local Group

Observer	v_0(km s^{-1})	ℓ	b	Reference
Mayall	300 ± 25	$93° \pm 6°$	$-14° \pm 4°$	**(M1)**
Humason and Wahlquist	292 ± 32	$106° \pm 6°$	$-7° \pm 4°$	**(H1)**
Byrnes	280 ± 23	$107° \pm 5°$	$-7° \pm 4°$	**(B2)**
de Vaucouleurs and Peters	315 ± 15	$95° \pm 6°$	$-8° \pm 3°$	**(D1)**
Yahil, Tammann, and Sandage	308 ± 23	$105° \pm 5°$	$-7° \pm 4°$	**(Y1)**

Galaxy should not be included in the sample, because these systems are dynamically bound to their primaries and hence reflect the motion of the primary. This fact implies that they do not move at random with respect to the center of the LG, and therefore they cannot be used to define the LG velocity centroid. Detailed discussion of these and other practical points can be found in (**Y1**).

The measured value of $\mathbf{v}_0$ is easily corrected for the effects of the Sun's peculiar vecocity, $\mathbf{v}_\odot$, to yield $\mathbf{v}_{\text{LSR}} = \Theta_0 \hat{\theta} + \mathbf{v}_G$, the velocity of the LSR with respect to the velocity centroid of the LG. The results from all the determinations listed in Table 6-4 are in quite good agreement, and they show that $\mathbf{v}_{LSR}$ has a magnitude of about $300 \pm 25 \text{ km s}^{-1}$ in the direction $\ell = 105^\circ \pm 5^\circ$, $b = -8^\circ \pm 4^\circ$. If we could safely assume that $\Theta_0 \gg v_G$, then we would infer that $\Theta_0 \approx 300 \text{ km s}^{-1}$. It should be noted, however, that this value is significantly larger than the maximum rotation velocities observed in external galaxies of the same morphological type as our Galaxy (see Table 8-3). This value would, moreover, imply so large a rotation rate for the system of globular clusters that it should be significantly flattened, contrary to observation. Thus the estimate of 300 km s^{-1} for Θ_0 is unlikely to be correct, and the actual value is probably smaller.

We arrived at a large value of Θ_0 by assuming that $v_G \ll \Theta_0$. Of course, we have no valid a priori reason for thinking that v_G is actually small. It could, in fact, be a rather large number, which would imply a smaller value of Θ_0 if v_G were properly oriented (although the question would then arise as to whether or not the LG is dynamically bound).

It is possible, at least in principle, to determine both Θ_0 and $\mathbf{v}_G$ from the observations by invoking dynamical hypotheses. An attempt to determine them has recently been made by D. Lynden-Bell and D. N. C. Lin (**L1**). Suppose we argue that, because our Galaxy and M31 dominate the mass of the LG, their momenta must be equal and opposite about the center of mass (which is also the velocity centroid and hence can be regarded to be at rest). Then, we can write

$$m_A \mathbf{v}_A + m_G \mathbf{v}_G = 0 \tag{6-37a}$$

or

$$\mathbf{v}_A = -\frac{m_G}{m_A}\mathbf{v}_G \equiv -\mu \mathbf{v}_G \tag{6-37b}$$

where the subscript A denotes "Andromeda." Now, the observed radial velocity of M31 with respect to the LSR is simply the projection of the relative velocity $\mathbf{v}_{\text{rel}} = \mathbf{v}_A - \mathbf{v}_{\text{LSR}} = \mathbf{v}_A - (\Theta_0 \hat{\theta} + \mathbf{v}_G)$ along the line of sight from the Sun to M31, which we specify with a unit vector $\mathbf{n}_A$; that is,

$$v_R(\text{M31}) = [\mathbf{v}_A - (\Theta_0 \hat{\theta} + \mathbf{v}_G)] \cdot \mathbf{n}_A = -\Theta_0(\hat{\theta} \cdot \mathbf{n}_A) - (1 + \mu)\mathbf{v}_G \cdot \mathbf{n}_A \tag{6-38}$$

Suppose that we have already determined $\mathbf{v}_0$ and hence $\mathbf{v}_{LSR}$. Then, by invoking equation (6-38) for given values of v_R(M31), μ, and $\mathbf{n}_A$ (of which only μ is uncertain), we have just enough information to determine Θ_0 and the three components of $\mathbf{v}_G$ separately.

From their analysis of the observational data, Lynden-Bell and Lin derive, formally, $\Theta_0 = 295 \pm 50$ km s^{-1}, and $\mathbf{v}_G = (34, 7, -16) \pm 25$ km s^{-1}, where $\mathbf{v}_G$ is expressed in (Π, Θ, Z) components consistent with the conventions described in §6-1. In essence, Lynden-Bell and Lin's results, taken at face value, imply that (1) the velocity of our Galaxy with respect to the center of mass of the LG is small and nearly radial outward along the Sun-center line; and (2) $\Theta_0 \approx 300$ km s^{-1}. It thus appears that we are again left with a large value of Θ_0, which is unsatisfactory for the reasons given earlier. But, it must be stressed that these results are actually extremely uncertain. Indeed, from a detailed statistical analysis, A. Yahil, G. A. Tammann, and A. R. Sandage (**Y1**) show that, because of the large size of the variance of the measured value of $\mathbf{v}_0$ and because of the large uncertainty in the other data (for example, the mass ratio μ) and in the dynamical hypotheses [for example, equations (6-36)] employed in effecting the decomposition of $\mathbf{v}_0$ into Θ_0 and $\mathbf{v}_G$, at the 90% confidence level one can conclude only that Θ_0 is restricted to the broad range $200 \lesssim \Theta_0 \lesssim 300$ km s^{-1}, which adds little to what we already know. At present, we can only hope that future work will yield more refined values for Θ_0 and $\mathbf{v}_G$. It is clear that, if we could reliably determine Θ_0 independent of observations of the LG, then $\mathbf{v}_G$ would follow immediately.

Summary and Prospects

All of the analyses discussed here show that Θ_0 lies between 200 and 300 km s^{-1}. From the hypothesis that the most extreme high-velocity stars are on the verge of escaping, we can show that $\Theta_0 \lesssim 275$ km s^{-1}, and, from the analysis of the space motions of RR Lyrae stars, we conclude that $\Theta_0 \gtrsim 225$ km s^{-1}. These results suggest that it is reasonable to adopt the compromise value $\Theta_0 = 250 \pm 25$ km s^{-1}, which is, in fact, the standard value that was endorsed by the International Astronomical Union (IAU) in 1964 and has been used in most studies since then. This choice is plausible because it lies within the range of rotation speeds observed in external galaxies of the same morphological type as our Galaxy. It is supported further by the fact that, at this value of Θ_0, half of the halo RR Lyrae stars have direct orbits and half have retrograde orbits around the galactic center. We shall henceforth assume $\Theta_0 = 250$ km s^{-1} in this book.

An improvement in our knowledge of Θ_0 will require further accurate observational work, and, even then, significant improvement will come only if we can introduce completely new and independent information into the analysis. A particularly interesting prospect in this connection is the possibility of a direct measurement, by radio observations, of the *angular rotation rate* $\omega_0 \equiv \omega(R_0) = (\Theta_0/R_0)$ of the LSR around the galactic center. A compact radio source (angular extent of the order of 0″.001) has been discovered at the position of the nucleus of our Galaxy [(**K1**), see also §9-4]. If we adopt $\Theta_0 \approx 250$ km s^{-1} and $R_0 \approx 9$ kpc, then $\omega_0 \approx 5 \times 10^{-3}$ seconds of arc per year. Given the present-day ability to make absolute position measurements of this small source with a precision of 0″.01 by the methods of radio astrometry, it is clear that the motion of the LSR will produce an observable proper motion of the galactic center over an interval of, say, ten years. Knowledge of an accurate value for ω_0 could lead to a significant improvement in present estimates of Θ_0 and of other related galactic-rotation parameters (see §8-2).

Absolute Space Motion of the LSR

Before leaving the subject of determinations of the motion of the LSR with respect to external reference frames, it is worthwhile to mention some other results even though they do not pertain directly to the rotation of the LSR around the galactic center and thus lie somewhat outside of the domain of the preceding discussion.

A direct measurement of the motion of the LSR with respect to the Universe as a whole comes from observations by Smoot, et al. (**S2**), (**S3**) and Cheng, et al. (**C1**) of a small anisotropy in the 3 K cosmic microwave background. This blackbody radiation field is a relic of the primeval fireball that occurred in the initial phases of the expansion of the Universe, and it should constitute an absolute inertial frame. Unless the anistropy is intrinsic to the cosmic radiation itself, it is most readily interpreted as resulting from a motion of the LSR relative to the inertial frame. The implied motion has a speed of about 365 ± 30 km s^{-1} in the direction $\alpha = 11^{\rm h}.7 \pm 0^{\rm h}.4$, $\delta = +5^\circ \pm 10^\circ$ ($\ell \approx 260^\circ$, $b \approx 55^\circ$). Correction for the motion of the LSR around the galactic center yields a peculiar velocity of our Galaxy relative to the inertial frame about 530 km s^{-1} in the direction $\ell \approx 270^\circ$, $b \approx 30^\circ$. The large size of this velocity is surprising given that the peculiar velocities of nearby galaxies are relatively small.

It is interesting to compare this result with measurements of the motion of the LSR relative to a large sample of galaxies. From two hundred sixty galaxies some 5 Mpc to 20 Mpc distant, G. de Vaucouleurs (**D2**) derived a velocity of 430 ± 60 km s^{-1} directed toward $\alpha = 13$h, $\delta = 83^\circ$. From about one hundred distant (70 Mpc to 130 Mpc) Sc galaxies, V. C. Rubin et al. (**R1**) found a velocity of 600 ± 125 km s^{-1} directed toward $\alpha = 2^{\rm h} \pm 1^{\rm h}$ and $\delta = 53^\circ \pm 11^\circ$. Both of these results differ strongly from the microwave background results and from one another; the directions of the velocity vectors disagree by up to 120°. If these results are taken at face value, they imply large motions of material within large volumes (radius $\sim$ 100 Mpc) with respect to the inertial frame of the Universe. Such a result may mean that dynamical interactions among clusters and superclusters of galaxies are more important than previously thought, or that the large-scale dynamics of the Universe is more complex than is usually considered in current "standard" cosmologies.

On the other hand, Yahil, Sandage, and Tammann (**Y2**) have measured a net motion of our Galaxy relative to a sample of galaxies within 80 Mpc at about half the speed given by the microwave results, but roughly in the same direction. Clearly, considerably more observational effort will be required to yield a definitive result, and future work on these questions should be rewarding.

6-6. SECULAR AND STATISTICAL PARALLAXES

In our discussion of methods of stellar distance determination in §3-1, we mentioned the methods of *secular* and *statistical parallaxes* but postponed their description. Having now considered the solar motion, we are in a position to study these approaches in detail.

Strategy

The most fundamental method of determining the distance to a star is to measure its trigonometric parallax, that is, the angular displacement on the sky produced when the star is viewed from opposite sides of the earth's orbit. We saw earlier that such determinations are limited to distances of 25 pc or less by the inherent errors of measurement of angular positions of stars on the sky. But it is obvious that, given that we can make *angular* measurements of a certain precision, we could extend the *linear* range of our distance determinations if we could use a longer baseline than the 2-a.u. diameter of the Earth's orbit. The Sun's motion of 20 km s^{-1}, relative to the average of nearby stars implies that the solar system translates in space 2×10^{-5} pc per year $= 4$ a.u. per year, and it is clear that, by allowing this movement to accumulate for, say, twenty years, we have in principle a much larger effective baseline to use in making distance determinations. The method of secular parallaxes is formulated in just such a way as to exploit this longer baseline. To begin, let us briefly sketch the basic idea behind this approach.

Suppose we measure the radial velocity and proper motion of an individual star. If we know the speed and direction of the solar motion, then we can immediately find the component of the star's space motion along the line of sight. The star's proper motion is more complex, being proportional to the difference of its transverse velocity and the components of the solar motion perpendicular to the line of sight divided by the star's distance. *If* we knew the star's transverse velocity and the solar motion, then, from the star's proper motion, we could deduce its distance. But, of course, for any particular star, the transverse velocity is not in general known in advance. Therefore, in the analysis of data from a single star, no advantage accrues from knowing the solar motion. But now suppose we make these observations for many stars well distributed over the sky. We can then argue that their average space motion corrected for the solar motion should be zero. (This follows immediately from the definition of the solar motion.) Therefore, any observed average proper motion for the group (over a definite time interval) must be merely the average angular change in the Sun's position relative to these stars produced by the solar motion itself (suitably projected and accumulated over the stated time interval), divided by the average distance to the stars. We know the distance traveled by the Sun and also the required projection factors, given the speed and direction of the solar motion and the direction to each star. Hence, from the measured average angular change, we can infer the unknown distance. Let us now formulate this method mathematically.

Basic Equations

To obtain reliable results from the secular-parallax method, one must apply it to a group of stars that are all nearly the same distance from the Sun. In practice, this requirement can be met if we select stars of a definite spectral type (and, hence, of a definite absolute magnitude) that lie within a restricted range of apparent magnitudes which, ideally, have been corrected for interstellar absorption. We can then argue that, because they have the same intrinsic and apparent brightnesses, the stars must all be at a fairly definite distance. In the past, some studies have been made using stars of all spectral types within some range of apparent magnitude, but, in view of the enormous spread in intrinsic stellar luminosities, it is not clear what the results of such studies mean. We shall henceforth assume that a homogeneous group of objects has been selected for the analysis.

Next, we must know the solar motion. We saw in §6-4 that the solar motion is quite different relative to different spectral types. Hence, if the group of stars to be analyzed is large enough, we should ideally determine the solar motion with respect to the group itself, using the methods described in §6-3. If radial velocity data are available, then we can determine both the speed and the apex of the motion. If only proper motions are available, then we can determine only the apex, and we must assume a value for the speed. In some cases, we cannot obtain an accurate enough result from the data for the group itself, and we must then simply assume both a speed and an apex for the solar motion. For spiral-arm and disk stars, we could adopt the standard or basic solar motions; for spheroidal-component stars, it would probably be better to use an apex near $\ell = 90°$, $b = 0°$, and a speed chosen according to, say, the velocity dispersion of the group under study [see Tables 6-3, 7-1, and 7-3].

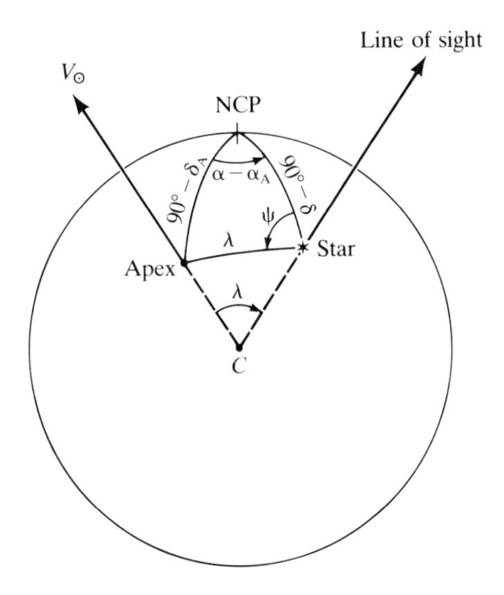

Figure 6-8. Relation of the angular distance λ and the angle ψ to the positions of the apex and a star on the celestial sphere.

Consider a star at position (α, δ) on the sky, as shown in Figure 6-8. Let the coordinates of the apex of solar motion be (α_A, δ_A). Let λ denote the angular distance between the star and the apex $(0° \leq \lambda \leq 180°)$, and ψ be the angle between the great circles joining the star to the north celestial pole and to the apex, taken positive from the north through the west. Using the standard formulae of spherical trigonometry, we can find λ from

$$\cos\lambda = \sin\delta\sin\delta_A + \cos\delta\cos\delta_A\cos(\alpha - \alpha_A) \tag{6-39}$$

and ψ (both magnitude and quadrant) from

$$\cos\psi\sin\lambda = \cos\delta\sin\delta_A - \sin\delta\cos\delta_A\cos(\alpha - \alpha_A) \tag{6-40a}$$

and

$$\sin\psi\sin\lambda = \cos\delta_A\sin(\alpha - \alpha_A) \tag{6-40b}$$

Clearly, we shall have $0 \leq \psi \leq 180°$ if $(\alpha - \alpha_A) \geq 0$ and $180° \leq \psi \leq 360°$ if $(\alpha - \alpha_A) \leq 0$.

In the equatorial system, the components of the star's observed proper motion are $(\mu_\alpha\cos\delta, \mu_\delta)$. We now choose a new set of reference axes as shown in Figure 6-9 and resolve the proper motions into an *upsilon component* υ along the great circle joining the star and the apex, and a *tau component* τ perpendicular to this great circle. We measure υ positive in the direction of the antapex. Then it is easy to see in Figure 6-9 that

$$\upsilon = \mu_\alpha\cos\delta\sin\psi - \mu_\delta\cos\psi \tag{6-41a}$$

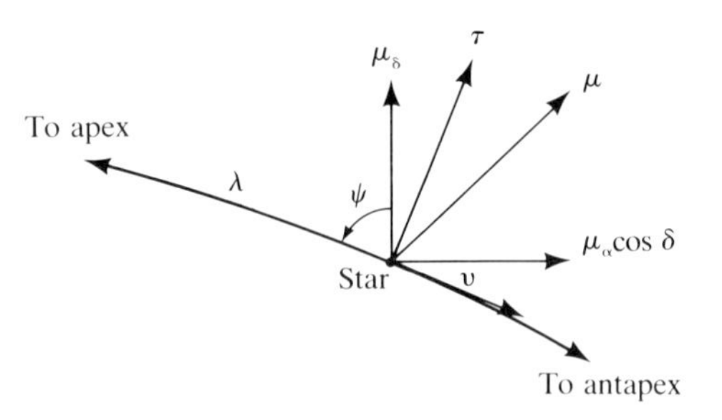

Figure 6-9. The proper motion μ of a star may be expressed in equatorial components ($\mu_\alpha \cos \delta$, μ_δ) or in components (υ, τ), measured along the great circle joining the star and the apex, and in the direction perpendicular to this circle, respectively. All vectors shown lie in the plane of the sky.

and that

$$\tau = \mu_\alpha \cos \delta \cos \psi + \mu_\delta \sin \psi \tag{6-41b}$$

where, as usual, the motions are expressed in seconds of arc per year. As the velocity vector $\mathbf{v}_\odot$ of the solar motion lies in the plane joining the star and the apex, it is evident that the upsilon component of the proper motion will contain a contribution from the solar motion, whereas the tau component, which is orthogonal to this plane, is independent of the solar motion. We now consider the analysis of each of these components in turn.

Upsilon Components

The observed upsilon component contains the upsilon motion υ_* of the star itself plus a contribution from the solar motion. From Figure 6-10, we can see that the component of the solar motion across the line of sight is just $S_\odot \sin \lambda$. Moreover, this component clearly lies in the plane defined by $\mathbf{v}_\odot$ and the line of sight to the star, and it must therefore lie along the great circle arc joining the star and the apex. Thus the contribution of the solar motion to the observed upsilon motion is $\upsilon_\odot = (\pi'' S_\odot \sin \lambda)/4.74$, and the total observed motion of the ith star is

$$\upsilon_i = \upsilon_{*i} + \frac{\pi_i'' S_\odot \sin \lambda_i}{4.74} \tag{6-42}$$

We have one such equation for each star of the group.

We now suppose that all stars in the group lie at about the same distance, so that we can replace π_i'' by $\bar{\pi}''$. We then solve for $\bar{\pi}''$ by the method of least squares. The procedure is essentially the same as that used in §6-3 to obtain the solar motion: (a) we take the moment of equations (6-42) against the coefficient of the unknown ($\bar{\pi}''$), and (b) we argue that, because the space

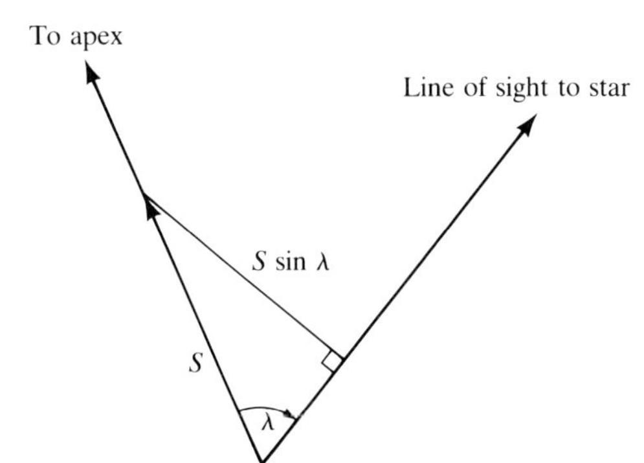

Figure 6-10. The velocity component of the solar motion across the line of sight at a star located an angular distance λ from the apex is $S_{\odot} \sin \lambda$, where $S_{\odot}$ is the speed of the solar motion.

motions of stars in the group are random with respect to their velocity centroid, the average of υ_{*i} will be zero. We thus obtain.

$$\begin{aligned}\sum_{i=1}^{N} \upsilon_i \sin \lambda_i &= \sum_{i=1}^{N} \upsilon_{*i} \sin \lambda_i + \frac{\bar{\pi}'' S_{\odot}}{4.74} \sum_{i=1}^{N} \sin^2 \lambda_i \\ &= \frac{\bar{\pi}'' S_{\odot}}{4.74} \sum_{i=1}^{N} \sin^2 \lambda_i \end{aligned} \tag{6-43}$$

Dividing by N, the number of stars in the group, and writing $\langle\ \rangle \equiv \sum/N$ to denote averages, we have

$$\bar{\pi}'' = \frac{4.74 \langle \upsilon \sin \lambda \rangle}{S_{\odot} \langle \sin^2 \lambda \rangle} \tag{6-44}$$

Notice that the derivation of a mean parallax from an analysis of upsilon components requires only proper-motion data and a knowledge of the apex and speed of the solar motion.

Tau Components

The tau components of proper motion are statistically independent of the upsilon components, and, by combining these data with observed radial velocities, we can obtain an independent estimate of the average distance to a group of stars. This method is known as the *statistical parallax method.* The observed space velocity of a star is $\mathbf{v} = \mathbf{v}_* - \mathbf{v}_{\odot}$, where $\mathbf{v}_*$ and $\mathbf{v}_{\odot}$ denote the peculiar velocity of the star and the Sun, respectively. Suppose we resolve $\mathbf{v}$ into components v_{υ} along the υ axis, v_{τ} along the τ axis, and v_R along the line of sight to the star, which is specified by the unit vector

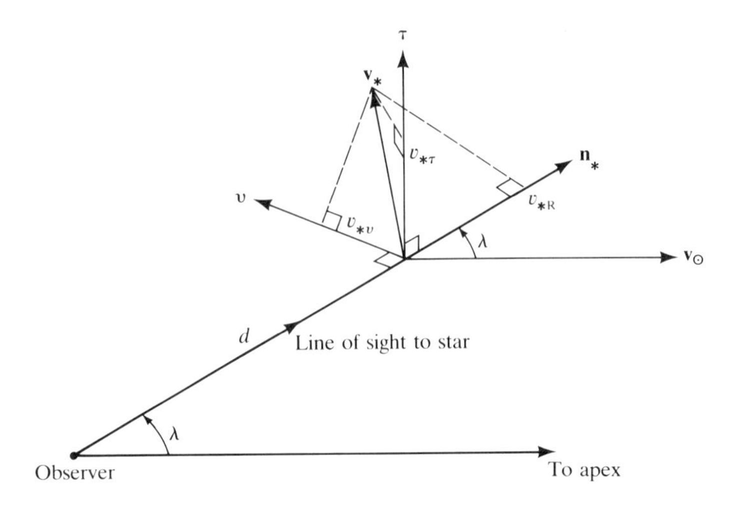

Figure 6-11. The coordinate system used to resolve a stellar velocity $\mathbf{v}_*$ into components v_{*R} along the line of sight to the star (unit vector $\mathbf{n}_*$), $v_{*\upsilon}$ along the υ axis, and $v_{*\tau}$ along the τ axis. The vectors $\mathbf{n}_*$, $\mathbf{v}_\odot$ and the υ axis all lie in the same plane, with the υ axis normal to $\mathbf{n}_*$ and on the opposite side of $\mathbf{n}_*$ from $\mathbf{v}_\odot$. The τ axis is normal to this plane.

$\mathbf{n}_*$ (see Figure 6-11). The υ axis lies in the plane defined by $\mathbf{n}_*$ and $\mathbf{v}_\odot$, and the τ axis is normal to this plane. It follows that v_τ has no contribution from $v_\odot$, hence

$$v_\tau = v_{*\tau} \tag{6-45a}$$

while

$$v_\upsilon = v_{*\upsilon} + S_\odot \sin \lambda \tag{6-45b}$$

and

$$v_R = v_{*R} - S_\odot \cos \lambda \tag{6-45c}$$

The observed τ components of the motion are produced by $v_{*\tau}$; thus

$$|\tau| = \frac{\pi''|v_{*\tau}|}{4.74} \tag{6-46}$$

Therefore, by averaging over all stars in the group, we can write

$$\overline{\pi}'' = \frac{4.74\langle|\tau|\rangle}{\langle|v_{*\tau}|\rangle} \tag{6-47}$$

In general, we have no way of knowing $\langle |v_{*\tau}| \rangle$. But, if the space velocities of stars in the group are oriented randomly, then, on the average, the projection of $|v_*|$ should be the same along each axis. We can thus expect

$$\langle |v_{*\tau}| \rangle = \langle |v_{*R}| \rangle = \langle |v_R + S_\odot \cos \lambda| \rangle \tag{6-48}$$

where we have made use of equation (6-45c). The last term of equation (6-48) can be evaluated directly from the observed radial velocities. We thus obtain finally

$$\bar{\pi}'' = \frac{4.74 \langle |\tau| \rangle}{\langle |v_R + S_\odot \cos \lambda| \rangle} \tag{6-49}$$

A more sophisticated analysis, which takes into account the ellipsoidal shape of the random velocity distribution (see §7-1) in calculating $\langle |v_{*\tau}| \rangle$, is described in (**T1**, Chapter 3.73).

Application

The methods described here have been applied to estimate the average distances, and hence the absolute magnitudes, of B stars and variable stars (for example, RR Lyrae, Cepheid, and late-type variables), which are found only beyond the range of the trigonometric parallax method. With accurate proper motions, these methods can yield reliable values of $\bar{\pi}''$ down to $0\overset{''}{.}002$ ($d = 500$ pc), and thus they extend greatly the volume of space within which we can obtain geometric estimates of stellar distances. The factor of twenty increase in the range of this method compared to that of direct trigonometric parallaxes is just what we would expect if we note that, in twenty years (a typical interval for proper motion measures), the Sun moves 80 a.u. in space, and that, on the average, roughly one-half of this change will be reflected in proper motions. This fact implies an effective baseline of 40 a.u., a factor of twenty larger than the 2 a.u. baseline used in trigonometric parallax work.

As a rough rule of thumb, if $S_\odot$ is greater than $\langle |v_{*R}| \rangle$, it is probably preferable to use the υ-component method to determine $\bar{\pi}''$, because then the solar motion is dominant over the stellar space velocities and hence produces the largest effects in the observations. Conversely, if $\langle |v_{*R}| \rangle$ is greater than $S_\odot$, it is probably preferable to use the τ-component method.

Finally, it should be noted that, when we apply the secular and statistical parallax methods to distant stars, the equations given here should, strictly speaking, be generalized to allow for the effects of differential galactic rotation. This generalization is straightforward (at least in principle), but it lies outside the scope of our discussion.

REFERENCES

(B1) Blaauw, A. and Schmidt, M. (eds.). 1965. *Galactic Structure.* (Chicago: University of Chicago Press).

(B2) Byrnes, D. V. 1966. *Pub. Astron. Soc. Pacific.* **78**:46.

(C1) Cheng, E. S., Saulson, P. R., Wilkinson, D. T., and Corey, B. E. 1979. *Astrophys. J. Letters.* **232**:L139.

(D1) de Vaucouleurs, G. and Peters, W. L. 1968. *Nature.* **220**:868.

(D2) de Vaucouleurs, G. 1978. In *Large Scale Structure of the Universe, IAU Symposium No. 79.* M. S. Longair and J. Einasto (eds.). (Dordrecht: Reidel). p. 205.

(E1) Eggen, O. J. 1964. *Roy. Obs. Bull.* No. 84, E111.

(F1) Fricke, W. 1949. *Astron. Nachr.* **278**:49.

(H1) Humason, M. L. and Wahlquist, H. D. 1975. *Astron. J.* **60**:254.

(I1) Isobe, S. 1974. *Astron. and Astrophys.* **36**:327.

(K1) Kellerman, K. I., Shaffer, D. B., Clark, B. G., and Geldzahler, B. J. 1977. *Astrophys. J. Letters.* **214**:L61.

(K2) Kinman, T. D. 1959. *Mon. Not. Roy. Astron. Soc.* **119**:559.

(L1) Lynden-Bell, D. and Lin, D. N. C. 1977. *Mon. Not. Roy. Astron. Soc.* **181**:37.

(M1) Mayall, N. U. 1946. *Astrophys. J.* **104**:290.

(R1) Rubin, V. C., Thonnard, N., Ford, W. K., and Roberts, M. S. 1976. *Astron. J.* **81**:719.

(R2) Russell, H. N., Dugan, R. S., and Stewart, J. Q. 1938. *Astronomy.* Vol. 2. (Boston: Ginn and Co.).

(S1) Sandage, A. R. 1969. *Astrophys. J.* **158**:1115.

(S2) Smoot, G. F., Gorenstein, M. V., and Muller, R. A. 1977. *Phys. Rev. Lett.* **39**: 898.

(S3) Smoot, G. F. and Lubin, P. M. *Astrophys. J. Letters.* 1979. **234**:L83.

(S4) Strömberg, G. 1924. *Astrophys. J.* **59**:228.

(S5) Strömberg, G. 1925. *Astrophys. J.* **61**:363.

(T1) Trumpler, R. J. and Weaver, H. F. 1953. *Statistical Astronomy.* (Berkeley: University of California Press).

(W1) Woolley, R. and Savage, A. 1971. *Roy. Obs. Bull. No. 170.*

(Y1) Yahil, A., Tammann, G. A., and Sandage, A. R. 1977. *Astrophys. J.* **217**:903.

(Y2) Yahil, A., Tammann, G. A., and Sandage, A. R. 1980. In *Physical Cosmology.* Les Houches Lecture Notes. R. Balain, J. Audouze, and D. N. Schramm (eds.). (Amsterdam: North Holland).

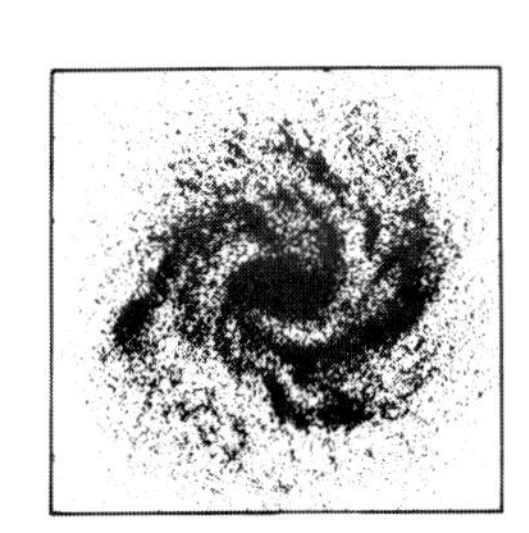

7 Stellar Kinematics: The Stellar Residual-Velocity Distribution

Having determined the solar motion with respect to the local standard of rest (LSR) and the rotation velocity of the LSR around the galactic center, we can now reduce the observed space motion of any star to the LSR and calculate its velocity relative to the center. Hence we can study both the statistical properties of the distribution of *stellar residual velocities* with respect to the LSR and the motions of various groups of stars with respect to the galactic center.

Historically, when this analysis was first done, two sharply distinguished groups of stars were recognized. Most stars in the field were found to have a smooth, random distribution of relatively small residual velocities with respect to the LSR. These stars are therefore moving on nearly circular orbits around the galactic center. In terms of Baade's picture of stellar populations, these stars would be considered to be Population I objects. In contrast, a second group, called the *high-velocity stars*, were found to have a large dispersion in all three components of the residual velocity and to lag systematically behind—that is, rotate more slowly than—the LSR. In terms of the old picture of stellar populations, they would be considered to be members of Population II.

As we said in Chapter 1, it was soon realized that this rotational lag implied that the extreme high-velocity stars (for example, RR Lyrae stars and subdwarfs) move on very eccentric (almost radial) orbits around the galactic center. We also described briefly the decisive role these stars played in the development of the Oort-Lindblad picture of galactic rotation.

We know today that the categorization of stars into only two kinematic groups as just described is an oversimplification, and our goal in this chapter will be to develop a fairly detailed description of the kinematic properties of stars as a function of age, type, and population group. As we saw in Chapter 4, the stars in the solar neighborhood are actually a mixture of many different kinds of objects, ranging from very old metal-poor, spheroidal-component (halo) stars created near the time of the formation of our Galaxy itself, to very young, metal-rich stars, newly born in spiral arms. It is reasonable to suppose that the present kinematic properties of any group of stars will reflect its dynamical history, and hence its age and the dynamical characteristics of our Galaxy during its lifetime. Other physical properties of stars—for example, their chemical composition—should also be correlated with their age and birthplace in our Galaxy, and hence ultimately with their kinematic properties. We shall see in this chapter that these expectations are justified, and we shall find that stars do indeed have strongly correlated kinematic and physical properties. In fact, we already cited these interrelations in our discussion of stellar population groups in §4-5, in particular in the delineation of a sequence of disk populations ranging from the spiral-arm population to the oldest disk population.

One of the major points we shall see in this chapter is that the variation of the kinematic properties of the various disk populations can be understood in terms of a progressive increase of the velocity dispersion with stellar age, which results from the randomizing effects of encounters of stars with interstellar cloud complexes and spiral arms. But, we shall also find that this mechanism cannot explain the kinematic properties of the spheroidal-component stars, and we now know that the properties of these stars contain fossil information about the dynamical state of our Galaxy in its distant past and that these properties are to be attributed to dynamical events in its early history.

7-1. KINEMATICS OF DISK STARS

Consider first the characteristics of the peculiar velocities of *disk stars* in the solar neighborhood. These characteristics can be deduced from analyses of observed radial velocities, of proper motions, or of space velocities. For ease of discussion, we shall consider only space velocities. As in Chapter 6, we shall again assume that all stars studied are close enough to the Sun that we can ignore the effects of differential galactic rotation (which can, of course, be included in the analysis if necessary).

The Velocity Ellipsoid

The primary result that emerges from an examination of the peculiar velocities of disk stars relative to the LSR is that they have very nearly a *random* (*Gaussian*) *distribution* in all three residual-velocity components (u, v, w). Examples are shown in Figure 7-1. An extremely elegant description of this state of affairs in terms of a *velocity ellipsoid* was proposed by K. Schwarzschild in 1907 (**S4**), (**S5**). [An alternative description in terms of

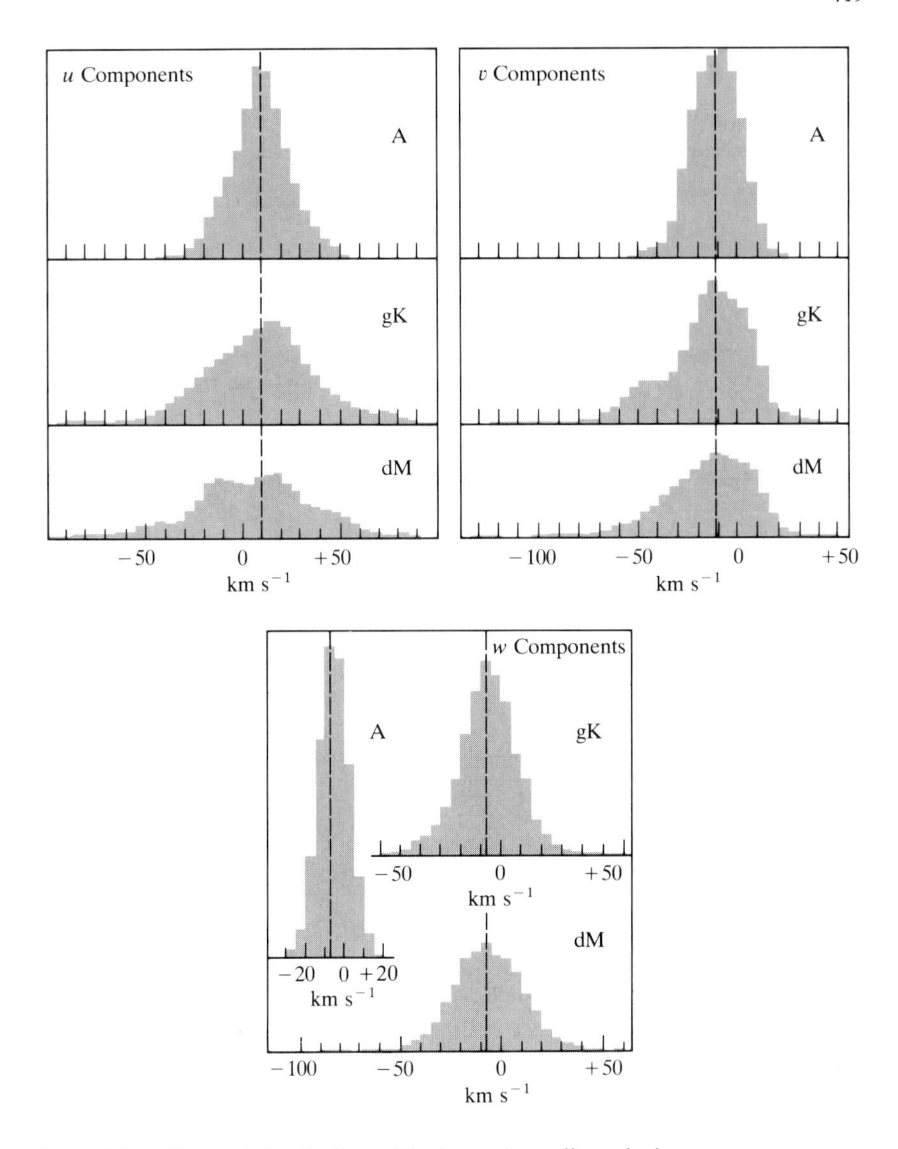

Figure 7-1. Observed distribution of the (u, v, w) peculiar-velocity components for A, gK, and dM stars. The vertical dashed lines show the solar motion $(u_\odot, v'_\odot, w_\odot)$ for these spectral types. Note the excess of stars with large negative velocities in the distribution of v components. [From (**B3**, Chapter 4), by permission. Copyright © 1965 by the University of Chicago.]

star streams can also be used (**T2**, Chapter 3.2), but this description can be reduced to an essentially equivalent superposition of one or more velocity ellipsoids around one or more centroids, and it will not be pursued further here.]

Suppose that we select a group of stars for analysis on the basis of, say, spectral type. We now choose an orthogonal set of axes and resolve each star's peculiar velocity into components (u_1, v_1, w_1) along them. What we find, as is obvious in Figure 7-1, is that the velocity distribution in each component is essentially Gaussian in shape, but that the dispersions are different along each axis. Any acceptable velocity distribution function must incorporate these features. (We also notice an asymmetry in v; we shall comment further upon this later). A distribution function having the required properties is

$$n(u_1, v_1, w_1)\, du_1\, dv_1\, dw_1 = \frac{\nu}{[8\pi^3\langle u_1^2\rangle\langle v_1^2\rangle\langle w_1^2\rangle]^{1/2}} \exp\left[-\left(\frac{u_1^2}{2\langle u_1^2\rangle} + \frac{v_1^2}{2\langle v_1^2\rangle} + \frac{w_1^2}{2\langle w_1^2\rangle}\right)\right] du_1\, dv_1\, dw_1 \tag{7-1}$$

Here $n(u_1, v_1, w_1)$ is the number of stars per unit volume with velocities on the range $(u_1, u_1 + du_1)$, $(v_1, v_1 + dv_1)$, and $(w_1, w_1 + dw_1)$; ν is the space density of the particular group of stars under study (that is, the total number of stars at any velocity per unit volume); and $\langle u_1^2\rangle^{1/2}$, $\langle v_1^2\rangle^{1/2}$, and $\langle w_1^2\rangle^{1/2}$ are the observed *velocity dispersions* along the three axes. The distribution function given by equation (7-1) acquired the name velocity ellipsoid by analogy of the form of the expression in the exponential with the formula

$$\frac{x^2}{a^2} + \frac{y^2}{b^2} + \frac{z^2}{c^2} = 1 \tag{7-2}$$

which is the equation for an ellipsoid with semiaxes a, b, and c.

It is easy to verify from equation (7-1) that

$$\langle u_1^2\rangle = \nu^{-1}\int_{-\infty}^{\infty} dw_1 \int_{-\infty}^{\infty} dv_1 \int_{-\infty}^{\infty} du_1 n(u_1, v_1, w_1) u_1^2 \tag{7-3}$$

and similarly for $\langle v_1^2\rangle$ and $\langle w_1^2\rangle$. Hence, in principle, we determine the velocity dispersions of the velocity ellipsoid simply by calculating averages of u_1^2, v_1^2, and w_1^2. But, thus far we have tacitly assumed that the peculiar velocities were resolved into components along the *principal axes* of the ellipsoid. In reality, the principal axes of the ellipsoid will not, in general, be along any particular set of coordinate axes we have chosen, say (u, v, w) as defined in Chapter 6. Therefore, a rotational transformation of the form

$$u_1 = \ell_1 u + m_1 v + n_1 w \tag{7-4a}$$

$$v_1 = \ell_2 u + m_2 v + n_2 w \tag{7-4b}$$

$$w_1 = \ell_3 u + m_3 v + n_3 w \tag{7-4c}$$

will be required to express velocity components along the principal axes in terms of the observed components. Here (ℓ_1, m_1, n_1) are the direction cosines of the u_1 axis in the (u, v, w) system, and, similarly, for (ℓ_2, m_2, n_2) and (ℓ_3, m_3, n_3).

If we substitute equations (7-4) into (7-1), we obtain a general expression for the velocity distribution function of the form

$$n(u, v, w)\,du\,dv\,dw = \frac{\nu}{[8\pi^3\langle u_1^2\rangle\langle v_1^2\rangle\langle w_1^2\rangle]^{1/2}} \times \exp[-(\alpha u^2 + \beta v^2 + \gamma w^2 + \delta uv + \varepsilon vw + \zeta uw)]\,du\,dv\,dw \quad (7\text{-}5)$$

where the coefficients $\alpha, \ldots, \zeta$ now depend on the velocity dispersions $\langle u_1^2\rangle$, $\langle v_1^2\rangle$, and $\langle w_1^2\rangle$ and the direction cosines $\ell_1, \ldots, m_3$. These coefficients are to be determined from the analysis of the observations. The six expressions that define $\alpha, \ldots, \zeta$, along with the requirements that the (u_1, v_1, w_1) axes be orthonormal, provide just enough relations to determine the twelve parameters $\langle u_1^2\rangle$, $\langle v_1^2\rangle$, $\langle w_1^2\rangle$, $\ell_1, \ldots, m_3$. Several general mathematical techniques for determining the coefficients $\alpha, \beta, \ldots, \zeta$ of the velocity ellipsoid from observational data have been developed and applied by Schwarzschild, Charlier, and others. We shall not describe these techniques in detail, but we shall summarize some of the important results that have been obtained for various classes of stars [see (**S6**, Chapter 5) for methods].

In practice, the analysis can be simplified by exploiting the empirical fact that one axis of the velocity ellipsoid is always found to be oriented perpendicular to the galactic plane, so that the other two axes must lie in the plane. It is also found empirically that the longest axis of the ellipsoid (that is, the direction of maximum velocity dispersion) points approximately in the direction of the galactic center. Therefore, to specify the orientation of the velocity ellipsoid, we really need determine only the galactic longitude along which the principal axis lies. This longitude is called the *longitude of the vertex*, ℓ_v. The transformation equations (7-4) in this case reduce to

$$u = u_1 \cos \ell_v + v_1 \sin \ell_v \quad (7\text{-}6a)$$

$$v = -u_1 \sin \ell_v + v_1 \cos \ell_v \quad (7\text{-}6b)$$

$$w = w_1 \quad (7\text{-}6c)$$

Given that one axis of the velocity ellipsoid is perpendicular to the galactic plane, the determination of the orientation of the other two axes becomes easy. For example, we could use the observed (u, v) velocity components of the stars under study to calculate the components along several trial axes with different values of ℓ_v and then compute the velocity dispersion along these axes. The axis giving the largest dispersion would be the major axis, and it would determine ℓ_v. Of course, one can also systematize the process and perform a least-squares fit to find ℓ_v, $\langle u_1^2\rangle$, and $\langle v_1^2\rangle$ simultaneously.

Correlation of Kinematic Properties with Stellar Types

Velocity ellipsoid parameters have been determined for a wide range of stellar types by several investigators. Typical results are summarized in Table 7-1, where we give the velocity dispersions $\langle u^2 \rangle^{1/2}$, $\langle v^2 \rangle^{1/2}$, and $\langle w^2 \rangle^{1/2}$ along the (Π, Θ, Z) axes defined in §6-1, and the longitude of the vertex ℓ_v. This information is sufficient to allow the velocity dispersions along the principal axes to be reconstructed, for, when we use the fact that $\langle u_1 v_1 \rangle \equiv 0$ from considerations of symmetry, it follows from equations (7-6) that

$$\langle u^2 \rangle = \langle u_1^2 \rangle \cos^2 \ell_v + \langle v_1^2 \rangle \sin^2 \ell_v \tag{7-7a}$$

$$\langle v^2 \rangle = \langle u_1^2 \rangle \sin^2 \ell_v + \langle v_1^2 \rangle \cos^2 \ell_v \tag{7-7b}$$

Equations (7-7) are readily solved for $\langle u_1^2 \rangle$ and $\langle v_1^2 \rangle$ when $\langle u^2 \rangle$, $\langle v^2 \rangle$, and ℓ_v are given.

An examination of the results in Table 7-1 immediately shows the following interesting features:

1. For all types, the dispersions obey the inequalities $\langle u^2 \rangle^{1/2} > \langle v^2 \rangle^{1/2} > \langle w^2 \rangle^{1/2}$. Roughly speaking, $\langle w^2 \rangle^{1/2} \approx 0.5 \langle u^2 \rangle^{1/2}$, while the ratio $\langle v^2 \rangle^{1/2} / \langle u^2 \rangle^{1/2}$ lies in the range from 0.55 to 0.75 for most stars. The fact that $\langle w^2 \rangle^{1/2} \neq \langle u^2 \rangle^{1/2}$ has important dynamical implications, as described in Chapters 13 and 14. As we shall show in §8-2, the numerical value of the ratio $\langle v^2 \rangle^{1/2} / \langle u^2 \rangle^{1/2}$ provides a useful constraint on the values of the parameters that describe the differential rotation of our Galaxy in the solar neighborhood (Oort's A and B constants).
2. Dwarfs of types earlier than F, the A and F giants, and all of the supergiants show markedly smaller velocity dispersions than dwarfs of types F5 and later and the late-type giants. From considerations of stellar evolution, we know that the former are all young stars. In contrast, the later types are a mixture of a few young stars and mostly old stars, and the fact that they have systematically larger velocity dispersions suggests the operation of a mechanism that leads to a progressive increase of the dispersion with time. We shall discuss this mechanism in greater detail soon.
3. The longitude of the vertex of the velocity ellipsoid shows that the principal axis points nearly in the direction of the galactic center for later spectral types, but it shows larger and larger departures from this direction for the early types. This departure is called the *vertex deviation.* We shall discuss the vertex deviation briefly later in this section, but, in subsequent dynamical work, we shall usually ignore it and assume that the velocity ellipsoid is aligned along the (Π, Θ, Z) axes, with the principal axis pointed toward the galactic center.

Table 7-1. Velocity-Ellipsoid Parameters for Various Types of Stars

Stellar type	Dispersions (km s^{-1}) $\langle u^2\rangle^{1/2}$	$\langle v^2\rangle^{1/2}$	$\langle w^2\rangle^{1/2}$	Longitude of vertex, ℓ_v(°)
		Disk-population stars		
Dwarfs				
B0	10	9	6	−50
A0	15	9	9	15
A5	20	9	9	19
F0	24	13	10	21
F5	27	17	17	13
G0	26	18	20	2
G5	32	17	15	14
K0	28	16	11	3
K5	35	20	16	11
M0	32	21	19	8
M5	31	23	16	−7
Giants				
A	22	13	9	27
F	28	15	9	14
G	26	18	15	12
K0	31	21	16	14
K3	31	21	17	4
M	31	23	16	7
Supergiants				
Classical Cepheids	13	9	5	–
O–B5	12	11	9	36
F–M	13	9	7	18
Other				
Carbon stars	48	23	16	–
Subgiants	43	27	24	–
Planetary nebulae	45	35	20	–
White dwarfs	50	30	25	–
RR Lyrae variables, $P < 0\overset{d}{.}45$	45	40	25	–
Long-period variables, $P > 300^d$	50	40	30	–
Long-period variables, $P < 300^d$	80	60	60	–
		Spheroidal-component stars		
Subdwarfs	100	75	50	–
RR Lyrae variables, $P > 0\overset{d}{.}45$	160	100	120	–

SOURCE: Adapted from (**B3**, Chap. 4), by permission.

4. The O and B stars are kinematically quite distinctive. These stars alone among all types have $\langle v^2 \rangle / \langle u^2 \rangle \approx 1$ and, moreover, show the largest vertex deviation. As we noted in Chapters 3 and 4, these stars are so young that they are virtually frozen into their birthplaces. In addition, their velocity distribution cannot have had time to respond to either the effects of encounters or to the general galactic potential field. Their velocity distribution must therefore reflect mainly the properties of the velocity field of the interstellar material from which they were formed.

Before we discuss items 2 and 3 in greater detail, there are a few more general points that we should note about the ellipsoidal velocity distribution.

First, we must realize that there are definite limitations to the use of this description. In particular, a velocity ellipsoid centered on the motion of the LSR does not describe the residual-velocity distribution of the high-velocity stars (comprising both old disk stars and spheroidal-component stars). These stars have an asymmetric velocity distribution with respect to the LSR, lagging behind it in the direction of galactic rotation by a significant fraction (nearly unity for the spheroidal-component stars) of the circular velocity around the galactic center. A velocity ellipsoid centered on the correct centroid does, in fact, give a fairly good description of each of these population groups although, in practice, there is difficulty in choosing a pure sample of such a group.

Second, even for relatively low-velocity stars, the velocity ellipsoid does not provide an exact description of the v component distribution for all stars of a specified spectral type taken together. As is easily seen in Figure 7-1, this distribution is skewed toward negative v. The skewness is again the result of the tendency of stars with nonzero $\langle u^2 \rangle$ (or $\langle \Pi^2 \rangle$) to lag behind the LSR as just mentioned (see also §6-4). If we choose stars according to spectral type alone, our sample inevitably includes stars having a range of ages. The progressive increase of residual velocities with age for disk stars implies that the older stars will both (a) lag behind the LSR and (b) have a larger $\langle v^2 \rangle$. Thus they will produce a long tail in the distribution function toward negative v. If we resolved the sample into individual age groups, we would see that this skewed distribution was actually composed of a superposition of velocity ellipsoids with progressively larger $\langle u^2 \rangle$, $\langle v^2 \rangle$, and $\langle w^2 \rangle$, centered on a sequence of progressively more negative values of $\langle v \rangle$.

Third, we must realize that we can legitimately consider the velocity ellipsoid to be only a convenient kinematical description that has been established observationally for disk stars *in the vicinity of the Sun.* Ideally, we would like to know the velocity distribution for all population groups at all positions in our Galaxy, but, of course, this can never be done directly. The form of the velocity distribution function has important dynamical implications, and, if we actually did know the velocity distribution everywhere in our Galaxy, we would have considerable insight into its dynamics. Indeed, as we shall see in Chapter 14, the mere fact that an ellipsoidal distribution exists even locally, when coupled with the equations of stellar dynamics, allows us to make reasonable guesses about the form of the velocity distribution and about the speed of galactic rotation elsewhere in our Galaxy. However, an a priori imposition of a Schwarzschild ellipsoidal velocity distribution at all points in our Galaxy is an unwarranted extrapolation, and dynamical theories that make such an assumption (as has sometimes been done) can be shown to be inconsistent.

Let us now consider the questions of the nature of the vertex deviation and the age variation of stellar residual velocity dispersion in somewhat greater detail.

The Vertex Deviation and Moving Groups

If we imagine our Galaxy to be an axially symmetric system in a steady state in which stars are distributed at random in their orbits, then, from considerations of symmetry alone, we would expect to find one axis of the velocity ellipsoid of stars in the galactic plane pointing exactly at the galactic center. This expectation is confirmed by analysis of an ellipsoidal velocity distribution with the equations of stellar dynamics (see Chapter 14). But, as was mentioned earlier, the longitude of the vertex ℓ_v often differs significantly from zero.

We can see in Table 7-1 that the vertex deviation is largest for stars of spectral type F0 and earlier, that is, for young stars. The vertex deviation of A dwarfs is shown quite plainly in Figure 7-2, where we also notice that the gK stars moving on orbits having low inclination to the galactic plane, which are young disk stars that have relatively low peculiar velocities, show a definite vertex deviation, while the relatively high-speed, older disk-population gK stars on highly inclined orbits show little, if any, vertex deviation. Further evidence that the vertex deviation is associated mainly with groups of stars having the smallest residual velocities is seen in Figure 7-3, which shows the (u, v) velocity components of F through M dwarfs within 20 pc of the Sun, taken from Gliese's catalog (see Table 2-6). Both the dF and dM stars have velocity distributions with major axes that are pointed almost toward $\ell = 0°$ at high speeds, but which show large deviations to higher longitudes at low speeds. In summary, it appears that the vertex deviation is associated primarily with very young, low-velocity stars.

The fact that the largest vertex deviations always occur for the youngest stars suggests that this phenomenon is produced by the local dynamics of the interstellar medium at the time of star formation, and perhaps is a manifestation of the velocity field of the gas in a spiral arm near the Sun (**B2**, 433) or of the initial spatial distribution of newly formed stars in the solar neighborhood (**B2**, 423). There is, in fact, considerable evidence that there are important local velocity fields associated with spiral-arm density waves. Recall also, as was mentioned in §4-1, that the brightest early-type stars

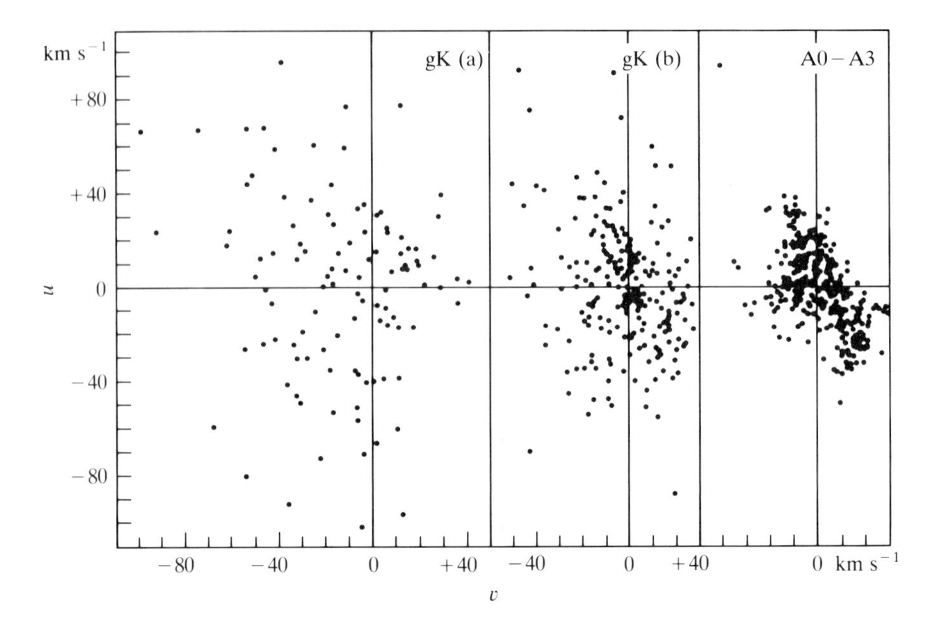

Figure 7-2. Distribution of (u, v) components of peculiar velocities of gK stars with (a) large $|w|$ velocities, hence orbits inclined to the galactic plane, and (b) small $|w|$ velocities, hence orbits essentially in the galactic plane. Also shown are (u, v) velocities of A0–A3 stars. [From (**B3**, Chapter 4), by permission. Copyright © 1965 by the University of Chicago.]

(all of them young objects) seen on the night sky do not lie in the galactic plane, but rather are found in *Gould's belt* in a plane inclined by about 16° to the plane of our Galaxy. A detailed kinematical analysis of these stars (**L1**) shows that they apparently form an *expanding group*. Their properties thus seem to be far from random and reflect strongly the velocity field associated with spiral structure.

Another phenomenon that may lead to the vertex deviation is the existence of *moving groups*. Reasonably compact *moving clusters* located in small areas of the sky have long been recognized. Indeed, as was discussed in §3-1, the analysis of their motions provides an important method of estimating stellar distances. Given that stars form in clusters and associations, and that cluster members have common space velocities, it is reasonable to expect that, when an association "dissolves" into the general field and its former members become spread out over a large area of the sky and thus are no longer easily distinguishable by their space concentration, the former members will nevertheless very nearly maintain their common motion and can therefore be recognized by having almost the same space-velocity components.

A number of moving groups of stars have, in fact, been suggested from time to time, but subsequent investigations have repeatedly called their reality into question. However, a recent reinvestigation of the question by Eggen has produced fairly convincing evidence for the existence of at least

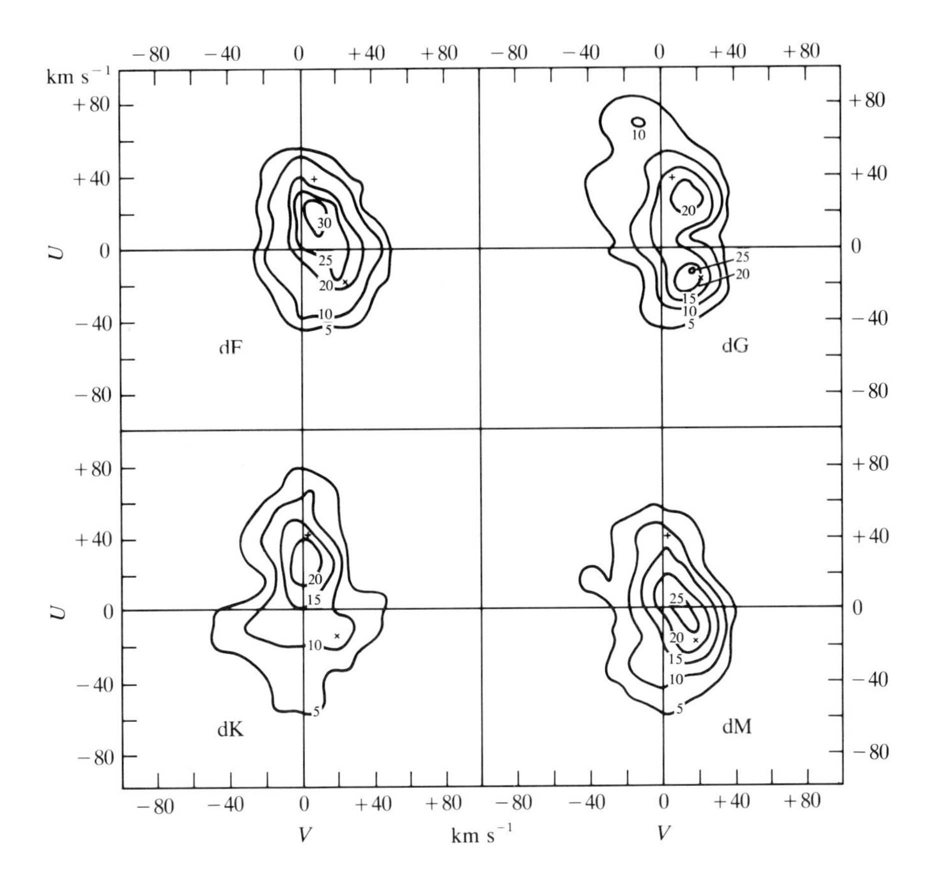

Figure 7-3. Distribution of (U, V) components of stellar velocities relative to the Sun for F–M dwarfs within 20 pc. Note vertex deviation of ellipsoidal distribution at low relative speeds and near lack of vertex deviation at high speeds. [From (**B3**, Chapter 4), by permission. Copyright © 1965 by the University of Chicago.]

half a dozen well-defined groups of low-velocity stars, the constituents of which have nearly identical space velocities and, taken together, yield coherent H-R diagrams [see (**B3**, Chapter 4), (**B3**, Chapter 6)]. Eggen has also identified several older moving groups of stars, some including extreme high-velocity objects.

The relevance of moving groups to the problem of the vertex deviation is that the young stars in a (u, v) diagram, such as Figure 7-2 or Figure 7-3, could possibly be members of a few distinct (if unrecognized) moving groups. If this were the case, then the number of truly independent points from which the characteristics of the velocity distribution are to be inferred becomes very small, and spurious estimates of the orientation of the velocity ellipsoid axes, and the velocity dispersion along these axes, may result. As a specific example, in the (u, v) diagram for A stars, Eggen has pointed out concentrations of points in the neighborhoods of the group velocities of the Sirius, Pleiades, Hyades, and Coma Bernices groups, along with two anonymous groups.

Clearly, the possible presence of moving groups in the (u, v) diagram is most serious for young stars, which presumably were cluster members in the recent past. If moving groups are actually responsible for the vertex deviation, then dynamically significant estimates of the velocity ellipsoid parameters may be obtainable only for later-type stars. Future research should shed some light on this issue because, if the observed vertex deviation is strongly influenced by the presence of a few moving groups, then the problem should vanish when a larger sample of stars is studied. If, however, the young stars really do have preferred directions of motion that are inconsistent with those expected for an axisymmetric galaxy (for example, as a result of an imposed spiral potential), then the effect will persist even with a larger set of data.

Variation of Kinematic Properties with Stellar Age

Empirical Evidence Probably the most important feature in the kinematic data presented in Table 7-1 is the steady increase of the residual-velocity dispersions with advancing spectral type. In fact, the effect is striking enough that P. P. Parenago (**P1**) suggested that there is actually an abrupt change in the kinematic properties of stars at about spectral type F5 on the main sequence (sometimes referred to as *Parenago's discontinuity*). It is easy to see from the data in Table 7-1 that, for main-sequence stars, the velocity dispersions in all three components are indeed about double those of early-type stars. An analogous but less striking effect can be seen in the solar motion (Table 6-3), where we find that the rotational lag of a group increases (larger values of $v'_{\odot}$) toward later spectral types; in contrast, $u_{\odot}$ and $w_{\odot}$ are roughly the same for all spectral types. We can also see similar increases in the velocity dispersions (especially $\langle v^2 \rangle^{1/2}$ and $\langle w^2 \rangle^{1/2}$) of the late-type giants relative to the A and F giants.

How are these effects to be explained physically? The simplest hypothesis is that there is a progressive increase with average age of the residual-velocity dispersions of groups of disk stars. Support for this hypothesis is found immediately when we compare groups of stars that we know must be young (for example, early-type stars, supergiants, and early-type giants) with objects that we believe to be old (for example, subgiants, planetary nebulae, and white dwarfs). The situation for the lower main sequence is less clear because these stars have lifetimes as long as the age of our Galaxy, or longer, and hence they comprise a random mixture ranging from very young to very old stars. But, if we note that the average age of all spectral types for which $\tau_{MS} \gtrsim \tau_G$ will be the same, namely about $\frac{1}{2}\tau_G$ (assuming a constant rate of star formation), we then understand why the velocity dispersions stop rising at the late G types and thenceforth remain practically constant for spectral types K and M. To infer the precise nature of the variation of the velocity dispersion with time, we must perform a more detailed analysis and actually segregate stars according to age groups.

It is clear from the foregoing remarks that we must expect to find groups of late-type stars, selected on the basis of, say, a rough spectral type or of their being in a specified range of intrinsic color, to be *kinematically inhomogeneous*—that is, to contain subgroups of stars with significantly different kinematic properties. That this conjecture is indeed true was shown particularly clearly for the F, G, and K stars by Nancy Roman (**R1**), (**R2**), (**R3**). She found that these types contain spectroscopically identifiable subgroups of differing line strengths (metallicities), and she showed that their kinematic and spectrophotometric properties are closely correlated. For didactic reasons, we shall defer a discussion of her work until the end of this section and focus first on a different example.

An example of kinematic inhomogeneity that is particularly interesting for our present purposes is provided by the M dwarfs. Among the M dwarfs, some stars (dMe) show emission lines in their spectra while others (dM) do not. The emission lines are indicators of strong chromospheric activity in these stars' atmospheres. From a study of the emission-line strengths observed in the spectra of M dwarfs found in associations and galactic clusters of known ages, it is found that the emission decreases fairly rapidly with increasing age. Hence the division of M dwarfs into the dMe and dM categories provides two clear samples: one from a very young disk population group and the other from a much older disk-population group. As can be seen in Figure 7-4, the dMe stars have much smaller peculiar velocities than

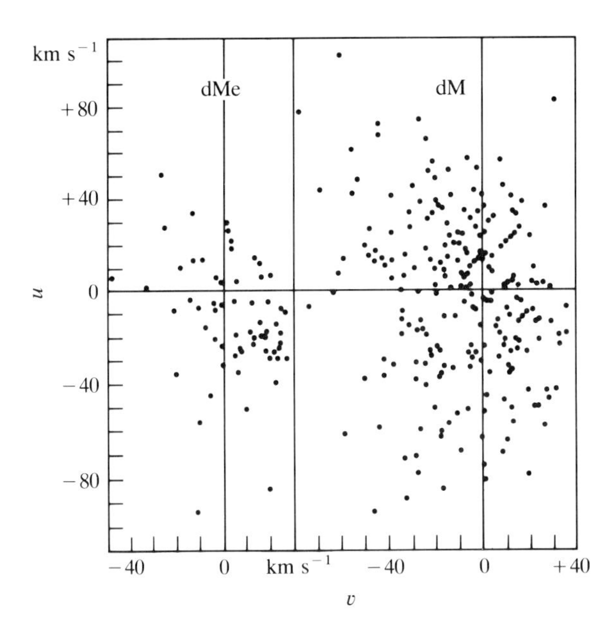

Figure 7-4. Distribution of (u, v) components of peculiar velocities of dM and dMe stars. Notice the considerable kinematic differences between these two groups of M stars. [From (**B3**, Chapter 4), by permission. Copyright © 1965 by the University of Chicago.]

the dM stars. The former show a total spread of about ± 40 km s^{-1} in the u component and about ± 30 km s^{-1} in the v component, while the latter show a spread of ± 80 km s^{-1} in u and range from about $+40$ km s^{-1} to -70 km s^{-1} in v. In addition to their larger random-velocity dispersions, the dM stars clearly have a skewed distribution in v, and they show a net rotational lag of about 15 km s^{-1} behind the LSR, while the dMe stars show essentially none. The dM stars have an rms speed with respect to the Sun of about 22 km s^{-1}, with a standard deviation of ± 30 km s^{-1} around this mean, while the dMe stars have a mean speed of about 11 km s^{-1}, with a standard deviation of ± 18 km s^{-1}. We have here an unequivocal example of the increase, with age, of the random velocity of a group of stars, and of the concomitant lag of the group behind the LSR. In terms of the population groups introduced in §4-5, the dMe stars are a mixture of stars from the spiral-arm and young disk populations. The dM stars have the characteristics of an intermediate disk population (which is probably to be translated to mean that they are a mixture of moderately young to rather old disk populations).

The data discussed thus far make the hypothesis that the peculiar velocity dispersion of disk stars increases with their age appear quite plausible; convincing evidence has emerged from recent work by J. Byl (**B4**), M. Mayor (**M1**), and R. Wielen (**W2**). Wielen's analysis is based on about 1000 stars contained in Gliese's catalog of stars within 20 pc of the Sun for which trigonometric parallaxes accurate to $\pm 10\%$, and accurate radial velocities and proper motions (hence space motions), are known. This sample can be directly plotted in an H-R diagram, and hence it can be divided into unambiguous age groups by choosing stars found in definite color intervals along the main sequence or near the positions of the subgiant or giant branches of clusters of known age. The details of how these divisions are made are described in (**W2**). For each main-sequence group, the average age is assumed to be about half the main-sequence lifetime of the appropriate stellar type (that is, we presume a constant rate of star formation), and the giants are assigned the ages of the clusters along whose giant branches they most nearly lie. The sample includes a large number of McCormick Observatory K and M dwarfs with known Ca II emission-line intensities, which were estimated visually from high-dispersion spectra and arranged on an empirical scale running from $+8$ for very intense emission to -5 for very weak or no emission. Mean ages for these emission-line stars can be derived statistically from their relative numbers by assuming a constant rate of star formation over the lifetime of our Galaxy, and these estimates can be checked using observed average emission-line strengths in clusters of known ages. (The two sets of ages agree well.)

For each age group, velocity dispersions and average rotational lags behind the LSR are readily derived, and they can then be plotted against mean age. The results, exhibited in Figure 7-5, show an unquestionable monotonic increase of the dispersion in each residual-velocity component, and of the

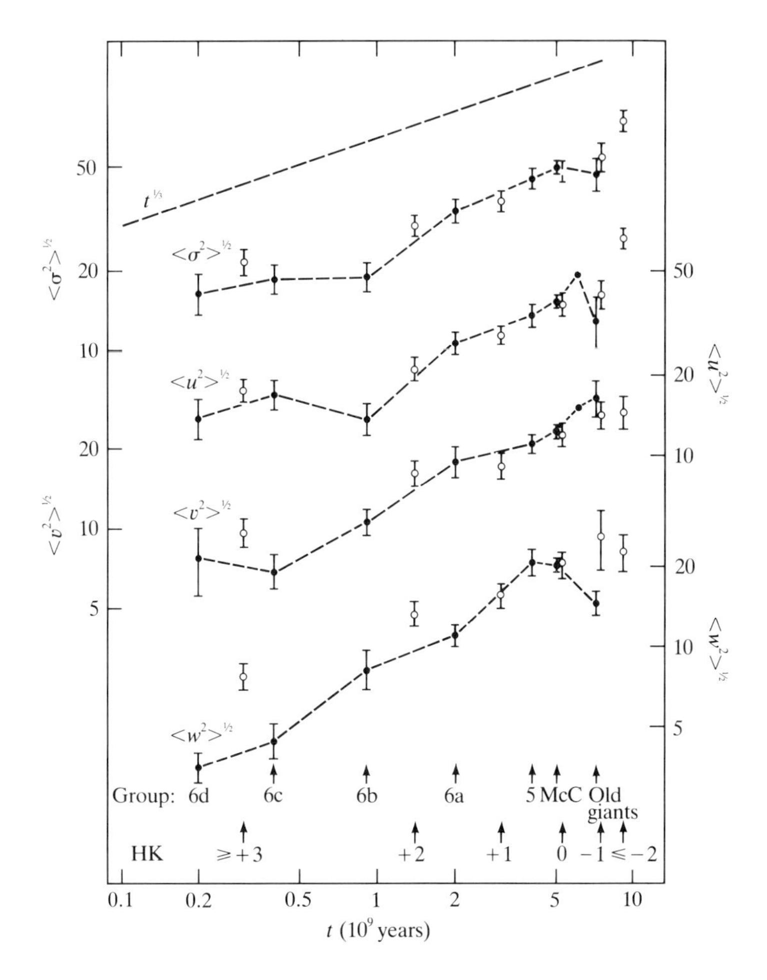

Figure 7-5. Variation of the peculiar velocities of disk stars with age. Curves give velocity dispersions in each coordinate and total rms speed. Solid dots give values for main-sequence age groups and old giants; open circles give values for McCormick K and M stars grouped by emission-line strength. Dashed line shows predicted growth resulting from Spitzer-Schwarzschild mechanism. [Adapted from (**W2**) by permission.]

rms speed, with increasing average stellar age. The dashed curve is the asymptotic result from theoretical arguments discussed next, and it is seen to be in excellent qualitative agreement with the observed behavior.

Theoretical Interpretation How can we explain the observed increase of the random velocities of disk stars with increasing age? There are essentially two classes of possible explanations. On the one hand, we could suppose that stars with different ages—that is, that were born at different epochs—were created with differing kinematic properties, the latter remaining unaltered thenceforth. While such a situation is not a priori impossible, it seems unlikely to be the cause of the observed behavior of the kinematic properties of the disk stars. Specifically, we see from Figure 7-5 that the observed velocity dispersions rise most rapidly for young stars and change relatively more slowly for old stars. On the hypothesis just proposed, we would have to require that the thickness of the layer of material from which stars form has decreased with time, the rate being most rapid at our present epoch. While it is possible that the gaseous layer from which stars form has become thinner with time, it is (cosmologically) unsatisfying to argue that the rate of contraction should suddenly have become large in the very recent past. We must therefore consider alternatives. Before we do, however, let us stress explicitly that we are addressing here only the origins of the kinematic properties of disk stars. We believe today that, in contrast, the major differences between the kinematic properties of spheroidal-component stars as a class versus disk stars as a class do, in fact, reflect the special dynamical conditions that existed in our Galaxy at the time the former group was formed.

An alternative class of explanations is based on the hypothesis that there exist mechanisms that progressively increase the velocity dispersions of disk stars from the initial values that they had at the time of their formation (which are about equal to the random velocities within the interstellar medium). Such an hypothesis is very appealing a priori because we know empirically that there are forces acting that lead to the disintegration of star clusters and stellar associations and to their dispersal into the general field. These explanations invoke accelerations imposed by time-fluctuating gravitational fields, which cause stars to deviate from the smooth orbits they would have had in a perfectly smooth, steady, axisymmetric galactic potential. The two basic ideas that have been proposed are (1) that stellar velocities are randomized through chance encounters of stars with other stars and with interstellar clouds, or (2) that stars are accelerated from time to time by interactions with the mass concentration in a passing spiral density wave. Let us briefly consider the results obtained from analyses of these processes, deferring the details of a dynamical discussion until Chapter 17.

When stars suffer encounters ("collisions") with other stars or interstellar clouds, they exchange energy and are deflected from their original orbits These encounters tend to randomize the velocity distribution, and they ultimately lead to *energy equipartition* between the collision partners. A calculation of the effects of star-star encounters shows that this mechanism is much too slow (by orders of magnitude) to account for the observed effect (see Chapter 17). But it was recognized by Spitzer and Schwarzschild (**S7**), (**S8**) that encounters with interstellar cloud complexes are, in comparison, vastly more effective because of the huge masses of the clouds. The clouds have enormously much more kinetic energy than a star, so in star-cloud collisions the stars tend to gain energy and move away at higher speeds.

A key point in the Spitzer-Schwarzschild analysis was the realization that, because of differential galactic rotation (see §8-1), complexes of interstellar clouds are, in effect, driven past the LSR and produce a time-fluctuating gravitational field (that is, collisions) quite efficiently. Spitzer and Schwarzschild solved the Fokker-Planck equation (see Chapter 18) to obtain a detailed quantitative description of the stellar velocity distribution. In particular, they showed that the time dependence of the rms speed of a star, $v_{\rm rms}$, is described approximately by the formula

$$v_{\rm rms}(t) \approx v_{\rm rms}(0)\left[1 + \left(\frac{t}{\tau_E}\right)\right]^{1/3} \tag{7-8}$$

where $v_{\rm rms}(0)$ is the initial rms speed fixed by the internal velocity dispersion in the interstellar medium (about 10 km s^{-1}), and τ_E is a characteristic *energy-exchange time* for the collision process. We shall give a heuristic argument that recovers equation (7-8).

By demanding that $v_{\rm rms}$ roughly double in about 1.5×10^9 years as observed, Spitzer and Schwarzschild determined $\tau_E \approx 2 \times 10^8$ years, and, from τ_E, they inferred the required mass of the cloud complexes using the results of the dynamical theory of encounters. The derived masses are about $\mathcal{M}_c \approx 5 \times 10^5 \mathcal{M}_\odot$, a value in accord with typical masses for the recently discovered *molecular cloud complexes* (**S9**) (see also §9-3). From their analysis of the then-available kinematic data, Spitzer and Schwarzschild concluded that the star-cloud encounter mechanism successfully explains the observed increase in velocity dispersion as a function of spectral type for most of the main sequence and for the giants. They felt, however, that their theory did not fully explain the velocity dispersion of the M dwarfs directly. But, recent work shows that their mechanism may, in fact, be able to account adequately for the kinematic properties of all disk stars. Finally, Spitzer and Schwarzschild concluded that their mechanism failed completely to explain the very large velocity dispersions of the spheroidal-component stars. This conclusion has remained unchanged by more recent work.

An investigation of the effects of interactions of stars with spiral density waves upon stellar kinematics has been carried out by B. Barbanis and L. Woltjer (**B1**). They conclude that the secular increase in the random velocities of disk stars can be understood as resulting from this process alone if either (a) the density in typical spiral arms a few billion years ago was perhaps three times its present-day value (and subsequently decreased), or (b) that the spiral arms always had nearly their present density but were nonstationary, that is, dissolving and reforming every few hundred million years. Of the two alternatives, the latter has greater basic appeal, because the former would imply arms that are much more massive than are commonly observed in external spiral galaxies of the same type as our Galaxy. A difficulty with this mechanism is that unlike the Spitzer-Schwarzschild mechanism, it cannot account for the observed increase in the z component of the velocity dispersion.

It should be noted that both the Barbanis-Woltjer mechanism and the Spitzer-Schwarzschild mechanism can be in operation simultaneously. In fact, they can even act in concert, because we know from observations of external galaxies (for example, M51) that both stars and interstellar gas moving with the local circular velocity overtake the spiral arms, running into them at supersonic speeds (compared to the sound speed in the interstellar gas), thus producing an interstellar gas shock along the inner edge of the arms. Empirically, we find that dust lanes lie along the inner edge of spiral arms, precisely along the position of this shock. Thus a star encountering a spiral arm will experience not only the forces produced by the mass concentration in the spiral wave itself but, in addition, the gravitational field it experiences will be very irregular as it runs the gauntlet through the interstellar gas and dust clouds piled up in the dust lanes.

R. Wielen has given (**W3**) a very interesting discussion of the kinematic consequences of these randomizing processes, and we can get both some intuitive insight and some useful numerical estimates from a brief consideration of his results. The effects of random gravitational encounters of stars with very massive objects can be approximated by a sequence of independent and essentially instantaneous impulses that are distributed stochastically. Each impulse changes the star's velocity $\mathbf{v}$ by an amount $\Delta\mathbf{v}_i$. Because they are random, we expect that, over a suitably long time interval t, the vector sum of these changes will average to zero, that is,

$$\langle \Delta\mathbf{v}_i \rangle = \lim_{t\to\infty} \frac{1}{t} \sum_{i=1}^{n} \Delta\mathbf{v}_i = 0 \tag{7-9}$$

where n is the number of collisions in time t. (The effects of *dynamical friction*—see Chapter 18—are thereby ignored). On the other hand, the sum of the squares of the velocity changes do not average to zero, but rather they tend to grow linearly with time,

$$\sum_{i=1}^{n} (\Delta v_i)^2 = Dt \tag{7-10}$$

where D is a *diffusion coefficient* that may, in general, depend up v and t. (Again, dynamical friction is ignored). Thus there is a diffusion of stars in velocity space; that is, the velocity vectors for a group of stars tend to scatter with an ever-increasing dispersion around their original values.

The diffusion in velocity space is described mathematically by the Fokker-Planck equation, which we shall discuss in Chapter 19. For our present purposes, it suffices to suppose that we can describe the process as an isotropic diffusion whose nature is unaffected by the details of a star's unperturbed orbit (a "force-free" diffusion) and that we can fix the value of the diffusion coefficient empirically from the observed data. To a rough approximation, we then expect the random velocity of a star to grow according to

$$\frac{d(v^2)}{dt} = D(v, t) \tag{7-11}$$

In particular, if $D \equiv D_0 =$ constant, then

$$v^2 = D_0 t + c \tag{7-12}$$

so that we expect a relationship of the form

$$v_{\mathrm{rms}}(t) = v_{\mathrm{rms}}(0)\left[1 + \left(\frac{t}{\tau}\right)\right]^{1/2} \tag{7-13}$$

to exist. This result is reminiscent of equation (7-8), but the exponent is different. From the theory of binary encounters (see Chapter 18), one finds that the diffusion coefficient is inversely proportional to v; that is, $D(v) = D_0/v$. Then, equation (7-11) yields

$$v^3 = \tfrac{3}{2}D_0 t + c \tag{7-14}$$

which implies a relationship of the form

$$v_{\mathrm{rms}}(t) = v_{\mathrm{rms}}(0)\left[1 + \left(\frac{t}{\tau}\right)\right]^{1/3} \tag{7-15}$$

This relationship recovers the result obtained by Spitzer and Schwarzschild.

Wielen has shown (**W3**) that the observed data can be fitted extremely well by equation (7-13), with $v_{\mathrm{rms}}(0) = 10$ km s^{-1} and $\tau \sim 2 \times 10^8$ years, or slightly less well by equation (7-15), with $v_{\mathrm{rms}}(0) = 10$ km s^{-1} and $\tau \sim 5 \times 10^7$ years. In the Spitzer-Schwarzschild theory, the latter value for τ would imply interstellar cloud masses of the order of $\mathscr{M}_c \sim 2 \times 10^6 \mathscr{M}_\odot$, which is a bit larger than, but of the same order as, the observed values for molecular clouds. The result that equation (7-13) apparently fits the observations better than equation (7-15) may possibly imply that the approximation of individual binary collisions on which the latter is based may break down; for example, perhaps multiple collisions occur simultaneously when a star runs into a spiral arm.

Wielen also made a detailed study (**W3**) of the diffusion of stellar orbits, applying the approximation of random perturbations to an underlying epicyclic motion around a circular orbit (see Chapter 13). The perturbations

Table 7-2. Kinematic Diffusion Effects on Orbits of Disk Stars

	RMS velocity changes (km s^{-1})				RMS position changes (kpc)			
Age (years)	Δu	Δv	Δw	$\Delta \sigma$	ΔR	ΔS	Δz	Δp
10^7	2	2	1.7	3	0.01	0.01	0.01	0.02
10^8	7	5	4	10	0.4	0.5	0.04	0.65
10^9	20	12	10	25	1	15	0.11	15
5×10^9	40	25	21	50	2	165	0.23	165
10^{10}	60	40	33	80	3	550	0.36	550

SOURCE: Adapted from (**W3**), by permission.

were presumed to produce an isotropic diffusion process of the kind just described. In his calculation, he used standard values of the galactic rotation constants (see §8-2) to compute the epicycle frequency in the plane and a representative z variation of the galactic potential to compute the oscillation frequency perpendicular to the plane (see Chapter 13). He finds the diffusion-induced rms increases in the velocity components Δu, Δv, Δw, and the speed $\Delta \sigma$ (all in kilometers per second) shown in Table 7-2. These velocity increments are to be summed statistically with the initial velocity dispersion (~ 10 km s^{-1} in each direction). He also calculates the rms deviations (in kiloparsecs) ΔR in the radial direction, $\Delta S = R\,\Delta\theta$ in the tangential direction, Δz in the direction perpendicular to the plane, and the total rms deviation in position Δp of a star away from its initial LSR. The particular results cited here are for the case of a velocity- and time-dependent diffusion coefficient, but the results for other cases are practically identical to these.

The results presented in Table 7-2 have several interesting implications.

1. Notice, first, that the computed limiting values of the ratios $\langle u^2 \rangle^{1/2}$:$\langle v^2 \rangle^{1/2}$:$\langle w^2 \rangle^{1/2}$ are about 1.0:0.64:0.53. These numbers are in remarkable agreement with those for the most representative group of old disk stars in Gliese's catalog, the McCormick K and M stars, for which one finds $\langle u^2 \rangle^{1/2}$:$\langle v^2 \rangle^{1/2}$:$\langle w^2 \rangle^{1/2} = 48{:}29{:}25$ (km s^{-1}) $= 1.0{:}0.61{:}0.52$. This agreement lends considerable credibility to the computed results. The fact that the residual-velocity distribution does not become isotropic even when dominated by randomizing influences reflects the presence of the constraints imposed by the general galactic potential field.

Given that the computations fit the observed data, we have used them (folded with an initial rms dispersion of 10 km s^{-1} in each component) as a guide in defining the kinematic properties of the archetypal population groups of Table 4-18. They also allow us to make reasonable estimates of what the kinematic properties of disk stars as old as 10^{10} years should be like if such stars exist. (Recall that current age estimates for the oldest known galactic cluster, NGC 188, are more like 6–8 × 10^9 years.) In particular, we see that the oldest disk stars could have rms speeds around 80 km s^{-1} with respect to the Sun, and hence, observationally they would be considered to be high-velocity stars. (Notice also that, because they lag the LSR by some 50 km s^{-1}, their velocities in the direction of galactic rotation will still be below the observed limit of about 65 km s^{-1}; see §7-2.)

An extremely important point that must be emphasized here is that the kinematic properties of the old disk stars do not merge continuously with the properties of the spheroidal-component stars, although they overlap them partially. Historically, this situation has caused considerable confusion in deciding to which population any given observed group of stars belongs, and the erroneous presumption that the two major groups (disk and spheroidal components) do merge has repeatedly caused great difficulties in the development of a coherent picture of stellar populations. In what follows, we shall attempt to make some fairly crisp distinctions among various kinematic groups on the basis of Table 4-18, but we stress that these can be only as secure as the results quoted in that table are realistic, and we freely admit that they may be subject to significant revision in the future.

2. Next, we can see that orbital diffusion effects imply a very large *spatial diffusion* of stars. The spread is particularly large in the tangential direction. After 10^{10} years, the number of revolutions a star 10 kpc from the galactic center will have made is uncertain by ± 9 revolutions around the average value of 40 revolutions. Such stars would have diffused to about ± 700 pc away from the plane ($\langle |z| \rangle = \beta_z$, the scale height), and they would scatter over a radial range of $\Delta R \approx \pm 3$ kpc. If we define an orbital "eccentricity" parameter as

$$e \equiv \frac{R_{\max} - R_{\min}}{R_{\max} + R_{\min}} \tag{7-16}$$

then the oldest disk stars in the solar neighborhood should have orbital "eccentricities" up to about 0.35. The random spatial diffusion of stars produced by these effects is so large that there is simply no hope of tracing the orbits of old stars backward in time, for example, to locate their initial birthplaces. (To do so would be an example of the classic problem of trying to integrate the diffusion equation in the wrong direction.)

3. Finally, we see that these diffusion mechanisms, which explain the kinematics of disk stars so well, are quite inadequate to explain the properties of the spheroidal-component stars. As we can be seen in Tables 6-3, 7-1, and 7-5, typical halo stars (subdwarfs, and RR Lyrae stars) found in the solar neighborhood have random velocities and a rotational lag far in excess of the values listed in Table 7-2, and, typically, they have orbital "eccentricities" approaching unity, again much too large to be explained by the results obtained here. These stars belong to a system having a very high kinematic "temperature," whose kinematic properties may have been fixed long before the disk even formed.

The results just described provide a reasonably satisfying empirical and theoretical rationale for the picture developed in §4-5 of a sequence of disk populations characterized by a progressive kinematic "heating," an increasing lag behind the LSR, and a progressively larger spatial diffusion as a function of stellar age. In terms of this picture, the data given in Tables 6-3, 7-1, and 7-5 become coherent and readily comprehended. (Note, in passing, that the value of $\langle v \rangle$ in Table 7-5 for the long period variables with $P < 200^{\mathrm{d}}$ seems too large for a pure disk-population group; it may perhaps include a mixture of halo stars.)

The basic kinematic elements of our picture of the disk populations are hardly new, having been suggested as early as 1924 by G. Strömberg (**S10**). Historically, they were prematurely pushed aside when it was recognized that they could not explain the observed properties of the highest-velocity stars. As was mentioned earlier, the picture was further muddied by the partial overlap of the kinematic properties of the older disk population and the spheroidal-component stars. But, in light of the observational and theoretical results just cited, it now appears appropriate to revive some of these early ideas.

Implications for Kinematic Heterogeneity of Spectral Groups Having developed what appears to be a satisfactory set of archetypal disk populations, let us return briefly to the problem of kinematic heterogeneity of spectral groups and attempt to interpret a few representative examples in terms of these archetypes.

As a first example, consider the K-type giants observed by A. N. Vyssotsky (**V1**) which we discussed earlier in connection with the vertex deviation. In Figure 7-2, the peculiar velocity components u and v are plotted for the two groups of gK stars: (1) those with small $|w|$ velocities, which therefore move on orbits that lie essentially in the galactic plane, and (2) those with large $|w|$ velocities, which move on orbits that are inclined to the plane and therefore penetrate to fairly large distances above and below it. The low-inclination group has a smaller spread in u components (± 40 km s^{-1}) and v components (from $+30$ to -50 km s^{-1}), and a smaller lag behind the LSR ($\langle v \rangle \approx -10$ km s^{-1}) than the high-inclination group, which scatters ± 80 km s^{-1} in u and from $+40$ to -80 km s^{-1} in v, and lags the LSR by about 20 km s^{-1}. In terms of the population types of Table 4-18, the low-inclination group would be considered to be dominantly composed of young disk-population stars, and the high-inclination group to be composed of intermediate disk-population stars. *Some* of the stars in the first group may, of course, really belong to the second, but they have been included in the former group simply because we have differentiated the two groups solely on the basis of the value of $|w|$ for each individual star, and we know that some individuals of the older group must have low, or zero, $|w|$, even though the group as a whole has a large value of $\langle w^2 \rangle^{1/2}$.

Table 7-3. Kinematic Properties of Strong-Line and Weak-Line F–K Stars

Spectral class	Group	Mean speed relative to LSR (km s^{-1})	Standard deviation from mean speed (km s^{-1})
F5–G5 (excluding G2–G3 giants)	Strong-line	28	14
	Weak-line	43	22
G5–K1 giants	Strong-line	25	13
	Weak-line	41	24
	λ4150 stars	42	42
	Weak CN band	96	47

SOURCE: Adapted from (**B3**, 76), by permission. Copyright © 1965 by the University of Chicago.

Certainly the most important example of the kinematic heterogeneity of classical spectral types is provided by the F, G, and K stars. By means of a careful spectral classification study, N. Roman (**R1**), (**R2**) was able to isolate within these classes several spectroscopically distinctive groups characterized by a progressive weakening of metal-line strengths. The kinematic properties of these groups are summarized in Table 7-3. It is worth stressing that these stars are all common field stars in the disk. The strong-line and weak-line late-F to early-G dwarfs constitute 85% of all stars of these spectral types in the solar neighborhood. Similarly, the strong-line and weak-line giants (plus the kinematically indistinguishable λ4150 stars) comprise about 85% of the late-G to early-K giants in the solar neighborhood, with the "ordinary" high-velocity giants (including the weak CN stars) contributing almost all of the remaining 15%. Only about 1% of the G–K giants near the Sun belong to the spheroidal component.

The striking result that emerged from Roman's work was the close correlation between the spectroscopic and kinematic properties of the subgroups. Specifically, the data show a progressive increase of the mean speed of the stars with respect to the Sun, and of their velocity dispersion around the mean, with progressively weaker spectrum lines. Both the strong-line dwarfs and giants have the kinematic characteristics of ordinary, young disk-population stars. The weak-line dwarfs, and the weak-line and λ4150 giants, are kinematically intermediate disk-population stars. The weak CN-band stars have an rms speed that implies that this sample must be composed primarily of the very old disk population.

Given that the spectroscopic line strengths indicate the metallicity of each group, it is clear that the oldest groups (as judged from their kinematic properties) of disk stars are also the most metal poor. We thus infer that the average metallicity of disk stars is an increasing function of time. These data alone do not yield a quantitative estimate of the rate of increase of metallicity, because the line-strength groups are only qualitative, but subsequent model-atmosphere analyses have given quantitative metal abundances and lead to the estimates of $d[\mathrm{Fe/H}]/dt$ quoted earlier in §3-9.

The trends described here are fully supported by the results obtained for high-velocity stars, that is, stars selected to have large space motions (usually $\gtrsim 65$ km s^{-1}) with respect to the Sun (**R3**), (**R4**), (**B3**, Chapter 16). This sample will naturally include a fair number of spheroidal-component stars (subdwarfs), but the latter are spectroscopically so distinctive that they can be easily excluded from the analysis. The overwhelming majority of a random sample of high-velocity stars have w components that imply that they remain within about ± 500 pc of the galactic plane and hence are indeed members of a disk population. Spectroscopically, what one finds is that the vast majority of the high-velocity stars are distinctly metal poor, although none of them have metallic lines as weak as those of stars in, say, M92. Both on the basis of their spectroscopic appearance and from their H-R diagram, these stars strongly resemble the stars in an old galactic cluster such as M67 or NGC 188. In particular, the giants in the sample are spectroscopically very similar to normal K giants, and they are significantly less luminous than the red giants in globular clusters. We conclude again that these stars are members of an old (perhaps very old), metal-poor, kinematically "hot" disk population whose overall properties harmonize well with the archetypes of §4-5.

Yet another set of data, obtained by J. L. Greenstein (**B3**, Chapter 17) for the F, G, K high-velocity stars, is summarized in Table 7-4. Again, we see the same trends of increasing random velocity dispersion σ and rotational lag $\langle v \rangle$ with decreasing line strength. (Both $\langle u \rangle$ and $\langle w \rangle$ are zero to within the statistical errors, as they should be.) The first three groups in the table are intermediate-to-old disk stars. The last two are members of the spheroidal component (note the marked jump in $\langle v \rangle$ for these stars). It should be stressed that the stars in this sample are highly selected. They were all included in the observing program because they were suspected of

Table 7-4. Space Motions and Velocity Dispersions of F and G High-Velocity Stars (km s^{-1})

Line-strength group	$\langle u \rangle$	$\langle v \rangle$	$\langle w \rangle$	$\langle \sigma \rangle$
Normal	$+30 \pm 15$	-65 ± 20	$+10 \pm 15$	40
Slightly weak	$+50 \pm 30$	-120 ± 40	-10 ± 30	90
Intermediate	-45 ± 30	-115 ± 40	-40 ± 30	95
Subdwarfs	-30 ± 40	-210 ± 65	-50 ± 45	115
Extreme subdwarfs	-20 ± 45	-215 ± 50	-35 ± 40	160

SOURCE: Adapted from (**B3**, 370), by permission. Copyright © 1965 by the University of Chicago

having high space velocities relative to the LSR, and none of the groups contains a representative sample of unselected F, G, K stars. This selection criterion explains why the rotational lag and rms speed σ are so large for even the normal line strength group. These stars populate the extreme fringe of the velocity distribution for normal young disk stars. Similarly, $\langle v \rangle$ for the weak-line and intermediate groups is almost twice as large as the value given in Table 4-18 for the oldest disk population, because stars with small $|v|$ were discriminated against from the outset.

All of the kinematic data discussed here are in basic accord with the population scheme developed earlier in this section and in §4-5.

Finally, it is worth remarking that, in the analysis of kinematic data, it is important always to be aware of the possibility of kinematic heterogeneity in any group of stars and to attempt to recognize it when it occurs. If we were to ignore a significant heterogeneity of some group, we would surely derive spurious estimates of the parameters that define the velocity ellipsoid, and, in addition, we would lose potentially valuable information concerning the age and population type dependences of the kinematic properties of stars in our Galaxy.

Let us now turn to an analysis of the kinematic properties of stars in the other major structural element of our Galaxy—the spheroidal component.

7-2. KINEMATICS OF SPHEROIDAL-COMPONENT STARS

Kinematics of the High-Velocity Stars

Oort's Analysis Even in the early work on stellar kinematics, it was recognized that the Schwarzschild velocity ellipsoid centered on the motion of the LSR provides an accurate description of stellar peculiar velocities only over a limited range of speeds, and that, above a certain critical speed (about 65 km s^{-1}), the velocity distribution is markedly different. The significance of this sharp change was recognized by Oort in 1927 and 1928. He realized that it revealed the rotation of our Galaxy (**O1**), (**O2**), (**O3**). This realization was a breakthrough in our understanding of our Galaxy, and it is of interest, even after fifty years, to review the data used in his pioneering study as well as some of the chain of reasoning that led to the inferences he made from them.

A plot of individual stellar velocity vectors relative to the LSR, projected onto the galactic plane, for the high-velocity stars in Oort's sample is shown in Figure 7-6. The ordinate and abscissa are the (Π, Θ) components relative to the galactic center, on the assumption that the rotation velocity of the LSR, Θ_0, is 300 km s^{-1} (about 50 km s^{-1} too large by current estimates). The square represents the velocity components of the LSR, and the circled dot represents the Sun. The smallest (dashed) circle centered on the Sun corresponds to a relative speed of 20 km s^{-1}; stars in this region were omitted

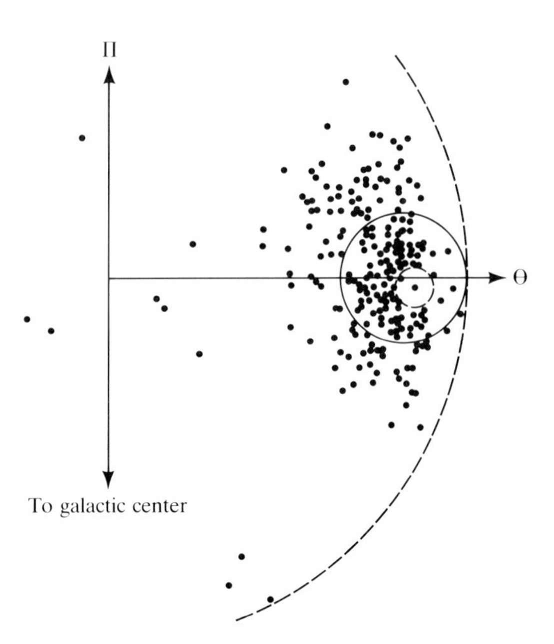

Figure 7-6. Distribution of Π and Θ velocities of high-velocity stars as derived by Oort in 1928. [Adapted from (**O3**).]

because the sample was incomplete owing to observational selection effects. The solid circle corresponds to stars moving with a speed of 65 km s^{-1} relative to the LSR. This speed is the critical one above which the velocity distribution changes fundamentally in character. The large dashed circle corresponds to a speed of 365 km s^{-1} with respect to the fundamental standard of rest (that is, the galactic center). At the time of Oort's study, no star was known to have a higher speed and thus fall outside this boundary.

From the data available to Oort, one can see the following:

1. The velocity distribution for speeds less than 65 km s^{-1} is approximately ellipsoidal with the long axis directed very nearly toward the galactic center. These are disk stars.
2. For speeds greater than about 65 km s^{-1} relative to the LSR, the velocity distribution is asymmetric around the direction to the galactic center (as identified by Shapley). This result is shown particularly clearly by Figure 7-7, where we see that the nature of the asymmetry is that the high-velocity stars preferentially lag behind the LSR along an axis perpendicular to the direction to the galactic center. Oort suggested that this effect could be readily understood if that axis lay in the direction of galactic rotation, and if the stars that are observed as high-velocity stars were part of a system (or systems) having a low systemic rotation rate around

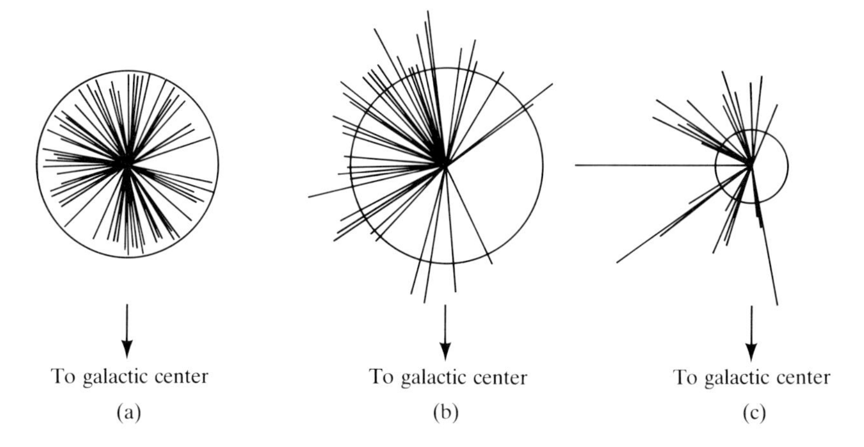

Figure 7-7. Velocity vectors in the (u, v) plane of stars with speed S, relative to the LSR, on the ranges (a) $S < 63$ km s^{-1}; (b) $63 \leq S \leq 100$ kms^{-1}; and (c) $S \geq 100$ km s^{-1}. The circle has a radius of 63 km s^{-1} in all three plots. Note that the velocity vectors in group (a) are essentially randomly distributed in direction, while those in groups (b) and (c) show a symmetry around the direction $\ell = 90°$ and a progressive tendency, with increasing S, to point away from this direction (the direction of galactic rotation). [Adapted from (**O1**).]

the galactic center, whose members would therefore lag far behind the Sun in its revolution on a circular orbit around the galactic center.

3. In addition to the asymmetry just discussed, Figure 7-7 shows that, while the directions of the motions of low-velocity stars are essentially random, at speeds $\gtrsim 65$ km s^{-1} relative to the Sun, no stars are found moving in the direction of galactic rotation. Oort suggested that stars moving with this speed in the direction of galactic rotation might have enough energy to escape from our Galaxy. This explanation is now believed to be incorrect (see §8-2), and the phenomenon is now thought to reflect instead a paucity of high-angular-momentum stars in the solar neighborhood. We shall return to this point later in this section.
4. The stars moving with high speeds relative to the LSR show, in general, large $|w|$ velocities. For the low-velocity stars, $\langle w^2 \rangle^{1/2}$ is of the order of 10 to 20 km s^{-1}, whereas, for the high-velocity stars, $\langle w^2 \rangle^{1/2}$ is of the order of 70 to 90 km s^{-1}. These stars must therefore move on orbits that have high inclinations to the galactic plane and that carry them far from the plane.

The major conclusion that emerged from Oort's analysis was that our Galaxy rotates around the center identified by Shapley, and that the bulk

Table 7-5. Asymmetric Drift (Rotational Lag) and Velocity Dispersions for Evolved Disk Stars and Spheroidal-Component Objects

Object	Number	Lag (km s^{-1}) $\langle\Theta\rangle - \Theta_0$	$\langle\Theta\rangle$	Dispersions (km s^{-1}) $\langle\Pi^2\rangle^{1/2}$	$\langle(\Theta - \langle\Theta\rangle)^2\rangle^{1/2}$	$\langle Z^2\rangle^{1/2}$
Evolved disk stars						
Long-period variables ($P > 300^d$)	167	−10	240	40	40	30
Planetary nebulae	110	−20	230	40	–	30
Long-period variables ($200^d \leq P \leq 300^d$)	143	−30	220	55	45	45
RR Lyrae variables ($\Delta S < 5$)	16	−50	200	60	50	35
Long-period variables ($150^d \leq P \leq 200^d$)	41	−105	145	95	90	60
Spheroidal-component stars						
Globular clusters	70	−165	85	145	–	–
Extreme subdwarfs	22	−185	65	170	90	65
RR Lyrae variables ($\Delta S > 5$)	33	−220	30	210	120	90

SOURCE: Adapted from (**B3**, 483), by permission.

of the stars in the solar neighborhood move on near circular orbits with a rotation speed of roughly 300 km s^{-1} (250 km s^{-1} by today's estimates). The high-velocity stars are then seen to be stars that have high velocities with respect to the Sun, but low rotation velocities around the galactic center. It follows that these stars move on very eccentric orbits around the center. Finally, as mentioned in Chapter 1, Oort also derived a reasonable first estimate of the mass of that part of our Galaxy inside the position of the Sun.

Modern Data The sample of stars available for Oort's discussion was small, and it was composed of a very inhomogeneous group of objects. Actually, only a small fraction of the stars plotted in Figure 7-6 are genuine spheroidal-component objects. Most of them are old disk stars comprising the "normal" high-velocity stars discussed in §7-1. It is, in fact, remarkable how Oort was able to draw such fundamentally sound conclusions from such fragmentary data.

Since the time of Oort's work, a great deal of observational material concerning the kinematics and physical properties of the high-velocity stars has been gathered. For example, in the 1940s, improved sets of data were published by G. Miczaika (**M2**) and W. Fricke (**F2**). By far the most complete material is contained in the extensive catalogs of space velocities, distances, spectroscopic and photometric data, and soon, for subdwarfs and high-velocity stars published by Eggen (**E1**), (**E2**) and Sandage (**S1**). Using spectrophotometric information [for example, $\delta(U - B)$ for subdwarfs or ΔS for RR Lyrae stars], we can now select from these data homogeneous samples of stars that truly represent the spheroidal component, and hence we can derive a fairly accurate picture of its kinematic properties near the Sun.

Typical velocity dispersions and rotational lags for representative spheroidal-component objects are listed in Tables 7-1 and 7-5. We see that they all have low rotation rates around the galactic center and an extremely large random velocity dispersion. Plots of two modern samples of subdwarfs and high-velocity stars in the (u, v) plane are shown in Figure 7-8 and 7-9. Stars with $\delta(U - B) \lesssim 0.15$ are dominantly old disk stars (and hence should be ignored for the present discussion), while those with large $\delta(U - B)$ are genuine halo stars. We can see in these figures that, to a good approximation, the residual velocities of the halo stars with respect to their centroid at $\langle\Theta\rangle \approx 50$ km s^{-1} (or $\langle v\rangle \approx -200$ km s^{-1}; see Tables 6-3 and 7-5) is ellipsoidal, with its principal axis directed toward the galactic center. Indeed, the mean ratios $\langle\Pi^2\rangle^{1/2}:\langle(\Theta - \langle\Theta\rangle)^2\rangle^{1/2}:\langle Z^2\rangle^{1/2} = 1.0:0.55:0.40$, obtained from the most reliable data (the extreme subdwarfs and RR Lyrae stars in Table 7-5), are rather similar to the corresponding values for old disk stars (1.0:0.61:0.52).

If the gravitational potential of our Galaxy is known, then an orbit for each high-velocity star can be computed from a knowledge of its (u, v, w)

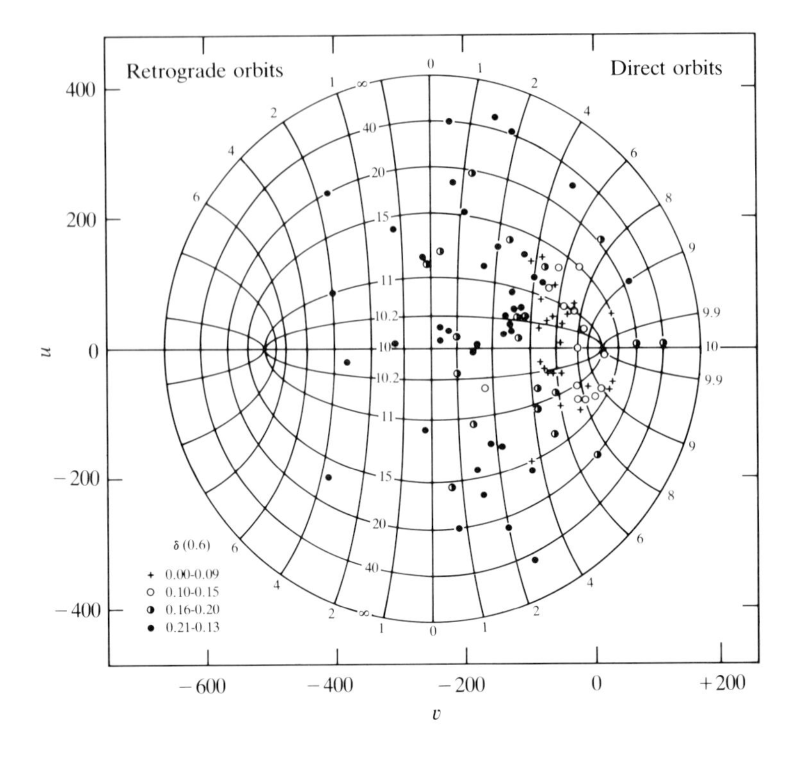

Figure 7-8. Bottlinger diagram for subdwarfs. Abscissa and ordinate give (u, v) components of peculiar velocities relative to the LSR in kilometers per second. Coding of points indicates degree of ultraviolet excess. Curves of constant apocenter (ovals) and pericenter (open) are marked in kiloparsecs; these were computed assuming $R_0 = 10$ kpc and $\Theta_0 = 250$ km s^{-1}. Stars to the right of the vertical line $R_{min} = 0$ travel on direct orbits; stars to the left travel on retrograde orbits. [From (**S1**), by permission. Copyright © 1969 by the University of Chicago.]

velocity components in the solar neighborhood (see Chapter 13 for a discussion of how this is done), and we can calculate, in particular, its *pericenter distance* (or *perigalacticon*) R_{min} and *apocenter distance* (*apogalacticon*) R_{max}. A convenient representation of the results is given by the *Bottlinger diagram*, which is shown in two different forms in Figures 7-8 and 7-9 (for two different samples of stars). We caution the reader that the potential fields used to construct these figures were chosen for analytical convenience and are very crude, so the results given there are not to be taken literally, but they should be regarded as illustrative only.

In Figure 7-8, curves of R_{min} and R_{max} are shown directly, while in Figure 7-9, the plot is given in terms of R_{max} and the orbital "eccentricity" parameter defined in equation (7-18). Because our Galaxy is not a point mass producing an inverse-square force field, the orbits of stars are not closed, and therefore

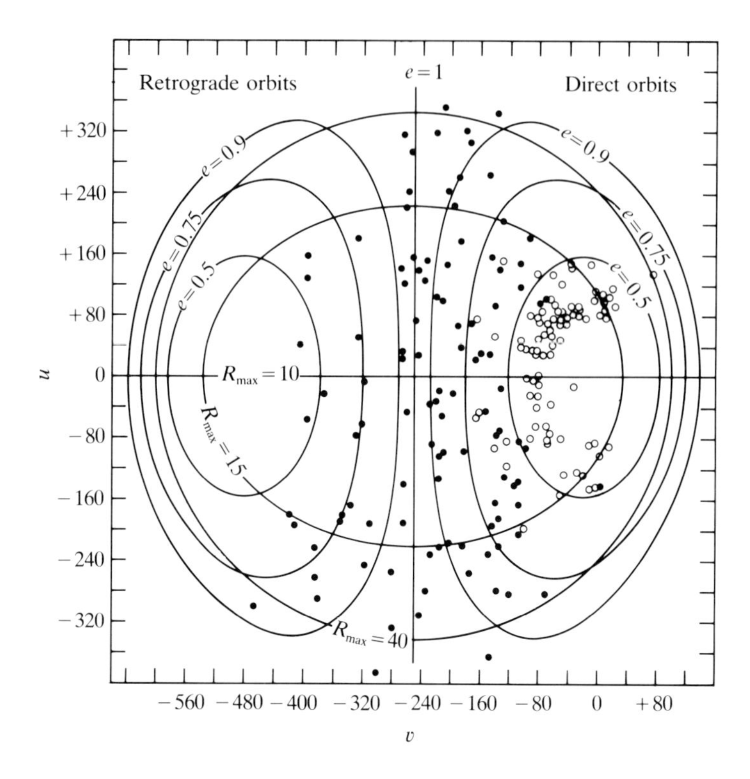

Figure 7-9. Bottlinger diagram for high-velocity stars with speeds relative to the LSR in excess of 100 km s^{-1}. Filled circles denote stars with $\delta(U-B) \geq 0\overset{m}{.}15$ and open circles those with $\delta(U-B) \leq 0\overset{m}{.}15$. Curves are marked with orbital eccentricities as defined by equation (7-16) and with apocenter distances in kiloparsecs. These were computed assuming $R_0 = 10$ kpc and $\Theta_0 = 250$ km s^{-1}. [From (**B3**, Chapter 6) by permission. Copyright © 1965 by the University of Chicago.]

the standard definition of orbital eccentricity becomes meaningless. The parameter e used here, though rough, does at least give a qualitative feeling for the degree of elongation of a star's orbit, $e \to 0$ for circular orbits and $e \to 1$ when $R_{max} \gg R_{min}$ ("plunging" orbits). If we want a more precise parameterization of a star's orbit, we could simply specify its total energy and angular momentum.

In Figures 7-8 and 7-9, we can see some noteworthy features. One is that the halo stars obviously do have very eccentric orbits and move through a large range of galactocentric distances. The majority of these stars penetrate to distances less than 4 kpc from the center; most of them have apocenter distances of less than 15 kpc, but some penetrate as far as 40 kpc from the center. Another interesting point is that there is a fairly large number of stars moving on retrograde orbits, that is, counter to the direction of galactic rotation. This is a natural consequence of the fact that the spheroidal component is kinematically so "hot," and so slowly rotating, that the velocity dispersion exceeds the system's rotation velocity. Hence stars can easily have negative tangential residual-velocity components that exceed the rotation speed of the centroid and thus have a net negative rotation velocity around the galactic center. Finally, it is clear from both Figures 7-8 and 7-9 that, if the subdwarf random-velocity distribution is symmetric about its centroid at $\langle\Theta\rangle \approx 50$ km s^{-1} (as we would expect it to be), then, for positive velocities, it overlaps the velocity domain of the disk stars. This is the phenomenon described in §7-1: the kinematics of disk stars overlap those of the spheroidal-component stars (or, more precisely, are overlapped by them) without merging continuously into them.

Correlations Between Kinematic and Spectrophotometric Properties

Some of the most important results that have emerged from recent studies of the high-velocity stars are the striking correlations that exist between their spectrophotometric and kinematic properties. We have already touched on these in our discussion of old disk stars, focusing there mainly on spectroscopic data. Let us now consider the large, homogeneous body of data presented by Eggen, Lynden-Bell, and Sandage [(**E3**), hereinafter referred to as ELS], who have studied the correlation of kinematic properties with the ultraviolet excess $\delta(U - B)$.

Empirical Results Consider first Figure 7-10, which gives a plot of $|w|$, the absolute value of the z velocity, versus $\delta(U - B)$. The right-hand ordinate is labeled with values of z_{max}, the maximum distance to which a star can penetrate away from the galactic plane according to the approximate galactic potential model used by ELS. At first sight, this plot appears to show a progressive, continuous increase of z_{max} with $\delta(U - B)$. In harmony with the then current pictures of stellar populations and galactic evolution, ELS suggested that this correlation could be interpreted as implying that metal-poor stars [large $\delta(U - B)$] were formed at almost any height above the plane up to 10 kpc, while progressively more metal-rich stars were formed in layers of progressively smaller thickness around the plane, in a sequence of successively more metal-rich and more highly flattened population groups. When more recent data (**S1**) are added to this figure, the impression just described is somewhat spoiled, for one finds several stars with $\delta(U - B) \approx 0.15$ and $150 \lesssim |w| \lesssim 300$.

An equally interesting alternative explanation is also possible, which harmonizes better with the ideas about stellar populations outlined in §4-5. As indicated by the boundaries drawn into the plot, one can consider the stars to fall into two distinct domains: (1) stars with $\delta(U - B) \lesssim 0.13$, all of

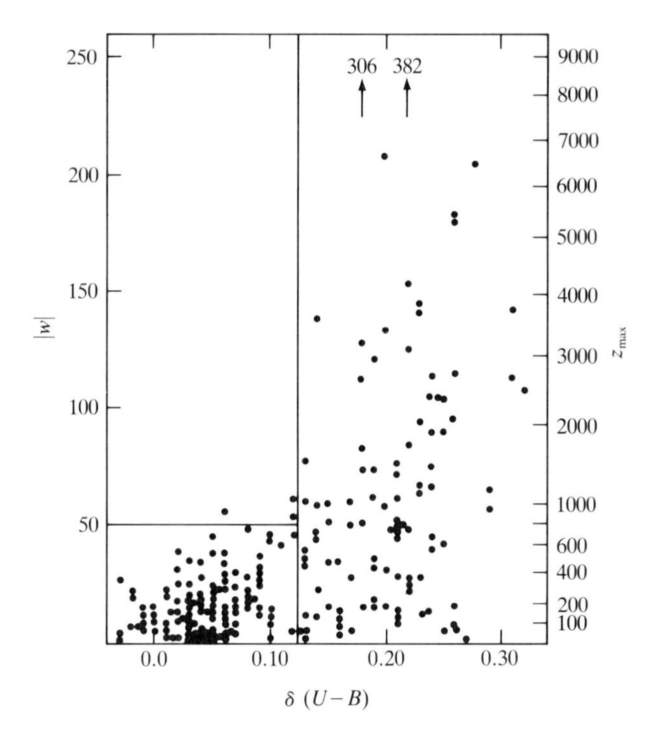

Figure 7-10 Correlation between $|w|$, the absolute value of the peculiar velocity (in kilometers per second) perpendicular to the galactic plane (*left-hand ordinate*), or $|z|_{\max}$, the maximum distance (in parsecs) that the star can penetrate away from the galactic plane (*right-hand ordinate*), and the ultraviolet excess $\delta(U - B)$ for high-velocity stars. [From (**E3**), by permission. Copyright © 1962 by the University of Chicago.]

which are intermediate-to-old disk stars having $|z_{\max}| \lesssim 700$ pc in accord with Table 4-18, and (2) stars with $\delta(U - B) \gtrsim 0.13$, most of which are halo stars and true members of the spheroidal component. It must be realized that disk stars with $\delta(U - B) \gtrsim 0.13$ could possibly exist, and hence it is possible that *some* of the stars in the lower part of the right-hand domain could conceivably be disk stars, not halo stars, but we obviously could not hope to discriminate from this diagram alone. If there are, in fact, disk stars mixed into the lower right-hand corner of the figure, then we would have a case of overlap of both the spectrophotometric and kinematic properties of the two groups, again without a continuous merging of the two.

Once the domains of disk stars and halo stars overlap in a plot such as Figure 7-10, unless some other discriminant can be used, one could hope to separate the two groups only on the basis of their relative frequencies in a statistically complete sample, knowing that halo stars contribute only about 1% of the F–K stars found in the solar neighborhood. Unfortunately, such a sample does not exist, and obtaining one would require an enormous observational effort. To guarantee completeness for a sample within, say, 25 pc, one would have to survey all stars in some area of the sky down to $V \approx 10$, choosing stars photometrically in order to avoid biasing the statistics on the basis of kinematics. That is, we want both the low- and high-velocity stars to be represented with their true frequencies, so we do not want to select on the basis of radial velocity or proper motion. The photometric screening would have to be thorough enough to discard reddened early-type stars, to separate dwarfs from giants, and to classify stars into metallicity groups. Then, the full complement of kinematic data would have to be measured and distances estimated (which could be done using spectroscopic parallaxes if the photometric classifications were precise enough). Clearly this is a large and time-consuming program. It would, however, be immensely rewarding, and it appears to be within the grasp of modern instruments. Work on this problem is currently under way under the leadership of B. Strömgren.

Consider now Figure 7-11, which is also from the work of ELS. At first sight, this diagram appears to show a monotonically increasing upper bound

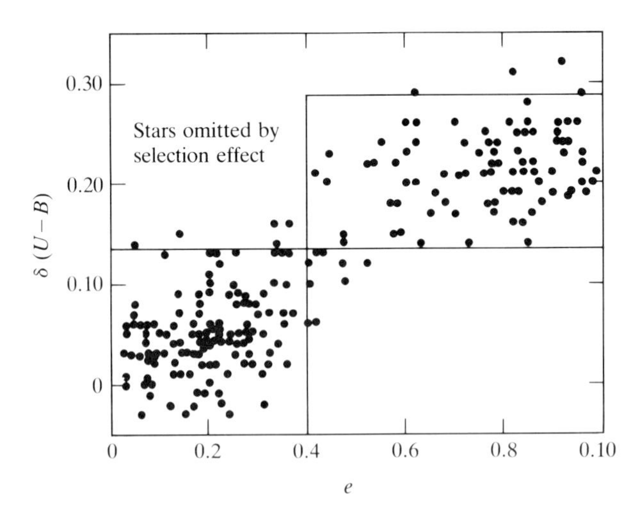

Figure 7-11. Correlation between ultraviolet excess $\delta(U - B)$ and the orbital eccentricity e [as defined by equation (7-16)] for high-velocity stars. In constructing the plot, it was assumed that $R_0 = 10$ kpc and $\Theta_0 = 250$ km s^{-1} for the Sun. [From (**E3**), by permission. Copyright © 1962 by the University of Chicago.]

to the orbital eccentricity as a function of $\delta(U - B)$, a result that ELS interpreted in terms of a rapid initial collapse and progressive circularization of the material of our Galaxy as it flattened toward a disk (see the following discussion). Note that the absence of stars in the upper left-hand region of this diagram results from a selection effect: stars in this region would be metal-poor stars moving on nearly circular orbits around the center, thus having a low space motion with respect to the Sun. In a statistically complete sample, we would expect to find a few such stars, but they are rare and hard to discover, and, because this sample was selected to have large space motions relative to the Sun, they are absent from this plot.

In terms of the population groups we defined in §4-5, we could reinterpret Figure 7-11 in the following way. Those stars with $\delta(U - B) \lesssim 0.13$ are old disk stars, and all move on orbits with $e \lesssim 0.4$. This limit is almost precisely the upper limit on e predicted by the calculations of the randomizing effects of time-fluctuating gravitational fields on disk-star velocities, as described in §7-1. Again, low-eccentricity disk stars with $\delta(U - B) \gtrsim 0.13$ could possibly exist, but they would be excluded from the sample by selection effects. Stars with $\delta(U - B) \gtrsim 0.13$ are halo stars and move on orbits with eccentricities ranging from some minimum value e_{min} all the way to the largest values plotted. The lower bound e_{min} could be as small as zero. In fact, we would expect it to be, given that there are a fair number of stars in Figure 7-9 with $v \approx -400$ km s^{-1}, which implies $|\Theta - \langle\Theta\rangle| \approx 200$ km s^{-1} if $\langle\Theta\rangle \approx 50$ km s^{-1}, and hence that some halo stars should have Θ's as large as $+250$ km s^{-1} if we assume symmetry around $\langle\Theta\rangle$. If e_{min} for halo stars is near zero, and if old disk stars with $\delta(U - B) \gtrsim 0.13$ do exist, then we would have an overlap of the spectroscopic and kinematic domains of the two groups in this diagram as well.

Finally, consider Figure 7-12, which gives the distribution of h, the angular momentum per unit mass (in units of 10^2 kpc km s^{-1}) versus $\delta(U - B)$ as plotted by ELS; a star moving on a circular orbit with $R \approx R_0$ has $h \approx 25$. Note that the upper right-hand corner of this diagram is devoid of stars because of precisely the same selection effects which left the upper left-hand corner of Figure 7-11 empty. Such stars would be hard-to-find, low-velocity subdwarfs which are undiscovered even though we expect them to exist. ELS interpreted this diagram as implying that there is a monotonic increase in the specific angular momentum of a star with increasing metal abundance, an interpretation in harmony with the then-current ideas of a sequence of progressively more flattened, rapidly rotating, metal-enriched populations spanning the range from the halo to the disk.

On the basis of the stellar population picture that we just used, we could reinterpret Figure 7-12 as showing that (1) the stars with $\delta(U - B) \lesssim 0.13$ are old disk stars, which all have a high specific angular momentum, typically on the range $17.5 \lesssim h \lesssim 27.5$; again, disk stars in the same range of h with larger values of $\delta(U - B)$ could possibly exist and would tend to be confused

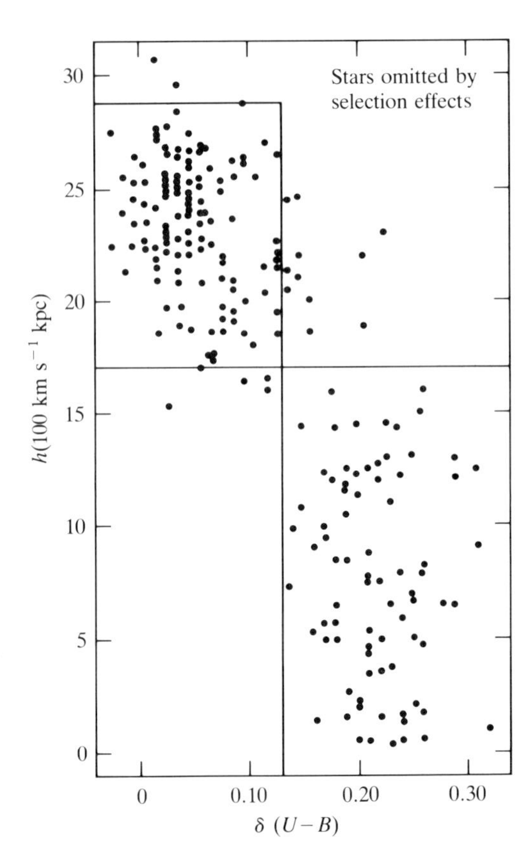

Figure 7-12. Correlation between angular momentum per unit mass h (in units of 10^2 kpc km s^{-1}) and ultraviolet excess $\delta(U - B)$ for high-velocity stars. In constructing the plot, it was assumed that $R_0 = 10$ kpc and $\Theta_0 = 250$ km s^{-1} for the Sun. [From (**E3**), by permission. Copyright © 1962 by the University of Chicago.]

with halo stars, and (2) the halo stars as plotted have specific angular momenta on the range from 0 to at least 22.5 and probably up to 25 in view of the selection effects just mentioned. It should be noticed that Figure 7-12 is misleading in one important respect: the angular momentum for all halo stars is plotted as if it were always positive, despite the fact that many of these stars move on retrograde orbits and hence have negative angular momenta. We shall return to this point later.

The upper bound of $h \approx 31$ for the angular momenta of the disk stars plotted in Figure 7-12 reflects the fact that no stars are found to be moving with peculiar velocities $\gtrsim 65$ km s^{-1} in the direction of galactic rotation. As we mentioned earlier, Oort suggested that this situation might be because such stars would escape from our Galaxy. But, as we shall see in §8-2, this interpretation is not correct, for the escape speed must be larger than the value implied by this limit. Rather, we apparently do not observe such stars simply because there is a genuine upper limit on the *angular momentum* of disk stars in the solar neighborhood. Why should this be the case? Calculations of the orbits of stars with $h \approx 31$ using a representative galactic potential show that they would reach apogalacticon at about 25 kpc. It has been suggested that a possible reason that we do not find such stars in the solar neighborhood might be that the stellar disk of our Galaxy has an "edge" at $R \sim 25$ kpc. This edge could be real (that is, the density of the disk drops sharply downward there), or it could be an effective edge marking the termination of the zone in which star formation has occurred in a gaseous disk that could extend to yet larger distances. But it may be that neither of these explanations is the answer: an equally plausible solution could be that, in the solar neighborhood, we do not observe disk stars from beyond 25 kpc even if they do exist because, to reach inward to 10 kpc, they would have to be on orbits with "eccentricities" $e \geq (25 - 10)/35 \approx 0.42$, which is above the maximum value of e produced for disk stars by the diffusion mechanisms described in §6-1. Although this agreement may be nothing more than a numerical coincidence, it is true that star formation has evidently occurred out to distances of 40 kpc or more in the halo (we observe stars from such distances), and it could equally well have occurred that far in the disk. It is also true that we would expect those disk stars to have an upper bound to their orbital eccentricities.

Dynamical Implications Let us briefly consider some of the dynamical implications of the kinematics of halo stars while the details of the data are freshly in mind, deferring a quantitative analysis until Chapter 19. ELS suggested that the kinematic data for the high-velocity stars, and what they interpreted as a continuous progression of metal content from halo stars to disk stars in a sequence of progressively more flattened subsystems, could be explained in terms of a rapid collapse of the primordial protogalactic material into the disk, accompanied by a rapid and essentially complete enrichment of the heavy elements during this collapse. In what follows, we shall try to summarize some of their arguments for this picture and to indicate how it could be modified in the light of the discussion in §4-5 and earlier in this chapter.

1. If we assume that, during the collapse, the metal abundance is an increasing function of time, and that the maximum heights above the galactic plane of stars formed at some particular epoch reflect the degree of flattening of the interstellar material when they formed, then, on the ELS picture, one would expect a correlation of decreasing $|z|_{\text{max}}$ with increasing metallicity. This correlation is consistent with the ELS interpretation of Figure 7-10. Noting that some metal-poor stars reach a maximum height of 10 kpc, while most metal-rich stars lie within 400 pc of the plane, ELS concluded that the collapse reduced the vertical extent of the material by about a factor of twenty-five. As we saw in Chapter 4, an alternative interpretation of the data is that the spheroidal component is roughly spherically symmetric with a radial composition gradient, whereas the disk consists of a sequence of populations which, on the average, become more metal poor as they grow older, and simultaneously diffuse to ever-larger distances away from the

plane. At any position in the halo, there is a large dispersion of metal abundances; in the solar neighborhood, its average metallicity is quite low compared to spiral-arm and young disk stars, but the spread is large enough to overlap the metallicity characteristic of the oldest disk stars. Because the metallicity ranges of the two groups overlap in the solar neighborhood, it is easy to confuse this overlap with an actual merging of the two in a single, continuous physical progression. But, in fact, the two groups are quite distinct kinematically, and, even though it may be difficult to ascertain to which category a given individual star belongs, it is possible to distinguish fairly clearly the properties of the two groups considered as entities.

2. A second suggestion made by ELS was that the very high eccentricities of halo stars are the result of the rapidity of the disk collapse. They hypothesized that these stars were formed at the onset of the collapse at large distances from the galactic center (near their present apocentric distances) in shells of material that had angular momenta equal to the present angular momenta of the stars. If the collapse occurs rapidly, so that a large amount of material originally outside a star streams quickly inward past it, then the gravitational forces that act on it strengthen significantly during a single rotation period (a few times 10^8 years), and there is an increase in the star's orbital eccentricity because it is more tightly bound by more mass even though its angular momentum remains unchanged. The infalling gas eventually suffers inelastic collisions and becomes circularized in orbits having the same original angular momentum, while the excess energy it gains in the collapse is radiated away. ELS suggested that the process just described could explain what they interpreted to be the progressive increase of orbital eccentricity with $\delta(U - B)$ shown in Figure 7-11. By demanding strict conservation of angular momentum in the collapsing gas, ELS estimated the radial collapse factor to be about ten.

There are, however, some important difficulties with this picture.

To begin, in this scheme some metal-poor stars should have been formed in material of higher angular momentum than the Sun, and today they should be found in the solar neighborhood leading the Sun in the direction of galactic rotation. ELS noted that there were no such stars in their sample, and, although a few have been found since (**S1**), they remain embarrassingly scarce.

Further, a collapse-induced increase of orbital eccentricity does not explain why high-velocity stars with negative angular momentum (retrograde orbits) are found fairly frequently in the sample. If strict angular-momentum conservation is invoked for each star, then there should be no stars with negative angular momenta (the original material had positive angular momentum), and there should be many more with higher-than-solar angular momentum than are observed. On the other hand, if it is argued that angular-momentum conservation should be abandoned for some stars, it cannot legitimately be invoked for any star, and the mapping procedure used by ELS to determine the scale and nature of the collapse breaks down. Yet other difficulties with this mechanism have been discussed by S. Isobe (**I3**).

Our present view of the kinematical situation is rather different. It is clear from both its smooth space distribution and its kinematic properties that the spheroidal component is dynamically relaxed (see Chapter 14). It is a system of very high internal velocity dispersion. The velocity distribution is roughly ellipsoidal, oriented with its principal axis directed toward the galactic center, and the ratios of the velocity dispersions in each direction are in accord with values produced by randomizing forces acting on a group of stars constrained by the galactic potential. Because the system's velocity dispersion is larger than its net rotation rate, it contains a large fraction of stars on retrograde orbits. At the same time, the dispersion is also large enough to allow a few stars to move with nearly circular orbits, and there is a significant overlap of the velocity domains occupied by disk stars and spheroidal-component stars, which causes confusion.

How did the spheroidal component become relaxed? As was true for the disk, star-star encounters are hopelessly ineffective. One strong possibility is that it underwent *violent relaxation* from rapid time changes in the gravitational potential during the collapse, as would surely be the case if the collapse occurred on the dynamical time scale (essentially a freefall time) suggested by ELS. Furthermore, there is no reason to suppose the collapse was smooth; large mass condensations could have arisen in a chaotic ("turbulent") collapse, yielding a hierarchy of interacting structures. Encounters of the more massive structures—for example, globular clusters (which may initially have been present in large numbers and subsequently destroyed)—with one another and with stars would lead to a disintegration of the clusters and a relaxation of the stars. The spheroidal component is at present relatively free of interstellar gas, and the present space density of globular clusters is quite low. There are, therefore, essentially no massive bodies with which halo stars can now suffer collisions. Hence, since the epoch of its formation, relaxation processes in the halo have occurred only at the glacial rate produced by star-star encounters, which suggests that its kinematic properties may have remained essentially unaltered from their initial values. (If, however, the galactic disk is sufficiently barrish near the center, then the kinematics of halo stars may evolve as these stars "bounce" off the moving force field of the bar.)

3. Finally, there is one further problem with the ELS picture: the suggestion that the present metal content of the disk was produced during a rapid disk collapse poses serious difficulties. Such a scenario manifestly requires a copious rate of star formation and extremely efficient mechanisms for ejecting and recycling stellar material. Recent studies (**T1**) of evolutionary models of our Galaxy that start with an initially metal-poor disk that forms stars and evolves as a closed system in such a way as to reproduce approximately

the present properties of the Galaxy in about 10^{10} years show that such models do not yield satisfactory results. They either produce too many low-mass, metal-poor stars, or they violate other reasonable physical and observational constraints. Models of this type fit the known constraints only if implausible assumptions are made about the time dependence of the rate of star formation for stars of various masses. In contrast, models (**O4**) in which the halo forms first, and the disk is formed later from primordial material enriched by metal-rich gas shed from old stars in the halo, encounter none of these problems. In this picture, the disk is metal rich because a substantial fraction of its matter (about one-fourth of the whole) is *secondary* material ejected from now-defunct massive halo stars. Such a picture is in accord with current stellar evolution estimates of the ages of halo and disk clusters, which indicate that the oldest disk cluster (NGC 188) is perhaps several billion years younger than typical halo globular clusters. We shall return to these problems in Chapter 19.

7-3. VELOCITY DISPERSIONS IN STELLAR SYSTEMS

Although we have fairly detailed information about the velocity distribution for various spectral classes and stellar populations in the vicinity of the Sun, for other stellar systems such as star clusters and galaxies, our knowledge is much less complete. These systems are faint, so observations of them are time consuming, and they are distant, which puts severe limits on the spatial resolution achievable using ground-based telescopes. Typically, we can measure only *velocity dispersions* along the line of sight for clusters, elliptical galaxies, and the spheroidal components of spirals, and *rotation curves* for rapidly rotating systems such as normal and barred spiral galaxies. As will be discussed in later chapters, these data can be used to derive estimates of the masses of these systems. In this section, we shall discuss only velocity dispersions (which are similar in character to the dispersions in the velocity ellipsoid) and defer a discussion of rotation curves until §8-4.

Methods

Essentially three methods can be used to determine the velocity dispersion in a stellar system: measurements of radial velocities of a sample of bright stars in the system, estimation of $\langle v_r^2 \rangle^{1/2}$ from proper motions of individual stars in the system, and analysis of the integrated spectrum of the system as a whole.

The first method is, in principle, straightforward and yields a great deal of information. But, in practice, it suffers limitations because it requires that numerous individual spectra be obtained, and, if standard techniques are employed, these will generally have to be at low dispersion (hence of limited accuracy), of only the brightest stars in the system (which may not be representative dynamically), and in relatively uncrowded regions (thereby excluding the cores of rich clusters). These constraints have been loosened somewhat by the work of J. Gunn and R. Griffin (**G1**), who have developed a technique that measures the wavelength shift (and hence radial velocity) in a stellar spectrum by cross-correlating it photoelectrically with a standard template, built into the spectrometer, of a spectrum of the same type. In any case, the method is inapplicable whenever individual stars cannot be measured, owing to either limited resolution or to the faintness of the system.

The second method has been applied to galactic clusters; obviously it can be used only for systems that are quite nearby and hence have easily measured and accurate proper motions. The method is nevertheless valuable, because the velocity dispersion in galactic clusters is quite small and cannot be obtained from individual radial-velocity measurements; their errors would swamp the quantity to be measured. The particular value of this method is that it yields information about *two* components of the velocity dispersion. If we know the distance to a cluster, we can convert observed proper motions to velocities across the line of sight. These velocities can, in turn, be resolved into components that lie along projected radii from the cluster center and are perpendicular to these radii. Averaging these components over all cluster stars, we obtain estimates of both $\langle v_r^2 \rangle^{1/2}$ and $\langle v_t^2 \rangle^{1/2}$, the velocity dispersions in the radial and tangential directions. These results are valuable because the ratio $R \equiv \langle v_r^2 \rangle^{1/2} / \langle v_t^2 \rangle^{1/2}$ gives information about the degree of isotropy of the velocity distribution, a matter of great interest to discussions of the dynamics of the system.

Neither of the preceding two methods can be applied to external galaxies where it is difficult to resolve any but the very brightest stars, and these only for relatively nearby systems. The third method is then advantageous, as it uses the light from vast numbers of stars simultaneously, and hence can be applied to faint systems. It can be implemented in several different ways. The basic idea is to exploit the fact that the integrated spectrum of the system is the superposition of individual stellar spectra, each having been Doppler shifted according to the star's radial velocity. Let us denote stellar types by a subscript k, and let $s_k(\Lambda)\,d\Lambda$ be the energy (or number of photons) received, per unit time, between the wavelengths $\lambda = e^\Lambda$ and $\lambda = e^{\Lambda + d\Lambda}$ from a star of stellar type k at rest in the system being observed. Supposed $f_k(v)$ is the fraction of all stars of type k which have velocities between v and $v + dv$. Then, because, to first order in (v/c), the Doppler shift causes light emitted by a star moving with speed v at $\log \lambda = \Lambda - (v/c)$ to be received at $\log \lambda = \Lambda$, the integrated (logarithmic) spectrum $i(\Lambda)$ observed will be given by the convolution integral

$$i(\Lambda) = \sum n_k \int_{-\infty}^{\infty} s_k\left(\Lambda - \frac{v}{c}\right) f_k(v)\,dv \tag{7-17}$$

where n_k is the number of stars of type k, and the sum extends over all types present in the system. Because we have no a priori knowledge of how $f_k(v)$ may vary from type to type, we have little choice but to assume a single Gaussian distribution

$$f_k(v) = (2\pi\langle v_r^2\rangle)^{-1/2} \exp - \left[\frac{(v - \langle v_r\rangle)^2}{2\langle v_r^2\rangle}\right] \tag{7-18}$$

for all types. The goal is to determine the *systemic velocity* $\langle v_r\rangle$ and the *dispersion* $\langle v_r^2\rangle^{1/2}$. In this case, equation (7-17) reduces to

$$i(\Lambda) = \int_{-\infty}^{\infty} \tilde{s}(\Lambda - u)\tilde{f}(u)\,du \tag{7-19}$$

where $\tilde{s}$ is a *composite intrinsic spectrum*

$$\tilde{s}(\Lambda) = \sum_k n_k s_k(\Lambda) \tag{7-20}$$

$$u \equiv (v/c),$$

and $f(u) \equiv (1/c)\tilde{f}(cu)$.

Suppose, now, that we know $\tilde{s}(\Lambda)$ and wish to find $\langle v_r\rangle$ and $\langle v_r^2\rangle^{1/2}$. One procedure is simply to perform the convolution in equation (7-19) for several trial values of $\langle v_r\rangle$ and of $\langle v_r^2\rangle^{1/2}$. By comparison of the broadened intrinsic spectrum with the observed integrated spectrum of the system, usually by direct superposition and visual inspection or by least-squares fitting of the plots, "best estimate" values of $\langle v_r\rangle$ and $\langle v_r^2\rangle^{1/2}$ can be obtained. Alternatively, one can make use of the *convolution theorem* of Fourier analysis, which states that, if $\mathscr{I}_{\text{obs}}(v)$, $\mathscr{S}(v)$, and $\mathscr{F}(v)$ are the Fourier transforms of $i(\Lambda)$, $\tilde{s}(\Lambda)$, and $\tilde{f}(u)$, respectively, then $\mathscr{I}_{\text{obs}}(v) = \mathscr{S}(v)\mathscr{F}(v)$. The transform $\mathscr{F}(v)$ can be calculated analytically for a given value of $\langle v_r\rangle$ and $\langle v_r^2\rangle^{1/2}$, and $\mathscr{I}_{\text{obs}}(v)$ and $\mathscr{S}(v)$ can be computed numerically from the observed $i(\Lambda)$ and assumed $\tilde{s}(\Lambda)$. One can then either (1) compute trial functions $\mathscr{I}(v)$ from $\mathscr{S}(v)$ and $\mathscr{F}(v)$ for different values of the velocity dispersion and fit (visually) to $\mathscr{I}_{\text{obs}}(v)$ to obtain an estimate of $\langle v_r\rangle$ and $\langle v_r^2\rangle^{1/2}$, or (2) compute $\mathscr{I}_{\text{obs}}(v)/\mathscr{S}(v)$ and fit this ratio with the analytical expression for $\mathscr{F}(v)$, choosing $\langle v_r\rangle$ and $\langle v_r^2\rangle^{1/2}$ to optimize the fit. Advantages of these Fourier transform techniques are that high-frequency noise can be filtered out of the data and that the procedure is fairly objective.

In the actual analysis of integrated spectra, a correct choice for the intrinsic spectrum $\tilde{s}(\Lambda)$ is critical but may be difficult, because we may not know precisely how to choose the relative numbers of the different kinds of stellar spectra forming the composite intrinsic spectrum. In the case of globular clusters, it was noted long ago by Morgan (**M3**) that the integrated spectrum shows a (surprisingly) low degree of compositeness; that is, in the usual blue-violet spectral region, the brightest stars all happen to be of about the same type. Detailed quantitative analyses of globular-cluster spectra (**I1**) have demonstrated that an excellent fit to the observed integrated spectrum is obtained by use of a single intrinsic stellar spectrum, and that the improvement obtained by allowance for compositeness is not worth the additional effort. In contrast, galaxy spectra are decidedly composite, and, if this fact is not taken into account, the velocity dispersion derived for a system can be somewhat overestimated (**W4**). (It is obvious that a composite spectrum, even at zero velocity dispersion, will be inherently more smeared out than that from a single type, so that, if compositeness is ignored, a spurious additional velocity dispersion will be required to produce the extra smearing inferred).

Results for Clusters

Velocity dispersions for the two galactic (open) clusters Pleiades and Praesepe based on proper motions have been published by B. F. Jones (**J1**), (**J2**). For Pleiades, assuming a distance of 125 pc, he obtains $\langle v_r^2 \rangle^{1/2} = 0.48$ km s^{-1} and $\langle v_t^2 \rangle^{1/2} = 0.29$ km s^{-1} for the cluster as a whole. Thus the velocity distribution is significantly anisotropic, with R, the ratio of radial to tangential velocity dispersions, about equal to 1.6, which shows that the orbits are elongated radially. His data for this cluster show that (1) for stars of a given mass (apparent magnitude), the ratio R increases outward from the cluster center (the largest value is 2.6); (2) at a given distance from cluster center, the velocity distribution is more nearly isotropic for low-mass than for high-mass stars; and (3) the velocity dispersions in both directions are larger for low-mass stars than they are for high-mass stars ($\langle v_r^2 \rangle = 0.41$ km s^{-1} and $\langle v_t^2 \rangle^{1/2} = 0.22$ km s^{-1} for $4 \leq V \leq 9$; $\langle v_r^2 \rangle^{1/2} = 0.53$ km s^{-1} and $\langle v_t^2 \rangle^{1/2} = 0.33$ km s^{-1} for $9 \leq V \leq 14$). For Praesepe, assuming a distance of 160 pc, Jones obtains $\langle v_r^2 \rangle^{1/2} = 0.52$ km s^{-1} and $\langle v_t^2 \rangle^{1/2} = 0.43$ km s^{-1}, hence $R \approx 1.2$ for the cluster as a whole. Although the ratio R is always greater than unity (and rises to 2.4 for stars most distant from the cluster center), there is no clear-cut variation of either the absolute sizes of the velocity dispersions or the ratio R with either stellar mass or distance from cluster center for Praesepe as there was for Pleiades.

Velocity dispersions for a number of globular clusters have been determined by G. Illingworth (**I1**) from integrated spectra, and for M92 by O. C. Wilson and M. F. Coffeen (**W5**), and for ω Cen by G. A. Harding (**H1**) from individual stellar spectra. Their results are given in Table 7-6 along with absolute magnitudes M_V, core radii r_c, and the ratio of the tidal radius r_t to r_c, which were defined in §5-2. Typical observational errors for the velocity dispersions are ± 1 km s^{-1}, and the dispersions at the center of the clusters are estimated to be about 4% larger than the values tabulated. Velocity

Table 7-6. Velocity Dispersions, Absolute Magnitudes, and Characteristic Radii for Selected Globular Clusters

NGC	M_V	$\langle v_r^2 \rangle^{1/2}$ (km s^{-1})	r_c(pc)	$\log(r_t/r_c)$
104 (47 Tuc)	−9.2	10.5	0.48	2.03
362	−8.4	7.5	0.55	1.70
1851	−8.4	7.9	0.37	1.83
2808	−9.2	14.2	0.66	1.75
5139 (ω Cen)	−10.1	9.8	3.8	1.36
6093 (M80)	−8.0	12.5	0.32	1.88
6266 (M62)	−9.2	13.7	0.54	1.63
6341 (M92)	−8.1	7.6	0.6	1.78
6388	−9.6	18.9	0.50	1.75
6441	−8.8	17.6	0.41	1.70
6715 (M54)	−9.5	14.2	0.67	1.83
6864 (M75)	−8.4	10.3	0.51	1.82

SOURCE: Data published in (**H1**), (**I1**), (**I2**), (**P2**), (**W5**)

dispersions for two globular clusters have been determined from proper motions by K. Cudworth (**C1**), (**C2**), who finds $\langle v^2 \rangle^{1/2} = 10$ km s^{-1} and 7–9 km s^{-1} for M15 and M92, respectively. In both clusters, the velocity distribution is essentially isotropic, in contrast to the situation for the galactic clusters studied by Jones.

Results for Galaxies

Velocity dispersions in elliptical galaxies and the bulges of spiral galaxies have been measured by several investigators. Typical results are summarized in Table 7-7; the information given there is meant only to be representative and not exhaustive. Although the internal errors of measurement quoted by the observers are relatively small, large systematic errors exist, as different sets of results for the same galaxy may differ by 30% in some cases.

An important source of error and uncertainty is the choice of composite spectrum $\tilde{s}(\Lambda)$. A priori, one knows neither what precise mix of MK spectral types furnishes most of the light observed, nor how strong-lined these stars are. Ideally, one would vary both the mix of stellar types represented in $\tilde{s}(\Lambda)$ and the line strengths of the spectra, and then adopt the lowest value of $\langle v_r^2 \rangle^{1/2}$ obtained as the true velocity dispersion. In practice, there is too much noise in the data for this procedure to lead to sensible results. Fortunately, it is found that, with the Fourier technique, the estimated value of $\langle v_r^2 \rangle^{1/2}$ is reduced by only a small amount ($\lesssim 10\%$) when one adopts for $\tilde{s}(\Lambda)$ a fully composite spectrum in place of the spectrum of a single giant of suitable type (**S2**). T. B. Williams (**W4**) finds, on the other hand, that if

Table 7-7. Velocity Dispersions and Absolute Magnitudes for Selected Galaxies

NGC	Hubble type	M_B	$\langle v_r^2 \rangle^{1/2}$ (km s^{-1})	Reference
		Disk galaxies		
4594	Sa	-22.7	263	**(W1)**
4254	Sc	-22.7	157	**(W1)**
4579	Sb	-22.6	195	**(W1)**
4450	Sab	-22.4	130	**(W1)**
5457	Scd	-22.0	80	**(W1)**
3627	Sb	-21.3	184	**(W1)**
3166	S0a	-20.8	114	**(W1)**
224 (M31)	Sb	-20.6	173	**(W1)**
3593	S0a	-19.3	117	**(W1)**
		Ellipticals		
5846	E0	-21.2	255	**(F1)**
741	E1	-22.4	275	**(S2)**
1172	E1	-19.6	100	**(S2)**
3379	E1	-20.2	135	**(W4)**
4374	E1	-21.2	285	**(F1)**
221 (M32)	E2	-15.3	55	**(W4)**
596	E2	-20.9	150	**(S2)**
1426	E2	-19.5	145	**(S2)**
4472	E2	-22.2	295	**(F1)**
5846A	E2	-18.5	200	**(F1)**
7626	E2	-21.6	240	**(S2)**
1395	E3	-21.1	220	**(S2)**
584	E4	-21.4	150	**(W4)**
1052	E4	-20.4	100	**(W4)**
4889	E4	-22.7	400	**(F1)**
4473	E5	-22.0	110	**(W4)**

dispersions are estimated by direct comparison of an artificially broadened spectrum with the galaxy spectrum, there is a tendency to overestimate the velocity dispersion unless a composite spectrum is used.

It was noticed by S. M. Faber and R. E. Jackson **(F1)** and confirmed by later studies **(S2)**, **(S3)**, **(W1)** that the velocity dispersions and luminosities of elliptical galaxies are approximately related by an expression of the form $L \sim \langle v_r^2 \rangle^2$. It is easily seen in Table 7-7 that the more luminous elliptical galaxies do indeed have larger velocity dispersions, but the table shows that the central velocity dispersions of disk galaxies are not strongly correlated with absolute magnitude. However, B. C. Whitmore et al. **(W1)** find that the velocity dispersions of disk galaxies are related to the *bulge* luminosities L_B

by $L_B \sim \langle v_r^2 \rangle^2$ although the normalization of this relationship is not the same as for the ellipticals. The velocity dispersion at the center of a spiral galaxy of bulge luminosity L_B tends to be lower than the central velocity dispersion of an elliptical at $L = L_B$.

REFERENCES

(B1) Barbanis, B. and Woltjer, L. 1967. *Astrophys. J.* **150**:461.

(B2) Becker, W. and Contopoulos, G. (eds.). 1970. *The Spiral Structure of Our Galaxy.* IAU Symposium No. 38. (Dordrecht: Reidel).

(B3) Blaauw, A. and Schmidt, M. (eds.). 1965. *Galactic Structure.* (Chicago: University of Chicago Press).

(B4) Byl, J. 1974. *Mon. Not. Roy. Astron. Soc.* **169**:157.

(C1) Cudworth, K. M. 1976. *Astron. J.* **81**:519.

(C2) Cudworth, K. M. 1976. *Astron. J.* **81**:975.

(E1) Eggen, O. J. 1962. *Roy. Obs. Bull. No. 51*, E79.

(E2) Eggen, O. J. 1964. *Roy. Obs. Bull. No. 84.* E111.

(E3) Eggen, O. J., Lynden-Bell, D., and Sandage, A. 1962. *Astrophys. J.* **136**:748.

(F1) Faber, S. M. and Jackson, R. E. 1976. *Astrophys. J.* **204**:668.

(F2) Fricke, W. 1949. *Astron. Nachr.* **278**:49.

(G1) Gunn, J. E. and Griffin, R. F. 1979. *Astron. J.* **84**:752.

(H1) Harding, G. A. 1965. *Roy. Obs. Bull. No. 99*, E65.

(I1) Illingworth, G. 1976. *Astrophys. J.* **204**:73.

(I2) Illingworth, G. and Illingworth, W. 1976. *Astrophys. J. Supp.* **30**:227.

(I3) Isobe, S. 1974. *Astron. and Astrophys.* **36**:333.

(J1) Jones, B. F. 1970. *Astron. J.* **75**:563.

(J2) Jones, B. F. 1971. *Astron. J.* **76**:470.

(L1) Lesh, J. R. 1968. *Astrophys. J. Supp.* **16**:371.

(M1) Mayor, M. 1974. *Astron. and Astrophys.* **32**:321.

(M2) Miczaika, G. 1940. *Astron. Nachr.* **207**:249.

(M3) Morgan, W. W. 1956. *Pub. Astron. Soc. Pacific.* **68**:509.

(O1) Oort, J. H. 1926. *Kapteyn Astron. Lab. Groningen Pub. No. 40.*

(O2) Oort, J. H. 1927. *Bull. Astron. Inst. Netherlands.* **3**:275.

(O3) Oort, J. H. 1928. *Bull. Astron. Inst. Netherlands.* **4**:269.

(O4) Ostriker, J. P. and Thuan, T. X. 1975. *Astrophys. J.* **202**:353.

(P1) Parenago, P. P. 1950. *Astron. Zhur. (USSR).* **27**:150.

(P2) Peterson, C. J. and King, I. R. 1975. *Astron. J.* **80**:427.

(R1) Roman, N. G. 1950. *Astrophys. J.* **112**:554.

(R2) Roman, N. G. 1952. *Astrophys. J.* **116**:122.

(R3) Roman, N. G. 1954. *Astron. J.* **59**:307.

(R4) Roman, N. G. 1955. *Astrophys. J. Supp.* **2**:195.

(S1) Sandage, A. 1969. *Astrophys. J.* **158**:1115.

(S2) Sargent, W. L. W., Schechter, P. L., Boksenberg, A., and Shortridge, K. 1977. *Astrophys. J.* **212**:326.

(S3) Schechter, P. L. and Gunn, J. E. 1979. *Astrophys. J.* **229**:472.

(S4) Schwarzschild, K. 1907. *Göttingen Nachr.* p. 614.

(S5) Schwarzschild, K. 1908. *Göttingen Nachr.* p. 191.

(S6) Smart, W. M. 1938. *Stellar Dynamics.* (Cambridge: Cambridge University Press).

(S7) Spitzer, L. and Schwarzschild, M. 1951. *Astrophys. J.* **114**:385.

(S8) Spitzer, L. and Schwarzschild, M. 1953. *Astrophys. J.* **118**:106.

(S9) Stark, A. A. and Blitz, L. 1978. *Astrophys. J. Letters.* **225**:L15.

(S10) Strömberg, G. 1924. *Astrophys. J.* **59**:228.

(T1) Thuan, T. X., Hart, M. H., and Ostriker, J. P. 1975. *Astrophys. J.* **201**:756.

(T2) Trumpler, R. J. and Weaver, H. F. 1953. *Statistical Astronomy.* (Berkeley: University of California Press).

(V1) Vyssotsky, A. N. 1951. *Astron. J.* **56**:62.

(W1) Whitmore, B. C., Kirshner, R. P., and Schecter, P. L. 1979. *Astrophys. J.* **234**:68.

(W2) Wielen, R. 1974. In *Highlights of Astronomy*, Vol. 3. XVth General Assembly of the IAU, G. Contopoulos (ed.). (Dordrecht: Reidel).

(W3) Wielen, R. 1977. *Astron. and Astrophys.* **60**:263.

(W4) Williams, T. B. 1977. *Astrophys. J.* **214**:685.

(W5) Wilson, O. C. and Coffeen, M. F. 1954. *Astrophys. J.* **119**:197.

8 The Rotation of Galaxies

The disk of our Galaxy is in a state of differential rotation around an axis through the galactic center. At any specified distance R measured from the center in the galactic plane, there is a particular velocity, called the *circular velocity*, such that a star moving with this velocity revolves around the center on a circular orbit on which the inward acceleration of the star is exactly equal to the gravitational force per unit mass exerted on the star by the rest of our Galaxy. If we can measure the run of circular velocity as a function of distance from the center, then we can obtain extremely valuable information about the gravitational forces acting in our Galaxy, and hence, about its dynamics.

As was discussed in Chapters 6 and 7, we know that the Sun is moving around the galactic center on a nearly circular orbit of radius R_0. From optical measurements of the radial velocities of globular clusters and of external galaxies, we obtained an estimate of $\Theta_0(R_0)$, the circular velocity of material in the solar neighborhood. We now wish to inquire how the circular velocity varies as a function of distance from the center. Although, as we shall see in §8-2, optical observations do yield some local information about the variation of the *rotation curve* $\Theta(R)$ near the Sun, they do not permit a determination of the run of the rotation velocity through the entire disk of our Galaxy, primarily because they are restricted by the effects of interstellar absorption to distances in the disk of only 2 or 3 kpc from the Sun. Moreover, as we saw in Chapter 7, although some stars move on very nearly circular orbits around the center, others move on eccentric (sometimes almost radial) orbits. The presence of such a wide mixture of kinematic properties within a given volume would complicate the analysis even if we were able to observe stars to large distances. However, along most lines of sight, our Galaxy is virtually transparent to the 21-cm radio line emitted by neutral hydrogen. Furthermore, as we mentioned in Chapter 1, clouds of interstellar gas have large mutual collision cross sections and are therefore rather strongly constrained to move on nearly circular orbits. Thus, by observation of the 21-cm line with radio telescopes, we can survey essentially the entire disk of our Galaxy and thereby determine the run of the rotation curve $\Theta(R)$ over large ranges of galactocentric distance R.

As we shall see, studies of 21-cm line emission by neutral hydrogen in our Galaxy enable us to derive reasonably reliable estimates of the galactic circular velocity $\Theta(R)$ for $R < R_0$. The situation is not so favorable beyond the Sun's distance from the galactic center. To obtain reliable information about the behavior of galaxian rotation curves at large radii, it is necessary to turn to studies of external galaxies. Optical studies of the velocities of giant emission regions can yield rotation velocities out to considerable radii, but the greatest strides in this area have been made over the last ten years as a result of the development of aperture-synthesis radio telescopes having sufficient resolving power to study the internal dynamics of nearby galaxies in some detail.

Until recently, our knowledge of rotation in external galaxies was confined to the spiral-arm population (young stars and gas). However, galaxian rotation curves obtained by measuring the frequencies of Doppler-shifted stellar absorption features like the H and K lines of Ca II have now become available for a number of elliptical and lenticular galaxies (whose light derives almost entirely from old evolved stars). We shall see in Chapter 14 that, combined with velocity-dispersion measurements derived from the same observational data, these rotation curves have helped open a new era in our understanding of the dynamics of spheroidal components.

8-1. KINEMATICS OF GALACTIC ROTATION

Qualitative Expectations

For the sake of simplicity, let us suppose that the material in the disk of our Galaxy is in strictly circular rotation around an axis through the galactic center, perpendicular to the galactic plane. Because the mass in our Galaxy is centrally concentrated, and because material outside the orbit of a star in the plane exerts almost no net force on it (strictly true only if it is distributed spherically symmetrically), we might expect stars to travel, to a rough first approximation, on nearly Keplerian orbits around the galactic center. In this case, the smaller the orbit of a star—that is, the closer it is to the center—the shorter is its period of revolution, and the greater its orbital angular velocity. Our Galaxy would then rotate *differentially*, with stars in the innermost parts completing an orbital revolution in a much smaller period than in the outer parts (see Figure 8-1). In reality, although our Galaxy does rotate differentially, the variation of its rotation velocity in the range of galactocentric distances over which it can be determined is not Keplerian. In fact, we shall see in §8-3 that the very innermost part of the disk ($R \lesssim 3$ kpc)

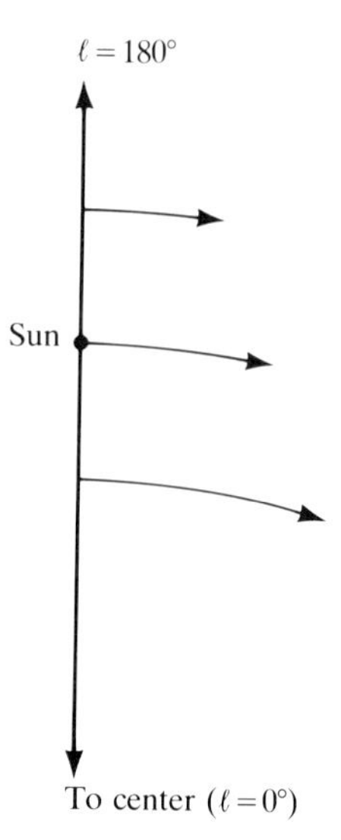

Figure 8-1. Rotation velocities in a differentially rotating Galaxy, assuming approximately Keplerian motion around the galactic center.

is almost in solid-body rotation with a rotation velocity that rises outward. In contrast, at larger distances the rotation velocity is, to a rough approximation, nearly constant, and it shows only a slow decline outward. However, in the bulk of the disk, the angular rotation rate decreases outward.

Let us now examine some qualitative features of the observable effects produced by differential galactic rotation. Consider first the radial velocities, relative to the Sun, of stars in such a system. As is clear from Figure 8-1, if we look toward $\ell = 0°$ or $\ell = 180°$, we will not observe systematic radial velocities induced by galactic rotation, for all stars in these directions are moving perpendicular to the line of sight. Similarly, in the directions $\ell = 90°$ and $\ell = 270°$, we again would not observe radial velocities for nearby stars, because stars in these directions are moving around the galactic center with the same circular orbital velocity as the Sun and therefore will show no net approach or recession. In the direction $\ell = 45°$, however, we would observe positive radial velocities, because stars in this direction move away from the Sun along the line of sight faster than the Sun approaches them (see Figure 8-2). At $\ell = 135°$, the situation is reversed: the Sun overtakes stars in this direction, so they appear to approach (negative radial velocities). Similar arguments can be made for the other quadrants, and we conclude that v_R for nearby stars in a differentially rotating Galaxy should have the dependence on galactic longitude shown in Figure 8-3. We shall derive below a quantitative expression for this variation.

For tangential velocities, which give rise to proper motions, there are two main contributions to the observed effects. First, a part of the observed proper motions will be produced by the relative motions of stars at different distances from the galactic center that result from the differential rotation

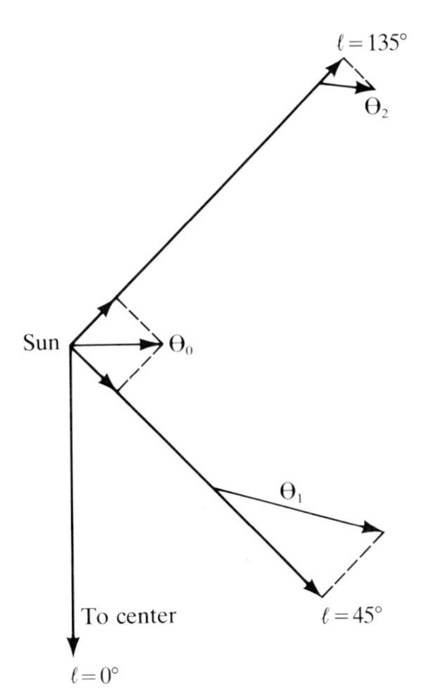

Figure 8-2. Effects of differential galactic rotation on the observed radial velocity v_R of stars in the solar neighborhood. Stars at $\ell = 45°$ have a rotation velocity $\Theta_1 > \Theta_0$, and hence they will appear to recede from us, whereas stars at $\ell = 135°$ have a rotation velocity $\Theta_2 < \Theta_0$, and hence they will appear to approach us.

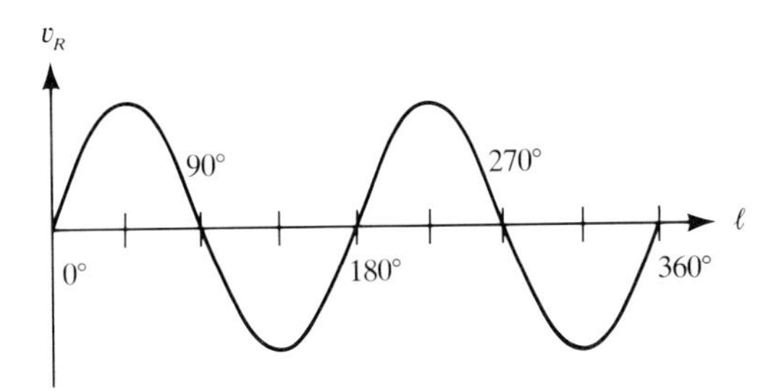

Figure 8-3. Variation with galactic longitude of the observed radial velocities of nearby objects in a differentially rotating galaxy.

of the galactic disk. As can be seen in Figure 8-4, stars seen in the direction $\ell = 0°$ rotate about the center faster than the Sun does, and therefore they tend to leave it behind. This net velocity leads to an increase in the observed galactic longitude of these stars and hence to a positive contribution to the proper motion μ_ℓ, which is measured positive in the direction of increasing ℓ. Similarly, the Sun tends to leave behind stars in the direction $\ell = 180°$, and again they appear to drift to larger galactic longitudes and hence to have positive proper motions. In the directions $\ell = 90°$ and $\ell = 270°$, because the stars are moving on the same circular orbit as the Sun, they have no differential tangential velocity, and, accordingly, μ_ℓ will be zero. In summary, differential rotation gives rise to proper motions in galactic longitude that are positive or zero. Superimposed on these differential effects is an additional

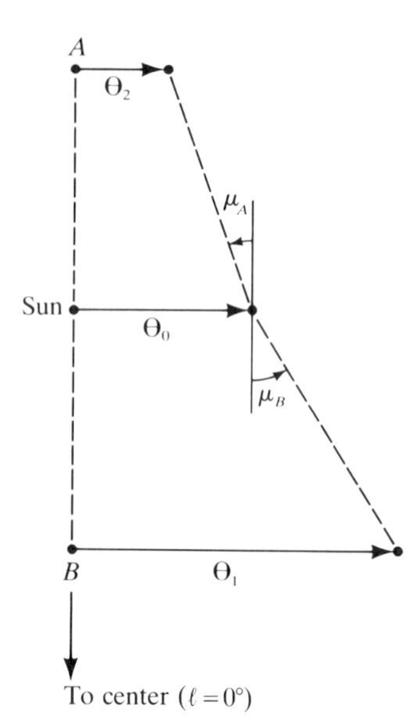

Figure 8-4. Proper motions produced by differential galactic rotation. Initially star A lies in the direction $\ell = 180°$ and star B in the direction $\ell = 0°$. At a later time, the galactic longitudes of both stars will be larger, and hence both stars have positive proper motions μ_ℓ.

contribution that results from the rotation of our Galaxy as a whole relative to a fixed frame. Suppose that our Galaxy rotated as a solid body and that no star moved differentially with respect to the Sun. Then, as is clear from Figure 8-1, the clockwise rotation of the Sun around the galactic center implies that all objects in our Galaxy will be seen to drift toward smaller galactic longitudes when their positions are determined relative to a truly inertial frame (for example, that established by external galaxies). The rate of apparent drift will be the same in all directions, hence there will be a constant (that is, independent of galactic longitude) negative term in the measured proper motions μ_ℓ. The final observed proper motions will be the sum of these two contributions.

General Rotation Formulae

Let us now write several useful formulae for the radial velocity v_R, the tangential velocity v_T, and the proper motion in galactic longitude μ_ℓ, in a differentially rotating Galaxy. We define the following quantities (see Figure 8-5):

$R \equiv$ the distance of a general point from the galactic center (in the galactic plane)

$R_0 \equiv$ the distance of the Sun from the galactic center (in the galactic plane)

$\Theta \equiv$ the linear rotation velocity of the Galaxy at radius R

$\Theta_0 \equiv$ the linear rotation velocity of the Galaxy at radius R_0

$\omega \equiv \Theta/R =$ the angular rotation rate of the Galaxy at radius R

$\omega_0 \equiv \Theta_0/R_0 =$ the angular rotation rate of the Galaxy at radius R_0

$d \equiv$ the distance from the Sun to an object at radius R from the galactic center

$v_R \equiv$ the observed radial velocity of an object, relative to the Sun

$v_T \equiv$ the observed tangential velocity of an object, relative to the Sun

$\ell \equiv$ the galactic longitude of an observed object

Finally, we introduce an auxiliary angle α as shown in Figure 8-5.

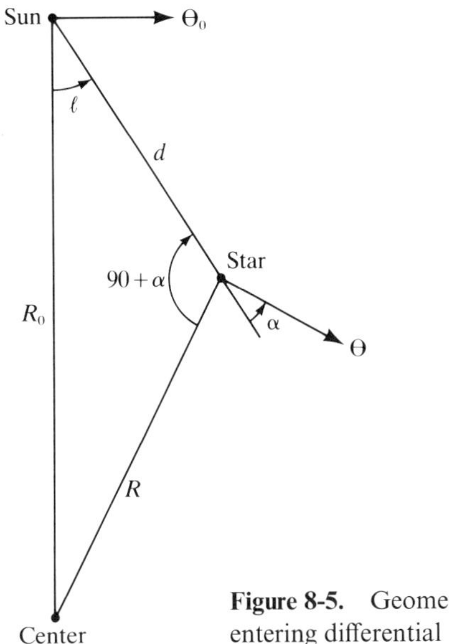

Figure 8-5. Geometric relationship among quantities entering differential rotation formulae.

From Figure 8-5, it is clear that a general expression for the radial velocity v_R is

$$v_R = \Theta \cos \alpha - \Theta_0 \sin \ell \tag{8-1}$$

Now, from the law of sines,

$$\frac{\sin \ell}{R} = \frac{\sin(90^\circ + \alpha)}{R_0} = \frac{\cos \alpha}{R_0} \tag{8-2}$$

Therefore,

$$v_R = \left(\frac{\Theta R_0}{R}\right) \sin \ell - \Theta_0 \sin \ell \tag{8-3}$$

or

$$v_R = (\omega - \omega_0) R_0 \sin \ell \tag{8-4}$$

Equations (8-3) and (8-4) are both general and assume only that the galactic rotation is strictly circular.

Let us now consider tangential velocities. From Figure 8-5, we easily find

$$v_T = \Theta \sin \alpha - \Theta_0 \cos \ell \tag{8-5}$$

where v_T is measured as positive in the direction of increasing ℓ. From Figure 8-6, we see that

$$R \sin \alpha = R_0 \cos \ell - d \tag{8-6}$$

so that

$$v_T = \left(\frac{\Theta}{R}\right)(R_0 \cos \ell - d) - \Theta_0 \cos \ell \tag{8-7}$$

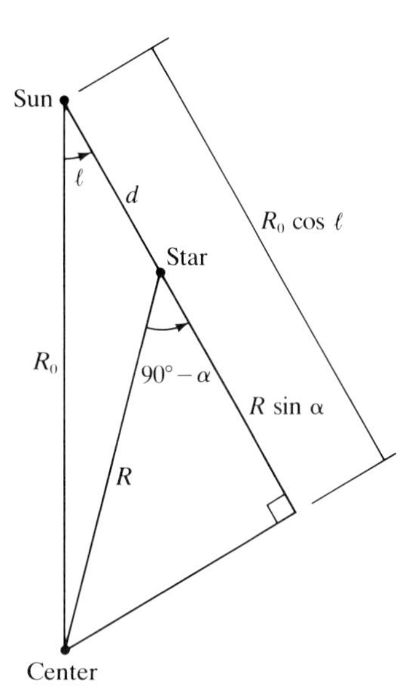

Figure 8-6. Geometric construction used in the derivation of equation (8-6).

or

$$v_T = (\omega - \omega_0)R_0 \cos \ell - \omega d \tag{8-8}$$

Again, equations (8-7) and (8-8) are valid in general for strictly circular rotation.

The equations just derived, particularly the expression for v_R, are of great value in the discussion of radio observations of the 21-cm line of interstellar atomic hydrogen, as will be described in detail in §8-3 and §8-5. Before leaving them, however, it is worthwhile to note a few additional points. Suppose that the angular rotation rate $\omega(R)$ is a monotonically decreasing function of R, and consider the variation of v_R as given by equation (8-4) as a function of distance along a line of sight at some prechosen galactic longitude ℓ. For $90^\circ \leq \ell \leq 180^\circ$, as we look to progressively larger distances d, it is obvious that we see regions lying at increasingly large distances $R \geq R_0$ from the galactic center, having progressively smaller values of $\omega \leq \omega_0$. It follows from equation (8-4) that we will measure increasingly negative values of v_R with increasing d, as shown schematically in Figure 8-7. In the quadrant $180^\circ \leq \ell \leq 270^\circ$, v_R will be positive and will increase monotonically with increasing distance d. On the other hand, if we look in some direction on the range $0^\circ \leq \ell \leq 90^\circ$, we first observe regions with smaller values of R, and thus larger values of ω, so that v_R for these regions is positive. At some point along its length, the line of sight will penetrate to a minimum radial distance $R_{min} = R_0 \sin \ell$ from the galactic center (see Figure 8-8). Here $d = R_0 \cos \ell$, and ω reaches its maximum value for the chosen value of ℓ, as does v_R. Beyond this *tangent point*, R increases again; therefore ω decreases, and v_R likewise decreases. Eventually, R will again equal R_0, and, because we assume that all objects at $R = R_0$ revolve around the galactic center with the same angular velocity ω_0 as the Sun, it is clear that they will remain fixed relative to the Sun and hence will have $v_R = 0$. Still farther along the line of sight,

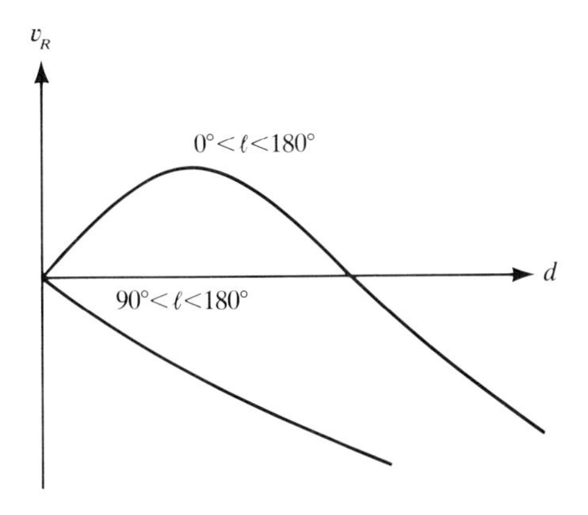

Figure 8-7. Schematic run of observed radial velocities as a function of distance for two ranges of galactic longitude; v_R passes through a maximum when $0^\circ < \ell < 90^\circ$, and it decreases monotionically for $90^\circ < \ell < 180^\circ$.

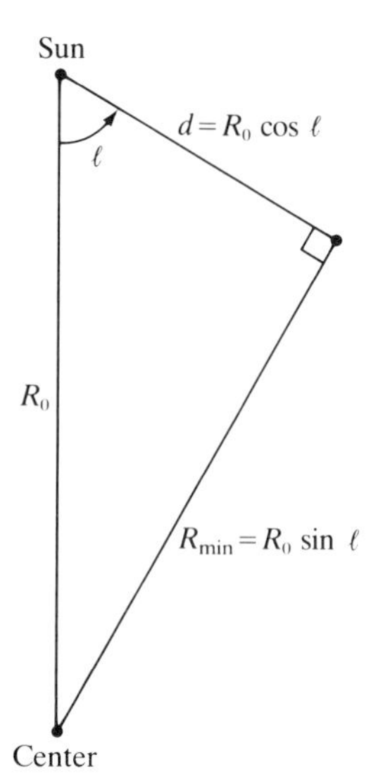

Figure 8-8. Geometric relationships among R_0, ℓ, and R_{min}.

we observe regions in which $R > R_0$; hence $\omega < \omega_0$ so that v_R finally becomes negative. The resulting variation of v_R with d is shown schematically in Figure 8-7. In the quadrant $270° \leq \ell \leq 360°$, the same variation of v_R, with the sign reversed, is obtained. We shall exploit these results extensively in the determination of the galactic rotation curve and hydrogen distribution from 21-cm observations.

Oort's Formulae

The results just obtained are general (within the framework of the assumptions made) and apply at arbitrary values of R from the galactic center and distances d from the Sun. But optical observations of stars in the disk of our Galaxy are generally restricted by interstellar absorption to relatively small distances (say, up to 2 kpc) from the Sun. It is therefore worthwhile to develop formulae that are valid for small d, and, in fact, we can obtain extremely convenient approximate forms of equations (8-4) and (8-8) that permit a direct analysis of the characteristics of differential galactic rotation in the neighborhood of the Sun. These expressions were first derived by Oort (**O1**), and they are named in his honor.

Consider first equation (8-4) for fixed ℓ. R_0 is fixed, and only the term $(\omega - \omega_0)$ can depend on distance. We can estimate the variation of $(\omega - \omega_0)$ by use of a first-order Taylor expansion

$$(\omega - \omega_0) \approx \left(\frac{d\omega}{dR}\right)_{R_0} (R - R_0) \tag{8-9}$$

Now

$$\frac{d\omega}{dR} = \frac{d}{dR}\left(\frac{\Theta}{R}\right) = \frac{1}{R}\frac{d\Theta}{dR} - \frac{\Theta}{R^2} \tag{8-10}$$

so that

$$\left(\frac{d\omega}{dR}\right)_{R_0} = \frac{1}{R_0}\left(\frac{d\Theta}{dR}\right)_{R_0} - \frac{\Theta_0}{R_0^2} \tag{8-11}$$

Therefore, to first order

$$v_R = \left[\left(\frac{d\Theta}{dR}\right)_{R_0} - \frac{\Theta_0}{R_0}\right](R - R_0)\sin \ell \tag{8-12}$$

Now, as can be seen in Figure 8-9, for $d \ll R_0$, the projection of R onto R_0 is almost equal to R, so that, to a good approximation,

$$R_0 - R \approx d \cos \ell \tag{8-13}$$

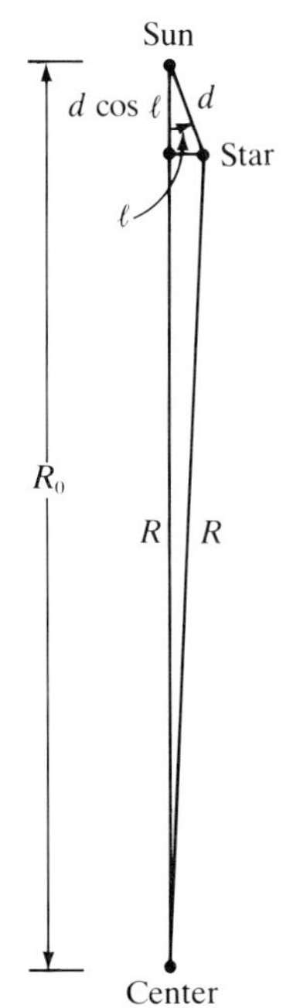

Figure 8-9. Geometric construction showing $R_0 - R \approx d \cos \ell$ when $d \ll R_0$.

which, when substituted into equation (8-12), yields

$$v_R \approx \left[\frac{\Theta_0}{R_0} - \left(\frac{d\Theta}{dR}\right)_{R_0}\right] d \sin \ell \cos \ell \tag{8-14}$$

Then, using the trigonometric identity $\sin \ell \cos \ell = \frac{1}{2} \sin 2\ell$, and defining

$$A = \frac{1}{2}\left[\frac{\Theta_0}{R_0} - \left(\frac{d\Theta}{dR}\right)_{R_0}\right] \tag{8-15}$$

which is called *Oort's constant A*, we obtain, to first order in d,

$$v_R = A\, d \sin 2\ell \tag{8-16}$$

In harmony with the results we obtained from the heuristic arguments given at the beginning of this section, equation (8-16) shows that, because of differential galactic rotation, the radial velocities of stars in the galactic disk will show a *double sine-wave* variation with galactic longitude (see Figure 8-3), with an amplitude that increases linearly with distance. Equation (8-16) is valid for all disk stars with $d \ll R_0$. Hence, if we can observe v_R and estimate d for a large sample of stars, we can estimate A and hence place constraints on the local values of ω_0 and $(d\omega/dR)_{R_0}$ [see equations (8-22) and (8-23)]. In most work, v_R is expressed in kilometers per second and d in kiloparsecs, so that A has units of kilometers per second per kiloparsec.

Now, consider equation (8-8) for the tangential velocity v_T. In the first term, we can write an expression for $(\omega - \omega_0)$ that is first order in the distance d directly from equations (8-9) and (8-13). In the second term, $\omega d = \omega_0 d + (\omega - \omega_0)d$, and, because $(\omega - \omega_0)$ is already of the first order in d, it is obvious that, if we are to retain only first-order terms, we can simply replace ωd with $\omega_0 d$. Thus equation (8-8) becomes

$$\begin{aligned} v_T &\approx \left[\left(\frac{d\Theta}{dR}\right)_{R_0} - \frac{\Theta_0}{R_0}\right](R - R_0) \cos \ell - \omega_0\, d \\ &\approx \left[\frac{\Theta_0}{R_0} - \left(\frac{d\Theta}{dR}\right)_{R_0}\right] d \cos^2 \ell - \left(\frac{\Theta_0}{R_0}\right) d \end{aligned} \tag{8-17}$$

or, using the trigonometric identity $\cos^2 \ell = \frac{1}{2}(1 + \cos 2\ell)$,

$$v_T = \frac{1}{2}\left[\frac{\Theta_0}{R_0} - \left(\frac{d\Theta}{dR}\right)_{R_0}\right] d \cos 2\ell - \frac{1}{2}\left[\frac{\Theta_0}{R_0} + \left(\frac{d\Theta}{dR}\right)_{R_0}\right] d \tag{8-18}$$

Now, using equation (8-15), and defining *Oort's constant B* as

$$B \equiv -\frac{1}{2}\left[\frac{\Theta_0}{R_0} + \left(\frac{d\Theta}{dR}\right)_{R_0}\right] \tag{8-19}$$

we finally obtain

$$v_T = d(A \cos 2\ell + B) \tag{8-20}$$

In terms of proper motions, we know from equation (3-2) that $\mu = v_T/4.74d$, where μ is expressed in seconds of arc per year, v_T in kilometers per second, and d in parsecs. Thus, if μ_ℓ is the proper motion in seconds of arc per year measured positive in the direction of increasing ℓ, then

$$\mu_\ell = \frac{A \cos 2\ell + B}{4.74} \tag{8-21}$$

In equation (8-21), both A and B have to be expressed in kilometers per second per parsec. The numerical values of the two constants in these units are a factor of 10^3 smaller than the values quoted later in this chapter, which are given in the customary units of kilometers per second per kiloparsec.

As expected from the qualitative analysis made at the beginning of this section, μ_ℓ is composed of two terms: the first arises mainly from differential motion effects, and the second reflects essentially the rotation of the coordinate system. We noted earlier that we expect the longitude-independent drift term to be negative, hence the constant B should be intrinsically negative. As we shall see in §8-2, $(d\Theta/dR)_{R_0} < 0$, but $\omega_0 = (\Theta_0/R_0) > |(d\Theta/dR)_{R_0}|$, hence B does turn out to be negative.

Equations (8-16) and (8-21) were obtained in 1927 by Oort (**O1**), who immediately showed that they were consistent with observation. He thereby demonstrated unequivocally the existence of differential galactic rotation, and he provided compelling evidence for the basic correctness of the Lindblad-Oort model of galactic rotation.

It is obvious from equations (8-15) and (8-19) that the local values of the angular rotation rate and the derivative $(d\Theta/dR)$ can be expressed directly in terms of the Oort A and B constants. We obtain

$$\omega_0 = \frac{\Theta_0}{R_0} = A - B \tag{8-22}$$

and

$$(d\Theta/dR)_{R_0} = -(A + B) \tag{8-23}$$

We shall use these two equations repeatedly in later work; students may even find it worthwhile to memorize these results.

Because the A and B constants provide an essentially complete *local* description of the kinematics of galactic rotation, and specifically because they can be used to evaluate ω_0 and $(d\Theta/dR)_{R_0}$, a considerable effort has been devoted to the determination of their numerical values. It is clear that the constant A can be derived from an analysis of radial velocity data for stars of known distances. In principle, both A and B can be inferred from proper-motion data, and, as can be seen from equation (8-21), the results thus obtained are independent of a distance scale. The latter is an important advantage because in many cases stellar distances are not known accurately. In practice, however, the observable effects are small [$\mu_\ell \approx$ (0.″003 cos 2ℓ − 0.″002) per year], and the proper-motion data available at present may contain systematic errors (see Chapter 2). A variety of different techniques have been developed and employed to infer the rotation constants, and, as we shall see in §8-2, one can develop a consistent set of values for the four parameters A, B, R_0, and Θ_0 by invoking certain theoretical relationships among them and by combining several different kinds of observational data.

The AR_0 Formula

One further result is useful. We have just seen that, along a given line of sight, v_R attains a maximum (absolute) value for objects that lie at the tangent point. For these bodies, the angle α of Figure 8-5 has value $\alpha = 0$, and therefore, by equation (8-1), we have for the maximum radial velocity along the line of sight at $\ell(-90^\circ < \ell < 90^\circ)$

$$v_{max} = \Theta(R_{max}) - \Theta_0 \sin \ell \tag{8-24}$$

where

$$R_{max} = R_0 \sin \ell \tag{8-25}$$

If we now restrict our attention to longitudes near 90°, when $R_{max} - R_0$ is small compared with R_0, we may expand $\Theta(R_{max})$ in powers of $(R_{max} - R_0) = -R_0(1 - \sin \ell)$ to find

$$\begin{aligned} v_{max} &= \Theta_0 - \left(\frac{d\Theta}{dR}\right)_{R_0} R_0(1 - \sin \ell) + \\ &\quad \frac{1}{2}\left(\frac{d^2\Theta}{dR^2}\right)_{R_0} R_0^2(1 - \sin \ell)^2 + \cdots - \Theta_0 \sin \ell \\ &= 2AR_0(1 - \sin \ell) + \frac{1}{2}\left(\frac{d^2\Theta}{dR^2}\right)_{R_0} R_0^2(1 - \sin \ell)^2 + \cdots \end{aligned} \tag{8-26}$$

The terms after the first one on the right-hand side of equation (8-26) can be safely ignored for a range of longitudes near $\ell = 90°$, both because $(d^2\Theta/dR^2)_{R_0}$ is small and because $(1 - \sin \ell)$ is small. Thus measurement of the dependence of v_{max} on ℓ allows us to determine the product AR_0 observationally, and the value of this product provides a useful constraint on possible values of A and R_0.

8-2. ESTIMATION OF THE LOCAL ROTATION CONSTANTS

In 1927, Oort both proved the existence of differential galactic rotation and derived a first estimate of the rotation constant A from a study of the radial velocities of supergiants, Cepheids, and O stars, using distances based on statistical parallaxes derived from proper-motion studies. He used these stars because they are highly luminous and hence can be observed to great distances, which allows the predicted effects to build to a measurable level. He obtained the numerical value $A = 31.7 \pm 3$ km s^{-1} kpc^{-1}. In 1928, from an analysis of proper motions, Oort found $A = 19$ km s^{-1} kpc^{-1} and $B = -24$ km s^{-1} kpc^{-1}.

Since Oort's pioneering efforts, numerous studies have been made to determine the values of A and B and the related values of Θ_0 and R_0. The status of this work through the early 1960s was discussed comprehensively by M. Schmidt (**B9**, Chapter 22); this review gives a clear description of both methods and results, and it merits the reader's attention. In 1964, the IAU adopted the values (derived by Schmidt) $A = 15$ km s^{-1} kpc^{-1}, $B = -10$ km s^{-1} kpc^{-1}, $R_0 = 10$ kpc, and $\Theta_0 = 250$ km s^{-1}. These values form an *internally consistent* set in the sense to be described below.

In the discussion that follows, we shall summarize briefly some of the recent work on the determination of the local rotation constants and present an alternative consistent set of values which, according to present evidence, may be preferable to the IAU values. It must be stressed, however, that both sets of numbers are to be regarded as provisional and subject to revision at any time, because the data upon which they rest are not yet of sufficient quality to yield definitive values.

Oort's Constant *A*

The rotation constant A may be found by the following methods:

1. From the amplitude of the variation with galactic longitude of the radial velocities of objects of known distances, using equation (8-16).
2. From the amplitude of the longitude dependence of proper motions, using equation (8-21).

3. From the definition

$$A = \frac{1}{2}\left[\frac{\Theta_0}{R_0} - \left(\frac{d\Theta}{dR}\right)_{R_0}\right] = -\frac{1}{2} R_0 \left(\frac{d\omega}{dR}\right)_{R_0} \qquad (8\text{-}27)$$

using an empirically determined rotation curve $\omega(R)$.

Before we discuss numerical results, each of these methods deserves comment. Note first that the value of A determined by method 1 is inversely proportional to the distance scale; thus it is vital to use as accurate a scale as possible. Moreover, when one compares different sets of results, this scaling must be borne in mind, and all results must be reduced to a common scale if necessary. At present, the distance scale obtained from absolute-magnitude calibrations of stars in galactic clusters (discussed in Chapter 3) is generally regarded as the most reliable one, and we shall adopt this scale as standard. Method 2 is not affected by errors in the distance scale, but it is subject to uncertainties produced by any systematic errors that may exist in the available proper motions. In method 3, one proceeds as follows. If the average radial velocity for any group of objects known to be at a common distance is measured, and if R_0 is taken as known, then $(\omega - \omega_0)$ follows from equation (8-4). The radial distance of this group from the galactic center can be derived from the law of cosines

$$R^2 = R_0^2 + d^2 - 2R_0 d \cos \ell \qquad (8\text{-}28)$$

when R_0 and the distance d are presumed known. If we can find $(\omega - \omega_0)$ versus R for several groups of objects over a reasonably wide range of R, then, by fitting a smooth curve to the results and evaluating its slope at $R = R_0$, we can estimate $(d\omega/dR)_{R_0}$ and hence A via equation (8-27). Inherent in this approach are any uncertainties in R_0 and in the distance scale.

Results from several recent determinations of the value of the constant A are listed in Table 8-1. The errors quoted are internal standard deviations, and the weights are those assigned in a very useful critical discussion of the subject by Oort and Plaut (**O3**). The weighted average value is, in round numbers, $A = 16 \pm 2$ km s^{-1} kpc^{-1}. This "best current estimate" is but little different from the IAU standard.

The value of A listed in Table 8-1 for Cepheids is not completely independent of the others, because the distance scale is based on absolute magnitudes derived from Cepheids found in galactic clusters. An independent estimate of Cepheid absolute magnitudes can be made by the statistical-parallax method. The result is a distance scale that is smaller by a factor of 0.8, which implies a value of $A = 19$ km s^{-1} kpc^{-1}. It is unfortunate that the galactic-cluster distance scale is not more consistent with the statistical-parallax scale, and it must be borne in mind that the measurement

Table 8-1. Measured Values of Oort's Rotation Constant A (km s^{-1}) kpc^{-1}

Object	Data analyzed	A	Weight	Reference
O and B Stars	Radial velocities	16.8 ± 0.6	2	**(B3)**
Optical interstellar lines	Radial velocities	$18.1 + 0.7$	1.5	**(B3)**
Cepheid variables	Radial velocities	15 ± 2	1.5	**(K4)**
Galactic clusters	Radial velocities	15 ± 3	1	**(J1)**
Fundamental catalog stars	Proper motions	15.6 ± 3	2	**(F3)**

of statistical parallaxes is delicate, difficult, and vulnerable to systematic errors in the proper motions. In any event, even if we had adopted the statistical-parallax scale, the mean value of A would have been 16.9 km s^{-1} kpc^{-1}, which is within the errors of the result quoted earlier.

Oort's Constant *B*

The only *direct* method by which the rotation constant B can be determinea is from an analysis of proper motions using equation (8-21). Such analyses have been attempted many times, but in the past the results have been discordant, which is not particularly surprising given the smallness of the effect to be measured and the difficulty met in eliminating systematic errors. Nonetheless, some recent high-quality determinations have begun to yield harmonious results. From a study of proper-motion data in the fundamental star catalogs GC, N30, FK3, and FK4, W. Fricke has recently derived **(F3)** the value $B = -10.9 \pm 3$ km s^{-1} kpc^{-1}. Two other determinations have been made using proper motions measured relative to external galaxies; these determinations are completely independent of the fundamental (meridian-circle) system. The result obtained by N. V. Fatchikhin **(F1)** is $B = -10.4 \pm 4$ km s^{-1} kpc^{-1}, while S. Vasilevskis and A. R. Klemola **(V4)** obtained $B = -12.5 \pm 4$ km s^{-1} kpc^{-1}. If we give the average of these last two determinations half the weight of the result found from the fundamental system **(O2)**, the mean value obtained from the direct method is $B = -11.1 \pm 3$ km s^{-1} kpc^{-1}.

An *indirect* estimate of the value of B can be obtained from the dynamical relation (to be proved in Chapter 13) that

$$\frac{-B}{A-B} = \frac{\sigma_\Theta^2}{\sigma_\Pi^2} \tag{8-29}$$

or

$$\frac{-B}{A} = \frac{1}{(\sigma_{\Pi}^2/\sigma_{\Theta}^2) - 1} \tag{8-30}$$

where σ_{Π}^2 and σ_{Θ}^2 are the dispersions of peculiar velocities in the solar neighborhood in the radial and tangential directions, respectively. If we use the values of $(\sigma_{\Theta}^2/\sigma_{\Pi}^2)$ listed in Table 7-1 for various types of stars, and if we adopt a value for the constant A, then we can estimate B. In practice, one uses the velocity dispersions for later-type stars, whose peculiar velocities presumably satisfy the criterion of being well mixed dynamically, which is the requirement for equation (8-30) to be valid. Thus, for F5 to M5 dwarfs, the average value of $(\sigma_{\Theta}^2/\sigma_{\Pi}^2) = 0.4$, which implies $(-B/A) = 0.67$, or $B = -10.8$ km s^{-1} kpc^{-1} for $A = 16$ km s^{-1} kpc^{-1}. Similarly, for A to M giants, $(\sigma_{\Theta}^2/\sigma_{\Pi}^2) = 0.42$, $(-B/A) = 0.71$, or $B = -11.4$ km s^{-1} kpc^{-1}. The basic difficulty with this approach is that the adopted ratio $(\sigma_{\Theta}/\sigma_{\Pi})$ may contain systematic errors produced by local irregularities in the stellar velocity distribution. The vertex deviation of the velocity ellipsoid (see §7-1) is evidence that such effects exist, but it is not known at present whether these effects seriously affect our estimates of $(\sigma_{\Theta}^2/\sigma_{\Pi}^2)$.

In summary, both of the methods described here suggest that a "best" value for the B constant is $B = -11 \pm 3$ km s^{-1} kpc^{-1}, again a value that differs but little from the IAU standard.

Distance to the Galactic Center, R_0

Knowledge of R_0, the distance from the Sun to the galactic center, bears upon the discussion of the local rotation constants, both because observational estimates can be made of the product AR_0 and because $\omega_0 = (\Theta_0/R_0)$ is related to the rotation constants A and B via equation (8-22). Direct estimates of R_0 can be obtained from studies of the space density of objects that are believed to be distributed symmetrically around the galactic center. Globular clusters are ideal for such work because, owing to their great luminosity, they can be observed at great distances. RR Lyrae variables are equally valuable because their characteristic light variation makes it possible to find them to a high level of completeness in any given field. In these studies, the basic problems encountered are the correct choice of the absolute magnitudes of the objects and the elimination of interstellar absorption effects.

One of the first accurate estimates of R_0 was made by W. Baade (**B1**) from a study of RR Lyrae variables in carefully selected fields of relatively low interstellar absorption near the galactic center; he obtained $R_0 = 8.2$ kpc. Subsequent measurements tended to yield larger values, and, in the IAU reference set, the estimate $R_0 = 10$ kpc was adopted. Considerable observa-

Table 8-2. Recent Determinations of R_0 (kpc)

Object	R_0	Reference
Globular clusters	8.5 ± 1.6	**(H1)**
RR Lyrae stars in galactic bulge	8.7 ± 0.6	**(O2)**
Main-sequence stars in galactic bulge	9.2 ± 2.2	**(V1)**
Disk OB stars (kinematical)	9.0 ± 1.6	**(B3)**

tional effort has recently been brought to bear on the determination of R_0, and the values obtained in three different attempts to measure R_0 directly are listed in Table 8-2.

In the first study cited, a variety of different distance indicators were used to estimate the distances to the clusters. These indicators were carefully compared and reduced to an internally consistent system. More than one hundred clusters were used to estimate the space centroid of the system (after making a detailed allowance for selection effects) and hence the value of R_0. The second study determined the distances to fairly sharp maxima in the space distribution of RR Lyrae stars in five fields near the galactic center. In this work, a careful analysis was made to eliminate the effects of incompleteness, absorption, random errors in individual stellar distance estimates, and the gradient in the large-scale space distribution. In the third study, an abrupt increase in the numbers of faint stars between $19 \lesssim m \lesssim 20$, observed with the 200-inch telescope, is used to estimate R_0 on the hypothesis that there are spheroidal-component stars near the evolutionary turnoff point for stars of the appropriate age. There is, of course, the possibility that this hypothesis may be in error, and the result must therefore be given a lower weight than the first two.

It is also possible to make an indirect kinematical estimate of R_0. The strategy here is to find luminous stars (for example, OB stars or Cepheids) that are at large distances from the Sun, but which lie at about the same distance from the galactic center as the Sun does. These stars should have zero radial velocity with respect to the Sun because they move around the galactic center with the same speed as the Sun; one can imagine them to be mounted with the Sun on a common, rigidly rotating hoop. From Figure 8-10, we see that such a star, observed at galactic longitude ℓ, is at distance $d = 2R_0 \cos \ell$. Therefore, if we can determine the longitude ℓ_0 at which stars distance d from the Sun appear to have zero radial velocity, we can obtain R_0 as

$$R_0 = \tfrac{1}{2} d \sec \ell_0 \tag{8-31}$$

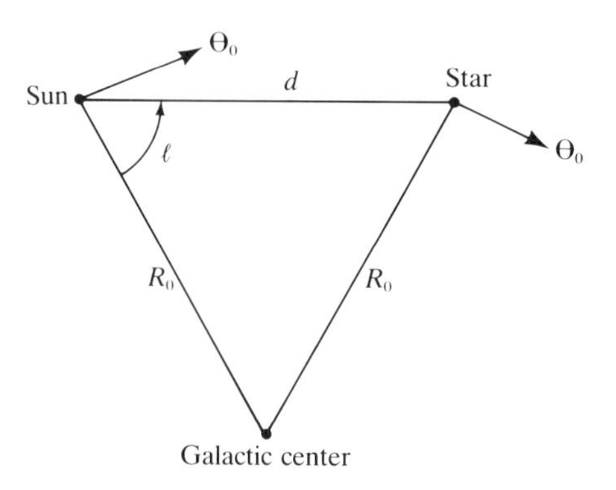

Figure 8-10. A star at distance R_0 from the galactic center will revolve with a circular velocity $\Theta(R_0) = \Theta_0$. Hence, it will remain fixed with respect to the Sun and will therefore have zero radial velocity. The distance from the Sun to such a star observed at galactic longitude ℓ is $d = 2R_0 \cos \ell$.

This method has two shortcomings, however. One is that it is vulnerable to the effects of local departures of stellar velocities from perfectly circular motion. The second is that it is extremely sensitive to small errors in the longitude ℓ_0. Thus it is easy to see from equation (8-31) that $(\Delta R_0/R_0) \approx \tan \ell_0 \, \Delta\ell_0$. Suppose $d \approx 3.5$ kpc for a typical group of stars; then $\cos \ell_0 \approx \frac{1}{5}$, $\sin \ell_0 \approx 1$, and $(\Delta R_0/R_0) \approx 5\,\Delta\ell_0$ (radians) or $0.1\,\Delta\ell_0$ (degrees). Thus even a small error in ℓ_0 (say, 2°) implies a serious error in R_0.

A weighted average of the results in Table 8-2 yields $R_0 = 8.8 \pm 0.7$ kpc, so that, for our "best" value, we may adopt in round numbers $R_0 = 9 \pm 1$ kpc, which again is fairly close to the IAU standard.

A useful consistency check on the results derived thus far is provided by the measured numerical value of the product AR_0. From a critical discussion of values of AR_0 obtained using equation (8-26) from 21-cm radio observations of the maximum radial velocities seen along various lines of sight, Schmidt concluded (**B9**, Chapter 22) that this product lies in the range $135 \lesssim AR_0 \lesssim 150$ km s^{-1}. This constraint is satisfied both by our "best" values, which yield $AR_0 = 145 \pm 25$ km s^{-1} and by the IAU standard values, which yield $AR_0 = 150$ km s^{-1}. Gunn et al. (**G1**), however, recently derived a lower value $AR_0 = 110 \pm 5$ km s^{-1} from similar data.

Rotation Speed Θ_0

As was discussed in Chapter 6, direct measurements of Θ_0, based on radial velocities of globular clusters or spheroidal-component stars in our Galaxy, or of external galaxies in the Local Group, yield values in the range $200 \leq \Theta_0 \leq 300$ km s^{-1}, and a "best" estimate of $\Theta_0 = 250$ km s^{-1} (which is also the IAU standard value). We shall see in §8-4 that this value is also compatible with observations of rotation curves in other galaxies. A dynamical method for estimating Θ_0 was also described in Chapter 6, but it was concluded that this approach does not yield reliable information. About the only other information that can be brought to bear on the matter follows from equation (8-22), which implies that $\Theta_0 = (A - B)R_0$. Using the values adopted earlier for A, B, and R_0, we find $\Theta_0 = 245 \pm 40$ km s^{-1}, which is consistent with our adopted value; the IAU standard data are also self-consistent. When a direct measurement of ω_0 by radio-astrometric methods becomes available, it will provide an extremely useful constraint on both Θ_0 and $(A - B)$.

The values of R_0 and Θ_0 derived earlier imply that the Sun revolves once around the galactic center in a period of

$$P_{\text{rev}} = \frac{2\pi R_0}{\Theta_0} = 2.2 \times 10^8 \text{ years} \tag{8-32}$$

which is a useful reference number. The IAU standard values give $P_{\text{rev}} = 2.45 \times 10^8$ years. Further, using the values of A and B quoted earlier, we can estimate that $(d\Theta/dR)_{R_0} = -5$ km s^{-1} kpc^{-1}, which suggests that the rotation curve of our Galaxy is declining in the solar neighborhood. As we shall see in §8-4, the rotation curves in disk galaxies characteristically show a relative maximum at distances of the order of 5 to 10 kpc from the center. From the negative value just derived for $(d\Theta/dR)_{R_0}$, we might conclude that the Sun is near, but slightly outside, the peak in the rotation curve of our Galaxy. However, we shall shortly see that the galactic rotation curve probably does not continue to decline at radii significantly larger than R_0.

8-3. DETERMINATION OF THE GENERAL ROTATION LAW OF OUR GALAXY

The first-order expansions developed in §8-2, with known values of A, B, and Θ_0, are adequate to describe differential galactic rotation only at galactocentric distances $R \approx R_0$. But what we desire is a specification of $\Theta(R)$ in the galactic disk at all values of R from the center. This information can be obtained only from observational study of the kinematics of objects in the disk over large portions of our Galaxy; however, the distance range of optical observations is severely limited by interstellar absorption, hence observations of stars cannot provide the needed data.

Fortunately, atomic hydrogen, which is a major constituent of the interstellar gas that permeates the galactic disk, possesses a hyperfine transition in its ground state, which gives rise to a radio spectral line having a wavelength of 21 cm. Because the wavelength of this line is so much larger than the typical size (about 10^{-4} cm) of particles in the interstellar medium, it suffers essentially no interstellar absorption, and, because the transition is forbidden, the line emission is virtually unattenuated by self-absorption. As a result, hydrogen 21-cm emission can be detected from material at all locations in the galactic plane, and observations of the rotation velocity of interstellar hydrogen enables us to determine $\Theta(R)$ for almost the whole galactic disk interior to the Sun.

21-cm Line Emission by Atomic Hydrogen

Emission in the 21-cm line of neutral atomic hydrogen is of such enormous importance for galactic astronomy that it is appropriate to outline briefly the most important aspects of this process. Much of the interstellar medium consists of neutral hydrogen atoms in their "ground states". Actually, two energy levels are accessible to the electron in this state of a hydrogen atom, depending on how its spin is oriented with respect to that of the proton which forms the nucleus of the atom. If the electron spin is aligned parallel to that of the proton (yielding a total spin quantum number $F = 1$), the energy of the electron is 6×10^{-6} eV higher than if the spins are antiparallel ($F = 0$). This energy difference is extremely small; it corresponds to a temperature $T = 0.07$ K (through $E = kT$) or to photons having frequency 1420.4 MHz and a wavelength of 21.105 cm. When an electron in a hydrogen atom reorients its spin, it absorbs or emits a 21-cm photon.

The temperature of the interstellar medium is invariably much higher than 0.07 K, so that more than enough energy is available in a typical atomic collision to excite any hydrogen atom in the lower level to the upper level. Once in the upper level, the atom will remain there until it is either de-excited to the lower level by a second collision or emits a 21-cm photon (either spontaneously or in response to stimulation by other 21-cm photons incident on it) and thereby radiatively decays to the lower level.

The probability for an excited atom to decay radiatively without stimulation by a strong 21-cm radiation field is small, because the transition between the energy levels is "forbidden"; that is, the decay involves coupling to the radiation field by a magnetic rather than an electric dipole. In fact, the average lifetime of an isolated excited atom is 3.5×10^{14} seconds, or about 10^7 years! The collisional de-excitation rate of the upper level is actually much more rapid than this: about once per 400 years. As we shall see later, this fact implies that collisions can establish equilibrium populations in the two levels, which means that there will be nearly three atoms in the upper level (which is threefold degenerate because there are three ways in which a total spin of $\hbar$ can be oriented) to every one in the lower level (**F2**). However, the number of photons emitted per unit time by the atoms in the upper level is not affected by the relatively rapid collisional de-excitation; only their distribution over frequency is slightly altered.

Because the collisional excitation and de-excitation rates are so much faster than the rate of radiative decay from the upper level, the atomic populations n_1 and n_2 in the two levels will be essentially the same as those expected in thermodynamic equilibrium, even though the 21-cm photon density will normally be much smaller than the thermal equilibrium photon density. Therefore, one can use the usual Boltzmann formula,

$$\frac{n_2}{n_1} = \left(\frac{g_2}{g_1}\right) \exp\left(\frac{-h\nu}{kT_S}\right) \tag{8-33}$$

where $(g_2/g_1) = 3$ is the ratio of the degeneracies of the two levels mentioned earlier. The quantity T_S appearing in equation (8-33) is called the *spin temperature* of the gas, and it is usually indistinguishable from the kinetic temperature that governs the speeds with which the atoms move. In a typical cloud, $T_S \approx 100$ K, so $(h\nu/kT_S) = 6.8 \times 10^{-4}$ and $\exp(-h\nu/kT_S) = 0.9993$ is extremely close to 1.

Atoms in the upper level can undergo either *spontaneous* or *induced* decay to the lower level, and atoms in the lower level can *absorb* radiation. These processes are described by the *Einstein probability coefficients* A_{21}, B_{21} and B_{12}, respectively, which are connected by the relations (**M2**, Chapter 4)

$$A_{21} = \left(\frac{2h\nu^3}{c^2}\right) B_{21} \tag{8-34}$$

$$g_1 B_{12} = g_2 B_{21} \tag{8-35}$$

Let ϕ_ν denote the *absorption profile* of the gas, normalized such that $\int \phi_\nu \, d\nu = 1$. Then ϕ_ν will be the fraction of atoms in either level that can either absorb or emit radiation of frequency ν. The number of spontaneous emissions per unit frequency interval $d\nu$ and solid angle $d\Omega$ will be $n_2 A_{21} \phi_\nu d\nu (d\Omega/4\pi)$, the number of induced emissions will be $n_2 B_{21} \phi_\nu I_\nu \, d\nu (d\Omega/4\pi)$, and the number of absorptions will be $n_1 B_{12} \phi_\nu I_\nu \, d\nu (d\Omega/4\pi)$. Here I_ν is the *specific intensity* of the radiation field. Each photon has energy $h\nu$, hence the transfer equation that expresses energy balance through an element of material of length ds is

$$\begin{aligned} \frac{\partial I_\nu}{\partial s} &= \left(\frac{h\nu}{4\pi}\right) [n_2 A_{21} \phi_\nu - (n_1 B_{12} - n_2 B_{21}) \phi_\nu I_\nu] \\ &\equiv \eta_\nu - \chi_\nu I_\nu \end{aligned} \tag{8-36}$$

where η_ν is the *emissivity* of the material, and χ_ν is the *absorption coefficient.*

If we define the *source function* as $S_\nu \equiv \eta_\nu / \chi_\nu$, and the *optical depth* (which measures the integrated absorption along the line of sight) as

$$\tau_\nu \equiv -\int \chi_\nu \, ds \tag{8-37}$$

then the transfer equation (8-36) becomes

$$\frac{\partial I_\nu}{\partial \tau_\nu} = I_\nu - S_\nu \tag{8-38}$$

Consider the simple case of constant S_ν in a medium of total optical depth τ_ν. The observed intensity is then

$$I_\nu(0) = S_\nu(1 - e^{-\tau_\nu}) + I_\nu(\tau_\nu)e^{-\tau_\nu} \tag{8-39}$$

Here $I_\nu(\tau_\nu)$ is the intensity incident at the far side of the material. As we shall integrate through the entire disk, we shall set $I_\nu(\tau_\nu) = 0$.

By use of the Einstein relations (8-34) and (8-35), along with equation (8-33), we can write

$$\chi_\nu \equiv (n_1 B_{12} - n_2 B_{21})\left(\frac{h\nu}{4\pi}\right)\phi_\nu = n_1\left(\frac{B_{12}h\nu}{4\pi}\right)(1 - e^{-h\nu/kT_S})\phi_\nu \tag{8-40}$$

and

$$S_\nu = \frac{n_2 A_{21}}{n_1 B_{12} - n_2 B_{21}} = \frac{2h\nu^3/c^2}{e^{h\nu/kT_S} - 1} = B_\nu(T_S) \tag{8-41}$$

where $B_\nu(T_S)$ is the Planck function, which describes the blackbody radiation from material at temperature T_S. Hence equation (8-39) can be rewritten

$$I_\nu = (1 - e^{-\tau_\nu})B_\nu(T_S) \tag{8-42}$$

Clearly, as the material becomes optically thick (that is, $\tau_\nu \to \infty$), the radiation field thermalizes, and

$$I_\nu(\text{thick}) \to B_\nu(T_S) \tag{8-43}$$

At radio frequencies, one generally has $(h\nu/kT) \ll 1$, so $\exp(h\nu/kT_S) \approx 1 + (h\nu/kT_S)$ and hence the Planck function $B_\nu(T_S)$ reduces to its *Rayleigh-Jeans limit*;

$$B_\nu(T_S) = \frac{2kT_S\nu^2}{c^2} \tag{8-44}$$

yielding, from equation (8-42),

$$I_\nu(0) \approx 2kT_S(\nu^2/c^2)(1 - e^{-\tau_\nu}) \tag{8-45}$$

Equation (8-45) has led radio astronomers to report surface brightness measurements in terms of a quantity called *brightness temperature*, T_B, which is defined by

$$T_B \equiv \left(\frac{c^2}{2k\nu^2}\right) I_\nu \tag{8-46}$$

With this definition of brightness temperature, one sees from equations (8-45) and (8-46) that, in general,

$$T_B = T_S(1 - e^{-\tau_\nu}) \tag{8-47}$$

Thus, for an optically thick source, $T_B = T_S$, and the observed brightness temperature of the source equals the spin temperature of the atoms, while at small optical depth

$$T_B \approx \tau_\nu T_S \tag{8-48}$$

Let us now calculate the optical depth in the 21-cm line. From equations (8-37) and (8-40) we obtain

$$\tau_\nu = \left(\frac{B_{12}h^2\nu^2}{4\pi k T_S}\right) N_1 \phi_\nu \tag{8-49}$$

where N_1 denotes the *integrated column density* of hydrogen in the lower level along the line of sight; that is,

$$N_1 = \int_0^\infty n_1(s)\, ds \tag{8-50}$$

Now, using the result that $N_H \approx 4N_1$ and equation (8-34), we find

$$\tau_\nu = \left(\frac{hc^2 g_2 A_{21}}{32\pi k g_1 \nu}\right)\frac{N_H}{T_S}\phi_\nu = \frac{CN_H\phi_\nu}{T_S} \tag{8-51}$$

where $C = 2.57 \times 10^{-15}$ cm^2 K^{-1}.

We must now inquire what shape the profile function has, and what is the characteristic distance over which we should integrate the column density N_H. The natural width of the 21-cm line is extremely small ($< 5 \times 10^{-13}$ Hz), so the line profile will be determined completely by the motions of the atoms in the gas; that is, it will have a Doppler profile

$$\phi_\nu = \left(\frac{1}{\pi^{1/2}\,\Delta\nu_D}\right) e^{-(\Delta\nu/\Delta\nu_D)^2} \tag{8-52}$$

where the *Doppler width* $\Delta\nu_D$ is given by

$$\Delta\nu_D = \frac{\nu_0 v_D}{c} \tag{8-53}$$

where v_D is a characteristic velocity of the hydrogen atoms along the line of sight. For the 21-cm line, $v_D = 1$ km s^{-1} implies a Doppler width of

4.74 kHz. If the atomic motions were strictly thermal, v_D would be about 1 km s^{-1}. In reality, all observed lines are much broader, and even the sharpest individual components have characteristic widths of the order of 5 km s^{-1}. This width is attributed to "turbulent" macroscopic motions of individual elements of gas within interstellar clouds and to random motions of individual clouds within cloud complexes.

The characteristic length over which the optical depth can accumulate is set by the effects of differential galactic rotation, which causes the absorption peak of atoms more than a certain maximum distance from the observer to be Doppler-shifted outside the frequency domain of the line profile of nearby atoms. It is this rotational shear that also produces the enormous line widths ($\pm$100 km s^{-1}) that are observed. [In fact, this situation is very closely similar to the Sobolev-theory limit for line transfer in an expanding atmosphere (**M2**, Chapter 14)]. From the size of Oort's constant $A \approx$ 15 km s^{-1} kpc^{-1}, it follows that a Doppler shift of 10 km s^{-1} can be produced over a distance of about 600 pc $\approx 2 \times 10^{21}$ cm, and, assuming a hydrogen density of around 0.5 atoms cm^{-3} over the entire pathlength, the column density $N_H \approx 10^{20}$ atoms cm^{-2}. If we adopt a typical temperature of the order of 100 K for the interstellar atomic hydrogen, set $v_D = 10$ km s^{-1} so that $\Delta\nu_D \approx 5 \times 10^4$ Hz, we find that, at line center, $\tau_\nu \approx 0.3$ (which is probably an overestimate because the line of sight may not pass through hydrogen clouds at every point along its length). Thus we see that, along most lines of sight, the 21-cm line will be just optically thin. Exceptions to this rule are the lines of sight to the galactic center ($\ell = 0°$) and, in the reverse direction, to the anticenter ($\ell = 180°$), along which the optical depth in the 21-cm line can approach unity because, in these two directions, the galactic rotation velocity is always tangential to the line of sight. Therefore, in the center and anticenter directions, and to a lesser extent in the directions $\ell = 90°$ and $\ell = 270°$ along which the systematic radial velocity can remain small over quite long pathlengths [recall equation (8-16)], the brightness temperature tends to saturate to the spin temperature. One concludes in this way that $T_S \approx 135$ K (**B14**).

It is instructive to calculate the rate N_ν at which photons will be received in bandwidth $\Delta\nu$ by a telescope of collecting area S from an optically thin source of optical depth τ_ν, spin temperature T_S, and angular size Ω. Dividing equation (8-45) by the energy $h\nu = hc/\lambda$ of each photon, one has

$$N_\nu = 2\tau_\nu\left(\frac{S}{\lambda^2}\right)\left(\frac{kT_S}{h\nu}\right)\Delta\nu\,\Omega \tag{8-54}$$

For example, the early observations of galactic 21-cm emission at Leiden (**V2**) had $S \approx 180$ m^2, $\Delta\nu \approx 12$ kHz, and $\Omega \approx 30$ square degrees, so, taking $T_S = 100$ K and $\tau_\nu = 0.3$, one finds that these observations were based on about 4×10^6 photons, or about 25 eV, per second (each photon carries 6×10^{-6} eV). However, if one considers that a modern aperture-synthesis telescope has $S \lesssim (100 \text{ m})^2$, $\Delta\nu \approx 1.5$ kHz (corresponding to $\Delta v = 1$ km s^{-1}), and $\Omega \approx (1 \text{ arcmin})^2$, so that the instrument collects no more than 250 photons, or only 1.5×10^{-3} eV, per second in each channel, it becomes clear that sensitive equipment and long integration times are required for surveys of neutral hydrogen at high spatial and velocity resolutions.

It is also worth remarking that we could have obtained a similar estimate of the flux emitted by an optically thin source by simply counting the number of excited atoms within our beam which decay to the lower level by the emission of a photon toward our collecting dish. However, even in an optically thin source ($\tau \approx 0.1$, say), thc vast majority of photons will be absorbed before they have gone even a small distance through the source. Thus equations (8-36), (8-33), and (8-35) tell us that the net absorption coefficient, χ_ν, whose integral along the line of sight is defined to be $-\tau_\nu$, is the very small difference of two large numbers. For every few photons which are net losses to the system, some 10^4 photons are absorbed and replaced by others formed by stimulated emission. Therefore, although the disk of our Galaxy is in one sense transparent at 21 cm, in another sense it is very opaque.

The column density (in atoms per square centimeter) of neutral atomic hydrogen being observed in some direction can be computed by integrating equation (8-51) over all frequency—recalling that ϕ_ν is normalized—to obtain

$$N_H = 3.88 \times 10^{14} \int_0^\infty T_S \tau_\nu \, d\nu \qquad (8\text{-}55)$$

The frequency units in equation (8-55) are hertz. If the more customary units of kilohertz are used, then the numerical factor should be 3.88×10^{17}, and if velocity units (kilometers per second) are used,

$$N_H = 1.82 \times 10^{18} \int_0^\infty T_S \tau(v) \, dv \qquad (8\text{-}56)$$

Now a great simplification results when the material is *optically thin*, for then, according to equation (8-48), we can replace the product $T_S\tau(v)$, which contains the only poorly known spin temperature, with the observed brightness temperature T_B and write

$$N_H(\ell, b) = 1.82 \times 10^{18} \int_0^\infty T_B(l, b, v) \, dv \qquad (8\text{-}57)$$

Thus we can find the total amount of atomic hydrogen in a column of 1 cm^2 cross section along the line of sight directly from observable intensities in 21-cm line profiles.

When we observe an external galaxy, we can determine its total atomic hydrogen content by integrating the H I column density over the surface area of the system. If we write an element of area as $dS = D^2\,d\Omega$, where D is the distance to the galaxy and $d\Omega$ is an element of solid angle on the sky, and if we write the angle- and velocity-dependent brightness temperature as $T_B(\theta, \phi; v)$, then

$$N_H(\text{total}) = 1.82 \times 10^{18} D^2 \int_0^\infty dv \int_\Omega T_B(\theta, \phi; v)\,d\Omega \qquad \text{(8-58)}$$

where Ω is the solid angle subtended by the whole galaxy. If we express D in megaparsecs and the angle integral of the brightness temperature in terms of the observed *flux* $S(v)$ measured in *flux units* (10^{-26} W m^{-2} Hz^{-1}) or *Janskys*, then a convenient formula for the total mass of atomic hydrogen contained by the galaxy is

$$(\mathscr{M}_H/\mathscr{M}_\odot) = 2.36 \times 10^5 D^2 \int_0^\infty S(v)\,dv \qquad \text{(8-59)}$$

Notice that both $\mathscr{M}_H$ and the total luminosity L of a galaxy are proportional to D^2, so that the ratio $(\mathscr{M}_H/L)$ is independent of the often uncertain distance to an external galaxy.

The basic observational data obtained from 21-cm observations in our Galaxy yield $T_B(\ell, b, v)$, the distribution of brightness temperature as a function of galactic coordinates (ℓ, b) and velocity v. These data can be displayed and analyzed in a wide variety of forms, each of which may have special advantages in certain classes of problems. Among the most widely used forms of data presentation are *contour maps* showing $T_B(\ell, v)$ at a given value of galactic latitude b, and *line profiles* $T_B(v)$ at a given value of (ℓ, b). A contour map of $T_B(\ell, v)$ in the galactic plane ($b = 0°$) is shown in Figure 8-11, and a schematic line profile is illustrated in Figure 8-12. Extensive atlases of H I contour maps and line profiles are referenced in Table 2-7. Notice that in Figure 8-11 we can clearly discern high-intensity ridges (for example, one that extends over about 50 km s^{-1}) that show a double sine-wave ($\sin 2\ell$) modulation in galactic longitude, which we know to be characteristic of differential galactic rotation. Thus it is immediately obvious that, if we can perform a suitable analysis of the data, we can hope to infer the rotation velocity of the hydrogen in our Galaxy. Let us now turn to this problem.

The Rotation Curve from 21-cm Line Observations

The basic assumption made in the derivation of the general rotation curve of our Galaxy from 21-cm observations is that gas clouds move in the galactic plane on strictly circular orbits around the galactic center. In this case, their

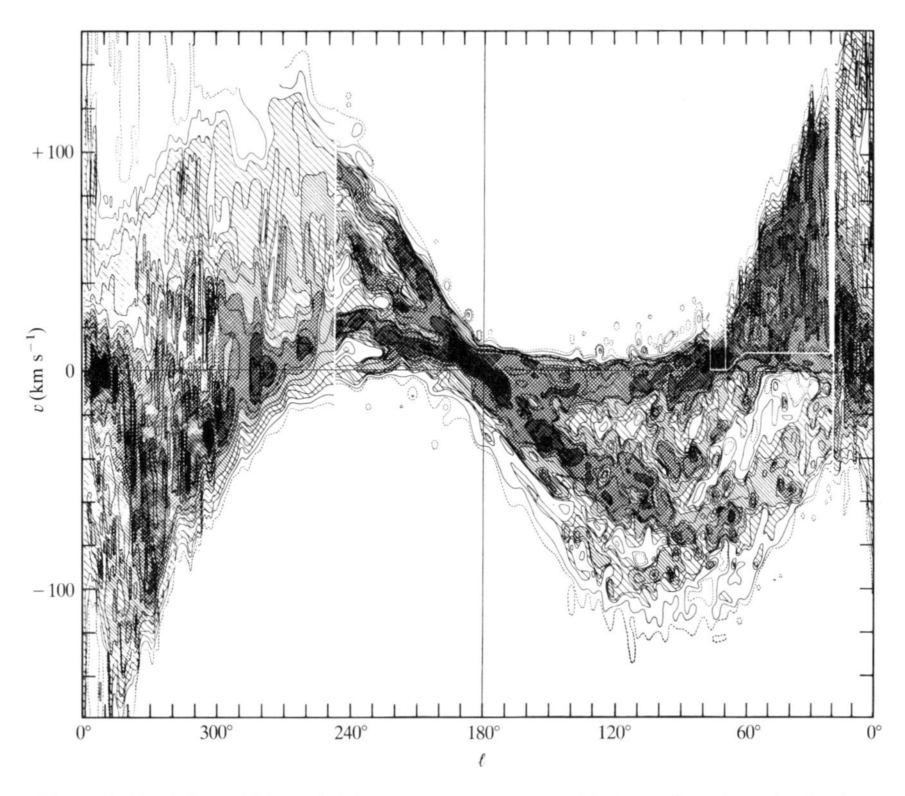

Figure 8-11. Map of 21-cm brightness temperature $T_B(\ell, v)$ as a function of galactic longitude and velocity. The intensity scale is qualitative, darker areas indicating higher temperatures. [From (**V3**, 143), by permission.]

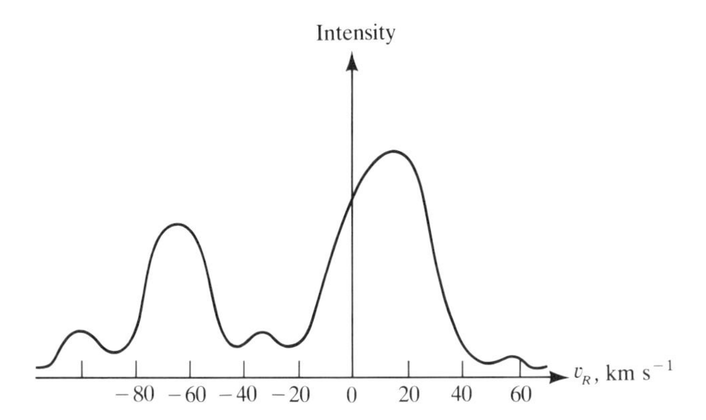

Figure 8-12. Schematic 21-cm line profile. At each point in the profile, the intensity of the radiation is proportional to the number of hydrogen atoms along the line of sight, while the frequency shift is proportional to the velocity of the material in the emitting region.

radial velocities are given by equation (8-4):

$$v_R = R_0[\omega(R) - \omega_0] \sin \ell \tag{8-4}$$

We shall assume that R_0 and ω_0 are known. In particular, for the present discussion of numerical results, we adopt the IAU standards $R_0 = 10$ kpc and $\Theta_0 = 250$ km s^{-1}. Along any given line of sight (which fixes ℓ), equation (8-4) allows us to associate an angular rotation rate $\omega(R)$ with each measured value of v_R. We do not, however, in general know the value of R at which this value of $\omega(R)$ applies. But, if we assume in addition that $\omega(R)$ is a monotonically decreasing function of R, then we recall from §8-1 that the maximum (absolute) radial velocity will occur at the tangential point where $R = R_{min} = R_0 \sin \ell$ (see Figures 8-7 and 8-8). Thus, by observing the maximum radial velocity along lines of sight at several galactic longitudes in the quadrants $0^\circ \leq \ell \leq 90^\circ$ and $270^\circ \leq \ell \leq 360^\circ$, one can deduce the run of $\omega(R)$ and hence $\Theta(R)$ for $R \leq R_0$. This analysis was first done by K. K. Kwee, C. A. Muller, and G. Westerhout (**K6**) in 1954. In the other two longitude quadrants, there is no distinctive maximum in the radial velocity curve (see Figure 8-7), and hence we cannot infer $\Theta(R)$ for $R > R_0$ from radio data.

In actual application of the method just described, the velocity maxima derived from the observed profiles must somehow be corrected for the effects of the smearing produced by the internal velocity structure ("turbulence") of individual H I clouds. Furthermore, the method fails in practice at longitudes within about 20° of the galactic center, both because strongly noncircular motions are found there and because the uncertainty in the line-of-sight position of any emitting region induced by the intrinsic width of the profiles becomes unacceptably large. The method is also weak in the range $75^\circ < \ell < 90^\circ$ and $270^\circ < \ell < 295^\circ$ because of the geometry of the situation: R, and thus v_R, changes slowly with increasing distance along the line of sight, which makes the maximum velocity ill defined and extremely sensitive to errors in measurement and to small, local, noncircular streaming motions. Thus the rotation curve for our Galaxy is most reliably determined from 21-cm data for $4 < R < 9$ kpc.

The rotation curve obtained by W. W. Shane and G. P. Bieger-Smith (**S4**) using the technique just outlined is shown in Figure 8-13. This estimate is currently the best one available of $\Theta(R)$ in the range from 4 to 9 kpc from the galactic center. The curve shown for $R \leq 4$ kpc is based on a dynamical model (**S5**) and may be quite unreliable. As is apparent in Figure 8-13, the rotation curve derived from 21-cm observations is not smooth, but it fluctuates by about ± 10 km s^{-1} with respect to the mean curve. There are at least three possible explanations for these irregularities: (1) actual local variations in the circular $\omega(R)$, (2) gaps in the observed hydrogen distribution, or (3) local streaming motions of the material relative to pure circular motion.

The true circular-velocity curve of our Galaxy is unlikely to have fluctuations of the kind observed in the derived 21-cm rotation curve, both because

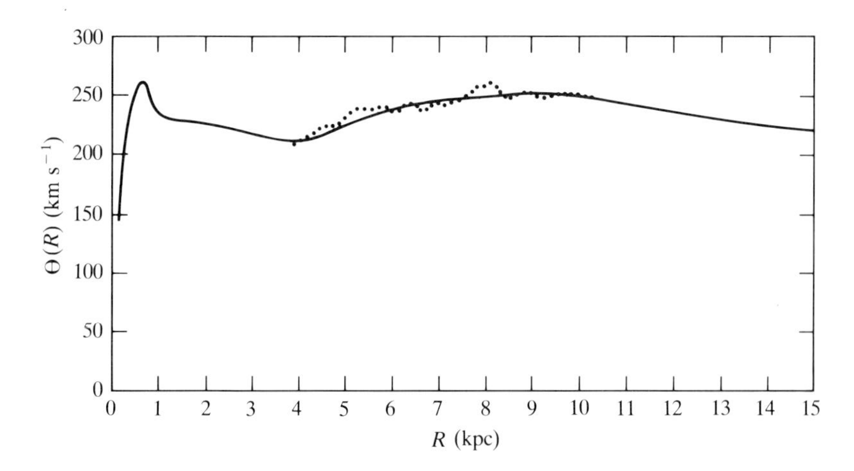

Figure 8-13. The rotation curve $\Theta(R)$ for the inner parts of our Galaxy as derived from 21-cm observations by W. W. Shane and G. P. Bieger-Smith (**S4**). Individual data points are plotted as dots, and the smooth curve is from dynamical models. [From (**B15**). Reproduced with permission from the *Annual Review of Astronomy and Astrophysics*, Volume 14. Copyright © 1976, Annual Reviews, Inc.]

such fluctuations would imply an implausible galactic mass distribution and because they are in any case kinematically implausible (see Figure 8-14). We therefore discard the first hypothesis. The second hypothesis was invoked in the older analysis by Kwee et al. (**K6**), who argued that, if there were regions where there happened to be no hydrogen at the tangent point, then we would obviously not receive radiation from that point, and we would therefore systematically underestimate the maximum $\omega(R)$ along that line of sight. On this hypothesis, the true rotation curve is to be drawn through the highest observed points. Subsequent work has shown, however, that this explanation cannot be correct. Low-intensity extensions of the 21-cm profile up to the presumed upper envelope velocity are never observed at the longitudes where the dips in the velocity curve occur. The complete absence of such features requires that the density contrast between the emitting and empty regions would have to be about 100 to 1, which is implausibly high. Furthermore, the empty regions would have to span a velocity range of 10 km s^{-1} or so along the line of sight, which would mean that regions that are of 4 to 5 kpc in extent and are preferentially oriented with respect to the Sun would have to be completely devoid of hydrogen. Such a distribution of the hydrogen in the disk is unacceptably artificial.

The only reasonable conclusion is that there are local irregularities in the velocity field of the gas, which have amplitudes of the order of 10 km s^{-1}. In fact, in at least one instance, rather good evidence has been presented

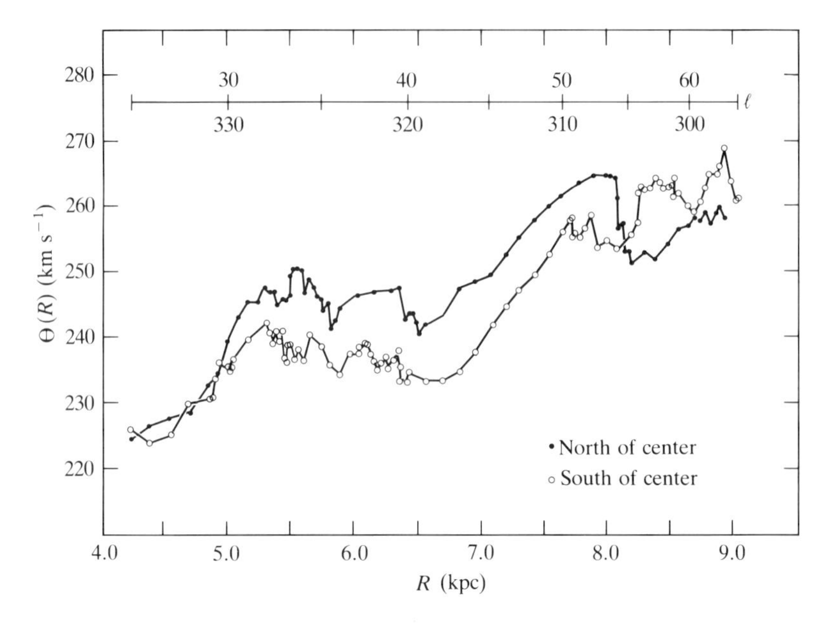

Figure 8-14. Comparison of the galactic rotation laws derived from Northern-Hemisphere ($25° \leq \ell \leq 60°$) and Southern-Hemisphere ($300° \leq \ell \leq 335°$) 21-cm observations, showing local fluctuations and large-scale asymmetry. [From (**K2**, 81).]

to show that a bump in the rotation curve is produced by systematic streaming motions of the gas adjacent to a major spiral arm (**B13**). Further support for this view is provided by the large-scale asymmetry between the rotation curves observed in the northern and southern hemispheres, as shown in Figure 8-14. We see there that the two curves have local irregularities at different places, and we also see that even smooth unperturbed curves drawn through the two sets of data differ systematically by about 8 km s^{-1} in the range $5 \leq R \leq 8$ kpc. In short, there is no alternative to accepting the existence of both small-scale irregularities and large-scale asymmetries in the velocity field, and hence noncircular motions, of the gas. Note that the observed fluctuations are only a few percent of the basic rotation speed, and therefore they are not likely to be too important dynamically. Indeed, the departures of the gas from pure circular motion in our Galaxy, which are probably the response of the gas to low-amplitude gravity waves, seem a good deal smaller than those often observed in external galaxies. Nevertheless, as we shall see in §9-1, even such small departures severely interfere with the mapping of the hydrogen distribution.

For $R < 4$ kpc, the situation becomes more complicated. At $R \approx 3.7$ kpc (assuming $R_0 = 10$ kpc), G. W. Rougoor and J. H. Oort (**R6**), (**R7**) discovered the *3-kpc expanding arm*, which rotates with a velocity of about 50 km s^{-1}.

The dynamics of this feature are not yet fully understood, but large-scale expansion of the gas may be a general phenomenon in the range $1 \lesssim R \lesssim 4$ kpc. Inside this region, Rougoor and Oort discovered a thin *nuclear disk*, which seems to be in pure circular rotation. The data indicate that there is an *inner disk* having a radius of about 300 pc and a rotation speed of about 200 km s^{-1}, surrounded by a *ring* having a radius of about 750 pc and a rotation speed of about 265 km s^{-1}. There is no sign of expansional motion in either the disk or the ring. The rotation velocities observed in the nuclear disk agree closely with the circular velocities derived from the mass distribution in the galactic nucleus, as inferred from infrared observations in our Galaxy itself and by comparison with the nucleus of M31 (**O2**).

The principal result we can obtain from the rotation curve of our Galaxy is an estimate of gravitational forces in the galactic plane. It is thus natural to ask whether the observed motions of H I clouds do indeed represent mainly these forces, or whether they reflect the action of nongravitational forces and hence differ systematically from the motions of stars. There are several lines of evidence that indicate that H I velocities are, in fact, consistent with those of stars and of other constitutents of the interstellar medium. Solutions for the solar motion with respect to local H I give results quite close to the standard solar motion relative to common stars (**K1**). The rotation curve derived from observations of radio recombination lines in H II regions is very similar to the H I rotation curve (**M1**), and, although H II regions do have individual peculiar velocities, there is no systematic difference between the motions of H I and H II regions in excess for a few kilometers per second (**B4**, 101), (**M3**). Similar results are found for interstellar Ca II lines. In sum, the basic result is that the H I and Ca II clouds and H II regions (and hence their exciting stars) all appear to move essentially together on nearly circular orbits like those of spiral-arm and young disk-population stars.

The Rotation Curve for $R > R_0$ from Optical Observations

Unfortunately, no very satisfactory method exists for determining the galactic rotation law beyond $R = R_0$. However, several approaches indicate that the galactic rotation curve does not decline interior to $R \approx 20$ kpc and possibly not even then.

If one knows the heliocentric distance d of an object in the outer galactic disk, one can determine its galactocentric distance R from a knowledge of its galactic longitude ℓ, R_0, and the law of cosines. Then, if one can further determine the object's radial velocity v_R, one can find $\omega(R)$ from equation (8-4). Combining results from a number of different objects, one can in this way build up information about the rotation curve out to $R \approx 15$ kpc.

An early application to this technique was that of R. P. Kraft and M. Schmidt (**K4**), who studied Cepheid variables. They determined the absolute magnitude of the stars from the period-color-luminosity relation, and then they obtained their distances from their apparent magnitudes and estimates of the interstellar extinction between them and the Sun. The rotation curve derived by Kraft and Schmidt, which extended to $R = 12$ kpc, was rather flat.

Recently, L. Blitz (**B10**) has measured the galactic rotation curve out to $R = 15$ kpc by observing young stars and molecular clouds associated with H II regions. Blitz estimates the distances to his H II regions by fitting their young stars to the ZAMS. He obtains the velocities of the H II regions from the Doppler shift of line emission at millimeter wavelengths from the molecular clouds (see §9-3). In this way, Blitz derives a rotation curve which *rises* by about 20 km s^{-1} between R_0 and 15 kpc.

The conclusion of Blitz that the galactic rotation velocity is still rising at the solar radius is very striking, for it implies that much of the matter in our Galaxy lies beyond the Sun, whereas most of the light is probably emitted from within the Sun. However, two entirely different studies confirm that there *is* a great deal of material beyond the Sun.

G. R. Knapp et al. (**K3**) have used observations of 21-cm emission near $\ell = 90°$ to derive a limit on how rapidly the galactic rotation curve might fall. When one looks out toward $\ell = 90°$, one's line of sight passes through material of ever greater negative radial velocity; the Sun is moving in that direction at speed Θ_0, whereas the material along that line of sight has part of its velocity directed across the line of sight. Consequently, we appear to be approaching material that lies along $\ell = 90°$ at $R > R_0$. If there were gas at, say, $R = 40$ kpc, its velocity would be almost entirely directed across the line of sight, with the result that we would be approaching this material at velocity $v_R \approx -\Theta_0$. Knapp et al. have searched for such high velocity gas in the hope of thus determining Θ_0; they found no gas with $-v_R > 160$ km s^{-1}. This dearth of material at $\ell = 90°$ with large negative radial velocity tells us that our Galaxy cannot possess a large neutral hydrogen disk like those surrounding many other spiral galaxies (see §9-2) and that the galactic rotation velocity falls slowly, if at all, outside the solar radius. The more narrowly the galactic H I is confined, the weaker is the constraint one can place on any decline in the rotation velocity from the absence of material with $-v_R > 160$ km s^{-1}. The approach adopted by Knapp et al. is to suppose that the surface density $\Sigma(R)$ of H I beyond R_0 falls off exponentially as $\Sigma(R) = \Sigma_0 \exp[(R_0 - R)/\lambda R_0]$. Observations of other galaxies indicate that the decline of Σ is unlikely to be steeper than this. The scale length λR_0 of the decline in Σ can be determined by the requirement that the total neutral hydrogen emission outside R_0 match the observed emission at $\ell = 90°$ integrated over all velocities; $\lambda = 0.4$ fits the data well. Then one finds that the absence of material at $-v_R > 160$ km s^{-1} requires either (1) $\Theta_0 \approx 250$ km s^{-1} and a rising rotation curve beyond R_0, or (2) $\Theta_0 \approx 220$ km s^{-1} and a flat rotation curve, or (3) $\Theta_0 < 200$ km s^{-1} and Θ declines beyond the Sun as $\Theta \sim R^{-1/2}$, as it must if the galactic mass is confined to $R < R_0$. We saw in Chapter 6 that Θ_0 is unlikely to lie much below 220 km s^{-1}. Therefore, one may conclude that the galactic rotation curve is not falling near the Sun, and it may well be rising.

Still another indication that the galactic rotation curve is flat or rising beyond the Sun, and thus that much of the galactic mass lies beyond R_0, comes from an analysis of the motions of distant globular clusters and dwarf galaxies by F. D. A. Hartwick and W. L. W. Sargent (**H2**). These systems move in, and are bound by, the gravitational field of our Galaxy, and hence it is intuitively clear that, the faster they are observed to be moving, the stronger must be our Galaxy's gravitational pull, and the greater our Galaxy's mass. Hartwick and Sargent find that the random motions of these systems do not decline with increasing galactocentric distance, and thus they conclude that the force field of our Galaxy, and hence the galactic rotation curve, do not decline sensibly out to $R \gtrsim 20$ kpc (see Chapter 21).

8-4. ROTATION LAWS OF OTHER GALAXIES

The rotation of external galaxies was first detected by V. M. Slipher (**S6**) and M. Wolf (**W2**) through measurements of line shifts in galaxian spectra. This detection was a full decade before Hubble proved conclusively that the spiral "nebulae" are, in fact, galaxies. Since that time, numerous measurements of rotation curves of galaxies have been made by means of optical observations of Doppler shifts in either the absorption-line spectrum of the stellar component or the emission-line spectrum of the ionized-gas component (H II regions), and at radio wavelengths by measuring Doppler shifts in 21-cm line emission from H I.

Historically, the early determinations of galaxian rotation were based on absorption-line observations of the central regions of disk galaxies. As we shall see later, velocity determinations based on the measurement of absorption-line spectra are very difficult even today, and it was only by Herculean efforts that the pioneers in this field were able to chart the rotation of the innermost parts of a few nearby disk galaxies from absorption spectra.

In order to obtain estimates of rotation beyond the central regions, spectra were obtained of emission regions. The emission spectra of these regions are easier to observe than are absorption spectra, because emission lines are typically narrow and can be detected on short-exposure spectra on which the continuum (against which absorption lines have to be measured) does not even register. However, with the equipment then available, and for the very nearby galaxies originally studied, it was not possible to obtain the velocities of many H II regions from the same spectrogram. Consequently, prodigious amounts of telescope time were required for the production of a single rotation curve, and the body of observational material in this area grew extremely slowly. One of the classic examples of this work is the study of rotation in M31 which H. W. Babcock (**B2**) published in 1939.

In the case of galaxies of relatively small angular size, the velocities of many H II regions can be determined simultaneously from a single exposure with a spectrograph that has a long slit, which can be judiciously placed across the galaxian image. During the 1960s, the Burbidges observed emission regions in this way to chart rotation in about a dozen relatively distant galaxies; references to these papers and to earlier rotation determinations will be found in (**B12**). Most recently, Fabry-Perot techniques have been developed which make possible the simultaneous measurement of the velocities of H II regions distributed over the entire two-dimensional field of a galaxy (**C1**).

The commissioning of sensitive 21-cm radio telescopes from about the middle 1960s, and especially the introduction around 1970 of large aperture-synthesis telescopes, permitted two important developments in the area of rotation measurements. (1) It became possible to follow galaxian rotation much farther out than is possible by observations of H II regions; as we shall see in §9-1, the density of neutral hydrogen in disk galaxies falls off less steeply than do the densities of stars or H II regions. (2) The velocity resolution (≈ 10 km s^{-1}) possible with H I observations is considerably better than that which had, up to that time, been obtainable optically. Also, the velocity field can be determined from 21-cm observations throughout the whole extent of a galaxian disk, rather than at the few isolated points where H II regions happen to be located. On the other hand, a major drawback of the early 21-cm observations was (and to some extent still is) that they offered only poor spatial resolution. Thus, because the angular resolution of a telescope having aperture D and operating at wavelength λ is $\approx(\lambda/D)$ rad, one needs $D \gtrsim 10(\lambda/\theta)$ if one is to obtain a reasonably sharp picture of the neutral-hydrogen structure of a galaxy having an apparent radius θ rad. For example, $\theta = 10$ arcmin for the nearby Sb galaxy M81, so one requires $D > 700$ m to obtain ten resolution elements along the major axis of this system. By contrast, much of the available 21-cm data derives from single-dish instruments about 100 m in diameter, which give substantially lower resolution. Only aperture-synthesis instruments can achieve effective apertures as large as 700 m, and these instruments are generally rather less sensitive than large single-dish instruments (which have larger collecting areas). We shall see later that the comparatively low spatial resolution characteristic of much 21-cm data gives rise to special problems of interpretation. When the Very Large Array becomes available for 21-cm line observations in a few years, it will become possible to observe neutral hydrogen with a spatial resolution comparable to that characteristic of ground-based optical studies.

Very recently, there has been a revival of interest in rotation curves obtained from optical absorption-line spectra because this is the only information available for early-type systems, which do not have prominent emission regions. Until the publication of the rotation curve of the innermost 3 kpc of NGC 4697 by F. Bertola and M. Capaccioli (**B5**) in 1975, only disk-galaxy rotation curves were available. However, it turns out that early-type galaxies have fundamentally different rotation properties from disk galaxies. This late date of appearance for the first rotation curve of an early-type galaxy reflects clearly the fact that absorption-line work is, as we have mentioned, substantially more difficult than the measurement of emission spectra. The lines whose positions and widths one has to determine are intrinsically broad, and they are often blended with neighboring weaker lines whose presence can shift the effective central wavelength of the combined feature. To make matters worse, there are no nearby very luminous early-type galaxies, so the objects one wishes to study are always rather faint. In the face of these difficulties, reliable velocity measurements can only be made with very high quality spectra, such as could not be readily obtained before the advent of modern sensitive photographic emulsions and two-dimensional linear light detectors.

R. Minkowski (**M4**) pioneered the study of the absorption-line spectra of early-type galaxies by estimating central velocity dispersions and central rotation gradients for about a dozen galaxies. However, at that time it was not possible to obtain high quality spectra of low surface-brightness regions beyond the cores of early-type galaxies, and therefore it was impossible to chart rotation in these systems outside the central region of "solid-body" rotation (see the material that follows on the geometry of disk galaxies). Also, there was until recently considerable uncertainty as to how one should interpret spectral information in terms of rotation velocities and velocity dispersions. This situation has changed radically in the last few years, and we now have rotation curves and velocity dispersions for about two dozen elliptical galaxies. Before we review these results, we shall discuss in more detail the methods and results of studies of the rotation curves of disk galaxies, which are in many ways simpler to comprehend than those for early-type systems.

Disk Galaxies

Geometry Suppose that we observe a galaxy having a well-defined disk in a plane inclined at an angle i to the plane of the sky. Assume that the system rotates about an axis that is perpendicular to the galaxian plane and thus makes an angle i with the line of sight. Choose polar coordinate systems (R, θ) in the plane of the galaxy and (ρ, ϕ) in the plane of the sky, locating the origin at the point where the rotation axis pierces the plane of the sky, and measuring both angles from the line of nodes. Then, by inspection of Figure 8-15, one can see that the radial velocity measured at (ρ, ϕ) will be

$$v_R(\rho, \phi) = v_0 + \Pi(R, \theta) \sin \theta \sin i + \Theta(R, \theta) \cos \theta \sin i + Z(R, \theta) \cos i \quad (8\text{-}60)$$

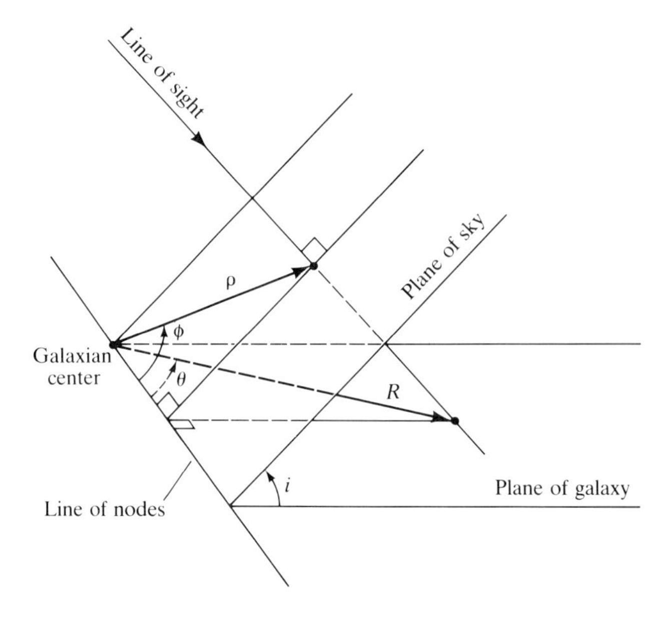

Figure 8-15. Geometric relationship of the coordinates (R, θ) in the plane of a galaxy to the coordinates (ρ, ϕ) in the plane of the sky. The galaxy is inclined at an angle i to the sky; the angles θ and ϕ are measured from the line of nodes, and the radial distances R and ρ, from the galaxian center.

where

$$R^2 = \rho^2(\cos^2 \phi + \sec^2 i \sin^2 \phi) \tag{8-61}$$

and

$$\tan \phi = \sec i \tan \theta \tag{8-62}$$

In equation (8-60), v_0 denotes the mean radial velocity of the system as a whole (that is, of its center of mass), while Π, Θ, and Z are the radial and tangential velocities in the plane and the velocity perpendicular to the plane, respectively. It is obvious that, in the most general case, we cannot hope to deduce three velocity components from only one measured component, and we must therefore impose further constraints in the analysis. In particular, if we assume *axial symmetry* about the center of the galaxy, then Π, Θ, and Z will be functions of R alone, and, at least in principle, we could infer all three from a knowledge of $v_R(\rho, \phi)$ over the whole disk. In practice, we shall impose even more restrictive assumptions.

To convert observed radial velocities to intrinsic velocity components, we must know the angle of inclination i. If we know that the material in the galaxy lies in a disk having an overall axial symmetry (which may be at least approximately true for spirals), then we can estimate i as $i = \cos^{-1}(b/a)$, where b and a are the semiminor and semimajor axes of the ellipse that most closely fits the projected shape of the system. This formula, which is valid for perfectly flat systems, underestimates i for nearly edge-on systems ($i \approx 90^\circ$) having a finite thickness.

It is obvious from equation (8-60) that we can obtain no information about the rotation of galaxies that are seen face on ($i \approx 0^\circ$). In general, we cannot choose an edge-on ($i \approx 90^\circ$) orientation either because, if the system has a dust layer, the line of sight will then penetrate only a small distance into the disk, and we shall measure only the radial velocity of the edge of the system (like the rim of a wheel). If, on the other hand, the system is dust free, at edge-on orientation the line of sight will penetrate essentially all the way through the disk and thus sample a wide range of velocities; this may make interpretation of the observations difficult. In practice, one must effect a compromise and choose systems that are sufficiently open to assure that each line of sight can be identified with a definite point on the disk, but are at the same time sufficiently inclined to have a substantial component of the rotation velocity projected along the line of sight.

Rotation is generally the dominant form of motion in disk galaxies, so let us set Π and Z in equation (8-60) equal to zero and investigate how the equal-velocity contours describing $v_R(\rho, \phi)$ would look in the plane of the sky. In the case of pure circular rotation, $\Theta = \Theta(R)$ will actually depend on R alone. Figure 8-16 shows a typical rotation curve $\Theta(R)$. Near the center, the rotation velocity increases outward; this region is called the region of *solid-body rotation* because $\Theta \propto R$ for a spinning solid body.

Figure 8-17 shows contours of constant v_R in the (ρ, ϕ) plane for a galaxy whose rotation curve is like that shown in Figure 8-16. Notice that the equal velocity contours in the region of solid-body rotation run straight and parallel to the apparent minor axis of the galaxy. The decline in the rotation curve of Figure 8-16 beyond R_{max} gives rise in Figure 8-17 to closed contours that are elongated along the apparent major axis of the galaxy. If the rotation curve of Figure 8-16 had not peaked but had gone on rising as some rotation curves do, or if it had risen to some constant value, there would be no closed contours in Figure 8-17. Notice that it is possible to determine from Figure 8-17 the directions of both the major and minor axes of the galaxy. The minor axis is the locus of points having the same velocity as the nucleus of the system (called the *systemic velocity*), whereas the major axis is defined by the two curves which run from the nucleus in the directions perpendicular to the local constant-velocity contours. In Figure 8-17, the major and minor axes determined in this way are identical with the apparent optical axes. However, if we had not set $\Pi = Z = 0$ in equation (8-60), the principal

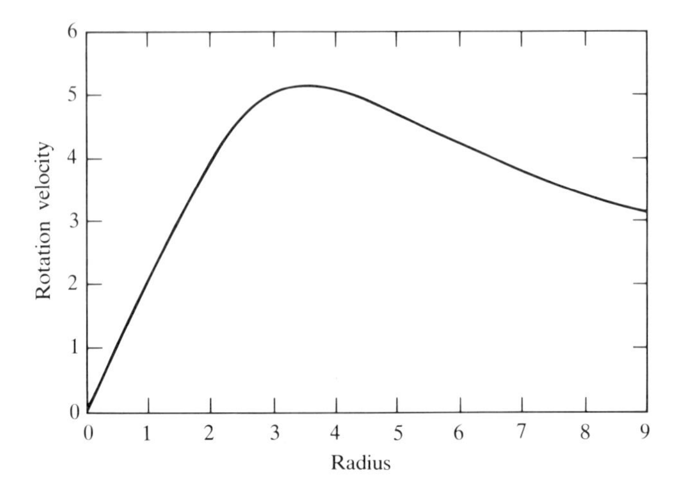

Figure 8-16. Model galaxian rotation-velocity curve; both the ordinate and the abscissa are in arbitrary units. [From (**V5**, Chapter 11), by permission.]

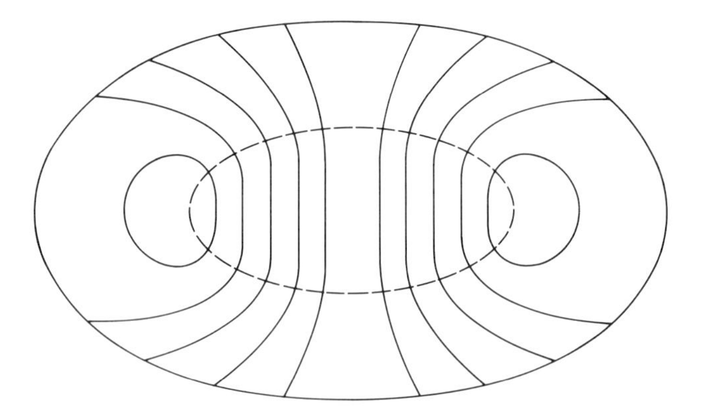

Figure 8-17. Contours of constant radial velocity for a galaxy rotating as shown in Figure 8-16. The region of solid-body rotation ($R < 3$ units) is enclosed by the dashed ellipse. [From (**V5**, Chapter 11), by permission.]

axes determined from the contour maps of v_R would not have coincided with those determined optically. Therefore, it is useful to distinguish between the *kinematic principal axes*, which are determined from velocity measurements, and the *optical principal axes*, which are derived from the brightness distribution. In fact, it has been found from 21-cm studies that, in many galaxies, the kinematic principal axes are appreciably displaced from the optical axes, by which we infer that the terms Π and Z in equation (8-60) cannot always be put equal to zero. But, before we discuss these details, let us decide how to derive the dominant term in equation (8-60), that is, the rotation term $\Theta(R)$.

Rotation Curves from Optical Data The determination of rotation curves from optical data is comparatively straightforward. One estimates the inclination i of the galaxy from the apparent principal axis lengths a and b of the system through $i = \cos^{-1}(b/a)$, and then one has, for the velocity v_k of the kth H II region, that

$$\Theta(R_k) = \frac{v_k - v_0}{\cos\theta_k \sin i} \tag{8-63}$$

where R_k and θ_k are given in terms of the coordinates ρ_k and ϕ_k of the H II region on the sky by equations (8-61) and (8-62). One may then plot Θ_k against R_k for each H II region and join neighboring points to form a smooth curve. Figure 8-18 shows a number of rotation curves derived in this way. Notice two points about this figure. (1) The curves tend to fall slowly, if at all, at large radii. This fact indicates that there is a good deal of matter in these galaxies that is located more than 10 kpc from the centers of the

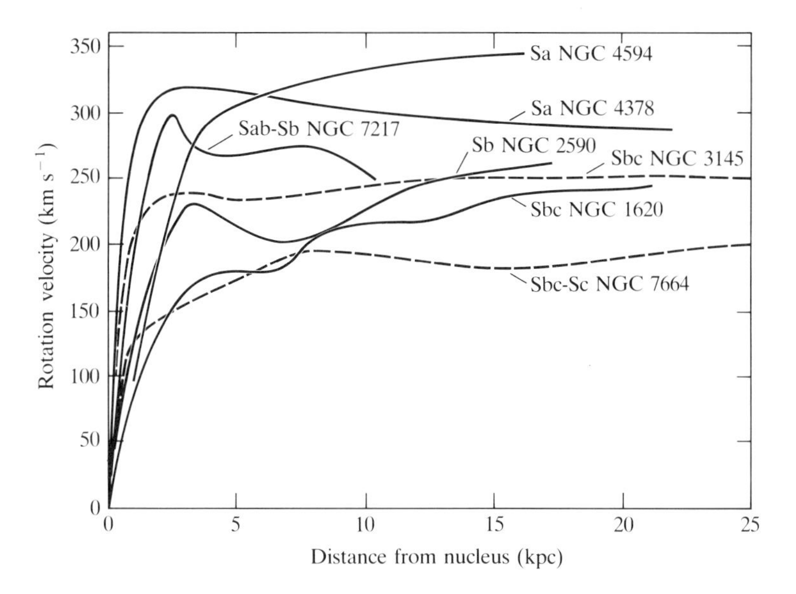

Figure 8-18. Rotation curves of spiral galaxies, obtained from optical measurements. [From (**R8**), by permission. Copyright © 1978 by the American Astronomical Society.]

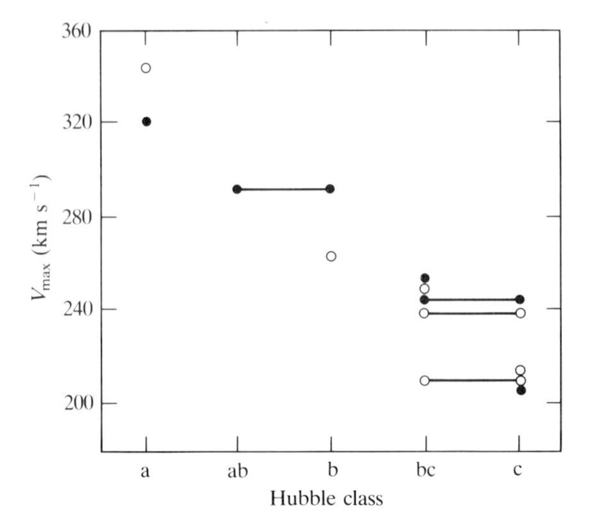

Figure 8-19. Early-type spirals tend to rotate faster than late-type spirals. Galaxies with uncertain classifications are shown as two points joined by a line. [From (**R8**), by permission. Copyright © 1978 by the American Astronomical Society.]

systems. A simple dynamical argument (see Chapter 12) shows that if $\Theta \approx$ constant, then the matter density ρ falls off as $\rho(R) \approx R^{-2}$, rather than following the steeper falloff characteristic of the light distribution (see §5-2). We shall see later that 21-cm velocity measurements reinforce this surprising conclusion. (2) The peak rotation velocity, Θ_{max}, is correlated with Hubble type in the sense that early-type spirals, Sa and Sb, have larger Θ_{max} than do Sc galaxies. The strength of this correlation emerges clearly from Figure 8-19, in which Θ_{max} is plotted against Hubble-type t (defined in Table 5-2).

Rotation Curves from 21-cm Data The interpretation of 21-cm observations of disk galaxies is more complicated than it is for the optical data. This is due to the fact that 21-cm observations lack the spatial resolution of optical measurements, so that interpreting 21-cm data can be likened to trying to make out the structure of a galaxy from a plate which is badly out of focus. Observations made with an aperture-synthesis telescope yield a series of *channel maps*; each channel map is a map of the intensity received at every point in the field in a certain narrow range of frequencies. By the Doppler effect, each such range of frequencies (that is, *channel*) corresponds to a range of velocities. A typical velocity range might be 20 km s^{-1} wide, and from 25 to 30 channels might be employed to ensure that all gas in the galaxy, no matter what its heliocentric velocity, will register in at least one channel. Figure 8-20 shows 25 channel maps obtained by the Westerbork

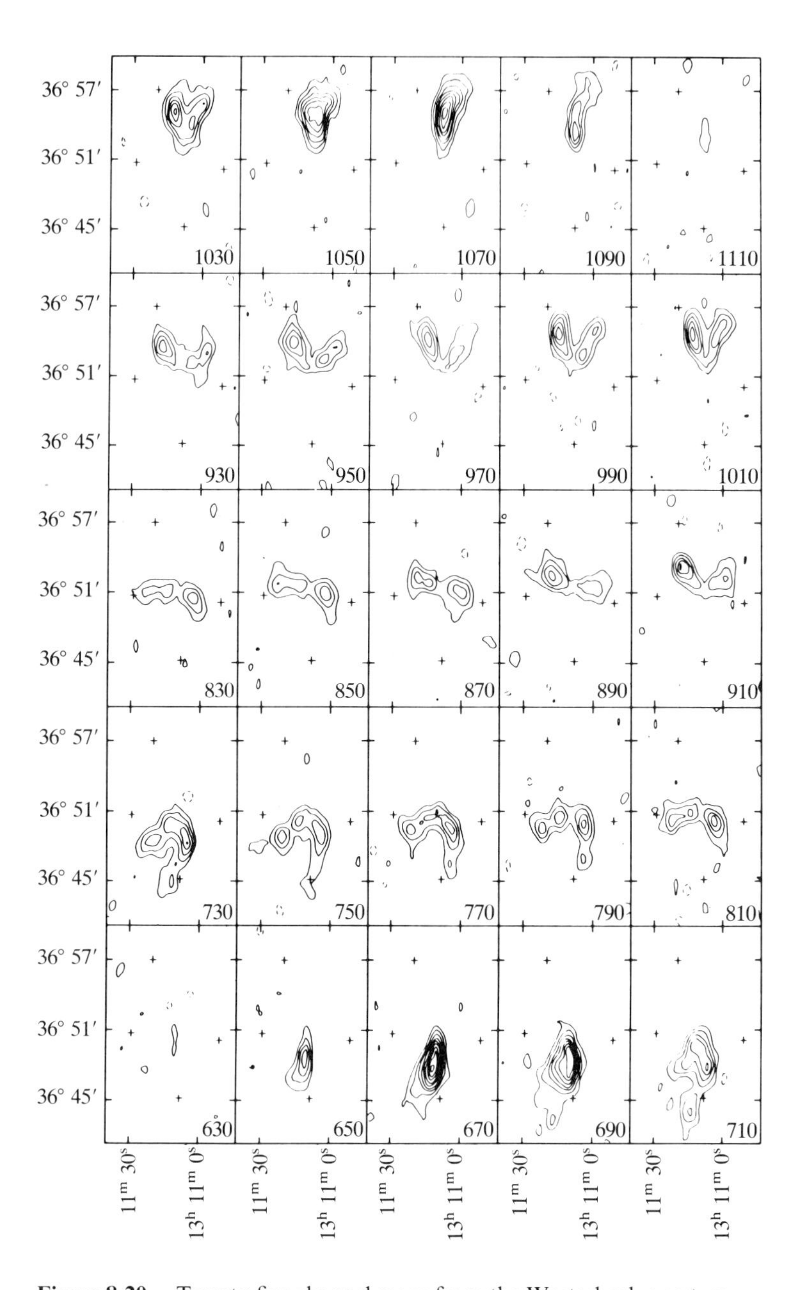

Figure 8-20. Twenty-five channel maps from the Westerbork aperture-synthesis telescope for the Sc galaxy NGC 5033. Each channel map gives the brightness temperature measured through a 129-kHz-wide filter centered on the frequency corresponding to the heliocentric velocity quoted in kilometers per second at bottom right. The sequence of maps should be read from top right to bottom left. Notice the asymmetry of many of the profiles; this is evidence that the disk of NGC 5033 is warped. [From (**B11**).]

aperture-synthesis telescope for the Sc galaxy NGC 5033. At the bottom right-hand corner of each map is given the heliocentric velocity corresponding to the middle of the channel associated with that particular map. Consider now how one might derive for this galaxy the H I surface density and a contour map of the lines of constant velocity like that shown in Figure 8-17.

The first step is to isolate the 21-cm line radiation from the continuum radiation. In this step, one examines each spectrum to decide which channels are line free and then subtracts the average intensity of these continuum channels from each of the channels that contain line radiation. The next step involves assigning a typical velocity to each point in the galaxy by identifying the frequency at which the line radiation has peak intensity, or some other suitably chosen central frequency. Finally, one integrates over frequency the total line emission from each point to determine, via equation (8-54), the column density at that point on the sky. A glance at Figure 8-21 shows that each of these steps is subject to an uncertainty that increases as the signal-to-noise ratio of the spectrum declines; an account of these difficulties will be found in (**B11**). When line centers and column densities have been assigned in this way to a grid of points on the sky, it is straightforward to construct H I density and H I velocity contour maps such as those shown for NGC 5033 and M81 in Figures 8-22, 8-23, and 8-24.

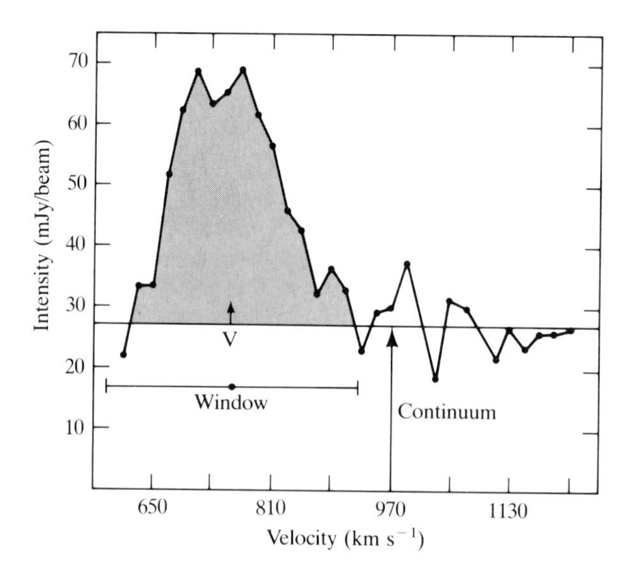

Figure 8-21. A typical velocity profile constructed from the channel maps shown in Figure 8-20. One possible estimate of the line center is marked V, and an estimate of the area of the profile associated with line emission is shaded. [From (**B11**).]

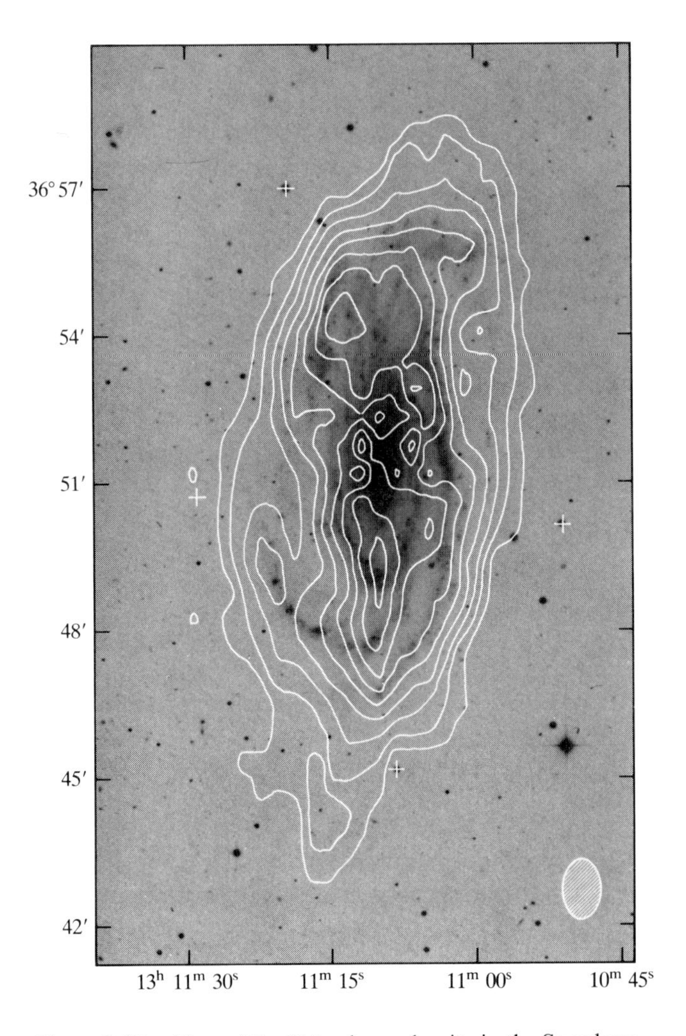

Figure 8-22. Map of the H I column density in the Sc galaxy NGC 5033 superposed on a photograph of the system. Notice that the gas appears to cover a much larger area than the optical galaxy. [From (**B11**).]

We must now consider how the rather poor spatial resolution typical of H I observations will be reflected in maps like those shown for NGC 5033 in Figures 8-22 and 8-23. When the beam is centered on a particular point (ρ, ϕ) on the sky, the system response at any frequency will depend on how much gas is moving with the corresponding velocity at all points throughout the beam. Therefore, the width of the velocity profile shown in Figure 8-21 does not indicate that there are elements of gas at a given point in the disk moving with radically different velocities. Rather, the width of the velocity

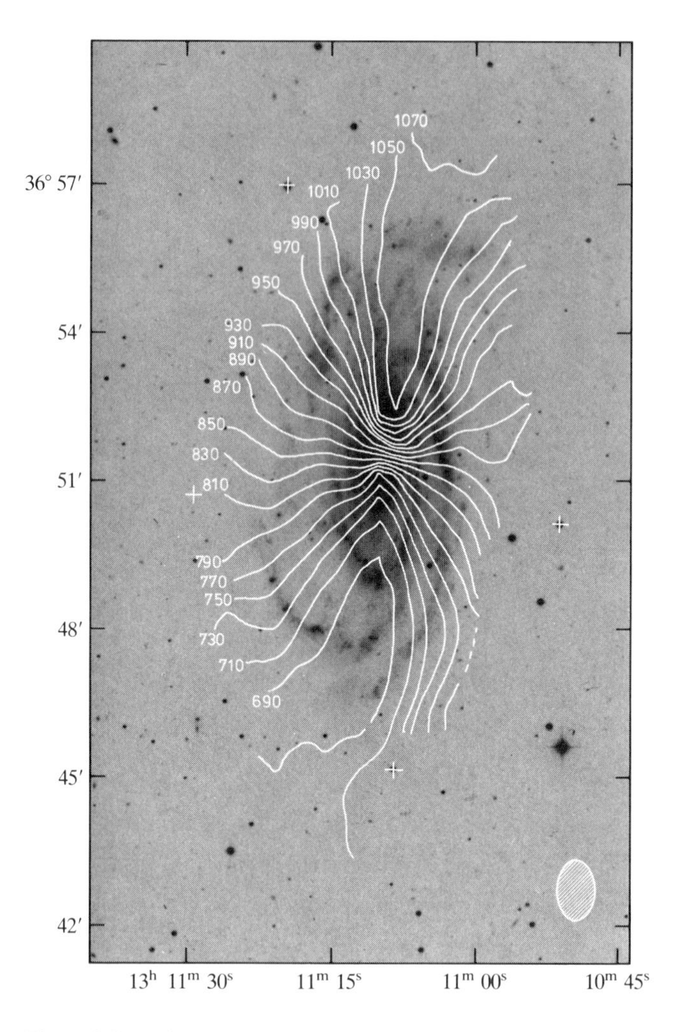

Figure 8-23. Contours of constant H I velocity in NGC 5033. Notice the curvature of the kinematic principal axes. [From (**B11**).]

profile arises primarily because there are radial-velocity gradients within the part of the disk that falls within the beam, so that, within the same beam, though at different places in the disk, gas is to be found over a range of velocities. The velocity assigned to (ρ, ϕ) is some sort of density-weighted average of the velocities within the beam. This average density may or may not be typical of the true velocity at that point, depending on the way the density variations in that region cause the different velocities within the beam to be weighted in the averaging procedure. Obviously, the column density assigned to (ρ, ϕ) will also be only a beam-averaged value.

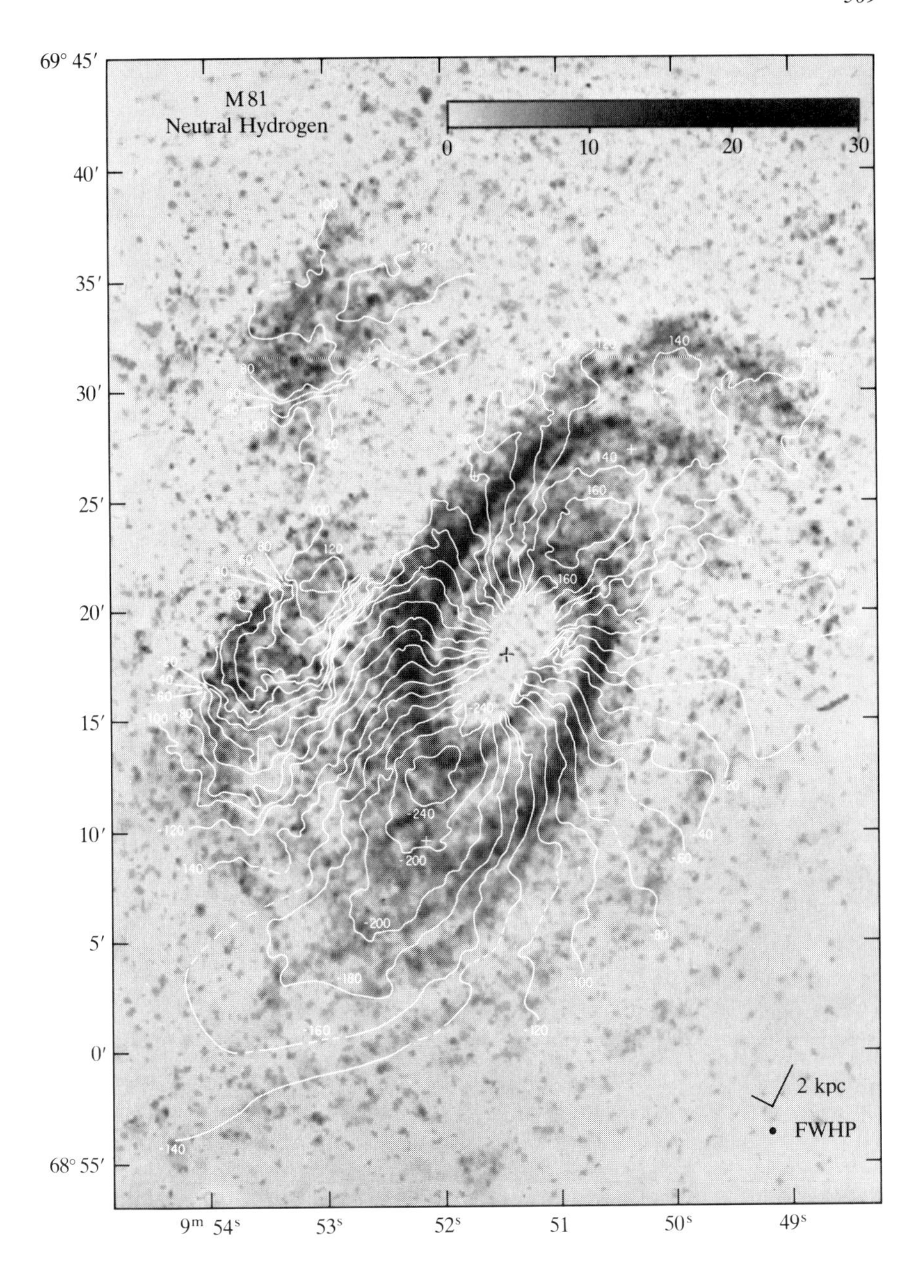

Figure 8-24. Radiograph made with the Westerbork aperture-synthesis telescope of the H I density distribution in M81 with radial velocity contours superimposed. The grey scale gives column densities in units of 10^{20} atoms cm^{-2}. Notice that in this high-resolution map, irregular motions associated with spiral arms are evident (see Figure 8-23). Notice also the central hole in the density distribution and the curvature of the kinematic principal axes. [From (**R5**), by permission.]

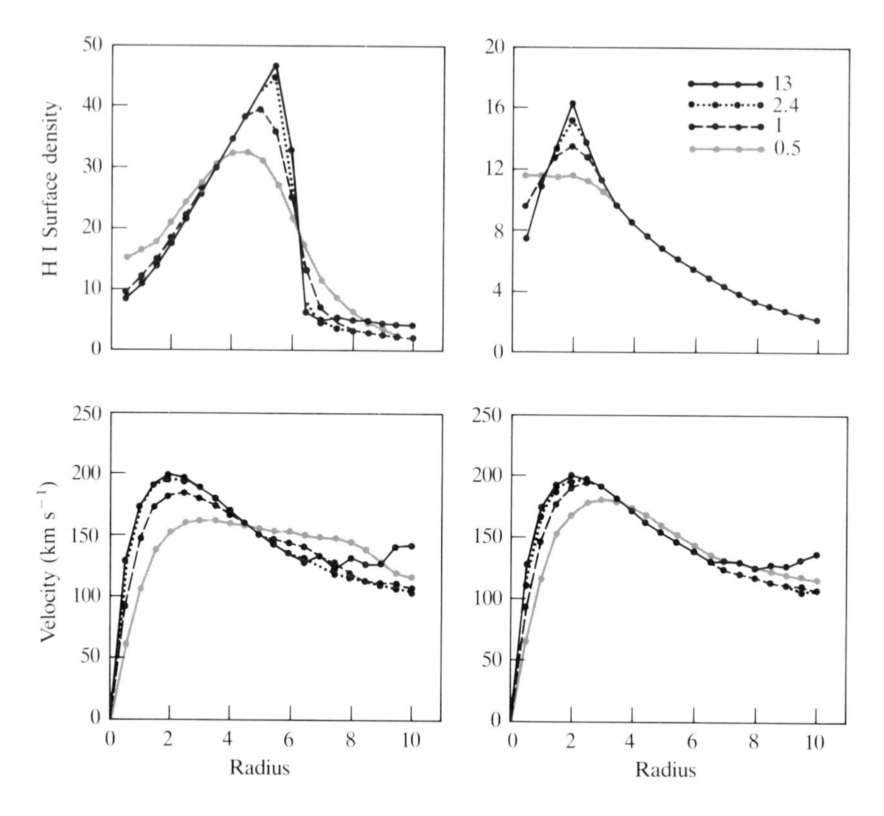

Figure 8-25. Radial distribution of $\sigma_H(R)$ and $v_c(R)$ derived for two model galaxies using "telescopes" of different resolutions. The resolution is defined in terms of the inverse of the half power beamwidth expressed in model scale lengths. [From (**B11**).]

Figure 8-25 illustrates how beam smearing alters radial H I surface density profiles and rotation curves. The curves plotted in the diagram are the H I surface-density curves and the H I rotation curves of two model galaxies, as reconstructed by the procedures described earlier from "data" obtained with radio telescopes, whose half-power beam widths are, respectively, $\frac{1}{13}$, $\frac{1}{2.4}$, $\frac{1}{1}$, and $\frac{1}{0.5}$ of the scale length of the model studied. Notice that both model galaxies have $R_{\max} = 2$ scale lengths, and that peaks in the gas density curves occur near 5 and 2 scale lengths, respectively. Therefore, if we compare the model shown in the two panels on the right of Figure 8-25 with our Galaxy, we have 1 kpc $\approx$ 1 scale length. Thus, from Figure 8-25, we see that if, for example, we wished to detect any central drop in the H I density of a galaxy like our Galaxy, we should require resolution at least as good as that furnished by the half-power beam width $\approx$ 1 kpc. And, even at this resolution, beam smearing effects would be far from negligible, particularly in relation to features like spiral arms, which might have structure on scales much smaller than 2 kpc.

In fact, data of 1-kpc resolution are at present available for only a minority of comparatively nearby systems. It is customary to measure the resolution of 21-cm studies of galaxies in terms of the ratio R_H/B of the Holmberg radius R_H (see §5-3) to the half power beam width B. [A better criterion would be the ratio r_s/B of the characteristic disk radius r_s, see equation (5-6), to B, because R_H is effectively defined in terms of the night-sky brightness, which clearly should not enter into the definition of radio resolving power. Unfortunately, r_s is known for only a few galaxies.] For example, M81 has $R_H = 10$ arcmin (Table 5-4). Therefore, 1-kpc resolution requires $B \approx$ 1 arcmin. As we saw earlier, only aperture-synthesis telescopes can achieve 1 arcmin resolution.

Low-resolution H I data are best interpreted by a model-fitting procedure in which one attempts to optimize the fit between predicted H I data derived from a suitably parametrized model galaxy and the actual observations. The model predictions can be fitted to the data either in their raw form (as channel maps or velocity profiles) or after reduction to H I density maps and H I velocity contour maps like those displayed in Figure 8-22 and 8-23. Unfortunately, one often finds that changes in the assumed radial H I density distribution and the assumed rotation curve can interact in such a way that equally satisfactory fits to the data can be obtained from more than one model. In such cases, it may be necessary to invoke additional criteria to select the "best" model. The average rotation curve $\Theta(R)$ is usually determined by fitting the data on the whole surface of the galaxy under study but giving highest weight to points on the major axis.

Once a best-fit rotation curve has been found, the theoretically predicted radial velocities induced by rotation alone can then be subtracted from the observed velocity field to give the *residual velocity* field. We shall discuss later the information one may extract from the residual velocity field about large-scale deviations of the motion of the gas from circularity and about small-scale noncircular motions produced, for example, by spiral arms or by tidal disturbance by satellite dwarf galaxies. In M31 and M33, for which excellent high resolution data are available, the velocities of the H II regions are in good systematic agreement with the H I velocities. Individual H II regions may depart from H I velocity at some point by 10–20 km s^{-1}, but this value is also the size of the velocity fluctuation from point to point within H II regions. Thus, as was true for our Galaxy, it appears that the neutral and ionized gas components (and presumably also the underlying stars that ionize the gas) of these galaxies move together.

Table 8-3 summarizes some important conclusions of a number of surveys of neutral hydrogen in disk galaxies. In Table 8-3, R_L denotes the limiting radius to which hydrogen has been detected, $R_{\max}$ denotes the radius at which the maximum rotation velocity is found, R_H is the Holmberg radius

Table 8-3. Properties of Galaxian Rotation Curves

Morphological type	$\langle R_L/R_H \rangle$	$\langle R_{max}/R_H \rangle$	$\langle \Theta_{max} \rangle$	Number
Sab	2	0.3	235 ± 15	1
Sb	2	0.51	260 ± 10	2
Sc	2.4	0.63	220 ± 15	2
Scd–Sd	2.6	0.83	130 ± 30	10
Sdm–Sm	2.5	1.1	90 ± 5	1
Im	4	2.1	55 ± 5	1

SOURCE: Data published in (**H4**) and (**K5**)

of the system, and Θ_{max} is the maximum observed rotation velocity. The number of systems used to calculate the mean values are also listed. The following conclusions may be drawn.

1. H I is often observed far beyond the optical radius of galaxies, and, as receivers of higher sensitivity become available, this conclusion is likely to be strengthened.
2. For early-type systems, the ratio (R_{max}/R_H) is small, and the rotation curve shows a clearly defined maximum well within the optically bright parts of the galaxy. In contrast, in later-type systems, the maximum in the rotation curve lies far out in the disk of the galaxy, and it is ill defined, often being part of a broad plateau.
3. The radio data confirm the conclusion we have already drawn from optical rotation curves (Figure 8-19), that Θ_{max} is essentially a monotonically decreasing function of advancing morphological type, but that the range in observed values of Θ_{max} is surprisingly small. The latter fact may provide an important clue concerning the origin of spiral structure.

In Figure 8-26, we show radio rotation curves for five thoroughly studied galaxies and for our Galaxy. From Figure 8-26, it is apparent that, although $\Theta(R)$ in some of the galaxies (for example, M81, NGC 3109, NGC 7217) shows a definite decline at large R, in many systems (for example, M31, IC 342), $\Theta(R)$ is essentially constant with increasing R out to the furthest observable point from the galaxian centers, and nothing like a Keplerian falloff is found. Thus radio observations confirm the conclusion we have already drawn from the optical data, that the rotation curves of many galaxies are remarkably flat at large radii. The particular importance of the neutral hydrogen observations is that they extend this result to very large radii, implying that these galaxies have a great deal of mass situated far out in their halos. The relative frequencies of occurrence of asymptotically flat or clearly declining rotation curves are poorly known. In some cases, different

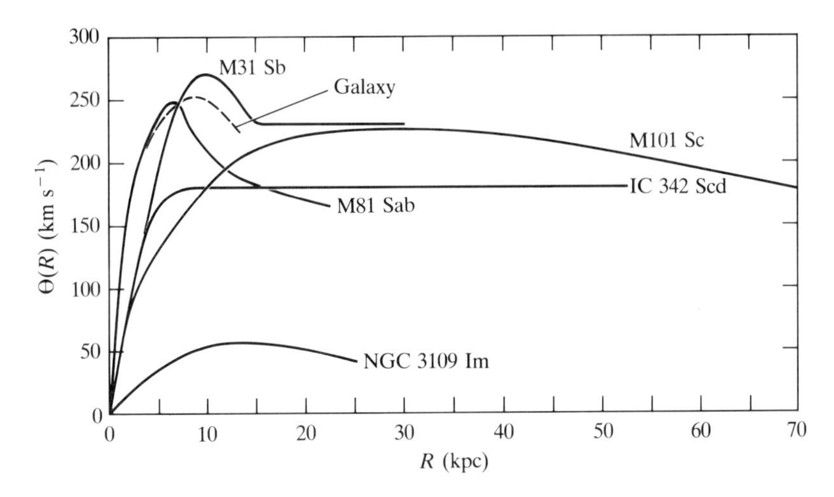

Figure 8-26. Rotation curves for spiral galaxies (including our Galaxy) from 21-cm line radio measurements. Notice that the rotation curves of M31 and IC 342 show no evidence for an outward decline and that we have no reliable information about our Galaxy's rotation curve beyond the solar radius. [Data from (**H3**, 331), (**H4**), (**R1**), and (**R5**).]

investigators draw conflicting conclusions about the same galaxy. M. S. Roberts found (**H3**, 331) that, out of fourteen galaxies he examined, at least seven showed an essentially constant rotation velocity for $R \lesssim R_H$. Of seventeen galaxies for which he compiled data, W. K. Huchtmeier (**H4**) found that at least seven showed a declining $\Theta(R)$ for large R, seven had apparently flat rotation curves, and in three cases the data were inconclusive.

Deviations from Circular Rotation We have seen that if galaxian disks are flat circular structures in which all the material (neutral hydrogen and H II regions) is on circular orbits about the galaxian centers, then the kinematical axes would form two perpendicular straight lines, and the whole velocity field would enjoy twofold rotation symmetry about the galaxian center (see Figure 8-17). Real H I velocity contour maps usually differ from this ideal configuration in one or more respects, and from this it follows that the structures of galaxian disks must be appreciably more complex than a simple plane, circularly symmetric model allows. Here we shall inquire how simple models of galaxian disks must be modified to account for these differences between the expected and observed H I velocity contour maps.

Symmetric Distortions We first discuss those observed H I velocity contour maps which are *twofold symmetric*, even though they deviate from the ideal model in other ways. The map of the velocity contours of M83 shown in Figure 8-27 is an example of a twofold symmetric map that differs markedly from the standard model. Notice how both the kinematic principal axes are

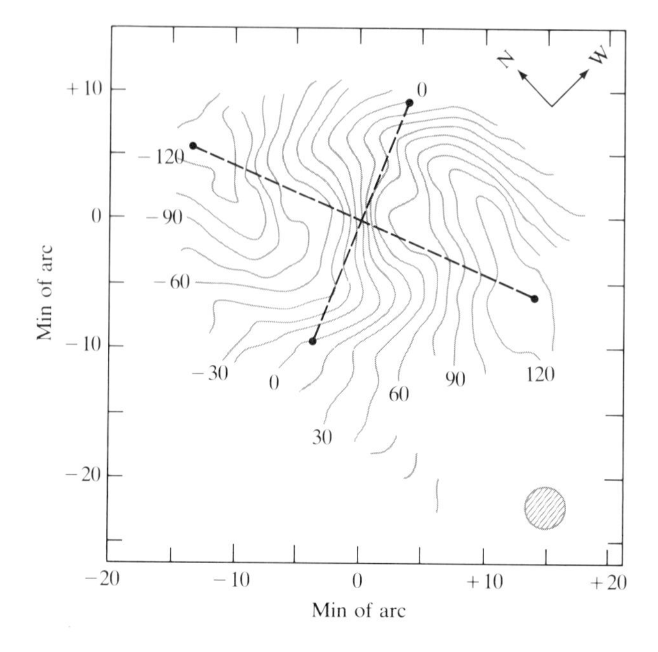

Figure 8-27. Contours of constant H I velocity for M83. Notice the severe s distortion of the kinematic principal axes, which nonetheless remain approximately perpendicular at equal true galactocentric distance. [From (**R2**), by permission. Copyright © 1974 by the American Astronomical Society.]

distorted into s shapes. Notice also how straight lines joining the nucleus of the galaxy to two points, one on the kinematic major axis and one on the kinematic minor axis, that are at roughly equal galactocentric distances, remain nearly perpendicular as the galactocentric distances of these points are varied. This type of velocity field results when the disk of a galaxy is *warped*; that is, the galaxian disk is not plane but is rather made up of a number of annular rings, each of which is tilted with respect to the others. Figure 8-28 depicts this state of affairs. An observer will obtain an H I velocity contour map to which each ring contributes an elliptical portion. If the material of each of the rings of Figure 8-28 is assumed to rotate on circular orbits within the plane of the annulus, the pattern of the contours within the ellipse associated with each ring will be like that within an elliptical annulus in Figure 8-17. Therefore, within this elliptical ring, the kinematic major and minor axes will be mutually perpendicular. But neighboring rings will have different orientations with respect to the observer's line of sight, with the result that the kinematic major axes of neighboring rings will not be exactly parallel. By suitably adjusting the orientation of the rings of material of different radii, one may obtain a warped disk model whose kin-

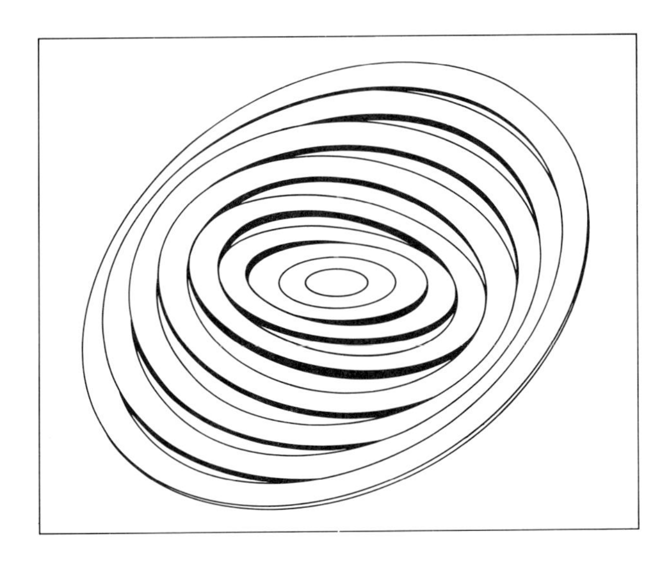

Figure 8-28. Tilted ring model of the warped disk of M83. [From **(R2)**, by permission. Copyright © 1974 by the American Astronomical Society.]

ematic major axis follows any predetermined curve. Figure 8-28 shows a warped disk model constructed to fit the observations of M83 shown in Figure 8-27.

There can be little doubt that warped disks like that of M83 are very common. Indeed, we shall see later that the disk of our Galaxy is known to be warped between 9 and 15 kpc from the galactic center, the disk of M31 is appreciably warped at large radii **(R1)**, and the third spiral galaxy in the Local Group, M33, has an extremely badly warped disk; that is, *all* the spiral galaxies in the well-studied Local Group have warped disks. Furthermore, the warping of the disk of M33 is so strong that some lines of sight from the Sun pass through the disk twice. Figure 8-29 is a schematic representation of the situation. Thus D. H. Rogstad et al. **(R4)** find that the velocity

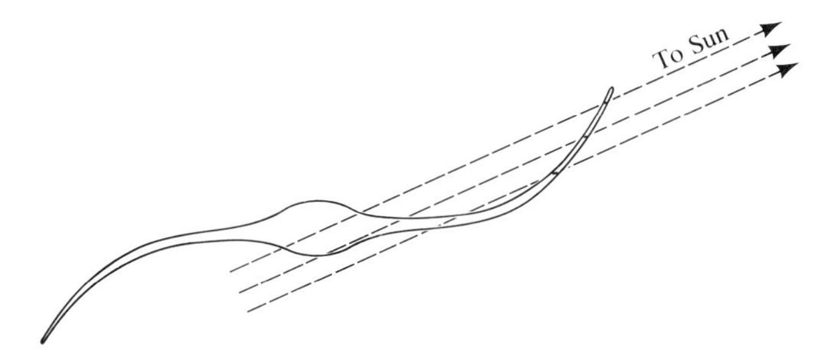

Figure 8-29. Some lines of sight from the Sun cut the warped disk of M33 twice.

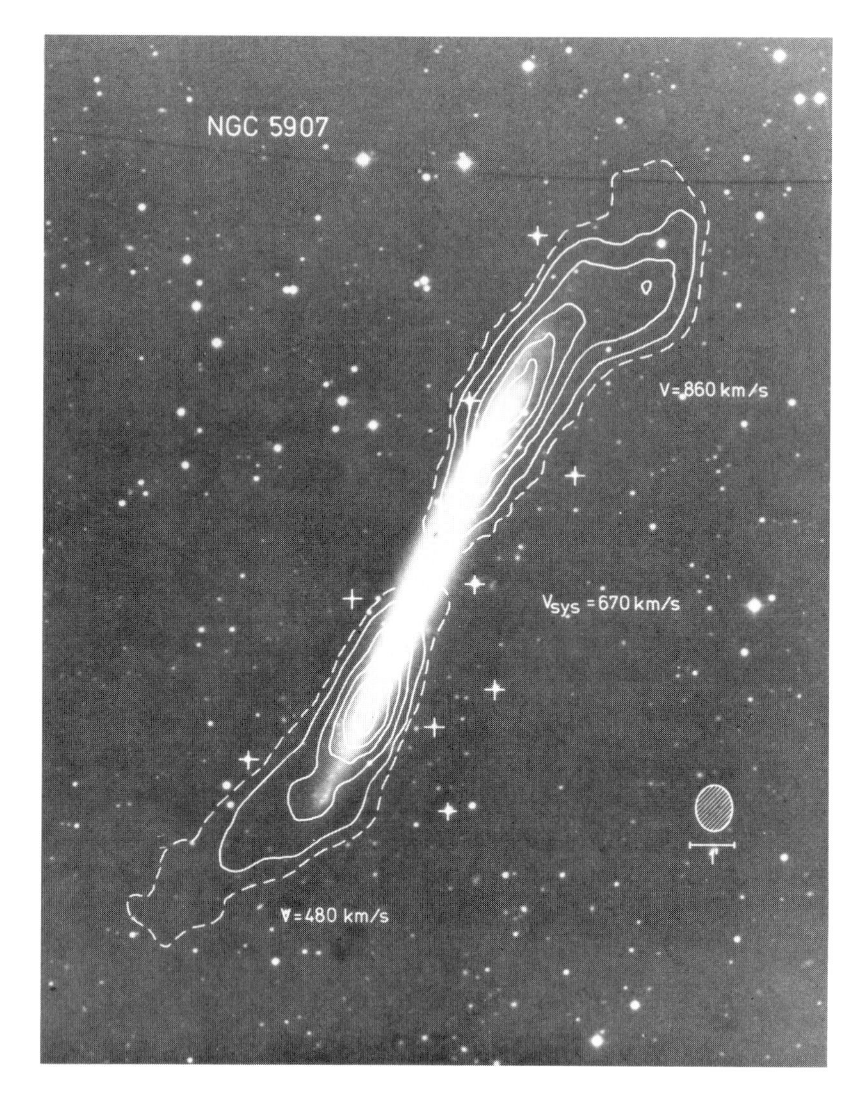

Figure 8-30. The warped neutral hydrogen disk of the nearly edge-on galaxy NGC 5907. Emission by gas that is moving near the systemic velocity of the galaxy has been supressed for clarity. NGC 5907 has no nearby neighbors that could have recently disturbed it tidally. [From (**S1**), by permission.]

profile associated with certain points in the disk of M33 show two peaks, rather than the single peak of the typical velocity profile displayed in Figure 8-21. One of these peaks they associate with the rapidly rotating gas in the main body of the disk, and the other they associate with the more slowly rotating material nearer the edge of the disk.

In addition to these inferred or *kinematic warps*, R. Sancisi (**S1**) has been able to show directly that the H I planes of many external galaxies are far from flat. For example, Figure 8-30 shows two channel maps from the Westerbork aperture-synthesis telescope of the near edge-on disk galaxy NGC 5907 superimposed on a photograph of the system. The velocities of the channels shown are such that only gas which is either at the top right or the bottom left of the galaxy registers. Clearly, the gas detected in this way does not lie in a plane. We shall see in Chapter 13 that the prevalence of warps in spiral galaxies is very hard to reconcile with the idea that most of the mass in these systems is concentrated in the optically bright disks. Thus the dynamical implications of galaxian warps are of great importance (**H5**).

Figure 8-31 shows the H I velocity contours of the barred spiral galaxy NGC 5383. This diagram is approximately twofold symmetric, but now the kinematic major and minor axes are by no means perpendicular. We have seen that this type of velocity field cannot be accounted for in terms of a warped disk model, but one can show that, if the lines of flow of the gas in a galaxy like NGC 5383 form oval shapes centered on the galaxian nucleus, then, for most observers, the kinematic minor axis of the galaxy will be inclined to the optical minor axis (**B7**). The kinematic major axis will, on the other hand, remain approximately parallel to the optical major axis (which is always perpendicular to the optical minor axis), with the result that most observers will conclude that the kinematic principal axes of an oval galaxian

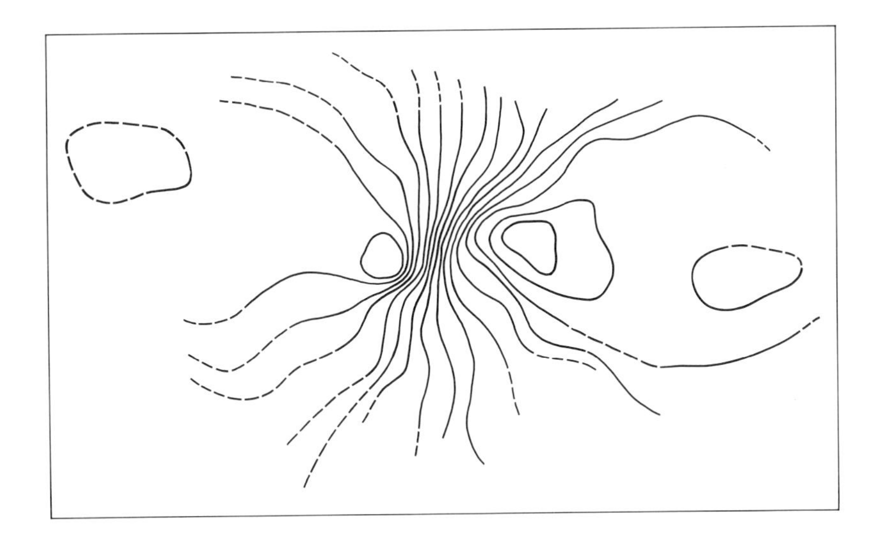

Figure 8-31. The neutral hydrogen velocity field of the SBb galaxy NGC 5383. Notice that, at the center, the kinematic principal axes are far from perpendicular. This is the signature of an oval disk. [From (**S2**) by permission.]

disk are not perpendicular. The H I velocity maps of many galaxies (for example, NGC 3198, NGC 4258, IC 342) suffer from this type of distortion, although the distortion is rarely as severe as it is in the case of NGC 5383 shown in Figure 8-31. It is possible that the motion of the gas in these disks is distorted into oval configurations by the gravitational field of some hot, triaxial component. We shall return to this important dynamical point in Chapter 13.

Asymmetric Distortions It is useful to divide those distortions which do not show twofold rotation symmetry about the galaxian center into three classes.

1. Large-scale asymmetries such as those exemplified by M101; the rotation curves one derives for M101 from the eastern and western kinematic principal axes have very different shapes. Undoubtedly, some large-scale asymmetries are to be ascribed to tidal disturbance by neighboring galaxies. Detailed models of some of the most spectacular systems of this type (for example, NGC 4676) have been successfully constructed by A. and J. Toomre (**T1**) and others.
2. Small-scale disturbances (possibly also of tidal origin) like those observed in M31 (**R1**); part of the disk of M31 has a very peculiar, even double-valued, velocity field, which may be due to disturbance by M32, the dwarf elliptical companion of M31. The gas in M81 may be similarly disturbed by NGC 3077. No entirely satisfactory model of this phenomenon is available, however.
3. Small-scale streaming motions produced by spiral arms; such streaming motions generally have amplitudes of about 10 km s^{-1}, and they have been observed in, for example, M81, M101, NGC 2403, and NGC 6946 (**R3**), (**R5**). For M81, the observed radial components of the streaming motion appear to be compatible with the predictions of a density wave theory; the tangential component is indeterminate because those motions cannot be separated observationally from rotational motions. In contrast, for M31, the observed motions seem to be incompatible with the density wave theory's predictions (**E1**). For yet other systems, for example, M33, no prominent streaming motions associated with spiral arms are observed at all (**W3**).

Early-Type Galaxies

In contrast to the relative wealth of information available about the rotation curves of spiral galaxies, the data concerning early-type galaxies, which lack H II regions or 21-cm line emission, and which therefore have to be studied by means of absorption-line observations, are fairly sparse. At the time of writing, no map of the systematic velocity field of an S0 or an E galaxy has become available, although we do now have rotation curves along various axes for a fair number of early-type systems. References to sources of information concerning the rotation of early-type galaxies can be found in (**B6**) and (**B8**). The few data that are available are of great importance because they indicate that elliptical galaxies are dynamically fundamentally different from all types of disk galaxies. Here we shall only summarize the available observations and very briefly indicate why they are so significant; a full discussion of these questions must be deferred until we develop the stellar dynamics of spheroidal components in Chapter 14.

In §5-3, we saw that there are photometric indications that elliptical galaxies may not be flattened, axially symmetric systems. If they were axially symmetric, then it would be a reasonable inference that rotation in these systems would take the form of rotation about the axis of symmetry, with the result that no velocity gradients would be detectible along their apparent minor axes. If, on the other hand, these systems are not axially symmetric, then there must be viewing directions along which there is "minor-axis rotation", that is, a velocity gradient along the apparent minor axis. In a few cases, this might be large enough to be detectible with modern equipment. T. B. Williams (**E2**, 187) has recently reported the detection of this effect at the center of the E3 galaxy NGC 596. Near the center of this system, the apparent rotation axis makes an angle of 30° with the minor axis of the isophotes, instead of being parallel to this line as would be the case if the galaxy were rotationally symmetric.

Minor axis rotation, such as that described earlier for NGC 596, is extremely hard to detect, because the rotation velocity of a giant elliptical galaxy tends to be small even along the apparent major axis where rotation should be most readily detected. G. Illingworth (**I1**) and P. L. Schechter and J. E. Gunn (**S3**) find that the peak major axis rotation velocities of twenty-four elliptical galaxies fall in the range from 0 to 170 km s^{-1}, with mean value near 60 km s^{-1}. By contrast, one finds the six Sa–Sc galaxies of Table 8-3 to have a mean true rotation velocity $\langle \Theta_{max} \rangle = 235$ km s^{-1}, so that, if one were to observe a sample of such galaxies at random orientations to the line of sight, one would find $\langle v_{max} \rangle \approx (\pi/4)\ \langle \Theta_{max} \rangle = 180$ km s^{-1}. In fact, the samples of Illingworth and of Schechter and Gunn may be biased toward highly inclined galaxies, which would have higher than average apparent rotation, so the difference between $\langle v_{max} \rangle \approx 60$ km s^{-1} for their samples and $\langle v_{max} \rangle = 180$ km s^{-1} for the Sa–Sc galaxies is very striking.

Why do elliptical galaxies generally rotate so much more slowly than spirals? One might speculate that ellipticals are simply less massive than spirals, with the result that their characteristic internal velocities are much smaller than those of spirals. But this speculation would be wrong for two reasons. (1) The luminosities of ellipticals whose rotation velocities have been measured are generally much higher than those of most spirals. The

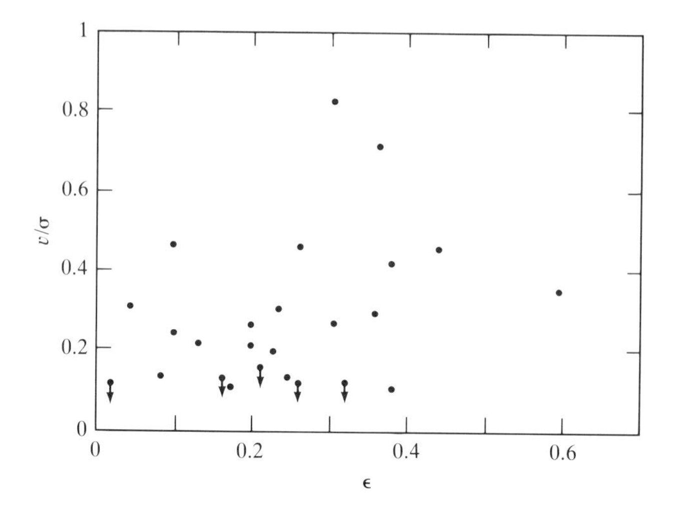

Figure 8-32. The shapes of elliptical galaxies are not produced by rotation alone. In this diagram we plot the peak line-of-sight rotation velocities of twenty-six elliptical galaxies, normalized by the appropriate central velocity dispersion, versus central ellipticity. No clear relation between these variables is apparent.

mean absolute magnitude of the twelve giant ellipticals in Illingworth's sample is $\langle M_B \rangle = -21.5$, whereas the mean absolute magnitude of the six Sa–Sc galaxies of Table 8-3 is $\langle M_B \rangle = -19.4$. Therefore, unless the mass-to-light ratios of ellipticals are much smaller than those of spirals (which is extremely unlikely), we have to conclude that the ellipticals are more, rather than less, massive than the spirals. (2) The typical internal velocities of giant ellipticals are directly observed to be large. The central line-of-sight velocity dispersions of the stars in these systems generally fall in the range $\sigma_0 \approx$ 130–250 km s^{-1}, which may be compared with the range $\sigma_0 \approx$ 75–215 km s^{-1} for central velocity dispersions found by T. B. Williams (**W1**) in four bright Sa–Sc galaxies. In fact, if we eliminate the effects of differences in the absolute sizes of galaxies by expressing the peak rotation velocity of each galaxy as a multiple of the central velocity dispersion of that system, then we find that most spirals have $(v_{max}/\sigma_0) > 1$, whereas, for most ellipticals, $(v_{max}/\sigma_0) < 1$.

Until recently, it was almost universally believed that elliptical galaxies were flattened into oblate spheroids by rotation, just as the figure of the Earth bulges at the equator on account of its rotation. If the equatorial flattening and the rotation of elliptical galaxies were causally related, it would follow that, in a plot of the true (deprojected) rotation speeds of elliptical galaxies (normalized relative to the central velocity dispersion to eliminate the effects of differences in absolute size) versus true ellipticity, all galaxies would lie on a single line. Actually, we do not know the inclinations i of elliptical galaxies, so we cannot correct the observed rotation velocity v_{max} to the true value Θ_{max}, nor can we correct the observed ellipticity ε to the true ellipticity ε_0. But, as we shall see in Chapter 14, for any inclination i, the observed quantities $\varepsilon(i)$ and $v_{max}(i)$ should lie along a unique curve. In Figure 8-32, we have plotted (v_{max}/σ_0) against ε for twenty-six ellipticals. Clearly, there is no tendency for these points to lie along a single curve. At any given apparent flattening ε, galaxies are to be found which rotate at entirely different rates. Therefore, not only do elliptical galaxies rotate slowly, but their shapes are evidently determined, at least in part, by something other than rotation. We shall investigate what this additional factor may be in Chapter 14.

Lenticular galaxies, in contrast to the ellipticals, have rotation properties similar to those of spirals. In Table 8-4, we list estimates of Θ_{max} for six lenticular galaxies. Comparing Tables 8-3 and 8-4, one can see that the peak rotation speeds of spiral and lenticular galaxies are very similar. This conclusion is an important one, particularly as the spiral rotation speeds refer to the gaseous component (neutral and ionized hydrogen), whereas the lenticular galaxy determinations are based on absorption-line measurements of the motion of the (relatively evolved) disk stars. We thus infer that both the stellar and gaseous components in galaxian disks have similar rotation curves and hence respond in the same way to the large-scale gravitational potential of the galaxy.

Table 8-4. Rotation Velocities of Lenticular Galaxies

NGC	Θ_{max}(km s^{-1})
128	260
3115	350
3998	240
4111	260
4459	160
4762	215

SOURCE: Data published in (**B6**)

REFERENCES

(**B1**) Baade, W. 1953. In *Symposium on Astrophysics.* (Ann Arbor: University of Michigan), p. 25.

(**B2**) Babcock, H. W. 1939. *Lick Obs. Bull.* **19**:41.

(**B3**) Balona, L. A. and Feast, M. W. 1974. *Mon. Not. Roy. Astron. Soc.* **167**:621.

(B4) Becker, W. and Contopoulos, G. (eds.), 1970. *The Spiral Structure of Our Galaxy*. IAU Symposium No. 38. (Dordrecht: Reidel).

(B5) Bertola, F. and Capaccioli, M. 1975. *Astrophys. J.* **200**:439.

(B6) Bertola, F. and Capaccioli, M. 1978. *Astrophys. J. Letters*. **219**:L95.

(B7) Binney, J. J. 1978. *Mon. Not. Roy. Astron. Soc.* **183**:779.

(B8) Binney, J. J. 1980. *Phil. Trans. Roy. Soc. London*. **296**:329.

(B9) Blaauw, A. and Schmidt, M. (eds.). 1965. *Galactic Structure*. (Chicago: University of Chicago Press).

(B10) Blitz, L. 1979. *Astrophys. J. Lett.* **231**:L115.

(B11) Bosma, A. 1978. *The Distribution and Kinematics of Neutral Hydrogen in Spiral Galaxies*. Thesis, University of Groningen, Netherlands.

(B12) Burbidge, E. M. and Burbidge, G. R. 1975. In *Galaxies and the Universe*. Sandage, A., Sandage, M., and Kristian, J. (eds.). (Chicago: University of Chicago Press).

(B13) Burton, W. B. 1966. *Bull. Astron. Inst. Netherlands*. **18**:247.

(B14) Burton, W. B. 1970. *Astron. and Astrophys.* **10**:76.

(B15) Burton, W. B. 1976. *Ann. Rev. Astron. and Astrophys.* **14**:275.

(C1) Comte, G., Monnet, G., and Rosado, M. 1979. *Astron. and Astrophys.* **72**:73.

(E1) Emerson, D. T. 1976. *Mon. Not. Roy. Astron. Soc.* **176**:321.

(E2) Evans, D. S. (ed.). 1979. *Photometry, Kinematics, and Dynamics of Galaxies*. (Austin: University of Texas).

(F1) Fatchikhin, N. V. 1970. *Soviet Astron.* **14**:495.

(F2) Field, G. B. 1958. *Proc. Inst. Radio Eng.* **46**:240.

(F3) Fricke, W. 1977. *Veroff. Astron. Rechen-Inst. Heidelberg*, No. 28.

(G1) Gunn, J. E., Knapp, G. R., and Tremaine, S. D. 1979. *Astron. J.* **84**:1181.

(H1) Harris, W. E. 1976. *Astron. J.* **81**:1095.

(H2) Hartwick, F. D. A. and Sargent, W. L. W. 1978. *Astrophys. J.* **221**:512.

(H3) Hayli, A. (ed.). 1975. *Dynamics of Stellar Systems*. IAU Symposium No. 69. (Dordrecht: Reidel).

(H4) Huchtmeier, W. K. 1975. *Astron. and Astrophys.* **45**:259.

(H5) Hunter, C. and Toomre, A. 1969. *Astrophys. J.* **155**:747.

(I1) Illingworth, G. 1977. *Astrophys. J. Letters*. **218**:L43.

(J1) Johnson, H. L. and Svolopoulos, S. N. 1961. *Astrophys. J.* **134**:868.

(K1) Kerr, F. J. 1969. *Ann. Rev. Astron. and Astrophys.* **7**:39.

(K2) Kerr, F. J. and Rodgers, A. W. (eds.). 1964. *The Galaxy and the Magellanic Clouds*. IAU Symposium No. 20. (Canberra: Australian Acad. Sci.).

(K3) Knapp, G. R., Tremaine, S. D., and Gunn, J. E. 1978. *Astron. J.* **83**:1585.

(K4) Kraft, R. P. and Schmidt, M. 1963. *Astrophys. J.* **137**:249.

(K5) Krumm, N. and Salpeter, E. E. 1977. *Astron. and Astrophys.* **56**:465.

(K6) Kwee, K. K., Muller, C. A., and Westerhout, G. 1954. *Bull. Astron. Inst. Netherlands.* **12**:211.

(M1) Mezger, P. G., Wilson, T. L., Gardner, F. F., and Milne, D. K. *Astron. and Astrophys.* **4**:96.

(M2) Mihalas, D. 1978. *Stellar Atmospheres.* 2nd ed. (San Francisco: W. H. Freeman).

(M3) Miller, J. S. 1968. *Astrophys. J.* **151**:473.

(M4) Minkowski, R. 1961. In *Problems of Extragalactic Research.* IAU Symposium No. 15, G. C. McVittie (ed.). (New York : Macmillan), p. 112.

(O1) Oort, J. H. 1927. *Bull. Astron. Inst. Netherlands.* **3**:275.

(O2) Oort, J. H. 1977. *Comments on Astrophys.* **7**:51.

(O3) Oort, J. H. and Plaut, L. 1975. *Astron. and Astrophys.* **41**:71.

(R1) Roberts, M. S. and Whitehurst, R. N. 1975. *Astrophys. J.* **201**:327.

(R2) Rogstad, D. H., Lockhart, I. A., and Wright, M. C. H. 1974. *Astrophys. J.* **193**:309.

(R3) Rogstad, D. H. and Shostak, G. S. 1971. *Astron. and Astrophys.* **13**:99.

(R4) Rogstad, D. H., Wright, M. C. H., and Lockhart, I. A. 1976. *Astrophys. J.* **204**:703.

(R5) Rots, A. H. 1975. *Astron. and Astrophys.* **45**:43.

(R6) Rougoor, G. W. 1964. *Bull. Astron. Inst. Netherlands.* **17**:381.

(R7) Rougoor, G. W. and Oort, J. H. 1960. *Proc. Nat. Acad. Sci.* **46**:1.

(R8) Rubin, V. C., Ford, W. K., and Thonnard, N. 1978. *Astrophys. J. Letters.* **225**:L107.

(S1) Sancisi, R. 1976. *Astron. and Astrophys.* **53**:159.

(S2) Sancisi, R., Allen, R. J., and Sullivan, W. J. 1979. *Astron. and Astrophys.* **78**:217.

(S3) Schechter, P. L. and Gunn, J. E. 1979. *Astrophys. J.* **229**:472.

(S4) Shane, W. W. and Bieger-Smith, G. P. 1966. *Bull. Astron. Inst. Netherlands.* **18**:263.

(S5) Simonson, S. C. and Mader, G. L. 1973. *Astron. and Astrophys.* **27**:337.

(S6) Slipher, V. M. 1914. *Lowell Obs. Bull.*, No. 62.

(T1) Toomre, A. and Toomre, J. 1972. *Astrophys. J.* **178**:623.

(V1) van den Bergh, S. 1974. *Astrophys. J. Letters.* **188**:L9.

(V2) van de Hulst, H. C., Muller, C. A., and Oort, J. H. 1954. *Bull. Astron. Inst. Netherlands* **12**:117.

(V3) van Woerden, H. (ed.). 1967. *Radio Astronomy and the Galactic System.* (London: Academic Press).

(V4) Vasilevskis, S. and Klemola, A. R. 1971. *Astron. J.* **76**:508.

(V5) Verschuur, G. L. and Kellerman, K. I. (eds.). 1974. *Galactic and Extragalactic Radio Astronomy.* (Berlin: Springer-Verlag).

(W1) Williams, T. B. 1977. *Astrophys. J.* **214**:685.

(W2) Wolf, M. 1914. *Vierteljahresschrift d. Astron. Gessel.* **49**:62.

(W3) Wright, M. C. H., Warner, P. J., and Baldwin, J. E. 1972. *Mon. Not. Roy. Astron. Soc.* **157**:337.

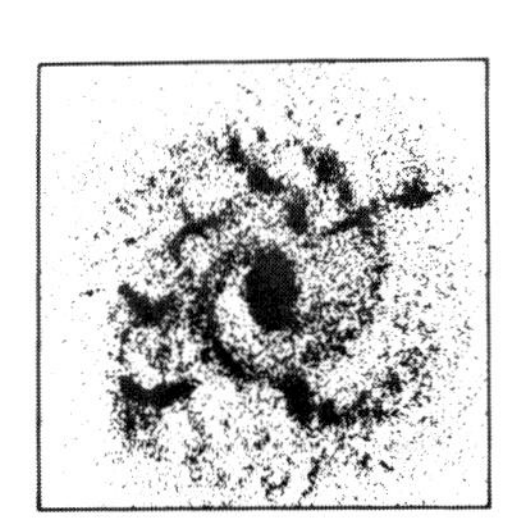

9 The Large-Scale Distribution of Gas in Galaxies

In the previous chapters, we were mainly concerned with the distribution of stars in our Galaxy and in external galaxies. In the first part of this chapter, we shall describe the distribution of gas in galaxies. As we described in §3-11, the interstellar medium is thought to comprise five components. These components are in approximate pressure equilibrium with one another, and, indeed, it is likely that material is constantly cycling from the cold clouds into the hot component while other material cools from the hot phase to join the cold clouds. The hot, low-density medium occupies most of the volume in the galactic disk ($\approx 70\%$). On the other hand, the clouds into which both the cold neutral material and the molecular material are concentrated contain most of the mass of gas in our Galaxy, because the density in these clouds is so much higher than it is in the surrounding hot regions.

A full discussion of the distribution of gas in galaxies would ideally include a description of the distribution of each of these five components of the interstellar medium. Measurements of radiation in the 21-cm line of neutral hydrogen detect principally the cold neutral clouds and their envelopes of warmer neutral material, because it is only in and around these clouds that the bulk of the gas is in the form of neutral atomic hydrogen. As was described in §8-4, fairly high-resolution 21-cm studies are now available for a substantial number of nearby galaxies; for a recent review of these observations, see (**V1**).

In addition to molecular hydrogen, which is difficult to detect, the material of the giant molecular clouds contains molecules, such as carbon monoxide (CO) and ammonia (NH_3), which emit millimeter line radiation when they pass from one rotation level to another, less energetic rotation level. This millimeter radiation serves as a tracer for the molecular clouds. During the last decade, it has become possible to build radio telescopes that operate at the high frequencies characteristic of these rotation energy-level transitions, and surveys of the galactic plane in the strongest lines emitted by giant molecular clouds are now available. Using these surveys, one can build a fairly complete picture of the disposition and kinematics of the molecular clouds in our Galaxy. Millimeter observations of galaxies other than our own are, by contrast, in their infancy, because the presently available millimeter telescopes have barely enough sensitivity to detect even the most readily detected species in the most nearby galaxies.

The hottest component of the interstellar medium is unfortunately very hard to map even within our own Galaxy. The existence of this component has only recently been conclusively established from studies of the diffuse X-ray background and the ultraviolet absorption spectra of nearby stars, and, at present, we lack the means to map its distribution through our Galaxy. Furthermore, we have no observational data pertaining to the possible existence of a similar hot component of the interstellar media of external galaxies. We shall not discuss this component further in this chapter, but the reader should be aware that this is necessitated by the absence of reliable data concerning its large-scale distribution in our Galaxy and is not a reflection of its probable importance for the structure and evolution of galaxies.

The last component of the interstellar medium, the warm ionized component, may be mapped by tracing the distribution of hydrogen recombination line radiation or of radio continuum radiation through our Galaxy and in external galaxies. Furthermore, this component is generally associated with one of the three cooler components, so that a knowledge of the distribution of these cooler components allows one to infer the distribution of the warm component.

9-1. THE DISTRIBUTION OF NEUTRAL HYDROGEN IN OUR GALAXY

Maps of Neutral Hydrogen Density and Spiral Structure

The 21-cm line of atomic hydrogen can be detected from essentially our entire Galaxy. In principle, it can therefore be used to determine the large-scale features of the hydrogen distribution and, in particular, to detect possible spiral structure in the gaseous component of our Galaxy. Recall that evidence for spiral structure in the stellar component of our Galaxy is limited by interstellar absorption to regions within 2 or 3 kpc of the Sun. If one assumes that the galactic neutral hydrogen is optically thin, then the intensity of the radiation at a given frequency is proportional to the column density of H I atoms that can radiate at the appropriate frequency displacement from line center, which is set by the Doppler shift produced by differential rotation and by the peculiar velocities of individual clouds.

In §8-3, we discussed how observed 21-cm profiles can be interpreted to give information about the rotation law $\Theta(R)$. But, it is obvious that they also contain information about the distribution and spin temperature of the H I in the galactic disk, and we now ask how this information can be recovered. It is clear from the outset that, because three independent variables (density, temperature, and line-of-sight velocity of the gas) can all influence the observed intensity, it will be necessary to make a number of simplifying assumptions in the analysis. In particular, the simplest case results if we assume that both the temperature distribution and the velocity field of the gas are known in advance.

The first comprehensive analysis of the hydrogen distribution in our Galaxy was made in 1957 by M. Schmidt (**S3**) and G. Westerhout (**W4**) using the then newly available 21-cm line profiles obtained with a 7.5-m dish at Kootwijk in the Netherlands. In this work, it was assumed that the hydrogen is everywhere optically thin, the spin temperature of the gas is everywhere the same, and the gas moves on perfectly circular orbits around the galactic center with velocities given by the rotation curve $\Theta(R)$ obtained from an earlier analysis of the data (**K11**). These assumptions imply that all observed intensity variations will necessarily be attributed completely to density variations in the material. This approach was certainly reasonable for a pioneering analysis, but it must be stressed that the third assumption in particular very strongly influences the results. Indeed, it should be emphasized that *any observed profile can equally well be reproduced by assuming a uniform medium and by choosing an appropriate velocity distribution along the line of sight.*

Having made the assumptions just stated, we use equation (8-4) to associate values of $\omega(R)$ with each radial velocity within a profile observed at a definite value of ℓ. From the known run of $\omega(R)$, we can then determine R (or, if needed, the distance d along the line of sight) and assign the density needed to produce the observed brightness temperature to the material at that position. In actual practice, there are complications. For example, corrections must be applied to the observed profiles to remove the effects of random motions within the clouds; these are often quite ill determined. What is worse, it is not always possible to assign a unique radius R to a given feature. For longitudes in the range $90^\circ \leq \ell \leq 270^\circ$ (which implies $R \geq R_0$), the procedure is unambiguous because there is only one value of R, or d, at which a given value of v_R, and hence $\omega(R)$, will occur (see Figure 8-7). The only difficulty that we may encounter is that the $\omega(R)$ curve might have to be estimated by extrapolation. However, when we look in toward the center (that is, for $-90^\circ \leq \ell \leq 90^\circ$, $R \leq R_0$) there are *two* values of R (or d) at which $\omega(R)$ will have the same value, and, because the H I is optically thin, radiation from either position can be seen. It therefore becomes necessary to decide whether a given feature is produced by H I at the nearer or more distant region. In the early work, this ambiguity was resolved by observing the radio signal at a number of latitudes above and below the galactic plane. Then, if it is assumed that the layer of neutral hydrogen has essentially the same thickness at all points radius R from the galactic center (as is suggested by the observations), it follows that nearby sources should have larger angular sizes than distant sources (see Figure 9-1). This effect allows us to make at least a statistically valid separation of individual

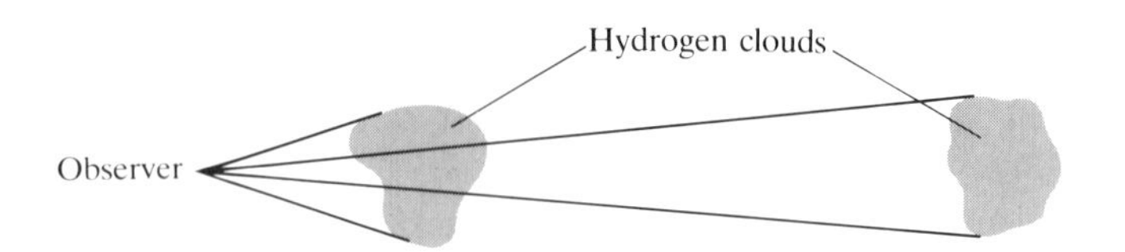

Figure 9-1. Method of distinguishing between nearby and distant hydrogen clouds. If the hydrogen layer in our Galaxy has a constant thickness, then, on the average, sources of large angular size can be presumed to be nearby, while those of small angular size can be presumed to be at greater distances.

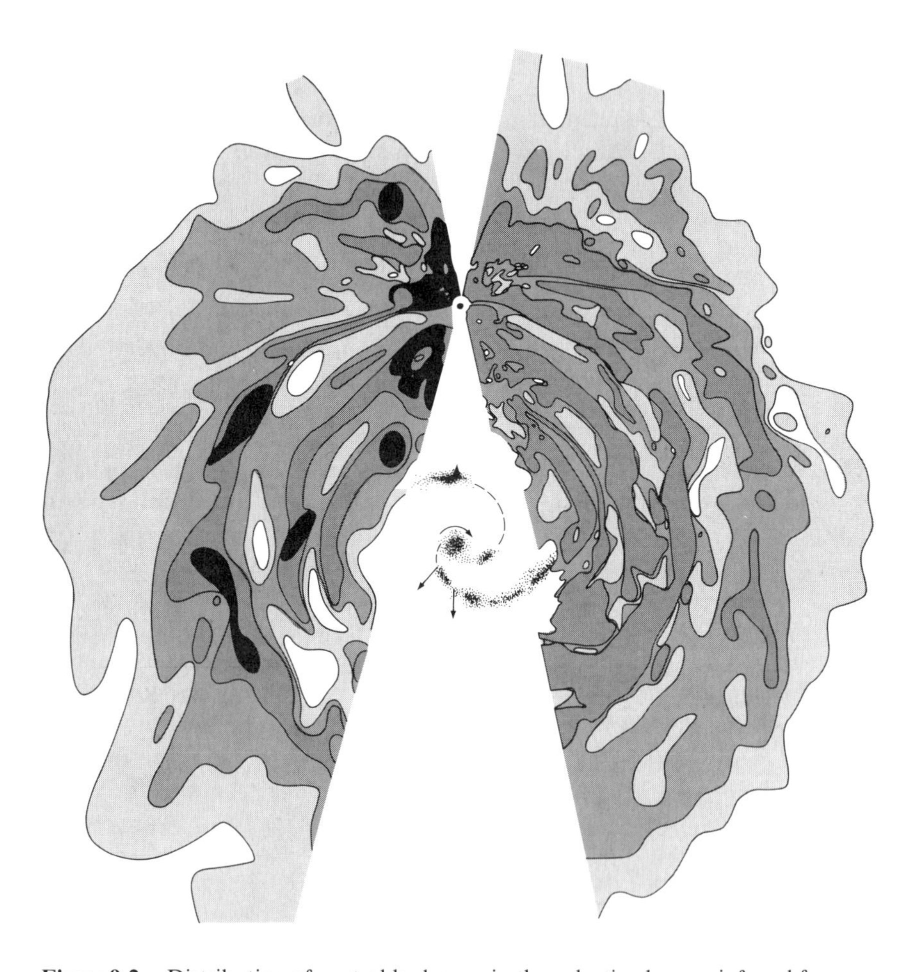

Figure 9-2. Distribution of neutral hydrogen in the galactic plane as inferred from early Dutch and Australian 21-cm observations by Kerr, Schmidt, and Westerhout. The grey scale indicates hydrogen density, with the darkest regions being the most dense. A schematic representation of the expanding 3-kpc arm and the nuclear disk as observed by Rougoor and Oort is shown at the center. The cone within $\pm 20^\circ$ of the Sun-center line is not mapped because there the mapping procedure breaks down. [From (**C1**, 121), by permission.]

features into the proper domains on the line of sight. Further details of the procedures actually used can be found in (**S3**), (**W4**).

The basic result of the work by Schmidt and Westerhout, supplemented by Southern-Hemisphere observations by F. J. Kerr and his collaborators (**K6**), is a now-classical H I density map, one version of which is shown in Figure 9-2 [see also the map published in (**O2**)]. The hydrogen appears to be arranged into more-or-less continuous armlike structures. The pattern is irregular and fragmented into a rather clumpy distribution, and although dominant trends can be seen, there is a large amount of splitting and interconnection among different features. The cone within $\pm 20°$ of the Sun-center line is not mapped, because here the circular velocities implied by circular motions are so small that the mapping procedure fails.

Much more refined analyses have been made using data that have higher spatial and frequency resolution (**W5**), (**W3**) and using alternative methods for identifying spatially related features in the profiles and for resolving the distance ambiguity. A useful approach that permits a relatively objective analysis of the data has been developed by H. F. Weaver [see, for example, (**B5**, 126) and (**K7**, 573)]. Here, again, circular motion with a known rotation curve is assumed. In this approach, the fact that prominent spatial structures in our Galaxy will produce distinctive features in plots of $T_B(\ell, v)$ is exploited fully. Consider, for example, spiral arms in a differentially rotating galaxy having an angular velocity $\omega(R)$ that is a monotonically decreasing function of R. Then, in a plot of $T_B(\ell, v)$, a structure such as the one shown in Figure 9-3a maps into the features shown in Figure 9-3b. Points on the inner parts of the spirals ($R < R_0$) transform

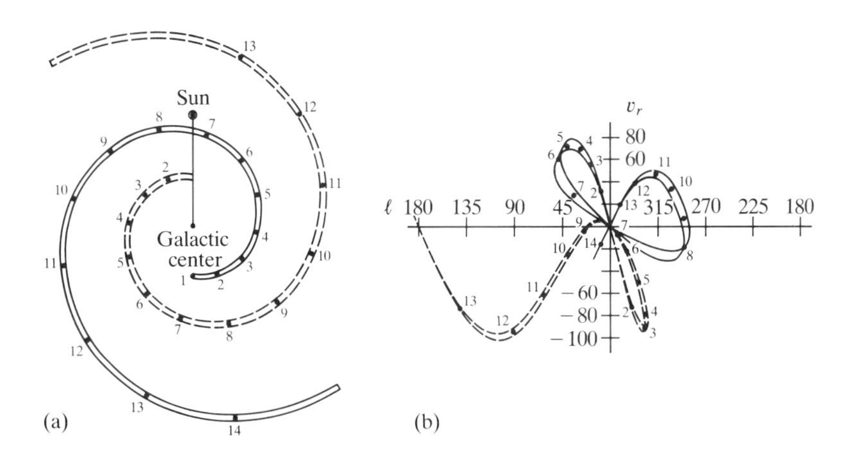

Figure 9-3. Mapping of an optically thin H I spiral-arm pattern (a) in a differentially rotating galaxy into the velocity-longitude diagram (b). Points are labeled with corresponding sets of numbers in both diagrams. [From (**K7**, 573), by permission.]

to points in quadrants II and IV of the (ℓ, v) plane and characteristically produce loops. Points on the outer parts $(R > R_0)$ transform into sinelike curves in quadrants I and III. If one can identify definite structures in the $T_B(\ell, v)$ diagram by choosing only the most intense features while being guided by considerations of continuity and knowledge of the shapes (for example, loops) expected in the diagram, then one can reasonably hope to delineate major features in the H I distribution by the inverse mapping procedure. Thus, starting with the observed distribution of brightness temperatures T_B greater than some cutoff value T_c at a given longitude—that is, $T_B(v, b|l, T_B \geq T_c)$, one can sum over a specified range of latitudes, say, $\pm b_{\text{max}}$, chosen to contain all the material in the plane, to obtain the function

$$I(v|\ell, T_B \geq T_c) \equiv \int_{-b_{\text{max}}}^{b_{\text{max}}} T_B(v, b|\ell, T_B \geq T_c)\, db \tag{9-1}$$

which provides a map in the (ℓ, v) plane of all features above the cutoff temperature. Then, in the (ℓ, v) plane, for each value of ℓ, one can plot points at the values of v that locate the maxima of $I(v|\ell, T_B \geq T_c)$. These maxima can then easily be joined into very clear ridge-line curves such as those shown in Figure 9-4. Finally, by performing the inverse mapping on these ridge lines, one obtains the space distribution of H I shown in Figure 9-5. This map is based on only very high-intensity features, and it should be about the best representation of major H I structures that is obtainable by use of the basic mapping procedure described thus far.

It has long been recognized that all of the H I maps derived in the manner just outlined are no more reliable than the basic kinematic assumptions

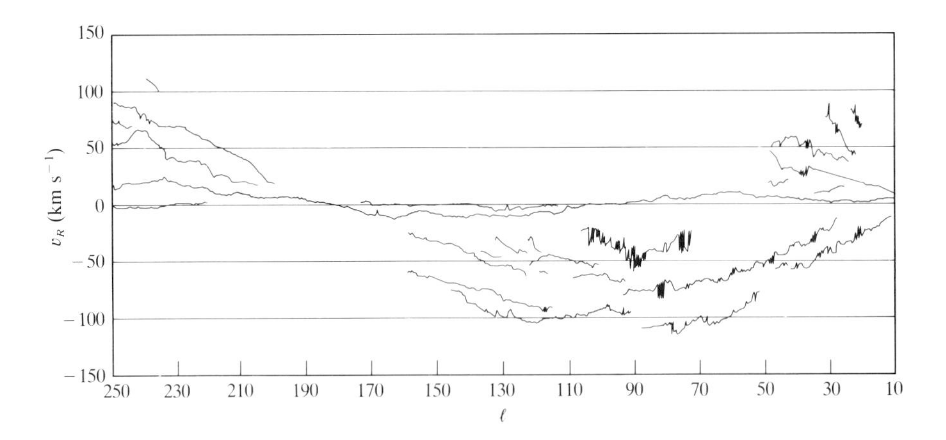

Figure 9-4. Loci of maximum observed 21-cm line brightness temperature in our Galaxy, plotted in the (ℓ, v) plane. [From (**K7**, 573), by permission.]

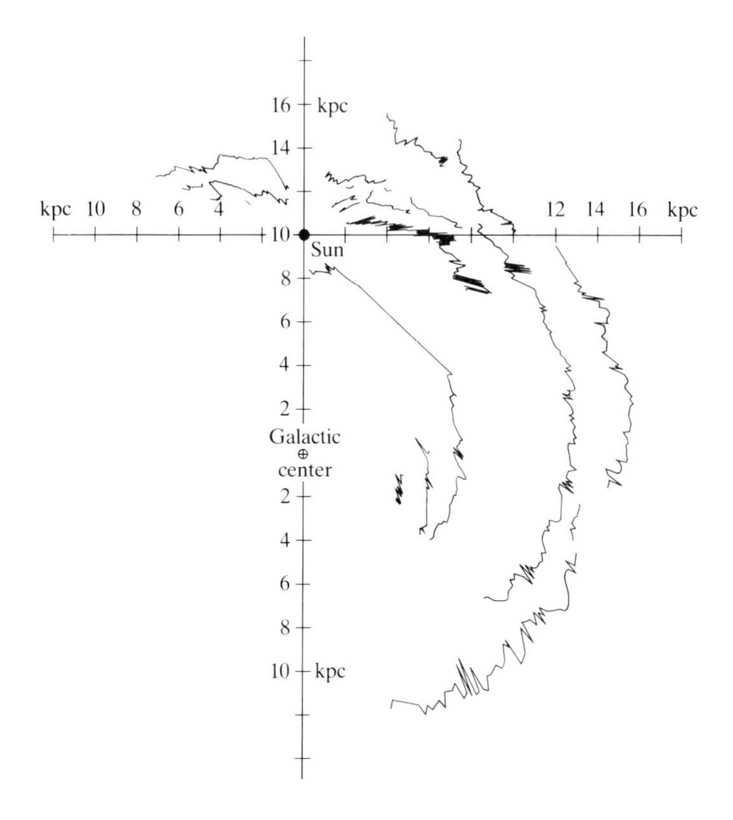

Figure 9-5. Spiral-arm pattern in our Galaxy derived from Figure 9-4 assuming circular rotation with the standard rotation curve. [From (**K7**, 573), by permission.]

invoked in their construction, and they are subject to more-or-less serious errors, depending on how large the departures of gas motions from strictly circular rotation are. Definite, if limited, evidence for noncircular motions of interstellar clouds has been available since the time of the first attempts at H I mapping. For example, G. Münch showed (**M4**) that interstellar lines observed in the spectra of Perseus-arm stars deviate by ± 20 km s^{-1} from the values expected from pure circular rotation, and J. J. Rickard found (**R1**) corresponding H I features for almost every interstellar line component. Furthermore, definite nonzero H I velocities have been observed in directions where they should not occur for circular motions ($\ell = 0°$, $\ell = 180°$), and positive H I velocities are observed in the direction $\ell = 90°$, whereas only negative velocities should be found for circular motion with $\omega(R)$ decreasing outward. More recently, it has become clear that local velocity irregularities of the order of ± 10 km s^{-1} away from circular motion are a widespread phenomenon throughout our Galaxy.

For some time it was hoped, nevertheless, that deviations of, say, a few percent from circular velocity would produce only minor mapping errors, and that the maps might be regarded as drawn on a "rubber sheet" which could be slightly deformed locally to account for the errors produced by irregularities in the velocity field. But it is now known from several studies **(B16)**, **(B17)**, **(T4)** that this hope was overoptimistic, and that in reality velocity irregularities affect the appearance of observed line profiles much more strongly than density fluctuations in the gas. For example, W. B. Burton has shown **(B17)**, (**K7**, 551), (**B18**) how features in the profiles in the directions $\ell = 90°$ and $\ell = 110°$ can easily be fit by assuming a uniform density medium with velocity perturbations of only a few kilometers per second away from pure circular motion. To fit these same features assuming strict circular motions would require density contrasts of a factor of 50 or 100 to 1, and furthermore, some of the features cannot be fit at all because they appear at radial velocities that are "forbidden" in such a kinematic model.

The fundamental problem that plagues attempts to construct maps of the density of neutral hydrogen in the galactic plane is that even small deviations of the velocity of an element of gas from the circular velocity can lead to large errors in the galactocentric azimuthal coordinate which one assigns to that element of gas. The galactocentric distance of an element of gas is, by contrast, only modestly affected by deviations from the circular velocity. To see why this is so, recall that, near the subcenter of a given line of sight (the point of closest approach of the line of sight to the galactic center), the radial component of the circular velocity changes only very slowly with distance along the line of sight from the Sun. Conversely, the inferred distance along the line of sight to an element of gas that lies near the subcenter changes rapidly with changes in that element's heliocentric velocity. However, the galactocentric distance of the element of gas is not sensitive to the element's position along the line of sight, because the line of sight is nearly perpendicular to the radius vector to an element of gas that is near the subcenter. Therefore, the uncertainty in the element's position is concentrated almost entirely in the galactocentric azimuthal coordinate.

One way we might hope to escape from the errors in galactocentric azimuth just discussed is to attempt to fit the observed $T_B(\ell, v)$ distribution to a dynamical model that predicts the deviations from circular velocity which are associated with given azimuthal density variations, for then one might hope to be able to obtain a unique solution that would allow each element of gas to be placed correctly. This improvement would be possible because the model would enable one to derive each element's component of noncircular velocity. An additional advantage of fitting a model galactic disk to the observed distribution of $T_B(\ell, v)$ is that one can then correctly solve the radiative transfer problem involved in deriving the theoretical $T_B(\ell, v)$ diagram from the assumed model of the hydrogen distribution in the disk. Figure 9-6 shows an example of such a modeling approach; panel (a) shows $T_B(\ell, v)$ as observed in the range $40° \leq \ell \leq 90°$, and panel (b) shows the

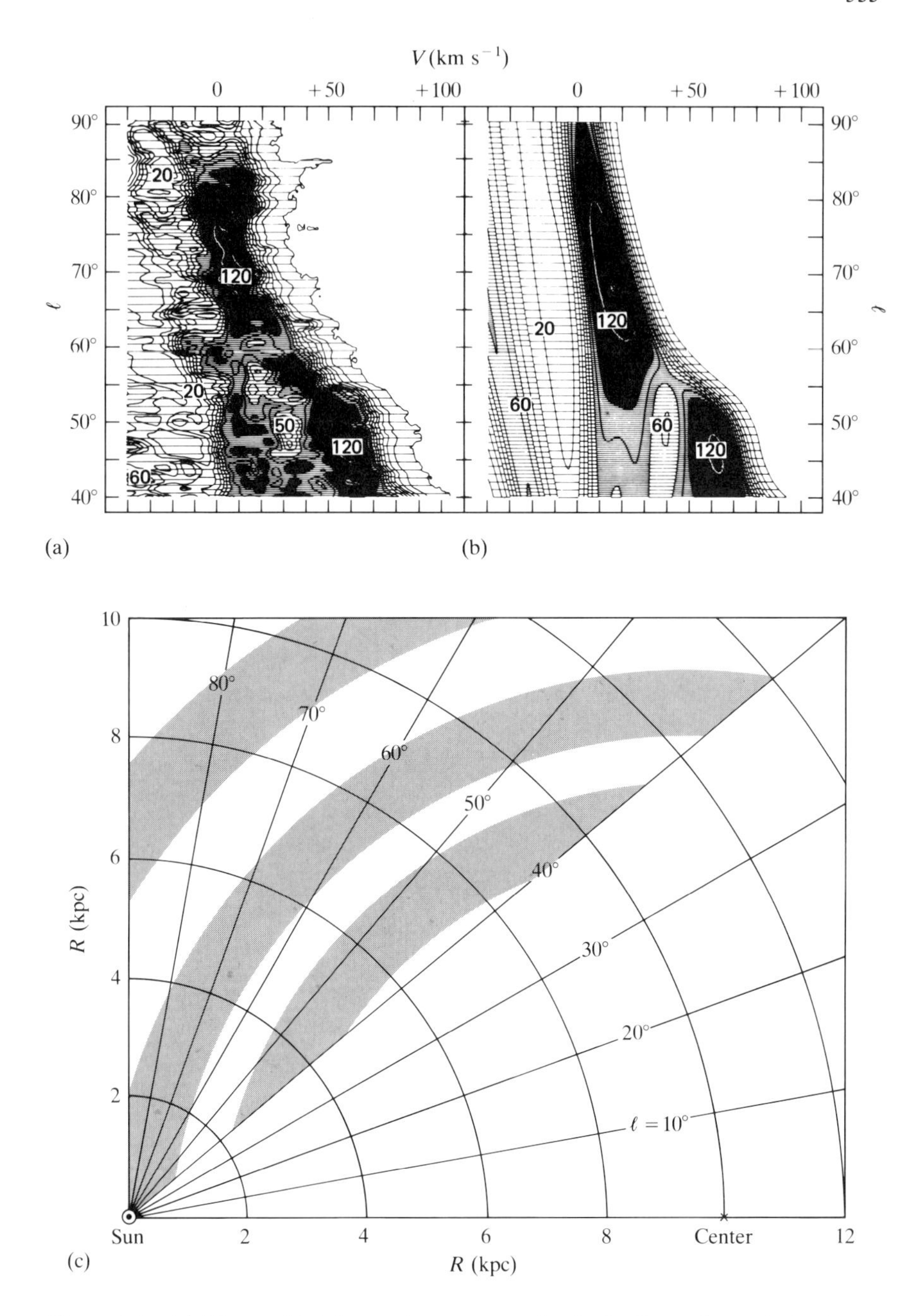

Figure 9-6. (a) Observed brightness temperature of galactic 21-cm line emission in the (ℓ, v) diagram. (b) Predicted brightness temperatures from hydrogen distribution in panel (c). (c) Model hydrogen distribution; shaded regions have above average density. The kinematical model used in calculating panel (b) allows for differential rotation and local streaming in a density-wave model. [From (**B16**), by permission.]

brightness-temperature map predicted from the spiral-like density distribution in panel (c), assuming circular motion plus local streaming velocities consistent with a spiral density-wave theory. The model gives a good fit to the data, shows the correct run of terminal velocity with longitude, and provides a good description of the radiation received from the Sagittarius arm, which is seen tangentially at $\ell = 50°$.

For some time, there were severe problems in matching the radio and optical maps of spiral arms in our Galaxy. For example, on comparing the classical radio map (similar to Figure 9-2) with optical spiral arms (see Figure 4-11), one finds that the optical and radio arms have markedly different slopes, and that the optical arms pass through regions that have very low hydrogen densities [see (**B12**, 153)]. Similarly, comparisons of the positions of galactic clusters and O associations by W. Becker (**B4**) and of individual O and B stars by A. Beer (**B7**), (**B8**) with the H I pattern revealed that there actually was a strong anticorrelation of these stellar objects with the hydrogen arms. This result is embarrassing, because these stars have such short lifetimes that they must have been formed at very nearly their present locations, and one would therefore expect them to be found in regions of high gas density where star formation is likely to occur. More modern maps show much better agreement between the optical data and the local radio-arm structure [see particularly the comparison by Weaver in (**B5**, 138)], but the basic problem remains that the radio maps cannot be any more accurate than the underlying kinematical assumptions.

In this connection, it is interesting that E. Fletcher showed (**F2**) that the discrepancies between the optical and early radio maps arose mainly from the inconsistency between the different methods used to construct them, positions of objects in the former being based on direct (spectroscopic) distances and in the latter being deduced from radial velocities and an assumed galactic rotation law $\Theta(R)$. In particular, he showed that if stellar radial velocities are used to construct an optical map by the same procedure used to make the radio map, then an almost perfect correlation between the stars and gas is achieved. Similarly, J. S. Miller (**M3**) showed that, in the Perseus arm, the apparent discrepancies between the positions of the H I arm and the stellar-arm tracers are produced entirely by small deviations of the gas motion from the assumed circular-rotation curve. In retrospect, we realize that these results simply tell us that the radio mapping procedure is strongly dominated by the kinematic model employed.

Radial Density Profile of Neutral Hydrogen

We have just seen that, although it is extremely difficult to estimate correctly the galactocentric azimuth of an element of gas from the (ℓ, v) coordinates of its 21-cm emission, there is reason to hope that the galactocentric radius of any element of gas is less subject to error. Therefore, if one integrates

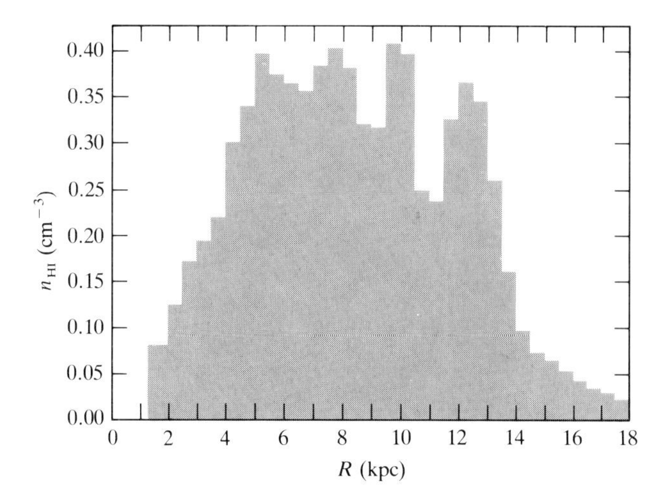

Figure 9-7. The radial density profile of neutral hydrogen in our Galaxy. [From (**B19**), by permission.]

the gas density shown in a plot like that of Figure 9-2 in annuli around the galactic center to obtain the radial density distribution, the azimuth errors should largely cancel one another, leaving one with a reliable profile. This profile is, moreover, directly comparable to those observed in other galaxies.

Figure 9-7 shows the radial density distribution of neutral hydrogen in our Galaxy that is obtained by a procedure which is equivalent to averaging a distribution like that of Figure 9-2 around annuli. Notice that this density distribution is not centrally peaked as one might have expected by analogy with the disk and bulge luminosity distributions discussed in Chapter 5, but actually it has a deep depression at the center ($R < 4$ kpc). Similar features are observed in other galaxies (see §9-2). Also, we shall see later that other components of the interstellar medium have central holes. However, no other component has the flat-topped profile characteristic of the neutral hydrogen distribution between 5 and 13 kpc from the galactic center. Beyond 14 kpc, the neutral hydrogen density shown in Figure 9-7 falls off quite steeply; note, however, that the density shown in this figure for the region outside the solar circle depends somewhat on the poorly known rotation curve at $R > R_0$.

It should also be borne in mind that, in the directions $\ell \approx 0°$ and $\ell \approx 90°$, the optical depth τ of the galactic disk at 21 cm differs significantly from zero. However, Figures 9-2 and 9-7 are based on the use of equation (8-57), which holds only if $\tau \approx 0$. In directions in which the optical depth differs significantly from zero, equation (8-57) gives underestimates of the gas density. Nevertheless, even when one takes full account of the effects of beam saturation by adopting the model-fitting approach described earlier, one obtains a radial density profile which differs only insignificantly from that shown in Figure 9-7 (**B18**).

z Distribution

Studies of the distribution of H I in the direction perpendicular to the galactic plane show that, for $R \leq R_0$, the gas is confined to a very thin, flat layer (**B5**, 183–189), (**J1**). The average full half-density thickness of the layer for 4 kpc $< R <$ 10 kpc is about 250 pc, and, within a radius of several kiloparsecs from the galactic center, the point of maximum density within the layer does not deviate from the galactic plane by more than $\pm$25 pc. The distance of the Sun above (that is, north of) the mean hydrogen plane is estimated to be 4 pc $\pm$ 12 pc. There is some evidence (**J1**) that the half-thickness of the H I layer shows local maxima near the positions of H I spiral arms and giant H II regions. Note that the approximate constancy of the thickness of the galactic H I layer over most of the disk implies that the space densities shown in Figures 9-2 and 9-7 are approximately proportional to the corresponding surface densities of neutral hydrogen.

For $R <$ 4 kpc, the half-thickness drops to about 100 pc and shows substantial fluctuations. At distances beyond 10 kpc from the galactic center, the hydrogen layer becomes thicker, and the half-thickness rises from about 300 pc for R = 11 kpc to about 650 pc near R = 15 kpc, and to perhaps 1000 pc for R = 20 kpc (**J1**).

At certain radii, the hydrogen layer shows a systematic deviation from the galactic plane, as illustrated in Figure 9-8. The distortion is perceptible at about R = 7 kpc, and it increases out to about R = 15 kpc to distances of the order of 1 kpc from the plane. In §8-4, we saw that there is evidence for similar distortions in many other galaxies. Several theoretical explanations of this warping have been proposed: (1) tidal distortion of the hydrogen layer by the Magellanic Clouds (**B15**), (**K4**); (2) distortion of the H I layer by the pressure field generated by a flow of intergalactic gas past the galaxy

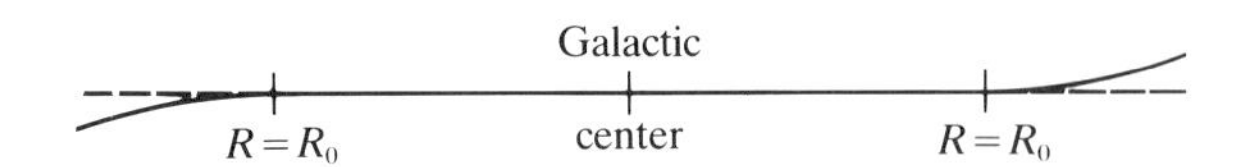

Figure 9-8. Cross-section of the hydrogen layer in our Galaxy, plotted in the plane through the galactic center and perpendicular to the Sun-center line. For $R < R_0$, the hydrogen lies in a very thin layer essentially in the galactic plane. At larger distances from the center, however, the hydrogen deviates systematically from the galactic plane as shown.

(**K1**); (3) the distortion of the disk represents some kind of free oscillation of the entire disk (**L5**), (**B11**) [reviews of these mechanisms will be found in (**H2**) and (**B11**)].

9-2. THE DISTRIBUTION OF NEUTRAL HYDROGEN IN OTHER GALAXIES

Optical observations show that spiral and irregular galaxies invariably contain dust and gas excited by young stars, and it is natural to expect 21-cm observations of H I in these systems to furnish important information about the form and dynamics of these objects. Now that it is possible to observe external galaxies with high resolution at radio wavelengths, we can map their H I distributions directly, without encountering the confusion produced by overlapping features along the line of sight or needing to use a kinematical model in the analysis as we do in the case of our Galaxy.

Lenticular galaxies and some ellipticals are also known to contain dust. Furthermore, there are good theoretical reasons for believing that large amounts of gas should have been shed by evolving stars in these systems over the lifetime of the Universe, and this gas might have remained in these galaxies to form an interstellar medium even if none was originally present. The question therefore arises as to whether or not early-type galaxies contain detectible amounts of neutral hydrogen. Recently, small quantities of H I have in fact been detected in a few early-type systems, although the available data on their H I properties are still very sparse, because the signals produced by their neutral hydrogen are generally exceedingly weak.

Low surface densities of neutral hydrogen are sometimes found in the form of bridges between neighboring galaxies, or as a more-or-less uniform background in the space between galaxies in small groups.

Spiral and Irregular Galaxies

Large-Scale Distribution of H I The distribution of neutral hydrogen in all late-type systems is quite unlike any of the light distributions discussed in Chapter 5. Thus H I distributions are typically neither centrally concentrated, nor smooth, nor necessarily symmetric about the galaxian major axes. Figure 9-9 shows the azimuth-averaged H I surface-density profiles of a selection of Sa–Im galaxies. Notice the similarity of the neutral hydrogen distribution of the Sab galaxy M81 to that shown in Figure 9-7 for our Galaxy. Notice also that several other profiles show marked central deficiencies of neutral hydrogen, but that this phenomenon is not universal. The Sc galaxy M33, for example, has a neutral hydrogen surface density that is constant from the center out to 5 kpc. The peak surface densities of neutral hydrogen range from $\approx 5 \times 10^{21}$ atoms cm^{-2} attained by M31 down to 5×10^{20} atoms cm^{-2} in M81 and the irregular galaxy DDO 125.

The observable H I in late-type galaxies can extend to appreciably greater radii than the optical luminosity as measured by the Holmberg radius R_H (see §5-3). Thus the H I in the Sbc galaxy NGC 5055 is still detectable at 40 kpc ($> 2R_H$) from the center; the surface density there is about 10^{20} atom cm^{-2}. In M33, by contrast, the surface density at R_H is already below this level. Some irregular galaxies—for example, NGC 3109 (**B9**, 264)—also have detectible neutral hydrogen that extends beyond $2R_H$. From the profiles shown in Figure 9-9, it will be seen that there is a general tendency for H I surface density to decline steeply near R_H, but that in some cases the H I density then flattens off at a detectable level. In those galaxies around which extensive H I envelopes have been detected, the level at which this flattening occurs is very close to the practical detection threshold of present-day aperture-synthesis telescopes (several times 10^{19} atoms cm^{-2}), so that it must not be concluded from a figure like Figure 9-9 that only some galaxies possess extensive low-surface-density hydrogen envelopes. Single-

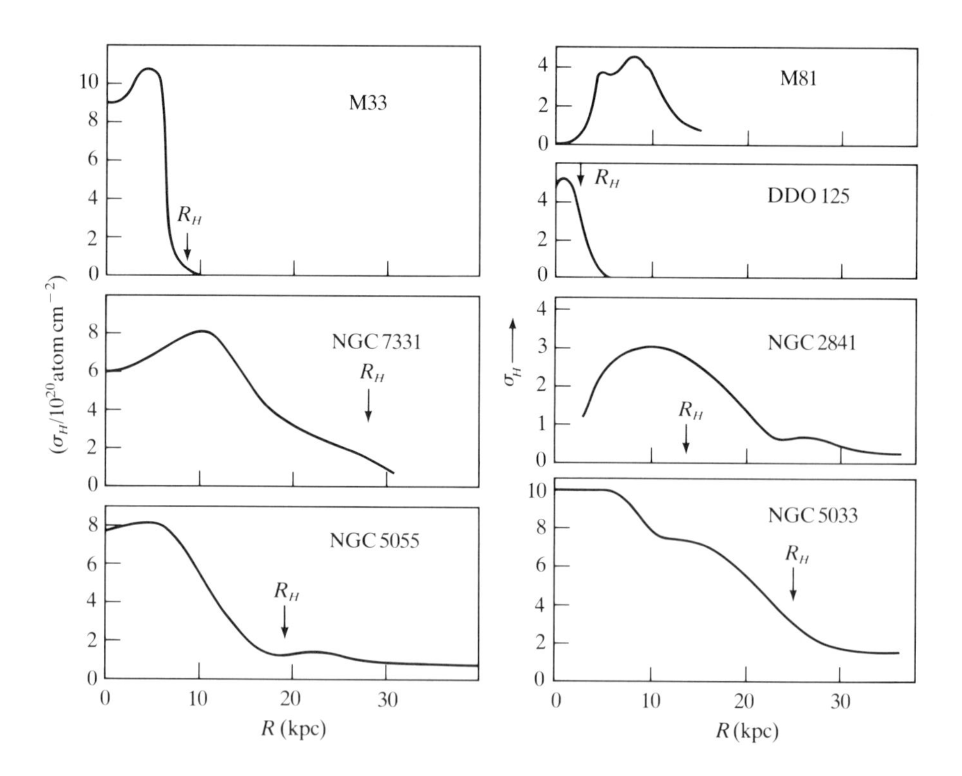

Figure 9-9. The azimuth-averaged, neutral hydrogen surface densities in eight Sa–Sm galaxies. [From data published in (**B14**) and from (**N1**), (**R3**), and (**T2**), by permission.]

dish measurements, which can have up to a factor-of-ten better sensitivity than is possible with aperture-synthesis telescopes, detect hydrogen well beyond R_H in nearly all galaxies.

The neutral hydrogen distributions of nearly all late-type galaxies have some large-scale asymmetric features. These asymmetries are particularly pronounced in the case of irregular galaxies, but they are still immediately apparent when one inspects the integrated column-density profiles of more massive late-type galaxies. For example, a very remarkable feature of the radiograph shown in Figure 8-24 of the neutral hydrogen column density in M81 are the two large concentrations of gas on the east (left) side of the figure. Figure 9-10 is a contour plot of this same neutral hydrogen density after deprojection to face-on orientation. From the diagram, it is apparent that the asymmetry of the H I distribution is not confined to the existence of the two large concentrations. In the region interior to these concentrations ($R < 13$ kpc), there is still 50% more H I to the east of the line of nodes than to the west. The nature and origin of these irregularities are something of a mystery, because one might have expected differential rotation to smear out any large condensations of neutral hydrogen that might form within one, or at the most two, rotation periods of the galaxy. One possibility is that these features are caused by tidal disturbance of the galaxy by neighboring systems.

Small-Scale Structure and Spiral Arms Figure 9-10 shows two noteworthy features of the H I distribution in M81 in addition to the large-scale asymmetry just mentioned: (1) on a small scale, the distribution appears to break up into a large number of small blobs, and (2) there is clear evidence that the neutral hydrogen is concentrated into spiral arms.

The patchiness of the H I distribution of Figure 8-24 is commonly observed in all types of late-type galaxy. Indeed, one finds that even in low-mass systems like the Magellanic Clouds (**M2**), (**H1**) and other low-luminosity systems (**T2**), the neutral hydrogen is concentrated into complexes that each contain around $5 \times 10^6 \mathcal{M}_\odot$ of H I and are a few hundred parsecs in diameter. Probably the fact that low-mass systems can contain only a few of these complexes ($\lesssim 50$ for an Im galaxy), whereas giant galaxies like M81 contain many hundred such complexes, contributes to the irregular appearance characteristic of low-mass, late-type systems.

The H I surface density distribution of M81 depicted in Figure 9-10 shows one of the most distinct and regular spiral patterns seen in 21-cm line radiation (**R3**), (**R4**). Out to $R = 10$ kpc, the H I distribution is dominated by a clear two-arm spiral pattern, which is well approximated by a logarithmic spiral with a pitch angle of about 15°. The ratio of the peak H I density in the arms to the average H I density near them is about a factor of two. The H I arms show an excellent correlation with optical arms; the H I ridge line tends to lie on the inner (concave) edge of the bluest part of the optical arm (which presumably is the region containing the highest density of young blue stars). A narrow dust lane, seen optically, lies along the inner

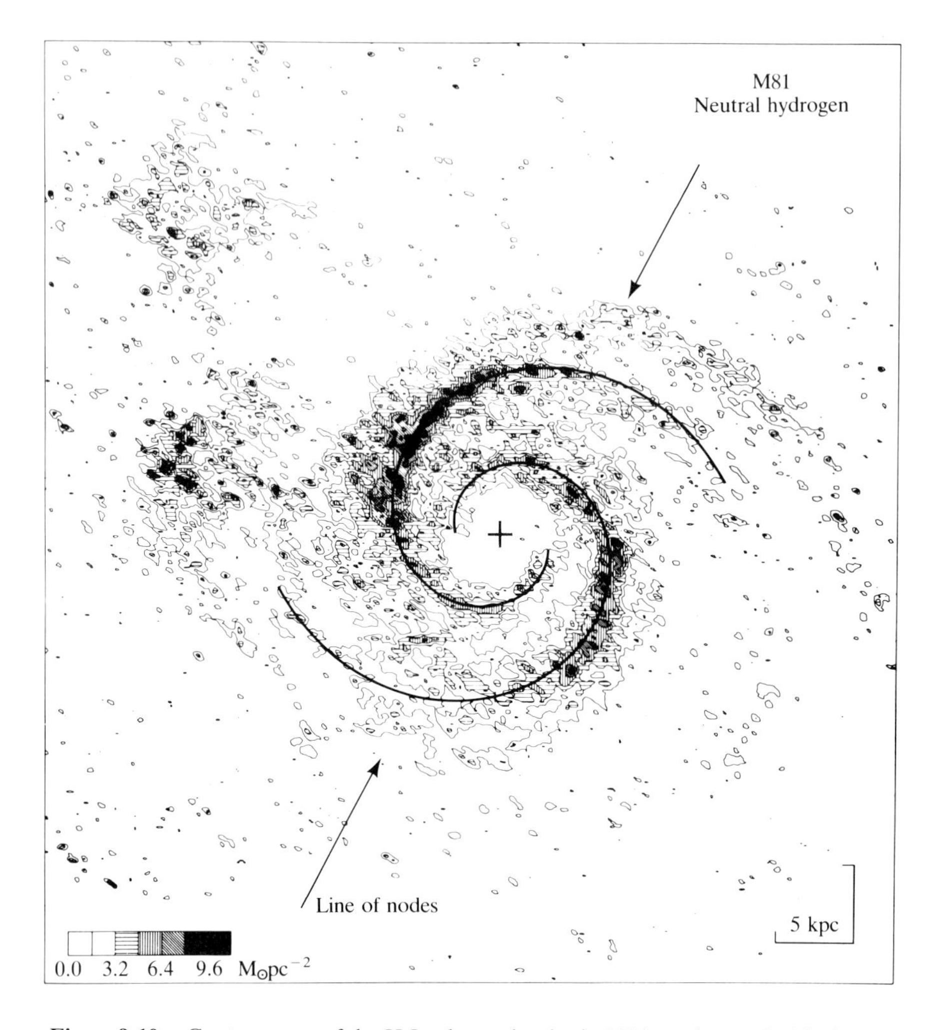

Figure 9-10. Contour map of the H I column density in M81 as observed with the Westerbork Aperture Synthesis Telescope, deprojected to a face-on view. Densities are given in $\mathscr{M}_{\odot}$ pc^{-2}. The spiral pattern has two principal arms that are well fitted by a logarithmic spiral with a pitch angle of 15°. [From (**R3**), by permission.]

edge of the eastern H I arm. The phase lag from the H I peak to the blue light peak is about 10° to 15°. Using an estimated rotation speed for the spiral pattern (chosen to be the corotation speed at $R \approx 10.5$ kpc), one deduces that the time required for an element of gas to pass from the maximum H I surface density to the highest density of young stars is 10^7 years. The ratio of hydrogen density to stellar mass density increases progressively along the arms; the H I arms can be traced far beyond the end of the optical arms. Outside $R \approx 10$ kpc, two outer-arm features develop. The northern arm is crisply defined and can be followed easily over several kiloparsecs. The southern arm is fragmented, and it joins into the two large H I concentrations to the east of the galaxy.

Early-Type Galaxies

Until very recently, few lenticular galaxies were known to contain appreciable amounts of neutral hydrogen, and no neutral hydrogen had been detected in an elliptical galaxy. Now H I has been detected in a large number of S0 galaxies, and gas has been observed in several elliptical galaxies (**B9**, 52). For example, H. van Woerden et al. have detected neutral hydrogen in sixteen of a random sample of fifty-five large-diameter southern S0 galaxies. These galaxies show a wide variation of the ratio $\mathcal{M}_H/L_B$ of the detected hydrogen mass to B luminosity; $0.03\mathcal{M}_\odot/\mathcal{L}_\odot < \mathcal{M}_H/L_B < 1.4\mathcal{M}_\odot/\mathcal{L}_\odot$. For comparison, Sc galaxies have, on the average, $\mathcal{M}_H/L_B \approx 0.3\mathcal{M}_\odot/\mathcal{L}_\odot$. A similar survey of elliptical galaxies by G. R. Knapp et al. (**K8**) led to only six neutral hydrogen detections out of a sample of thirty-eight systems observed.

Generally, elliptical galaxies are extremely gas poor ($\mathcal{M}_H/L_B < 0.03$ $\mathcal{M}_\odot/\mathcal{L}_\odot$). Nothing is known with any certainty about whether the gas in gas-containing ellipticals forms a rotating disk or some other structure. There are, however, indications that the gas is not confined to the central regions of the galaxies and that the line profiles produced by these galaxies are broader than those characteristic of spiral galaxies (**B10**), (**K8**).

De Vaucouleurs and his collaborators have defined an index of neutral hydrogen richness, analogous to a color index, by the relation

$$\mathrm{H\ I} = 16.6 - 2.5 \log S_H + B_T^0 \tag{9-2}$$

In this equation, S_H is the 21-cm flux density (corrected for optical depth effects) integrated over all velocities and expressed in units of watts per square meter, and B_T^0 is the total blue apparent magnitude in the sense of the *Second Reference Catalogue* (see §5-3). The number 16.6 in equation (9-2) has been chosen to make H I $\approx$ 1 for an average spiral. Notice that H I is effectively the ratio (expressed as a magnitude) of the galaxian luminosity to the total 21-cm flux. The H I index is negative for extremely gas-rich systems and positive for galaxies that are poorer in hydrogen. Figure 9-11 shows how H I correlates with Hubble-type t. As we have seen, late-type galaxies are richer in neutral hydrogen than early-type systems. Notice also that, for $t \lesssim 2.5$, there is little correlation between H I and t because, although S0 and E galaxies are generally gas poor, they do show large individual fluctuations from the mean hydrogen richness. Finally, there is a tendency, shown only schematically in the figure, for late-type giants to be poorer in H I than late-type dwarfs.

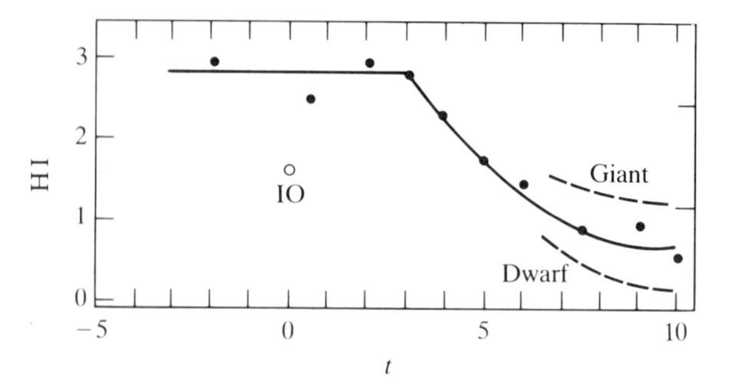

Figure 9-11. Late-type galaxies tend to have more neutral hydrogen per unit luminosity than do early-type systems. This figure shows the correlation of the hydrogen index H I [defined by equation (9-2)] with Hubble-type t (see Table 5-2). [From (**T1**, 43), by permission.]

Gas in Groups of Galaxies

A number of lines of evidence indicate that the space between galaxies in groups of galaxies may not be empty. The possible existence of diffuse gas in groups of galaxies is important because the majority of galaxies are probably in some type of group (**T3**).

In the Local Group, clear evidence for intergalactic material is provided by studies of the *Magellanic Stream* (**W2**), (**M1**). This feature consists of a series of six elongated neutral hydrogen clouds that are aligned with each other to form a swath of neutral hydrogen emission that stretches over more than 60° in the southern sky. One end of this swath merges with the diffuse H I envelope around the Magellanic Clouds (Figure 9-12). This structure has been variously interpreted as tidal debris torn from the Large Magellanic Cloud (LMC) by the tidal field of our Galaxy (**L6**), or as a kind of vapor trail that has condensed out of a hypothetical hot, gaseous intergalactic halo of our Galaxy in the wake of the LMC (**M1**).

The so-called *high-velocity clouds* that are observed in 21-cm line radiation—both in the galactic plane, with velocities differing by more than ≈ 50 km s^{-1} from those of a simple differential rotation, and also at high latitudes—have often been interpreted as intergalactic clouds. However, it now seems likely that most of them are actually features in the galactic disk. Indeed, K. Y. Lo and W. L. W. Sargent (**L3**), who have searched without success for similarly massive ($\mathscr{M} \geq 10^8 \mathscr{M}_\odot$) neutral hydrogen clouds in nearby groups of galaxies, argue that the intergalactic interpretation of the high-velocity clouds is unlikely to be correct, because it would imply that the Local Group is a rather special system. G. L. Verschuur (**V2**) has recently reviewed the status of this problem.

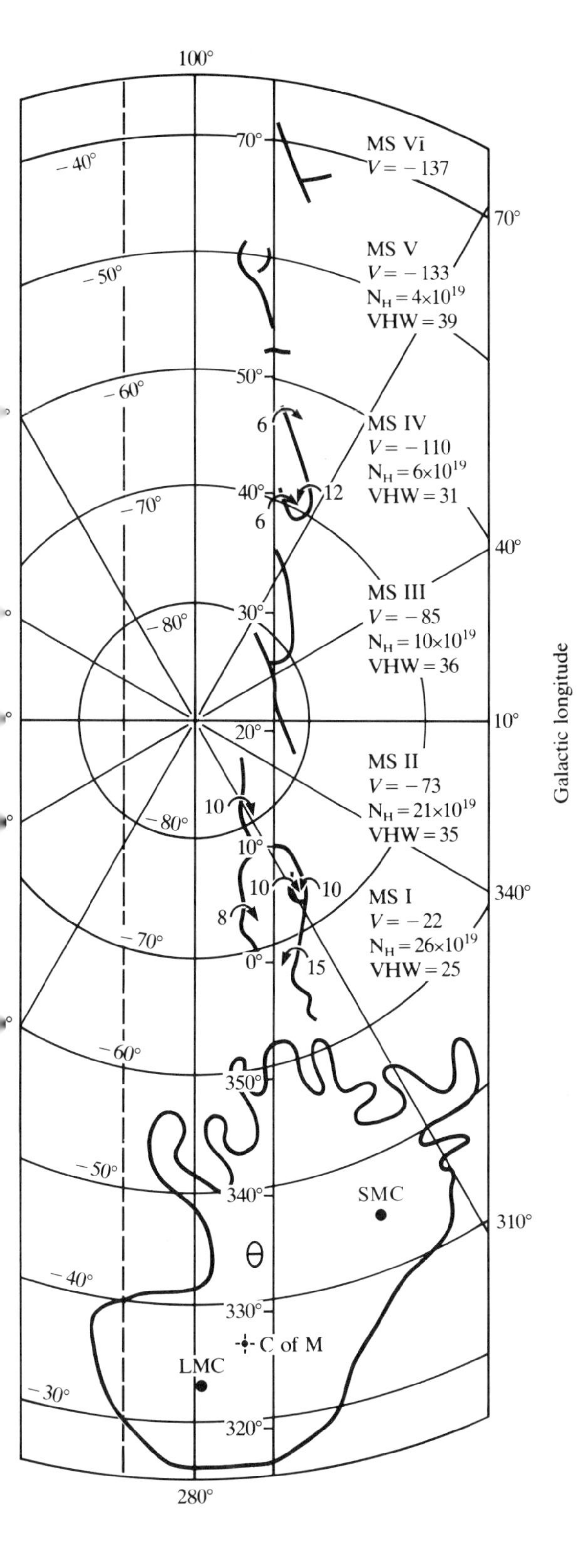

Figure 9-12. The Magellanic Stream. The contour surrounding the Magellanic Clouds is the 5×10^{19} atoms cm^{-2} H I surface density contour. [From (**M1**), by permission. Copyright © 1977 by the American Astronomical Society.]

Clearer evidence for intergalactic material around galaxies comes from the high-sensitivity neutral hydrogen surveys mentioned earlier. Studies of, for example, the M81/M82/NGC 3077 group show that complexes of neutral hydrogen having detectable surface densities ($>10^{18}$ atoms cm^{-2}) can envelop several adjacent galaxies (**B9**, 274).

A further important line of evidence for widespread intergalactic gas in groups of galaxies comes from the existence of absorption features in the ultraviolet, optical, and radio spectra of distant active galaxies and quasars. For many years, the origin of the absorption features that are commonly seen in the spectra of high-redshift quasars was hotly debated. Many of the absorption lines in these systems could be unambiguously identified with transitions in ions such as C IV and Si II. These identifications made it possible to establish that the redshift of many of the absorption systems were much smaller than the redshifts of the associated quasars; that is, the material causing the absorption must be moving at relativistic speed with respect to the emission source in the quasar against which the absorption is seen. A natural interpretation of this phenomenon is that the absorbing material is not associated with the quasar, but just happens to lie along the line of sight from the Sun to the quasar, and that the difference in redshift between the quasar emission spectrum and the absorption system represents the cosmological recession of the quasar from the absorbing material. However, this interpretation of the nature of the absorbing material requires an uncomfortable assumption: one finds that there are so many absorbing systems along the typical line of sight to, say, a quasar with redshift $z = 2.5$ that, if it is material in normal galaxies that is responsible for the absorption system, then these galaxies must be of the order of 100 kpc in radius (**B1**)!

The great linear size of galaxies implied by the intervening-matter hypothesis led many people to believe that the absorbing matter was associated with the quasar and that the large differences often observed between the redshifts of the emission and absorption systems indicate that the absorbing material has been ejected from the quasar at relativistic velocity. Actually, it has long been the case [see Chapter 5 and, for example, (**B3**)] that the available photometric evidence suggests that the tenuous outer regions of luminous galaxies merge smoothly with one another, so one has to consider seriously the evidence of the quasar absorption spectra for large gaseous halos around galaxies. Furthermore, two facts now strongly suggest that the majority of absorbing clouds are *not* associated with quasars but are clouds that just happen to intersect the line of sight from the Sun to the quasar: (1) One finds that high-redshift quasars have on the average more absorbing systems in their spectra than do low-redshift quasars, just as one would expect if the absorption systems arise from chance passages through intervening clouds of lines of sight of varying length (**S2**). (2) In a few cases, galaxies are actually seen at the same heliocentric redshift as an absorbing feature in the spectrum of an adjacent quasar (**B9**, 287). Instances in which a galaxy has been identified at the same redshift as an absorption system in a quasar spectrum are very rare, because, until recently, only either extremely distant or very thick absorbing clouds could be detected, whereas normal galaxies can be seen only at relatively small cosmological distances. The situation should change dramatically now that it is possible to observe the ultraviolet spectra of nearby quasars.

To summarize, several lines of evidence indicate that there is gaseous material around galaxies out to radii of *at least* 100 kpc. However, the typical intergalactic separation in groups of galaxies is less than or of the order of 200 kpc. Therefore, it is likely that at least a substantial fraction of the intergalactic space in groups of galaxies is filled with some sort of intergalactic medium. Just what this medium consists of is still a matter for speculation. It seems probable that, like the interstellar medium, any intergalactic medium will comprise more than one phase, and our present knowledge is probably restricted by an ability to detect only some of the phases.

9-3. MOLECULAR CLOUDS IN OUR GALAXY

Our conception of the interstellar medium has changed dramatically since observations of molecular species at millimeter wavelengths became possible in the early 1970s (**W6**). In this section, we shall summarize evidence that suggests that a large fraction of the galactic gas is concentrated in dense clouds that consist principally of molecular material and dust. These clouds are of particular importance because there are indications that they are the sites of new star formation. However, many questions concerning this component of our Galaxy remain open, and our discussion will of necessity be incomplete. In particular, it is still very uncertain exactly how much molecular material there is, and even fundamentals of the structure and life cycle of the clouds have yet to be worked out.

Molecular Material near the Sun

In the solar neighborhood, studies of the ultraviolet absorption spectra of nearby stars have detected H_2 molecules in diffuse interstellar clouds. These clouds generally have less than 2 mag of visual extinction, and the material at their cores is only partially converted to molecular form (**S8**). The chemistry of these clouds is evidently a complex balance between the formation of molecular species at particle densities $n \gtrsim 1000\ \text{cm}^{-3}$ and their subsequent photodissociation by ultraviolet photons (**K9**). No simple relation holds between the column densities of different molecular species and the total visual extinction.

E. B. Jenkins and D. B. Savage (**J2**) do, however, find that the sum N_H of the neutral and molecular hydrogen column densities along various lines of sight

$$N_H \equiv N_{HI} + 2N_{H_2} \tag{9-3}$$

are well correlated with visual extinction. They find that a plot of N_H against $E(B - V)$ for the lines of sight to ninety-five stars obeys the mean relation

$$\frac{N_H}{E(B - V)} = 7.5 \times 10^{21} \text{ atoms cm}^{-2} \text{ mag}^{-1} \tag{9-4}$$

which corresponds, via equation (3-75), to

$$\frac{N_H}{A_V} = 2.3 \times 10^{21} \text{ atoms cm}^{-2} \text{ mag}^{-1} \tag{9-5}$$

An inventory of the gaseous material within 1 kpc of the Sun is given in Table 9-1. Considerable uncertainty is attached to the quoted density of molecular gas, because the concentration of this material into discrete clouds makes it hard to be certain that one has sampled these in an unbiased way or has corrected properly for any bias toward more-or-less massive clouds. The figure quoted in Table 9-1 is that derived from the *Copernicus* observations of ultraviolet absorption by H_2.

Table 9-1. Inventory of Gas Within 1 kpc of the Sun

Ionized gas	$0.003\mathcal{M}_\odot/\text{pc}^3$
Neutral atomic gas	$0.031\mathcal{M}_\odot/\text{pc}^3$
Molecular gas	$0.007\mathcal{M}_\odot/\text{pc}^3$

SOURCE: (**F1**, 215)

Dark Molecular Clouds

Within a radius of a few hundred parsecs from the Sun, there are a number of well-studied clouds having extinctions that are too large ($A_V > 2$ mag) for ultraviolet observations to be possible. In these cases, one is obliged to infer the presence of the dominant molecular species, H_2, either from optical extinction or from millimeter observations of molecules such as CO that (unlike H_2) have observable radio-frequency lines. Carbon monoxide is the most widely observed molecule in our Galaxy, because its spectrum contains intense lines, at wavelengths of 1.3 mm and 2.6 mm, that arise from the $J = 2 \rightarrow 1$ and the $J = 1 \rightarrow 0$ changes in the rotation states of these molecules. Furthermore, besides the common isotope $^{12}C^{16}O$

(^{12}CO), there are the rarer forms ^{13}C^{16}O (^{13}CO) and ^{12}C^{18}O (C^{18}O), whose abundances are approximately 40 and 500 times smaller than that of the common isotope. Each of these species emits at a slightly different wavelength, and careful comparison of the profiles of the various lines furnishes valuable information about the structure of the clouds observed.

We shall see that, in some respects, the millimeter emission of CO may be used as a tracer of the molecular gas through the disk in much the same way as the 21-cm line of neutral hydrogen. There are, however, a number of important differences between CO and 21-cm observations that arise from differences in the physics involved in the production of the two types of line. One important distinction is that, whereas the hyperfine transition that gives rise to the 21-cm line of neutral hydrogen is a forbidden transition, the changes in the rotation transitions that give rise to the 1.3 mm and 2.6 mm lines of CO are allowed. This difference implies that much smaller column densities of CO molecules are required to establish a given optical depth in these lines than the neutral-hydrogen column density required to establish the same optical depth at 21 cm. In fact, the intrinsic strength of the rotation transitions of CO effectively compensates for the low abundance of CO molecules relative to hydrogen, so that the strongest CO lines saturate at similar column densities of all atoms ($\approx 2 \times 10^{19}$ atoms cm^{-2}/km s^{-1}) as does the 21-cm line (see Table 9-2 and §8-3). Furthermore, the clouds into which the molecular material is concentrated are so dense and cold that, not only do the strong lines saturate, but even some of the weaker ones saturate as well.

An observation of a saturated line yields the kinetic temperature of the emitting material near the surface at which optical depth unity is attained, but it provides no indication of the column density of emitting atoms. In Chapter 8, we saw by contrast how easy it is to infer a neutral-hydrogen column density from 21-cm line observations, because the 21-cm line is rarely fully saturated. This disadvantage of optically thick lines has led

Table 9-2. Carbon Monoxide Rotational Lines

Species	Relative abundance	Transition	$\sigma_{\tau=1}(^{12}\mathrm{CO})$*	$\sigma_{\tau=1}(\mathrm{H}_2)$
^{12}C^{16}O	1	$J = 2 \rightarrow J = 1$	9.6×10^{15}	7×10^{19}
		$1 \rightarrow 0$	2.2×10^{16}	2.8×10^{20}
^{13}C^{16}O	1/40	$2 \rightarrow 1$	2.2×10^{17}	2.8×10^{21}
		$1 \rightarrow 0$	8.8×10^{17}	1.1×10^{22}
^{12}C^{18}O	1/500	$2 \rightarrow 1$	2.8×10^{18}	3.5×10^{22}
		$1 \rightarrow 0$	1.1×10^{19}	1.4×10^{23}

* For each species $\sigma_{\tau=1}$, (^{12}CO) is the column density in atoms cm^{-2} of ^{12}C^{16}O molecules at $T = 20$ K, and distributed over a 1 km s^{-1} velocity range, that will saturate the given line; $\sigma_{\tau=1}(\mathrm{H}_2)$ is the corresponding hydrogen column density from equation (9-7).

SOURCE: Data published in (**G1**)

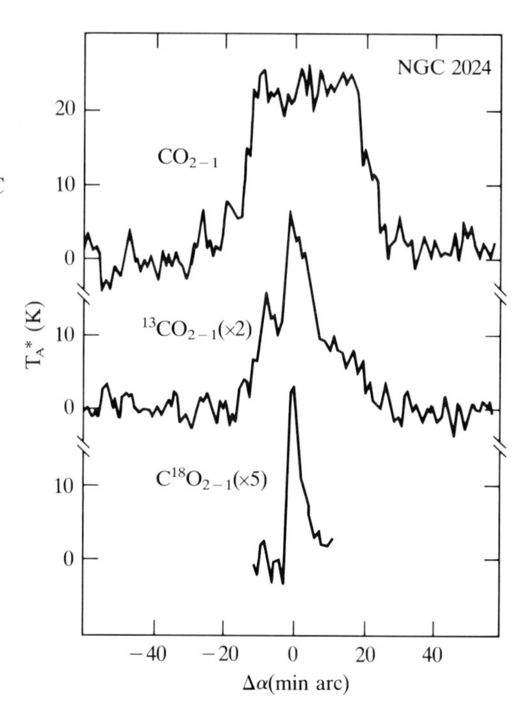

Figure 9-13. The dark cloud NGC 2024 traced in the 1.3-mm line radiation of three species of carbon monoxide. The top profile suggests that the region that is optically thick in the ^{12}CO line is spatially resolved and that the gas has a kinetic temperature near 20 K. Only a smaller portion of the cloud is optically thick in the ^{13}CO line, and the $C^{18}O$ line may be optically thin everywhere. Notice that the peak antenna temperatures differ by much less than the abundances of the various species. [From (**P1**), by permission. Copyright © 1979 by the American Astronomical Society.]

astronomers to study successively less intense CO lines in the hope of mapping the galactic molecular material in an optically thin line and thus obtaining estimates of column densities and total masses of this material.

T. G. Phillips et al. (**P1**) have assembled observations of several dark clouds in most of the CO lines listed in Table 9-2. They find that the antenna temperature in the 1.3-mm ^{12}CO line is invariably very similar to that in the 2.6-mm ^{12}CO line. From this observation, it follows that the cloud must be optically thick in both these lines because, if it were thin in the less opaque line (the 2.6-mm line), the latter would yield a lower antenna temperature. Figure 9-13 shows the spatial profiles of the dark cloud in NGC 2024 when observed in the 1.3-mm lines of ^{12}CO, ^{13}CO, and $C^{18}O$. The peak intensities of these radiations are in the ratio $1:\frac{1}{2}:\frac{1}{5}$, from which it follows that the cloud must be optically thick in the 1.3-mm line of ^{13}CO because, if it were thin in this line, it would be thin also in the corresponding line of $C^{18}O$, which would then be fainter than the ^{13}CO line by a factor of $500/40 > 10$, compared to the observed value of 2.5. Comparison of the 1.3-mm and 2.6-mm ^{13}CO antenna temperatures of several clouds (**P1**) confirms that ^{13}CO is usually optically thick. If either of these lines were optically thin, the 1.3-mm temperature would be nearly a factor of four higher than the 2.6-mm temperature, whereas they are, in fact, comparable. Nevertheless, the ^{13}CO line temperatures are invariably at least a factor two lower than the ^{12}CO temperatures. This behavior suggests that the unit-optical-depth surface of ^{13}CO has either a smaller radius or a lower brightness temperature than that of ^{12}CO.

In order to determine at what overall column density these various lines become optically thick, it is necessary to know the ratio of hydrogen molecules to CO molecules in these clouds. Unfortunately, it is extremely difficult to predict theoretically the abundance of CO relative to H_2, because one does not know what fraction of the carbon in the interstellar medium should be in atomic form or locked up in dust grains. Therefore, one is forced to an empirical determination. On the one hand, one has to calibrate the relationship between CO and H in relatively thin clouds, because optical and ultraviolet observations are impossible for very dense clouds, and, on the other hand, it is important that the clouds studied be sufficiently dense that their CO abundances are typical of dense clouds in which molecules might form easily and be less subject to photodissociation than in less-dense and well-shielded condensations. These considerations effectively restrict one to discussion of the correlation between optical extinction A_V and CO line strengths for clouds having $2 < A_V < 5$. R. L. Dickman (**D1**) has made such a study using ^{13}CO 2.6-mm line strengths and extinctions (estimated from star counts) for thirty-eight dark clouds. After converting his values of A_V to total hydrogen column densities N_H through equation (9-5), he concludes that

$$N(H_2) = (5.0 \pm 2.5) \times 10^5 \, N(^{13}CO) \tag{9-6}$$

The terrestrial abundance of ^{13}C relative to ^{12}C is 1/89, but it has been argued (**W1**) that the interstellar value lies near 1/40. If we accept this latter value, we find that

$$N(H_2) = (1.25 \pm 0.6) \times 10^4 \, N(^{12}CO) \tag{9-7}$$

With this ratio, about 12% of the carbon in molecular clouds is tied up in CO.

If we use these abundances to convert the CO column densities in the fourth column of Table 9-2 to hydrogen column densities, we obtain the numbers given in the fifth column of the table. All these column densities are per kilometers per second of the width of the line profile, so that a typical cloud having $T \approx 10$ K and a profile 5 km s^{-1} broad requires $N(H_2) \approx 10^{22}$ cm^{-2} to saturate the 2.6-mm ^{13}CO line. A cloud whose temperature exceeds 20 K requires even larger column densities to saturate its lines, because the saturation density is roughly proportional to T^2 (**G1**). The physical densities implied by these column densities are very high ($\gtrsim 5 \times 10^4$ cm^{-3}), and the associated freefall times (see Chapter 15) are extremely short ($\lesssim 10^6$ yr). At present, we do not understand what happens to the material of these clouds over longer periods of time. However, the slow rate at which material is incorporated into new stars indicates that most of the matter in molecular clouds must remain in the interstellar medium for much longer than a cloud freefall time.

Large-Scale Distribution of Molecular Material in Our Galaxy

The conclusion we have just reached, that most dense clouds are optically thick in the 2.6-mm line of ^{13}CO, is of great importance, for it has frequently been assumed that this line is optically thin, and then this assumption has been used to derive, from the ^{13}CO observations, estimates of the mass of molecular gas in the disk. As we have already emphasized, one cannot readily estimate masses from observations of optically thick lines. Therefore, before we can address the crucial question of what is the total mass of molecular material in our Galaxy, we must either (1) await the completion of a galactic survey in a line which is optically thin, or (2) estimate the masses of individual clouds or complexes with the help of dynamical models of some kind. We shall describe one dynamical estimate of the masses of molecular complexes, but, at the present time, one must regard the latter as very un-

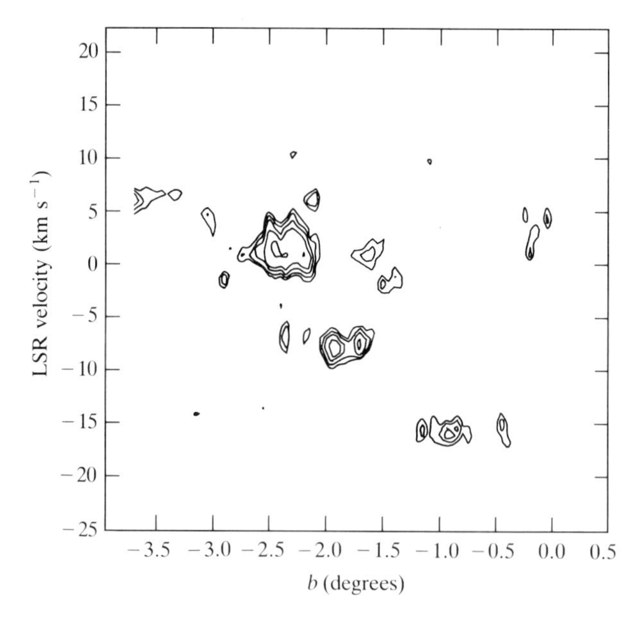

Figure 9-14. Observations of the 2.6-mm line of $^{12}C^{16}O$ in the anticenter direction indicate that CO in our Galaxy is distributed in discrete clouds. The contours in this diagram are drawn at antenna temperatures of 0.4, 0.9, 2, 4, and 9 K. The resolution element has dimension ($0.2^\circ \times 0.7$ km s^{-1}). [From (**S9**).]

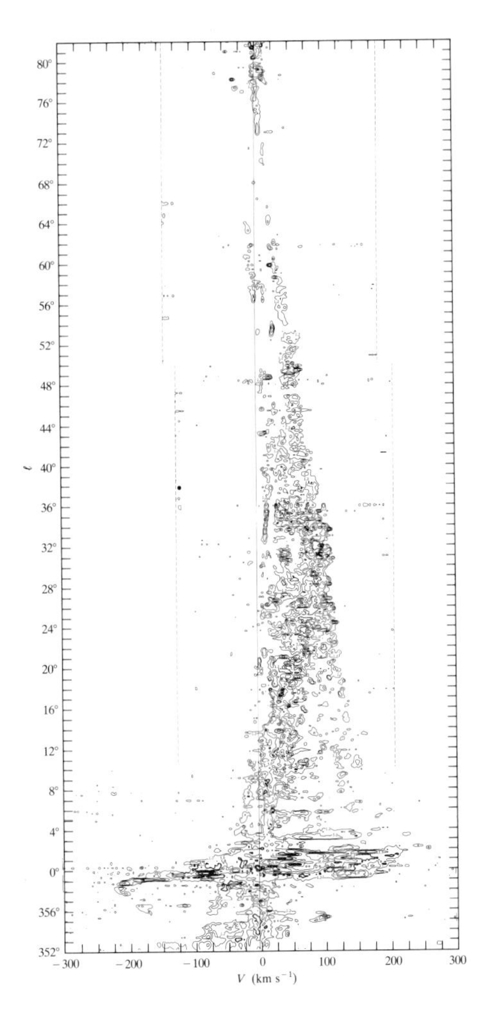

Figure 9-15. The galactic plane surveyed in the 2.6-mm line of ^{12}CO. Notice how strongly clumped the emission is and the scarcity of high-velocity emission in the interval ($4° < \ell < 18°$), as well as the brightness and large velocity range of the emission from the center region. [From (**B19**), by permission.]

certain. Fortunately, we shall see that it is possible to derive a good deal of information about the distribution of the molecular clouds through our Galaxy from measurements of ^{12}CO and ^{13}CO lines, even if they are optically thick.

Figure 9-14 shows the distribution of 2.6-mm ^{12}CO line antenna temperature in the anticenter direction. Notice that the CO emission breaks up into isolated islands in the (v, b) plot, each one of which can be identified with an interstellar cloud. Some of these clouds are known from optical observations as dark nebulae. Notice also that the ^{12}CO profile of the largest cloud is rather flat topped and steep sided on account of its optical thickness to ^{12}CO line radiation. Figure 9-15 shows the 2.6-mm ^{12}CO emission from the first quadrant ($0° < \ell < 90°$) of our Galaxy. Notice again how very clumpy the emission is; no emission is detectable in many areas that are adjacent to points showing strong emission. In this regard, the CO emission is unlike the smoother distribution of H I emission shown in Figure 9-4. Two further points are noteworthy in Figure 9-15: there is very strong emission from the galactic center region ($|\ell| < 3°$), and there is a conspicuous lack of high-velocity ($v > 60$ km s^{-1}) material in the interval ($4° < \ell < 18°$). We shall see that this gap in the CO emission can be interpreted as an absence of molecular material in the region 4 kpc $> R >$ 500 pc, similar to that observed in the distribution of H I (see Figure 9-4).

A plot of the 2.6-mm ^{13}CO antenna temperature through the disk closely resembles that shown in Figure 9-15 for the ^{12}CO distribution (S7). This important fact rules out a widely discussed model of the distribution of molecular material in our Galaxy, namely, the model in which the ^{13}CO is optically thin and is emitted by clumps that are large enough (diameter $>$ 10 pc) to be resolved even at the galactic center, for, according to this hypothesis, an enhancement in the ^{12}CO antenna temperature should be associated with a change in the ratio of the ^{12}CO and ^{13}CO antenna temperatures. Actually, this ratio $\approx$5.5 and shows no systematic variation with ^{12}CO antenna temperature as the telescope beam is swept in either ℓ or v (S7). This similarity of the ^{12}CO and ^{13}CO distributions can, on the other hand, be understood if we imagine that the brightness variations in Figure 9-14 are caused by variations in the number of small, unresolved cloudlets falling within the field of the telescope. Furthermore, this picture of the overall distribution of CO-emitting gas breaking up into a large number of compact, optically thick cloudlets is consistent with the studies of relatively well-resolved dark clouds near the Sun discussed earlier.

If this model of the distribution of the emitting molecular gas is correct, the strong temperature variations shown in Figure 9-15 must be caused by the individual cloudlets being strongly clumped into *cloud complexes* in the same way that stars cluster into galaxies. Figure 9-16 shows a histogram of the diameters of the complexes that one sees at R = 5.5 kpc from the galactic center. Complexes smaller than $\simeq$12 pc in diameter are probably very numerous, but they would not have been detected in the survey from which Figure 9-16 was constructed. This distribution of the sizes of complexes has a property that is remarkably reminiscent of the galaxy luminosity function discussed in §5-3: if we assume that the mass of a complex is simply proportional to its volume, then we find that, although there are more small complexes than big ones, the big complexes contain most of the molecular material in our Galaxy. Figure 9-17 illustrates this result. It is likely that more than half the molecular material in the disk is concentrated into complexes having diameter $D \gtrsim 80$ pc.

Two dynamical arguments have been used to obtain estimates of the mass of giant molecular cloud complexes: (1) the virial theorem (Chapter 18) can be used to estimate the masses from knowledge of the widths of the velocity profiles of individual complexes; and (2) one can determine the minimum

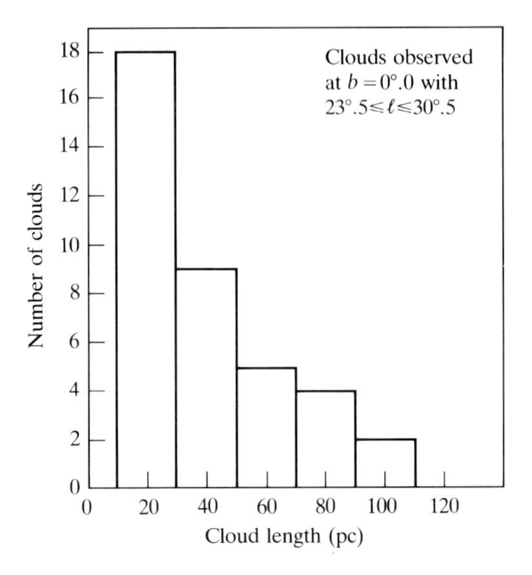

Figure 9-16. The distribution of the diameters of molecular cloud complexes seen in the longitude range $23.5° < \ell < 30.5°$. [From (**S6**), by permission.]

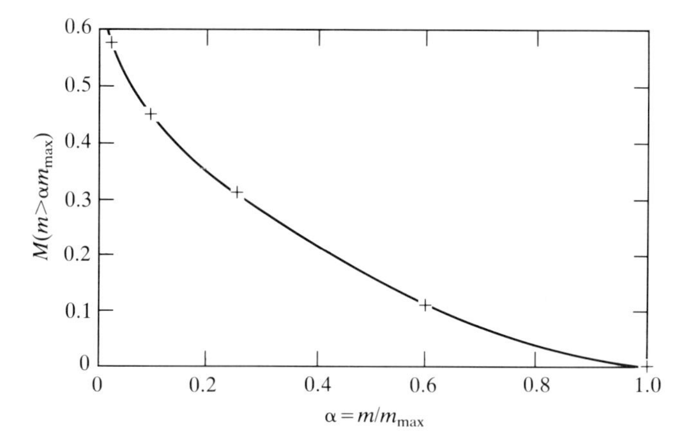

Figure 9-17. Most of the galactic molecular gas is probably concentrated into a few thousand giant complexes. This diagram shows the fraction M of the entire mass that is contained in complexes whose masses are greater than a fraction α of the mass of the most massive complex. [Adapted from (**S6**), by permission.]

mass a complex must have to avoid being torn apart by the tidal field of our Galaxy (**S10**). Neither of these arguments is conclusive because one cannot be certain that the complexes are long-lived objects that are in virial equilibrium and stable against tidal disruption by our Galaxy, nor can one be certain that their overall line widths are not produced by motions within their tightly bound subcomponents. In any event, both criteria lead to $\mathscr{M} \approx 10^6 \mathscr{M}_\odot$ for the largest complexes. Thus it is possible that giant molecular complexes are, in fact, extremely massive objects.

It has been estimated **(S6)** that there are about 4000 giant molecular complexes in our Galaxy. Thus, if we adopt $\mathscr{M} \approx 10^6 \mathscr{M}_\odot$ for the mass in each complex, we conclude that the total mass in them is around $4 \times 10^9 \mathscr{M}_\odot$. For comparison, the 21-cm surveys that we discussed in §9-1 indicate (**B12**, Chapter 5), (**K5**) that the total mass of neutral hydrogen in our Galaxy is about $5 \times 10^9 \mathscr{M}_\odot$. We emphasize again that substantial uncertainty attaches to the mass of molecular material in our Galaxy, but this result suggests that a large portion of the gas in our Galaxy may be concentrated into a few giant molecular complexes. As was mentioned in Chapter 7, it is likely that the tendency of moving cloud complexes to scatter gravitationally and thus accelerate disk stars is responsible for the observed correlations between the kinetic and chemical properties of stars in the solar neighborhood.

An important consequence of the compactness of the cloudlets that are responsible for most of the CO emission is that the CO antenna temperature in any range of ℓ and v must be proportional to the total mass of molecular material in that range, precisely as was assumed to be the case for the 21-cm (ℓ, v) distribution of Figure 9-4. Therefore, we can apply the same techniques used to convert Figure 9-4 to Figure 9-7 for the radial distribution of neutral hydrogen. Figure 9-18 shows the radial distribution of CO that one derives

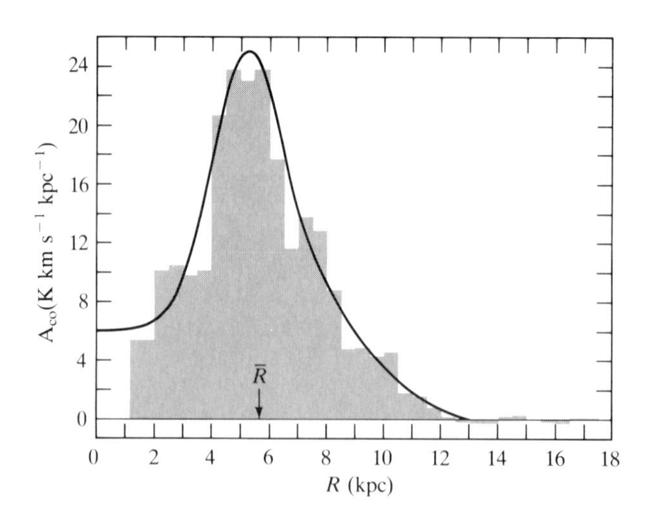

Figure 9-18. The radial distribution of molecular clouds as deduced from CO emission-line observations. This figure should be compared with the H I distribution shown in Figure 9-7. [From (**B19**), by permission.]

in this way from Figure 9-15. Notice that the distribution of molecular material is much less flat than is the neutral hydrogen distribution. Indeed, much of the molecular material appears to be concentrated into a ring having a mean radius $\approx$ 5.5 kpc and width $\approx$ 3 kpc, which has no analogue in the neutral hydrogen distribution. Both the molecular and the neutral hydrogen distributions fall steeply interior to 4 kpc. The molecular material, unlike the neutral hydrogen, is largely concentrated inside the solar circle.

The molecular material is found to be more highly concentrated toward the galactic plane than is H I. The molecular clouds have a mean half-thickness at half maximum intensity of about 60 pc compared to about 130 pc for H I (**S6**). The thickness of the molecular cloud layer, which seems to increase slightly from $R = 3$ kpc to $R = 8$ kpc from the galactic center, should be related to the random velocities of the clouds. A. A. Stark (**S9**) finds most cloud complexes move with random velocities characterized by velocity dispersion $\sigma \approx 8 \text{ km s}^{-1}$, although the most massive complexes tend to move more slowly, $\sigma \approx 3 \text{ km s}^{-1}$.

Distributions of Molecular Clouds and Young Stellar Objects

Evidently, a prerequisite for the formation of stars in an annulus of radius R about the galactic center is the presence of gas. We have seen that the two forms of gas that we can detect through our entire Galaxy have rather different radial distributions, and it is interesting to enquire whether either of these distributions resembles the distribution in the disk of young stellar objects.

There are several ways in which one might endeavor to plot the distribution of young stellar objects.

1. One can plot the distribution of the H II regions that are excited by young blue stars by detecting the radio-frequency recombination lines such as H166α.
2. Alternatively, one can plot the positions of supernova remnants as shown by their radio continuum radiation.
3. Or, one can plot the distribution over galactic longitude of γ rays. It is uncertain how γ rays in the energy band 35–100 MeV are produced, but the intensity of this class of γ ray is concentrated toward the galactic plane and toward the center direction, which suggests that they are produced by energetic disk objects of some sort, very possibly supernovae.

Figure 9-19 shows estimated radial distributions of these three classes of source. Substantial uncertainties attach to two of these distributions on account of the lack of kinetic data from radio-continuum and γ-ray observations. However, it is apparent that all of the distributions of Figure 9-19

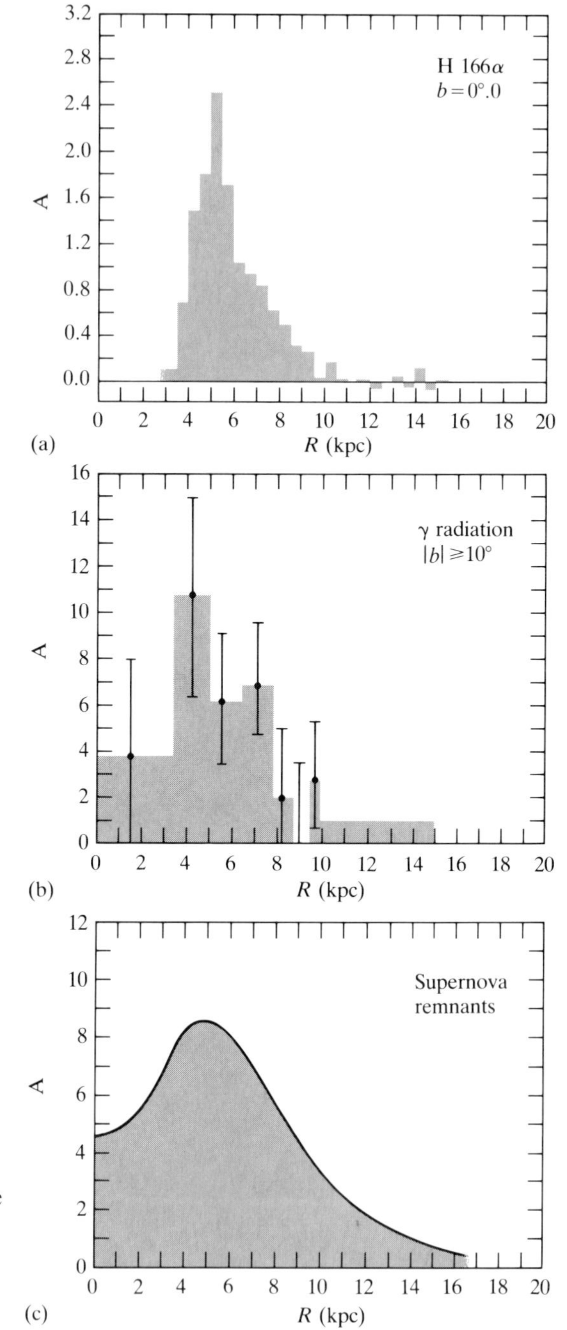

Figure 9-19. Tracers of young stellar objects seem to have similar distributions to the molecular cloud distribution shown in Figure 9-18. Shown are (a) the distribution of H II regions that is derived from radio recombination line studies (**L4**), (b) the inferred distribution of 35–100 MeV γ-ray sources (**S11**), and (c) the distribution of supernova remnants that has been derived from radio continuum radiation studies (**K10**).

more closely resemble the CO distribution shown in Figure 9-18 than the H I distribution of Figure 9-7. This fact strongly suggests that molecular clouds are the sites of star formation in our Galaxy. Indeed, in the Orion nebula, we see that molecular sources and young stars are closely allied. In other nebulae, for example, M8, extremely opaque globules are detected optically against the bright background of the emission nebula. It is possible that these globules [often referred to as *Bok globules* (**B13**)] are the optical counterparts of the dense cores of molecular emission discussed earlier.

9-4. THE GALACTIC CENTER

The galactic center is hidden from us at optical wavelengths by very strong interstellar absorption in the disk. However, over the past two decades, radio and infrared observations of the region $R < 4$ kpc have enabled us to study the galactic center in considerable detail. In particular, we have been able to obtain data concerning the disposition and motions of gas at the center of our Galaxy which are much more complete than those available for other galaxies. Unfortunately, the rapid observational progress that has characterized the last two decades has not been matched by progress in our theoretical understanding of the nature of the center region. Even the gross outlines of the structure of the region $R < 4$ kpc remain very obscure, and the task of synthesizing the wealth of available observational data into a single coherent picture remains one of the outstanding challenges of modern galactic astronomy.

21-cm Observations

Figure 9-20 shows a contour plot of H I brightness temperature at $b = 0^\circ$ in the longitude-velocity (ℓ, v) plane. A glance at this figure will convince the reader that no simple model will be able to account for the distribution of neutral hydrogen at the galactic center. Indeed, if the neutral hydrogen at the galactic center were arranged in a disk that rotated around the galactic nucleus, the (ℓ, v) plot would be symmetric on rotation by 180° about the point $(\ell = 0^\circ, v = 0)$, which Figure 9-20 clearly is not. In particular, it is clear that there is much more neutral hydrogen at positive longitudes and velocities (top right in the figure) than at negative longitudes and velocities (bottom left). Furthermore, even if the H I were concentrated into clouds which, like the globular clusters, moved in a random fashion around the galactic center, then one would still expect the (ℓ, v) plot to be symmetric on rotation by 180°.

Figure 9-20 does, nevertheless, show large-scale structure. One very conspicuous feature of the galactic neutral hydrogen distribution is the thin tongue of 21-cm emission that extends at $-4^\circ < \ell < 0^\circ$ from velocity

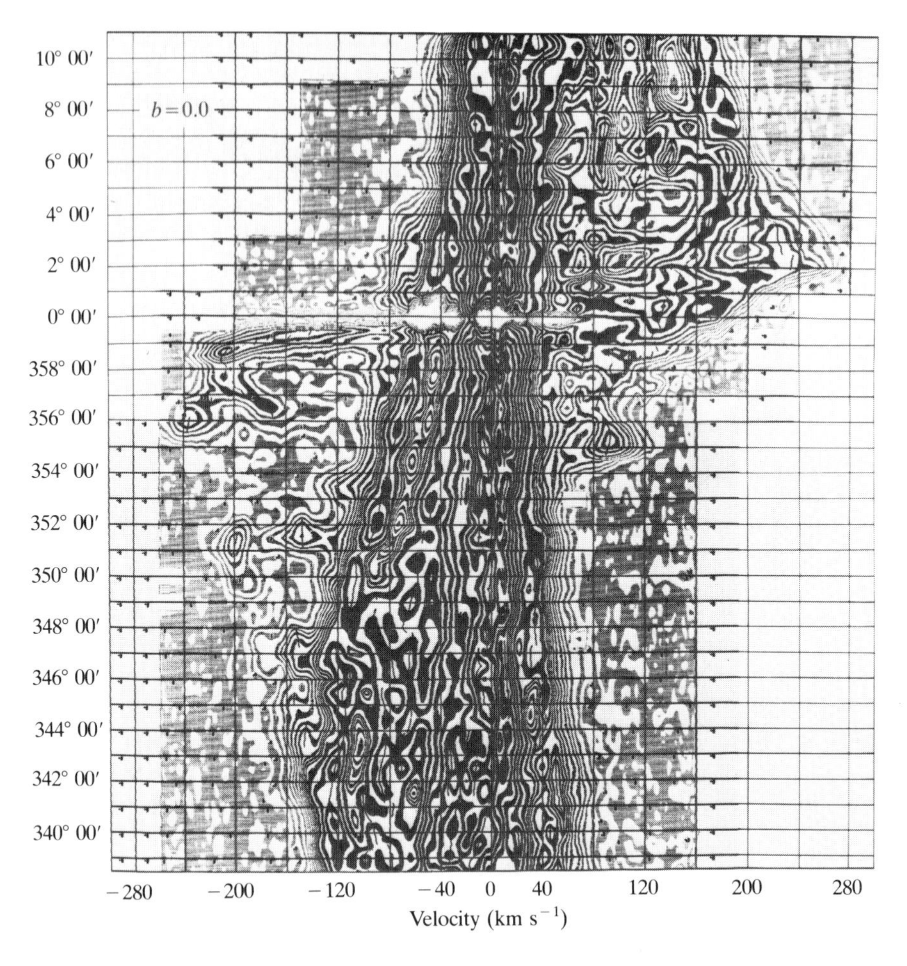

Figure 9-20. Contours of equal 21-cm brightness temperature at $b = 0°$. Notice that the figure is not symmetrical on rotation by 180° about ($\ell = 0°$, $v = 0$). [From **(S5)**.]

$v \approx -90$ km s^{-1} out to $v \approx -240$ km s^{-1}. If one imagines [for example, **(R5)**] that a similar extension exists at positive longitude and velocities ($0° < \ell < 4°$, $+90$ km s^{-1} $< v <$ $+240$ km s^{-1}), then it is possible to interpret this feature as arising from a disk of rapidly rotating gas. The radius of the disk may be estimated as $R \approx 700$ pc from the maximum extension of the feature in longitude.

Two other large-scale features in Figure 9-20 have aroused considerable interest. The first of these is the long ridge of emission that runs in an almost straight line from ($\ell = 0°$, $v \approx 50$ km s^{-1}) down toward ($\ell \approx -15°$, $v \approx -120$ km s^{-1}). On the far side of the $\ell = 0°$ line, this ridge may be followed to at least $\ell = 5°$. Figure 9-20 shows a strong depression in the

H I brightness temperature along the ridge near $\ell = 0^\circ$. This depression arises because, at $\ell = 0^\circ$, the material whose 21-cm emission gives rise to the ridge is seen in absorption against the high-temperature continuum radiation from the galactic nucleus. The ridge terminates near longitude $\ell = 20^\circ$, which corresponds to $R \approx 3.5$ kpc. This termination radius has caused the feature to be known as the *3-kpc arm*. Actually, it does not follow from Figure 9-20 that this feature arises from any kind of spiral arm, or that it arises from material at $R \approx 3$ kpc. However, there does appear to be some kind of structure near the galactic center whose 21-cm emission extends at least over $-15^\circ < \ell < 5^\circ$ in longitude and $-120 \text{ km s}^{-1} < v < -40 \text{ km s}^{-1}$ in velocity. Notice that the fact that we see this feature in absorption against the galactic center at nonzero radial velocity ($v \approx -53 \text{ km s}^{-1}$) necessarily implies that the material that gives rise to it is not in circular rotation about the galactic center. Indeed, the material of the 3-kpc arm clearly has a component of motion away from the galactic center toward the Sun.

A second feature in Figure 9-20 is the emission at negative longitudes ($-5^\circ < \ell < 0^\circ$) and positive velocities ($50 \text{ km s}^{-1} < v < 200 \text{ km s}^{-1}$). The presence of 21-cm radiation in this region of the (ℓ, v) plane is surprising, because any rotational contribution to the radial velocity of material at negative longitudes should have a negative sign. Notice also that there is no depression in the 21-cm brightness temperature along the line $\ell = 0^\circ$ at positive velocities like that seen at ($\ell = 0^\circ$, $v = -53 \text{ km s}^{-1}$) caused by absorption by the material of the 3-kpc arm against the nuclear continuum sources. These observations suggest that the material responsible for the 21-cm emission at ($\ell < 0$, $v > 0$) lies behind, and, like the material of the 3-kpc arm, is expanding away from the galactic center. This feature is frequently referred to as *the expanding arm at 135 km* s^{-1}. However, it must be remarked that, if its material were very clumpy, it might lie on our side of the center and still not be seen in absorption against the center sources. In this case, it would be falling into the center rather than expanding away from it.

Many other ridges and peaks have been identified in Figure 9-20 and in similar diagrams that are obtained by surveying the 21-cm brightness temperature at values of galactic latitude other than $b = 0^\circ$; reviews of these features can be found in (**O1**) and (**S5**). Until we have a successful model of the distribution of H I in the center region, it is hard to assess the significance of most of these features. What is important at this stage is that, interior to 4 kpc, there is probably a rapidly rotating neutral hydrogen disk having radius $R \approx 700$ pc, and there are additional large-scale ($R \approx 1$ kpc) features which have components of velocity that are directed away from the galactic center. We do not know whether these motions represent ejection of gas by the nucleus, or whether they arise from our particular view of a noncircular pattern of circulation around the center, because we have no way of estimating components of velocity transverse to the line of sight. Oort (**O1**) has discussed the arguments for and against these two possibilities.

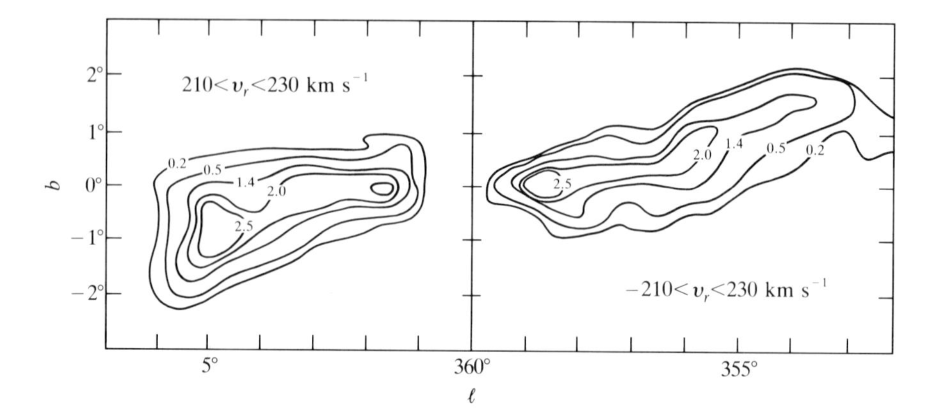

Figure 9-21. These channel maps of the galactic center at $v = \pm 220$ km s^{-1} show that neutral hydrogen within about 1.5 kpc of the galactic center is rotating rapidly about a tilted rotation axis. The measurements have an angular resolution of 24 arcsec. [From (**S5**).]

The neutral hydrogen inside $R = 4$ kpc is not strongly confined to the plane. The half thickness of the material interior to $\ell = \pm 10°$ ($R < 1.7$ kpc) is about 1°, and isolated features have been seen at latitudes of up to 10° (**S1**). Figure 9-21 shows that the rotation axis of the central H I disk is tilted with respect to the galactic plane, as is the plane of much of the "expanding" neutral hydrogen that surrounds the central disk (**C2**).

Molecular Clouds at the Galactic Center

About 300 pc from the center, the density of molecular cloud complexes (which is, in general, small interior to $R \approx 4$ kpc; see §9-3) rises dramatically. Indeed, if we accept the estimates of the masses of molecular cloud complexes discussed earlier, then we conclude from 2.6-mm CO observations that most of the gaseous material interior to $R = 300$ pc ($\ell = 2°$) is in molecular form (**L2**). T. M. Bania (**B2**) estimates that there is more than $10^8 \mathscr{M}_\odot$ of H_2 interior to 300 pc, which may be compared with an estimated $4 \times 10^6 \mathscr{M}_\odot$ of neutral hydrogen in the entire 700-pc radius H I disk (**O1**).

The distribution and kinematics of the central molecular material are broadly similar to those of neutral hydrogen at $R \lesssim 4$ kpc. Figure 9-22 shows the distribution in the (ℓ, v) plane at $b = 0°$ of 2.6-mm line emission by ^{12}CO. One again notices in this figure an asymmetry on rotation of the diagram by 180° about ($\ell = 0°, v = 0$), which is similar to that displayed by Figure 9-20 for H I. Once again, we find that there is more gas at positive longitudes and velocities than at negative longitudes and velocities. Notice

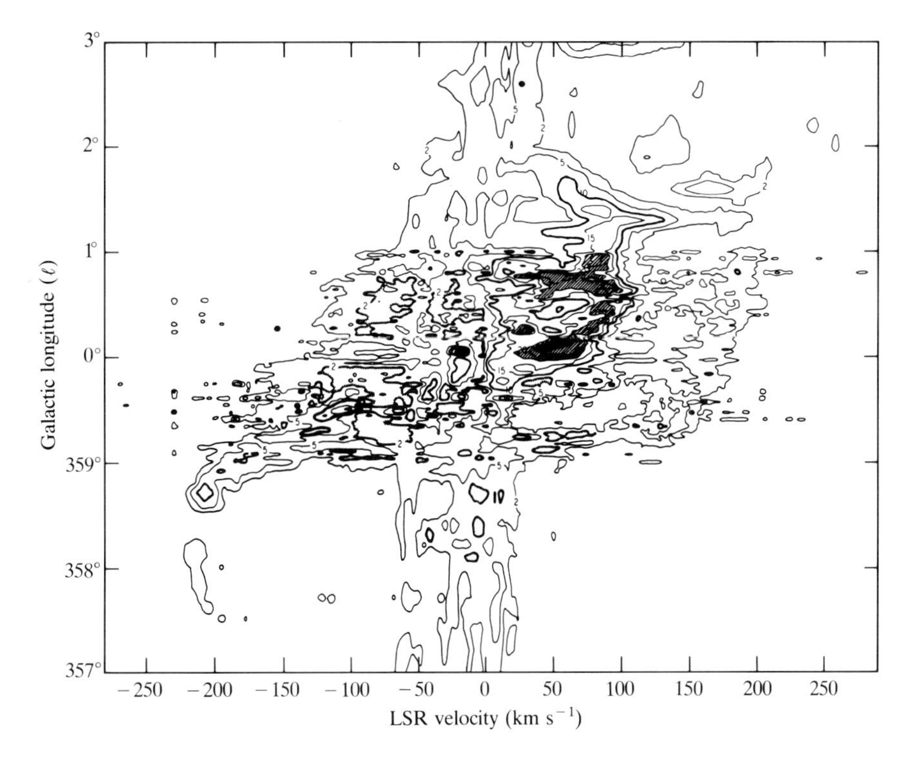

Figure 9-22. Contours of equal 2.6-mm ^{12}CO brightness temperature at $b = 0°$. Notice similarities with the H I distribution shown in Figure 9-20. The emissions of the 40 km s^{-1} cloud and of the Sagittarius B2 complex are shown shaded. [From (**O1**). Reproduced with permission from the *Annual Review of Astronomy and Astrophysics*, Volume 15. Copyright © 1977 by Annual Reviews, Inc.]

also that Figure 9-22 shows that there are two giant molecular complexes at ($\ell = 10'$, $v \approx 40$ km s^{-1}) and at ($\ell \approx 45'$, $v \approx 70$ km s^{-1}). The first of these clouds (often referred to as the "40 km s^{-1} cloud") may possibly be associated with the radio continuum source, *Sagittarius A*, that is thought to mark the exact center of our Galaxy. The other cloud is associated with at least seven H II regions and is known as the *Sagittarius B2 complex*.

Just as the observation of a 21-cm absorption feature in the spectrum of part of the 3-kpc arm suggests that the 3-kpc arm lies in front of, and has a component of motion away from, the center, so measurements of the absorption features due to the hydroxyl radical (OH) and formaldehyde (H_2CO) in the spectra of diffuse radio-continuum sources have been used to suggest that the Sagittarius B2 complex lies behind the center and is expanding away from it. In fact, it has even been suggested, on the basis of OH and H_2CO absorption measurements (**K2**), (**S4**), that the Sagittarius B2 complex is a member of an expanding ring of molecular clouds. However, it must be remembered that the emitting molecular material is likely to be very clumpy, so the Sagittarius B2 complex could lie on our side of the center and still not be visible in absorption.

The layer of molecular clouds has roughly the same half-thickness at 100 pc from the center as at $R = 5$ kpc. We shall see in Chapter 14 that this fact indicates that the velocity dispersion of the clouds must be appreciably higher at $R = 100$ pc than at $R = 5$ kpc.

Infrared and Radio-Continuum Observations

The highest-resolution data for the center region are provided by observations at radio and infrared wavelengths. Figure 9-23 shows the large-scale distribution of 3.75-cm radiation from the nuclear region. The peak bright-

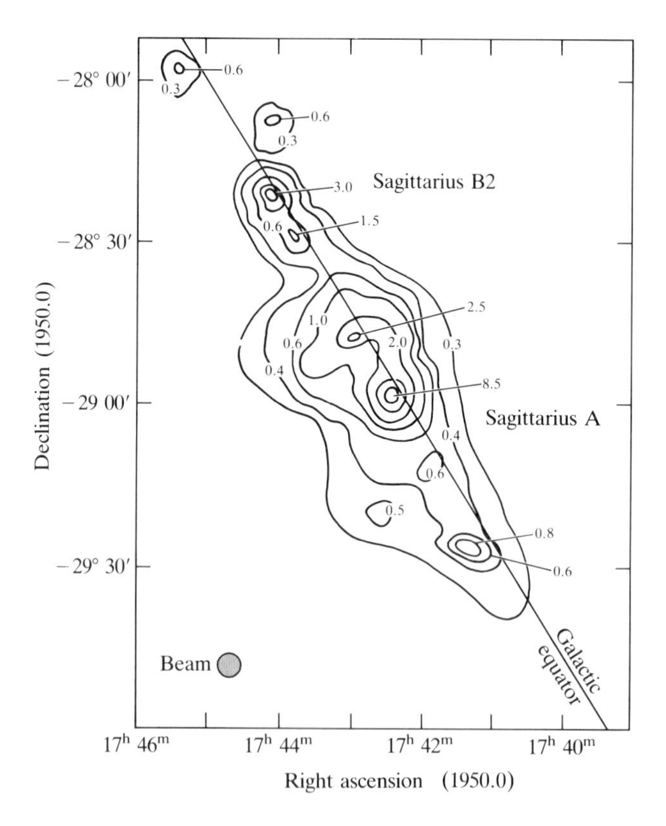

Figure 9-23. The large-scale distribution of 3.75-cm radio-continuum emission from the galactic center. [From (**O1**). Reproduced with permission from the *Annual Review of Astronomy and Astrophysics*, Volume 15. Copyright © 1977 by Annual Reviews, Inc.]

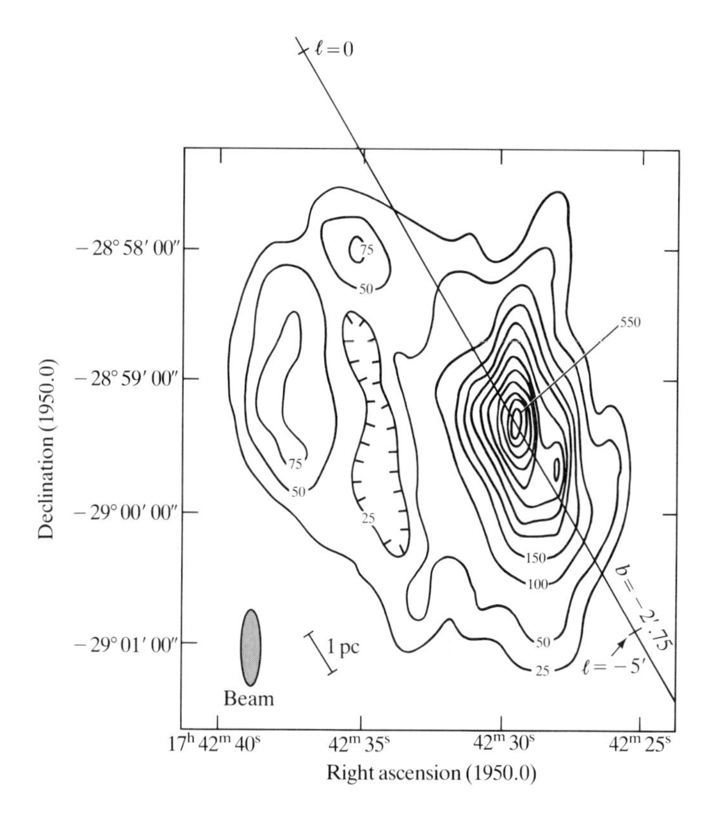

Figure 9-24. The 6-cm radio-continuum emission of Sagittarius A on a scale fifteen times larger than the scale of Figure 9-23. Sagittarius A West (the right-hand peak, which marks the true galactic center) has galactic coordinates $\ell = -3'.34$, $b = -2'.75$). The features appear more extended in the N–S direction than they really are because the beam shape (bottom left) is highly elliptical. [From (**O1**). Reproduced with permission from the *Annual Review of Astronomy and Astrophysics*, Volume 15. Copyright © 1977 by Annual Reviews, Inc.]

ness in this figure is emitted by the source Sagittarius A, which is located at one end of an arc of enhanced radiation. In Figure 9-24, which shows the 6-cm radiation of Sagittarius A on a scale fifteen times larger than Figure 9-23, one sees that Sagittarius A actually comprises two sources. Sagittarius A East is probably a supernova remnant, while very long baseline radio interferometry has shown that Sagittarius A West contains an ultracompact nonthermal source of diameter $\theta \lesssim 0.001''$, that is, $D \lesssim 10$ a.u. (**K3**). It is known that many active galaxies have very compact central sources, but for no external galaxy is the limit on the size of such a source so stringent as a mere 10 a.u. The radio luminosity of the galactic center

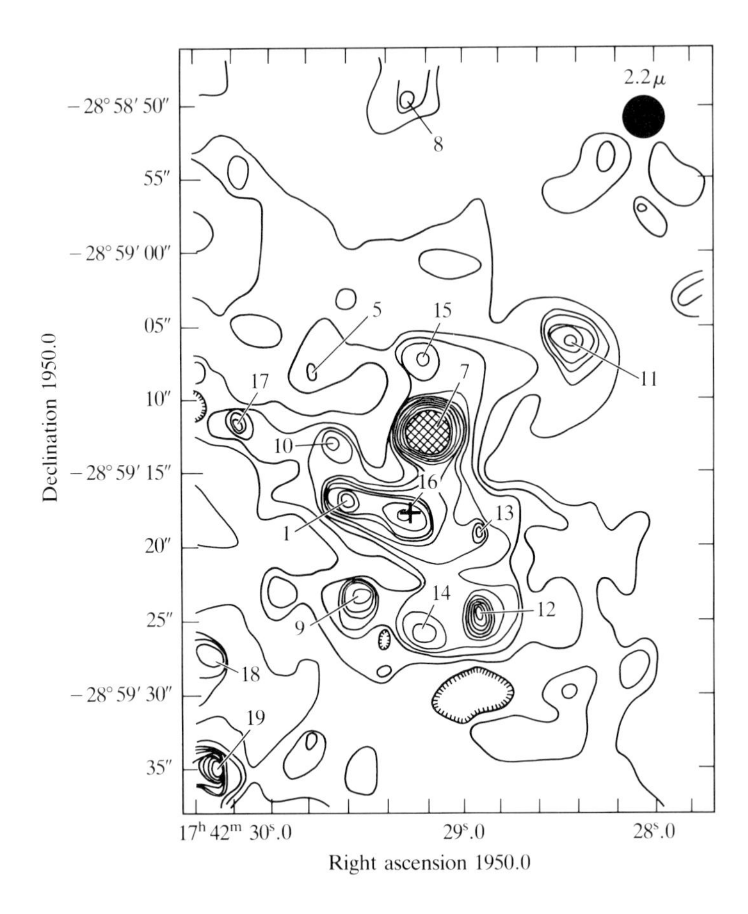

Figure 9-25. The galactic center region in 2.2μ infrared radiation. The position of Sagittarius A West is marked by a cross. The cross hatching at the center of source 7 represents thirty-five contour levels. Notice how the (probably stellar) infrared sources cluster around Sagittarius A West. [From (**B6**) by permission. Eric E. Becklin and Gerry Neugebauer, Palomar Observatory, California Institute of Technology. Copyright © 1975 by the American Astronomical Society.]

region is only about 10^{33} erg s^{-1} ($\frac{1}{4}L_{\odot}$), which is some 13 orders of magnitude smaller than that of the radio nuclei of some galaxies. Yet, the ultracompact source in Sagittarius A West does have an extremely high brightness temperature ($T \approx 10^{10}$ K), which implies that the physical processes occurring in this region are of an exotic nature.

Figure 9-25 shows that the position of the core of Sagittarius A West is located at the center of a ring of 2.2μ infrared sources and within 5 arcsec of the brightest galactic-center infrared source. Infrared radiation at 2.2μ is probably largely stellar in origin (arising mostly from K giants rather than interstellar dust), which implies that Sagittarius A West is at the very center of our Galaxy, where the star density is highest. In fact, the radial distribution of near-infrared radiation at the galactic center closely resembles the distributions of both infrared and optical radiation from the center of M31 (**B6**). From the spectrum of the galactic infrared radiation and comparisons with observations of M31, one derives extinctions of 3 mag and about 27 mag at 2μ and in the V band, respectively, for the line of sight to the center. Correcting for this extinction and adopting a mass-to-light ratio of eight in solar units, one concludes that the stellar mass within 50 pc is about $5 \times 10^8 \mathscr{M}_{\odot}$, a value that is consistent with the rotation rate of the H I nuclear disk.

Infrared observations have also furnished an important piece of kinematic information concerning the galactic nucleus. E. R. Wollman and his collaborators (**W7**) have used observations of a line at wavelength 12.8μ that arises from a fine-structure transition in once-ionized neon (NeII) to show that Sagittarius A West is surrounded by a rotating mass of ionized gas. The gas can be detected out to about 15 arcsec (0.7 pc) on either side of Sagittarius A West and shows a systematic velocity shift with galactic longitude. The magnitude of the rotation velocity deduced from this velocity shift is not large (only about 150 km s^{-1}), but this systematic effect is associated with a large line width (about 200 km s^{-1}), which suggests that noncircular motions help to support the gas against the gravitational attraction of the nucleus. From these velocities, Wollman et al. obtain a tentative upper limit for the mass within 1 pc of the galactic center of $4 \times 10^6 \mathscr{M}_{\odot}$. J. H. Lacy et al. (**L1**) have presented evidence that the central mass of gas is very clumpy and rotates about an axis that is strongly inclined to the general galactic rotation axis. The dynamics of this material are not well understood at present.

REFERENCES

(**B1**) Bahcall, J. 1975. *Astrophys. J.* **200**:1.

(**B2**) Bania, T. M. 1977. *Astrophys. J.* **216**:381.

(**B3**) Baum, W. A. 1955. *Pub. Astron. Soc. Pacific.* **67**:328.

(**B4**) Becker, W. 1961. *Z. für Astrophys.* **51**:155.

(**B5**) Becker, W. and Contopoulos, G. (eds.). 1970. *The Spiral Structure of Our Galaxy.* IAU Symposium No. 38 (Dordrecht: D. Reidel).

(**B6**) Becklin, E. E. and Neugebauer, G. 1975. *Astrophys. J. Letters.* **200**:L71.

(**B7**) Beer, A. 1961. *Mon. Not. Roy. Astron. Soc.* **123**:191.

(**B8**) Beer, A. 1961. *Mon. Not. Roy. Astron. Soc.* **123**:216.

(**B9**) Berkhuijsen, E. M. and Wielebinski, R. (eds.). 1978. *Structure of Properties of Nearby Galaxies*, IAU Symposium No. 77. (Dordrecht: D. Reidel).

(**B10**) Bieging, J. H. 1978. *Astron. and Astrophys.* **64**:23.

(**B11**) Binney, J. J. 1978. *Mon. Not. Roy. Astron. Soc.* **183**:779.

(B12) Blaauw, A. and Schmidt, M. (eds.). 1965. *Galactic Structure*. (Chicago: University of Chicago Press).

(B13) Bok, B. J. 1977. *Pub. Astron. Soc. Pacific*. **89**:597.

(B14) Bosma, A. 1978. Ph.D. Thesis, University of Groningen, Netherlands.

(B15) Burke, B. F. 1957. *Astron. J.* **62**:90.

(B16) Burton, W. B. 1970. *Astron. and Astrophys.* **10**:76.

(B17) Burton, W. B. 1972. *Astron. and Astrophys.* **19**:51.

(B18) Burton, W. B. 1976. *Ann. Rev. Astron. and Astrophys.* **14**:275.

(B19) Burton, W. B. and Gordon, M. A. 1978. *Astron. and Astrophys.* **63**:7.

(C1) Chiu, H.-Y. and Muriel, A. (eds.). 1970. *Galactic Astronomy*. (New York: Gordon and Breach).

(C2) Cohen, R. J. and Davies, R. D. 1976. *Mon. Not. Roy. Astron. Soc.* **175**:1.

(D1) Dickman, R. L. 1978. *Astrophys. J. Supp.* **37**:407.

(E1) Emerson, D. T. 1978. *Astron. and Astrophys.* **63**:L29.

(F1) Fichtel, C. E. and Stecker, F. W. (eds.). 1977. *The Structure and Content of the Galaxy and Galactic Gamma Rays* (Washington, D.C.: NASA).

(F2) Fletcher, E. 1963. *Astron. J.* **68**:407.

(G1) Goldreich, P. and Kwan, J. 1973. *Astrophys. J.* **189**:441.

(H1) Hindman, J. V. 1967. *Australian J. Physics*. **20**:147.

(H2) Hunter, C. and Toomre, A. 1969. *Astrophys. J.* **155**:747.

(J1) Jackson, P. D. and Kellman, S. A. 1974. *Astrophys. J.* **190**:53.

(J2) Jenkins, E. B. and Savage, D. B. 1974. *Astrophys. J.* **187**:243.

(K1) Kahn, F. D. and Woltjer, L. 1959. *Astrophys. J.* **130**:705.

(K2) Kaifu, J., Kato, T., and Iguchi, T. 1972. *Nature*. **238**:105.

(K3) Kellerman, K. I., Shaffer, D. B., Clark B. G., and Geldzahler, B. J. 1977. *Astrophys. J. Letters*. **214**:L61.

(K4) Kerr, F. J. 1957. *Astron. J.* **62**:93.

(K5) Kerr, F. J. 1969. *Ann. Rev. Astron. and Astrophys.* **7**:39.

(K6) Kerr, F. J., Hindman, J. V., and Gum, C. S. 1959. *Australian J. Phys.* **12**:1270.

(K7) Kerr, F. J. and Simonson, S. C. (eds.). 1974. *Galactic Radio Astronomy*, IAU Symposium No. 60. (Dordrecht: D. Reidel).

(K8) Knapp, G. R., Kerr, F. J., and Williams, B. A. 1978. *Astrophys. J.* **222**:800.

(K9) Knapp, G. R. and Jura, M. 1976. *Astrophys. J.* **209**:782.

(K10) Kodaira, K. 1974. *Publ. Astron. Soc. Japan*. **26**:255.

(K11) Kwee, K. K., Muller, C. A., and Westerhout, G. 1954. *Bull. Astron. Inst. Netherlands*. **12**:117.

(**L1**) Lacy, J. H., Baas, F., Townes, C. H., and Geballe, T. R. 1979. *Astrophys. J. Letters.* **227**:L17.

(**L2**) Liszt, H. S., Burton, W. B., Sanders, R. H., and Scoville, N. Z. 1977. *Astrophys. J.* **213**:38.

(**L3**) Lo, K. Y. and Sargent, W. L. W. 1979. *Astrophys. J.* **227**:756.

(**L4**) Lockman, F. J. 1976. *Astrophys. J.* **209**:429.

(**L5**) Lynden-Bell, D. 1964. *Mon. Not. Roy. Astron. Soc.* **129**:299.

(**L6**) Lynden-Bell, D. and Lin, D. N. C. 1977. *Mon. Not. Roy. Astron. Soc.* **181**:59.

(**M1**) Mathewson, D. S., Schwarz, M. P., and Murray, J. D. 1977. *Astrophys. J. Lett.* **217**:L5.

(**M2**) McGee, R. X. and Milton, J. A. 1966. *Australian J. Phys.* **19**:343.

(**M3**) Miller, J. S., 1968. *Astrophys. J.* **157**:1215.

(**M4**) Münch, G. 1957. *Astrophys. J.* **125**:42.

(**N1**) Newton, K. 1980. *Mon. Not. Roy. Astron. Soc.* **190**:689.

(**O1**) Oort, J. H. 1977. *Ann. Rev. Astron. and Astrophys.* **15**:295.

(**O2**) Oort, J. H., Westerhout, G., and Kerr, F. J. 1958. *Mon. Not. Roy. Astron. Soc.* **118**:379.

(**P1**) Phillips, T. G., Huggins, P. J., Wannier, P. G., and Scoville, N. Z. 1979. *Astrophys. J.* **231**:720.

(**R1**) Rickard, J. J. 1968. *Astrophys. J.* **152**:1019.

(**R2**) Rogstad, D. H., Wright, M. C. H., and Lockhart, R. A. 1976. *Astrophys. J.* **204**:703.

(**R3**) Rots, A. H. 1975. *Astron. and Astrophys.* **45**:43.

(**R4**) Rots, A. H. and Shane, W. W. 1975. *Astron. and Astrophys.* **45**:25.

(**R5**) Rougoor, G. W. and Oort, J. H. 1960. *Proc. Natl. Acad. Sci. USA.* **46**:1.

(**S1**) Saraber, M. J. M. and Shane, W. W. 1974. *Astron. and Astrophys.* **36**:365.

(**S2**) Sargent, W. L. W., Young, P. J., Boksenberg, A., Carswell, R. F., and Whelan, J. A. J. 1979. *Astrophys. J.* **230**:49.

(**S3**) Schmidt, M. 1957. *Bull. Astron. Inst. Netherlands.* **13**:247.

(**S4**) Scoville, N. Z. 1972. *Astrophys. J. Letters.* **175**:L127.

(**S5**) Sinha, R. P. 1979. Ph.D. Thesis, University of Maryland.

(**S6**) Solomon, P. M. and Edmunds, M. (eds.). 1979. *Giant Molecular Clouds in the Galaxy.* (Oxford: Pergamon).

(**S7**) Solomon, P. M., Scoville, N. Z., and Sanders, D. B. 1979. *Astrophys. J. Letters.* **232**:L89.

(**S8**) Spitzer, L. and Jenkins, E. B. 1975. *Ann. Rev. Astron. and Astrophys.* **13**:113.

(**S9**) Stark, A. A. 1979. Ph.D. Thesis, Princeton University.

(S10) Stark, A. A. and Blitz, L. 1978. *Astrophys. J. Letters.* **225**:L15.

(S11) Strong, A. W. 1975. *J. Phys. A.* **8**:617.

(T1) Tinsley, B. M. and Larson, R. B. (eds.). 1977. *The Evolution of Galaxies and Stellar Populations.* (New Haven: Yale University Observatory).

(T2) Tully, R. B., Bottinelli, L., Fisher, J. R., Gouguenheim, L., Sancisi, R., and van Woerden, H. 1978. *Astron. and Astrophys.* **63**:31.

(T3) Turner, E. L. and Gott, J. R. 1976. *Astrophys. J. Supp.* **32**:409.

(T4) Tuve, M. A. and Lundsager, S. 1972. *Astron. J.* **77**:652.

(V1) van der Kruit, P. C. and Allen, R. J. 1978. *Ann. Rev. Astron. and Astrophys.* **16**:103.

(V2) Verschuur, G. L. 1975. *Ann. Rev. Astron. and Astrophys.* **13**:257.

(W1) Wannier, P. G., Penzias, A. A., Linke, R. A., and Wilson, R. W. 1976. *Astrophys. J.* **204**:26.

(W2) Wannier, P. G. and Wrixon, G. T. 1972. *Astrophys. J. Lett.* **173**:L119.

(W3) Weaver, H. and Williams, D. R. W. 1973. *Astron. and Astrophys. Supp.* **8**:1.

(W4) Westerhout, G. 1957. *Bull. Astron. Inst. Netherlands.* **13**:201.

(W5) Westerhout, G. 1973. *Maryland-Green Bank Galactic 21-cm Line Survey*, 3rd ed. (College Park: University of Maryland).

(W6) Wilson, R. W., Jefferts, K. B., and Penzias, A. A. 1970. *Astrophys. J. Letters.* **161**:L43.

(W7) Wollman, E. R., Geballe, T. R., Lacy, J. H., Townes, C. H., and Rank, D. M. 1976. *Astrophys. J. Letters.* **205**:L5.

Epilogue

In this volume, we have reviewed what galaxies *are*. Next we turn to a study of *how* they function. There is much to explain. How do the motions of stars on their orbits through our Galaxy lead to the observed motions of the stars of the solar neighborhood? How are these motions determined by the large-scale structure of our Galaxy? Why do stars clump in the galactic disk? Why do elliptical galaxies have the brightness profiles they do? Why are these systems not associated with disk components like those that encircle the bulges of spiral galaxies? Why are luminous elliptical galaxies redder than faint ellipticals, and why do H II regions at different radii in the disks of spirals have different excitations? How do the properties of galaxies change as the Universe ages? Do galaxies grow fainter or brighter, greater or smaller with the passage of time? Do galaxies retain their integrity as individual systems, or do they sometimes merge?

We shall not be able to provide satisfactory answers to all of the questions posed here, but we shall see that many of the facts we have assembled in this volume lock together nicely like pieces of a giant jigsaw puzzle. Other facts remain obstinately incompatible with each other, or they form small isolated islands that cannot yet be connected to the main body of our understanding. One is forced to suspect that some key pieces of the puzzle have yet to be unearthed. Only when we have found them will the grand design of the whole become apparent. Nevertheless, we shall see in the next volume that careful examination of the pieces that are already in our possession strongly suggests what that grand design might be.

A Model of Galactic Evolution

The figures shown on the opening page of each chapter are computer-generated plots of results from n-body calculations by Frank Hohl, of NASA Langley Research Center, of a rotating disk composed of 10^5 point masses representing stars. The particles in the disk move according to Newton's laws of motion under the influence of their mutual gravitational interactions.

The calculation was carried out in two steps. Initially, the stars are distributed into a dynamically balanced, uniformly rotating disk. The disk is "cold," in the sense that the stars have a very small velocity dispersion around their rotation velocity, the dispersion being chosen to satisfy a local stability criterion. The disk is dynamically unstable globally and after about two rotations develops a central bar structure and becomes rather "hot," in the sense that the stars have a large velocity dispersion. The resulting distribution is then artificially smeared in longitude to give an axisymmetric hot disk, which is used to carry out the second step of the calculation.

In the numerical experiment shown here, the hot disk is "cooled" by arbitrarily removing the random velocities of 30% of the stars per rotation. This cooling induces the development of spiral structure and condensations reminiscent of those observed in late-type spirals. Each figure is labeled with a time measured in units of the rotation period of the initial cold, balanced, uniformly rotating disk. The figures shown here are at times $t = 0$, $t = 3.0$, $t = 4.0$, $t = 5.0$, $t = 6.0$, $t = 7.0$, $t = 8.0$, $t = 9.0$, and $t = 11.0$, where time is measured in rotation periods of the original uncooled disk.

Glossary of Symbols

Symbols used in this text are listed here, along with a brief description of their meaning and the page number on which each first appears. Standard mathematical symbols, dummy indices and variables, and notations used only in one place are not included.

a	Semimajor axis in an orbit, 79
a	Semimajor axis of a galaxy, 289
A	Oort constant for differential galactic rotation, 474
A_λ	Interstellar absorption at wavelength λ (in magnitudes), 182
A_{21}	Einstein spontaneous emission coefficient, 485
$A(m)$, $A(m,b)$, $A(m, S)$	Number of stars in apparent magnitude range $m \pm \frac{1}{2}$ or $m \pm \frac{1}{2}$dm, 201
b, b^{I}, b^{II}	Galactic latitude, 40
b	Blue magnitude in Strömgren system, 61
b	Semiminor axis of a galaxy, 289
b_A	Galactic latitude of apex of solar motion, 385
$b(\mathcal{M}, t)$	Stellar birthrate function, 231
$(b - y)$	Color index in Strömgren system, 61

B	Blue magnitude in Johnson-Morgan system, 56
B	Oort constant for differential galactic rotation, 475
B_{12}	Einstein absorption probability, 485
B_{21}	Einstein induced emission coefficient, 485
$B.C.$	Bolometric correction, 97
$(B - V)$	Color index in Johnson-Morgan system, 59
$(B - V)_0$	Intrinsic (unreddened) $(B - V)$, 185
$B_\nu(T)$	Planck function, 92
c	Velocity of light, 47
c	$\log_{10}(r_t/r_c)$, 307
c_1	Color index difference in Strömgren system, 61
d	Distance from Sun to a star, 44
d_1, d_2	Diameters of eclipsing stars, 90
$\dot{d}$	Radial velocity of a star, 390
D	Diffusion coefficient for velocity in phase space, 434
D	Distance to a galaxy, 490
$D(r)$	General relative density function, 210
$D_S(r)$	Relative density function for stars of spectral type S, 209
e	Orbital "eccentricity" parameter, 437
$E(B - V)$	Color excess in $(B - V)$, 185
$E(U - B)$	Color excess in $(U - B)$, 185
f	Integrated radiation flux, 54
f_λ, f_ν	Spectral energy flux, 53
$f(q)$	Number of galaxies with apparent axial ratios in range $(q, q + dq)$, 336

$f(\pi)$	True distribution of parallaxes, 219
$f_k(v)$	Velocity distribution of stellar type k, 457
$f_{obs}(\pi')$	Observed distribution of parallaxes, 219
$\tilde{f}(u)$	Renormalized velocity distribution, 458
$\mathscr{F}_\lambda$, $\mathscr{F}_\nu$	Absolute monochromatic flux emitted by star, 95
[Fe/H]	Logarithmic iron abundance relative to solar value, 121
$F(m, \pi)$	Number of stars of apparent magnitude m and parallax π, 220
$\mathscr{F}(v)$	Fourier transform of $\tilde{f}(u)$, 458
$\mathscr{F}_\nu$(BB)	Monochromatic flux from blackbody, 92
g	Stellar surface gravity, 94
g_i	Statistical weight of atomic level, 485
G	Newtonian gravitation constant, 79
$G(m, \pi \mid \mu \geq \mu_0)$	Number of stars of apparent magnitude m, parallax π, having proper motions $\geq \mu_0$, 219
h	Planck's constant, 92
h	Angular momentum per unit mass, 451
H, H_0	Hubble constant, 326
i	Angle of inclination of orbital plane to sky, 84
$i(\Lambda)$	Observed spectrum from galaxy or star cluster, 457
I_λ, I_ν	Specific intensity, 182
$\mathscr{I}_{obs}(v)$	Fourier transform of $i(\Lambda)$, 458
k	Boltzmann's constant, 92
k_λ, k_ν	Extinction coefficient, 182
ℓ, ℓ^{I}, ℓ^{II}	Galactic longitude, 40
ℓ_A	Galactic longitude of apex of solar motion, 385

ℓ_v	Galactic longitude of vertex of velocity ellipsoid, 421
ℓ_1, ℓ_2	Apparent brightnesses of eclipsing stars, 90
$\ell(m)$, $\ell(m, b)$	Light received from stars in apparent magnitude range $m \pm \frac{1}{2}$, 201
(ℓ/H)	Ratio of mixing length to scale height, 164
L	Stellar luminosity, 93
$\mathscr{L}$	Luminosity in solar units, 224
L_t	Total luminosity of a galaxy, 373
L_*	Characteristic luminosity in galaxy luminosity function, 352
$\mathscr{L}(m)$	Light received from all stars with apparent magnitudes $\leq m$, 207
m	Apparent magnitude, 38
m_{pg}	Photographic apparent magnitude, 55
m_{pv}	Photovisual apparent magnitude, 55
m_ν	Monochromatic magnitude at frequency ν, 62
m_1	Color index difference in Strömgren system, 61
M	Absolute magnitude, 58
$\mathscr{M}$	Mass (stellar, solar, and so on), 79
M_{bol}	Bolometric absolute magnitude, 58
M_V	Visual absolute magnitude, 58
M_0	Mean absolute magnitude in luminosity function, 226
M_*	Absolute magnitude corresponding to L_*, 352
$\overline{M(m)}$	Mean absolute magnitude of stars of apparent magnitude m, 240
n	Number density, 182
N_{H}	Hydrogen column density, 489
N_{HI}	Neutral atomic hydrogen column density, 546

N_{H_2}	Molecular hydrogen column density, 546
$N(m)$, $N(m, b)$	Number of stars down to limiting apparent magnitude m, 183
$N(\beta)$	Number density of galaxies with true axial ratios on range $(\beta, \beta + d\beta)$, 337
O	Orange magnitudes, 341
p	Pressure, 135
P	Period of orbital revolution, 79
$q = (b/a)$	Apparent axial ratio of an elliptical galaxy, 329
$q(m, r)$	Number of stars at distance r that appear at magnitude m, 240
Q	Stellar pulsation constant, 158
Q	Reddening-free parameter in (UBV) system, 186
r	Distance from Sun, 205
r_c	Core radius (of a galaxy or globular cluster), 307
r_e	Effective radius in de Vaucouleurs law, 311
r_s	Characteristic length in disk brightness profile, 323
r_t	Tidal radius (of a galaxy or globular cluster), 307
r_0	Structural length in Hubble law, 311
R	Stellar radius, 86
R	Radial distance from galactic center, 260
R	Radial-tangential velocity-dispersion ratio, 457
R	Radial coordinate in plane of a galaxy, 499
R_{max}	Maximum radial distance in galactic plane of star from galactic center, 437
R_{max}	Radius of maximum galaxian rotation velocity, 511

$R_{\min}$	Minimum radial distance in galactic plane of star from galactic center, 437
R_V	Ratio of total absorption in V band to color excess in $(B - V)$, 188
R_0	Sun's distance from galactic center, 257
s	Pathlength, 485
$\tilde{s}(\Lambda)$	Composite intrinsic spectrum of galaxy or star cluster, 458
$s_k(\Lambda)$	Energy distribution of star of type k, 457
S_{esc}	Escape speed from galaxy, 403
S_λ	Receiver sensitivity, 59
S_ν	Source function, 485
$S_\odot$	Speed of solar motion, 385
$\mathscr{S}(\nu)$	Fourier transform of $\tilde{s}(\Lambda)$, 458
t	Hour angle, 33
t	Hubble type, 296
T	Absolute temperature, 92
T_B	Brightness temperature, 486
T_{eff}	Effective temperature, 93
T_R	Radiation temperature, 109
T_S	Spin temperature, 485
u	Ultraviolet magnitude in Strömgren system, 61
u	Shape parameter of isodensity surfaces, 334
$u = \Pi - \Pi_{\text{LSR}}$	Velocity radially outward in galactic plane relative to LSR, 383
u_1	Component of star's peculiar velocity in rectangular coordinate system, 420
$u_\odot$	u component of Sun's peculiar velocity relative to LSR, 383
$(u - b)$	Color index in Strömgren system, 61

U	Ultraviolet magnitude in Johnson-Morgan system, 56
U_*	u velocity of a star relative to Sun, 383
$(U - B)$	Color index in Johnson-Morgan system, 59
$(U - B)_0$	Intrinsic (unreddened) $(U - B)$, 185
v	Space velocity, 77
v	Violet magnitude in Strömgren system, 61
v	Orbital velocity, 84
$v = \Theta - \Theta_{LSR}$	Tangential (rotational) velocity in galactic plane relative to LSR, 383
v_R	Radial velocity, 46
v_r	Velocity radially outward from cluster center, 457
v_{rms}	Root-mean-square speed of a star experiencing collisions, 433
v_S	Space motion; speed, 381
v_T	Tangential velocity, 77
v_t	Velocity in tangential direction relative to cluster center, 457
v_τ	Tau component of space motion of a star relative to Sun, 414
v_υ	Upsilon component of space motion of a star relative to Sun, 414
v_1	Component of star's peculiar velocity in rectangular coordinate system, 420
$v_\odot$	v component of Sun's peculiar velocity relative to LSR, 383
$v'_\odot$	v component of solar motion, 385
$\mathbf{v}_A$	Vector velocity of M31 relative to center of mass of Local Group, 406
$\mathbf{v}_G$	Vector velocity of Galaxy relative to center of mass of Local Group, 405

$\mathbf{v}_{LSR}$	Vector velocity of LSR relative to velocity centroid of Local Group, 406
$\mathbf{v}_{rot}$	Vector rotation velocity of LSR around galactic center, 405
$\mathbf{v}_0$	Vector velocity of Sun relative to velocity centroid of Local Group, 405
$\mathbf{v}_\odot$	Vector peculiar velocity of Sun relative to LSR, 405
V	Visual magnitude in Johnson-Morgan system, 56
V_*	v velocity of a star relative to Sun, 383
$w = Z - Z_{LSR}$	Velocity perpendicular to galactic plane relative to LSR, 383
w_1	Component of star's peculiar velocity in rectangular coordinate system, 420
$w_\odot$	w component of Sun's peculiar velocity relative to LSR, 384
W_*	w velocity of a star relative to Sun, 384
x	Linear distance along sun-center (of Galaxy) line, 257
x	Equatorial rectangular component of position of star relative to Sun, 390
$\dot{x}$	Equatorial rectangular component of velocity of star relative to Sun, 390
X	Fractional abundance by weight of hydrogen in stellar material, 99
X_*	x component of star's velocity in equatorial rectangular system, 390
$X_\odot$	x component of Sun's velocity in equatorial rectangular system, 390
y	Yellow magnitude in Strömgren system, 61
y	Linear distance perpendicular to sun-center line in direction of galactic rotation, 257

y	Equatorial rectangular component of position of star relative to Sun, 390
$\dot{y}$	Equatorial rectangular component of velocity of star relative to Sun, 390
Y	Fractional abundance by weight of helium in stellar material, 99
Y_*	y component of star's velocity in equatorial rectangular system, 390
$Y_\odot$	y component of Sun's velocity in equatorial rectangular system, 390
z	Zenith distance, 32
z	Distance perpendicular to galactic plane, 249
z	Equatorial rectangular component of position of star relative to Sun, 390
$\dot{z}$	Equatorial rectangular component of velocity of star relative to Sun, 390
Z	Fractional abundance be weight of "metals" in stellar material, 99
Z	Velocity perpendicular to galactic plane, 382
Z_*	z component of star's velocity in equatorial rectangular system, 390
$Z_\odot$	z component of Sun's velocity in equatorial rectangular system, 390
α	Right ascension, 34
α_A	Right ascension of apex of solar motion, 385
$\dot{\alpha} d \cos \delta$	Transverse velocity in right ascension of a star, 391
β_S	Scale height for z distribution of stars of spectral type S, 250
$\beta(t)$	Total birthrate function, 133
δ	Declination, 34
δ_A	Declination of apex of solar motion, 385

$\dot{\delta}d$	Transverse velocity in declination of a star, 391
$\delta(U - B)$	Ultraviolet excess, 118
ΔS	Spectral type difference between H-line and K-line classifications, 156
$\Delta\lambda$	Wavelength shift or wavelength interval, 46
$\Delta\nu_D$	Doppler width, 487
$\Delta(B - V)$	Blanketing effect in $(B - V)$, 118
$\Delta(U - B)$	Blanketing effect in $(U - B)$, 118
$\Delta_S(\rho)$	Fictitious relative density function, 214
$\varepsilon = 1 - (b/a) = 1 - q$	Ellipticity of a galaxy, 331
ε_λ	Line-blocking coefficient, 116
η	Integrated line-blocking coefficient, 119
η_ν	Emissivity, 485
θ	Position angle, 41
θ	Polar angle in spherical coordinates, 92
θ	Stellar angular diameter, 95
θ	Angular coordinate in plane of a galaxy, 499
$\hat{\boldsymbol{\theta}}$	Unit vector in plane of galaxy in direction of galactic rotation, 405
Θ	Tangential (rotational) velocity in galactic plane, 382
Θ_{esc}	Escape velocity from galaxy, 13
Θ_{max}	Maximum galaxian rotation velocity, 512
$\Theta_\odot$	Rotation velocity of Sun around galactic center, 13
Θ_0	Circular rotation velocity, 383
λ	Wavelength, 47
λ	Angular distance from star to cluster convergent point, 76
Λ	$\log_e \lambda$, 457

μ	Proper motion, 41
μ	$\log_e 10$, 206
μ_B	Magnitude (blue) per square second of arc, 303
μ_ℓ	Proper motion in galactic longitude, 467
μ_α	Proper motion in right ascension, 41
μ_δ	Proper motion in declination, 41
ν	Frequency, 53
ν, $\nu(r, M, S)$, $\nu(r, \ell, b, M, S)$	Space density of stars at distance r, galactic coordinates (ℓ, b), absolute magnitude M, and spectral type S, 209
$\nu(\mathscr{M})$	Space density of stars of mass $\mathscr{M}$, 232
$\nu(R)$	Space density of globular clusters, 261
$\xi(\mathscr{M})$	Initial mass function, 230
π, π''	Parallax, 43
π'	Observed parallax (including errors), 219
Π	Velocity radially outward from center in galactic plane, 382
ρ	Density, 135
ρ	Apparent distance to star (before allowance for extinction), 208
ρ	Radial coordinate in plane of sky, 499
σ	Stefan-Boltzmann constant, 92
σ	Magnitude dispersion in luminosity function, 226
σ	Dispersion of Gaussian fit to seeing profile, 315
σ_Θ	Dispersion of residual velocities in tangential direction, 479
σ_Π	Dispersion of residual velocities in radial direction, 479
Σ_e	Surface brightness of a galaxy at $r = r_e$, 311

Σ_S	Surface density in disk of stars of spectral type S, 251
Σ_S	Characteristic disk surface brightness, 323
Σ_0	Central surface brightness, 307
$\Sigma(r)$	Surface brightness of a galaxy, 311
$\Sigma_{app}(r)$	Apparent brightness distribution (smeared by seeing), 314
$\Sigma_t(r)$	True brightness distribution, 314
τ	Tau component of proper motion, 412
τ_E	Energy-exchange time, 433
τ_G	Age of our Galaxy, 230
τ_{ms}	Main-sequence lifetime of a star, 137
τ_λ, τ_ν	Optical depth, 182
τ_1	Optical half-thickness of galactic disk, 183
υ	Upsilon component of proper motion, 411
ϕ	Azimuthal angle in spherical coordinates, 92
ϕ	Angular coordinate in plane of sky, 499
ϕ_ν	Absorption profile, 485
Φ_0	Coefficient luminousity function, 226
$\Phi(M)$	General luminousity function, 210
$\Phi(M, S)$	Luminosity function for stars of spectral type S, 209
$\Phi_m(M, S)$	Luminosity function for stars of apparent magnitude m, 239
χ_ν	Absorption coefficient, 485
$\Psi(M_V)$	Initial luminousity function, 230
ψ	Auxiliary angle, 411
ω	Solid angle, 92
ω	Angular rotation rate around galactic center, 469
$\omega_0 = (\Theta_0/R_0)$	Angular rotation rate of LSR around galactic center, 408

Ω	Solid angle, 485
$\oplus$	Earth symbol, 79
☾	Moon symbol, 81
*	Star symbol, 97
$\odot$	Sun symbol, 58

Index